MCWASP XV: International Conference on Modelling of Casting, Welding and Advanced Solidification Processes

IOP Conference Series: Materials Science and Engineering
Volume 861

Jonkoping, Sweden
22 - 23 June 2020

ISBN: 978-1-7138-1813-7
ISSN: 1757-8981

Printed from e-media with permission by:

Curran Associates, Inc.
57 Morehouse Lane
Red Hook, NY 12571

Some format issues inherent in the e-media version may also appear in this print version.

Institute of Physics
Dirac House, Temple Back
Bristol BS1 6BE UK

Phone: 44 1 17 929 7481
Fax: 44 1 17 920 0979

techtracking@iop.org

Additional copies of this publication are available from:

Curran Associates, Inc.
57 Morehouse Lane
Red Hook, NY 12571 USA
Phone: 845-758-0400
Fax: 845-758-2633
Email: curran@proceedings.com
Web: www.proceedings.com

TABLE OF CONTENTS

KEYNOTES

DOES MCWASP STILL FOLLOW MOORE'S LAW? FORTY YEARS OF ADVANCES IN MICROSTRUCTURE MODELING 1
M Rappaz

COMPARING MESOSCOPIC MODELS FOR DENDRITIC GROWTH 11
D Tourret, L Sturz, A Viardin, M Založnik

REDUCED ORDER MODELS FOR WELDING AND SOLIDIFICATION PROCESSES 25
P F Mendez

THE VIRTUAL CORE - MODELLING AND OPTIMIZATION OF CORE MANUFACTURING AND APPLICATION 48
J Thorborg, S Kumar, I Wagner, J C Sturm

ADDITIVE MANUFACTURING

SOLIDIFICATION MICROSTRUCTURE SELECTION MAPS FOR LASER POWDER BED FUSION OF MULTICOMPONENT ALLOYS 64
P Mohammadpour, A B Phillion

RESOLVING THE EFFECTS OF LOCAL CONVECTIVE HEAT TRANSFER VIA ADJUSTMENT OF THERMO-PHYSICAL PROPERTIES IN PURE HEAT CONDUCTION SIMULATION OF LASER POWDER BED FUSION (L-PBF) 72
M Bayat, V K Nadimpalli, S Mohanty, J H Hattel

COLUMNAR-TO-EQUIAXED TRANSITION IN A LASER SCAN FOR METAL ADDITIVE MANUFACTURING 80
L Yuan, A S Sabau, D StJohn, A Prasad, P D Lee

THERMOMECHANICAL MODELING OF ADDITIVELY MANUFACTURED STRUCTURAL PARTS - DIFFERENT APPROACHES ON THE MACROSCALE 88
J Thorborg, P Esser, M Bayat

STUDY OF THERMOELECTRIC MAGNETOHYDRODYNAMIC CONVECTION ON SOLUTE REDISTRIBUTION DURING LASER ADDITIVE MANUFACTURING 96
A Kao, T Gan, C Tonry, I Krastins, K Pericleous

CAPTURING MARANGONI FLOW VIA SYNCHROTRON IMAGING OF SELECTIVE LASER MELTING 104
S J Clark, C L A Leung, Y Chen, L Sinclair, S Marussi, P D Lee

EFFECT OF EVAPORATED GAS FLOW ON POROSITY AND MICROSTRUCTURE OF IN718 PARTS PRODUCED BY LPBF-PROCESSES 110
J Jakumeit, C Huang, R Laqua, J Zielinski, J H Schleifenbaum

MELT POOL MORPHOLOGY IN DIRECTED ENERGY DEPOSITION ADDITIVE MANUFACTURING PROCESS 118
Y Chen, S Clark, A C L Leung, L Sinclair, S Marussi, R Atwood, T Connoley, M Jones, G Baxter, P D Lee

PART-SCALE MECHANICAL MODELLING OF LPBF INCLUDING MICROSTRUCTURAL EVOLUTION EFFECTS 124
D De Baere, M Bayat, S Mohanty, J H Hattel

CONTINUOUS CASTING

COLUMNAR-EQUIAXED TRANSITION IN CONTINUOUS CASTING BASED ON A MICRO-MACRO SOLIDIFICATION MODEL WITH LONG-RANGE SOLUTAL MIXING 132
M Uddagiri, J Spee, S Hubig, I Steinbach

MODELLING VISCOPLASTIC BEHAVIOR OF SOLIDIFYING SHELL UNDER APPLIED ELECTROMAGNETIC BREAKING DURING CONTINUOUS CASTING 140
A Vakhrushev, A Kharicha, M Wu, A Ludwig, G Nitzl, Y Tang, G Hackl, J Watzinger, C M G Rodrigues

SIMULATION OF LONGITUDINAL SURFACE DEFECT IN STEEL CONTINUOUS CASTING 147
M L S Zappulla, B G Thomas

INTEGRATED NUMERICAL SIMULATION OF THE INDUSTRIAL CONTINUOUS CASTING PROCESS 155
J Fainberg, E Hepp, W Schäfer, C Krafthöfer

A MESHLESS THERMOMECHANICAL TRAVELLING-SLICE MODEL OF CONTINUOUS CASTING OF STEEL 163
B Mavrič, T Dobravec, R Vertnik, B Šarler

PERFORMANCE OF TURBULENCE MODELS FOR FLOW PREDICTION IN A MOULD OF CONTINUOUS STEEL CASTER 171
J Gregorc, A Kunavar, B Šarler

DEFECTS

MODELLING OF LOCALISED SHRINKAGE PRESSURE IN A DIRECTIONALLY SOLIDIFYING DOMAIN 179
B Nenchev, S Gill

MULTIPHASE-FIELD MODELLING OF HYDROGEN PORE EVOLUTION DURING ALLOY SOLIDIFICATION 186
A Zhang, Z Guo, Q Wang, S Xiong

ADVANCED MODELING OF SHRINKAGE POROSITY IN CASTINGS 194
P Esser, C Schankies, V Khalajzadeh, C Beckermann

THREE-DIMENSIONAL NUMERICAL SIMULATION OF SOLIDIFICATION MICROPOROSITY AND MICROSTRUCTURE OF ALUMINUM ALLOYS 202
C Gu, A A Luo

ELECTROMAGNETIC COUPLING

MODELING MIXED COLUMNAR-EQUIAXED SOLIDIFICATION OF SN-10WT%PB ALLOY UNDER FORCED CONVECTION DRIVEN BY TRAVELLING MAGNETIC STIRRING 209
Z Zhang, M Wu, H Zhang, E Karimi-Sibaki, A Ludwig, A Kharicha

IN SITU OBSERVATION OF DIRECTIONAL SOLIDIFICATION IN GA-IN ALLOY UNDER A TRANSVERSE DC MAGNETIC FIELD 217
S He, N Shevchenko, S Eckert

COMPARISON OF TWO-PHASE AND THREE-PHASE MACROSCOPIC MODELS OF EQUIAXED GRAIN GROWTH IN SOLIDIFICATION OF BINARY ALLOY WITH ELECTROMAGNETIC STIRRING 225
T Wang, S Semenov, E Wang, Y Delannoy, Y Fautrelle, O Budenkova

HEAT TRANSFER AND FLUID FLOW

PREDICTING THE INTERFACIAL HEAT TRANSFER COEFFICIENT OF CAST MG-AL ALLOYS USING BECK'S INVERSE ANALYSIS 233
M U Bilal, N Hort

A MESHLESS SOLUTION OF A OF LID-DRIVEN CAVITY CONTAINING A HETEROGENEOUS POROUS MEDIUM 241
V Hatić, Z Rek, K Mramor, B Mavrič, B Šarler

PERMEABILITY TENSOR FOR VARIOUS COLUMNAR DENDRITE STRUCTURES 249
Y Mitsuyama, T Takaki, S Sakane, Y Shibuta, M Ohno

INGOT AND SHAPE CASTING

MODELLING OF AN INDUSTRIAL DIE CASTING PROCESS FOR THE PRODUCTION OF ALUMINUM AUTOMOTIVE PARTS 257
J Ou, C Wei, D Maijer, S Cockcroft, L A, Y Zhang, Z Chen, Z Zhu

OPTIMISATION OF THE FILLING PROCESS IN COUNTER-GRAVITY CASTING 265
M Papanikolaou, P Saxena, E Pagone, K Salonitis, M Jolly

PREDICTION OF SOLIDIFICATION STRUCTURES IN A 9.8 TONNE STEEL INGOT 273
B Gerin, H Combeau, M Založnik, I Poitrault, M Cherif

MODELLING AND SIMULATION OF INGOT SOLIDIFICATION WITH THE OPEN-SOURCE SOFTWARE CODE_SATURNE 281
C Demay, M Ferrand, S Belouah, V Robin

MULTISCALE SOLIDIFICATION SIMULATION OF SR-MODIFIED AL-SI-MG ALLOY IN DIE CASTING 289
B Zhou, G Laschet, J Eiken, H Behnken, M Apel

MULTI-SCALE SIMULATION OF HYBRID LIGHT METAL STRUCTURES PRODUCED BY HIGH PRESSURE DIE CASTING 297
J Jakumeit, H Behnken, R Laqua, J Eiken, J Brachmann

A VERIFICATION OF THERMOPHYSICAL PROPERTIES OF A POROUS CERAMIC INVESTMENT CASTING MOULD USING COMMERCIAL COMPUTATIONAL FLUID DYNAMICS SOFTWARE 305
C A Jones, M R Jolly, A E W Jarfors, M Irwin, R Svenningsson, J Steggo, J Eriksson

MECHANICAL BEHAVIOUR

X-RAY TOMOGRAPHY, DIGITAL VOLUME CORRELATION AND FE MODELLING: A SYNERGISTIC COMBINATION TO STUDY THE PROCESSING-STRUCTURE-PROPERTY RELATIONS IN DUCTILE IRON 313
T Andriollo, Y Zhang, S Fæster, J Thorborg, N S Tiedje, V Kouznetsova, J Hattel

IMPACT OF LOCAL SI SEGREGATION ON STRAIN LOCALIZATION IN DUCTILE CAST IRON 321
C L Xu, T Andriollo, F Lin, Y B Zhang, J Hattel, N Tiedje

IN SITU ANALYSIS OF CAST IRONS MECHANICAL BEHAVIOUR USING SYNCHROTRON X-RAY TOMOGRAPHY AND 3DXRD 326
T Sjögren, S Hall, L Elmquist, E Dartfeldt, E Larsson, M Majkut, J Elfsberg, P Skoglund, J Engqvist

MODELLING MACROSEGREGATION MODIFICATION IN DC CASTING OF ALUMINIUM ALLOYS IN SHEET INGOTS ACCOUNTING FOR INLET MELT FLOW, EQUIAXED GRAIN MORPHOLOGY AND TRANSPORT 333
A Pakanati, M M'Hamdi, H Combeau, M Založnik

NUMERICAL INVESTIGATION OF AN IN-SITU OBSERVED FLOW REGIMES DURING SOLIDIFICATION OF AN NH4CL – 70 WT%H2O SOLUTION 341
M Wu, M Stefan-Kharicha, A Kharicha, A Ludwig

IMPACT OF CRYSTAL SEDIMENTATION AND VISCOPLASTIC SEMI-SOLID DYNAMICS ON MACROSEGREGATION 349
C M G Rodrigues, A Ludwig, M Wu, A Kharicha, A Vakhrushev

MICRO- AND MACRO-SEGREGATION

MODELLING THE EVOLUTION OF MNS INCLUSIONS AND MACROSEGREGATION DURING SOLIDIFICATION USING A THREE-PHASE MODEL 357
D Cai, J Li, H Dong, J Li

SENSITIVITY OF MACROSCOPIC TRANSPORT CALCULATIONS TO UNCERTAIN MICROSCALE RELATIONSHIPS DURING METAL ALLOY SOLIDIFICATION 364
J Coleman, M J M Krane

MICROSEGREGATION MODEL INCLUDING SORET EFFECT: AN APPLICATION FOR SN-BI ALLOY 371
J E Santos Filho, R Heringer

STRUCTURE FORMATION ON MICROSCALE

CELLULAR AUTOMATON MODELLING OF M7C3 CARBIDE GROWTH DURING SOLIDIFICATION OF FE-C-CR ALLOY 379
H W Zhang, S Zhang, Y C Wang, Y Z Hao, M Miao, K Nakajima, H Lei, Q Wang, J C He

EFFECT OF AL-V-B GRAIN REFINER ON REFINING ALUMINIUM ALLOYS: ESTIMATION FROM AB INITIO CALCULATIONS 387
S Ma, N Zong, H Dong, T Jing

INFLUENCE OF CRYSTAL FRAGMENTATION ON THE FORMATION OF MICROSTRUCTURE AND MACROSEGREGATION DURING DIRECTIONAL SOLIDIFICATION UNDER FORCED CONVECTION CONDITION 395
H Zhang, M Wu, Z Zhang, A Ludwig, A Kharicha

THREE-DIMENSIONAL NEEDLE NETWORK MODEL FOR DENDRITIC GROWTH WITH FLUID FLOW 403
T Isensee, D Tourret

COMPARISON OF SOLIDIFICATION KINETICS OF COMPACTED AND LAMELLAR CAST IRONS 411
J Lacaze, U de la Torre, J Sertucha, A Regordosa

IN SITU OBSERVATION OF AUSTENITE COARSENING INDUCED BY MASSIVE-LIKE TRANSFORMATION DURING SOLIDIFICATION IN FE–C ALLOYS 419
H Yasuda, T Suga, K Ichida, T Narumi, K Morishita

3D CELLULAR AUTOMATON MODELLING OF SILICON CRYSTALLIZATION INCLUDING GRAINS IN TWIN RELATIONSHIP 427
A Pineau, G Guillemot, G Reinhart, G Regula, N Mangelinck-Noël, Ch-A Gandin

MOLECULAR DYNAMICS SIMULATION OF HETEROGENEOUS NUCLEATION FROM CONCAVE CAVITY AT SURFACE OF GRAIN REFINER 435
T Fujinaga, Y Shibuta

THE INTEGRATION OF STRUCTURAL MECHANICS INTO MICROSTRUCTURE SOLIDIFICATION MODELLING 441
P Soar, A Kao, G Djambazov, N Shevchenko, S Eckert, K Pericleous

CALPHAD-BASED PHASE-FIELD STUDY OF THE INTERPLAY BETWEEN SPHEROIDAL GRAPHITE GROWTH AND CHEMICAL SEGREGATION IN DUCTILE CAST IRON 449
J Eiken

EFFECTS OF BI ADDITION ON SI FEATURES, TENSILE PROPERTIES AND WEAR RESISTANCE OF HYPEREUTECTIC AL-15SI ALLOY 457
C L Pereira, L F Gomes, J E Spinelli

APPLICATIONS OF PATTERN RECOGNITION FOR DENDRITIC MICROSTRUCTURES 465
J Strickland, B Nenchev, H B Dong

MODELLING OF INTERMETALLIC LAYERS FORMATION DURING SOLID-LIQUID JOINING OF DISSIMILAR METALLIC MATERIALS 471
V Glück Nardi, T Greß, B Tonn, W Volk

CELLULAR AUTOMATON MODELLING TO PREDICT MULTI-PHASE SOLIDIFICATION MICROSTRUCTURES FOR FE-C PERITECTIC ALLOYS 479
J Ogawa, Y Natsume

PHASE FIELD MODELLING OF DENDRITIC SOLIDIFICATION BY USING AN ADAPTIVE MESHLESS SOLUTION PROCEDURE 487
T Dobravec, B Mavrič, B Šarler

A CELLULAR AUTOMATON APPROACH FOR THE PREDICTION OF GRAIN SIZE IN GRAIN REFINED ALLOYS 494
A Jacot

THERMOMECHANICS AND THERMOPHYSICAL PROPERTIES

TOWARD THE DEVELOPMENT OF A THERMAL-STRESS MODEL OF AN INDUSTRIAL COUNTER PRESSURE CASTING PROCESS 502
C Wei, J Ou, P Fan, F Mehr, D Maijer, S Cockcroft, Y Zhang, Z Chen, L A, Z Zhu

CASTINGS DISTORTION OF GAS TURBINE ENGINE PARTS DURING SOLIDIFICATION AND COOLING 510
A Inozemtsev, A Dubrovskaia, K Dongauzer, A Alapanova, A Maleev

MOLECULAR DYNAMICS STUDY OF COMPOSITION DEPENDENCE OF SOLID-LIQUID INTERFACIAL ENERGY OF FE-NI BINARY ALLOY 518
K Ueno, Y Shibuta

OBSERVATION OF SEMISOLID DEFORMATION BY USING 4D-CT AND 3DXRD 525
T Narumi, T Nakata, H Yasuda

MODELLING OF SHEAR BANDS DURING SOLIDIFICATION 533
A Ludwig, C M G Rodrigues, M Wu, A Kharicha

APPLICATION OF MULTI-PHASE-FIELD LATTICE BOLTZMANN METHOD TO SEMI-SOLID DEFORMATION 541
N Yamanaka, S Sakane, T Takaki

CONCURRENT AND COUPLED RESOLUTION OF FLUID FLOW AND SOLID DEFORMATION IN SOLIDIFICATION PROCESSES 548
S Zhang, G Guillemot, C-A Gandin, M Bellet

ULTRASONICS

THE ULTRASOUND EFFECTS ON THE FORMATION OF THE SOLIDIFICATION STRUCTURE OF A356 INGOTS PROCESSED VIA A 2-ZONE INDUCTION MELTING FURNACE 556
A Dong, L Nastac

PROGRESS IN THE DEVELOPMENT OF A CONTACTLESS ULTRASONIC PROCESSING ROUTE FOR ALLOY GRAIN REFINEMENT 564
K A Pericleous, C Beckwith, V Bojarevics, G Djambazov, A Dybalska, W D Griffiths, C Tonry

WELDING

IN SITU X-RAY OBSERVATIONS OF TRANSIENT STATES IN ARC WELD POOLS 572
F Wu, K V Falch, M Drakopoulos, W U Mirihanage

PHASE-FIELD STUDY ON MICROSTRUCTURE FORMATION IN MAR-M247 DURING ELECTRON BEAM WELDING AND CORRELATION TO HOT CRACKING SUSCEPTIBILITY 578
B Böttger, M Apel, T Jokisch, A Senger

PREDICTING SOLIDIFICATION CRACKING SUSCEPTIBILITY OF STAINLESS STEELS USING MACHINE LEARNING 586
S Feng, H B Dong

Author Index

Preface

The international conference on Modelling of Casting, Welding and Advanced Solidification Process (MCWASP) traces back to 1980, when the first conference on "Modeling of Casting and Welding Processes" was held in New Hampshire, USA. Since the previous MCWASP XIV, held in Japan 2015, the community has eagerly been awaiting and preparing to celebrate the 40th year anniversary with the 15th edition of MCWASP scheduled for 21-26. June 2020 at the venue Djurönäset in the archipelago of Stockholm, Sweden. The planning was made for a full 5-day conference consisting of presentations of recent scientific findings, and with plenty of time for personal contacts, exciting discussions and unforgettable experiences. However, as the new decade started, the spread of the COVID-19 pandemic led to closed borders and international travel restrictions, and in March 2020 the decision was made to cancel the physical meeting. Nevertheless, the organizing committee continued the process of reviewing and publication of the papers contributed to the conference, and to proceed online. With this in mind, we are very happy to deliver the proceedings of the 40-year anniversary edition of this important conference series, the first MCWASP conference to be held online 22-23. June 2020 hosted from Jönköping, Sweden.

In the proceedings, the first section contains the papers from the four keynote speakers. These speakers were invited to show the diversity in the topics of the conference, to provide a tribute to the history of MCWASP as well as perspectives on the future of scientific and industrial challenges within the scope of the conference. Michel Rappaz provides an exciting recap of the developments within microstructure modelling as well as computational power during the 40-year history of the MCWASP conference, with a keynote entitled "Does MCWASP still follow Moore´s law?". Damien Tourret continues with a keynote entitled "Comparing mesoscopic models for dendritic growth", showing a benchmark of different multi-scale modeling approaches for dendritic growth, discussing their relative strengths, limitations, and choices of parameters to achieve quantitative predictions. Patricio Mendez' paper "Reduced order models for welding and solidification processes" shows that complex problems can indeed be treated with generality, accuracy, and simplicity. Jesper Thorborg completes the Keynote section with a paper entitled "The virtual core – modeling and optimization of core manufacturing and application" showing the details behind modeling the lifecycle of a sand core and its influence on casting quality.

In planning for the physical meeting, the committee wanted to find ways of increasing the presence and significance of the Poster presentations. A new session concept called Teaser sessions was developed. The intention was that ten authors of Poster papers each day would have the opportunity to present their work in a three-minute prerecorded video shown on the big screen in the conference hall. This would enable them to make use of audio and video to attract interest to their work and provide inspiration for vivid discussions at the Poster session later the same day. In order to distribute the contributions between Oral and Poster/Teaser contributions, 45 contributions were assigned Oral presentations and 45 Poster presentations.

In the online version of the conference, the distribution between Oral and Posters was maintained. Now all authors can present their work on Online Presentation pages, containing their contribution in a digital format with creative use of digital communication methods.

In total, the conference contains 94 contributions, of which 73 are accompanied with papers in the current proceedings. Among the contributions, the emerging field of *Additive Manufacturing* is well represented. There are many contributions on the more traditional MCWASP topic *Structure formation at microscale*, demonstrating the significant developments in numerical techniques as well as computational power that has taken place since the previous MCWASP conference in 2015. The other topics represented are *Continuous casting*, *Defects*, *Electromagnetic coupling*, *Heat transfer and fluid flow*, *Ingot and shape casting*, *Mechanical behaviour*, *Micro- and macro-segregation*, *Thermomechanics and thermophysical properties*, *Ultrasonics* and *Welding*. Authors report advances in aspects such as in-situ experiments and numerical techniques, including machine learning and pattern recognition.

It can be concluded that the traditional MCWASP themes still attract significant research interest. The complexity of these advanced solidification phenomena is still a challenge for the community and although significant progress has been made, many challenges still remain for the future.

Jönköping, May 2020

Jakob Olofsson

Chairman MCWASP XV

School of Engineering
Jönköping University
Sweden

Peer review statement

All papers published in this volume of *IOP Conference Series: Materials Science and Engineering* have been peer reviewed through processes administered by the proceedings Editors. Reviews were conducted by expert referees to the professional and scientific standards expected of a proceedings journal published by IOP Publishing.

IOP Conf. Series: Materials Science and Engineering **861** (2020) 012001 doi:10.1088/1757-899X/861/1/012001

Does MCWASP still follow Moore's law? Forty years of advances in microstructure modeling

M Rappaz

Ecole Polytechnique Fédérale de Lausanne, Institute of Materials, Station 12, CH-1015 Lausanne, Switzerland

E-mail: michel.rappaz@epfl.ch

Abstract. Over the past forty years, the series of MCWASP conferences has been the witness of the fantastic developments in the field of modeling of welding and solidification processes. This has been made possible thanks to the exponential increase in computing power but also to the emergence of new numerical methods and tools. While the "size" of computer simulations reported in MCWASP proceedings continues to follow Moore's law describing the evolution of computers power, the present contribution focuses on the evolution of microstructure modeling. In this respect, a major turning point is the introduction of the phase field (PF) method in the mid-90's. While PF is still the method of choice for the direct simulation of microstructure formation, it is also very CPU intensive. Therefore, it did not replace and rather complement other coarser methods such as analytical micro-macro models, Cellular Automata coupled with Finite Elements (so-called CAFE approach), modified-CAFE methods, mesoscopic model based on a dendrite envelope or Dendritic Needle Network (DNN) model. Some of these models as well as PF are coupled with fluid flow models based either on standard Navier-Stokes or Lattice-Boltzmann solutions. In parallel to these developments, wonderful experimental tools have become available to complement and validate numerical simulations: in situ X-ray radiography and tomography, EBSD, atom probe, FIB, etc. Besides a review of the past forty years, the present contribution will outline future directions of research in the area of microstructure modeling.

1. Introduction

In 1980 was held in New Hampshire the first "Modeling of Casting and Welding Processes" conference, with Harold Brody and Diran Apelian as chairmen [1]. If 1-Dimensional (1-D) simulations of heat flow involving melting or freezing date back to 1959 [2], the casting and welding community was clearly behind mechanical engineering, physics or meteorology in the field of numerical modeling. At the end of this first conference, the participants concluded that "*...activity in the field [of solidification processes] would merit a second conference in 2-3 years.... the format of this conference contributed to its success and that future conferences should be limited to a group of about 100 individuals*".

Over the past forty years, this first New Hampshire conference has been followed by fourteen more editions, including the present one in Stockholm, Sweden [3-16]. Section 2 traces back the initial developments of the conference series, from the first one held in 1980 up to the fifth one held in Davos ten years later, which was then renamed with the actual brand name "MCWASP". The period from MCWASP VI to VIII (section 3) has seen the emergence of two important methods in the field of microstructure modeling, namely Phase Field (PF) and Cellular Automata (CA). From the turn of the millennium, the MCWASP conference series has seen many applications of these two methods to various problems, but also the development of other original meso-scale approaches (section 4). Finally, in section 5, it will be shown that the development of microstructure modeling is strongly linked with the evolution of the available computing power, but also the usage of adaptive meshing techniques and massively parallel computations on GPUs. For simplicity, papers published in MCWASP proceedings are only referenced with the volume number of the volume in which they appear ([1, 3-15]).

2. From the first Modelling of Casting and Welding Conference to the first MCWASP

At the time of the first Modeling of Casting and Welding Processes conference in 1980 [1], the focus of the conference was mainly on heat and fluid flow, with one session dedicated to stresses and deformation. However, M. Glicksman, in charge of organizing a half-day session on analytical and computer modeling of solidification kinetics and structure, contacted W. Kurz to present an invited talk. The title of Kurz and Fisher's presentation "*Directional growth of eutectic and dendrites at the limit of stability*" reminds us that the marginal stability criterion for the growth of a dendrite tip, which gives the right order of magnitude for the tip radius and undercooling (while missing the question of anisotropy), was proposed only a few years before by Langer and Müller-Krumbhaar. While most of the presentations in this session were focused on experimental results and theoretical interpretations, Gilmer described an Ising model for the growth of a crystal-melt interface. A few talks were also concerned with microsegregation modeling. Interestingly, what is maybe missing nowadays in MCWASP conferences, a full session was dedicated to outline future directions and needs in the field.

The second conference, dedicated to Chvorinov for his 80th birthday, was essentially focused on the same topics [3]. Rapid solidification, a topic extensively studied in the 1980's which has seen a regain of interest nowadays with additive manufacturing processes, was the subject of an entire session. Flood and Hunt presented their first Columnar-to-Equiaxed Transition (CET) model, followed in 1984 by the publication of Hunt's CET analytical criterion [17]. This criterion was further extended to unsteady-state conditions in the third proceeding by the same authors in [4].This conference could have already been renamed MCWASP since it focused on several advanced solidification processes.

Besides the first appearance of color maps grouped at the end of the proceeding, the IVth conference proceeding [5] has published the first micro-macroscopic models of equiaxed dendritic solidification for grain structure predictions. This approach, initially developed in Stefanescu's group and at EPFL, was able to produce grain structure or undercooling maps (see figure 1). It should be pointed out that Oldfield pioneered such micro-macro approaches for cast iron solidification more than twenty years earlier [18]. Interestingly, a steel hammer casting was used for a Round Robin session to test and validate several heat flow simulations performed by different groups. This test was certainly key to demonstrate the importance of materials properties, initial and boundary conditions for the accuracy of temperature predictions, a prerequisite for any accurate microstructure modeling.

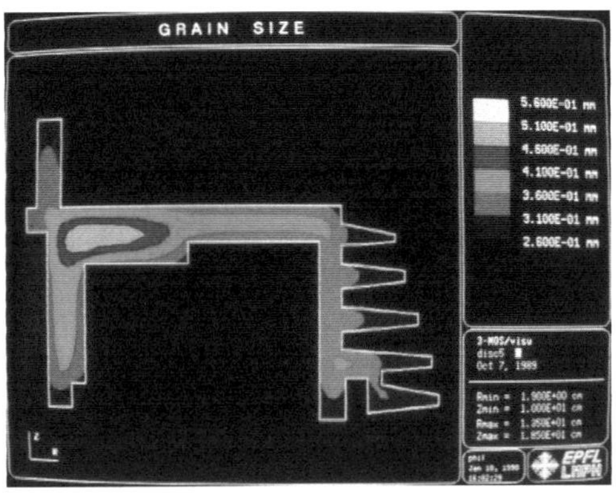

Figure 1. Grain-radius map of equiaxed grains in a 2-Dimensional (2-D) section of an axisymmetric Al-Si brake cylinder, as calculated with a micro-macro approach. As there were no color printer at that time, this is a screen photograph taken with an argentic camera, thus explaining the distortion of the frame. (From the work of Desbiolles *et al.* presented in [5]).

In 1990, the Vth conference [6] certainly has set up a landmark in microstructure and defect modeling with a full session dedicated to this topic, with a keynote delivered by G. Lesoult from Nancy. Organized for the first time outside the USA in Davos, Switzerland, the conference was renamed with its present label: "*Modeling of Casting, Welding and Advanced Solidification Processes*", or *MCWASP* in short. It was also the first reporting in MCWASP proceedings of the much more refined micro-macro models of Beckermann and co-workers, using a two-phase formalism approach, also developed at that time by Bennon and Incropera [19] and Voller and Prakash [20]. The cover page of this proceeding, reproduced from the communication of Brown and Spittle, shows four cross sections of Al-Cu sand castings demonstrating nicely the influence of the Cu-composition on the CET (figure 2(a)). These authors were developing a 2-D stochastic model of equiaxed and columnar grain growth based on a Monte-Carlo-type approach (figure 2(b)). Although this method developed for grain growth in the solid state does not encapsulate the physics of solidification, it really paved the direction to go, i.e. produce simulated grain structures or microstructures which can be directly compared with metallographic cross sections or in-situ solidification observations.

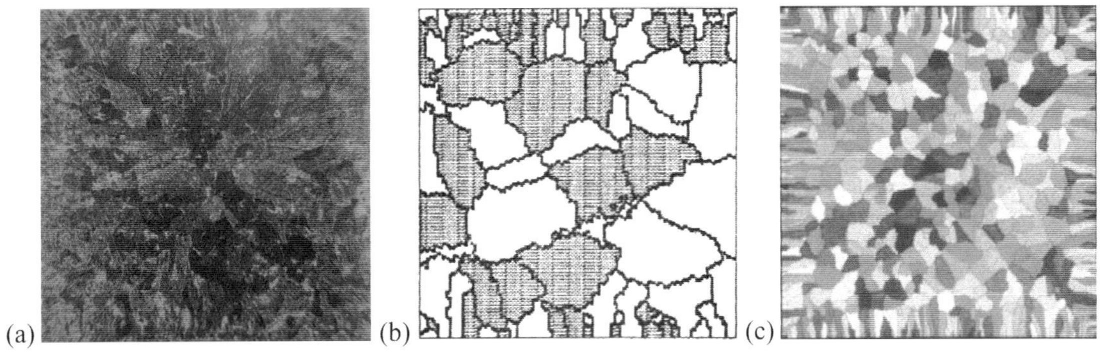

Figure 2. (a) Sand-casting of Al-0.75wt,%Cu and (b) first 2-D stochastic model based on a Monte-Carlo method for grain structure predictions in a square ingot (from Brown and Spittle in [6]). (c) First 2-D Cellular Automaton (CA) prediction for a similar configuration (from Gandin *et al.* in [7]).

3. From MCWASP VI to MCWASP VIII

Clearly for the VIth MCWASP conference [7], the novelties in the field of microstructure modeling were the appearance for the first time of two new methods, namely Phase Field (PF) and Cellular Automata (CA), which have become now standard techniques in the field. The history of the PF method development can be found in the recent review of W. Kurz *et al.* [21], including the first 1978 notes of J. Langer outlining the bases of this technique. R. Kobayashi was the first to compute a full 3-D thermal dendrite (or seaweed when there is no anisotropy) using the PF method [22]. The method was introduced in the MCWASP community by W. Boettinger and co-workers in [7] (see figure 3). A few years later, Warren and Boettinger extended the PF method to solutal dendrites [23] (see also in [8]) and the number of PF contributions has risen sharply since then (e.g. a dozen publications related to PF in [15]).

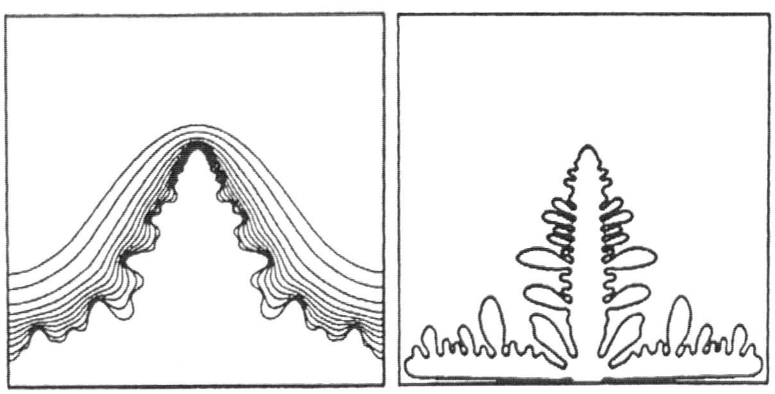

Figure 3. First 2-D phase-field simulation reported in MCWASP proceedings [7] by Boettinger *et al.* The left and right figures show the temperature and the phase-field maps, respectively.

Cellular Automata, first developed for an isothermal volume (see figure 2(c)), was initially referred to as a "stochastic approach" of grain structure modeling, even though the only stochastic aspects are related to the location and crystallographic orientation of heterogeneous nuclei. Once this is known in a volume (or surface) of a casting, the capture of CA cells is deterministic, based on an analytical dendrite tip (and arms) growth kinetics law. The CA technique was then coupled with Finite Elements (FE) heat flow calculations, leading to the so-called CAFE model, the principle of which, like PF, was later applied to other materials science problems (e.g. [24] for fracture). Applications made with the 2-D and 3-D versions of CA can be found in [8] and [9], respectively. Although the CAFE model does not consider the solutal interactions at the detailed level of dendrite tips/arms (i.e. hard- instead of soft-impingement), it has become an industrial tool to predict grain structures in casting and welding processes.

In parallel to PF and CAFE, substantial progresses were made in analytical micro-macro modeling to predict mushy zone evolution and associated macrosegregation. For predicting freckles formation (e.g. Combeau and Lesoult, Poirier and Heinrich in [7]), the solid is assumed to be fixed while the Navier-Stokes equation for the liquid phase is modified to include a Darcy-type term describing the drag effect exerted by the solid already formed in the mushy zone. During this period, Beckerman, keynote speaker in [7], released the assumption of a fixed solid and developed two-phase approaches which account for different velocities of the liquid and solid phases, with proper interactions between the phases derived from averaging over a representative volume element. This allows to include the effect of grain sedimentation or floatation in macrosegregation predictions.

It should also be pointed out that inverse modeling, first appearing in [6] and introduced in the field by Beck and Zabaras, was applied by Dantzig and co-workers to the optimization of solidification conditions in [7], and of the shape of the Round Robin hammer casting in [8]. In the same MCWASP VII proceeding of the conference held in London, a benchmark for mold filling calculations was proposed by Campbell based on a comparison with in situ X-ray radiographies.

MCWASP VIII established a new, and yet today unsurpassed record: that of the thickest proceeding (1267 pages)! It was dedicated to K. Brimacombe, who passed away one year before in 1997. Thanks to the progresses made over past MCWASP conferences for cast iron modeling, a full session with a keynote by Svensson was organized on this topic. In the same way, a session focused on "Direct microstructure simulation" included a keynote by Steinbach, while most contributions were based on PF. Adaptative grids, massively parallel PF computations start to appear (e.g. Provatas *et al.*), together with multi-PF approaches in particular for eutectic simulation (e.g. Seesselberg and Tiaden) as well as the coupling between PF and fluid flow computations (e.g. Diepers *et al.*, Tong *et al.*, figure 4(a)).

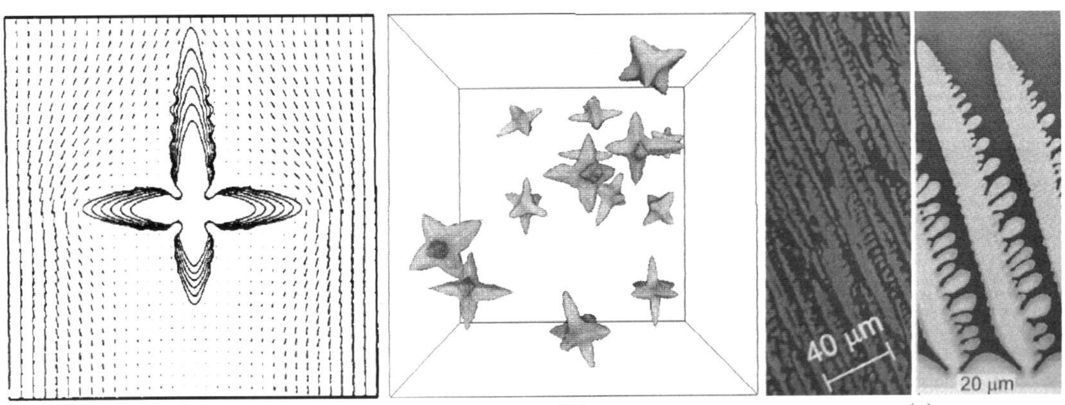

Figure 4. (a) 2D PF simulation of an equiaxed dendrite in a forced flow [25]. (b) 14 equiaxed grains randomly distributed with a random orientation, as predicted from a mesoscopic model considering thermal interactions [26]. (c) Metallographic section of dendrites during e-beam remelting of Al-Si and modified-CA predictions (from Dilthey and Pavlik in [9]).

A new mesoscopic model based on PF was developed by Steinbach *et al.* for a thermal dendrite: it considers the dendrite envelope without the full details of the internal arms structure to speed up the computations, while a confocal envelope at some fixed distance is used to compute the undercooling (figure 4(b)). The CA model is modified by Dilthey and Pavlik to include solute diffusion around growing dendrites, with an estimation of the curvature based on the local fraction of solid distribution (figure 4(c)). Although these so-called "modified-CAFE" are usually not validated and are thus semi-quantitative, they provide an interesting view of the internal dendritic structure of the grains at a scale which, at that time, could hardly be reached with PF.

4. From MCWASP IX to MCWASP XIV

At the turn of the millennium, MCWASP IX [10] organized in Aachen is more oriented toward industry, with interesting keynotes setting one step out of the field: one on meteorology and climate research modeling by Kaas, and another one on discrete particle modeling by Tsuji. There is no major innovation in the field of microstructure modeling, except for the introduction of an orientation-parameter in PF by Warren *et al.* to predict the impingement of a family of grains and the extension of the method to multiphases and multicomponents, sometimes coupled with thermodynamic databases, by Nestler *et al.* or Steinbach *et al.*

A full session dedicated to PF is organized in 2003 at MCWASP X [11]. One of the key actors in this area, A. Karma, had developed quantitative methods with the introduction of an anti-trapping term to compensate for the artificially increased thickness of the diffuse interface [27]. A full session is devoted also to CA and modified-CA applications.

But more importantly in MCWASP X [11], the first in-situ radiographies of directional solidification of Al-Cu thin specimens are reported by Mathiesen and Arnberg, with unprecedent resolution thanks to the use of X-ray synchrotron radiation (figure 5(a)). In the same proceeding, Bernard *et al.* use X-ray tomography to deduce the 3-D morphology of eutectic regions in Al-Cu specimens, and calculate with this information and solving Stokes equation the permeability of these alloys just above the eutectic temperature.

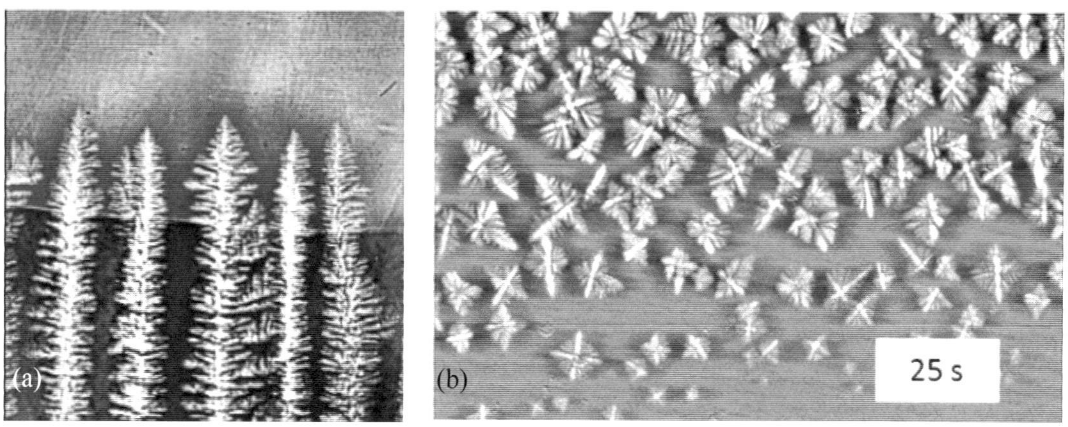

Figure 5. (a) First high-resolution X-ray radiography reported in MCWASP by Mathiesen and Arnberg for directionally solidified Al-30wt.%Cu in [11]. (b) X-ray radiography of nucleation and growth in Al-15%Cu inoculated with Al5Ti1B in a small gradient (Prasad *et al.* in [15]).

In situ X-ray observations (radiography, tomography, topography) see such a fantastic development that a full session devoted to this topic is organized in the following MCWASP conference organized in Opio-France [12]. Among the main contributions, one can mention the tomography observation of pores formation by Salvo *et al.*, radiography of the CET in Al-Ni by Reinhart *et al.*, dendrite fragmentation by Mathiesen *et al.*, dendrite extension past a re-entrant corner by Yasuda *et al.* The PF method continues to be more and more extensively used in 3-D situations (Puztai *et al.*), for eutectic growth (Parisi and Plapp, figure 6(a)), for ternary eutectics (Böttger *et al.*, Nestler *et al.*) and for Mg alloys (Eiken *et al.*).

In MCWASP XII, held on a cruise boat leaving Vancouver for Alaska [13], and even more in MCWASP XIII, held in Schladming-Austria [14], in situ X-ray observation techniques are increasingly used. At the microscale, tomography is used to characterize the 3-D shape formation of pores and intermetallics by Lee *et al.* in [13], while Yasuda *et al.* in [14] use radiography to observe the γ-δ transition in Fe-C. In situ radiography is also used at a larger scale, e.g. for flow imaging by Reilly *et al.* in [13], for freckles formation in Ga-In by Shevchenko *et al.* or for particle pushing by Kareh *et al.* in [14]. This last group also presents a Discrete Element Method (DEM) to explain their observations

Micro-macro modeling including the transport of grains in [13] and of (grains + inoculant particles) in [14] is further developed by the group of Založnik and Combeau. Using a granular approach, Phillion *et al.* in [13] have extended to 3-D the solidification model for randomly oriented globular grains previously reported in MCWASP XI [12] by Vernède *et al.* This granular approach was completed by Sistaninia *et al.* (in [14]) in order to ultimately model in situ hot tearing tomography observations performed by Terzi *et al.* [28, 29]. In the same proceeding, the same experiment was modelled by Zaragoci *et al.* using a level-set method.

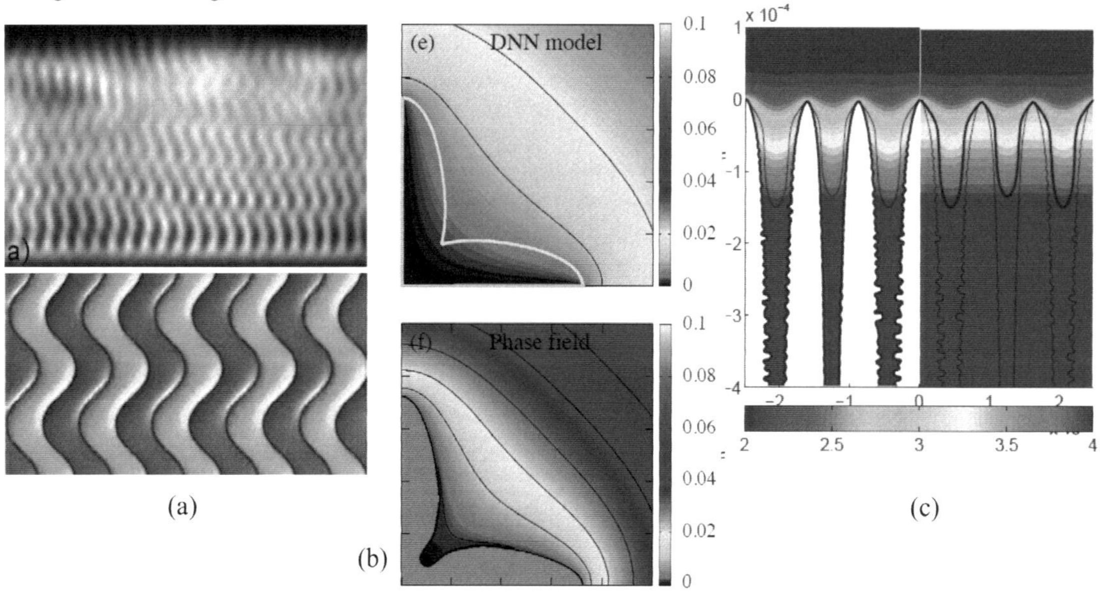

Figure 6. (a) Stable zig-zag structure observed with a long-distance microscope in CBr_4-C_2Cl_6 (top) and 3-D PF simulations (bottom) by Parisi *et al.* in [12] (the top and bottom figures are taken from [30] and [31] respectively). (b) Comparison of DNN and PF calculations for an equiaxed dendrite showing the solutal field by Tourret and Karma in [14]. (c) Comparison of the solutal fields between PF (left) and mesoscopic model (right) predictions for columnar dendrites by Založnik *et al.* in [15].

While PF modeling continues to be applied to various situations, e.g. twinned dendrites by Salgado *et al.* in [13], Dendrite Orientation Transition (DOT) observed in Al-Zn alloys by Friedli *et al.* in [13] and [14], MCWASP XIII sees the first appearance of the Dendrite Needle Network model of Tourret and Karma (figure 6(b)), a model that, like the mesoscopic model, bridges the gap between PF and CAFE. As the name indicates, the DNN model replaces a dendritic network of primary, secondary and higher other branches by sharp needles that interact through solutal diffusion fields. The tip velocity of each needle is determined by combining a standard solvability condition and an additional solutal flux balance made over the dendrite tip region. During the same conference, Wu *et al.* present for the first time the coupling of a CA model with a fluid flow model based on the Lattice Boltzmann Method (LBM) for the prediction of dendrite and bubble formation during solidification. Figure 7 is a good illustration of the evolution of the CAFE model over the past twenty years (to be compared with figure 2(c)): the 3-D CA model is coupled with FE solution of heat, mass, momentum and solute transport in order to predict the evolution of the grain structure and macrosegregation for a benchmark experiment developed in Grenoble (from Carozzani *et al.* in [14]).

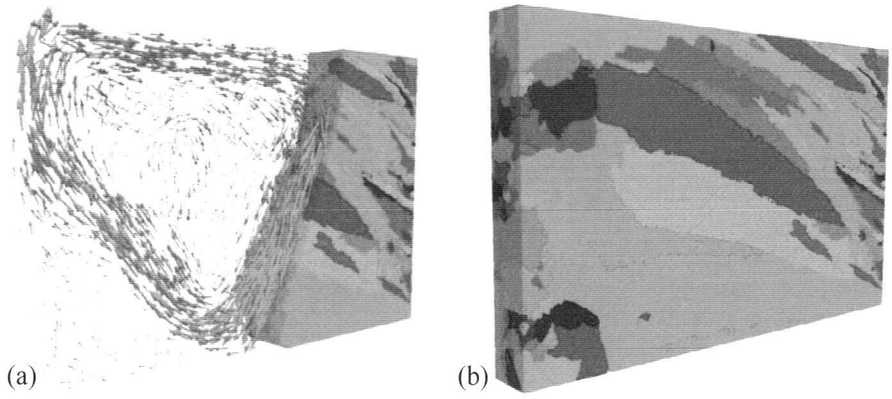

Figure 7. Simulation of a macrosegregation benchmark experiment for Sn - 3%Pb, using a 3-D CA coupled with FE solution of heat-and-mass transfers, (a) at t = 3000 s when the cavity is partially liquid and (b) t = 5000 s after the alloy has solidified. From Carozzani *et al.* in [14].

It is interesting to point out that the proceeding of MCWASP XIII [14] is for the first time published as an open source publication of IOP in a bundle format (paper and electronic version). The proceeding of the following MCWASP conference held in Awaji-Japan [15] was also published as an open source but only under an electronic format, demonstrating the evolution of the publication modes. This last conference honoured two key members of the MCWASP community: Itsuo Ohnaka, who very sadly died in 2016, only six months after giving an invited talk at Awaji " *How to solve complex problems in foundry plants - future of casting simulation*", and Jon Dantzig, who was chairman of the second conference with John Berry.

MCWASP XIV confirms the evolution seen over the two previous conferences, in particular large scale computations, and combined in situ X-ray observations and simulations [15]. Takaki's group uses massively parallel PF computations on GPUs to simulate dendrite growth, without convection (figure 8) and with convection using LBM. Adaptive meshing techniques are used for PF simulations performed by Bollada *et al.* or by Guo and Xiong. The effect of Sr on the solidification of Al-Si is modelled in 3-D by Eiken and Apel, while Akamatsu *et al.* used both PF and the boundary-integral method to investigate the effect of the anisotropy of α-β interfaces during eutectic growth. The DNN model is extended to 3-D and compared with in situ experiments on Al-Cu by Tourret *et al.* The mesoscopic model is compared with PF by Založnik *et al.* (figure 6(c)), while Mondoux *et al.* analyze the percolation in 3-D of a family of smooth granular grains. No doubt that many of these contributions will be extended and included in the present MCWASP proceeding [16].

Figure 8. 3-D PF simulations showing the competition of dendrites at a converging grain boundary in directional solidification by Sakane *et al.* in [15]. The grain on the left has a ⟨100⟩ direction well oriented with the vertical thermal gradient G, while that on the right is misoriented by 30 deg and is slightly more undercooled due to a larger velocity.

5. Moore's law evolution of MCWASP conferences

The development of numerical models over the past forty years was only possible thanks to the fantastic evolution of computers' power. To show that, it is sufficient to consider the first workstation that we bought at EPFL in 1985 to develop the micro-macro model producing the result of figure 1. This Apollo DN 550 workstation had a performance of 0.3 Mips, with 1.5 Mbytes of central memory and 75 Mbytes of disc! Yet, its cost was 135'000 CHF (about the same amount in US$ at present conversion rate). In 2020, the performance-over-price ratio of computers is about 10^8 to 10^9 better as compared with the DN 550, while Moore's law over 35 years gives a factor 10^7!

In 2002, Voller and Porté-Agel [32] made an interesting count of the number of computation nodes reported in the proceedings of the MCWASP conference series since the first edition (see figure 9). They concluded that the mean number of nodes was evolving with an exponential law: $1000 \times 2^{0.67(\text{year-1980})}$, as a function of the year following the first conference (held in 1980, proceedings published in 1981 [1]). In other words, the number of nodes was doubled every 1.5 years, exactly like Moore's law for the computers power evolution.

Celebrating the 40th anniversary of the MCWASP conference series, we should see in 2020 computations with about 100 billion nodes if this law is correct! Taking the previous proceeding [15], the largest PF simulations reported by Takaki's group are 1.6×10^9 nodes (for figure 8), while the extrapolation of Voller and Porté-Agel predicts 10.6×10^9 nodes. However, in their 2013 publication [33], the same group reports simulations involving $4'000^3$ nodes, a value 16 times larger than the extrapolated Moore's law of figure 9. Therefore, without making further detailed accounting of the number of nodes, we can quite confidently conclude that the "size" of the microstructure simulations continues to follow the evolution of computers power, thanks in particular to the use of massively parallel calculations performed on GPUs.

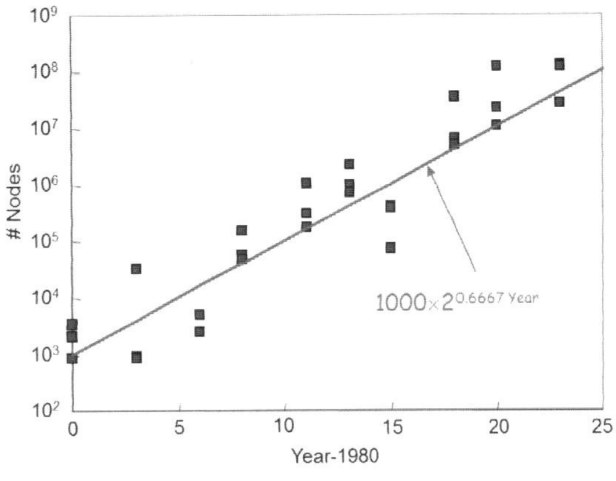

Figure 9. Number of nodes reported in the first 10 MCWASP proceedings as a function of the years following the first conference. The figure has been redrawn from the original analysis made by Voller and Porté-Agel [32]. The linear fit in this log-lin scale shows that the "size" of the simulations follows Moore's law, i.e. the number of nodes in the largest simulations reported in MCWASP conferences doubles every 1.5 years.

6. Conclusion

The past forty years have been very exciting from the point of view of microstructure modeling, thanks to the development of new approaches and methods as well as to the fantastic evolution of computation power. However, the complexity of casting and welding processes are such that their modeling requires multiple approaches in order to address the ten orders of magnitude length scales involved. In between ab initio calculations at the atomistic scale and FE/FV solutions of average conservation equations at a component scale, PF, CAFE, DNN, granular models or mesoscopic approaches allow to predict microstructures, defects and grain structures at intermediate scales. In parallel to modeling, new experimental tools have become available, in particular in situ X-ray observations techniques, but also EBSD, EDX, atom probe, FIB, STEM, etc., sometimes used for 3-D characterizations.

Still, many materials properties and behaviors are key in obtaining valuable modeling results and/or understanding some of the microstructure/defects forming during solidification. Among those: solid-liquid interfacial properties, grain boundary energies, diffusion coefficients, thermal diffusivity, attachment kinetics contribution, high temperature behavior of the solid, partial ordering of the liquid phase, etc. Atomistic calculations will probably contribute increasingly in the future to calculate some of these properties.

Finally, as stated by the mathematician S. Karlin (1924-2007), "*The purpose of a [mathematical] model is not to fit the data, but to sharpen the questions*". In 2020, year of the XVth MCWASP conference, well-controlled experiments, analysis tools and modeling at all scales can be used jointly to sharpen many remaining questions in the field of solidification.

References

[1] *Modeling of Casting and Welding Processes* 1981, ed H D Brody and D Apelian (The Metallurgical Society of AIME, Warrendale, PA USA)
[2] Murray W D and Landis F 1959 Numerical and machine solutions of transient heat-conduction problems involving melting or freezing *Trans. ASME* C **81** 106
[3] *Modeling of Casting and Welding Processes II* 1983, ed J A Dantzig and J T Berry (The Metallurgical Society of AIME, Warrendale, PA USA)
[4] *Modeling and Control of Casting and Welding Processes* 1986, ed S Kou and R Mehrabian (The Metallurgical Society, Warrendale, PA USA)
[5] *Modeling and Control of Casting and Welding Processes IV* 1988, ed A F Giamei and G J Abbaschian (The Minerals, Metals and Materials Society, Warrendale, PA USA)
[6] *Modeling of Casting, Welding and Advanced Solidification Processes V* 1990, ed M Rappaz, M R Özgü and K W Mahin (The Minerals, Metals and Materials Society, Warrendale, PA USA)
[7] *Modeling of Casting, Welding and Advanced Solidification Processes VI* 1993, ed T S Piwonka, V Voller and L Katgerman, (The Minerals, Metals and Materials Society, Warrendale, PA USA)
[8] *Modeling of Casting, Welding and Advanced Solidification Processes VII* 1995, ed M Cross and J Campbell (The Minerals, Metals and Materials Society, Warrendale, PA USA)
[9] *Modeling of Casting, Welding and Advanced Solidification Processes VIII* 1998, ed B G Thomas and C Beckermann (The Minerals, Metals and Materials Society, Warrendale, PA USA)
[10] *Modeling of Casting, Welding and Advanced Solidification Processes IX* 2000, ed P R Sahm, P N Hansen and J G Conley (The Minerals, Metals and Materials Society, Warrendale, PA USA)
[11] *Modeling of Casting, Welding and Advanced Solidification Processes X* 2003 ed D M Stefanescu J A Warren M R Jolly and M J M Krane (The Minerals Metals and Materials Society Warrendale, PA USA)
[12] *Modeling of Casting, Welding and Advanced Solidification Processes XI* 2006, ed C A Gandin and M Bellet (The Minerals, Metals and Materials Society, Warrendale, PA USA)
[13] *Modeling of Casting, Welding and Advanced Solidification Processes XII* 2009, ed S L Cockcroft and D M Maijer (The Minerals, Metals and Materials Society, Warrendale, PA USA)

[14] *Modeling of Casting, Welding and Advanced Solidification Processes XIII* 2012, ed A Ludwig, M Wu and A Kharicha (IOP Publishing, Bristol, UK)
[15] *Modeling of Casting, Welding and Advanced Solidification Processes XIV* 2015, ed H Yasuda (IOP Publishing, Bristol, UK, open access https://iopscience.iop.org/article/10.1088/1757-899X/84/1/011001)
[16] *Modeling of Casting, Welding and Advanced Solidification Processes XV* 2020, eds J Olofsson, A Dahle and J Hattel (IOP Publishing, Bristol, UK, open access, this proceeding)
[17] Hunt J 1984 Steady state columnar and equiaxed growth of dendrites and eutectics *Mater. Sci. Eng.* **65** 75
[18] Oldfield W 1966 A quantitative approach to casting solidification - Freezing of cast iron *ASM Trans* **59** 945.
[19] Bennon W D and Incropera F P 1987 A continuum model for momentum, heat and species transport in binary solid liquid-phase change systems *Int. J. Heat Mass Tran.* **30** 2161
[20] Voller V R, Brent A D and Prakash C 1989 The modelling of heat, mass and solute transport in solidification systems *Int. J. Heat Mass Tran.* **32** 1719
[21] Kurz W, Fisher D J Rohit Trivedi R 2019 Progress in modelling solidification microstructures in metals and alloys: dendrites and cells from 1700 to 2000 *Int. Mater. Rev.* **64** 311-54
[22] Kobayashi R 1994 A numerical approach to three-dimensional dendritic solidification *Exp. Math.* **3** 59
[23] Warren J A and Boettinger W J 1995 Prediction of dendritic growth and microsegregation patterns in a binary alloy using the phase-field method *Acta Metal. Mater.* **43** 689
[24] Shterenlikht A and Howard I C 2006 The CAFE model of fracture—application to a TMCR steel, in *Fatigue Fract. Eng. M.* **29** 770
[25] Tong X, Beckermann C, Karma A and Li Q 2001 Phase-field simulations of dendritic crystal growth in a forced flow *Phys. Rev.* **63** 061601
[26] Steinbach I, Beckermann C, Kauerauf B, Li Qand Guo J 1999 Three-dimensional modeling of equiaxed dendritic growth on a mesoscopic scale *Acta Mater.* **47** 971
[27] Karma A 2001 Phase-Field Formulation for Quantitative Modeling of Alloy Solidification *Phys. Rev. Lett.* **87** 115701
[28] Terzi S, Salvo L, Suéry M *et al.* 2009 In situ x-ray tomography observation of inhomogeneous deformation in semi-solid aluminium alloys *Scripta Mater.* **61** 449
[29] Sistaninia M, Terzi S, Phillion A B, Drezet J-M and Rappaz M 2013 3-D granular modeling and in situ X-ray tomographic imaging: A comparative study of hot tearing formation and semi-solid deformation in Al–Cu alloys *Acta Mater.* **61** 3831
[30] Akamatsu S, Bottin-Rousseau S and Faivre G 2004 Experimental evidence for a zigzag bifurcation in bulk lamellar eutectic growth *Phys. Rev. Lett.* **94** 175701
[31] Parisi A and Plapp M 2008 Stability of lamellar eutectic growth *Acta Mater.* **56** 1348
[32] Voller V R and Porté-Agel D 2002 Moore's Law and Numerical Modeling *J. Comput. Phys.* **179** 698
[33] Takaki T, Shimokawabe T, Ohno M, Yamanaka A and Aoki T 2013 Unexpected selection of growing dendrites by very-large-scale phase-field simulation *J. Cryst. Growth* **382** 21

IOP Conf. Series: Materials Science and Engineering **861** (2020) 012002 doi:10.1088/1757-899X/861/1/012002

Comparing mesoscopic models for dendritic growth

D Tourret[1], L Sturz[2], A Viardin[2] and M Založnik[3]
[1] IMDEA Materials Institute, Getafe, Madrid, Spain
[2] Access e.V., Aachen, Germany
[3] Université de Lorraine, CNRS, IJL, F-54000 Nancy, France

E-mail: damien.tourret@imdea.org

Abstract. We present a quantitative benchmark of multiscale models for dendritic growth simulations. We focus on approaches based on phase-field, dendritic needle network, and grain envelope dynamics. As a first step, we focus on isothermal growth of an equiaxed grain in a supersaturated liquid in three dimensions. A quantitative phase-field formulation for solidification of a dilute binary alloy is used as the reference benchmark. We study the effect of numerical and modeling parameters in both needle-based and envelope-based approaches, in terms of their capacity to quantitatively reproduce phase-field reference results. In light of this benchmark, we discuss the capabilities and limitations of each approach in quantitatively and efficiently predicting transient and steady states of dendritic growth. We identify parameters that yield a good compromise between accuracy and computational efficiency in both needle-based and envelope-based models. We expect that these results will guide further developments and utilization of these models, and ultimately pave the way to a quantitative bridging of the dendrite tip scale with that of entire experiments and solidification processes.

1. Introduction

In metallic alloys obtained by solidification processing, dendritic microstructures are common [1]. From a fundamental standpoint, dendritic growth theory and modeling stands as a challenge to combine phenomena across a wide range of scales, from microscopic capillarity at dendritic tips to macroscopic transport of heat and solute in the melt [2,3]. Due to this multiscale aspect, numerous approaches have emerged over the years that aim at bridging length scales in solidification modeling [4-6]. Each of these approaches operates within a different range of scales, which makes them valuable tools from a technological innovation perspective, since a key hurdle on the way to effective ICME (Integrated Computational Materials Engineering) implementations relies upon our ability to couple models at different length scales [7,8]. Yet, most modeling approaches to dendritic growth have thus far been developed separately, with little effort to compare them and to discuss their capabilities and limitations on a quantitative basis.

In this article, we focus on models of solidification that operate from the microscopic to an intermediate, or mesoscopic, scale between that of the microscopic solid-liquid interface pattern and that of the macroscopic solidification process. At this intermediate scale, interactions between grains occur that determine microstructural features such as grain size, shape, morphology, and crystal texture. Mesoscopic models are more recent than their microscopic and macroscopic counterparts, such that a rigorous quantitative assessment of their relative advantages and limitations remains to be performed. We also focus on moderate solute supersaturation, $\Omega \leq 0.25$, which is a common regime for most processes that do not involve rapid solidification.

Here, we focus on three prominent models for dendritic crystal growth — namely phase-field, dendritic needle network, and mesoscopic grain envelope models. Using a set of benchmark simulations for isothermal equiaxed growth in a binary alloy at given undercoolings (i.e. at different solute supersaturations), we compare the predicted growth velocities of the dendrite tips in both steady and transient growth regimes. The objective of this comparison is to highlight the advantages and limitations of the different methods, and to suggest how to select model and numerical parameters in order to achieve quantitative predictions while retaining profitable computational efficiency. While this study is still ongoing, the preliminary results presented here already provide useful insight into the capabilities of the models and the choice of parameters.

2. Models

2.1. Phase-field

The phase-field (PF) method is a powerful tool for the resolution of free-boundary problems. The method can handle moving interfaces of arbitrary morphological complexity. Interfaces and boundaries are not tracked explicitly, but instead different phases or grains are represented by a "phase field" that varies continuously through an interface of finite width W (figure 1(a)).

The free energy of the system is formulated, typically using a Ginzburg-Landau functional [9,10], which describes the energetic contribution of interfaces, transformations, and local state (e.g. temperature, concentration, stresses, etc.) [11-16]. The time evolution of the fields derives directly from the functional derivatives of the constructed free energy functional, using the so-called Cahn-Hilliard equation [17] for a conserved field, such as the solute concentration, or a time-dependent Ginzburg-Landau (also referred to as Allen-Cahn [18]) equation for non-conserved fields. A number of PF models have been proposed, including some that consider a grand potential functional in place of a free energy description [19,20], but the basic recipe typically follows these same common traits for every phase-field model. Hence, what makes the distinction between different PF models is essentially the choice of the number of phase fields (or order parameters) necessary to describe the system and the transformation(s), the mathematical/physical formulation of the functional, the interpolation functions used for different properties between different phases, and the matching of the model parameters to physical parameters relevant to the phenomena one aims to describe.

In the context of solidification, the PF method has been particularly successful, in part due to the presence of well-defined sharp interface problems to be matched by the PF formulations [12,14]. In particular, "thin-interface" asymptotic analyses have permitted using a diffuse interface width W much higher than the actual width of the solid-liquid interface [21,22]. The introduction of a so-called "anti-trapping current" has allowed quantitative simulations of alloys by eliminating spurious numerical solute trapping for wide interfaces [22,23]. In the model considered here, the temperature is assumed homogeneous and constant in time, solid-state diffusion and kinetic undercooling are neglected, a binary alloy is considered within its dilute limit where its phase diagram can be linearized with a constant interface solute partition coefficient k, and solute transport is limited to diffusion in the liquid with a constant diffusivity D. These assumptions result in the following evolution equations [21-24]

$$a_s(\mathbf{n})^2\frac{\partial\psi}{\partial t} = \nabla\cdot[a_s(\mathbf{n})^2\nabla\psi] + \varphi\sqrt{2} - \lambda\sqrt{2}\,(1-\varphi^2)U + \sum_{m=x,y,z}\left[\partial_m\left(|\nabla\psi|^2 a_s(\mathbf{n})\,\frac{\partial a_s(\mathbf{n})}{\partial(\partial_m\psi)}\right)\right] \quad (1)$$

$$[1+k-(1-k)\,\varphi]\frac{\partial U}{\partial t} = D'\,\nabla\cdot[(1-\varphi)\nabla U] + [1+(1-k)U)]\,\frac{(1-\varphi^2)}{\sqrt{2}}\frac{\partial\psi}{\partial t} + \nabla\cdot\left\{[1+(1-k)U]\,\frac{(1-\varphi^2)}{2}\frac{\partial\psi}{\partial t}\frac{\nabla\psi}{|\nabla\psi|}\right\}, \quad (2)$$

where the dimensionless solute field is

$$U = \frac{1}{1-k}\left[\frac{c/c_l^0}{(1-\varphi)/2 + k\,(1+\varphi)/2} - 1\right], \quad (3)$$

with c the solute concentration field, c_l^0 the equilibrium solute concentration of a flat interface at the considered temperature T_0. In these equations, space is scaled with the diffuse interface width W and time with the relaxation time τ_0 at T_0. Non-dimensional values of the liquid diffusion coefficient $D' = D\tau_0/W^2 = a_1a_2W/d_0$ and the coupling factor $\lambda = a_1W/d_0$, with $a_1 = 5\sqrt{2}/8$, and $a_2 = 47/75$ from thin-interface asymptotic analysis [21]. We use the standard cubic symmetry of the surface tension with $a_s(\mathbf{n}) = (1 - 3\varepsilon_4) + 4\varepsilon_4 (n_x^4 + n_y^4 + n_z^4)$, where n_x, n_y, and n_z are the components of the unit vector $\mathbf{n}$ normal to the interface, and ε_4 is the strength of the surface tension anisotropy. The classical phase field φ, with $\varphi = +1$ in the solid and $\varphi = -1$ in the liquid phase, was replaced by a preconditioned phase field ψ, defined by $\varphi = \tanh(\psi/\sqrt{2})$. This change of variable stabilizes the numerical resolution of the equations, thus allowing the use of a coarser grid spacing with negligible loss in accuracy [25]. The model is solved using finite differences on a homogeneous grid of cubic elements and an explicit Euler time stepping scheme. The code is written in cuda (Compute Unified Device Architecture) programming language, which allows its acceleration through multi-threading on Nvidia graphics processing units (GPUs).

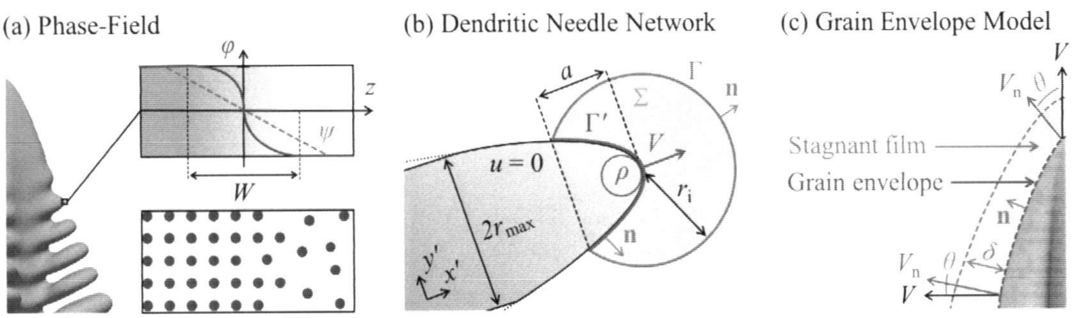

Figure 1. Schematic illustration of the phase field (a), dendritic needle network (b), and envelope (c) models, illustrating characteristic model parameters, namely the PF diffuse interface width W, the DNN integration radius r_i and bounding radius r_{max}, and the envelope stagnant-film thickness δ.

2.2. Dendritic Needle Network

The DNN approach aims at bridging the scale of the dendrite tip radius ρ and the larger scale of transport phenomena, e.g. the solutal diffusion length $l_D = D/V$, with V the crystal growth velocity. The model, rigorously derived in the limit of small Péclet numbers, $P = \rho/(2l_D) \ll 1$, represents a dendritic crystal as a network of thin needles that approximate the primary stems and higher order sidebranches of the grain. In earlier descriptions of the model, in two dimensions, the needle-like branches were treated as infinitely sharp [26]. In later versions of the model, branches were given a parabolic shape, which lead to the same governing equations, while also allowing a three-dimensional extension of the model [27].

The DNN model was shown to quantitatively reproduce primary dendritic spacings measured in directional solidification experiments in different Al-based alloys [27-29]. Implementations have been developed that include fluid flow in the liquid [30,31] and the interaction between nucleation and growth [32,33].

The evolution of the network during solidification is described by the growth dynamics of each needle tip and the generation of new sidebranches. For a binary alloy, typical assumptions of the model include a linear phase diagram with an interface partition coefficient k, negligible fluid flow in the liquid phase (at least in the vicinity of the dendrite tips [30]), no solute transport in the solid phase, and negligible release of latent heat during solidification. Under these conditions, for an isothermal domain at a given solute supersaturation Ω, in three dimensions, the needle tip radius, ρ, and its velocity, V, are calculated by combining the following set of equations [27]:

a solvability condition at the needle tip scale $\approx \rho$

$$\rho^2 V = 2Dd_0/\sigma \, , \tag{4}$$

a solute conservation at intermediate scale much larger than ρ and much smaller than l_D

$$\rho V = D\mathcal{F} \, , \tag{5}$$

and solute diffusion in the bulk liquid at a scale $\approx l_D$

$$\partial_t u = D\nabla^2 u \, , \tag{6}$$

with

$$u = \frac{c_l^0 - c}{(1-k)\, c_l^0} \tag{7}$$

the normalized solute concentration, c_l^0 the liquid equilibrium concentration and d_0 the chemical capillary length, both at the fixed temperature T_0, and σ the tip selection parameter. The flux intensity factor $\mathcal{F}$ is a measure of the solute flux towards the needle in the vicinity of its tip (figure 1(b)), and captures intra- and intergranular solutal interaction. It is defined as

$$\mathcal{F} = \frac{1}{2\pi a} \int_{\Gamma'} \frac{\partial u}{\partial n} \, \mathrm{d}\Gamma' \approx \frac{1}{2\pi a} \left(\int_{\Gamma} \frac{\partial u}{\partial n} \, \mathrm{d}\Gamma + \frac{V}{D} \int_{\Sigma} \frac{\partial u}{\partial x'} \, \mathrm{d}\Sigma \right), \tag{8}$$

where a is a chosen integration distance behind the tip, typically a few tip radii, and Γ' is the resulting surface of the paraboloid up to a distance a from the tip (figure 1(b)). In equation (8), the last equality assumes that the field u is Laplacian in a moving frame of velocity V, i.e. $D\nabla^2 u = -V\,\partial u/\partial x'$, at the scale of the integration domain, with x' the principal growth direction of the branch. This assumption allows calculating the normal solute flux along the surface Γ′ using the normal solute flux through any surface Γ that intersects the parabolic tip at a length a behind the tip, with an enclosed volume Σ (figure 1(b)) [27]. In earlier model implementations [26-30], the outer surface Γ was set as a square or a cuboid for their ease of implementation onto a square numerical grid. In later works, a circular or spherical shape of the integration domain is preferred [31,32], which facilitates the implementation of grains of arbitrary orientations regardless of the numerical grid.

With c_∞ the alloy nominal solute concentration, the supersaturation $\Omega = (c_l^0 - c_\infty)/[(1-k)c_l^0]$ serves as boundary condition far from the interface, where $u \rightarrow \Omega$, while thermodynamic equilibrium neglecting capillary correction is applied along the dendritic branches, with $u = 0$.

We solve the equations numerically on a 3D finite difference mesh with a homogeneous spacing Δx, using an explicit time stepping scheme with a constant time step Δt. At each time step, we integrate the value of $\mathcal{F}(t)$ using equation (8) with a spherical integration domain of radius r_i centered on the tip of the needle [30], which provides the value of the tip velocity $V(t)$ and radius $\rho(t)$ from equations (4) and (5).

Far behind its tip, we bound the maximum thickness of a branch to a cylinder of radius r_{max} (figure 1(b)). While it does not affect the paraboloid shape in the tip vicinity, this allows a tip to slow down and stop with $V \rightarrow 0$ without numerical instabilities that would otherwise be induced by the constancy of $\rho^2 V$. Sidebranches normal to the growth direction (here assuming ⟨100⟩ growth directions) are generated at positions $\delta = (N_{br}+\delta_{br})\times\rho_s$ behind the tip in all four normal directions, when the tip has advanced by a distance $N_{br}\times\rho_s$ since the last branching event. A random fluctuation δ_{br} is generated for every branching event within a range of amplitude Δ_{br}. The average distance between sidebranches N_{br} and its fluctuation Δ_{br}, both expressed in units of the theoretical steady-state tip radius ρ_s, are user input parameters. Both truncating and branching methods are similar to those presented and discussed in previous works [26-32].

2.3. Mesoscopic Grain Envelope Model

The Mesoscopic Grain Envelope Model (GEM) [34,35] bridges the scales from the dendrite tip, over the internal grain structure, to the mesoscopic scale of an ensemble of grains that interact by diffusion and convection of solute. The core idea of the GEM is that solutal interactions at the mesoscopic scale can be accurately simulated without a detailed representation of the ramified structure of the solid-liquid interface. Instead, the description of a dendritic grain can be simplified and represented by its envelope, an artificial smooth surface that links the tips of the actively growing dendrite branches (figure 1(c)). Individual branches are not represented by the model, although, by definition, the growth of the envelope depends on the growth speed of the branch tips. The growth of the tips is controlled by the solute flux that they reject into the liquid and is therefore determined by the local solute concentration of the surrounding liquid. In the envelope model a constitutive model of tip kinetics is used that gives the tip growth speed as a function of the solute concentration field adjacent to the envelope.

The model was shown to be capable of quantitatively predicting grain envelope shapes, growth velocities, and internal solid fraction for a single grain growing into an infinite melt [34,36,37] and for transient interactions among several grains [35,38]. The envelope model has been successfully used for simulations of *in situ* synchrotron X-ray imaging experiments [38-40], for modeling spacing adjustments and growth competition in columnar growth [37,41], and for characterizing grain shapes and growth kinetics under the influence of solutal interactions used in upscaling to macroscopic models [42]. The model was also extended to include fluid flow and solute advection [40,43].

Two key assumptions are made: (i) that the phenomena controlling the growth of a dendrite tip are universal and are therefore valid for tips of any order (primary, secondary, tertiary,...) and (ii) that the characteristic time of the interactions at the small (tip) scale is much smaller than that of the interactions at the large (grain) scale. In this work, we further assume an isothermal domain at a given solute supersaturation, Ω, in three dimensions, and we formulate the model as follows. The normalized concentration at the tip interface is that of thermodynamic equilibrium, neglecting capillary corrections: $u = 0$. A stagnant-film formulation of the Ivantsov solution is written that relates the normalized concentration difference, u_δ, between the liquid at the tip and that at a finite distance δ from the tip [44] to the growth Péclet number of the tip, $P = \rho V/(2D)$, as

$$u_\delta = P \exp(P) \left\{ \mathrm{E}_1(P) - \mathrm{E}_1\left(P\left[1 + \frac{2\delta}{\rho} \right] \right) \right\}. \tag{9}$$

Equation (9) is used to solve for P. The concentration u_δ is obtained from the concentration field in the liquid around the envelope, which is resolved numerically as explained below. The tip selection criterion $\rho^2 V = 2Dd_0/\sigma$, is a supplementary relation that gives the tip radius, $\rho = d_0/(\sigma P)$, needed in Equation (9), and the tip speed, $V = 2\sigma P^2 D/d_0$. Dendrite tips are assumed to grow in predefined growth directions. In the present study, the cubic crystal dendrites are given six possible growth directions perpendicular to one another. The normal envelope growth velocity, V_n, is calculated from the local tip speed, V, by the relation $\mathbf{V}_\mathrm{n} = V\,\mathbf{n}\cos(\theta)$, where θ is the angle between the outward normal to the envelope, $\mathbf{n}$, and the preferential growth direction that forms the smallest angle with $\mathbf{n}$ (figure 1(c)). To track the envelope on a numerical grid we use a phase-field-like interface tracking method [45], combined with a surface reconstruction method for better accuracy [36]. Combining these methods, one can avoid explicit tracking of the envelope, which is defined throughout the whole domain with a level set (iso-surface) of a continuous indicator field. Details are given in references [36,45].

Solute transport at the mesoscopic scale, here by diffusion, is described by volume-averaged equations that are valid in the whole domain, both inside and outside the envelopes. Solidification inside the envelope uses the Gulliver-Scheil assumptions of thermodynamic equilibrium at the solid-liquid interface, negligible diffusion in the solid, and instantaneous diffusion in the liquid [46]. The resulting liquid fraction field, g_l, in the interior of the envelopes, was shown to realistically represent the distribution of solid and liquid within an actual dendritic grain envelope [34,36,37]. The liquid inside the envelope is at the normalized equilibrium concentration, $u_l = 0$. These assumptions lead to the conservation equation for the solute in the liquid phase

$$g_l \partial_t u_l = D\nabla \cdot (g_l \nabla u_l) + \partial_t g_l \,. \tag{10}$$

The solution of equation (10) gives the liquid concentration u_l outside the envelope and g_l inside the envelope. Outside the envelope, $g_l = 1$ and equation (10) reduces to a single-phase diffusion equation. Inside the envelope, $u_l = 0$ is the known equilibrium concentration and equation (10) gives the evolution of the liquid fraction inside the envelope. The concentration of the solid growing inside the envelope at constant Ω is $u_s = 1$.

The supersaturation across the stagnant film, u_δ, needed to calculate the envelope growth velocity (equation (9) and additional relations), is obtained from the concentration field in the liquid around the envelope, which is fully resolved numerically (equation (10)). This means that the locally valid Ivantsov solution for diffusion around a dendrite tip is matched to the solution of the mesoscopically valid solute transport equation at a distance δ from the envelope (figure 1(c)).

The solute conservation equation (10) and the equation of the evolution of the indicator field used for envelope tracking are solved by a finite volume solver with the OpenFOAM® toolbox. Second and higher order discretization schemes are used for the spatial operators. Implicit Euler time-stepping is used for the solute conservation equations and explicit time-stepping for the envelope indicator field equation. Adaptive time-stepping is employed and is useful during fast transients, such as the initial transient after nucleation. During other growth stages the time step is essentially limited by the characteristic diffusion time, $\Delta x^2/D$, at the scale of the mesh cell, Δx. In the present work uniform meshes with cubic cells are used. The mesh size needs to be adapted for an accurate description of the diffusion layer adjacent to the envelope, i.e. $\Delta x \lesssim 0.2D/V$, where V is a typical primary tip speed. A special search and interpolation algorithm is designed to provide the concentration at the distance δ from the envelope, needed for the calculation of the envelope velocity [36].

3. Benchmarks

Here, we compare the predictions of the three methods described above with a similar benchmark test case for undercooled isothermal equiaxed growth of a binary alloy. The phase field simulation uses well-converged numerical parameters, i.e. small enough grid spacing and diffuse interface width W, such that the PF results will be used as a reference for the discussion of the effect and choice of parameters in the other two methods.

We set a solid seed (about as small as the method allows) in the corner of a cubic domain initialized at a given solute supersaturation Ω. Thus, we simulate the growth of 1/8 of an equiaxed grain, using no-flux (mirror symmetry) Neumann boundary conditions on every face of the cube. We choose a value of the domain size L (edge length of the cube) that is large enough to reach (or closely approach) a steady growth velocity, but also small enough to make simulations computationally achievable within a few days (less than two weeks for the longest) using quantitative phase-field using the current single-GPU-parallelized implementation. Hence, we selected $L/l_D \approx 2$, 4, 6, 8, and 10 for $\Omega = 0.05$, 0.10, 0.15, 0.20, and 0.25, respectively.

In the PF simulations, we use a small interface energy anisotropy $\varepsilon_4 = 0.007$ and a solute partition coefficient $k = 0.1$. In the DNN and GEM simulation we consider a tip selection constant $\sigma = 0.0264$. This value was estimated from preliminary PF simulations of isothermal growth for $0.05 \leq \Omega \leq 0.25$ in a domain with 512^3 grid points with tip velocities and radii measured as soon as the diffusion field gets affected at the side of the domain opposite to the dendrite. This value agrees with results from joint directional solidification simulation with similar ε_4 and k. We noted a typical uncertainty of the order of 10 to 15% depending upon the procedure chosen to extract the dendrite tip radius [47]. Here, we follow the method described in reference [48] (figure 5 therein), i.e. by 2D cross-section least-square fitting to a second order parabola over a fitting range that spans a distance of one tip radius behind the tip. A deeper analysis in terms of selection parameter is underway, such that the value used here only provides a reasonable estimation of σ to use consistently in DNN and GEM simulations.

The objective is to compare predicted growth velocities, not only in the steady-state, which should closely match the Ivantsov solution, but also and most importantly the initial transient, as the velocity tends towards the free dendrite steady-state, and the final transient, as the velocity decreases to zero when the dendrite approaches the no-flux boundary wall.

For a given solute supersaturation Ω, the theoretical Péclet number $P_s = \rho_s V_s/(2D)$ at steady-state (hence marked with a subscript "s") is given by the Ivantsov solution [49]. It directly gives the ratio between the steady-state diffusion length D/V_s and dendrite tip radius ρ_s. In this paper, most results are presented scaled with respect to the steady-state tip radius ρ_s and velocity V_s. For each value of Ω considered here, table 1 summarizes the correspondence between these theoretical steady-state values and material parameters in physical units, such as the liquid solute diffusivity D and the interface capillarity length d_0. Within the selected supersaturation range, there is an order of magnitude of variation in the theoretical steady-state values of the Péclet and tip radius, and two orders of magnitude of variation of the tip velocity.

In the phase-field simulations, the finite difference grid element size is set equal to the diffuse interface width with $\Delta x = W \approx \rho_s/10$, which was found to yield well-resolved converged results. The initial spherical seed has a radius $r_0 = 2.5\Delta x$, with a preconditioned phase field $\psi = r - r_0$ with $r = (x^2 + y^2 + z^2)^{1/2}$ and a solute field $U = -\Omega$ everywhere, which is equivalent to setting $c = c_\infty$ (i.e. $u = \Omega$) in the liquid and $c = kc_\infty$ in the solid.

In the DNN simulations, we initialize a solid seed with $u = 0$ in the corner of the domain and set $u = \Omega$ in the liquid. The seed consists of three quarters of paraboloids along the three normal grid directions with initial tip velocity $V = 0$, radius $\rho = \rho_s$, and length L_0 larger than r_i and r_{max}. In particular, we study the effect of mesh spacing, Δx, and truncation radius, r_{max}. During the final transient, the integration sphere may be truncated if it exceeds the limits of the domain.

In the GEM simulations, we initialize a spherical envelope seed with $u = 0$ and $g_l = 1 - \Omega$ in the corner of the domain and set $u = \Omega$ in the liquid. The size of the envelope seed is given by the smallest radius of curvature that can be accurately represented by the envelope tracking method on a mesh with spacing Δx [36,45]: $r_0 = 4.46\Delta x$. The mesh spacing and time step needed for well-resolved converged results were determined by systematic convergence studies. Remaining mesh dependence, due to the initial seed size, which can be reduced on finer grids, is perceptible only during the initial growth transient and is discussed in section 3.2.

Table 1. Theoretical steady-state Péclet number P_s, tip radius ρ_s, and velocity V_s, for different Ω.

Ω	$P_s = \dfrac{\rho_s V_s}{2D}$	$\dfrac{\rho_s}{d_0} = \dfrac{1}{\sigma P_s}$	$\dfrac{V_s d_0}{D} = 2\sigma P_s^2$
0.05	0.0131	2895	$9.03 \cdot 10^{-6}$
0.10	0.0341	1111	$6.14 \cdot 10^{-5}$
0.15	0.0624	607	$2.06 \cdot 10^{-4}$
0.20	0.0988	383	$5.15 \cdot 10^{-4}$
0.25	0.1446	261	$1.10 \cdot 10^{-3}$

4. Results and discussions

The results presented here were obtained using model and numerical parameters that gave a good fit to the time evolution of the growth velocities obtained from PF simulations, while conserving a substantially lower computational cost. Since we used a wide range of computers (e.g. CPU/GPU) and methods (e.g. explicit/implicit time stepping), we did not perform a systematic performance analysis and focused on the potential of each method to yield quantitative predictions. Yet, in the following subsection, we provide values used for grid spacing and time steps, which provide a consistent picture of the relative computational workload to each method.

4.1. Steady-state growth

First, we compare steady-state growth predictions by all models for the range of solute supersaturation $0.05 \le \Omega \le 0.25$. Steady-state growth Péclet numbers predicted by PF, DNN, and GEM, are compared to the theoretical Ivantsov solution in figure 2. Values reported in figure 2 are averaged in time during steady-state growth, in order to integrate numerical oscillations. Such oscillation may occur as a dendrite tip crosses successive grid points and are due to the coarse grids used in DNN and GEM. The fluctuations in tip velocity and radius (see e.g. figure 4 discussed later) lead to a negligible fluctuation in Péclet number (within the symbol size used in figure 2). In summary, all models are capable of solving numerically the Ivantsov problem, and they thus appropriately predict the steady growth state.

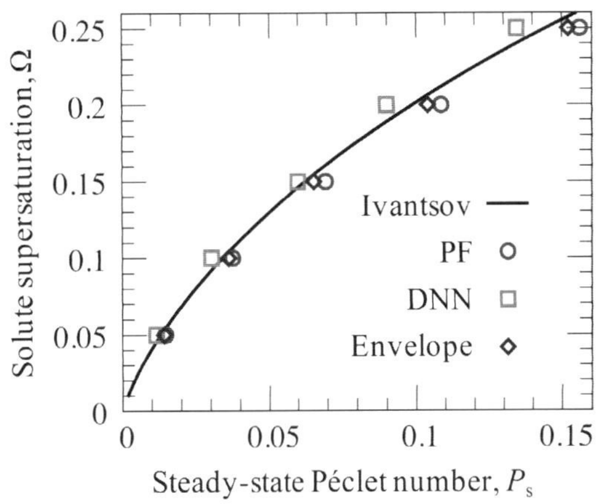

Figure 2. Predicted steady-state Péclet number for different imposed supersaturation Ω compared to the analytical Ivantsov solution [49].

Model parameters used for these simulations are summarized in table 2. In order to provide a consistent illustration of the computational cost for all three methods, we also summarize the numerical parameters, namely the grid element size and time step, for each Ω in table 3.

Table 2. Model parameters used with the three methods.

	PF	DNN		GEM
Ω	W/d_0	r_i/ρ_s	r_{max}/ρ_s	$\delta/(D/V_s)$
0.05	289	2.6	3, 10	0.06
0.10	110	2.4	3	0.20
0.15	60	1.6	3	0.40
0.20	38	1.9	3	0.60
0.25	26	0.6	2	0.80

Table 3. Normalized mesh element size and time step used with the three methods.

Ω	$\Delta x/(D/V_s)$			$\Delta t/(D/V_s^2)$		
	PF	DNN	GEM	PF	DNN	GEM
0.05	0.0026	0.0065	0.02	10^{-6}	$7\cdot10^{-6}$	10^{-4}
0.10	0.0068	0.017	0.04	$7\cdot10^{-6}$	$4\cdot10^{-5}$	$4\cdot10^{-4}$
0.15	0.0125	0.031	0.06	$2\cdot10^{-5}$	10^{-4}	$9\cdot10^{-4}$
0.20	0.0198	0.049	0.08	$6\cdot10^{-5}$	$4\cdot10^{-4}$	10^{-3}
0.25	0.0289	0.072	0.10	10^{-4}	$8\cdot10^{-4}$	$2\cdot10^{-3}$

The shape of the predicted grain when the length of the primary trunk reaches one half of the domain size is shown in figure 3 for all three models for $\Omega = 0.05$ (top) and $\Omega = 0.25$ (bottom). For the sake of illustration, the DNN microstructure is shown for a simulation performed without sidebranches and no thickness bounding (i.e. with $r_{max} = \infty$) for $\Omega = 0.05$ and with sidebranches and $r_{max} = 2\rho_s$ for $\Omega = 0.25$. In the latter case, the effect of a finite r_{max} becomes apparent from the cylindrical shape of the branches. While at this stage the velocity has already started decreasing due to the effect of boundaries, the shape of the microstructures is barely distinguishable from that at steady-state, for all three models.

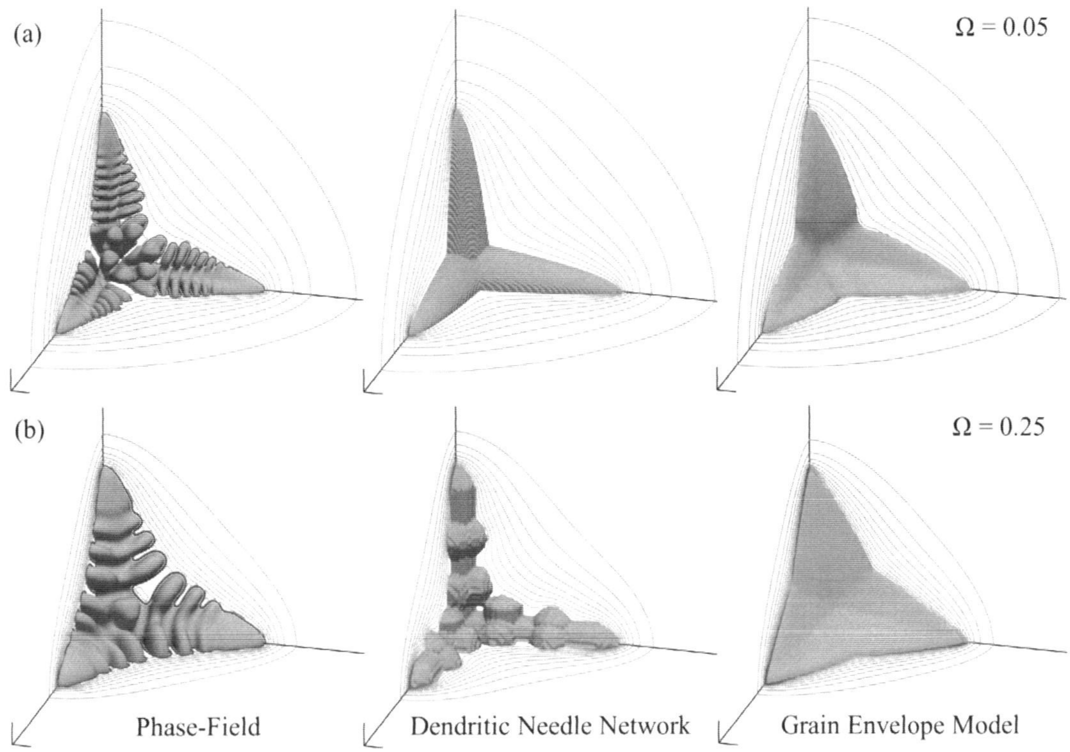

Figure 3. Grain shape (gray) and iso-concentration lines, from $u = 0$ (red) to $u = 0.9\Omega$ (blue) with steps of 0.1Ω, along the ($x = 0$), ($y = 0$), and ($z = 0$) planes when the length of the primary dendritic trunk reaches one half of the simulation domain for $\Omega = 0.05$ (a) and 0.25 (b) simulated using phase-field (left), DNN (middle), and Envelope (right) methods. The total size of the domain, illustrated by the visible bottom left corner of each simulation, is $2\,D/V_s \approx 221400\,d_0$ for $\Omega = 0.05$ and $10\,D/V_s \approx 9060\,d_0$ for $\Omega = 0.25$.

4.2. Transient growth

Next, we compare the predicted transient evolution of grain growth velocities. We focus on $\Omega = 0.05$ and $\Omega = 0.25$, which are illustrative of results obtained at intermediate $0.05 < \Omega < 0.25$. The evolution of the velocity from initial to final transient appears in figure 4 for $\Omega = 0.05$ (a) and $\Omega = 0.25$ (b). The velocity is shown as a function of the tip location normalized by the domain length. This way, we filter out the possible different onset times for the final decrease of the velocity — such differences arise when the steady-state is maintained at a different velocity. The right-hand-side panels, showing the evolution of the primary trunk length versus time, illustrate that the oscillations appearing in the left-hand-side velocity plot have a negligible effect on the length of the dendrite, once integrated over time. These results show that, for appropriately chosen sets of parameters, both DNN and GEM models can reproduce PF results with reasonable accuracy.

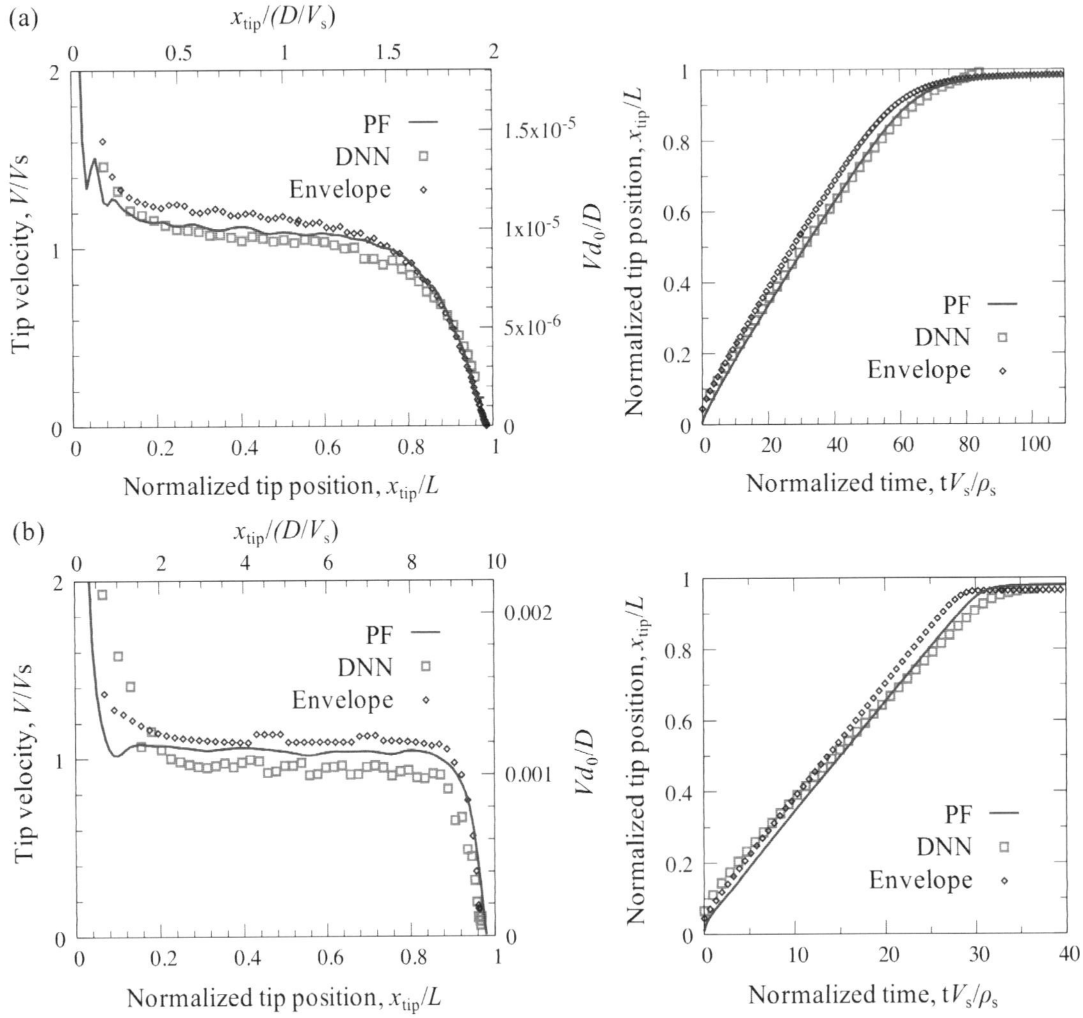

Figure 4. Evolution of the primary dendrite tip velocity from the initial to the final transient state (left) and primary trunk length versus time (right) for $\Omega = 0.05$ (a) and $\Omega = 0.25$ (b). In order to allow comparing models all together at different supersaturations, time and space are scaled with respect to the theoretical steady-state velocity V_s and tip radius ρ_s. The velocity evolution is plot as a function of the tip location normalized by the domain length. Note that left-hand-side velocity plots have been smoothed to reduce numerical oscillations and improve readability (right-hand-side length plots are raw).

DNN simulations were performed with $\Delta x = 0.25\rho_s$, $r_i = 2.6\rho_s$, $r_{max} = 3\rho_s$ for $\Omega = 0.05$, and with $\Delta x = 0.25\rho_s$, $r_i = 0.6\rho_s$, $r_{max} = 2\rho_s$ for $\Omega = 0.25$ (see table 2). We chose a resolution providing a compromise between accuracy and simulation time. The chosen grid size is smaller than the typical value used in DNN simulations, i.e. $\Delta x \approx \rho_s$ or higher. Other parameters are close to typical parameters used in the DNN model, i.e. fulfilling $r_i \ll l_D$ and r_{max} chosen as to not bound the needle thickness within the flux integration domain. Interestingly, we notice very little influence of sidebranching on the velocity of the primary tip, both in steady and transient states. It is also worth noting that the integration radius r_i can be taken lower than the tip radius, as long as the integration domain contains a sufficient number of integration points.

GEM simulations were performed with a stagnant film thickness of $\delta = 0.06\ D/V_s$ and $0.80\ D/V_s$, and with a mesh spacing of $\Delta x = 0.02\ D/V_s$ and $0.10\ D/V_s$, for $\Omega = 0.05$ and 0.25, respectively. The δ and the mesh used at high Ω are close to those already recommended by Souhar *et al.* [36] for steady-state equiaxed growth. At low Ω, significantly smaller δ were needed to accurately match both the initial and the final transient of the tip evolution. The same conclusion was recently made by Olmedilla *et al.* [38] when comparing the GEM to experiments on transient equiaxed growth at comparably low undercoolings. The mesh spacing is dictated by the diffusion length at high Ω ($\Delta x \approx 0.1\ D/V_s$ here), however at low Ω the small stagnant-film thickness, δ, becomes the limiting factor. For an accurate calculation and interpolation of u_δ the mesh spacing should be $\Delta x < 0.5\delta$.

4.3. *Effect of DNN parameters*

Figure 5 shows the tip velocity evolution predicted by DNN simulations for different combinations of grid spacing Δx and needle truncation radius r_{max}, for a flux integration radius $r_i = 2.6 \sim 2.7\ \rho_s$ and $\Omega = 0.05$.

Decreasing the grid spacing Δx for identical values of r_{max} and r_i results mainly in (i) less fluctuation of the velocity and (ii) higher velocities, in better agreement with PF results. On the other hand, the effect of increasing the radius of the stem truncation r_{max} for identical values of Δx and r_i is to decrease the steady-state tip velocity.

The effect of the initial length of the seed branches needs to be discussed. The initial length L_0 of the branches differs among the simulations because the simulations are initialized with paraboloids of length $L_0 = r_{max} + \Delta x$. The difference in initial length most notably affects the growth kinetics during the initial and final transient stages. All four simulations nonetheless stabilize close to the theoretical steady-state velocity given by the Ivantsov solution. This plateau region, however, lasts for a shorter time for the simulations with larger r_{max} and hence longer L_0. At this low Ω, in order for PF simulations to be achievable within a few days, we limited the domain size to $2l_D$. For this reason, the amount of time spent in steady-state is limited, and for high L_0 values, the final transient sets in before steady-state is fully maintained. In the last stage of the final transient, a velocity plateau may be observed (not represented here but visible by interrupted plots) due to a numerical restriction set to the tip velocity when the distance to the wall becomes lower than r_i. This limitation might be accommodated by either (i) truncating the integration domain, (ii) integrating boundary conditions within the integration of $\mathcal{F}$, or (iii) using an adaptive r_i that decreases when approaching a boundary.

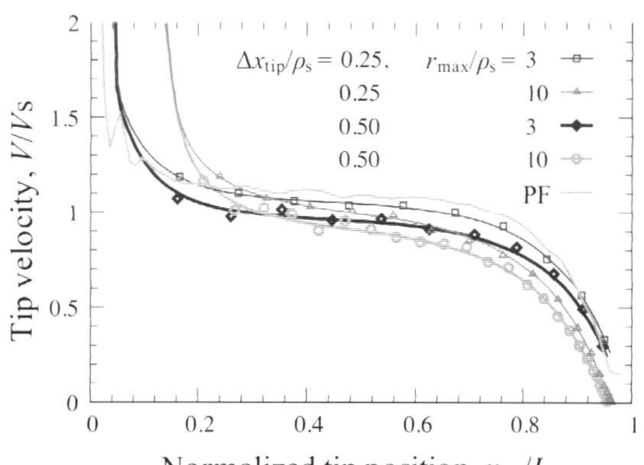

Figure 5. Evolution of the primary dendrite tip velocity for $\Omega = 0.05$ from DNN simulations, showing the effect of grid spacing Δx and stem truncation radius r_{max}. In all runs, the radius of the integration sphere r_i/ρ_s was kept to 2.6 or 2.7. For readability, symbols show raw data whereas lines show smoothed data (filtering numerical oscillations as the tip progresses between successive grid points).

4.4. *Effect of envelope parameters*

The main model parameter of the GEM is the stagnant film thickness, δ. Table 2 summarizes the δ that gave the best match to the tip evolution predicted by phase field. The best δ clearly increases with Ω. The dependence of the transient growth velocity upon the stagnant film thickness δ at $\Omega = 0.05$ is shown in figure 6(a) for a fixed grid spacing $\Delta x/(D/V_s) = 0.02$. The most striking observation is the effect on the steady-state velocity (already pointed out by Souhar *et al.* [36]), which affects the whole evolution. The de- and accelerations during the initial and final transient do not significantly depend on δ.

Figure 6(b) shows the influence of the mesh spacing on the transient tip growth at $\Omega = 0.05$. The influence of the mesh spacing is twofold. In addition to its direct impact on the solution of the PDEs, Δx also limits the smallest radius of the initial nucleus, r_0, and thus the initial conditions (see section 2.4). Typically, $r_0 \geq 4.46\ \Delta x$. The direct impact of Δx, excluding that of r_0, is shown by four simulations in which r_0 was kept fixed and only the mesh spacing was varied from $0.013\ D/V_s$ to $0.04\ D/V_s$. The initial radius was $r_0 = 0.18\ D/V_s$ which is the smallest r_0 that can be resolved on the coarsest of these four grids. The differences between the four solutions are strictly limited to the initial transient, and perfectly overlap at larger arm lengths. The reason for this is that the thin diffusion layer around the grain during the initial stages of growth requires a fine mesh resolution to be accurately resolved. Once the tip decelerates and the diffusion layer grows wider, coarse meshes are sufficient.

However, the most important influence related to the mesh resolution is actually that of the initial nucleus size. This is shown through a comparison of the evolution of tip velocity for $r_0 = 0.18\ D/V_s$ and $0.09\ D/V_s$ with the same $\Delta x = 0.02\ D/V_s$ (open-red versus full-black square symbols in figure 6(b)). We note a much slower deceleration during the initial transient at smaller r_0. Furthermore, the steady growth stage is replaced by a smooth transition to the final transient, which is not significantly affected.

In summary, accurate simulation of transient equiaxed growth requires a stagnant film thickness that roughly scales as $\delta \sim \Omega D/V_s$. This does not contradict, but goes beyond prior results [36] that established a recommendation for steady-state growth: $\delta \sim D/V_s$. If compared, the new recommendation is slightly less accurate in steady-state (with a difference of $< 10\%$) but is more accurate during transients.

The mesh spacing and the initial nucleus radius have a significant influence on the results only during the initial growth transient, but not during steady-state and the final transient. Note that because of the relatively small domain size ($L = 2\ D/V_s$) in the case presented here ($\Omega = 0.05$), the impact of the initial transient on the whole evolution is prominent. This is not the case at higher Ω, where larger domains are used and the initial transient is a smaller portion of the growth. Generally, $\Delta x \sim 0.1D/V_s$ is sufficiently fine for steady-state and for the final growth transient, but should be refined if good accuracy during initial transients is required. Furthermore, $\Delta x < 0.5\delta$ is an additional constraint that is important at low Ω, where smaller $\delta/(D/V_s)$ must be used.

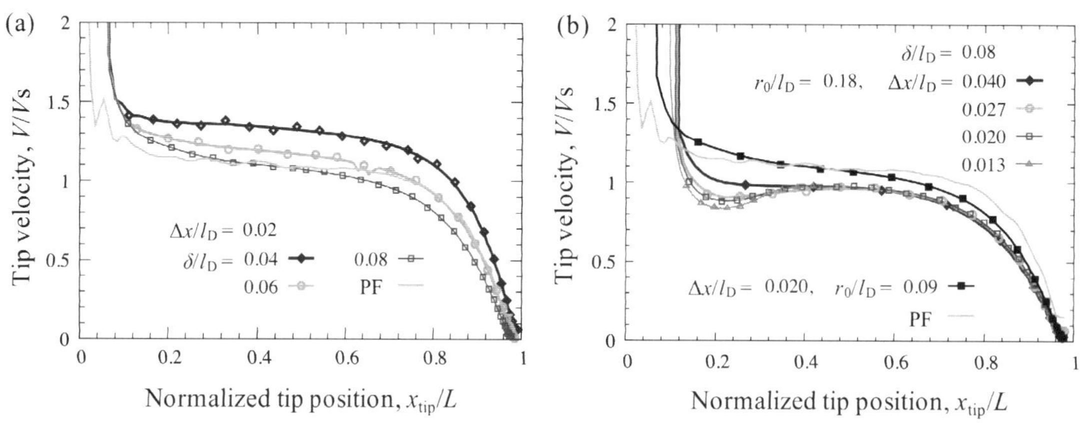

Figure 6. Evolution of the primary dendrite tip velocity for $\Omega = 0.05$ from GEM simulations, showing the effect of the model and numerical parameters. (a) The effect of the stagnant film thickness, δ; all simulations with a mesh $\Delta x/(D/V_s) = 0.02$. (b) The effect of the mesh spacing Δx, and of the size of the initial nucleus, r_0; all simulations with $\delta/(D/V_s) = 0.08$.

5. Summary and perspectives

We have compared the results of phase-field, dendritic needle network, and grain envelope models for a simulation of the equiaxed growth of an isothermally undercooled binary alloy. Using well-converged PF results as the benchmark reference, we discussed the effect and choice of model and numerical parameters in the DNN and GEM methods upon their accurate prediction of steady and transient dendrite growth velocities.

As expected, all models can appropriately predict steady-state growth conditions prescribed by microscopic solvability and Ivantsov solutions. Moreover, in the transient regimes, both GEM and DNN can accurately describe growth velocities, if appropriate model and numerical parameters are selected. The final transients (solutal interactions between grains in practical applications) are typically accurate with standard parameters. The early-stage growth transient, on the other hand, requires a fine tuning of model parameters in order to achieve quantitative predictions, such that the initial seed size and mesh spacing need to be finer than typical values recommended in previous works, in both DNN and GEM approaches.

While the present study provides already some new insight and guidance for the use of these relatively new modeling techniques, these results remain preliminary, and the study is ongoing. Work in progress on the current benchmarks include: (i) the extraction of scaling laws for initial and final transient growth stages, (ii) further systematic exploration of model parameters (e.g. r_i for DNN), (iii) comparison of envelope shapes in equiaxed growth, which is important in the scope of grain interactions, and (iv) exploring the advantages and limitations of models at higher supersaturations. Perspectives for future benchmarks include: (i) benchmarks for directional solidification, in particular in the scope of spacing selection, and (ii) configurations with convection in the liquid phase.

From the results of these studies, we expect that new numerical techniques will improve the predictive capabilities of the mesoscopic models. Such techniques could include, for instance, the introduction of adaptive parameters to ensure better treatment of transient regimes, e.g. using adaptive r_i in DNN or δ in GEM that could depend upon the growth acceleration/deceleration. We also expect that the present work, and those to follow, will provide useful guidance to potential users and developers of these promising tools for bridging length and time scales in the modeling of dendritic solidification.

Acknowledgements

D.T. acknowledges support from the European Union's Horizon 2020 research and innovation programme under a Marie Skłodowska-Curie Individual Fellowship (Grant Agreement 842795). L.S. and A.V. acknowledge support by the German Ministry of Economy via the Deutsche Zentrum für Luft- und Raumfahrt, DLR (Contract 50WM1743). M.Z. acknowledges support by the French State through the program "Investment in the future" operated by the National Research Agency (ANR) and referenced by ANR-11 LABX-0008-01 (LabEx DAMAS).

References

[1] Flemings MC 1974 *Solidification Processing* (New York: McGraw Hill)
[2] Langer JS 1980 *Rev. Mod. Phys.* **52** 1
[3] Trivedi R and Kurz W 1994 *Acta Metall. Mater.* **42** 15
[4] Boettinger WJ, Coriell SR, Greer A, Karma A, Kurz W, Rappaz M and Trivedi R 2000 *Acta Mater.* **48** 43
[5] Asta M, Beckermann C, Karma A, Kurz W, Napolitano R, Plapp M, Purdy G, Rappaz M and Trivedi R 2009 *Acta Mater.* **57** 941
[6] Karma A and Tourret D 2016 *Curr. Opin. Solid St. Mater. Sci.* **20** 25
[7] Allison J 2011 *JOM* **63** 15
[8] Panchal JH, Kalidindi SR and McDowell DL 2013 *Comput. Aided Design* **45** 4
[9] Ginzburg V and Landau L 1950 *Soviet Phys. JETP* **20** 1064
[10] Halperin B, Hohenberg P and Ma S 1974 *Phys. Rev.* B **10** 139

[11] Chen LQ 2002 *Annu. Rev. Mater. Res.* **32** 113
[12] Boettinger WJ, Warren JA, Beckermann C and Karma A 2002 *Annu. Rev. Mater. Res.* **32** 163
[13] Moelans N, Blanpain B and Wollants P 2008 *Calphad* **32** 268
[14] Steinbach I 2009 *Model. Simul. Mater. Sci. Eng.* **17** 073001
[15] Plapp M 2016 *J. Indian Inst. Sci.* **96** 179
[16] Tonks MR and Aagesen LK 2019 *Annu. Rev. Mater. Res.* **49** 79
[17] Cahn JW and Hilliard JE 1958 *J. Chem. Phys.* **28** 258
[18] Allen SM and Cahn JW 1979 *Acta Metall.* **27** 1085
[19] Plapp M 2011 *Phys. Rev.* E **84** 031601
[20] Choudhury A and Nestler B 2012 *Phys. Rev.* E **85** 021602
[21] Karma A and Rappel WJ 1998 *Phys. Rev.* E **57** 4323
[22] Echebarria B, Folch R, Karma A and Plapp M 2004 *Phys. Rev.* E **70** 061604
[23] Karma A 2001 *Phys. Rev. Lett.* **87** 115701
[24] Tourret D and Karma A 2015 *Acta Mater.* **82** 64
[25] Glasner K 2001 *J. Comput. Phys.* **174** 695
[26] Tourret D and Karma A 2013 *Acta Mater.* **61** 6474
[27] Tourret D and Karma A 2016 *Acta Mater.* **120** 240
[28] Tourret D, Clarke AJ, Imhoff SD, Gibbs PJ, Gibbs JW and Karma A 2015 *JOM* **67** 1776
[29] Tourret D, Karma A, Clarke AJ, Gibbs PJ and Imhoff SD 2015 *IOP Conf. Ser.: Mater. Sci. Eng.* **84** 012082
[30] Tourret D, Francois MM and Clarke AJ 2019 *Comput. Mater. Sci.* **162** 206
[31] Isensee T and Tourret D 2020 *IOP Conf. Ser.: Mater. Sci. Eng.* ***current issue***
[32] Sturz L and Theofilatos A 2016 *Acta Mater.* **117** 356
[33] Geslin P and Karma A 2016 *TMS 2016 Annual Meeting, Supplemental Proceedings, Symposium Frontiers in Solidification* 29
[34] Steinbach I, Beckermann C, Kauerauf B, Li Q and Guo J 1999 *Acta Mater.* **47** 971
[35] Steinbach I, Diepers HJ and Beckermann C 2005 *J. Cryst. Growth* **275** 624
[36] Souhar Y, De Felice VF, Beckermann C, Combeau H and Založnik M 2016 *Comput. Mater. Sci.* **112** 304
[37] Viardin A, Založnik M, Souhar Y, Apel M and Combeau H 2017 *Acta Mater.* **122** 386
[38] Olmedilla A, Založnik M and Combeau H 2019 *Acta Mater.* **173** 249
[39] Delaleau P, Beckermann C, Mathiesen RH and Arnberg L 2010 *ISIJ Int.* **50** 1886
[40] Olmedilla A, Založnik M, Cisternas Fernández M, Viardin A and Combeau H 2019 *IOP Conf. Ser.: Mater. Sci. Eng.* **529** 012040
[41] Li Y, Olmedilla A, Založnik M, Zollinger J, Dembinski L and Mathieu A 2019 *Proceedings 5th International Conference on Advances in Solidification Processes* (ICASP-5)
[42] Torabi Rad M, Založnik M, Combeau H and Beckermann C 2019 *Materialia* **5** 100231
[43] Delaleau P 2011 *Mesoscale modeling of dendritic growth during directional solidification of aluminium alloy* Ph.D. thesis NTNU, Trondheim, Norway
[44] Cantor B and Vogel A 1977 *J. Cryst. Growth* **41** 109
[45] Sun Y and Beckermann C 2007 *J. Comput. Phys.* **220** 626
[46] Scheil E 1942 *Z. Metallkd.* **34** 70
[47] Karma A, Lee YH and Plapp M 2000 *Phys. Rev.* E **61** 3996
[48] Clarke A, Tourret D, Song Y, Imhoff S, Gibbs P, Gibbs J, Fezzaa K and Karma A 2017 *Acta Mater.* **129** 203
[49] Ivantsov GP 1947 *Dokl. Akad. Nauk. SSSR* **58** 567

IOP Conf. Series: Materials Science and Engineering **861** (2020) 012003 doi:10.1088/1757-899X/861/1/012003

Reduced order models for welding and solidification processes

P F Mendez

Department of Chemical and Materials Engineering, University of Alberta, Edmonton, AB, Canada

E-mail: pmendez@ualberta.ca

Abstract. This paper reviews the foundations of modelling of materials processing and methodologies of analysis outside the traditional discretization of the domain (e.g. finite elements, finite volume, etc.). It starts with a review of the mathematical framework with a focus on dimensional analysis and characteristic values. The methodology of minimal representation and correction factor is discussed with emphasis on computational methods to obtain the minimal representation and on the methodology of blending to obtain the correction factors. Welding on a thin plate with surface heat losses is used as a case study, and the results obtained are validated against published data for many materials, processes, and researchers.

1. Introduction

Currently, the word "modeling" is interpreted, nearly universally to mean "numerical simulation"; however, there are other modeling techniques that can be better choices in frequent circumstances. Numerical simulations are convenient when the cost of the prototype is high; for example, it is convenient to simulate numerically an airplane design before building a prototype, since a prototype can cost hundreds of millions of dollars, and no simulation is that costly. Numerical simulations are also essential for problems with complex geometries that cannot be reduced to simple shapes. Conversely, design rules of the form of engineering formulae, tables, and graphs are convenient when the cost of the prototype is low, as is often the case of mechanical design, electrical circuit design, and even airplanes and rocket conceptual design.

It is not surprising to realize that welding procedures seldom involve numerical modeling, since trial and error is often less costly than specialized software and personnel. The vast majority of welding procedures are determined empirically and tested to pass codes or standards. The large amount of theoretical knowledge and advanced simulations has had very little reach to the engineers in charge of decisions; instead, experienced engineers seem to arrive to the right process parameters guided by their intuition, which enables them to discard secondary phenomena, and focus only on the dominant factors. It is surprising, though, that in one hundred years of welding, the field has not developed design rules like most other engineering disciplines. Indeed, most of the field of materials processing lags far behind other engineering disciplines in its progress towards design rules. Efforts in the field of materials processing include [1–4]. Relevant efforts in the field of transport phenomena include [5–16]. For welding, current scientific knowledge and simulations seldom reach the practitioners because it is typically costlier than "just try another weld." If this knowledge could be synthesized and generalized in a way amenable to practitioners, its application could be less costly than trial and error, at least in some cases.

Many fields of engineering have their knowledge synthesized and generalized in formulae, tables, and graphs, such as aerodynamics. Materials processing is more difficult to synthesize and generalize because it involves many more physical phenomena than aerodynamics; not only inertial and viscous forces are present, but also conduction, convection, electromagnetic, capillarity, and many more, and they are tightly coupled [17–19]. The mathematical foundation for dealing with multicoupled, multiphysics problems is lacking, and current progress is reviewed in this paper. Because of constraints in space, this paper refers to the appropriate references where derivations, examples, and comprehensive literature surveys are treated in great detail.

2. Structure of governing equations

The mathematical representation of a problem typically involves equations (often differential) capturing the physical phenomena considered relevant. The list of phenomena to include can be long, and judgement is often needed to decide what phenomena to include or not in the governing equations. Including more physical phenomena can increase the complexity of the problem significantly, while omitting a phenomenon, runs the risk of missing an unexpected dominant factor. The methodologies described here are more tolerant than discretization methods to an excess of physical phenomena, reducing the risk of omitting dominant factors.

In the detailed analysis of governing equations, it is important to discriminate the different type of quantities involved. Parameters are quantities that are fixed for a given problem, but vary between two problems of the same type. Independent variables are quantities that can vary freely and independently from each other for a given problem. Dependent variables are functions of both the parameters and independent variables.

As a case study, consider a 2D point heat source moving in steady-state on an infinite domain as shown in figure 1. This problem was first solved by Wilson in 1904 [20], and it was solved independently again by Rosenthal [21–23] in the 1940s, with a focus on welding. Comprehensive reviews of these solutions are in [24–28].

This problem is a good representation of welding under a broad range of practical conditions, often termed the "Rosenthal thin plate solution." It was validated with experiments in [29]. This formulation involves important simplifications: constant thermophysical properties (including no phase change), infinitesimal size of heat source, steady-state in motion in a straight line, infinite length and width of plate, and negligible gradients along the thickness of the plate. The small effect of size of the heat source, thickness, temperature-dependent properties, and melting was verified in [28].

In an Eulerian formulation, the heat source is considered fixed and the substrate moving along the x-axis in the negative direction. The governing equation and boundary conditions for thin plates with surface heat losses is:

$$\frac{\partial^2 T}{\partial x^2}+\frac{\partial^2 T}{\partial y^2}=-\frac{U}{\alpha}\frac{\partial T}{\partial x}+\frac{h+h'}{kd}(T-T_0) \tag{1}$$

$$\frac{\partial T}{\partial r}=-\frac{q}{2\pi r k d} \qquad \text{as } r=\sqrt{x^2+y^2}\to 0 \tag{2}$$

$$T=T_0 \qquad \text{as } r=\sqrt{x^2+y^2}\to \infty \tag{3}$$

In this case, $\{X\}=\{x,y\}$ is the set of independent variables (r is a notation shorthand, not an independent variable), $\{U\}=\{T(x,y)\}$ is the set of dependent variables (only temperature in this case), and $\{P\}=\{U,\alpha,k,q/d,(h+h')/d,T_0\}$ is the set of parameters. Some parameters appear grouped because they do not occur separately in the governing equation. To be rigorous, the notation $T=T(x,y,U,\alpha,k,q/d,(h+h')/d,T_0)$ should be employed; however, most engineers use $T=T(x,y)$, which is displays explicitly only the independent variables and omits the parameters. The exact solution of equation (1) with the boundary conditions of equations (2) and (3) is:

$$T(x,y)=T_0+\frac{q}{2\pi k d}\exp\left(-\frac{Ux}{2\alpha}\right)K_0\left[r\sqrt{\left(\frac{U}{2\alpha}\right)^2+\frac{h+h'}{kd}}\right] \tag{4}$$

IOP Conf. Series: Materials Science and Engineering **861** (2020) 012003 doi:10.1088/1757-899X/861/1/012003

where K_0 is the modified Bessel function of second kind and zero order. The sets of independent and dependent variables are the same in the solution. This equation provides the temperature value for each point in the substrate. The singularity at the origin ($r = 0$) is a consequence of assuming the heat source is concentrated in an infinitesimal area.

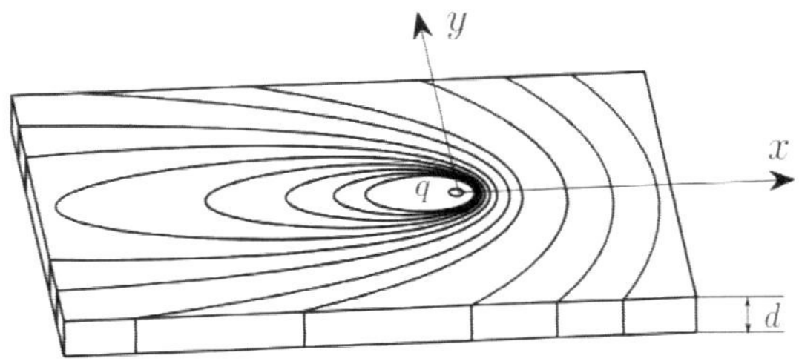

Figure 1. Isotherms for a point heat source of intensity q on a thin substrate of thickness d. The domain is $-\infty < x < \infty, -\infty < y < \infty$ and gradients in z are negligible [49].

The definitions and example above can be generalized to state that the governing equations of a problem (e.g. equations (1)-(3)) have the following form [30]:

$$\sum_j A_j g_j(\{X\},\{U\},\{P\}) = 0 \tag{5}$$

where g_j are functions of the independent variables (and associated differential expressions), the dependent variables, and the parameters. The parameters are defined as positive, adjusting the sign of the associated function g_j when necessary. The coefficients A_j have the form of a power law based only on the parameters:

$$A_j = K_j \prod_k P_k^{a_{jk}} \tag{6}$$

where K_j is a dimensionless numerical constant, but the exponents a_{jk} result in units for A_j. Coefficients that do not have a power-law form can often be reduced to the form of equation (5) [30]. In general terms, the solution to the governing equations is a set of expressions for the dependent variables: $u_j(\{X\},\{P\})$.

2.1. *Characteristic values*

The dependent variables are functions of the independent variables and the parameters. Often in engineering problems, given a set of parameters, the problem in practice is to determine particular values of the dependent variables over the domain, such as a a maximum temperature, minimum width, or average velocity. These values are called “characteristic values.”

An important property of characteristic values is that they do not depend on the independent or dependent variables, they depend only on the parameters: $u_c(\{P\})$. The characteristic values often can be expressed in the form of power laws based on the set of parameters, with exceptions discussed in [30].

Another type of characteristic value is when a variable (independent or dependent) has a value determined by factors outside the model considered. For example, the point heat source of equation (4) often considers a characteristic value of temperature given by phenomena such as melting or phase transformations, unrelated to the considerations on which the governing equations are built. Based on this characteristic temperature, other characteristic values can be defined, such as width of an isotherm. In this case, the externally imposed characteristic value becomes a new parameter. For the case of the moving heat source on a thin plate, the externally imposed characteristic temperature T_c is a new problem parameter.

Figure 2 illustrates relevant characteristic values for moving heat source problems. These characteristic values are associated with the isotherm $T = T_c$, and represent centerline heating and cooling rates, start and end of the isotherm on the centerline, width of the isotherm and lateral temperature gradient, and maximum temperature away from the centerline. Based on these characteristic values, other characteristic values of practical use can be derived immediately, such as $t_{8/5}$, thickness of HAZ, solidification time, aspect ratio of the isotherm, and more [31, 32].

The characteristic values of differential expressions can be approximated using the characteristic values of the original function and basic knowledge of the shape of the function. If $u(x)$ is the original function and we consider the first derivative $v(x) = du/dx$, the exact characteristic value of $v(x)$ in

$$v_{c,e} \approx v_c = \left(\frac{du}{dx}\right)_c = C\frac{\Delta u}{\Delta x} \tag{7}$$

with the values of C tabulated for many relevant cases and shapes of functions in [30].

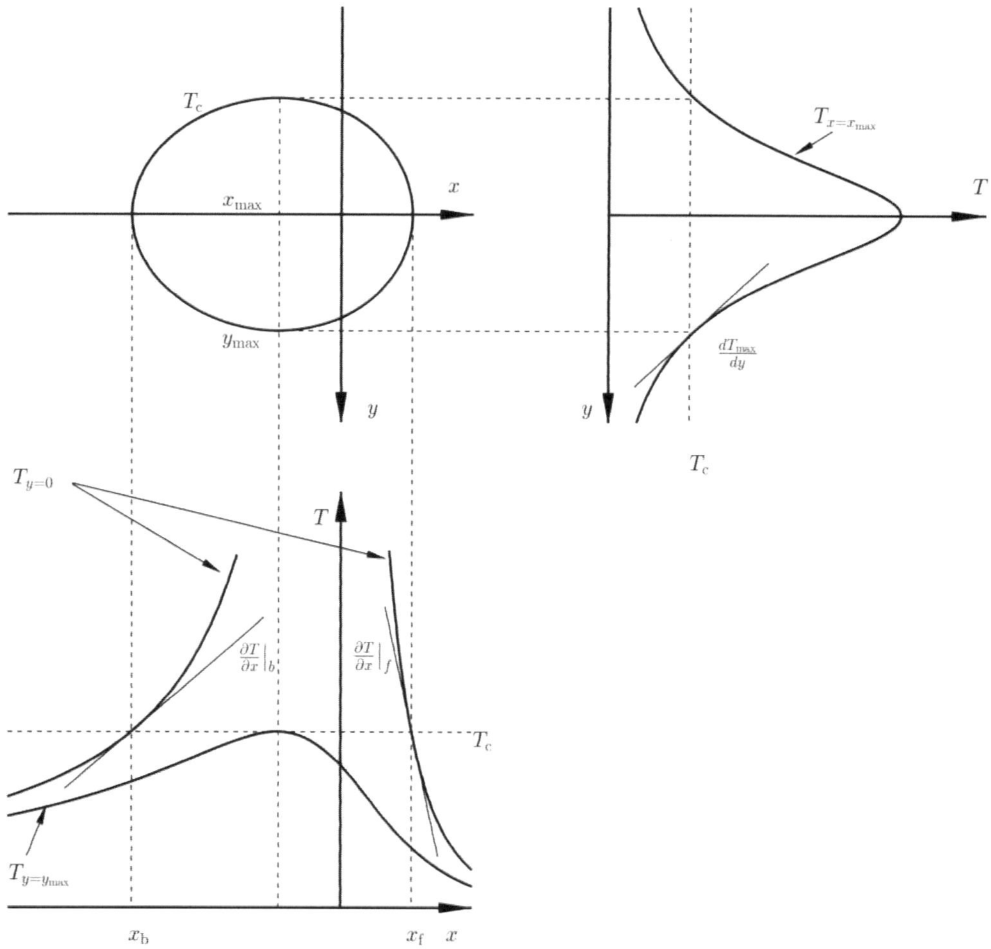

Figure 2. Characteristic values for moving heat sources associated with isotherm $T = T_c$ [31].

2.2. Dimensional analysis

Dimensional analysis states that a problem can be described using dimensionless groups, and those groups are fewer than the number of parameters plus variables of the system. Foundational work includes [33–39].

The most common formulation states that if a system involves n parameters plus variables, and the number of independent units involved is k, then the system can be represented with $m = n - k$ independent dimensionless groups. Independent units cannot be reduced; for example, if a problem involves the units of m, kg, and s, and involves Newton's laws, then N is not an independent unit because it is a combination of the former units. If temperatures always use a scale with an arbitrary zero (e.g. Celsius or Fahrenheit), then $m = n - k - 1$ [40]. In some problems, the number of dimensionless groups can be even smaller (for example when the governing equations do not involve the exchange of energy between mechanical work and heat). Another way to determine the number of independent dimensionless groups is through the normalization of equations.

Because systems of units are based on power laws, the dimensionless groups also have the form of power laws.

For the example of a 2D point moving heat source, the number of independent parameters is 6, with one dependent and two independent variables, for a total of 9 magnitudes with dimensions. A possible reference set of units is $\{R\}$ = {m, kg, s, °C} (k =4). Because all temperatures in this problem have an arbitrary zero, the number of dimensionless groups in this problem is 4, two of them can be associated with independent dimensionless variables (x^*, and y^*), one with the dependent dimensionless variable ($T^*(x^*,y^*)$), and one with the surface heat losses (h^*).

$$x^* = \frac{Ux}{2\alpha} \tag{8}$$

$$y^* = \frac{Uy}{2\alpha} \tag{9}$$

$$T^* = \frac{2\pi kd(T-T_0)}{q} \tag{10}$$

$$h^* = \frac{4\alpha^2(h+h\prime)}{kdU^2} \tag{11}$$

resulting in the following dimensionless equivalent of equation (4):

$$T^*(x^*,y^*) = \exp(-x^*)K_0\left(r^*\sqrt{1+h^*}\right) \tag{12}$$

In general, equation (5) can be expressed in dimensionless form as

$$\sum_j B_j(\{\Pi_P\})h_j(\{X^*\},\{U^*\},\{\Pi_P\}) = 0 \tag{13}$$

where $\{X^*\}$ is the set of the set of dimensionless independent variables, $\{U^*\}$ is the set of the set of dimensionless dependent variables, and $\{\Pi_P\}$ is the set of dimensionless groups that do not include variables. All these dimensionless groups together constitute the set of dimensionless groups $\{\Pi\}$ = $\{\{X^*\},\{U^*\},\{\Pi_P\}\}$. This is the type of normalization used in inspectional analysis. The coefficients B_j have the form of a power law based on $\{\Pi_P\}$:

$$B_j = K_j \prod_k {\Pi_P}_k^{b_{jk}} \tag{14}$$

where K_j and B_j are dimensionless. The solution of equation (13) is the set $\{U^*\}$, made by functions

$$u_j^*(\{X^*\},\{\Pi_P\}) \tag{15}$$

Because characteristic values do not depend on the variables, only on the parameters, a dimensionless characteristic value will have the functional dependence only on the dimensionless groups that depend on the parameters:

$$u_{c_j}^*(\{\Pi_P\}) \tag{16}$$

It is important to keep in mind that not always there is a one-to-one correspondence between variables and their dimensionless counterpart. In cases of self-similarity, the number of dimensionless independent variables is smaller than the number of independent variables with units.

2.3. Normalization of functions

Normalization of functions is a key concept in the scaling of differential equations; its ultimate goal is to capture the order of magnitude of functional expressions, thus allowing for efficient approximations. This goal is accomplished by considering only the characteristic values of the variables without knowing the exact (or numerically approximated) solution to the governing equations over the whole domain. The approach of considering only the characteristic values is in contrast with the integration of the differential equations (e.g. finite elements or finite differences methods), which involves *all* (or very many) values of the variables.

Similarly to dimensional analysis, normalization involves the generation of dimensionless functions and dimensionless groups; however, the dimensionless entities generated by normalization are designed to capture the order of magnitude of terms in the governing equations, which is not the case for dimensional analysis.

Using the concept of characteristic value, the function g_j in equation (5) can be normalized as:

$$g_j(\{X\},\{U\},\{P\}) = (g_j)_c(\{P\})(g_j)^{\circ}(\{X^*\},\{U^*\},\{\Pi_P\}) \qquad (17)$$

where the notation (g) indicates a dimensionless magnitude obtained by normalization by a characteristic value. The characteristic values $(g_j)_c$ are typically obtained from equation (7), and the whole methodology is described in detail in [30]. If the variables are also normalized by their characteristic values, the general expression of a differential equation with normalized functions is:

$$\sum_j A_j (g_j)_c(\{P\})(g_j)^{\circ}(\{X^{\circ}\},\{U^{\circ}\},\{\Pi_P\}) = 0 \qquad (18)$$

Equation (18) is similar to equation (13), except that $(g_j)_c$ are not necessarily power laws of the parameters. Another similarity is that each term (or group of terms) represents a physical phenomenon.

Identifying physical phenomena through normalization is convenient, since there are typically far fewer physical phenomena parameters. A problem with a large number of parameters often confuses practitioners, yet experts typically know to focus on the phenomena.

An important difference with equation (13) is that when the characteristic value is chosen as a maximum value in the domain, at the point $\{X_c^{\circ}\}$ where the maximum occurs, then $(g_j)^{\circ} = O(1)$. When the point $\{X_c^{\circ}\}$ in the domain corresponds to the maximum of a dominant term, equation (18) can be turned into an algebraic equation [41, 42]. In this case, the coefficient of the dominant term is approximately 1, while other coefficients, being O(1), can be approximately 1, or much smaller than 1.

For the example of a moving heat source, the normalized expression of equation (1) is:

$$\left(\frac{\partial^2 T}{\partial x^2}\right)_c \left(\frac{\partial^2 T}{\partial x^2}\right)^{\circ} + \left(\frac{d^2 T}{dy^2}\right)_c \left(\frac{\partial^2 T}{\partial y^2}\right)^{\circ} = -\frac{U}{\alpha}\left(\frac{\partial T}{\partial x}\right)_c \left(\frac{\partial T}{\partial x}\right)^{\circ} + \frac{h+h\prime}{kd}(T_c - T_0)(T - T_0)^{\circ} \qquad (19)$$

This normalization presents the challenge that the characteristic value for temperature and its derivatives is not given by a maximum, since this function has an asymptote of $T \rightarrow \infty$ at the origin.

2.4. Balances and regimes

In equations (18) and (14), not all terms are of the same order of magnitude, which means that not all the physical phenomena involved are of the same relevance. In the asymptotic extreme, two terms (dominant terms) are much larger than the rest (secondary terms). Some terms will be positive and some will be negative; arbitrarily, we can call the positive terms "input" and the negative terms "output." The two dominant terms must have opposite sign to balance the equation; thus, there will be a "dominant input" and a "dominant output."

A regime is a combination of parameters and independent variables yielding the same dominant input and the same dominant output. The division between regimes is not sharp, but meaningful separations can be established where a dominant term becomes of the same magnitude as the largest secondary term. The combinations of parameters and independent variables near the separations are "intermediate regimes," while combinations far away from the separations are asymptotic regimes. In the asymptotic regimes, a balance between the two dominant terms alone typically yields good estimates of unknown characteristic values. Regimes delimited by values of the independent variables are also called "regions," and can be mapped with much physical insight [43].

For the example of welding on a thin plate with surface heat losses, when the characteristic values are associated with a characteristic temperature T_c (e.g. maximum width of isotherm T_c), the conditions $T^* = T_c^*$ and $y_m^* = \max(y^*)$ reduce the four dimensionless degrees of freedom

(x^*, y^*, T^*, h^*) to only two: T_c and h^*. These two degrees of freedom establish four possible regimes in the combinations of high and low T_c and h^*. In this problem, T_c^* is also expressed as the Rosenthal number:

$$\mathrm{Ro} = \frac{q}{2\pi k d(T_c - T_0)} = \frac{1}{T_c^*} \tag{20}$$

A large Rosenthal number indicates relatively fast heat sources, and a relatively small Ro indicates slow heat sources. The four regimes will be called Regime III for high Ro and small h^*, Regime IV for small Ro and small h^*, Regime IIIa for high Ro and high h^*, Regime IV for small Ro and high h^*, illustrated in figure 3 [44].

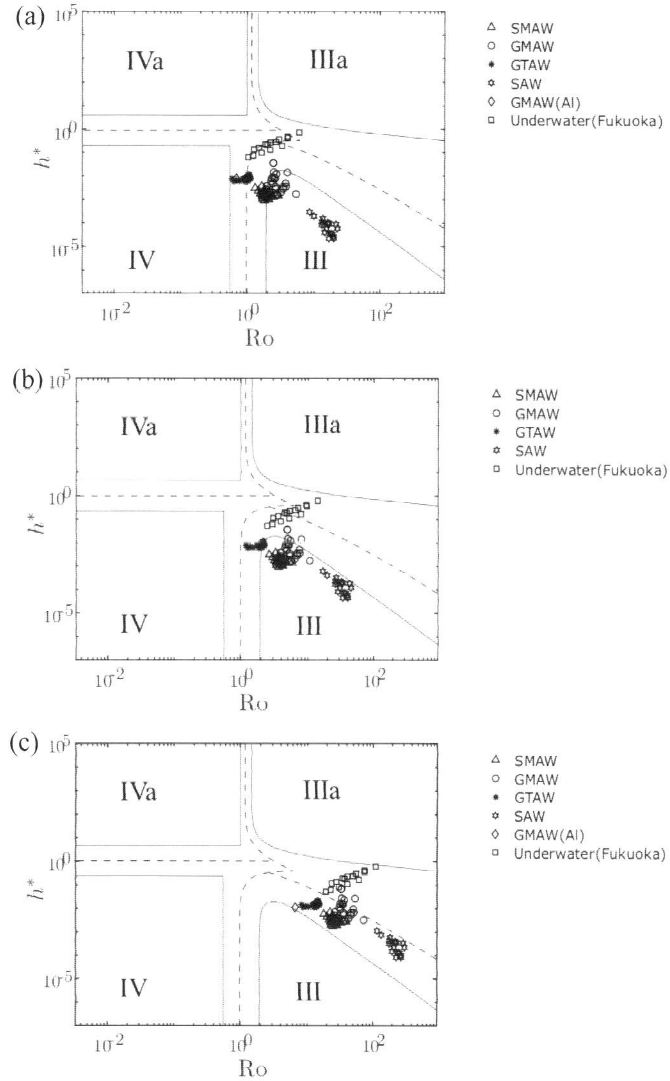

Figure 3. Process map for welding on a thin plate with surface heat losses illustrating the range of the four regimes present. The lines were determined using y_c^*. Thin lines indicate an error of 10% between the minimal representation and the exact solution. The thicker lines correspond to points in which the minimal representations of contiguous regimes yield the same prediction. The points marked correspond to typical welding procedures [44].
(a) T_c=melting, (b) T_c=630 °C, (c) T_c=yield temperature.

The characteristic value addressed in this figure 3 is the maximum width of an isotherm, but the map is essentially the same for all other characteristic values. The horizontal axis is the Rosenthal number in logarithmic scale (indicating fast or slow heat sources), and the vertical axis corresponds to h^* in logarithmic scale, indicating negligible or high heat losses. The points in this diagram correspond to typical welding procedures mostly from [45]. Characteristic values associated with melting temperature in typical welding conditions often fall in Regime III and the transition into Regime IV, at the temperatures typical of austenite decomposition (630 °C), characteristic values also fall in Regime III, transitioning into Regime IIIa. Calculations related to the shear zone in Friction Stir Welding, often falls in Regime IV [46], and calculations associated with residual stresses, underwater wet welding, and in-service welding tend to fall in the transition between Regime III and Regime IIIa [47]. The lines in this process map will be explained later in the "Blending" section.

In Regime III, for $x < 0$ (cooling), dominant terms that balance each other in the governing equation (equation (1)), are the first term on the right (advection in x, "input") and the second term on the left (conduction in y, "output"). This balance will be analyzed in detail in the section on Minimal Representation.

3. Minimal representation and correction factors

The approach formalized here is termed Minimal Representation and Correction Factors (MRCF), and is well suited for multiphysics, multicoupled problems. It has been used intuitively in all engineering disciplines for long. It focuses on characteristic values, and consists on a prediction based on a simplified problem ("minimal representation") and correction factors based on the phenomena that were neglected in the simplified problem:

$$\begin{array}{ccccccc} \text{reality} & \approx & \text{(prediction)} & = & \text{(simple formula)} & \times & \text{(correction factor)} \\ \mathrm{u_{c,e}}(\{\mathrm{P}\}) & \approx & \widehat{\mathrm{u_{c}}}_{\mathrm{j}}^{+}(\{\mathrm{P}\}) & = & \widehat{\mathrm{u_{c}}}_{\mathrm{j}}(\{\mathrm{P}\}) & & \mathrm{f_j}(\{\Pi_\mathrm{P}\}) \end{array} \tag{21}$$

where $u_{c,e}(\{P\})$ is the "real" magnitude of a characteristic value (e.g. from a good measurement), $\widehat{u_c}_j^{\,+}(\{P\})$ is a prediction based on the minimal representation of Regime j, $\widehat{u_c}_j(\{P\})$, corrected with the correction factor $f_j(\{\Pi_P\})$. The characteristic values can have units and depend on the set of arameters $\{P\}$, but the correction factors are dimensionless, and depend on the set of dimensionless groups based only on the parameters. $\{\Pi_P\}$. The notation $\widehat{u_c}_j$ indicates that the characteristic value u_c is estimated at the asymptotic regime j, and $\widehat{u_c}_j^{+}$ indicates that the characteristic value u_c is estimated at the asymptotic regime j.

The MRCF approach differs from numerical simulations in important aspects. Numerical simulations yield a numerical estimate of the solution over the whole domain for a given set of parameters $u(\{X\},\{P\})$, while MRCF yields an estimate of the characteristic value as a function of the problem parameters $u_c(\{P\})$. The dependence of the characteristic value on the parameters is explicit, in contrast with numerical simulations, for which a new simulation needs to be run for a new set of parameters. While numerical simulations aim to capture reality as faithfully as possible, MRCF focuses on the dominant physics. Problems with complex geometry are perfectly suitable for numerical modeling, but are outside the scope of MRCF; conversely, problems involving multiple coupled physical phenomena are challenging for numerical analysis, but well suited for MRCF. One similarity stands out, and it is that both numerical models and MRCF can be accurate to within measurement error.

3.1. MRCF steps

In broad terms, MRCF consists of 6 steps:

- *Step 1)* List all physics considered relevant: Adding more physics to MRCF is typically much simpler than doing it to numerical methods, and problems of convergence or excessive number of nodes are avoided.
- *Step 2)* Identify dominant factors: There are two approaches to identifying the dominant factors. First, when an exact solution exists, the dominant factors are obtained directly from asymptotic

analysis; this is the case of the moving heat source example in this paper. Another approach, when an exact solution is not available, is to postulate dominant factors, and confirm their validity in Step 4.

- *Step 3)* Solve the approximate problem considering only dominant factors: This is the "Minimal Representation" at the core of this methodology, discussed in a specific section below.
- *Step 4)* Check for self-consistency: If the dominant factors were postulated in Step 2, and a simplified solution was obtained in Step 3, when that solution is used to calculate the magnitude of the secondary phenomena, it should indicate that it is indeed negligible, or at least smaller than the phenomena considered dominant. Testing a proposed set of dominant factors can have three results: "consistent", indicating that was considered small is indeed small, "inconsistent", when what was neglected is evaluated as large when using the simplified solution, and "incompatible", when the proposed dominant factors cannot provide an approximate answer to the problem. The latter is common in systems involving multiple coupled equations, such as [42]. Self-consistency is part of the technique of dominant balance in applied mathematics [14, 15, 48–53].
- *Step 5)* Compare predictions to "reality:" Reality are characteristic values obtained from field measurements, controlled experiments, or validated simulations. In this comparison, the minimal representation must be of the same order of magnitude of the reality, and most follow the same trends as different parameters are considered.
- *Step 6)* Create correction factors: In a proper formulation, systematic discrepancies between reality and the minimal representation are due to the neglected phenomena. Correction factors of explicit form can be developed empirically based on the comparison of Step 5, reducing the error of predictions to the order of the irreducible scatter in the original data. Methodologies to obtain correction factors are described later in their own section.

3.2. Minimal representation

The minimal representation is often obtained manually, and recent progress has contributed algorithms to implement the process in a computer.

3.2.1. Manual asymptotics. Manual asymptotics are a common way of obtaining a minimal representation, often without even realizing it is being done. Four approaches are common, the first three: "Informed" Dimensional Analysis, Inspectional Analysis, and Order of Magnitude (or Balancing) involve the governing differential equations, and the fourth: Asymptotics of Closed Form Expressions, involves the solutions to the governing equations, when available.

In "Informed" Dimensional Analysis, knowledge of the physical meaning of dimensionless groups and their magnitude is used to simplify the problem. For example, when it is anticipated that surface heat losses are negligible, the 2D moving heat source problem can be simplified by assuming $h^* = 0$. Examples of the application of this technique are [54–58]

In Inspectional Analysis, the order of magnitude of terms in the governing equations is assessed, and the smaller terms are neglected. For example, for fast moving heat sources, the characteristic length in x is much longer than the characteristic length in y, and it can be stated that

$$\left(\frac{\partial^2 T}{\partial x^2}\right)_c = O\left(\frac{\Delta T}{x_c^2}\right) \ll \left(\frac{\partial^2 T}{\partial y^2}\right)_c = O\left(\frac{\Delta T}{y_c^2}\right) \tag{22}$$

resulting in the elimination of the first term in equation (1). Examples of the application of this technique are [29, 59–61]

In Order of Magnitude techniques, balances between terms are established intuitively, to be confirmed for self-consistency later. Examples of the application of this technique are [1, 5, 6]. Because too many balances are possible, not all balances can be explored. In equation (19), a balance between the second term (transverse conduction) and the third term (advection) yields the correct relationship between the characteristic length and width of an isotherm in fast moving heat sources:

$$y_c = C' \sqrt{\frac{\alpha x_c}{U}} \tag{23}$$

where C' is related to the factor C from equation (7) for the differential expressions involved.

In Asymptotic of Closed-Form Expressions, it is necessary to possess a closed form solution to a problem. The minimal representation in this case comes from asymptotic analysis of this solution (typically performed in dimensionless form). For example, the solution expressed in equation (12) can be used to estimate the maximum width of an isotherm when the convective effects are negligible ($h^* \approx 0$). In this case, there are two minimal representations [28]; in dimensionless form:

$$\widehat{y^*_{\text{max}}}(\text{Ro}) = \sqrt{\frac{\pi}{2e}}\text{Ro} \qquad \text{Regime III (large Ro,small h}^*) \tag{24}$$

$$\widehat{y^*_{\text{max}}}(\text{Ro}) = 2\exp(-\gamma)\exp(-\text{Ro}^{-1}) \qquad \text{Regime IV (small Ro,small h}^*) \tag{25}$$

where $\gamma \approx 0.5772$ is the Euler Mascheroni constant. Examples of the application of this technique are [24, 26, 27, 29, 31,62–64].

3.2.2. Computational asymptotics

Computational Asymptotics are techniques to obtain a minimal representation through computer-implemented algorithms. There are two general approaches: *Statistical Analysis* (data mining), and *Analysis of Governing Equations*.

The Statistical Analysis approach is driven by existing data, and tries to uncover predictive equations. One approach in this category is SLAW [65], in which it is assumed that a target characteristic value is determined by a power law expression of the parameters. The power law expression can be determined as a constrained linear regression in a logarithmic space, and the constraint is that the units of the proposed power-law must match the units of the target magnitude. To obtain a minimal representation, this methodology first performs a regression against all parameters, and subsequently eliminates parameters from the regression in a systematic way, starting by those which cause the smallest error increase in predictions. This process of backwards elimination continues until the error caused by the elimination of a parameter jumps suddenly, indicating that the parameter eliminated was part of the minimal representation. An exponent rounding algorithm is applied to deliver exponents as ratios of integers (with a denominator typically equal or smaller to 4). In its current implementation, this technique requires the vast majority of the input data to belong in the same regime. The advantage of this technique is that it only needs data to be applied, without need to explore the governing equations. Because of the constrain in reproducing the units of the target variable, this technique can predict the functional form of parameters for which no variations were tested (e.g. gravity). One challenge in the application of this technique is that it is not always known if the data set to be analyzed belongs to a single regime.

The Analysis of Governing Equations technique is a computational implementation of the Order of Magnitude techniques, and was pioneered as an early application of artificial intelligence in [52]. The computer implementation allows for the exhaustive exploration of all balances. In a more recent algorithmic implementation (OMS algorithm [41, 42]), it was necessary to develop a foundation for the estimation of the order of magnitude of each term in a differential equation [30], and also, a mathematical foundation to obtain an estimate for any chosen balance of terms. The latter task was implemented using linear algebra based on the fact that the characteristic values of each term of equation (19) can be expressed in the form of a power law; similarly to SLAW, described above, power laws in logarithmic space become linear expressions.

3.3. Correction factors

3.3.1. Errors of minimal representations. Once minimal representations have been obtained for one or more regimes, a comparison with "reality" is needed to validate and calibrate the predictions. The focus of the comparisons is a chosen characteristic value, for example the maximum width of an isotherm y_{max}, which does not depend on the independent variables {X}, but will vary from test to test as the parameters {P} change.

The discrepancies in the comparisons can be attributed to many factors. The assessment of "reality" involves experimental or simulation errors, which can be random or systematic. Random errors in GTAW welding experiments were assessed to be of the order of ±15% [66]. The mathematical simplifications in obtaining minimal representations can also result in systematic error, easy to correct with one, or few, tests. There are also errors in the values of materials properties used. This error manifests itself as a systematic error for comparisons with experiments in a given material, and as a random error for experimental comparisons against a variety of materials. For comparisons against simulations, this error is small or inexistent when the minimal representation and the simulation use the same materials properties. Comparisons against simulations typically show much smaller scatter than comparisons experiments.

The most interesting source of discrepancy between the minimal representation and reality is in the physical phenomena that were neglected to arrive to a minimal representation. Each neglected phenomenon induce systematic error, and is captured by a dimensionless group. The effect of the neglected phenomena can be recovered with high accuracy and simplicity with equation (21) by developing correction factors based on the comparison against reality. For the example of welding a thin plate, if surface heat losses are neglected, their effect can be accounted for by regressions against the dimensionless group h^*.

Correction factors based on neglected physical phenomena are mathematically rigorous, and can be developed based on abundant data available in the literature or from industrial practice. In contrast, blind correction factors in which a prediction is multiplied by a factor just to match an observation is not rigorous, and often fails when comparing under different conditions. Similarly, using materials properties, boundary conditions, or solver parameters as adjustable variables for fitting in numerical solutions is not rigorous, and likely to fail. In general, nonrigorous fitting falls into tautological traps of no use in practice. For example, if a numerical model has two fitting parameters and it is used to represent an approximately linear trend, then the model was not needed, since representing a linear trend with two fitting parameters is a trivial exercise. Similarly, validating a numerical model with a single experiment (or small number) is invalid when the model involves fitting parameters.

One advantage of the MRCF approach is that it can be validated against dozens of experiments over decades of literature; this is challenging or impractical with numerical models. Numerical models in which the number of adjustment variables is of the order of the number of comparisons of reality are unscientific and should not be accepted.

3.3.2. Properties of correction factors. The functional form of correction factors is typically unknown, but it has properties of much use. In the asymptotic limit of a particular regime (when the neglected phenomena are indeed negligible), the correction factor is constant of the order of magnitude of 1:

$$f_j(\{\Pi_P\}) \to\approx 1 \qquad \text{for Regime } j \qquad (26)$$

When the minimal representation is based on an exact solution, as in our example of welding on a thin plate, $\widehat{u_c}_j = \widehat{u_{c,e}}_j$, and the correction factor tends to exactly 1. Following equation (21), the functionality of correction factor $f_j(\{\Pi_P\})$ (correaponding to regime j) in regime k can also be determined, with minimal representation $\widehat{u_c}_k$:

$$f_j(\{\Pi_P\}) \rightarrow \frac{\widehat{u_c}_k}{\widehat{u_c}_j} \qquad \text{for Regime } k \tag{27}$$

where $f_j(\{\Pi_P\})$ typically tends to zero or infinity in Regime i.

4. Blending

While asymptotic cases are well captured by minimal representations, intermediate cases can pose an additional problem, as they are typically not captured by either asymptotic behavior. However, almost always, intermediate cases yield results that are intermediate between asymptotic cases.

The technique to calculate cases that are intermediate between asymptotic extremes is called "blending." For the case in which two asymptotic extremes are given by a single dimensionless group, there is a systematic treatment of intermediate cases [67]. A remarkable property of blending the ability to approximate complicated functions with small errors and much simplicity. The practical simplicity makes blending the frequent choice in engineering applications over perturbation analysis. For this reason, blending is commonly used in engineering expressions such as correlations for convection coefficients [68].

Blending was first studied specifically in [69, 70], dealing with monotonic functions that cross once, such as power laws. When one of the functions to be blended is not a power law, three complications can arise; first: that the two functions do not cross, second: that even if they cross, the asymmetry of the two functions can result in high errors when using equation (28), and third: that the asymptotic behavior chosen is not defined over the whole range of Π, thus blending cannot be performed over the whole domain.

Consider that a single dimensionless group Π is the source of two regimes (i for $\Pi \rightarrow 0$ and j for $\Pi \rightarrow \infty$), with two minimal representations $\widehat{u_c}_i(\{P\})$ and $\widehat{u_c}_j(\{P\})$. The analysis is best explained in dimensionless form, involving $\widehat{u_c^*}_i(\Pi)$ and $\widehat{u_c^*}_j(\Pi)$.

4.1. Blending of monotonic functions that cross

When $\widehat{u_c^*}_i(\Pi)$ and $\widehat{u_c^*}_j(\Pi)$ are monotonic over Π and they cross, the following blending approach is often effective [69]:

$$u_c^*(\Pi) \approx \widehat{u_c^*}^+(\Pi) = \left[\widehat{u_c^*}_i^n(\Pi) + \widehat{u_c^*}_j^n(\Pi)\right]^{1/n} \tag{28}$$

where the $^+$ notation indicates that the minimal representation involves a correction, and the blending exponent n is determined by comparisons "reality" to minimize the maximum error over the whole range of Π [62]. When the minimal representations are both power laws, past experience suggests that equation (28) has an accuracy typically within 10%, and the absolute value of n is not far from unity. Correction factors can be developed from equation (28):

$$u_c^*(\Pi) \approx \widehat{u_c^*}^+(\Pi) = \widehat{u_c^*}_i^+(\Pi) = \widehat{u_c^*}_i(\Pi)\left\{1 + \left[\frac{\widehat{u_c^*}_j(\Pi)}{\widehat{u_c^*}_i(\Pi)}\right]^n\right\}^{1/n} = \widehat{u_c^*}_i(\Pi) f_i(\Pi) \tag{29}$$

For the case of aspect ratio AR (length to width of an isotherm) in equation (12) with no convection, blending involves two power laws [32]:

$$\widehat{\mathrm{AR}}_{\mathrm{III}}(\mathrm{Ro}) = \sqrt{\frac{\pi e}{8}}\mathrm{Ro} \qquad \text{Regime III (large Ro, small } h^*) \tag{30}$$

$$\widehat{\mathrm{AR}}_{\mathrm{IV}}(\mathrm{Ro}) = 1 \qquad \text{Regime IV (small Ro, small } h^*) \tag{31}$$

where Regime III corresponds to high Rosenthal number (relatively fast welds), and Regime IV corresponds to low Ro (slow welds). Blending in this case can be done using equation (28):

$$\widehat{\mathrm{AR}}^{+}(\mathrm{Ro}) = \left[1 + \left(\sqrt{\frac{\pi \mathrm{e}}{8}}\mathrm{Ro}\right)^{n}\right]^{1/n} \tag{32}$$

with an optimal value $n = 1.972$ and an error always below 3.3% for all possible Ro values. Equation (29) yields the following correction factors for minimal representations of equations (30) and (31):

$$f_{\mathrm{AR_{III}}}(\mathrm{Ro}) = \left[1 + \left(\sqrt{\frac{8}{\pi \mathrm{e}}\frac{1}{\mathrm{Ro}}}\right)^{n}\right]^{1/n} \quad \text{Regime III (large Ro, small } h^{*}) \tag{33}$$

$$f_{\mathrm{AR_{IV}}}(\mathrm{Ro}) = \left[1 + \left(\sqrt{\frac{\pi \mathrm{e}}{8}}\mathrm{Ro}\right)^{n}\right]^{1/n} \quad \text{Regime IV (small Ro, small } h^{*}) \tag{34}$$

4.2. *Blending of monotonic functions that do not cross or are not power laws*

Without losing generality, consider the case in which $\widehat{u^*_{\mathrm{c}i}}(\Pi)$ is not a power law, but $\widehat{u^*_{\mathrm{c}j}}(\Pi)$ is, these functions can be expressed as:

$$\widehat{u^*_{\mathrm{c}i}}(\Pi) = v^*_i(\Pi)p^*_i(\Pi) \tag{35}$$

$$\widehat{u^*_{\mathrm{c}j}}(\Pi) = p^*_j(\Pi) \tag{36}$$

where $p^*_i(\Pi)$, and $p^*_j(\Pi)$ are power laws in Π, and $v^*_i(\Pi)$ is not a-power law such that $v^*_i(\Pi) \to 1$ when $\Pi \to \infty$ (Regime j). In this case, the function $\widehat{u^*_{\mathrm{c}j}}(\Pi) = p^*_j(\Pi)$ can be replaced by $v^*_i(\Pi)p^*_j(\Pi)$, with the same asymptotic behavior. Now equation (28) can be applied resulting in

$$\widehat{u^+_{\mathrm{c}}}(\Pi) = v^*_i(\Pi)\left[p^{*n}_i(\Pi) + p^{*n}_j(\Pi)\right]^{1/n} \tag{37}$$

This methodology is also useful for obtaining simpler expressions when blending functions that are not power laws, even if they cross. For the case of maximum width in equation (12) with no convection, the asymptotic behavior for $\widehat{y^*}_{\mathrm{max_{III}}}(\mathrm{Ro})$ is a power law (equation (24)), while $\widehat{y^*}_{\mathrm{max_{IV}}}(\mathrm{Ro})$ (equation (25)) is not. The blending of these two equations based on equation (37) is

$$\widehat{y^*}^{+}_{\mathrm{max}}(\mathrm{Ro}) = \exp(-\mathrm{Ro}^{-1})\left[(2\exp-\gamma)^{n} + \left(\sqrt{\frac{\pi}{2\mathrm{e}}}\mathrm{Ro}\right)^{n}\right]^{1/n} \tag{38}$$

with an optimal value $n = 1.407$ and an error always below 6.8% for all possible Ro values. Equation (38) yields the following correction factors for minimal representations of equations (24) and (25):

$$f_{y_{\mathrm{max_{III}}}}(\mathrm{Ro}) = \exp(-\mathrm{Ro}^{-1})\left\{1 + \left[\sqrt{\frac{8\mathrm{e}}{\pi}}\frac{\exp(-\gamma)}{\mathrm{Ro}}\right]^{n}\right\}^{\frac{1}{n}} \quad \text{Regime III} \tag{39}$$

$$f_{y_{\mathrm{max_{IV}}}}(\mathrm{Ro}) = \left\{1 + \left[\sqrt{\frac{8\mathrm{e}}{\pi}}\frac{\exp(-\gamma)}{\mathrm{Ro}}\right]^{-n}\right\}^{1/n} \quad \text{Regime IV} \tag{40}$$

4.3. *Alternative blending of monotonic functions that cross*

When equation (28) yields too much error (often when one or both asymptotic behaviors are not power laws), a useful alternative approach developed by Yi Lu as part of his doctoral studies at the University of Alberta is:

$$\widehat{u^+_{\mathrm{c}}}(\Pi) = \left[\widehat{u^*_{\mathrm{c}i}}^{\pm 1}(\Pi) + \widehat{u^*_{\mathrm{c}j}}^{\pm 1}(\Pi) + a\Pi^{\pm b}\right]^{\pm 1} \tag{41}$$

where $a > 0$ and b are determined by optimization minimizing the maximum error over all values of Π. The value of 1 in the exponents could also be replaced by a third adjusting parameter c and $1/c$; however fixing it to 1 reduces the complexity of the expression, and experience has shown that the two adjusting parameters a and b are enough to give acceptable errors.

This approach requires the asymptotic behaviors to be monotonic and to cross one each other. It can also be extended to apply in the case of monotonic functions that do not cross or are not power laws. When $\widehat{u_{c}^{*}}_i(\Pi)$ and $\widehat{u_{c}^{*}}_j(\Pi)$ are power laws, the exponent b in equation (41) is intermediate between the exponent of the two power laws.

For the case of centerline cooling rate in equation (12) with no convection, the asymptotic behavior for $\hat{\dot{T}}^*_{b_{III}}(\mathrm{Ro})$ is a power law, while for $\hat{\dot{T}}^*_{b_{IV}}(\mathrm{Ro})$ is not [32]:

$$\widehat{\dot{T}^*}_{b_{III}}(\mathrm{Ro}) = \frac{1}{\pi \mathrm{Ro}^3} \qquad \text{Regime III (large Ro,small h}^*\text{)} \tag{42}$$

$$\widehat{\dot{T}^*}_{b_{IV}}(\mathrm{Ro}) = \frac{1}{2}\exp\left(\frac{1}{\mathrm{Ro}} + \gamma\right) \qquad \text{Regime IV (small Ro,small h}^*\text{)} \tag{43}$$

where the dimensionless time is

$$t^* = \frac{U^2 t}{2\alpha} \tag{44}$$

These two asymptotic behaviors do not cross, and the first step in blending these equations is to apply equations (35) and (36), to obtain a blending of power laws that do cross. In this case, equation (41) with negative exponents yields a lower error than equation (28), resulting in the following blended expression

$$\widehat{\dot{T}^*_{b}}^{+}(\mathrm{Ro}) = \frac{\exp\left(\frac{1}{\mathrm{Ro}}\right)}{\pi \mathrm{Ro}^3 + 2\exp(-\gamma) + a\mathrm{Ro}^{-b}} \tag{45}$$

with optimal values of a =3.839 and b =-2.108, and an error below 5.8% for all possible Ro values [32]. Equation (45) yields the following correction factors:

$$f_{\dot{T}_{b_{III}}}(\mathrm{Ro}) = \exp\left(\frac{1}{\mathrm{Ro}}\right)\left[1 + \frac{2}{\pi}\exp(-\gamma)\mathrm{Ro}^{-3} + a\mathrm{Ro}^{-b-3}\right]^{-1} \qquad \text{Regime III} \tag{46}$$

$$f_{\dot{T}_{b_{IV}}}(\mathrm{Ro}) = \left[1 + \frac{1}{2}\pi\exp(\gamma)\mathrm{Ro}^3 + \frac{1}{2}\exp(\gamma)a\mathrm{Ro}^{-b}\right]^{-1} \qquad \text{Regime IV} \tag{47}$$

4.4. Blending of functions that change sign or are not defined over the whole domain

Consider the case in which $\widehat{u_{c}^{*}}_i(\Pi)$ is a power law, but $\widehat{u_{c}^{*}}_j(\Pi)$ is not, and it is not defined or changes sign for values of Π below or above a certain critical value Π_c. A new function $\widehat{v_{c}^{*}}_j(\Pi)$ can be defined as

$$\widehat{v_{c}^{*}}_j(\Pi) = \widehat{u_{c}^{*}}_j[(\Pi^{\pm 1} + a^{\pm 1})^{\pm 1}] \tag{48}$$

with positive exponents and when the problems with the domain are below Π_c and negative exponents and $a < \Pi_c$ when they are above. The value of a is always positive, and is adjusted by optimization; the value of 1 in the exponents was chosen for simplicity.

For all values of a, the asymptotic behavior of $\widehat{v_{c}^{*}}_j(\Pi)$ is the same as $\widehat{u_{c}^{*}}_j(\Pi)$ in Regime j, and function $\widehat{v_{c}^{*}}_j(\Pi)$ can replace $\widehat{u_{c}^{*}}_j(\Pi)$ in any of the blending techniques described above. The correction factor is applied to $\widehat{v_{c}^{*}}_j(\Pi)$, not $\widehat{u_{c}^{*}}_j(\Pi)$.

For the case of maximum temperature at a distance y_c from the centerline in equation (12) with no convection, the asymptotic behavior of $T^*_{\max}$ depends on y^*_c, not Ro. Regime III results in a power law, while Regime IV involves a change in sign above $y^*_c = 1$ [32].

$$\widehat{T^*}_{\max_{III}}(y^*_c) = \sqrt{\frac{\pi}{2e}\frac{1}{y^*_c}} \quad \text{Regime III (large Ro,small h}^*\text{)} \tag{49}$$

$$\widehat{T^*}_{\max_{IV}}(y^*_c) = \ln\left(\frac{1}{y^*_c}\right) \quad \text{Regime IV (small Ro,small h}^*\text{)} \tag{50}$$

In this case, the asymptotic behavior for Regime IV is replaced by $\ln(1/y^*_c + 1/a)$, and the standard blending approach of equation (28) is applied, yielding the following blended expression:

$$\widehat{T^*}_{\max} = \left[\left(\sqrt{\frac{\pi}{2e}\frac{1}{y^*_c}}\right)^n + \ln\left(\frac{1}{y^*_c} + \frac{1}{a}\right)^n\right]^{1/n} \tag{51}$$

with optimal values of n =-2.013 and a =0.3350, and an error below 2.1% for all possible Ro values [32]. Equation (51) yields the following correction factors:

$$f_{T^*_{\max_{III}}}(y^*_c) = \left\{1 + \left[\sqrt{\frac{2e}{\pi}}\, y^*_c \ln\left(\frac{1}{y^*_c} + \frac{1}{a}\right)\right]^n\right\}^{1/n} \quad \text{for Regime III} \tag{52}$$

$$f_{T^*_{\max_{IV}}}(y^*_c) = \left\{1 + \left[\sqrt{\frac{\pi}{2e}}\frac{1}{y^*_c \ln\left(\frac{1}{y^*_c} + \frac{1}{a}\right)}\right]^n\right\}^{1/n} \quad \text{for Regime IV} \tag{53}$$

The correction factor of equation (53) applies to $\ln(1/y^*_c + 1/a)$, not to equation (50).

4.5. Blending with more than one dimensionless group

When more than one dimensionless group is involved, for example Ro and h^* for the 2D moving heat source, the complexity of blending increases significantly. New constraints appear in the choice of dimensionless groups over which to develop blending. With multiple dimensionless groups, there will be correction factors of first, second, and higher orders, as they capture interactions between dimensionless groups. A systematic methodology for blending over multiple dimensionless groups does not currently exist, and is the focus of research efforts.

For the case study of welding on a thin plate with surface heat losses, figure 3 illustrates the four regimes to be blended. The thin lines delimiting each regime correspond to an error of 10% between the minimal representation and Rosenthal's solution. In these regions, using just the minimal representation is accurate within 10%, which is narrower than the repeatability of welding experiments [66]. The thicker lines indicate a conventional limit of regimes; these lines correspond to the parameter combination for which the minimal representation of two contiguous regimes yield the same prediction. The predictive error of minimal representations over these lines is not necessarily constant.

The minimal representations for regimes III and IV is given by equations (24) and (25). The correction factors of equations (39) and (40) allow to extend the prediction of Regime III into Regime IV, and vice versa, for very small values of h^*. One of the complexities of blending in this case, is that for any small, but finite value of h^*, a transition into Regime IIIa is expected at high enough Ro numbers; however, neither equations (39) nor (40) account for this case (because they assume h^*=0).

5. Validation of MRCF for the thin plate with surface heat losses

The techniques described above have been applied successfully to many problems including stress analysis in pipeline welding [71], solidification of chromium carbides [72], laser cladding [73], solid fraction in solidification and phase transformations [74–76], thermocapillary flows in the weld pool [42, 59] micro-electron beam welding [77–79], non-linear heat transfer at the electrode extension [80], friction stir welding [46], weld penetration in arc welding at high currents [66], application of solid lubricant sprays [81], ceramic to metal joining [82, 83], additive manufacturing and joining in semi-solid state [84–87], semi-solid die casting [88, 89] and velocities in the plasma arc [90].

For the case study of welding on a thin plate with surface heat losses, the minimal representations and correction factors were developed considering the exact Rosenthal solution as "reality." Expressions accounting for surface heat losses require blending with more than one dimensionless group, which is work in progress. The expressions presented here, valid when the surface heat losses are negligible, are within 10% of the exact solution for any value of Ro.

The exact solution required large simplifications of the physics of the problem to be tractable; then the interesting question is: If "reality" is defined by experiments or higher order models instead of the exact solution, do the MRCF expressions presented still provide useful predictions? Because the MRCF expressions are virtually identical to Rosenthal's exact solution, validation of the former is the same as validation of the latter, which was already carried out in [29]. A specific validation for the specific characteristic values analyzed here was still performed, based on data available in the literature from the span of decades. None of the comparisons that follow involves any fitting parameters associated to the experimental or numerical data used.

5.1. *Aspect ratio of isotherm*

For the characteristic value of aspect ratio, the validation of the MRCF expressions is illustrated in figure 4. It shows no obvious systematic error, and a scatter that increases with AR (i.e. high Ro), where the effect of surface heat losses might start playing a role. The accumulation of points near AR=1 is because for all small Rosenthal numbers, the isotherm approaches a circle with an aspect ratio of 1.

5.2. *Maximum temperature away from the centerline*

For maximum temperature away from the centerline, the validation of the MRCF expressions is illustrated in figure 5. This comparison shows no obvious systematic error, and a relatively small scatter over more than one order of magnitude of data.

5.3. *Cooling rate*

For the characteristic value of cooling rate, the validation of the MRCF expressions in the absence of surface heat losses is illustrated in figure 6 over almost two orders of magnitude. The comparison shows a slight systematic error, overpredicting fast cooling rates, and underpredicting slow cooling rates. The underwater wet welding data of [97] stands out in discrepancy. This is because surface heat losses are significant in underwater wet welding, but are not considered in equation (45).

IOP Conf. Series: Materials Science and Engineering **861** (2020) 012003 doi:10.1088/1757-899X/861/1/012003

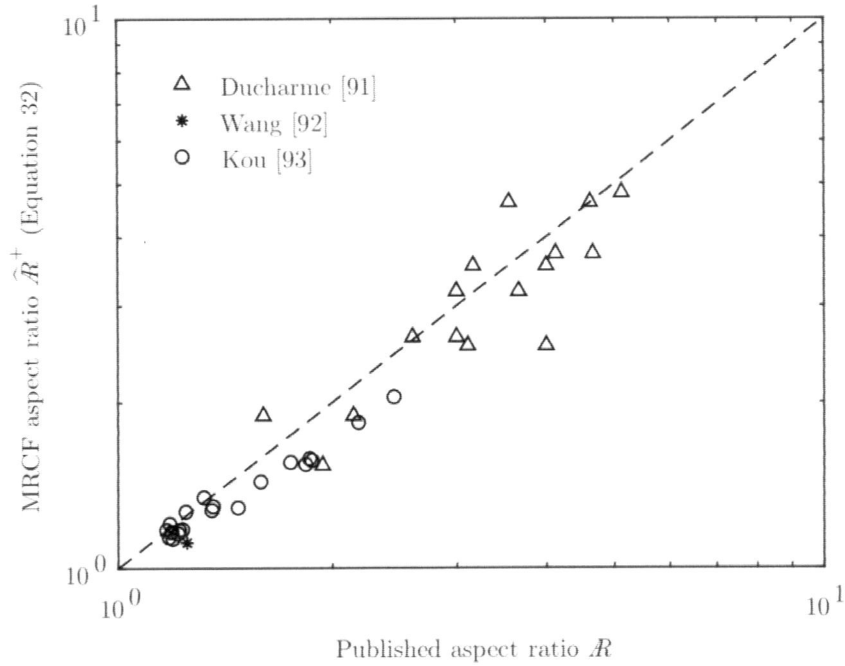

Figure 4. Validation of MRCF expressions for isotherm aspect ratio [32].

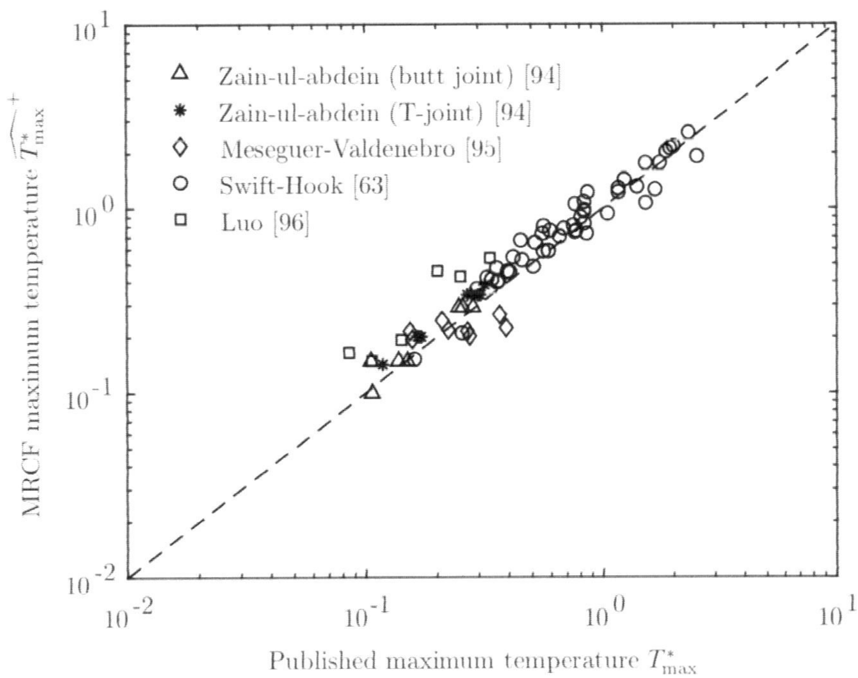

Figure 5. Validation of MRCF expressions for maximum temperature away from the centerline [32]. Published data was normalized with equation (10). MRCF maximum temperature from equation (51) [32].

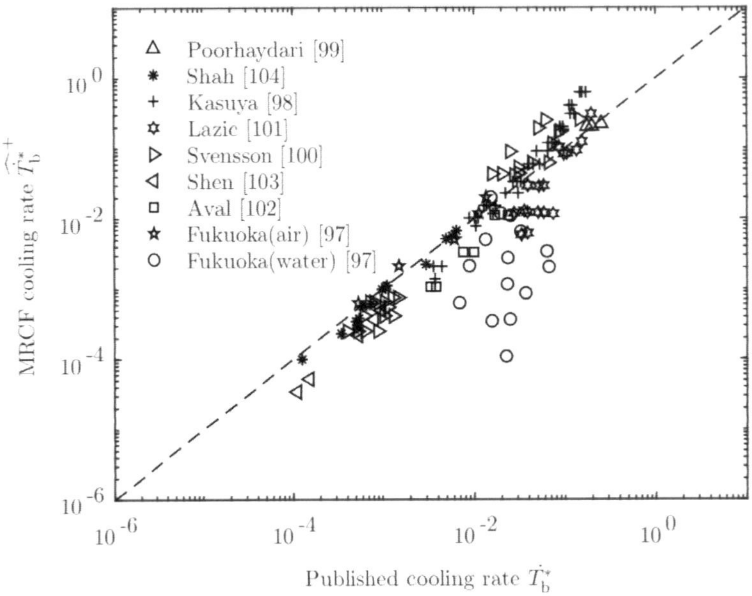

Figure 6. Validation of MRCF expressions for cooling rate [32] neglecting surface heat losses. Published data was normalized with equations (10) and (44). MRCF cooling rate from equation (45) [32].

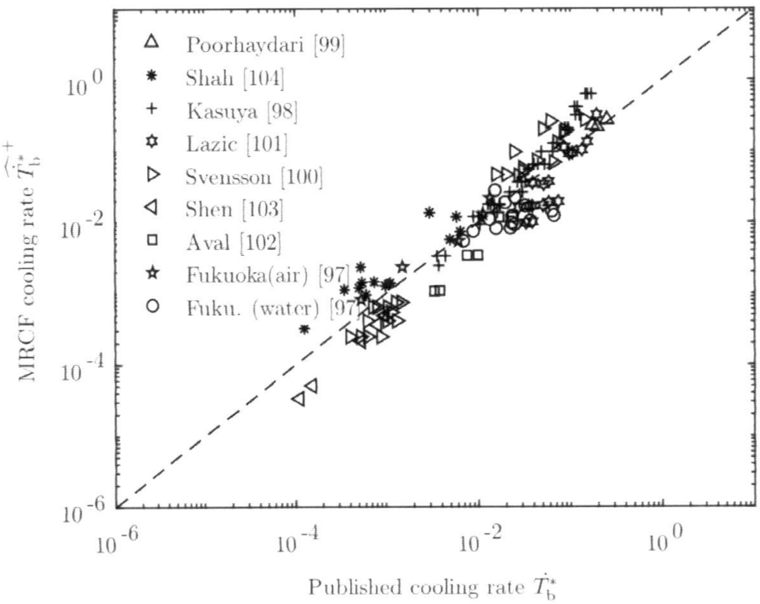

Figure 7. Validation of MRCF expressions for cooling rate [32] accounting for surface heat losses. Published data was normalized with equations (10) and (44). MRCF cooling rate not included in this paper, but indistinguishable from exact solution [47].

Figure 7 illustrates a comparison of experimental data against MRCF in which blending is attempted over both Ro and h^{*}. The blending equation is too involved to include here, and it is still the focus of research efforts. At the scale of this figure, the blending predictions are undistinguishable from the exact solution. What is remarkable in this case is that underwater wet welding data from [97] is part of the same cloud of points with all other data; also, some of the systematic error (e.g. [98]) is eliminated. This good match has important practical implications, including the ability to include practical predictions in codes and standards, as well as procedure developments for underwater wet welding, in-service welding, and other situations in which surface heat losses are relevant, such as the calculation of residual stresses.

5.4. Discussion of validation results

The narrow scatter and lack of significant systematic error shown in figures 4 to 7 is typical of MRCF. This comparison against multiple sources, under a broad range of materials and conditions, and without the use of ad-hoc fitting parameters does not exist for numerical simulations of welding.

There is no published evidence that numerical simulations, in the absence of ad-hoc calibration are more accurate than properly built reduced models, such as those presented here. This is contrary to the common belief that reduced order models result in unacceptable errors. If ad-hoc calibrations are included in the MRCF predictions presented, the error would nearly vanish.

In contrast with the correction factors presented, ad-hoc calibrations often have no physical basis; they can vary from a "fudge factor" correcting the overall expression, to artificially imposed materials properties to match experiments, for example, unrealistic viscosity values of molten metal. Ad-hoc calibrations have the drawback that they assign systematic experimental errors to the calibration factors, thus making the results unsuitable to different experimental conditions. For the case in which the experimental conditions are not expected to vary, such as when the same machine is used with the same material, ad-hoc calibrations are satisfactory. These is the case in many industrial applications. However, for the narrow range of validity required standard design of experiments often proves adequate; this is one of the drivers for the low adoption of numerical modeling in welding.

The reduced models presented here can also be extended to capture through rigorous correction factors the phenomena neglected, such as 3D conduction effects, finite heat source, variable material properties, latent heat, etc. Because of its explicit expression, owed to the blending techniques, predictions can be made very quickly, without problems of convergence.

6. Summary

It is possible to provide simple, accurate, rigorous, and physically meaningful predictions of complex materials processing systems. The approaches described focus on characteristic values and their dependence on problem parameters. The use of dimensional analysis and asymptotic techniques allows researchers to reduce problems to their minimal representation, in which all secondary factors are neglected. The discarded physics can be recovered with accuracy and simplicity through correction factors based on dimensionless groups and blending expressions. The scope of methodologies described is for problems with relatively simple geometries, but there are no restrictions about type and number of physical phenomena, linearity, or coupling involved in the problem. The expressions obtained are suitable for real-time control systems, codes and standards, handbooks, education, and design. A 2D moving heat source is chosen as an example through this article, and it is validated against experiments and simulations from the literature without any use of ad-hoc fitting parameters. The references cited contain detailed examples and literature surveys to enable the implementation of techniques to new problems.

Acknowledgements

This line of work has been supported by the US DOE (Basic Energy Sciences), NSF (CAREER Award), NSERC (Discovery Grant), as well as many grants targeting specific problems. Many of the findings are the result of the effort and talent of many graduate and undergraduate students. Special thanks to doctoral candidates Y. Lu and M.R. Grams for the help with figures and proofreading this manuscript.

Appendix

Table A1. Notation

Thin plate	**Units**	
d	m	Plate thickness
k	$Wm^{-1}K^{-1}$	Thermal conductivity of plate
q	W	Intensity of the heat source (accounting for thermal efficiency)
h, h'	$Wm^{-2}K^{-1}$	Effective convection coefficient on front and back surfaces
Ro	–	Rosenthal number (equation (20))
$T(x,y)$		Temperature in the plate
T_c		Characteristic temperature of interest (e.g. melting)
T_0	C	Initial temperature in the plate (e.g. preheat)
U	$m\ s^{-1}$	Heat source travel velocity
x, y	m	Coordinates (figure 1)
x_b	m	Trailing edge of isotherm on centerline (figure 2)
x_f	m	Leading edge of isotherm on centerline (figure 2)
x_{max}	m	x-coordinate of maximum width of isotherm (figure 2)
y_{max}	m	y-coordinate of maximum width of isotherm (half width) (figure 2)
α	m^2s^{-1}	Thermal diffusivity of plate
γ	-	Euler-Mascheroni constant ≈5.772
$\widehat{\dot{T}}_f$	$K\ s^{-1}$	Heating rate at x_f (figure 2)
T_b	$K\ s^{-1}$	Cooling rate at x_b (absolute value) (figure 2)
AR	–	Aspect ratio of isotherm (length/width)
MRCF		
$\{X\}$	e.g. x, y, z, t	Set of independent variables
$\{U\}$	e.g. $T(x, y)$	Set of dependent variables
$\{P\}$	e.g. k, α, T_c	Set of parameters
$\{\Pi\}$	e.g. Ro, x^*, y^*	Set of independent dimensionless groups
$\{\Pi_P\}$	e.g. Ro, h^*	Set of independent dimensionless groups not involving variables
K_j	e.g. $2, \pi, \gamma$	Numerical constants
A_j	Coefficients of governing equations (equation (5))	
B_j	Coefficients of dimensionless governing equations (equation (13))	
g_j	Functional form of terms of governing equation (equation (5))	
h_j	Functional form of terms of dimensionless governing equation (equation (5))	
Subscripts and superscripts		
c		Approximate characteristic value (equation (7))
c,e		Exact characteristic value (equation (7))
*		Dimensionless magnitude (equation (13))
∘		Normalized magnitude (equation (17))
$\frown$		Minimal representation (equation (21))
$\frown$ +		Corrected or blended minimal representation (equations (21) or (28))

References

[1] Dantzig J A and Tucker C L 2001 *Modeling in Materials Processing* (Cambridge University Press)
[2] Kou S 1996 *Transport phenomena and materials processing* (Wiley)
[3] Geiger G H and Poirier D R 1973 *Transport phenomena in metallurgy* Addison-Wesley series in metallurgy and materials (Addison-Wesley Pub. Co.)
[4] Szekely J and Themelis N J 1971 *Rate phenomena in process metallurgy* (Wiley)
[5] Krantz W B 2007 *Scaling Analysis in Modeling Transport and Reaction Processes: A Systematic Approach to Model Building and the Art of Approximation* (Wiley)
[6] Bejan A 2004 *Convection heat transfer 3rd ed* (Wiley)
[7] Ruckenstein E 2003 *Ind. Eng. Chem. Res.* **42** 2525–9
[8] Sides P J 2002 *Chem. Eng. Ed.* **36** 232–5
[9] Deen W M 1998 *Analysis of transport phenomena* (Oxford University Press)
[10] Astarita G 1997 *Chem. Eng. Sci.* **52** 4681–98
[11] Zlokarnik M 1991 *Dimensional Analysis and Scale-up in Chemical Engineering* (Springer)
[12] Chen M M 1990 Scales, similitude, and asymptotic considerations in convective heat transfer. *Annual Review of Heat Transfer* vol 3 1st ed ed Tien C L (Hemisphere) pp 233–291
[13] Ruckenstein E 1987 *Adv. Chem. Eng.* **13** 11–112
[14] Kline S J 1986 *Similitude and Approximation Theory* (Springer)
[15] Denn M M 1980 *Process Fluid Mechanics* 1st ed Prentice-Hall International Series in the Physical and Chemical Engineering Series (Prentice-Hall)
[16] Aris R 1976 *Chem. Eng. Ed.* **10** 114–24
[17] Mendez P and Eagar T 2001 *J. Mater. Process. Tech.* **117** CD–ROM, Section B9: Modeling
[18] Mendez P F 2011 *Sci. Technol. Weld Joi.* **16** 348–56
[19] Mendez P F, Tello K E and Gajapathi S S 2012 Generalization and Communication of Welding Simulations and Experiments Using Scaling Analysis. *9th Int. Conf. Trends Weld. Res.* (ASM International) pp 249– 258
[20] Wilson H A 1904 *P. Camb. Philos. Soc.* **12** 406–23
[21] Rosenthal D and Schmerber R 1938 *Weld. J.* **17** 2s–8s
[22] Rosenthal D 1941 *Weld. J.* **20** 220s–34s
[23] Rosenthal D 1946 *Trans. ASME* **68** 849–66
[24] Rykalin N N 1951 *Calculation of Heat Flow in Welding* (Mashgis)
[25] Carslaw H S and Jaeger J C 1959 *Conduction of heat in solids* 2nd ed (Clarendon Press)
[26] Myers P S, Uyehara O A and Borman G L 1967 *Welding Research Council Bulletin* 1–46
[27] Grong Ø 1994 *Metallurgical Modelling of Welding* 1st ed (Institute of Materials)
[28] Lu Y, Wang Y and Mendez P F 2020 *Int. J. Heat Mass. Tran. (Preprint)*
[29] Christensen N, Davies, V de L and Gjermundsen K 1965 *British Weld. J.* **12** 54–75
[30] Mendez P F 2010 *J. Appl. Mech.-T. ASME* **77** 061017
[31] Wang Y, Lu Y and Mendez P F 2019 *Int. J. Heat Mass. Tran.* **135** 1118–29
[32] Lu Y and Mendez P F 2020 *Int. J. Heat Mass. Tran. (Preprint)*
[33] Fourier J B 1822 *Théorie Analytique de la Chaleur* (F. Didot)
[34] Maxwell J C 1871 *P. London Math. Soc.* **III** 224
[35] Rayleigh J W S 1877 *The theory of sound* (Macmillan)
[36] Buckingham E 1914 *Phys. Rev.* **4** 345–76
[37] Tolman R C 1914 *Phys. Rev. Lett.* **3** 244 – 55
[38] Bridgman P W 1916 *Phys. Rev. Lett.* **8** 423 – 31
[39] Bridgman P W 1922 *Dimensional Analysis* 1st ed (Yale University Press)
[40] Washio T and Motoda H 1999 Extension of Dimensional Analysis for Scale-types and Its Application to Discovery of Admissible Models of Complex Processes. *Int. Workshop on Similarity Methods* pp 129–47
[41] Mendez P F and Eagar T W 2013 *J. Appl. Mech.-T. ASME* **80** 011009

[42] Mendez P F and Stier N 2013 OMS: A computer algorithm to develop closed-form solutions to multicoupled multiphysics problems. *Mathematical Modelling of Weld Phenomena 10* (Seggau, Austria: T. U. Graz) pp 219–54
[43] Mendez P F, Argaez M A R, Alvarez A D, Apaoblaza Chaer D A, Antonio A M and Sanchez A V 2019 Multiphysics in the plasma arc. *IIW Annual Assembly* Doc. 212-1609-19
[44] Lu Y, Grams M R and Mendez P F 2020 *In preparation*
[45] Electric L 2000 *The Procedure Handbook of Arc Welding* 14th ed (The James F. Lincoln Arc Welding Foundation)
[46] Mendez P F, Tello K E and Lienert T J 2010 *Acta Mater.* **58** 6012–26
[47] Lu Y and Mendez P F 2020 *In preparation*
[48] Kruskal M D 1963 *Asymptotology Mathematical Models in Physical Sciences* (Notre Dame, IN) pp 17–48
[49] Segel L A 1972 *SIAM Rev.* **14** 547–71
[50] Van Dyke M 1975 *Perturbation Methods in Fluid Mechanics* (The Parabolic Press)
[51] Bender C M and Orszag S A 1978 *Advanced Mathematical Methods for Scientists and Engineers* 1st ed (McGraw-Hill)
[52] Yip K M K 1996 *Artif. Intell.* **80** 309–48
[53] White R B 2005 *Asymptotic analysis of differential equations* (Imperial College Press)
[54] Murray P E and Scotti A 1999 *Sci. Technol. Weld Joi.* **4** 112–7
[55] Murray P E 2000 *Sci. Technol. Weld Joi.* **5** 221–6
[56] Murray P E 2002 *Weld. J.* **81** 125s–131s
[57] Fuerschbach P W and Knorovsky G A 1991 *Weld. J.* **70** 287s–297s
[58] Fuerschbach P W 1996 *Weld. J.* **75** 24s–34s
[59] Rivas D and Ostrach S 1992 *Int. J. Heat Mass. Tran.* **35** 1469–79
[60] Schlichting H 1987 *Boundary-layer theory* 7th ed McGraw-Hill (McGraw-Hill)
[61] Kou S and Wang Y H 1986 *Metall. Trans. A* **17A** 2265–70
[62] Mendez P F, Lu Y and Wang Y 2018 *J. Heat Trans.-T. ASME* **140** 081301
[63] Swift-Hook D T and Gick A E F 1973 *Weld. J.* **52** 492s–499s
[64] Heller K, Kessler S, Dorsch F, Berger P and Graf T 2017 *Int. J. Heat Mass. Tran.* **106** 958–69
[65] Mendez P F and Ordóñez F 2005 *J. Appl. Mech.-T. ASME* **72** 648–57
[66] Duman U 2009 *Modeling of Weld Penetration in High Productivity GTAW* Ph.D. Colorado School of Mines
[67] Wang Y, Lu Y, Grams M, Cesaro A and Mendez P F 2019 Asymptotics and Blending in the Modeling of Welding. *Mathematical Modelling of Weld Phenomena 12* (Seggau, Austria: T. U. Graz) pp 907–32
[68] Bergman T L, Incropera F P, DeWitt D P and Lavine A S 2011 *Fundamentals of heat and mass transfer* (Wiley)
[69] Churchill S W and Usagi R 1972 *AIChE J.* **18** 1121–8
[70] Churchill S W and Usagi R 1974 *Ind. Eng. Chem. Fund.* **13** 39–44
[71] Grams M R and Mendez P F 2020 A Quantitative Index to Assess the Influence of Joint Fit-Up on Pipeline Weld Root Discontinuities. *Proc. 13th International Pipeline Conference* (IPC 2020)
[72] Barnes N, Clark S, Seetharaman S and Mendez P F 2018 *Acta Mater.* **151** 356–65
[73] Wood G and Mendez P F 2016 First Order Prediction of Bead Width and Height in Coaxial Laser Cladding. *Mathematical Modelling of Weld Phenomena 11* (Seggau, Austria: T. U. Graz) pp 701–34
[74] Kamyabi-Gol A, Clark S J, Gibbs J W, Sridhar S and Mendez P F 2016 *Acta Mater.* **102** 231-40
[75] Kamyabi-Gol A and Mendez P F 2015 *Metall. Mater. Trans. A* **46** 622–38
[76] Gibbs J W, Schlacher C, Kamyabi-Gol A, Mayr P and Mendez P F 2015 *Metall. Mater. Trans. A* **46** 148–55
[77] Gajapathi S S, Mitra S K and Mendez P F 2011 *Int. J. Heat Mass. Tran.* **54** 5545–53

[78] Gajapathi S S, Mitra S K and Mendez P F 2012 *Sci. Technol. Weld Joi.* **17** 429–34
[79] Gajapathi S S, Mitra S K and Mendez P F 2012 *Sci. Technol. Weld Joi.* **17** 435–40
[80] Lehnhoff G and Mendez P F 2011 *Int. J. Heat Mass. Tran.* **54** 2651-60
[81] Mendez P F and Powell A C 2008 *Scripta Mater.* **59** 784–7
[82] Park J W, Mendez P F and Eagar T W 2002 *Acta Mater.* **50** 883–99
[83] Park J W, Mendez P F and Eagar T W 2005 *Scripta Mater.* **53** 857–861
[84] Mendez P F, Rice C S and Brown S B 2003 *J. Mater. Sci. Lett.* **22** 1047–9
[85] Mendez P F, Rice C S and Brown S B 2002 *Weld. J.* **81** 180s–187s
[86] Rice C S, Mendez P F and Brown S B 2000 *JOM* **52** 31–33
[87] Mendez P F and Brown S B 1999 *US Patent* 1–13
[88] Rice C S and Mendez P F 2001 *Adv. Mater. Process.* **159** 49–52
[89] Rice C S R, Mendez P F, Brown S B, Myojin S, Mendez P F, Rice C S R and Myojin S 2001 *US Patent* 1–10
[90] Mendez P F, Ramirez M A, Trapaga G and Eager T W 2001 *Metall. Mater. Trans. B* **32** 547–54
[91] Ducharme R, Williams K, Kapadia P, Dowden J, Steen B and Glowacki M 1994 *J. Phys. D Appl. Phys.* **27** 1619–27
[92] Wang X and Li R 2014 *J. Intell. Manuf.* **25** 1301–13
[93] Kou S, Kanevsky T and Fyfitch S 1982 *Weld. J.* **61** 175s–181s
[94] Zain-ul abdein M, Nélias D, Jullien J F and Deloison D 2010 *Mater. Sci. Eng. A* **527** 3025–39
[95] Meseguer-Valdenebro J L, Serna J, Portoles A, Estrems M, Miguel V and Martínez-Conesa E 2016 *T. Indian I. Metals* **69** 783–91
[96] Luo C, Cao Y, Zhao Y, Zhao L and Shan J 2018 *Weld. J.* **97** 214s–228s
[97] Fukuoka T and Fuku S 1994 *B. Mar. Eng. Soc. Japan* **22** 86–92
[98] Kasuya T and Yurioka N 1993 *Weld. J.* **72** 107s–115s
[99] Poorhaydari K, Patchett B M and Ivey D G 2005 *Weld. J.* **84** 149s–155s
[100] Svensson L E, Gretoft B and Bhadeshia H K D H 1986 *Scand. J. Metall.* **15** 97–103
[101] Lazić V N, Sedmak A S, Živković M M, Aleksandrović S M, Čukić R D, Jovičić R D and Ivanović I B 2010 *Therm. Sci.* **14** 235–46
[102] Aval H J, Farzadi A, Serajzadeh S and Kokabi A H 2009 *Int. J. Adv. Manuf. Tech.* **42** 1043–51
[103] Shen S, Oguocha I N A and Yannacopoulos S 2012 *J. Mater. Process. Tech* **212** 286–94
[104] Shah A K, Kulkarni S D, Gopinathan V and Krishnan R 1995 *Weld. J.* **74** 297–304

The virtual core - modelling and optimization of core manufacturing and application

J Thorborg, S Kumar, I Wagner and J C Sturm

MAGMA GmbH, Kackertstraße 11, 52072 Aachen, Germany

E-mail: J.Thorborg@magmasoft.de

Abstract. Development and manufacturing of high quality cast parts with tight dimensional requirements depend to a large extent on the stability and application of the sand cores. The core-sand is based on a complex granular multi-material system, where curing of an organic or inorganic system bond the grains into a porous core. The chosen combination of sand and binder determines the thermo-mechanical properties of the core. The paper provides an overview of current modelling capabilities along the life cycle of sand cores. A detailed analysis of the production steps is presented as well as the resulting properties linked to the application in the casting process. Several questions and some answers are given to the understanding of core related defects and how simulation can bridge the gap between core production and application of cores in the casting process. Today, process simulation of sand core production is an accepted tool for core box design and prediction of robust production conditions. The first modelling step is core shooting where expanding air is the driving force for sand flow from a core shooting machine into the core box. Coupled two phase flow modelling of air and granular material is required to represent the process sufficiently. The final process step in core manufacturing is core curing and moisture transport where porous media gas flow modelling is applied. Phase transformations as well as the relevant transport phenomena and the complete heat balance in case of thermally controlled processes need to be considered. During casting the deformation of the bonded sand material is generally governed by thermal expansion, phase transformation and the location of core prints. For long thin walled cores, buoyancy forces due to density differences between the cast material and the bonded sand material can play an important role due to creep effects in the organic binder systems. For inorganic systems moisture content and moisture transport during casting are affecting the mechanical behaviour, such as bending strength. The paper presents a detailed overview of a recent implemented soil plasticity model which is calibrated for different binder systems and sand types. The data are based on comprehensive mechanical tests to provide temperature, moisture and time dependent input data for the shown examples. This includes new steps and future challenges in modelling the anisotropy of printed sand cores and digitalization of the foundry in general.

1. Introduction

Sand cores play an important role in shaping complex casting designs, where a stable core design can be a challenge to achieve both during the manufacturing of the core as well as during the performance of the core in the casting process. The complexity and high quality standards of the cast parts today require a stable process with limited variations in the final part quality. To achieve this, the focus has traditionally been directly on the cast part itself, the alloy and variations in the process conditions. But in recent years it has been increasingly important to pay more systematic attention to the sand core to understand and control its quality and performance.

Simulation and optimization of the sand core manufacturing and the integration into classical casting process simulation have developed in parallel to the increasing focus on the quality of the sand core. Today it is possible to "virtually" analyse the flow of air and sand mixture inside the core box for different process conditions like shooting pressure and the interaction with different shooting nozzle geometries and layouts of the vents. The subsequent hardening and curing of the considered binder system is possible to evaluate by simulating the core gassing and purging. Pressure gradients govern the flow of gases through the porous media, where the permeability is dependent on the simulated porosity distribution. Similar models and approaches are used to simulate the transport of gases due to pyrolysis and binder degradation during the casting process. The results give a detailed knowledge of the gas pressure distribution and the risk of gas defects in the interface between the liquid metal and the sand core.

Recent developments have focused on the moisture content of inorganic cores to consider vaporization, transport and condensation of water during the casting process. The influence of temperature and hardening time on moisture content is possible to analyse and correlate to the mechanical strength of the sand core. Similar to predicting the risk of gas defects in the interface between the liquid metal and the organic bonded core it is possible to predict gas defects related to water vapor. These new functionalities follow the growing interest in industry to switch to inorganic cores where it is possible.

Simulation capabilities of the mechanical performance of sand cores in the casting process have been extended intensively in recent years. Constitutive models from soil plasticity and creep models to describe time effects at elevated temperatures have been tested and evaluated to describe the hot distortion of sand core. Various mechanical tests have been selected to provide input from different stress states to require mechanical data. Thermal expansion and expansion due to phase transformation in silica sand are measured and provide a significant contribution to the loading of the core material. Deformation due to small buoyancy forces can be considered in the simulation, which can have significant influence on the behaviour of long fragile cores. The new developments make it possible to analyse and optimize the location of core prints to reduce the risk of unwanted deformations and cracks due to high stress concentrations.

The objectives of the first part of the paper are to review the current status of simulating the life cycle of a sand core based on proper physical models and measured input data, including an extension in the field of moisture transport and associated risk of gas defects. The second part of the paper is presenting recent developments and new implementations used to simulate the mechanical performance of the sand core in the casting process. These activities are important steps in further digitalizing of the sand core material and for providing a virtual simulation tool for the day-to-day optimization of sand cores in the casting process. The integration of simulating core manufacturing and casting process simulation is illustrated in figure 1.

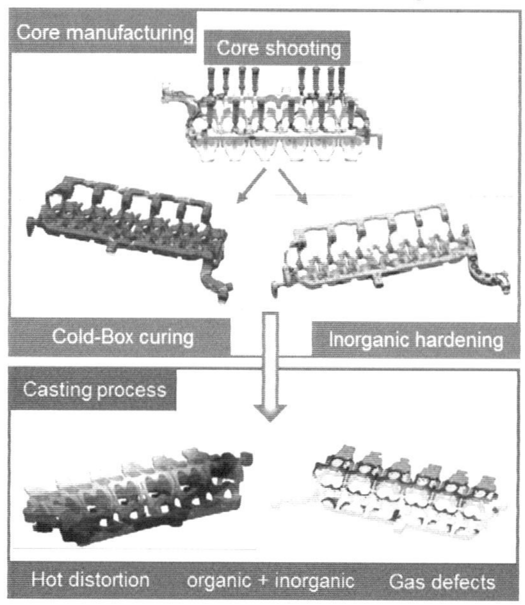

Figure 1. Life cycle of the sand core from core manufacturing to application in the casting process.

2. Core manufacturing

2.1. Core shooting

Flow of liquid metal starts the manufacturing of cast parts followed by a phase change to the final solid state during solidification and cooling. Similarly, manufacturing of sand cores starts by the flow of binder-coated sand particles driven by pressurized air followed by a state change to a porous solid state when the particles settle and the binder system is activated. The flow of metal and sand particles is obviously quite different in nature and proper analysis of the sand flow requires a two-phase flow simulation where the interaction between the binder coated particles and the air is properly described. For the Eulerian-Lagranian method the binder coated sand is treated as discrete particles, while for the considered Eulerian-Eulerian formulation, presented in this section, the binder coated sand and air are treated as two separate continuum phases. For the considered formulation, the mass and momentum conservation equations for the sand/binder and air phases can be written as, [1,2].

$$\frac{\partial}{\partial t}(\varepsilon_s\rho_s) + \nabla\cdot(\varepsilon_s\rho_s\vec{v}_s) = 0 \tag{1}$$

$$\frac{\partial}{\partial t}(\varepsilon_g\rho_g) + \nabla\cdot(\varepsilon_g\rho_g\vec{v}_g) = 0 \tag{2}$$

$$\frac{\partial}{\partial t}(\varepsilon_s\rho_s\vec{v}_s) + \nabla\cdot(\varepsilon_s\rho_s\vec{v}_s\vec{v}_s) = -\varepsilon_s\nabla P_g - \nabla P_s + \varepsilon_s\rho_s\vec{g} + \nabla\cdot\bar{\bar{\tau}}_s + \beta(\vec{v}_g - \vec{v}_s) \tag{3}$$

$$\frac{\partial}{\partial t}(\varepsilon_g\rho_g\vec{v}_g) + \nabla\cdot(\varepsilon_g\rho_g\vec{v}_g\vec{v}_g) = -\varepsilon_g\nabla P_g + \varepsilon_g\rho_g\vec{g} + \nabla\cdot\bar{\bar{\tau}}_g + \beta(\vec{v}_s - \vec{v}_g) \tag{4}$$

where ε represents a phase volume fraction, ρ is density, $\vec{v}$ is velocity, P is pressure, $\bar{\bar{\tau}}$ is a viscous stress tensor and β is a frictional coefficient that describes the momentum exchange between the phases. The subscripts s and g are used to indicate the sand/binder and air (gas) phases, respectively.

The interaction between the particles and by that the dissipation of the kinetic energy depends on the sand fraction. The three primary mechanisms of viscous dissipation are: kinetic dissipation, kinetic plus collisional dissipation and frictional dissipation [3,4]. The correlation between the different mechanisms and the sand fraction in the particle flow is shown in figure 2.

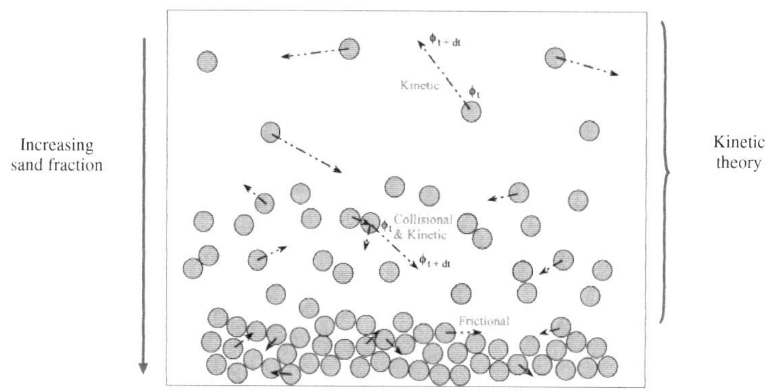

Figure 2. Different mechanisms of viscous dissipation depending on the sand fraction.

The two phase flow formulation ensures a separate flow behaviour of the two phases governed by the right physical coupling according to the illustration in figure 2. The flow of the two phases is possible to visualize separately with tracer particles as shown in figure 3.

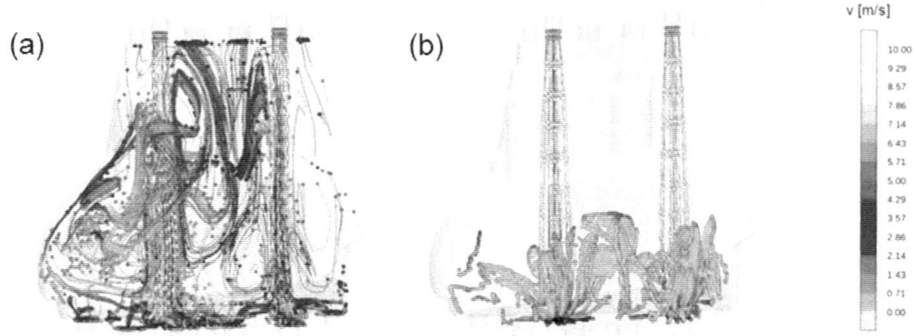

Figure 3. Visualization of the flow of the two phases, (a) air and (b) sand/binder mixture.

Solving the governing equations of the discretized system, with proper constitutive information to describe the kinetic behaviour, makes it possible to analyse the complex flow of sand particles from pressurization in the hopper, to flow through the shooting plate and the shooting nozzle, to the final settlement inside the core box. The distribution of settled sand particles determines the density distributions which can be used as initial condition for the subsequent curing and hardening phase. Further details of the implementation and validation based on measurements can be found in [1].

2.2. Core hardening

The binder is activated after the core shooting process to bond the particles into a stable sand core with sufficient mechanical properties for the casting process. For the cold box system the hardening process is governed by the flow of amine gas, which is flushed through the pore space of the sand core. The gas flow through the core is primarily dependent on the core geometry, the nozzle design, the gas pressure and the permeability of the particle/binder mixture. The curing itself depends on many factors such as binder composition and amount, solvent type and amount, binder wetting of the sand grains, morphology of the sand grains and temperature. For inorganic cores the hardening process is governed by the flow of hot air and moisture transport of vaporized water. The model also takes into account condensation in colder regions of the core. Vaporization and condensation need to be fully coupled to the thermal simulation to get realistic temperature distribution in the heated core box.

The transport of a catalyst gas or moisture during the hardening process can be described by using models for flow through porous media. The conservation of a chemical species k during the gassing of the core can be described by

$$\frac{\partial}{\partial t} C_k + \nabla \cdot \left(\rho_k \vec{U} \hat{C}_k\right) = \nabla \cdot \left[(1 - \varepsilon_s)\rho_k D_k \nabla \hat{C}_k\right] + R_k \tag{5}$$

where C_k is the mass concentration and $\hat{C}_k$ is the pore space fraction of the gas component, D_k is the mass diffusivity of the component, $\vec{U}$ is Darcy's velocity, and R_k represents in-situ mass transformations due to e.g. condensation/evaporation of the gas component in the core.

The gas phase velocity U is related to the pressure gradient in the pore space of the core by using Darcy's law

$$U = -\frac{K}{\mu_s} \nabla P_g \tag{6}$$

where K is the permeability of the core and μ_s is the gas phase viscosity. Combining this equation with equation (5) and taking advantage of the fact that the pore space is always completely occupied by the gas phase, it is possible to derive an equation which describes the variation of the gas pressure. The local permeability in the core is possible to relate to the simulated density distribution from the core shooting process. Details on the experimental investigation and the reaction source for vapor can be found in [1].

The transport of gas through the porous media can be used to evaluate the distribution of amine gas in the cold box core to optimize the core box design, the time of gassing and the subsequent purging of dry air. The results visualize the local gas effectivity to highlight critical areas with insufficient core stability. The result of simulating the gas effectivity of a cold box core is shown in figure 4(a), which corresponds well to the observations on the real part in figure 4(b). Results from simulating the moisture content during hardening/drying of an inorganic core is shown in figure 4(c). The transparent zone represents the hardened shell and the wet condensation zone is visualized in the centre of the bulky zone.

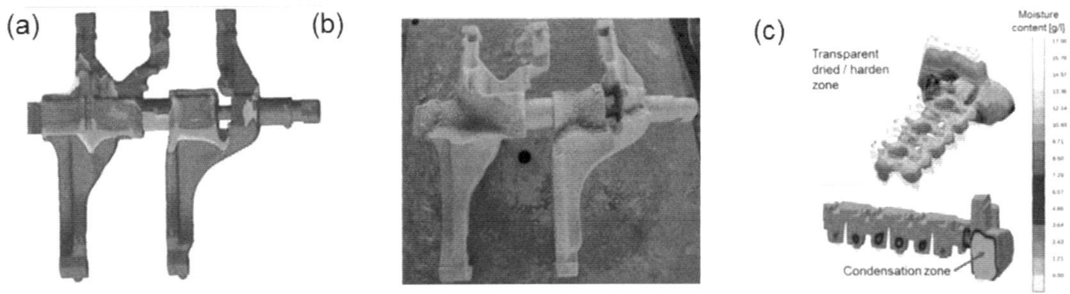

Figure 4. (a) Simulation of gassing effectivity for a cold box core, (b) defects observed on the real core and (c) moisture content in an inorganic core.

3. Cores in the casting process

3.1. Binder degradation and gas defects

After manufacturing and storing the core, it is finally used in the casting process where the surface of the core is subjected to a thermal shock when the liquid metal enters the cavity and hits the surface of the core. The thermal impact on the core immediately degrades the cold box binder and gases evolve in the core. The gases have to escape out of the core at core prints to avoid critical pressure levels which may cause gas defects in the casting.

The thermal stability of the binder is possible to analyse by evaluating the kinetic parameters for polymer decomposition which are based on the weight loss for isothermal conditions or constant heating rates. The degree of binder left in the core is possible to describe by this information using an Arrhenius expression to describe the reaction rate of the combined decomposition, [5]

$$R_k = \frac{dM_k}{dt} = M_B r_k(T) = M_B f_k exp\left\{\frac{-E_k}{RT}\right\} \tag{7}$$

where f_k and E_k are experimentally determined reaction frequencies and activation energies, respectively.

Evolution of gases and the associated thermal expansion of the gases increase the pressure locally which drives the transport of gas through the porous core. Simulating the degradation and the gas flow makes it possible to analyse the risk of gas defects caused by high gas pressure at the interface between the core and the liquid metal.

The core gas pressure is shown as an example in figure 5, where the red box in the simulation result in figure 5(b) indicates a critical level of gas pressure at the interface to the liquid metal. The pressure builds up because the gas transport in this section of the core is limited by the narrow passage indicated by the yellow box. As shown in this example it is possible to check the core design and check if the gas is able to escape through core prints or vents. The gas defect detected on the real part is shown in the photo in figure 5(a).

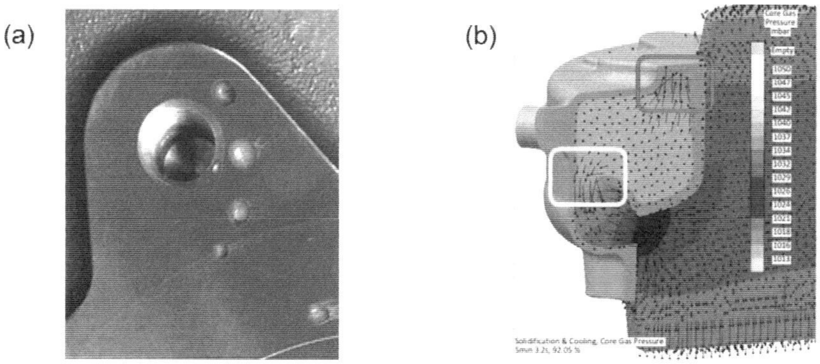

Figure 5. (a) Gas defect on the surface of the machined part and (b) result showing high gas pressure in the core indicated by the red box because there is only a narrow passage (yellow box) for gas flow out of the critical area.

3.2. Inorganic cores – vaporization and condensation of moisture and gas defects

The physical effects of vaporization and condensation during the casting process can affect the quality of metal castings where inorganic cores are used. Any left-over water in the sand core from insufficient hardening/drying can lead to water vapor during the casting process, which either condensates in colder regions of the core, escapes to the outside or enters the mould cavity with a high risk of causing gas defects. Therefore, a proper understanding of transport of water vapor in inorganic sand cores during the casting process is very important. The implemented model considers a time-dependent three-dimensional flow analysis of mixed water vapor and air through a porous medium taking vaporization and condensation effects into account. The governing mass conservation equation for the system is

$$\Phi \frac{\partial \rho}{\partial t} + \nabla \cdot (\rho U A) = S_p w \tag{8}$$

where Φ is the sand porosity, S_p is the specific sand area and w is the rate of mass transfer due to condensation or vaporization. Following the approach as described in [2], w is expressed as

$$w = M_T \rho_k \left(\frac{P_k^{sat}}{P} - C_{v_k} \right) \tag{9}$$

where M_T is the mass transfer coefficient, ρ_k is the density of water vapour, P_k^{sat} is the saturation vapour pressure for water vapor, P is the pressure of the gas mixture, C_{v_k} is the volume fraction of the water vapour in the pore space, [6]. Darcy's law, similar to equation (6), is used to relate the gas phase velocity to the pressure gradient. Vaporization and condensation are finally described by the two equations

$$\partial_t C_k + \nabla \cdot (\rho_k U C_{v_k}) = S_p w \tag{10}$$

$$\partial_t C_k' = -S_p w \tag{11}$$

Depending on the manufacturing of the inorganic core the properties and the distribution of moisture can be very different through the thickness of the core during the casting process. Ideally, the core would be completely dry, however in practice a small amount of unbonded water is typically left in the core. At elevated temperature the water vaporizes and depending on core design and possibility to vent the water vapor out of the mould there is a risk of vapor to enter the mould cavity. Depending on the pressure balance at the interface between water vapor and liquid metal there is a risk of gas defects. This risk is possible to analyse by highlighting areas at the interface where the gas pressure in the vapor is higher than the pressure in the liquid metal. Results on a water jacket core are shown in figure 6.

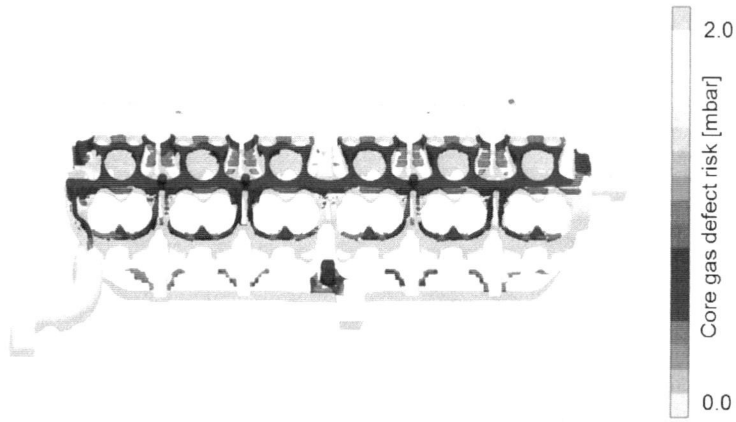

Figure 6. Gas pressure balance at the interface between the casting and core. The risk of gas defects is indicated by gauge pressure of the gas.

4. Mechanical behaviour of sand cores

The bonded sand core material is a brittle porous material at room temperature, with small bridges of binder between the grains, see figure 7. At elevated temperature the stiffness and strength of the core material typically decrease when the binder bridges degrade due to oxidation or increased moisture content. Mechanical loading can lead to pore collapse or shear failure of the bonded structure. The main objective of the work presented in this section has been to evaluate the distortion of bonded sand material during the casting process and use the information to improve the prediction of deformations of the sand cores. Both cold box and inorganic binder have been considered in this work and several tests have been performed to characterize the thermal and mechanical behaviour of the material.

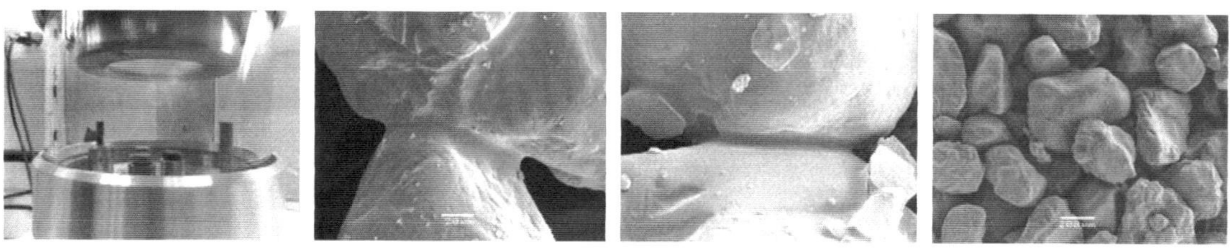

Figure 7. SEM images of the binder bridges between the sand grains.

4.1. Soil plasticity – modelling brittle material behaviour

The bonded sand material is fundamentally behaving differently in compression and tension due to the bonding of the granular structure of the quartz sand (or other foundry sand types). It is therefore not sufficient to use a standard J_2-plasticity formulation, which is only formulated in the deviatoric stress and strain space, [7, 8]. Two soil plasticity models have been evaluated to model the mechanical behaviour of the bonded sand, the Cam-Clay model and the Drucker Prager cap model [9]. Both models have a pressure dependent yield surface, with a high strength for high confining pressure and a lower strength for tensile conditions. The Cam-Clay model is not presented here.

The classical Drucker Prager cap model is fundamentally defined by two regions which define the yield surface of the model, see figure 8. Similar to the Cam-Clay model one region describes how the material dilates during plastic deformation and another region defines how the material is compressed during plastic deformation. The first region is described by the failure envolope, which is defined by the line with the slope M and the pressure offset, P_t. The second region is described by the cap, which limits the deformation in compression. To improve the convergence behaviour of the implemented model smaller modifications can be done to smooth the interface between the cap and the failure line and the failure line and the pressure axis. Smoothing the two interfaces reduces the unwanted influence of the vertexes on the yield surface. The yield surface can be described by the two yield conditions:

$$\Phi^a(\sigma) \equiv \sqrt{J_2(s(\sigma))} + \eta[p(\sigma) - p_t] = 0 \tag{12}$$

$$\Phi^b(\sigma, a) \equiv \frac{1}{\beta^2}[p(\sigma) - p_t + a]^2 + \left[\frac{q(\sigma)}{M}\right]^2 - a^2 = 0 \tag{13}$$

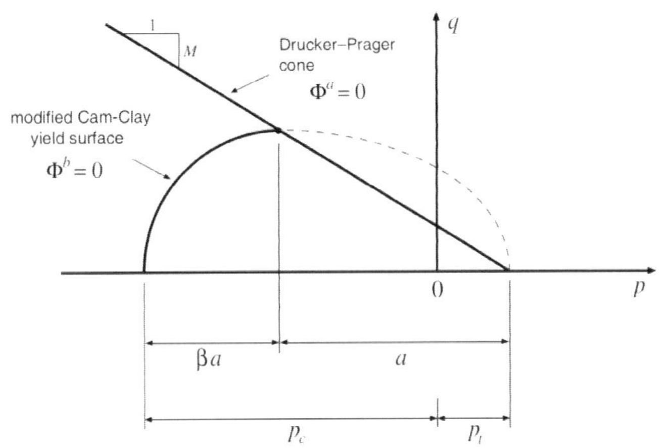

Figure 8. Sketch of the classical Drucker Prager cap model. The yield surface is defined by the failure envelope and the cap. The slope M defines the failure envelope and the ellipse defines the cap.

4.2. Soil plasticity – mechanical tests

Different mechanical tests are needed to provide sufficient input for the Drucker Prager cap model. The tests have to provide strength information for different stress states in terms of positive and negative pressure. The most general but also comprehensive and expensive test is the tri-axial test, where different load levels can be applied for different confining pressures. This is mainly relevant to get a good description of the compression cap. The failure line can be determined from three conventional mechanical tests, which have different stress states and by that different locations on the pressure axis. The 3-point bending test provides input to the tensile region, while the two cylinder tests provide information to the compressible region, see figure 9 below, [10]. The die compaction test can be used to collect information about hardening during compression of loosely bonded sand.

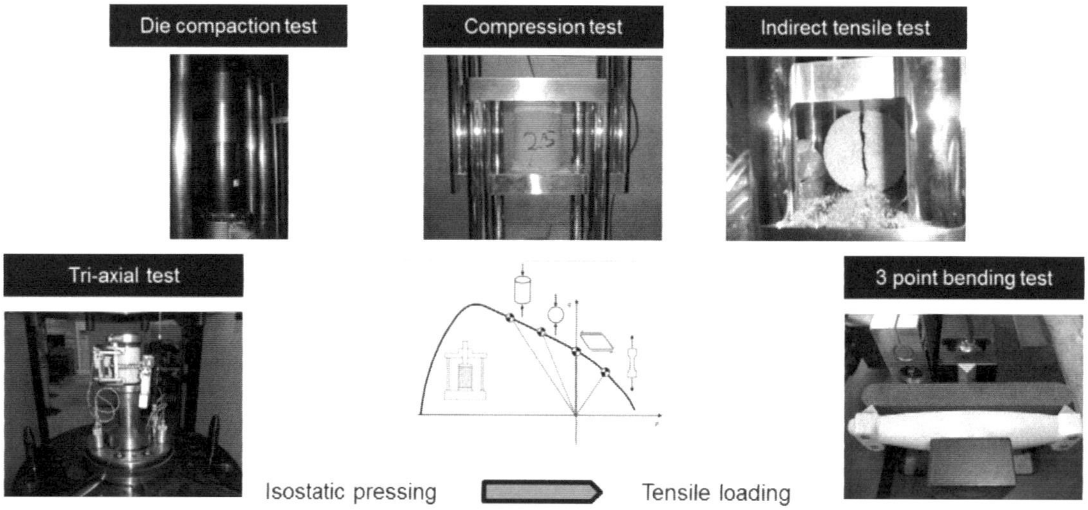

Figure 9. Different mechanical tests and the relationship to the definition of the yield surface.

The load information is easily used to calculate the P and Q values for the three simple tests, compression, indirect tension and 3-point bending while the P and Q information from the tri-axial test can be used directly. The formulas used to calculate P and Q are listed in the table below.

Table 1. Expressions to calculate strength as function of pressure for different stress states.

	σ	P	Q
Compression test	$\frac{4F}{\pi d^2}$	$-\frac{1}{3}\sigma$	σ
Indirect tensile test	$\frac{2F}{\pi dh}$	$-\frac{2}{3}\sigma$	$\sqrt{13}\sigma$
3 point bending	$\frac{3FL}{2bh^2}$	$\frac{1}{3}\sigma$	σ

4.3. Soil plasticity – extracting mechanical data

To extract data for the Drucker Prager cap model the core material was generally tested at different temperature levels from room temperature to 400 °C. The results were collected in P-Q diagrams for the different temperature levels and average values were used for property extraction and for comparing the two binder systems, figure 10. The graphs show some trends which can give some indications of how strong and stable the different binder systems are at different temperature levels.

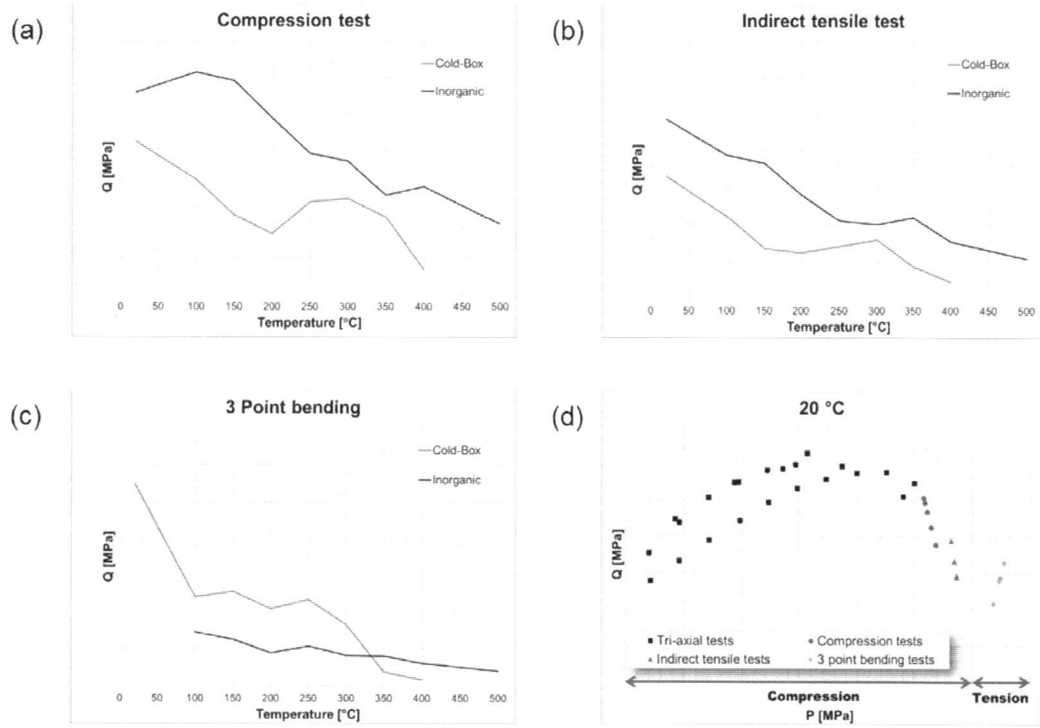

Figure 10. Response of the mechanical tests as function of temperature (a) compression test, (b) indirect tensile test, (c) 3-point bending and (d) the collection of stress state, including the tri-axial measurements.

For the compression tests, figure 10(a) and the indirect tensile tests, figure 10(b), the inorganic system is generally stronger than the cold box system. Curing of the cold box system increases the strength above 200 °C, followed by a rapid decrease in strength above 250 °C dure to binder degradation. The inorganic system has a steady decrease in strength as function of temperature. For 3-point bending the inorganic system has a steady low strength, figure 10(c), probably due to a more brittle behaviour when compared to the cold box cores. The cold box cores start out stronger, again with a small increase in strenght around 200 °C due to curing, followed by a rapid decrease in strength above 250 °C. The P-Q data are colleced in one graph for the cold box binder and combined with the triaxial test data at room temperature, figure 10(d), to give an impression about the input needed for the Drucker Prager cap model.

4.4. Creep model(s) – modelling time dependent response

The cold box system has a tendency to deform as function of time, especially for tensile load conditions. This behaviour is also observed for cores in production. Different rehological models can be used to describe different time dependent behaviour. For the considered cold box system the Maxwell element was used and the strain contribution was added to the plastic strain component from the soil plasticity model to comprise a combined response, goverend by both yielding and creep. The creep response from the rheological model is made pressure sensitive by using a pressure dependent viscosity. For the Maxwell element the inelastic strain rate of the material is described directly by a dashpot. A Hooke element is attached in a serial connection to describe the elastic response, see figure 11. The combined response of the Maxwell element is an instantaneous elastic response followed by a delayed viscous response of the dashpot.

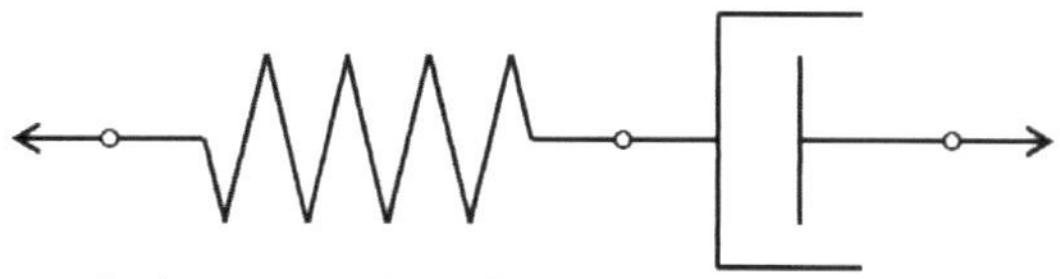

Figure 11. The Maxwell element to describe the creep behaviour in the cold box binder.

The strain tensor in the dashpot can be generalized in two different ways, either in the same way as the elastic strain tensor in Hookes's law, which includes both a deviatoric and a volumetric contribution or the deviatoric part only

$$\dot{\varepsilon}_{ij}^{vi} = \frac{1}{\eta}\{(1+\upsilon)\sigma_{ij} - \upsilon\delta_{ij}\sigma_{kk}\} \quad or \quad \dot{\varepsilon}_{ij}^{vi} = \frac{1}{\eta}s_{ij} = \frac{1}{\eta}\left\{\sigma_{ij} - \frac{1}{3}\delta_{ij}\sigma_{kk}\right\} \qquad (14)$$

where, σ_{ij}, s_{ij}, E and η are the stress in the Maxwell element, the stiffness of the spring and the viscosity of the dashpot, respectively.

It is possible to derive a modified Young's modulus for Hooke's law, which still relates the elastic strain to the stress, but also includes the contribution from the dashpot. The strain decomposition equation can easily be extended to include the plastic strain contribution from the soil plasticity model(s), which gives a corresponding change to the stress update equation.

4.5. Creep model(s) – mechanical tests

In the current work, tensile tests have been used to directly extract information about the viscosity. In addition two other tests, 3-point bending and buoyancy trials have been used to calibrate the pressure sensitivity of the viscosity.

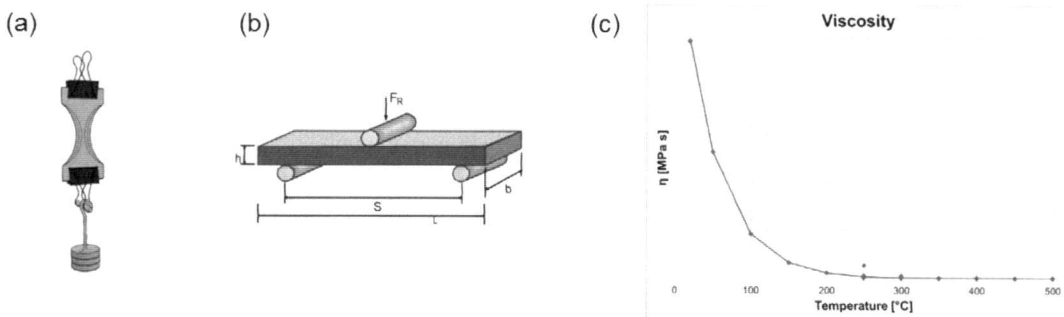

Figure 12. Creep tests - constant load applied at different temperature levels. (a) Tensile test, (b) 3-point bending test and (c) extracted viscosity for the dashpot.

The measurements confirmed that only the organic binder showed a clear tendency for continuous creep during loading in tension. This behaviour was also observed for 3-point bending, where constant loads in mechanical tests showed creep effects and submerged bars in liquid aluminum similarly showed time dependent deformation due to buoyancy. The tested inorganic binder system did not show a significant tendency to creep in any of the load cases.

4.6. Thermal expansion

In addition to the mechanical tests used for calibrating input data for the constitutive models, dilatometer measurements are needed to determine the thermal expansion.

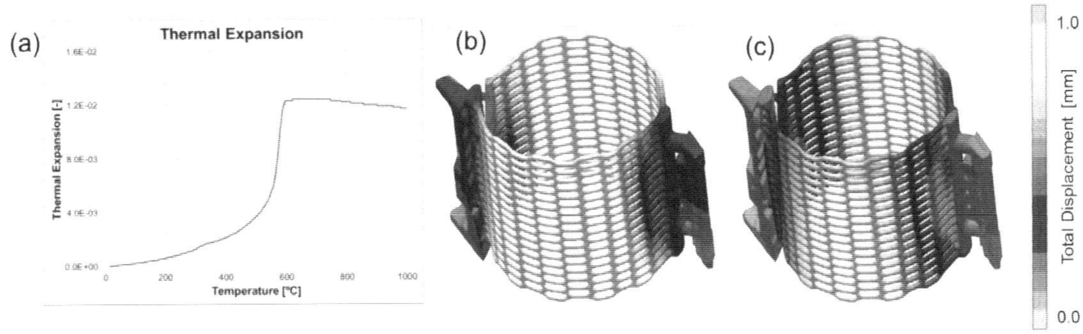

Figure 13. (a) Thermal expansion of quartz sand showing the α-quartz to β-quartz phase transformation. (b) Expansion of an E-motor housing water jacket core using quartz sand to the left and Bauxite sand to the right.

The measurements show a large expansion for quartz sand in general plus a significant amount of expansion for the α-quartz to β-quartz transformation around 570 °C. The thermal expansion of quartz sand has a significant influence on the deformation of the sand core. Other sand types showing less thermal expansion should be selected in cases where expansion becomes critical for the performance of the sand core.

4.7. Buoyancy – organic cores

For cold box cores in general the deformation is not only governed by thermal expansion. Also, density differences between the core material and the metal leads to buoyancy forces, which leads to time dependent deformations in long slender cores. Several tests were performed in a demonstrator mould where several bars with different dimensions were submerged into liquid metal. Aluminium and lead alloys were used to promote different levels of buoyancy forces at different temperature levels to evaluate the difference of bending coming from thermal expansion and buoyancy forces. As seen in figure 14 below the bending of the sand cores is significantly larger for the lead alloy due to the bigger difference in density.

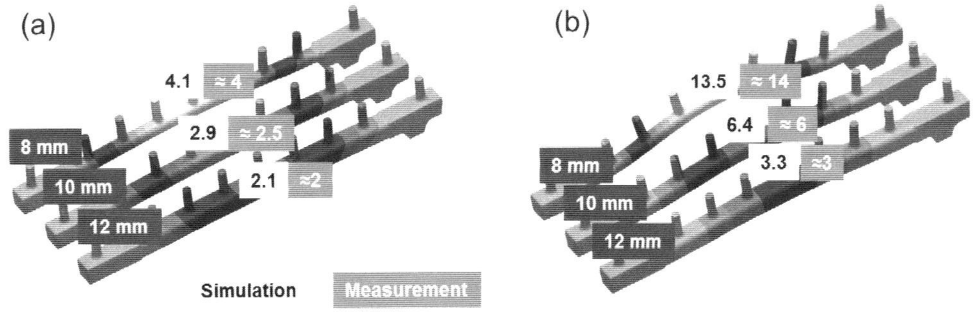

Figure 14. Bending of cold box cores due to different levels of buoyancy forces for (a) aluminium and (b) lead, respectively.

The results were compared to optical measurements, where in-situ deformations were measured on the tip of the small domes sticking out of the melt. The domes are seen on the bending bars in figure 14.

4.7.1. Marine application example. Buoyancy and creep effects can lead to critical levels of deformation in cast parts during production when organic binder systems are used. A core package used for marine application Al-castings is shown in figure 15. Requirements in the design are a free end of the relatively long cylinder shaped cores. For the cold box system this free end allows the core to deform as function of time due to the small buoyancy forces, see the upwards deformation in figure 15(a). Results from the core simulation were used to pre-shape the core box to compensate the deformation in the final part and it was possible to obtain straight castings within the required tolerances. To illustrate the different behaviour of the inorganic system, where no creep is taking place at elevated temperature, a similar deformation result is shown in figure 15(b). The result shows that the cores do no deform in a similar upwards direction and only thermal expansion gives a volumetric contribution to the deformation components. This gives an almost straight casting without any pre-shaping of the core box.

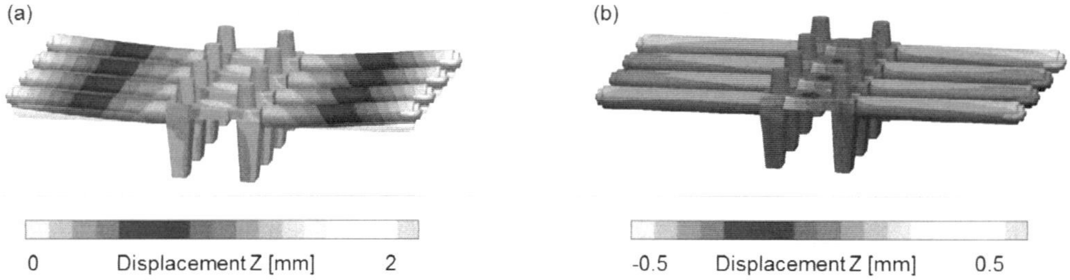

Figure 15. Evaluation of bending due to buoyancy and creep. (a) cold box – bending of the fee ends due to creep and (b) inorganic – no creep and hence no bending

4.7.2. Automotive oil gallery example. Castings used in automotive applications often include long slender cores to shape interior channels and thin-walled designs, e.g. water jacket cores and oil gallery cores. Similar to the example above this can lead to critical levels of deformations due to buoyancy and creep when cold box cores are used. An automotive example is shown in figure 16, where bending of a thin-walled oil gallery core is measured at room temperature by sectioning the engine block and scanning the interior parts optically. The predicted deformations are shown in figure 16(a) and compared to the measurements in figure 16(c). The comparison shows an overall good agreement between the predicted results and the measurements. The illustration in figure 16(b) indicates how the footprint of the core on the cast part is governed by the sand material before a sufficiently strong shell of cast material solidifies around the core. This might be obvious, but it is nevertheless important information which is needed to be considered for a proper description of the mechanical interface between the sand core and the cast material.

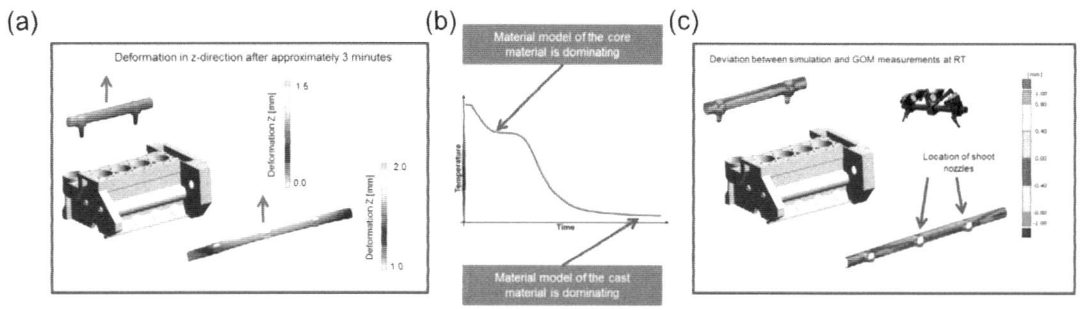

Figure 16. Deformation of a thin-walled oil gallery cores due to buoyancy and creep. (a) Simulation result, (b) governing deformation contributions at different temperature levels and (c) results compared to optical measurements at room temperature.

4.8. Moisture content – inorganic cores

The influence of moisture content on the mechanical behaviour is possible to take into account for the simulation of hot distortion of inorganic cores. Wet inner regions of partially hardened cores and regions with high moisture content due to condensation are modelled by removing the elements from the simulation domain or by reducing the strength and stiffness in the elements as function of increasing moisture content. To illustrate the influence of different moisture content on the bending strength several sand cores with different hardening time are shown in figure 17. The diameter of the wet centre increases with decreasing drying time, in the curves going from left to right. Taking the different moisture content into account in the simulation shows how the overall different stiffness of the bars lead to a faster increase in the stress level on the lower surface of the bars, which resembles their lower bending stiffness and strength.

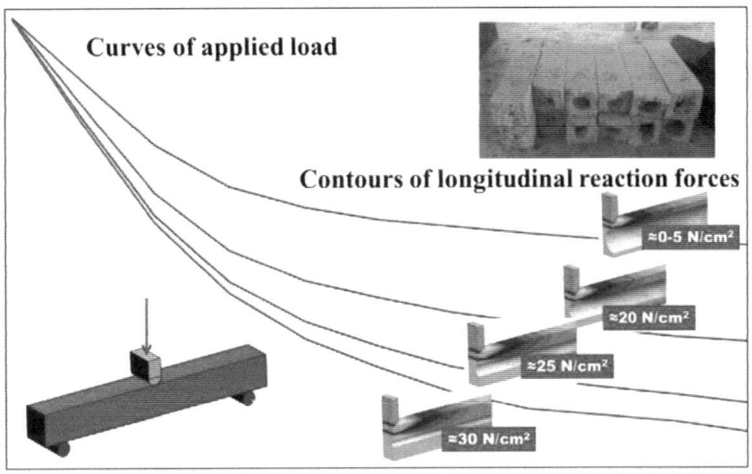

Figure 17. Inorganic bending bars with different drying time and corresponding different measured bending strength.

5. Outlook

Printed sand cores have gained an increasing interest in the last years. The design freedom of 3D printing in general makes the technology very attractive for highly complex geometries where other methods are not applicable or would require assembly of several subcomponents. The process is based on a powder bed technology where a re-coater distribute layers of sand followed by a printing step where the print head bond the sand grains using the jetting process to form the final geometry of the core, see figure 18. Compared to the traditional manufacturing of sand cores the method is known to have directional dependencies in the mechanical properties and the density is typically reported to be lower compared to methods where the sand is compacted during manufacturing.

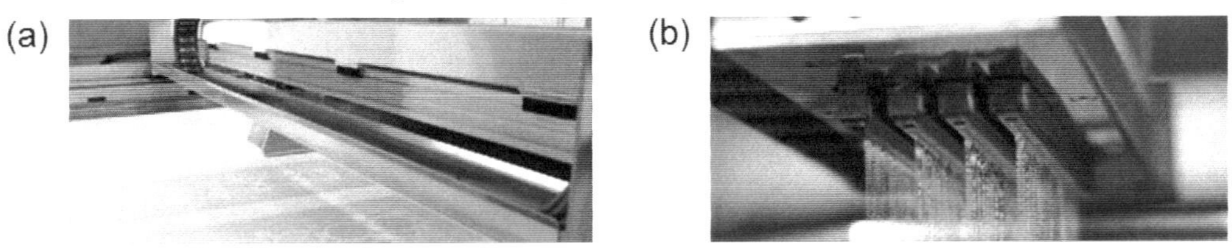

Figure 18. (a) Powder bed and print head. (b) Binder nozzles for the jetting process. Images shown on www.exone.com

The surface quality of the printed sand cores shows visible signs of the layered process. It is therefore necessary to have a post treatment cleaning step after the printing process to ensure a sufficient quality of the sand core surface, see figure 19.

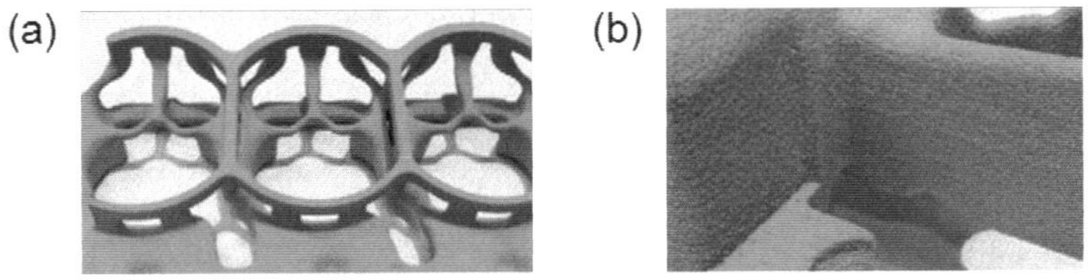

Figure 19. (a) 3D printed water jacket core and (b) surface quality after preparation of the layered surface. Examples from www.voxeljet.com

Measurement of bending strength and thermal expansion coefficients indicate that simulation of the mechanical behaviour of 3D printed sand cores will require an anisotropic formulation to take the print direction into account. Especially the z-direction is reported to have a lower strength compared to the x-direction of printing.

6. Conclusions

Simulation of the sand core manufacturing steps from core shooting, curing and hardening, to the application of the sand core in the casting process is today state of the art in designing and analysing the stability of the sand core. The two-phase flow simulation of the core shooting step gives a detailed knowledge about the flow inside the core box, which is useful knowledge to ensure a core box design where the sand is homogeneously distributed. The subsequent curing and hardening analysis highlight critical areas using the gas effectivity result for the cold box system and the moisture content result for the inorganic binder system. The moisture content result makes it possible to ensure that the hardened dry surface is sufficiently thick to avoid moisture from the inner part of the sand core to move to the interface of the casting during the casting process. Results from the core manufacturing process are used for the subsequent casting process to analyse the degradation of binder and transport of gases and moisture to evaluate if the pressure at the interface between the sand core and the liquid metal reaches critical levels to form gas defects.

Today, it is possible to simulate hot distortion of sand cores during the casting process. The material is described by porous soil plasticity models with a creep extension to include the time dependent deformation of e.g. the cold box system. Various standard mechanical tests are needed at different temperature levels to provide input for different stress states to the yield surface description. Dilatometer measurement of the thermal expansion describes the expansion of the sand core during heating, which is needed to analyse the dimensional stability of different sand types. In addition to the thermal strain it is possible to include the deformations from buoyancy effects due to density differences between the sand core and liquid metal.

The paper presents how the full lifecycle of the sand core is integrated into useful models and predictive tools which can be used to analyse the core design and stability before the sand core is put in production. The tools and results form the basis of the important digital twin used in industry today, where digitalization and quantification of material behaviour including sand cores become increasingly important to meet quality requirements.

References

[1] Schneider M 2008 Experimental Investigation, Physical Modeling and Simulation of Core Production Processes *AFS Trans.* 08-058(04) 1–14

[2] Rogers C A 2006 Virtual tool for the manufacture and use of foundry cores and molds *Proc. of MCWASP – XI* ed C A Gandin and M Bellet (Warrendale: TMS) pp 1123-30

[3] Garzó V and Dufty J W 1999 Dense fluid transport for inelastic hard spheres *Phys. Rev. E - Stat.*

Physics, Plasmas, Fluids, Relat. Interdiscip. Top. **59** 5895–911
[4] Bocquet L, Losert W, Schalk D, Lubensky T C and Gollub J P 2002 Granular shear flow dynamics and forces: Experiment and continuum theory *Phys. Rev. E - Stat. Physics, Plasmas, Fluids, Relat. Interdiscip. Top.* **65** 1–19
[5] Samuels G and Beckermann C 2011 Measurement of Gas Evolution from PUNB Bonded Sand as a Function of Temperature 65th AFSA Chicago
[6] Kubo K and Pehlke R D 1986 Heat and moisture transfer in sand molds containing water *Metall. Trans. B* **17** 903–11
[7] Tvergaard V 2001 Plasticity and creep in structural materials (Technical University of Denmark. Department of Mechanical Engineering, Solid Mechanics)
[8] Simo J C and Hughes T J R 1998 *Computational inelasticity* (New York: Springer)
[9] Desai C S and Siriwardane H J 1984 Constitutive laws for engineering materials, with emphasis on geologic materials (Englewood Cliffs: Prentice-Hall)
[10] Brewin P R, Brewin P R, Coube O, Doremus P and Tweed J H 2008 *Modelling of Powder Die Compaction* (London: Springer)

IOP Conf. Series: Materials Science and Engineering **861** (2020) 012005 doi:10.1088/1757-899X/861/1/012005

Solidification microstructure selection maps for laser powder bed fusion of multicomponent alloys

P Mohammadpour and A B Phillion

Dept. of Materials Science and Engineering, McMaster University, Hamilton, Canada

Email: andre.phillion@mcmaster.ca, mohamp3@mcmaster.ca

Abstract. Solidification Microstructure Selection (SMS) maps provide a simple yet effective approach to predict the non-equilibrium solidification microstructure and grain morphology during Additive Manufacturing. In this study, SMS maps have been created for the Inconel 625 (IN625) alloy processed by Laser Powder Bed Fusion (LPBF). Toward this end, theoretical solid growth models, a model of the Columnar to Equiaxed Transition (CET), interface response theory, thermal simulation results and computational thermodynamics are utilized. The predicted microstructures are compared both qualitatively and quantitatively to experimentally-obtained micrographs. The theoretical analysis was also compared to the earlier analytical calculation for Al-10Si-0.5Mg alloy to show how differences in thermophysical properties affect the microstructural predictions. The theoretical predictions are shown to be in good agreement with the experimental results in terms of the resulting microstructure and dendrite arm spacings. A discussion on the use of SMS maps, formed over a broad range of thermophysical conditions, to help guide industry in improving LPBF microstructure, is provided.

1. Introduction

Laser Powder Bed Fusion (LPBF) is a mature Additive Manufacturing (AM) process that uses metal powder as a raw material, along with a laser heat source, to fabricate end-use parts with complex geometries. In metallic components, the mechanical properties are significantly affected by the microstructure—grain size, morphology, and crystallographic orientation. Although metal LPBF has been recently adopted on an industrial scale, the knowledge to optimize the microstructure and consequently mechanical properties of the final product remains inadequate. In this regard, the advanced investigation between processing parameters, solidification microstructure, and mechanical properties is necessary [1].

LPBF is characterized by extremely high solidification rates, V, due to the high laser scan speed, V_s, on the order of 0.1-1.5 m/s, and micron-sized melt pools, which thus leads to high thermal gradients, G, on the order of mega K/m. The laser power needs to be high enough to avoid a lack of fusion during laser melting, but not too high to produce keyhole-induced defects [2]. Large values of G and V change the diffusion length and finally lead to non-equilibrium effects at the solid/liquid interface during solidification [3]. Due to this complicated heat transfer condition at small length scales during LPBF, *in situ* microstructural observation is challenging. This makes numerical methods—including Phase-Field method (PF) and Cellular Automata (CA) as well as analytical approaches *i.e.* solidification microstructure selection (SMS) maps [4,5]—an effective alternative for investigating the relationship between solidification parameters and microstructure. Although numerical simulation of solidification is a powerful method for predicting microstructure, these methods are quite computationally expensive,

IOP Conf. Series: Materials Science and Engineering **861** (2020) 012005 doi:10.1088/1757-899X/861/1/012005

especially for exploring structure-processing relationships [5]. Unlike PF and CA, SMS maps provide faster and more comprehensive predictions of microstructure in non-equilibrium solidification [4] and, being computationally efficient, allow for investigation over a wide range of compositions and growth velocities in a short time [4].

SMS maps are graphical charts that show the stable solidification microstructure and grain morphology under specific solidification conditions. They are useful for controlling solidification morphology through manipulating the processing parameters in order to obtain AM components with specific properties. Two types of SMS maps can be defined. The first type identifies the stability region of different solidification structures for a set of growth velocities and alloy compositions at a constant thermal gradient. For LPBF, the relevant microstructures include planar, dendritic, eutectic, and banded morphologies. The second type identifies the critical combination of interface velocity and thermal gradient at which the Columnar to Equiaxed Transition (CET) occurs for the dendritic morphology in an alloy with constant composition [3, 4, 6].

The analytical prediction of solidification morphologies is based on a set of growth models that describe the behaviour of the solid/liquid interface during non-equilibrium solidification. They were first proposed by Kurz *et al.* for various types of solid growth during directional solidification and laser treatment applications [3,7]. Utilizing Hunt's analytical model for the CET [8], later extended by Gauman *et al.* [6] to describe this transition during non-equilibrium solidification, columnar and equiaxed morphologies were also included. These analytical models, combined with thermal and thermodynamic data, can be used to create SMS maps [4].

This study aims to improve the AM design space for the popular multi-component Ni alloy Inconel 625 (IN625) processed by LPBF by combining knowledge of microstructure development gained through analytical predictive approaches and experimentation. The SMS maps were created for non-equilibrium solidification of this alloy. The microstructure of LPBF processed material, specifically the primary-dendrite-arm-spacing, was then characterized to validate the SMS maps both qualitatively and quantitatively. Finally, SMS map results for IN625 have been compared with previously published results for AlSi10Mg alloy [4]. These maps can be used to show how microstructure and properties are influenced by changing in solidification parameters and composition.

2. Creation of SMS maps

2.1. Description of material and phase formation

IN625 is a nickel-based superalloy which is strengthened mainly by solid solution strengthening via niobium, molybdenum, and chromium within the nickel-based (γ-FCC) matrix [9]. The nominal chemical composition of IN625 based on the ASTM F3056 standard is shown in Table 1 [10]. This alloy shows outstanding corrosion and oxidation resistance in elevated temperature/corrosive atmosphere conditions, as well as excellent yield strength, creep strength and fatigue strength due to the formation of intermetallic precipitates including γ" -Ni_3(Nb; Al; Ti), δ -Ni_3(Nb; Mo), laves-$(Ni; Cr; Fe)_2$(Nb; Mo; Ti) and MC carbides in the γ-FCC matrix [9,11].

Table 1. Chemical composition of Inconel 625 [10]

Element	C	Mn	Si	P	S	Cr	Co	Mo	Nb	Ti	Al	Fe	Ni
Min (wt%)	-	-	-	-	-	20.00	-	8.00	3.15	-	-	-	Remainder
Max (wt%)	0.10	0.50	0.50	0.015	0.015	23.00	1.00	10.00	4.15	0.40	0.40	5.00	

Figure 1 shows an isopleth section of the Ni-Cr-Nb-Fe- Mo phase diagram with 21 wt% Cr, 5 wt% Fe, 9 wt% Mo, and 0.8 wt% Co calculated via the Thermo-Calc software using the TCNI9 database [12]. The composition of IN625 is shown by the dashed line. As can be seen, equilibrium solidification of this alloy results in a single solid phase, γ-FCC. However, based on the Scheil solidification path [12] and previous experiments, the microstructure of IN625 parts fabricated via LPBF contain multiple phases: γ" -Ni_3Nb, laves, and (on rare occasions) δ -Ni_3Nb in addition to γ-FCC(Ni-Cr) [9, 13-14]. As can be seen in figure 1, the γ" -Ni_3Nb and laves phases must have formed through a eutectic reaction.

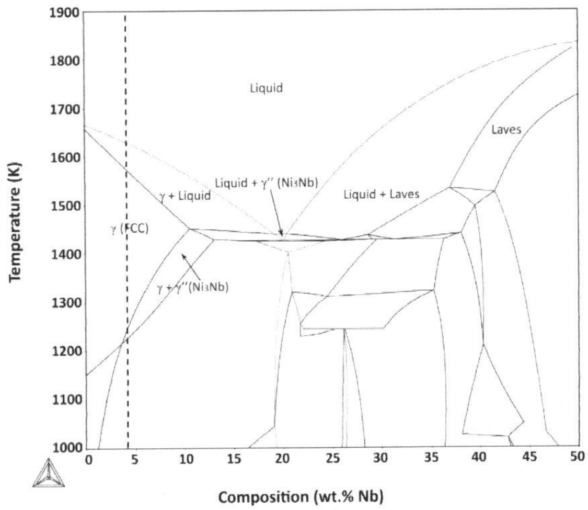

Figure 1. Isopleth section of Ni-Cr-Nb-Fe-Mo phase diagram with 21 wt% Cr, 5 wt% Fe, 9 wt% Mo, and 0.8 wt% Co - calculated via the Thermo-Calc software [12].

2.2. Analytical calculation

The aforementioned Kurz' and his colleagues' non-equilibrium growth models and interface response theory, i.e. the relationship between the solid/liquid interface temperature and velocity, as well as Gaumann's CET model, have been employed to create the type (1) and type (2) SMS maps. These models and calculation procedures have been documented in the literature [3,7,8] and most-recently summarized in [4] and are not completely repeated, but they are briefly discussed in this paper.

The general idea is based on calculating the undercooling of the S/L interface for all possible solidification growth morphologies while taking into account the effect of interface velocity on thermodynamic parameters in order to estimate the interface temperature over a range of C_0, G, V, and finally, to predict the microstructure under non-equilibrium solidification conditions. The stable morphology is then the one with the highest interface temperature. The interface temperatures for planar, dendritic, and eutectic growth, i.e. the relevant morphologies for the IN625 system, are shown below,

$$T_{PL}=T_m-\sum^i C_{0,i}\frac{m_i^v}{k_i^v}-\frac{V}{\mu_{k,i}} \tag{1}$$

$$T_D=T_{liq}-\Delta T_{tip} \text{ with } \Delta T_{tip}=\sum^i\left(\frac{k_i^v\Delta T_{0,i}^v Iv(Pe_d)}{1-(1-k_i^v)Iv(Pe_d)}+C_{0,i}(m_i-m_i^v)+\frac{V}{\mu_{k,i}}\right)+\frac{2\Gamma}{R}+\frac{GD}{V} \tag{2}$$

$$T_E=T_{eut}-\Delta T_{eut} \text{ with } \Delta T_{eut}=K_1\delta V+\frac{K_2}{\delta} \text{ and } \delta^2 V=\frac{K_2}{K_1} \tag{3}$$

where T_{pl}, T_D, and T_E are the planar, dendritic, and eutectic interface temperatures, ΔT_{tip} and ΔT_{eut} are the dendrite tip and eutectic solidification undercooling, T_{eut}, T_{liq}, and T_m are equilibrium eutectic, liquidus, and pure metal melting temperatures, V is the solidification velocity, D_l is the diffusion coefficient in the liquid, R the dendrite tip radius, Γ the Gibbs-Thomson coefficient, $\mu_{k,i}$ the kinetic attachment, δ the lamellar spacing, and C_0, m_i^v and k_i^v the initial composition, velocity-dependent liquidus slope, and velocity-dependent partition coefficient. The i subscript refers to the alloying element. The terms m_i^v and k_i^v are calculated via Aziz' model [15] for solute redistribution during non-equilibrium solidification as $m_i^v=m_i(1+(k_i-k_i^v[1-ln(k_i^v/k_i)])/1-k_i)$ and $k_i^v=(k_i+Pe)/(1+Pe)$ where Pe is the Péclet number, m_i and k_i are the equilibrium liquidus slope and partition coefficient for element i.

The general equation of the CET model is defined as

$$G=\frac{1}{1+n}\sqrt[3]{\frac{-4\pi N_0}{3\ln(1-\phi)}}\Delta T_n\left(1-\frac{\Delta T_n^{n+1}}{\Delta T^{n+1}}\right) \tag{4}$$

where ΔT_n and N_0 are the nucleation undercooling and nucleation density of the equiaxed grains, n is a constant and ϕ is the volume fraction of the equiaxed grains.

Previous studies of LPBF processed IN625 have shown that the microstructure contains the γ-FCC(Ni-Cr) phase as a matrix with the texture in < 100 > as well as γ”-Ni_3Nb and laves precipitates. Potentially, each of these phases could form during non-equilibrium solidification and as such the possible solidification microstructures that must be considered include: Planar- γ-FCC(Ni-Cr), dendrite-γ-FCC(Ni-Cr), planar- γ”-Ni_3Nb, primary- γ”-Ni_3Nb, planar-laves, primary-laves, eutectic (γ- γ”), eutectic (γ-laves) and the banded structure. Of these nine different microstructures, only four are probable, Planar- γ-FCC(Ni-Cr), dendrite- γ-FCC(Ni-Cr), eutectic (γ- γ”), and eutectic (γ-laves), given the alloys' low Nb content. Thus, only these four microstructures will be considered within the calculation of SMS maps.

To perform the necessary calculations, material physical properties are needed. Thermodynamic data, including the equilibrium liquidus slope (m), partition coefficient (k), liquidus temperature (T_{liq}), and solidus temperature (T_S) are calculated with Thermo-Calc software using TCNI9 database [12]. Other needed properties were taken from the available literature. Furthermore, the thermal gradient of the LPBF melt pool must be known. Although sophisticated methods exist, we have used the Rosenthal solution [16]—an analytical solution to the heat conduction equation developed for welding that is able to estimate the temperature field around a moving heat source—to match the desire for a computationally efficient prediction of microstructure during LPBF. The general Rosenthal solution formula, equation (5), was utilized to predict the temperature history:

$$T=T_0+\frac{\lambda P}{2\pi k r}\exp\left[-\frac{V_s(x+r)}{2\alpha}\right] \qquad (5)$$

where T_0 is the temperature far from the heat source, r is the radial distance from the moving point, λ is absorptivity, k and α are thermal conductivity and diffusivity, and P is laser power. The thermal gradient for LPBF processing of IN625 is estimated to be 10^6 to 10^{10} K/m within a meltpool when utilizing P=200 W and V_s=1100 mm/s as scanning power and velocity during single track LPBF processing.

3. LPBF single track experiments

3.1. Laser powder bed fusion experiment

In order to verify the microstructures predicted by the SMS maps, a single-track LPBF experiment has been performed utilizing the EOSINT M280 machine equipped with a 400 W Ytterbium fiber laser. The IN625 powder had a size distribution of 15-45μm and was processed using a power of P=200 W and scan velocity of V_s=1100 mm/s.

3.2. Characterization

After fabrication via AM, samples were prepared for metallography. First, samples were cut transverse cross-sectionally, mounted, polished and then etched chemically using aqua regia 3 HCl: 1 HNO3. Then, the melt pool microstructure was investigated by a KEYENCE optical microscope (OM) and a JEOL JSM-7000F Scanning Electron Microscope (SEM) equipped with an Oxford AZtecHKL Electron Backscatter Diffraction (EBSD) detector.

4. Results

4.1. Qualitative comparison of the SMS maps of LPBF IN625 with experimental findings

Figure 2(a) shows the interface responses of the four solidification microstructures that are expected to form during LPBF processing of IN625 as a function of interface velocity at the thermal gradient of 2×10^7 K/m. Based on the maximum interface approach, dendrite-γ-FCC(Ni-Cr) has the most stable morphology in the processing range of LPBF, as expected. The information contained in figure 2(a) can be used to create the SMS Maps. Figures 2(b) and 2(c) show the type 1 and type 2 maps, respectively, as well as the LPBF processing conditions. As can be seen in these maps, the dendritic γ-FCC(Ni-Cr) and banded-FCC(Ni-Cr) solidification microstructure, as well as fully columnar grains, are expected to form during LPBP processing of IN625. Figure 3(a) shows an OM micrograph of the transverse cross-

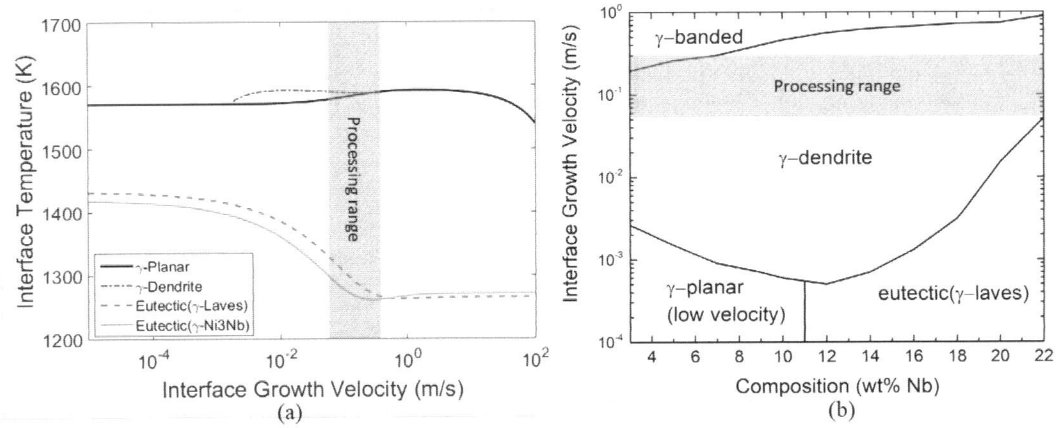

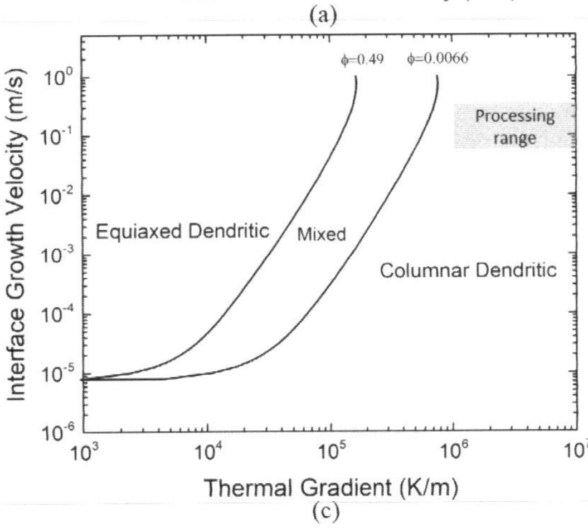

Figure 2. (a) The variation of interface temperature with solidification velocity for possible growth morphologies of the IN625 alloy at G=of 2×10^7K/m, (b) Type (1) SMS map for the IN625 alloy at $G = 2\times10^7$K/m over a range of composition for Nb, (c) Type (2) SMS map for the IN625 alloy.

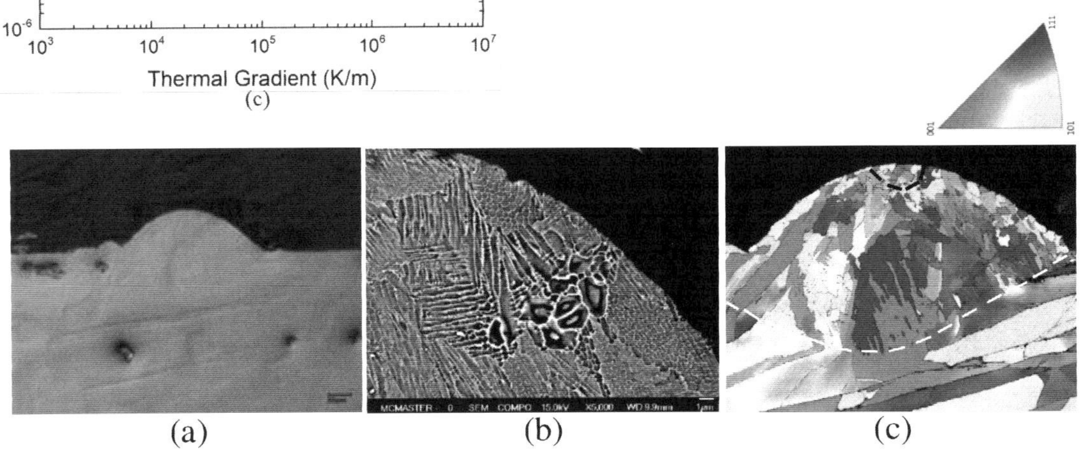

Figure 3. (a) OM (b) SEM micrographs, and (c) EBSD image of the transverse cross-section of IN625 single track melt with powder at scan velocity= 1100 mm/s and laser power= 200 W.

section of single-track LPBF processing of IN625 for the case with a scan velocity of 1100 mm/s and laser power 200 W; the hump represents the deposited layer. Figures 3(b) and 3(c) provide the corresponding high-resolution SEM micrograph and the EBSD orientation map of the melt pool. As can be seen in figure 3(b), the microstructure consists of dendritic γ-FCC(Ni-Cr) with a small amount of γ”-Ni_3Nb and laves participates within the inter-dendritic region. As a result of the non-equilibrium solidification process, the dendritic structure contains mainly primary arms with very short secondary arms that result in the structure appearing to have a cellular morphology. Further, at the base of the melt pool, figure 3(c), the columnar grains appear to have the same orientation as the substrate, providing evidence that they grew epitaxially towards the centre of the deposit. The epitaxial growth is related to the partial melting of the substrate while the elongated morphology is a result of the high thermal gradient and localized directional heat extraction at the edge of the melt pool.

Comparing figure 3(b) to the SMS map prediction in figure 2(b), it can be seen that while dendritic γ-FCC(Ni-Cr) is correctly predicted by the SMS map, the formation of the γ"-Ni_3Nb and laves participates is not anticipated from the analytical solution. On one hand, the phase diagram indicates that the γ"-Ni_3Nb and laves precipitates form as a solid-state transformation, and thus it is reasonable to not see them on the SMS map since only solidified microstructures can be predicted via the SMS map approach. On the other hand, Scheil solidification predicts their formation as a eutectic reaction. However, as the amount of Nb (4 wt%) in IN625 is much smaller than the eutectic composition (10wt% Nb for eutectic (γ- γ") and 20wt%Nb for eutectic (γ-laves)) it is unlikely that the eutectic reaction is taking place. In fact, the SMS Type 1 map (figure 2(b)) shows that these phases precipitate as a eutectic only for a high amount of Nb (larger than 11wt%) and at interface velocities much lower than the LPBF processing range. Based on these two points, it is likely that the γ"-Ni_3Nb and laves precipitates formed during solid-state phase transformation. The banded γ-FCC(Ni-Cr) was also expected to form ahead of the columnar dendrites. However, it was not observed in the microstructure. This is likely because, as shown in figure 3(c), equiaxed dendrites have formed instead ahead of the columnar dendrites. Finally, comparing figure 3(c) to the SMS map prediction in figure 2(c), it would appear that while columnar solidification is correctly predicted by the SMS map towards the base of the melt pool, the formation of equiaxed grains was not anticipated by the analytical solution. These equiaxed grains have nucleated ahead of the columnar front due to a drop in the thermal gradient at the end of melt pool's solidification. The calculation shown in figure 2(c) indicates that a CET will occur when the thermal gradient is less than approx. 5×10^5 K/m. Thus, it is likely that the thermal gradient near the top of the melt pool is significantly reduced as compared to what is predicted via the Rosenthal equation.

4.2. Quantitative comparison of primary arm spacing between experiment and calculation

Based on knowledge of the thermal gradient, Kurz and Fisher's model [17] can be used to quantify the Primary Dendrite Arm Spacing (PDAS) during dendritic growth,

$$\lambda=4.3\left(\frac{\Delta T}{G}\right)^{\frac{1}{2}}\left(\frac{D_l\Gamma}{Vk\Delta T_0}\right)^{\frac{1}{4}} \tag{6}$$

where λ is the primary arm spacing, k the equilibrium partition coefficient, and ΔT_0 and ΔT the equilibrium and non-equilibrium solidification range. The thermophysical parameters are listed in table 2, while the non-equilibrium solidification interval is given by $\Delta T= m^v C_0 (k^v-1)/k^v$.

Table 2. Thermophysical properties used for SMS maps and PDAS calculations in IN625[12,18]

Parameter		Unit
Initial composition, C_0	21Cr-9Mo-5Fe-4.1Nb-0.8Co	wt%
Liquidus slopes for: $m_{Nb,\gamma}$, $m_{Nb,laves}$, $m_{Nb,Ni3Nb}$	1199, 1356, 737	$Kwt\%^{-1}$
Partition coefficients for: $k_{Nb,\gamma}$, $k_{Nb,laves}$, $k_{Nb,Ni3Nb}$	0.5, 0.48, 0.37	-
Gibbs-Thomson coefficient for: Γ_γ, Γ_{Laves}, Γ_{Ni3Nb}	10^{-7}, 2×10^{-7}, 2×10^{-7}	mK
Diffusion coefficient in the liquid, D_l	3×10^{-9}	m^2s^{-1}
Material property, n	3.4	-
Nucleation undercooling of equiaxed grains, ΔT_n	1.5	K
Nucleation density of equiaxed grains, N_0	2×10^{15}	m^{-3}
Equilibrium solidification range, ΔT_0	75	K

Figure 4(a) shows a comparison between the experimentally measured PDAS and the values calculated with equation (6). The calculated values are shown by the solid and dotted lines, for two thermal gradients. The experimentally-measured points are given by black squares. As expected, the calculation shows that an increase in the growth velocity and thermal gradient decreases the PDAS. For the range of conditions [19] seen in LPBF processing, the calculated PDAS at $G=10^7$ K/m and $G=2\times10^7$ K/m are seen to change from 0.75 to 1.08 µm and 0.5 to 0.76 µm respectively for different growth velocities. These values match quite closely the experimental points, with locally-averaged measurements of 0.4 and 0.7 µm at two different positions within the melt pool, given uncertainties in materials properties especially D_l and the potential for Marangoni-induced convection. The PDAS is seen to decrease from the melt pool edge to the center due to the increase in the solidification velocity and the decrease in the thermal gradient.

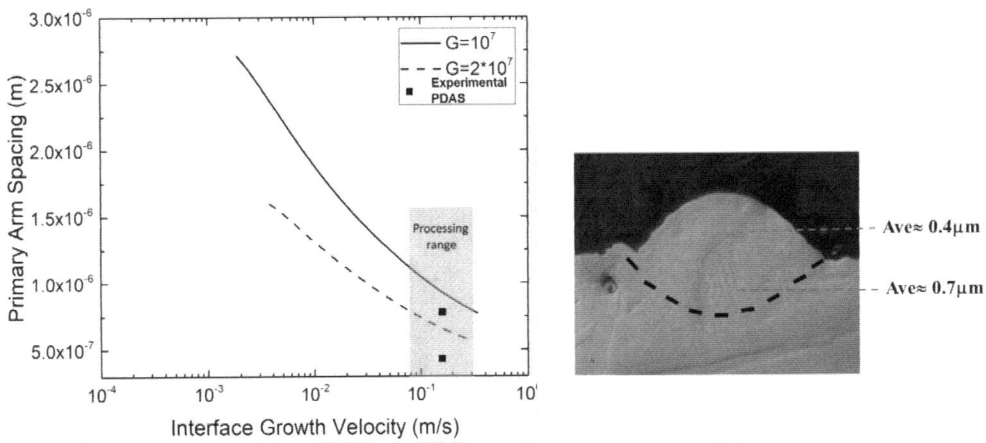

(a) (b)

Figure 4. (a) Predicted PDAS over a range of growth velocity at the thermal gradient of $G=10^7$ and 2×10^7 K/m (b) experimental PDAS.

4.3. Comparison of IN625 and AlSi10Mg SMS maps

SMS maps can provide guidance to improve LPBF processed microstructure and understand phase formation. To give an example, the differences between the SMS maps of IN625, and AlSi10Mg [4], figure 5, are striking, owing to the different thermal and thermodynamic properties and compositions.

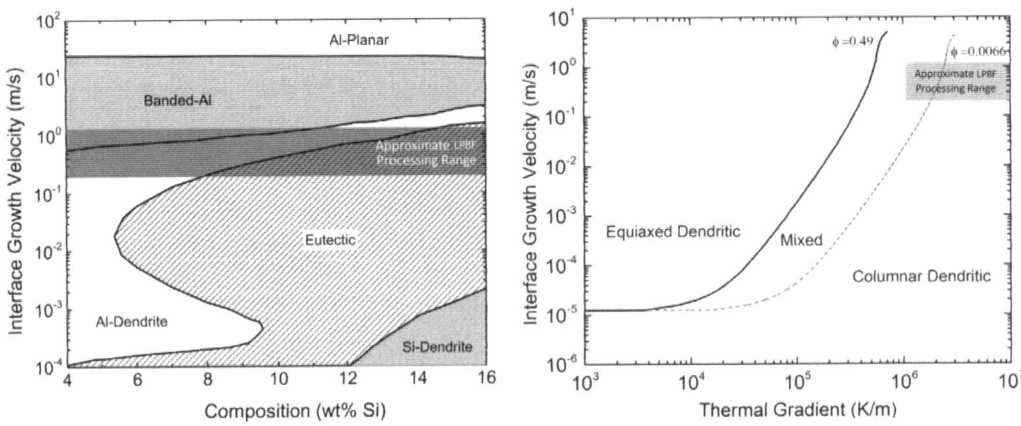

Figure 5. SMS maps for the AlSi10Mg alloy (a) Type (1) at $G = 10^6$K/m and (b) Type (2) over a range of Si compositions [4].

AlSi10Mg has a near-eutectic composition (12wt% Si), which leads to the formation of the eutectic microstructure. However, in IN625, the composition is far from the eutectic composition and as a result, the only predicted phase is dendritic γ-FCC(Ni-Cr). Furthermore, the grain morphology for LBPF processed AlSi10Mg is expected to be both columnar and equiaxed, whereas only the formation of columnar grains is predicted for IN625. Different parameters can affect the critical *G* in the columnar to equiaxed transition, such as type and composition of alloying elements, thermodynamic parameters (liquidus slope and partition coefficient), nucleation density, and nucleation undercooling. The overall contribution of these parameters produces a lower critical *G* for CET in IN625.

5. Conclusion

The modes of solidification and grain morphology formed during LPBF processing are dependent on the *G*, *V*, and initial composition of the material. *G* and *V* are a function of not only processing parameters but also the position within the melt pool. Due to the high *G* and *V* during LBPF processing, the formation of planar, eutectic, columnar dendrite, bands and equiaxed dendrite are expected. The SMS maps predict that columnar dendritic and banded microstructure form in LPBF microstructure of IN625. However, columnar to equiaxed transition has been observed in the experimentally micrograph of this alloy; this can be related to the drop in the thermal gradient in the center of the melt to the lower value than the predicted value by the Rosenthal solution. Since the heat source is moving, the amount of *G* and *V* are changing in the melt pool, which leads to the uneven distribution of solidification modes, grain size and PDAS in the single melt pool. The microstructure of LPBF processed IN625 mostly contains columnar dendrites with a PDAS ranging from 0.4 to 0.7 μm, in agreement with the analytical calculation results in terms of both the mode of solidification and primary arm spacing.

Aknowledgement
The authors acknowledge funding from the National Sciences and Engineering Council of Canada, and experimental support from McMaster's Additive Manufacturing Research Group.

References

[1] Babu S S, Love L, Dehoff R, Peter W, Watkins T R and Pannala S 2015 *MRS Bull.* **40** 1154-61
[2] DebRoy T, Wei H L, Zuback J S, Mukherjee T, Elmer J W, Milewski J O, Beese A M, Wilson-Heid A, De A and Zhang W 2018 *Prog. Mater. Sci.* **92** 112-224
[3] Kurz W 2001 *Adv. Eng. Mater.* **3(7)** 443-52
[4] Mohammadpour P, Plotkowski A and Phillion A B 2020 *Addit. Manuf.* **31** 100936
[5] Dantzig J A 2000 *JOM* **52(12)** 18-21
[6] Gäumann M, Trivedi R and Kurz W 1997 *Mater. Sci. Eng., A* **226** 763-9
[7] Kurz W l, Giovanola B and Trivedi R 1986 *Acta Metall.* **34(5)** 823-30
[8] Hunt J D 1984 *Mater. Sci. Eng.* **65(1)** 75-83
[9] Li S, Wei Q, Shi Y, Zhu Z and Zhang D 2015 *J. Mater. Sci. Technol.* **31(9)** 946-52
[10] F3056 2015 *ASTM International*
[11] Dinda G P, Dasgupta A K and Mazumder J 2009 *Mater. Sci. Eng. A* **509(1-2)** 98-104
[12] Andersson J O, Helander T, Höglund L, Shi P and Sundman B 2002 *Calphad* **26(2)** 273-312
[13] Keller T, Lindwall G, Ghosh S, Ma L, Lane B M, Zhang F, Kattner U R, Lass E A, Heigel J C, Idell Y and Williams M E 2017 *Acta Mater.* **139** 244-53
[14] Marchese G, Lorusso M, Parizia S, Bassini E, Lee J W, Calignano F, Manfredi D, Terner M, Hong H U, Ugues D and Lombardi M 2018 *Mater. Sci. Eng. A* **729** 64-75
[15] Aziz M J 1982 *J. Appl. Phys.* **53(2)** 1158-68
[16] Rosenthal D 1941 *Weld J* **20(5)** 34-220
[17] Kurz W and Fisher D J 1981 *Acta Metall.* **29(1)** 11-20
[18] Lee Y, Nordin M, Babu S S and Farson D 2014 *Metall. Mater. Trans. B* **45(4)** 1520-9
[19] Promoppatum P, Yao S C, Pistorius P C and Rollett A D 2017 *Engineering* **3(5)** 685-94

IOP Conf. Series: Materials Science and Engineering **861** (2020) 012006 doi:10.1088/1757-899X/861/1/012006

Resolving the effects of local convective heat transfer via adjustment of thermo-physical properties in pure heat conduction simulation of Laser Powder Bed Fusion (L-PBF)

M Bayat[1], V K Nadimpalli[1], S Mohanty[1] and J H Hattel[1]

[1] Department of Mechanical Engineering, Technical University of Denmark, building 425, Lyngby, Denmark

E-mail: mbayat@mek.dtu.dk

Abstract. Numerical models, especially when validated against experimental findings, are valuable tools that can be used as a cheap and reliable alternative to the expensive and time-consuming experimental investigations. In the current work, a multiphysics model of laser powder bed fusion (L-PBF) has been developed for stainless steel SS 316-L and validated against experimental findings. By configuring the model, the effects of the inverse Marangoni flow on the melt pool morphology and the heat and fluid flow conditions were analysed. It was shown that for high surface tension sensitivities, the total heat flux through the melt pool boundaries increases. The enhanced cooling will lead to a smaller melt pool size with larger depth to width ratio and causes a relatively uniform temperature field, which reduces the role of the conduction heat transfer. Furthermore, a Nusselt number was derived to quantify the role of the Marangoni effect on the convection heat transfer from the melt pool. This expression was used to calculate an effective thermal conductivity required for a modified pure conduction heat transfer calculation. It was shown that, via using a combination of the derived Nusselt number and the concept of the effective thermal conductivity, the average melt pool temperature determined from the modified pure conduction model gets very close to that of the more advanced CFD model.

1. Introduction

L-PBF is a branch of the Metal Additive Manufacturing (MAM) processes, where parts are manufactured layer-by-layer. The L-PBF process has found various functional applications in industries like automotive, energy, aerospace, and medical technology sectors [1], primarily due to its large flexibility in design freedom for manufacturing complex parts.

In L-PBF, tailoring the final microstructure and having a consistent melt track (melt pool profile) is essential for controlling the part quality. A numerical model which is validated against experimental findings can provide deep insights into the effects of the thermo-physical properties on the melt pool profile and its size and at a very low cost. However, by sole use of *ex-situ* and *in-situ* experimental analysis, one will miss many details regarding the spatio-temporal variation in the heat and fluid flow conditions. Even with the most advanced online monitoring systems, it is not possible to either determine the correct temperature [2] or fluid velocity at a time resolution even close to those the numerical models can predict. This once more necessitates the implementation of the numerical models as an alternative to the costly and time-consuming experimental investigations.

Computational Fluid Dynamics (CFD) has recently attracted attention for modelling of the L-PBF process, as it can predict the thermo-fluid conditions and the melt pool profiles with a very good accuracy. Depending on their complexity, the CFD models can take different physical phenomena into consideration, such as the surface tension, the thermo-capillary effect, the recoil pressure [3], etc. The author group has recently developed multi-physics models involving the aforementioned phenomena, to predict the lack-of-fusion and keyhole-induced porosities using the same technique [4-5].

Recent studies from the authors have shown that the pure conduction models overestimate temperature and the melt volume, since they neglect the fluid flow [6]. On the other hand, it must be stated that the CFD models are extremely CPU-heavy, as they solve for many more unknown variables such as the velocity components and pressure. This will in turn incur a lot of computational costs on the one who does these calculations. Accordingly, this is not favored by the industries who look for fast-responding models. However, while using the pure conduction heat transfer models, one can get more reliable results at a much lower computational time, if the effective thermal properties like thermal conductivity, are used. By using the fluid dynamics and heat transfer theories, it is possible to determine these effective thermal properties to improve the output from the pure heat conduction models.

In this work we developed a high-fidelity finite volume (FVM) numerical model for the L-PBF process of 316-L stainless steel in the commercial software package, Flow-3D. The predicted melt pool profile is compared with the experimental melt pool micrograph where a very good agreement is achieved. By only keeping thermo-capillarity, the model is modified and then used in a parametric study to investigate the effect of the inverse Marangoni (with positive surface tension sensitivity) flow on the heat and fluid flow conditions inside the melt pool. Furthermore, a Nusselt number is found to quantify the role of convection in the transfer of heat through the melt pool boundaries. This Nusselt number is then used to find the effective thermal conductivity in a modified pure conduction heat transfer model, in order to get more reliable melt pool temperatures.

2. Modelling and experiments

2.1. The CFD model

Partial differential equations of balance of mass and linear momentum are solved to find the coupled fields of velocity and pressure. For an incompressible flow we have

$$\vec{\nabla}\cdot\vec{V}=0, \tag{1}$$

$$\rho\left[\frac{\partial}{\partial t}(\vec{V})+\vec{V}\cdot\vec{\nabla}(\vec{V})\right]=-\vec{\nabla}p+\vec{\nabla}\cdot\left[\mu\left(\vec{\nabla}\vec{V}+\vec{\nabla}\vec{V}^{T}-\frac{2}{3}\vec{\nabla}\cdot\vec{V}\right)\right]-\frac{K_C(1-f_l)^2}{C_K+f_l^3}\vec{V}-\rho\vec{g}\beta\Delta T_{ref}, \tag{2}$$

where V is the fluid velocity vector and p is pressure, respectively in equation (2). K_c and C_K are two constants related to Darcy flow, used in order to freeze and free the fluid flow upon solidification and melting. β is the thermal expansion coefficient in equation (2), where the Boussinesq approximation is used. f_l the liquid fraction function which in this work is assumed to be a linear function of temperature.

The heat transfer equation is solved to find the temperature field

$$\rho\left[\frac{\partial}{\partial t}(h)+\vec{V}\cdot\vec{\nabla}(h)\right]=\vec{\nabla}\cdot\left[k\left(\vec{\nabla}T\right)\right]. \tag{3}$$

h is the fluid enthalpy and ρ and k are respectively the bulk density and thermal conductivity. In the L-PBF or welding process, there are three interfacial forces that have a significant impact on the liquid motion.

$$F_{int}=\left[P_{recoil}+\sigma(T)\kappa\right]\vec{n}+\gamma\left[\vec{\nabla}T-\vec{n}\left(\vec{\nabla}T\cdot\vec{n}\right)\right]. \tag{4}$$

P_{recoil} is the recoil pressure which along with the surface tension force, $\sigma\kappa$, act normal to the interface, while the Marangoni effect, the second term in equation (5), acts tangential to the interface. γ in equation (4) is the surface tension sensitivity to temperature. Moreover, to track the free surface of the fluid, the VOF method with the split Lagrangian scheme is used. The exposed surface of the metal transfers heat via convection and conduction to the surroundings.

$$-k\frac{\partial T}{\partial n}=h(T\text{-}T_{surr})+\varepsilon\eta\left(T^4\text{-}T_{surr}^4\right)\text{-}q''_{max}\,exp\left(\text{-}2\frac{r^2}{r_b^2}\right)+q''_{evap} \quad (5)$$

h, η, and q_{evap} are convective heat transfer coefficient, Stephan Boltzmann constant and the evaporative heat flux expressed in equation (5), respectively. The third term on the right hand side of equation (5) belongs to the laser where r_b is the radius of which the heat flux reaches e^{-2} of its maximum. In this work the heat source is modelled with ray-tracing but with a constant absorptivity.

2.2. Experiments and setup

The experiments were performed on an in-house made L-PBF machine developed at Technical University of Denmark (DTU) and with 316-L stainless steel powder [7]. The experimental setup consists of a 250 W single mode fibre laser with a nominal effective beam radius (r_b) of 50 μm and 1070 nm wavelength, an F-θ scan lens in addition to an in-house made galvanometer and laser controller to change the scanning velocity. The laser can be used in continuous wave (CW) and pulsed modes up to a frequency of 100 kHz. In the current work, only the continuous mode was utilized. Single track specimens were made on the top of a flat sheet of steel (SS316) with different scanning speeds and input powers where a custom single-layer powder handling unit was used. Then the samples were cut and analysed with light optical microscopy to find the melt pool cross-sectional profile. The scanning speed of the laser in the experiments was set to 210 mm.s^{-1}.

3. Results and discussion

3.1. Experimental validation of the multi-physics model

Figure 1(a) shows the 3D view of the temperature contour along with the surface deformation at t = 6.3 ms. According to figure 1(a), the peak temperature is about 3500 K which is slightly higher than the boiling temperature for 316-L stainless steel. At this relatively high localized temperature, the recoil pressure is active and overcomes the capillary force, causing a depression at the location of the hotspot, see figure 1(a). Figure 1(b) shows the comparison between the experimental and numerical melt pool profile from the y-z view. Figure 1(c) shows the solid fraction contour along with the velocity vectors from a longitudinal view. According to figure 1(c), in the depression zone, the recoil pressure will drive and pump the liquid material to the back of the melt pool. Right behind the zone of which the recoil pressure is active, the positive Marangoni effect exists which drives the liquid from the hot to the cold part of the melt pool. This effect will act as the second mechanism which drives the material to the cold tail of the melt pool, subsequent to the recoil pressure. At the very back of the melt pool, the liquid metal starts solidifying and as the temperature starts to decrease, the surface tension increases, which will form a hump at the tail of the melt pool, see figure 1(c). In this respect, the hump formation is caused by, the movement of the cold liquid metal and it's swelling at the back of the melt pool, in addition to the dominant capillary force at the cold tail of the melt pool.

3.2. The inverse Marangoni effect

To be able to investigate the sole effect of the inverse Marangoni flow on the heat transfer from the melt pool boundaries, the recoil pressure and the buoyancy effects are removed from the fluid dynamics part of the developed multi-physics model. Moreover, the heat transfer via radiation and convection to the ambient are removed where in this way, it is guaranteed that all the heat input from the laser goes to the underneath solid domain. In this respect, a parametric study was performed with six different positive surface tension sensitivities (γ in equation (4)) in addition to a pure conduction case. The model specification for the parametric study are given in table 1.

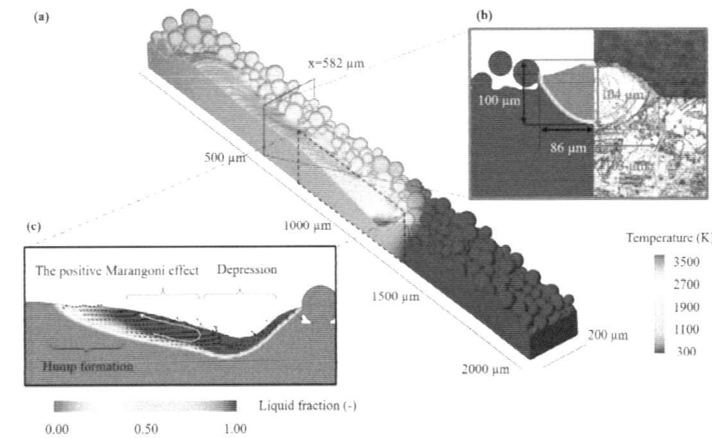

Figure 1. (a) The 3D view for the temperature contour at t=6.3 ms. (b) Comparison between numerical and experimental melt pool profile. (c) An x-z view of the melt pool liquid fraction contour.

Table 1. Model specifications for the parametric study.

Case #	Model ID	Power (W)	Speed (m.s^{-1})	D_b (μm)	γ (N.m^{-1}.K^{-1})
1	conduction				0.00
2	gamma_01				2.5e-5
3	gamma_02				5.0e-5
4	gamma_03	80	500	30	2.5e-4
5	gamma_04				5.0e-4
6	gamma_05				7.5e-4
7	gamma_06				1.0e-3

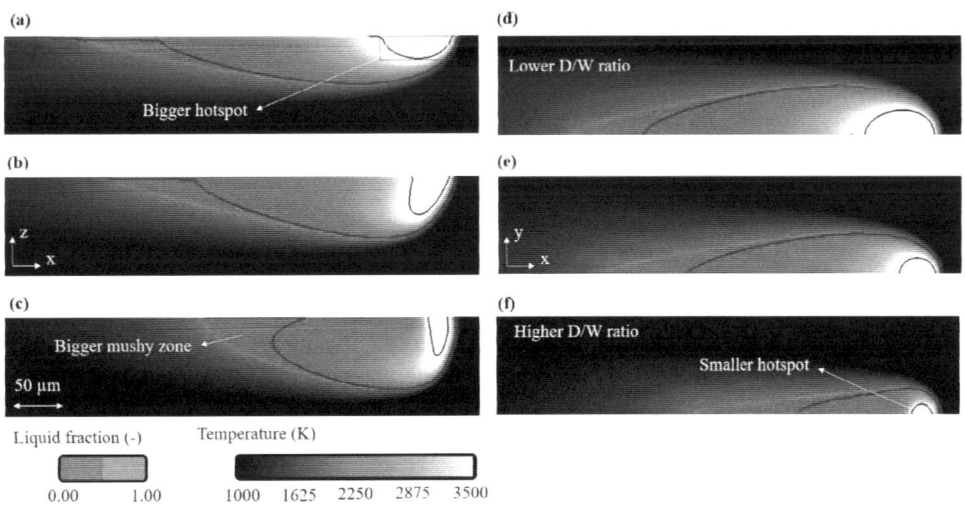

Figure 2. Longitudinal view of the temperature contour along with solid/liquid lines for (a) conduction, (b) gamma_02 and (c) gamma_05 at t=1.26 ms. (d), (e) and (f) respectively show the corresponding contours of (a-c), from the top view.

Figure 2 shows the temperature contours and the melt pool boundaries from the longitudinal and top views at t=1.26 ms and for chosen cases listed in table 1. One can notice the relatively larger size of the melt pool in the pure conduction case where there is no fluid flow involved. According to figure 2(a) and (d), in the pure conduction case, the hotspot region is larger as compared to the gamma_02 case shown in figure 2(b) and (e). In the pure conduction case, the melt pool will grow equally in width and depth, see figure 2(a) and (d). However, when the fluid flow is added to the model, the melt pool profile becomes deeper and at the same time smaller in width, see figure 2(c) and (f). It is noted that the higher the γ, the deeper the melt pool will be. This is in agreement with the experimental findings of Mills *et al.* [8], who observed a better weld penetration in high-sulphur content steels (with higher γ), as compared to low-sulphur samples. Details of the melt pool morphology are given in table 2.

Table 2. Melt pool size and morphology details for the parametric study.

Case	Length (μm)	Width (μm)	Depth (μm)	D/W ratio	L/W ratio
conduction	387.54	107.25	53.62	0.50	3.61
gamma_01	379.71	96.23	62.03	0.64	3.95
gamma_02	362.03	89.86	68.99	0.77	4.03
gamma_03	293.91	68.41	78.26	1.14	4.30
gamma_04	284.93	62.61	81.74	1.31	4.55
gamma_05	268.12	57.97	82.03	1.42	4.63
gamma_06	268.12	53.33	78.26	1.47	5.03

According to both figure 2 and table 2, it was observed that the depth/width and length/width ratio increase with higher Marangoni coefficients. Also it is noted that the melt pool shrinks in size for higher Marangoni coefficients. Interestingly based on figure 2(c) and (f), the mushy zone (the area shown between the red and blue lines) extends to the end and top of the melt pool, for higher fluid flows. This means that the melt pool starts solidifying from the top, even though the top boundary is set to adiabatic. Moreover, one can visually see in figure 2 that the hotspot region, despite getting smaller in size, becomes directed towards the bottom of the melt pool, for higher Marangoni coefficients. Details regarding the calculated average thermal properties of the melt pool are listed in table 3.

Table 3. Calculated averaged thermal and fluid dynamics related properties belonging to the melt pool.

γ	$T_{avg,liq}$	V_{liq}	δ_T	$V_{avg,mag}$	***Ma***	***Ec***	***Pe***	V_{mush}/V_{melt}
N.m^{-1}.K^{-1}	K	m^3	(-)	m.s^{-1}	(-)	(-)	(-)	(-)
2.50E-05	2151.78	7.523E-13	71.88	1.55E-01	8.17E+01	5.69E-08	1.92E+00	0.124
5.00E-05	2116.63	7.114E-13	63.70	2.45E-01	1.49E+02	1.54E-07	2.99E+00	0.139
2.50E-04	2037.56	5.236E-13	43.37	6.46E-01	5.52E+02	1.32E-06	7.18E+00	0.168
5.00E-04	2015.50	5.043E-13	37.59	8.05E-01	1.02E+03	2.20E-06	8.85E+00	0.170
7.50E-04	2007.32	4.852E-13	34.76	9.10E-01	1.47E+03	2.88E-06	9.89E+00	0.170
1.00E-03	1989.97	4.850E-13	32.09	1.00E+00	1.85E+03	3.70E-06	1.09E+01	0.173

According to table 3, a higher Marangoni effect will lead to a higher average liquid velocity inside the melt pool, as expected. Moreover, the melt pool volume reduces as we increase the Marangoni coefficient γ, according to table 3. V_{mush}/V_{melt} in table 3 shows the ratio of the mushy zone volume to the total melt pool volume. According to table 3, for γ=1.0e-3, more than 17% of the melt pool is composed of the mushy zone. At the same time, it was observed that increasing γ leads to a decrease in the average liquid temperature inside the melt pool. δ_T is the dimensionless variance of temperature inside the melt pool, in table 3, called the temperature non-uniformity factor, which decreases for higher Marangoni flows. Also the definition of the Peclect number can be helpful to understand the relative importance of convection compared to conduction in heat transfer. Table 3 shows that convection becomes more important with increasing Marangoni effect, as the calculated Peclect number (denoted Pe in table 3) becomes five times bigger for γ=1.0e-3, compared to γ=2.5e-5.

Figure 3 shows the velocity magnitude contour along with the velocity vectors in the x-z plane and at the centre of the melt pool. According to figure 3(a), the inverse Marangoni effect drives the liquid on the top surface from the cold tail of the melt pool and to the hot centre where the surface tension is higher. This observation is exactly the opposite of the validation case shown in figure 1(c), belonging to a negative surface tension sensitivity which used to drive the liquid from the hot central zone to the cold outer zones. In figure 3(b) and its blow-up it is observed that two circulations are formed on the front and at the back of the laser spot location. However, due to the low liquid volume in the melt pool front, the front circulation is found to be much smaller in size. By increasing the Marangoni coefficient, it is noted that a strong vertical jet of liquid forms right at the place where these two vortices meet each other see figure 3(c). The mentioned jet of the liquid metal, thus gets heated at the hotspot location and will play the role of a driller which digs in the cold material underneath – resulting in higher depth/width ratios.

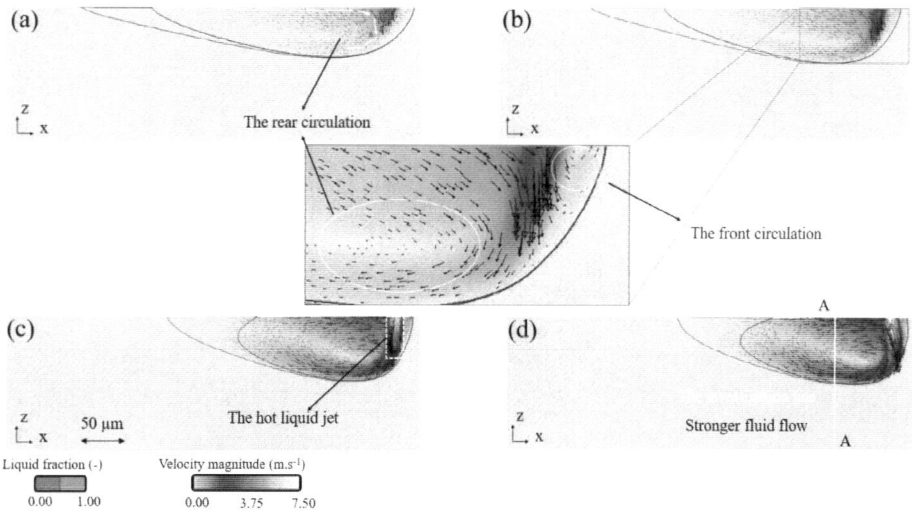

Figure 3. The velocity magnitude contours at the x-z cross-section and at the centre of the melt pool for (a) gamma_01, (b) gamma_02, (c) gamma_04 and (d) gamma_06 cases at t=1.26 ms (after reaching the quasi-steady condition).

To have a better understanding regarding the fluid flow and the temperature field inside the melt pool, temperature and x-direction velocity (denoted u) fields are plotted against the z-coordinate in figure 4(a) and (b), respectively, at the investigation line A-A which is shown in figure 3(d).

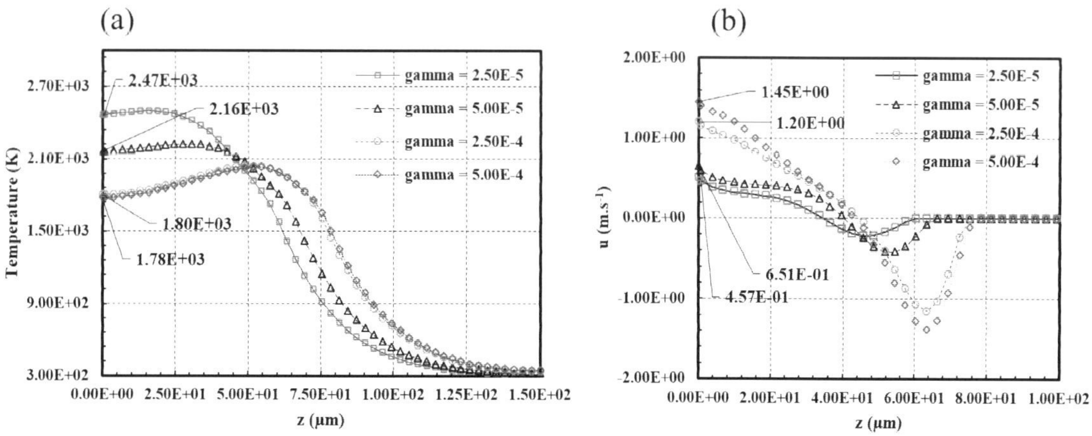

Figure 4. (a) Temperature and (b) u-velocity profile along the vertical investigation line A-A (shown in figure 3(d)) for different Marangoni coefficients.

The temperature profile along the z-direction is displayed in figure 4(a) for different Marangoni coefficients. One can see that by choosing a lower Marangoni coefficient, both the peak and the average temperatures increase, based on figure 4(a). It is interesting to note that the temperature is higher at the bottom part of the melt pool, compared to its upper part, for higher Marangoni numbers, see figure 4(a). This means that the return flow belonging to the rear vortex is initially warm, which is driven by the heated downward vertical jet mentioned earlier. This in addition to a lower temperature predicted at the top part of the melt pool, once more underline the rapid cooling in the high-Marangoni cases which will ultimately cause solidification of the liquid metal from the upper back of the melt pool, which was discussed earlier. Figure 4(b) shows the vertical distribution of the x-direction velocity along the A-A line. It is observed that the peak u-velocity increases with increasing Marangoni coefficients, as expected. Moreover, to fulfil the mass conservation, the returning flow is also stronger for the higher Marangoni number cases, see figure 4(b). Moreover, a higher viscous stress is predicted for higher Marangoni numbers, despite a more uniform thermal field.

3.3. *The effective thermal conductivity*

As the convection and radiation to the ambient are turned off, in the quasi-steady condition, one can divide the total input heat flux into a convective and a conductive part. This can be formulated by

$$q''_{tot}=q''_{cond}+q''_{conv}=k_l \nabla T+h_{sl}\Delta T, \tag{6}$$

k_l and h_{sl} are the stationary liquid conductivity and the convection heat transfer coefficient from the melt pool to the solid. We assume that the conductive and convective heat fluxes in equation (6), can be found via the average melt pool temperature difference,

$$q''_{cond}=k_l\frac{\left[T_{avg,liq}-T_s\right]}{l_{ch}}, \tag{7}$$

$$q''_{conv}=h_{sl}\left[T_{avg,liq}-T_s\right], \tag{8}$$

where h_{sl} is unknown in equation (8). One can find the solid/liquid heat transfer coefficient by substituting the conductive and convective heat fluxes from equations (7) and (8), into equation (6). We can solve for the unknown heat transfer coefficient via finding the corresponding Nusselt number

$$Nu=\frac{h_{sl}l_{ch}}{k_l}=\frac{q''_{tot}l_{ch}}{\left[T_{avg,liq}-T_s\right]k_l}-1. \tag{9}$$

When the heat transfer coefficient is determined, one can use the following expression found by Ladani *et al.* [9] to find the effective thermal conductivity for a pure conduction heat transfer problem

$$k_{eff}=k_l+h_{sl}\cdot l_{ch}, \tag{10}$$

where l_{ch} in equations (7-10) is the characteristic length of the melt pool, which is the diameter of a hemisphere which has an equivalent volume to that of the melt pool [6].

Figure 5(a) shows the plot of the calculated Nusselt number as a function of the Marangoni number, both of which are determined from the CFD models belonging to the parametric study.

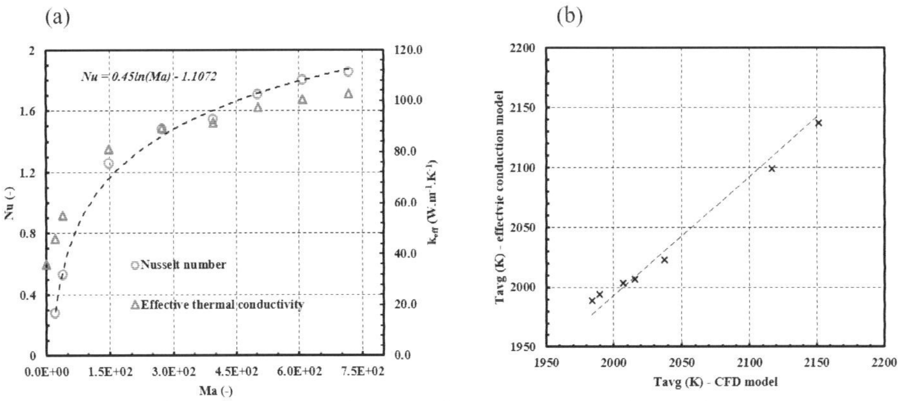

Figure 5. The plot of the calculated (a) Nusselt number versus the Marangoni number. (b) The plot of the average melt pool temperature from the effective conduction model versus the one predicted by the CFD model.

According to figure 5(a), the Nusselt number increases with the Marangoni number, which underlines the fact that the convective heat transfer is getting stronger. Moreover, figure 5(a) shows the effective thermal conductivity determined from equation (10) and as expected, it increases for higher Marangoni numbers. Figure 5(b) shows the plot of the average melt pool temperature calculated from the effective conduction model against the one calculated from the CFD model. Accordingly a very good match is observed between the average melt pool temperature calculated from the CFD model and the conduction model using the effective thermal conductivity. As mentioned earlier, the effective thermal conductivity is determined based on the Nusselt number shown in figure 5(a).

4. Conclusion

In this work a multi-physics numerical model was developed for the L-PBF process of 316-L stainless steel and the predicted melt pool profile was validated against the experimental findings. The predicted melt pool profile was in a very good agreement with the experimental micrograph. The multi-physics model was then modified to enable the investigation of the sole effect of inverse Marangoni on the heat and fluid flow conditions inside the melt pool and also its profile, size and morphology. Thus a parametric study was performed to investigate the effect of the inverse Marangoni flow and the results suggest that the melt pool shrinks as the surface tension sensitivity to temperature increases. Furthermore it was found that the temperature inside the melt pool is relatively uniform for higher Marangoni flows. This makes the conduction heat transfer contribution weaker as the temperature gradients become smaller. In addition, a Nusselt number was developed that quantifies the effect of the convection heat transfer through the melt pool boundaries. Moreover, based on the calculated Nusselt number, an effective thermal conductivity was found. Utilizing a combination of the developed Nusselt number and the concept of the effective thermal conductivity, the average melt pool temperature will get very close to the predictions by the more advanced CFD model and at a much lower computational cost.

References

[1] Debroy T, Mukherjee T, Milewski J O, Elmer J W, Ribic B, Blecher J J and Zhang W 2019 *Nat. Materials* **18** 1026-32
[2] Hooper P A 2018 *Addit. Manuf.* **22** 548-59
[3] Zhao Y, Koizumi Y, Aoyagi K, Wei D, Yamanaka D and Chiba A 2018 *Addit. Manuf.* **26** 202–14
[4] Bayat M, Mohanty S, and Hattel J H 2019 *Int. J. Heat Mass Transf.* **139** 95–114
[5] Bayat M, Thanki A Mohanty S, Witvrouw, Yang S, Thorborg J, Tiedje N S and Hattel J H 2019 *Addit. Manuf.* **30** 100835
[6] M Bayat M, Mohanty S, and Hattel J H 2019 *Int. J. Heat Mass Transf.* **139** 213–30
[7] Nadipalli V K, Andersen S A, Nielsen J S, and Pedersen D B 2019 in *Proc. of the Joint Special Interest Group meeting between euspen and ASPE Advancing Precision in Additive Manufacturing,* ed. A. Bernard *et al.* (The European Society for Precision Engineering and Nanotechnology), pp 66-69
[8] Mills K C, Keene B J, Brooks R F, and Olusanya A 1984 (National Physical Laboratory, Teddinton, Middlesex, UK)
[9] Ladani L, Romano J, Brindley W, and Burlatsky S 2017 *Addit. Manuf.* **14** 13-23

Columnar-to-equiaxed transition in a laser scan for metal additive manufacturing*

L Yuan[1], A S Sabau[2], D StJohn[3], A Prasad[3] and P D Lee[4]

[1] University of South Carolina, Columbia, SC, 29201, US
[2] Oak Ridge National Laboratory, Oak Ridge, TN 37830, US
[3] The University of Queensland, St Lucia QLD 4072, Australia
[4] University College London, London WC1E 6BT, UK

E-mail: langyuan@cec.sc.edu

Abstract. In laser powder bed fusion additive manufacturing (LPBFAM), different solidification conditions, e.g., thermal gradient and cooling rate, can be achieved by controlling the process parameters, such as laser power and laser speed. Tailoring the behaviour of the columnar to equiaxed transition (CET) of the printed alloy during fabrication can facilitate the production of highly customized microstructures. In this study, effective analytical solutions for both thermal conduction and solidification are employed to model solidifying melt pools. Microstructure textures and solidification conditions are evaluated for numerous combinations of laser power and laser speed under bead-on-plate conditions. This analytical-based high-throughput tool was demonstrated to select specific process parameters that lead to desired microstructures. Two selected process conditions were examined in detail by a highly parallelized microstructural solidification model to reveal both nucleation and grain growth. Both numerical solutions agree well with experiments that are performed based on bead-on-plate conditions, indicating that these numerical models aid evaluation of the nucleation parameters, providing insights for controlling CET during the LPBFAM processing.

1. Introduction

The Laser Powder Bed Fusion Additive Manufacturing process (LPBFAM) is one of the most used additive manufacturing processes for direct materials consolidation. One of the major advantages is that process parameters, e.g. laser power, speed, spot size and scanning paths and build plate temperature, can be adjusted locally over wide ranges independently, leading to dramatically different solidification conditions, e.g. thermal gradient and solidification speed. In turn, tailoring the solidification environment can result in location specific microstructures, thus, customized material properties for optimal performance [1–3].

Effectively controlling the columnar to equiaxed transition (CET) is one of the basic requirements to fabricate materials with deliberate microstructures. Although CET has been well studied for solidification processes, to eliminate columnar grains and promote equiaxed grains in metal additive

*__Notice__: *This manuscript has been authored in part by UT-Battelle, LLC, under contract DE-AC05-00OR22725 with the US Department of Energy (DOE). The United States Government retains and the publisher, by accepting the article for publication, acknowledges that the United States Government retains a non-exclusive, paid-up, irrevocable, world-wide license to publish or reproduce the published form of this manuscript, or allow others to do so, for United States Government purposes. The Department of Energy will provide public access to these results of federally sponsored research in accordance with the DOE Public Access Plan (http://energy.gov/downloads/doe-public-access-plan).*

manufacturing (AM) is rather challenging. In the melt pool, columnar grains typically form on account of the prevailing local directional solidification conditions, where grains grow epitaxially from the preexisting grains with no energy barrier for nucleation. Means to promote CET during AM can be generally attributed to the following four catagories: 1) control process parameters [4, 5]; 2) modify alloy composition [6]; 3) add nano/micro particles [7, 8]; and 4) interact with external fields [9, 10]. Recent research through both numerical analysis [4] and experimental observations [11] shows that thermal conditions via controlling process parameters nevertheless play an important role to facilitate other factors, such as variation of alloy composition and grain refiner, to promote nucleation. Due to high computational cost for physics-based numerical models, an effective means based on analytical solutions that can correlate solidification conditions from a wide range of process parameters to CET is preferred to assist the selection of process conditions for desired microstructure morphologies.

Classic solidification theory suggests that sufficient constitutional supercooling (CS) ahead of the solidification front is required to activate potent heterogenous nucleant particles, which essentially triggers a nucleation event, eventually resulting in CET. Among the theories, the Interdependence Model describes the establishment of Nucleation Free Zone (NFZ) due to the existence of CS whose length governs the distance at which a new nucleation event is triggered. On this basis, the model predicts the average grain size based on the location of the next nucleation event [12, 13]. This model has also been successfully applied to understand the grain refinement during AM [11]. To theoretically predict the CET, the analytical solution proposed by Hunt [14] has been widely adopted to generate the CET map based on thermal gradient and solidification speed [15]. Gaumann *et al.* extended Hunt's model with the Kurz–Giovanola–Trivedi (KGT) model to account for rapid solidification and applied it to achieve single crystals during a laser deposition process [16].

In this study, analytical solutions for both temperature and CET are resolved to effectively screen a wide range of laser process parameters, namely, laser power and laser speed, for bead-on-plate conditions during the LPBFAM process. The selection of bead-on-plate conditions is due to its simplicity and the identical solidification principle during the fabrication process. A process map based on the predictions is generated to guide the experimental design. To further evaluate the predictions, selective process parameters that result in different grain morphologies are predicted by a highly parallelized microstructural solidification model to reveal both nucleation and grain growth. Predictions from both methods are compared with experiments, providing effective tools to examine CET during LPBFAM processing.

2. Model description

Analytical solutions have been widely applied to evaluate process conditions and microstructure morphologies during solidification processes. They are effective methods to screen a large range of process parameters in comparison to physics-based high-fidelity numerical models. With necessary calibrations, they can provide acceptable accuracy that reflects the general material's behavior. Two analytical solutions for the temperature profile in a moving distributed heat source and the CET event, respectively, are implemented here to assess process conditions. Meanwhile, to examine details about the solidification microstructures, a highly parallel solidification model was integrated with the thermal analytical solutions to predict the nucleation events and grain growth. A summary of both analytical solutions and the solidification model is provided below.

2.1. Analytical solutions

Assuming only heat conduction occurs during laser scanning, a steady-state temperature profile can be reached for a moving heat source. Rosenthal has provided a solution for a moving point heat source [17]. For a Gaussian laser beam, the distribution of laser intensity is described as:

$$q(r) = \frac{2AP}{\pi R^2}\exp\left(-\frac{2r^2}{R^2}\right) \tag{1}$$

where P is the laser power, A is the laser absorption coefficient, R is the laser beam radius and r is the distance to the laser center. By integrating the heat flux based on the distributed heat source on the surface of a semi-infinite plate and assuming the laser is moving along the x direction, the temperature profile at quasi-steady state can be expressed as [17, 18]:

$$T(x,y,z) = T_0 + \frac{2AP}{\pi^{1.5}\rho c_p} \times \frac{\int_0^{\infty} \exp\left(-\frac{2(x+Vt)^2 + 2y^2}{R^2 + 8\alpha t} - \frac{z^2}{4\alpha t}\right)}{\sqrt{\alpha t}(R^2 + 8\alpha t)} dt \quad (2)$$

where (x, y, z) is the location coordinate relative to the heat source, T_0 is the build plate temperature, ρ is the density, c_p is the specific heat, α is the thermal diffusivity, V is the laser speed and t is the time.

Equation (2) provides a 3D quasi-steady thermal profile within the melt pool. The solidification conditions, both thermal gradient and solidification rate, can be extracted for the prediction of CET. The extended model proposed by Gaumann *et al.* [16] is applied here to evaluate the thermal conditions. The dendrite tip velocity, v, is approximated by:

$$v = a\,(\Delta T)^n \quad (3)$$

where a and n are alloy dependent constants. The fraction solid of equiaxed grains, ϕ_E, then, can be expressed as:

$$\phi_E = \frac{4\pi n_0}{3}\left(\frac{1}{G(n+1)}\left(1 - \frac{\Delta T^{n+1}}{\left(\frac{v}{a}\right)^{\frac{n+1}{n}}}\right)\left(\frac{v}{a}\right)^{\frac{1}{n}}\right)^3 \quad (4)$$

where n_0 is the nucleation density, G is the thermal gradient, v is the solidification speed, ΔT is the tip undercooling. When $\phi_E < 0.0066$, the structure is fully columnar. When $\phi_E > 0.66$, a fully equiaxed structure is expected [14]. For solidification under a high temperature gradient, the nuclei density plays an important role and the nucleation undercooling can be safely neglected [16]. Therefore, equation (4) can be simplified to:

$$\frac{G^n}{v} = \frac{1}{a}\left(\frac{1}{n+1}\left(\frac{4\pi n_0}{3\phi_E}\right)^{\frac{1}{3}}\right)^n \quad (5)$$

With given G and v, the higher the value obtained from equation (5), the easier the growth of columnar grains.

2.2. *Solidification microstructure model*

To predict the solidification microstructure based on thermal conditions, a parallel solidification code, adapted from the open source code µMatIC [19–21], was developed to predict the grain structure during rapid solidification. In this model, at every time step, the local quasi-steady temperature profiles are directly calculated from the thermal analytical solution and drive the nucleation event and grain growth. The code takes a master-worker approach, where a master node allocates subdomains to each worker (computational node) for heavy computation based on the status of the thermal profiles. For instance, if the temperature in the subdomain is fully solid or below the solidus temperature, no calculation will be assigned to workers to reduce both calculation time and memory burden. This allows a significant increase of computational efficiency for large geometries on high performance computers. In the model, the nucleation of new grains is modelled assuming a Gaussian distribution of nucleation density (n):

$$\frac{dn}{d(\Delta T)} = \frac{n_0}{\sqrt{2\pi}\sigma_n}\exp\left[-\frac{1}{2}\left(\frac{\Delta T - \Delta T_n}{\sigma_n}\right)^2\right] \quad (6)$$

where σ_n is the standard deviation of the Gaussian distribution, ΔT_n is the mean nucleation undercooling. and grain growth is then simulated based on the modified KGT model according to [16] as shown in equation (3). The implementation details can be found in prior publications [19–21].

3. Results and discussion

The nickel-based superalloy, IN625, was studied both numerically and experimentally. The thermal analytical solution does not take into account the temperature dependent material properties nor the latent heat during solidification. As such, the material's properties and laser absorption coefficient were calibrated by comparing with bead-on-plate experiments in LPBFAM. In addition, only conditions where the laser operates in the conduction melting mode are considered. The applied simulation parameters and general process conditions are listed in table 1. Figure 1 shows the predicted thermal profiles and the melt pool size for 75W and 100mm/s laser conditions. Melt pool shape and the dimensions can be consistently repeated with the current set of parameters in comparison with experiments for views of both cross section and top surface (Note that figure 1(d) is the cross-section of the laser scan and shows the melt pool depth and width; figure 1(e) shows the top view at the end of laser scan which reveals the length and width of the melt pool). Note that the accuracy of the prediction may vary as the laser conditions change. However, the general behaviors of thermal conditions remain valid to evaluate the process conditions and the potency of the CET.

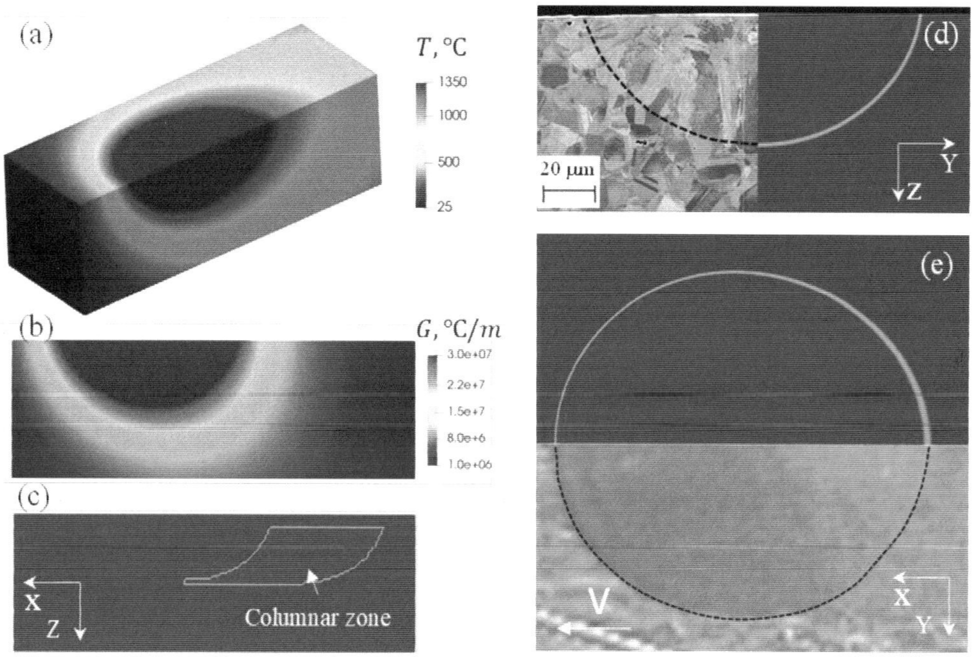

Figure 1. Analytical thermal predictions for laser parameters at 75W and 100mm/s: (a) steady state 3D thermal profiles; (b) thermal gradient at x-z cross section; (c) indication of CET; (d) comparison of melt pool size in the z-y direction; and (e) comparison of melt pool along the substrate surface where the laser was turned off.

Equation (5) was calculated based on the local G and v to estimate CET. To determine the tip growth velocity, v, both a and n are estimated by fitting equation (3) to the dendrite/cellular tip undercooling calculated with the Marginal Stability growth model [24]. Using the thermal dynamic data calculated by ThermoCalc to obtain the liquidus slopes and partitioning coefficients and assuming a constant Gibbs–Thomson coefficient of 1.0×10^{-7} m K and equal diffusion coefficients for the alloying elements of 3.0×10^{-9} m^2/s., a and n were calculated at $3.97 \times 10^{-6} m/s°C^{3.1}$ and 3.1, respectively. The detailed approach can be found in [25]. Since the nucleation density n_0 largely determines the critical value for CET, $n_0 = 2 \times 10^{15}$ m^{-3} was taken as a reference value to demonstrate the potential columnar zone, as shown in figure 1(c). Note that changing n_0 will change the size of columnar zone from the analytical solution. To effectively evaluate the solidification conditions for each case, average values for cooling rate, R, and G^n/v are calculated for the semisolid region located at the tail of the melt pool (the region after the maximum depth). The values of two cases, P=75W, V=100mm/s and P=75W, V=500mm/s, are shown in table 2, together with measured average primary dendrite arm spacing (PDAS). Case 2 had a higher cooling rate and higher G^n/v value, indicating that it should yield(?) finer PDAS and higher potency to grow columnar grains. The microstructures from bead-on-plate experiments for both cases are shown in figure 2. Qualitatively, Case 2 showed less equiaxed zone in comparison with Case 1. In addition, by measuring the average PDAS, Case 2 had the average value of 0.38 μm while Case 1 was 0.68 μm. The analytical predictions are consistent with experimental observations.

Table 1. Simulation Parameters for IN625[22, 23]

Parameter	Unit	Value
Density, ρ	kg/m^3	8440
Specific heat, c_p	J/kg/ °C	600
Liquidus temperature, T_{sol}	°C	1290
Liquidus temperature, T_{liq}	°C	1350
Thermal diffusivity, α	m^2/s	4.94e-6
Laser absorption coefficient, A		0.33
Laser beam radius, R	mm	0.05
Build plate temperature, T_0	°C	25

Table 2. Representative values from the analytical solutions

	P, W	V, mm/s	R_{avg}, K/s	G^n/v	PDAS_Exp, μm
Case 1	75	100	1.02×10^6	6.66×10^{26}	0.68±0.07
Case 2	75	500	5.47×10^6	2.43×10^{27}	0.38±0.06

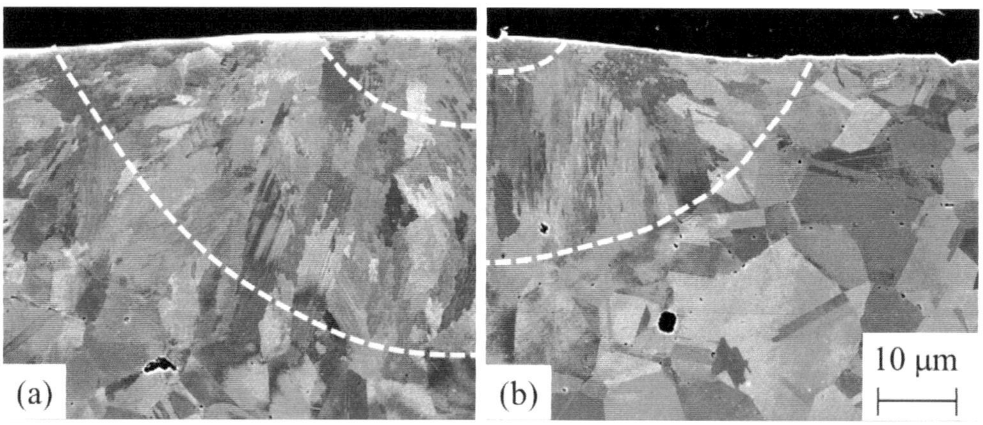

Figure 2. Microstructures at the y-z cross-section for (a) Case 1: P=75W, V=100mm/s; and (b) Case 2: P=75W, V=500mm/s. Within the melt pool, the top line indicates the CET line and the bottom line indicates the melt pool boundary.

Based on the average R and G^n/v values, a wide range of process parameters can be examined. Assuming both laser power and speed can vary from 50W to 500W and from 0.05 m/s to 2 m/s, respectively, a design of experiment was performed based on the Latin Hypercube sampling method [26] to fill the parametrical space. To avoid both insufficient melting and keyholing [27], the correlation of P and V were constrained by:

$$100 < \frac{P}{\sqrt{V}} < 350 \tag{7}$$

Both the lower and upper bounds in equation (7) were determined by bead-on-plate experiments on IN625. The combinations of *P* and *V* are shown in figure 3(a), which includes 270 different process conditions. The average *R* and G^n/v, calculated from the analytical solutions, is then displayed in figure 3(b). The *R* and G^n/v map will assist the selection of process conditions for potentially different solidification features. For example, to produce potential columnar grains structures with less residual stress or with less potency to solidification cracking, process conditions with high G^n/v values and low cooling rate are preferred, respectively. In figure 3(b), the conditions located in left-upper corner most likely can satisfy the requirements. By selecting the $\log_{10}(G^n/v)$ value of 26.5 and average cooling rate of 2.5×10^6 °C/*s* as the lower bounds, the conditions are highlighted as boxed dots in figure 3(a), which can be used to specify the experimental conditions.

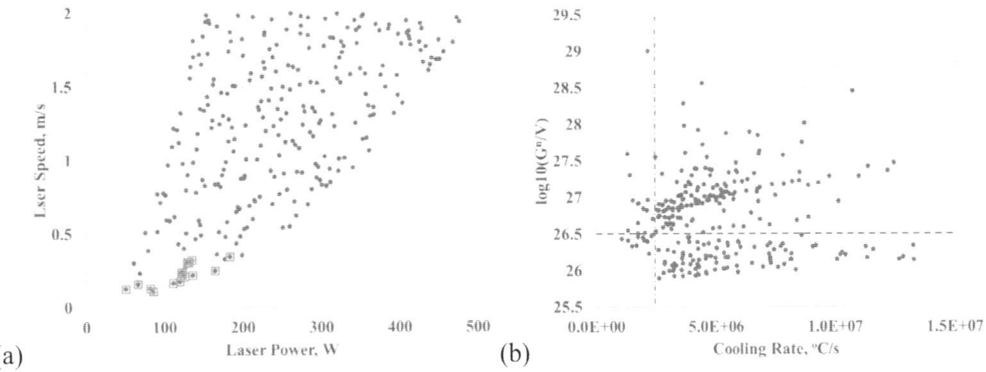

Figure 3. (a) Design of Experiment for laser power and laser speed: the boxed dots indicate the conditions with low average cooling rate and low CET potency. (b) Distribution based on average cooling rate and G^n/v.

Both cases in table 2 were simulated by the solidification microstructure model using the temperature profiles calculated by the analytical solution. In the current simulation mode, segregation information is neglected. Therefore, only thermal undercooling drives the nucleation and growth. A grid size of $2.5\times 10^{-7} m$ was chosen based on the PDAS. The selection of the nucleation parameters will determine the morphologies and size of the grain. Since the relative observations between the two cases were of interest in the current study, a nucleation undercooling of 10 °C with standard deviation of 2 °C and a maximum nucleation density of 1×10^{17} were used for both simulations. The calculation domain was set as a cube with a length of 5.0×10^{-5} m, which covered the entire melt pool for the display of the steady state grain structures. To mimic the initial grain conditions, the domain starts with fine grains. As the laser scanned and melted the substrate, remelting and solidification would occur to form the final microstructures. The predicted grain structures are shown in figure 4.

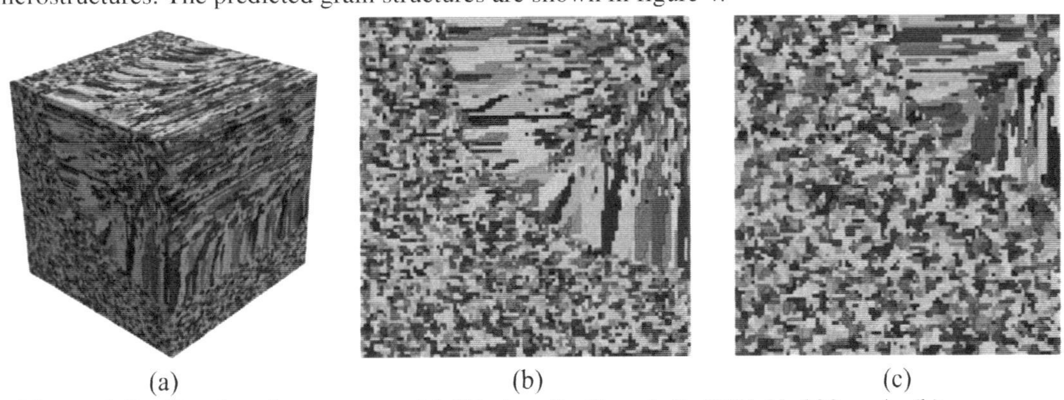

Figure 4. Predicted grain structures: (a) 3D view for Case 1: P=75W; V=100mm/s; (b) y-z cross-sectional view for Case 1; (c) y-z cross-sectional view for Case 2: P=75W, V=500mm/s.

With the current set of simulation parameters, the results reflect the general trend as shown in figure 2 that Case 1 tends to have a larger nucleation zone. Epitaxial growth dominates the bottom of the melt pool, where the columnar grains grow from the original fine grains.

High thermal gradients and high cooling rates are expected in the LPBFAM process. As shown in figure 5, following the Interdependence Model, with increasing thermal gradient, the magnitude of the constitutional supercooling is largely suppressed in terms of the Interdependence Model, shown in figure 5(a) by the maximum difference between T_E (the equilibrium temperature) and T_{A_highG} (actual temperature with high thermal gradient). With a high cooling rate, both the segregation and diffusion length of the solute are compressed, as shown in figure 5(b). The contribution of CS for nucleation may be neglected, where thermal supercooling dominates the solidification process. This also explains that columnar grains are preferred at the bottom of the melt pool besides epitaxial growth, where there is insufficient undercooling for nucleation at the initial stage of solidification. Therefore, limiting thermal gradient and grain growth speed will benefit nucleation, and in turn, facilitate the formation of equiaxed grains. The further development and implementation of the current solidification model can physically evaluate the theorical predictions and provide insights to the CET formation.

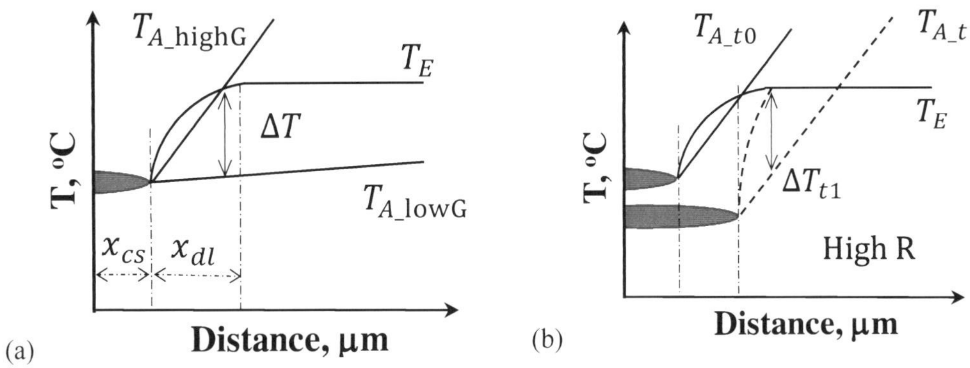

Figure 5. Schematic representation of the constitutionally supercooled zone as described by the Interdependence model: (a) with low and high thermal gradients; and (b) with high cooling rate at two time steps, $t0$ and $t1$.

4. Conclusions

Analytical solutions for both thermal conduction and solidification are applied to a broad range of laser power and speed on bead-on-plate conditions during LPBFAM. The average cooling rate, R and the CET criteria based on G^n/v are calculated for the solidifying region to evaluate the microstructure's texture. Qualitatively, the analytical solutions match experimental observations with necessary calibrations with the material properties. By combining different R and G^n/v, the process conditions that produce the desired grain size, morphology and potential defects can be selected. The extension of such a high throughput toolset provides effective means to examine full sets of key process parameters, including laser spot size, scanning paths and substrate temperature, to guide experimental design and locally control microstructure morphology.

Numerical simulations for two process conditions that generate different grain structures are simulated by a highly parallel microstructural solidification model to reveal both nucleation and grain growth. The predictions reflect the general solidification behaviours, as shown in the experiments with identical conditions, during solidification with both high thermal gradient and high cooling rate. The Interdependence Model suggests that thermal supercooling dominates the nucleation behaviour. Although the microstructure model did not include the alloy composition directly, it can be a valuable tool to directly examine process conditions and nucleation behaviours to guide design the microstructure.

Acknowledgements
This research was partially conducted for the project "ExaAM: Transforming Additive Manufacturing through Exascale Simulation", which was supported by the Exascale Computing Project (17-SC-20-SC), a collaborative effort of the U.S. Department of Energy (DOE) Office of Science and the National Nuclear Security Administration. This research used resources of the Oak Ridge Leadership Computing Facility, which is a DOE Office of Science User Facility supported under Contract DE-AC05-00OR22725. The research was performed under the auspices of the US DOE by Oak Ridge National Laboratory under contract No. DE-AC0500OR22725, UT-Battelle, LLC. PDL acknowledges support by the UK-EPSRC (EP/P006566/1).

References

[1] DebRoy T *et al.* 2018 *Prog. Mater. Sci.* **92** 112–224
[2] Bourell D L 2016 *Annu. Rev. Mater. Res.* **46** 1–18
[3] Gu D D, Meiners W, Wissenbach K and Poprawe R 2012 *Int. Mater. Rev.* **57** 133–64
[4] Haines M, Plotkowski A, Frederick C L, Schwalbach E J and Babu S S 2018 *Comput. Mater. Sci.* **155** 340–9
[5] Liang Y-J, Cheng X, Li J and Wang H-M 2017 *Mater. Des.* **130** 197–207
[6] Zhang D, Qiu D, Gibson M A, Zheng Y, Fraser H L, StJohn D H and Easton M A 2019 *Nature* **576** 91–5
[7] Li X P *et al.* 2017 *Acta Mater.* **129** 183–93
[8] Lin T-C, Cao C, Sokoluk M, Jiang L, Wang X, Schoenung J M, Lavernia E J and Li X 2019 *Nat. Commun.* **10** 4124
[9] Todaro C J *et al.* 2020 *Nat. Commun.* **11** 142
[10] Ning F, Hu Y, Liu Z, Cong W, Li Y and Wang X 2017 *Procedia Manuf.* **10** 771–8
[11] Bermingham M J, StJohn D H, Krynen J, Tedman-Jones S and Dargusch M S 2019 *Acta Mater.* **168** 261–74
[12] StJohn D H, Qian M, Easton M A and Cao P 2011 *Acta Mater.* **59** 4907–21
[13] Prasad A, Yuan L, Lee P D and StJohn D H 2013 *Acta Mater.* **61** 5914–27
[14] Hunt J D 1984 *Mater. Sci. Eng.* **65** 75–83
[15] Acharya R, Bansal R, Gambone J J and Das S 2014 *Metall. Mater. Trans. B* **45** 2279–90
[16] Gäumann M, Bezençon C, Canalis P and Kurz W 2001 *Acta Mater.* **49** 1051–62
[17] Dye D, Hunziker O and Reed R C 2001 *Acta Mater.* **49** 683–97
[18] Rosenthal D 1941 *Weld. J.* **20** 220–34
[19] Yuan L and Lee P D 2012 *Acta Mater.* **60** 4917–26
[20] Yuan L and Lee P D 2010 *Model. Simul. Mater. Sci. Eng.* **18** 55008
[21] Wang W, Lee P D and McLean M 2003 *Acta Mater.* **51** 2971–87
[22] Arısoy Y M, Criales L E and Özel T 2019 *Opt. Laser Technol.* **109** 278–92
[23] Mills K C, Youssef Y M, Li Z and Su Y 2006 *ISIJ Int.* **46** 623–32
[24] Trivedi R and Kurz W 1994 *Int. Mater. Rev.* **39** 49–74
[25] Lian Y, Gan Z, Yu C, Kats D, Liu W K and Wagner G J 2019 *Mater. Des.* **169** 107672
[26] Florian A 1992 *Probabilistic Eng. Mech.* **7** 123–30
[27] Hann D B, Iammi J and Folkes J 2011 *J. Phys. D. Appl. Phys.* **44** 445401

IOP Conf. Series: Materials Science and Engineering **861** (2020) 012008 doi:10.1088/1757-899X/861/1/012008

Thermomechanical modeling of additively manufactured structural parts - different approaches on the macroscale

J Thorborg[1], P Esser[1] and M Bayat[2]

[1] MAGMA GmbH, Kackertstraße 11, 52072 Aachen, Germany
[2] DTU Mechanical Engineering, Technical University of Denmark Building 425, 2800 Kgs. Lyngby, Denmark

E-mail: J.Thorborg@magmasoft.de

Abstract. Modeling the metal additive manufacturing process can be done on different length scales depending on the field of interest. However, due to the moving heat source and the narrow heat affected zone, a fine mesh and fine time resolution are required to analyze detailed information on various length scales. In this paper, the focus is on macro-scale temperature, deformation and residual stresses. The inherent strain method is applied layer by layer on the full part, to include the thermal contraction and inelastic strains generated by the moving heat source. The inherent strain components are mapped from the mesoscale simulations, where different boundary conditions can be used to resemble different regions from the macroscale. As an alternative to the inherent strain method and to include the thermal simulation on macroscale without simulating the moving heat source, a layer-by-layer approach is proposed and applied in the paper. The price to pay is the loss in detail of the thermal load. However, for the suggested layer-by-layer approach, the energy corresponds to the energy delivered from the laser. The proposed and implemented methods are applied to two examples and compared to measurements. Finally, large structural parts are simulated to evaluate the application on new trends in the automotive industry.

1. Introduction

Application of additive manufacturing has steadily been growing for many years and today the process is generally used to build complex parts in a wide range of alloys. The design freedom of the parts has been the main driver of the process. The possibility to build complex parts without expensive tooling has made the process interesting for prototyping and small series, especially for cases with a focus on service load analysis and an associated topology optimization. In laser powder bed fusion (LPBF), the building process of the parts comes with a price, in terms of steep thermal gradients around the moving heat source generated by the laser. Already solidified material is re-melted and affected by heating when new layers are added. High stress and inelastic strain can develop during the building process, which sometimes lead to large deformations when the part is cut off from the base plate and the support structure is removed. Heat treatment of the parts is often used to relax the stress level and to minimize these deformations. To analyze the influence of the process parameters on stability and the part quality, the process simulation is increasingly used today.

This paper presents the implementation of a general thermomechanical model, which can be used to simulate the moving heat source on a mesoscale, where the source term is modeled with the well-known Goldak model [1]. Furthermore, a layer-by-layer thermomechanical model is implemented to simulate the material response on a macroscale to provide results on the component level. Finally, the inherent strain method is implemented and applied to make a first qualitative evaluation of this 'simplified' approach.

The different functionality is implemented in MAGMASOFT® as an add-on to the integrated stress solver, MAGMAstress. The governing equations for the thermal and mechanical problems are discretized and integrated with standard linear hexagonal finite elements. All material data are temperature dependent. For the thermal solver it is possible to model convection and radiation on the outer surfaces. The constitutive law for the mechanical simulation is a standard time independent plasticity formulation, which is used during the printing process. During heat treatment a unified creep formulation is used to include time and rate effects at elevated temperature to model the stress relaxation.

Different parts from small academic size up to a large structural part from the automotive sector are simulated on different scales to evaluate the application of the integrated modeling approach. The results are compared to measurements which were performed during the evaluation of the implementation plus some comparison to findings in literature.

2. Mesoscale – moving heat source

On the mesoscale, each single laser path is simulated in detail. The energy distribution is described by a Goldak heat source model, see figure 1. The source term is directly used in the thermal simulation, where the laser parameters and the process conditions determine the moving heat source size and location. The state and temperature dependencies of the material data is taken into account to have a reasonable description of the variation in thermo-physical and mechanical properties of the powder and the solid material. For the presented results the state of the material is changed from powder to solid when the temperature is exceeding the solidus temperature. For the powder state the thermal conductivity and the surface emissivity are significantly reduced [2, 3], and the elements are not mechanically activated. For this type of simulation it means, that a new layer of powder material is activated in the thermal domain, when the re-coater has distributed the powder. The mechanical domain is dynamically growing inside the layer when the local temperature exceeds the solidus temperature and the thermal strain integration starts when the temperature drops below solidus again.

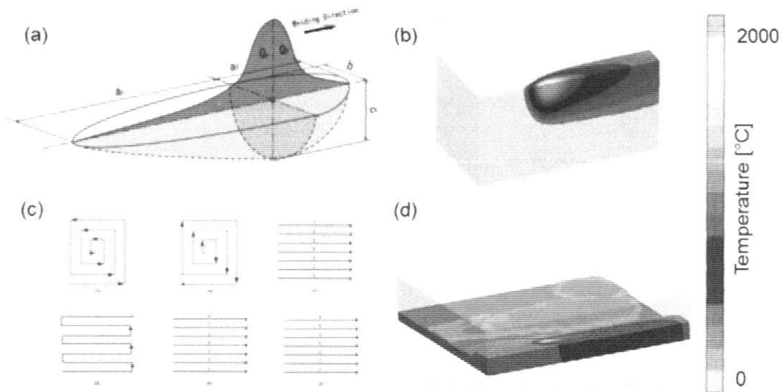

Figure 1. (a) Goldak's double-ellipsoidal heat source model, (b) temperature distribution around the moving source term, (c) typical scanning paths and (d) temperature distribution in a patch while being built.

The results from mesoscale simulations can be useful to analyze various local effects of hatching strategies and laser parameters and the influence on temperature and stress distributions. However, due to the moving heat source and the narrow heat affected zone a fine mesh and fine time resolution are required to analyze detailed information on very small length scales. If the focus is mainly on macroscale deformations and residual stresses it is unrealistic to resolve the details of the moving heat source even on smaller parts. To perform large scale simulations on macroscale it is often necessary to switch to a layer-by-layer strategy, where heated layers or layers of predefined inherent strain components are activated in the calculation domain step by step. The advantage of such approaches is obviously a significantly reduced calculation time due to the simplified description of the actual process, where the moving heat source is replaced by the activation of one or several full layers. Two quite different layer-by-layer approaches will be evaluated in the paper:

1. The inherent strain method, which is a pure mechanical model.
2. The thermo-mechanical method, where the thermal simulation is based on releasing the energy from the laser in one or several layers, followed by a mechanical simulation where the thermal strain is based on the simulated temperature distributions.

Measurements of deformation and stresses from two parts will be used to validate the two layer-by-layer approaches.

The heat transfer equation is solved on both mesoscale and macroscale in the thermomechanical case

$$\rho c_p \frac{\partial T}{\partial t} = \frac{\partial}{\partial x_i}\left\{k \frac{\partial T}{\partial x_i}\right\} + \dot{Q} \tag{1}$$

where ρ is the density, c_p is the specific heat capacity, T the temperature and $\dot{Q}$ the source term.

The static equilibrium equations are solved in all cases, with either the thermal strain or the inherent strain components as the main load input, it should be noticed that for the inherent strain approach the thermal strain is part of the inherent strain components

$$\sigma_{ij,i} + p_j = 0 \tag{2}$$

where σ_{ij} is the stress tensor and p_j is the volumetric force vector.

Time independent plasticity is used for the building process as the constitutive description, with the J_2 von Mises yielding condition

$$f = \sigma_e^2 - \sigma_y^2 = 0, with\ \sigma_y = \sigma_y(\varepsilon_e^{pl}, T) \text{ and } \sigma_e = \sqrt{3J_2} \tag{3}$$

Using the consistency condition to determine the load step

$$\dot{f} = \frac{\partial f}{\partial s_{ij}} \dot{s}_{ij} + \frac{\partial f}{\partial \varepsilon_e^{pl}} \dot{\varepsilon}_e^{pl} + \frac{\partial f}{\partial T} \dot{T} = 0 \tag{4}$$

where f is the yield function, σ_e the von Mises stress, σ_y the yield stress and ε_e^{pl} the effective plastic strain used as hardening the parameter.

A Unified creep formulation is used for the heat treatment simulation to include the time effects in the results

$$\dot{\varepsilon}_{ij}^{in} = A\, exp\left(\frac{-Q}{RT}\right)\left(\frac{\sigma_e}{\sigma_{ref}}\right)^m n_{ij} \tag{5}$$

where A is a reference strain rate, Q the activation energy, R the gas constant, T the absolute temperature, σ_{ref} the hardening stress, m the strain rate sensitivity exponent and n_{ij} is the outwards normal to the yield surface. For heat treatment the inelastic strain tensor makes a contribution to the well-known strain decomposition.

$$\varepsilon_{ij}^{tot} = \varepsilon_{ij}^{el} + \varepsilon_{ij}^{in} + \delta_{ij}\varepsilon^{th} \tag{6}$$

3. The inherent strain method

The inherent strain method is a well-known approach for numerical modeling of deformation and stress in the field of classical welding, [4]. The method distributes measured or modelled permanent strain in a simplified macroscopic calculation, including the loading from the moving heat source.

For the finite element method, the inherent strain components are added to the elastic strain tensor in the strain decomposition equation

$$\varepsilon_{ij}^{tot} = \varepsilon_{ij}^{el} + \varepsilon_{ij}^{inh} \tag{7}$$

with three non-zero normal components $\varepsilon_{ij}^{inh} = \delta_{ij}\varepsilon_{ij}^{*}$

Inserting this in Hooke's law yields the standard expression for the stress tensor

$$\sigma_{ij} = \frac{E}{1+\nu}\left\{\varepsilon_{ij}^{tot} - \delta_{ij}\varepsilon_{ij}^{*} + \delta_{ij}\frac{\nu}{1-2\nu}(\varepsilon_{kk}^{tot} - \varepsilon_{kk}^{*})\right\} \tag{8}$$

In addition to the inherent strain components, the usual inelastic strain tensor is included in the implementation.

3.1. U-profile

To validate the application of the inherent strain method on simple geometries, two AlSi10Mg U-profiles were built on a GE M2 machine and then simulated. The distance between the legs of the U-profile was 4mm and 8mm, respectively. Again two inherent strain simulations were performed assuming the inherent strain components to be -1.3%, which is estimated from earlier work on the mesoscale.

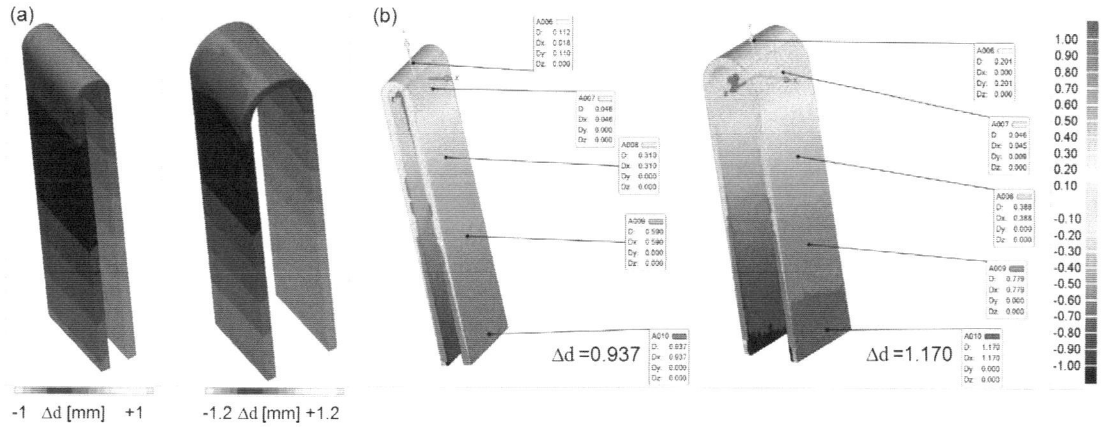

Figure 2. Comparison between the numerical results and the optical measurements for a 4mm and 8mm U-profile. (a) The simulation results and (b) the measurements. Δd is the deviation in measured displacement and the un-deformed geometry.

The results from the inherent strain method show same trends as the measurements. The deviation, Δd, between the measured displacement and the un-deformed geometry of the U-profile after machining, increases for the increased distance between the legs from approximately +-1.0mm to +-1.2mm. The main reason for the bending is the elastic spring-back during maching, when the residual stress state is released. Stresses mainly build up in the top layers of the U-profile when the final layers are printed and contract.

4. The thermomechanical layer-by-layer method

The layer-by-layer inherent strain method is based on a selective mapping of results from mesoscale to macroscale, where the choice of mapping becomes critical for the accuracy of the results. For complicated parts the definition and selection of the inherent strain components might not be so clear and the consequence could be a lack in quality in the prediction of deformations and stresses. Furthermore, the inherent strain approach does not include a thermal simulation on macro scale, which of course saves some calculation time, but it also means that it is not possible to evaluate the temperature related effects like the cooling conditions from support structure and powder leading to e.g. critical locations of hot spots.

To include the thermal simulation on macroscale without simulating the moving heat source, it is proposed in literature to release the energy in bigger regions at the same time, see e.g. [5] for a more detailed description or [6] for an alternative approach. The obtained thermo-mechanical hatch-by-hatch, patch-by-patch or layer-by-layer (often with multiple layers combined) approaches offer a simplified thermal solution, which can also be used as input for a subsequent mechanical simulation. The price to pay is the loss in details of the thermal load. However, for the suggested thermo-mechanical layer-by-layer approach, the energy is flashed in one or several layers, consistent with the power of the laser. The time of energy release is estimated from the process conditions to ensure a reasonable peak temperature in the layer, which resembles the peak temperature of the mesoscale simulation of the moving heat source.

The delivered energy from the laser in one layer is estimated from the build time and the effective power of the laser

$$E = P \cdot \eta \frac{A}{h \cdot v} \quad and \quad t_{flash} = \frac{L_{spot}}{v} \tag{9}$$

where P is the power of the laser, η effective absorbtivity of the laser, A the area of the layer, h the hatch distance and v the laser velocity.

If the source term used in the thermal simulation would be based on the full time it takes to build a layer, the energy density would be very small and the peak temperature in the layer would be too low. It is therefore proposed to flash the total energy for one layer in the same time it takes the laser spot to pass a generic material point. A similar suggestion is presented in [5].

Figure 3. (a) Illustration of a time line for the moving heat source and for the release of energy layer-by-layer. (b) The time interval for releasing the energy is estimated from the spot size and the velocity of the laser.

In this way it is possible to get a peak temperature in one layer which is similar to the peak temperature in the moving heat source calculation on mesoscale. The main reason for using this approach is to ensure thermal gradients between the layers, which are similar to the gradients in the mesoscale simulation. This is of course important for the mechanical simulation where the thermal gradients govern the mechanical response of the material. Nevertheless, when several layers are flashed at the same time it is typically necessary to scale up the time to allow energy to conduct away and reduce the peak temperature to a reasonable level.

4.1. Bridge example

The thermomechanical layer-by-layer method is applied on a bridge example to compare the residual stress results to the measurements presented in a recent publication from, [7]. A uniform mesh with element size 0.225mm is used and each powder layer in the simulation is based on 10 physical powder layers. The laser energy is flashed in 0.33ms followed by approximately 15s of cooling. The time corresponds well to the process time for building and distributing powder for the 10 physical layers.

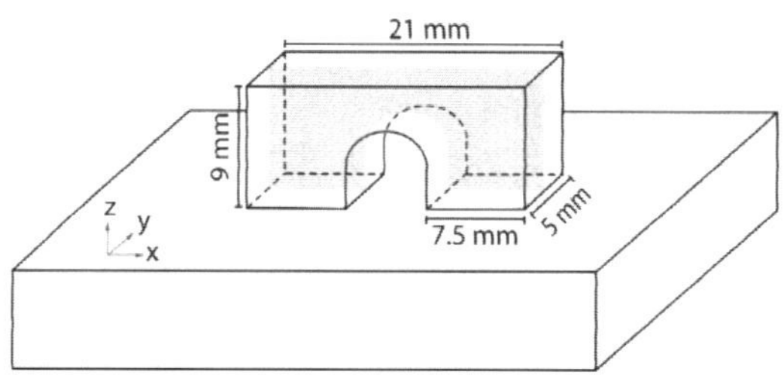

Figure 4. Dimensions of the bridge example [7].

The material is Ti-6Al-4V and temperature dependent properties were used for the simulations similar to the data presented in the publication. Three simulations were performed, two identical thermal layer-by-layer simulations with different integration of the thermal strain in the powder and one simulation based on the inherent strain method using -2% for the inherent strain components. The activation of elements in the mechanical macroscale model is different from the activation of elements in the mesoscale model. In the mesoscale model the elements become active in the mechanical simulation when the material switches from powder to solid. In the macroscale model the elements are activated in a layer-wise manner in both the thermal and the mechanical models. The immediate activation in the mechanical simulation leaves a choice of thermal strain integration in the powder during heating. In the first simulation thermal strain is integrated in the powder during heating, while in the second simulation it is assumed that the powder does not build up thermal strain as a solid and the integration of thermal strain is only performed during cooling. The measured residual normal stress components are presented in figure 5 together with the results from the 3 simulations.

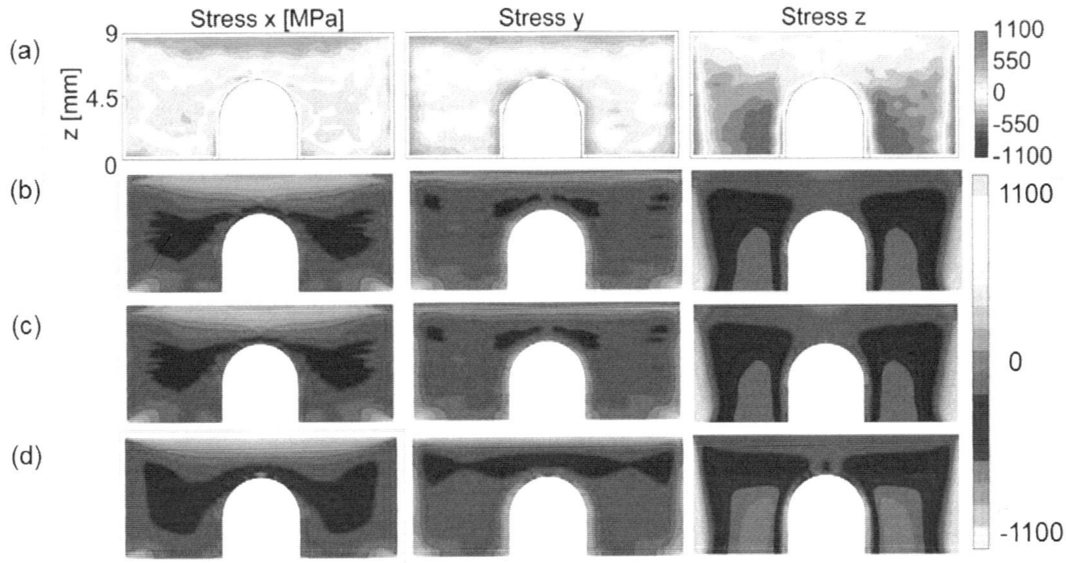

Figure 5. (a) Measurements, [7], (b) and (c) thermomechanical simulation, without and with thermal strain in the powder while heating, respectively and (d) the inherent strain method.

Generally, the results from the three simulations are quite similar and overall the agreement to the measurements is also quite good. The compression zone in the z component matches quite well with the measurements and the outer tensile zones in all 3 directions are overall in good agreement in size and location. It is interesting to see that even the inherent strain method predicts similar stress levels as the layer-by-layer thermomechanical simulation and the measurements, even by assuming the simple uniform inherent strain components.

It is furthermore interesting to notice the similar results for the two layer-by-layer thermomechanical simulations, where the thermal strain in the powder is only integrated during heating in the first simulation. However, by having a closer look at the sequence of the added layers, it is seen that in the case where no thermal strain is integrated in the powder, the powder is stretched by the solid material below when it is heated. The stretching of the powder leads to accumulated plastic strain in the powder, which to a certain extent corresponds to the thermal expansion which is integrated in the powder in the first simulation. The plastic stretching of the powder in the second simulation leads to a quite different level of residual plastic strain in the two simulations. But having the different thermal strain level and the general strain decomposition in mind it is interesting to see how the stress levels nevertheless are quite similar.

$$\varepsilon_{ij}^{tot} = \varepsilon_{ij}^{el} + \varepsilon_{ij}^{pl} + \delta_{ij}\varepsilon^{th} \tag{10}$$

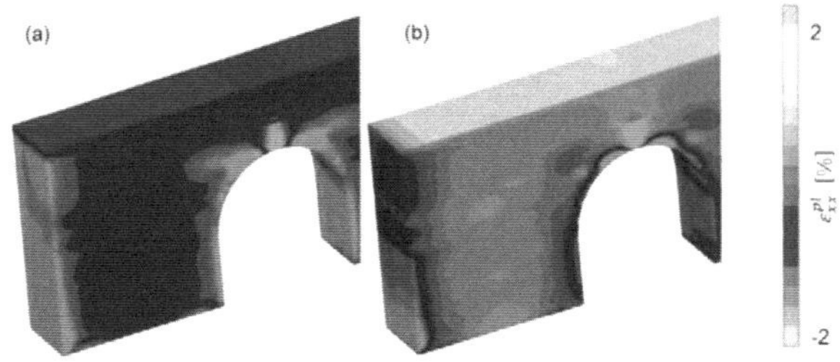

Figure 6. Plastic strain component in the x-direction, (a) with and (b) without thermal strain in the powder while heating, respectively.

5. Heat treatment

High thermal gradients during the building process often lead to high stress levels, which increase the risk of cracks and deformations during machining and service loading. Heat treatment is a useful way of relaxing the stress level in addition to optimize the mechanical properties. To analyze the effect of heat treatment it is possible to simulate the stress relief at elevated temperature. A unified creep model is applied during a 4 hours heat treatment process at 650 °C followed by an air cooling step.

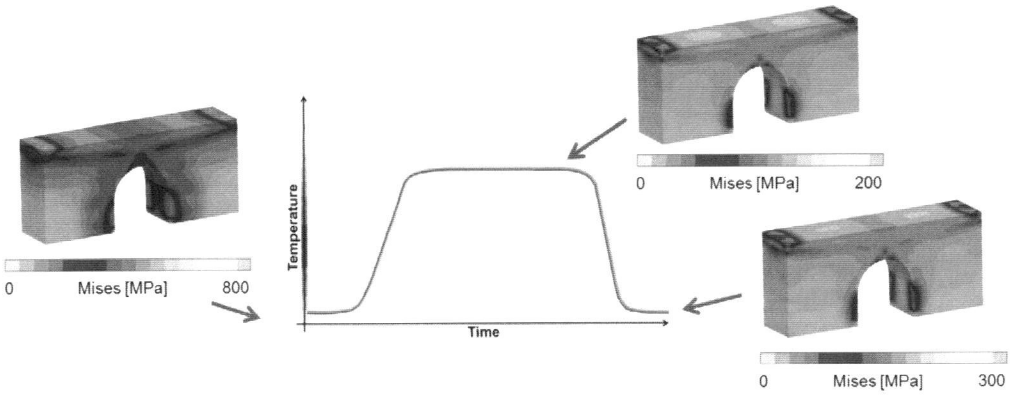

Figure 7. Stress relaxation in the bridge example during heat treatment.

The stress level is relaxed from a peak level of approximately 800 MPa at the end of building to approx. 200 MPa at the elevated temperature level. After cooling to room temperature the stress level increases to a peak level of 300 MPa. With the implemented functionality it is possible to quantitatively see the effect of different heat treatment temperature and time levels.

6. Automotive structural part

3D printing of large structural parts is possible even though it is slow and expensive compared to conventional production. However, for prototyping and small series it is becoming interesting to evaluate the process on even large automotive parts. To show the current state in printing large structural aluminum parts an experimental shock tower from the research project CustoMat3D is shown below. The main focus of the project has been to develop new aluminum alloys for additive manufacturing, with high fatigue strength, which should be suitable for load carrying automotive components, like the topology optimized demonstrator part.

To challenge the implemented model the part is simulated. Similar to the production the part is split in two, see figure 8 below. The two simulation domains include the support structure highlighted in yellow and the base plate in dark grey. The mesh is uniform with 2mm side length. The support structure is included in the simulation. The material data corresponds to the AlSi10Mg alloy at room temperature, with uniform inherent strain components equal to -1.3%.

The presented results are based on the inherent strain method. The results are obtained in approximately 8 hours for the small part and 16 hours for the large part. Both simulations were performed on 32 cores. The approach makes it possible to check and make relative comparisons of different designs and support structures in a reasonable calculation time.

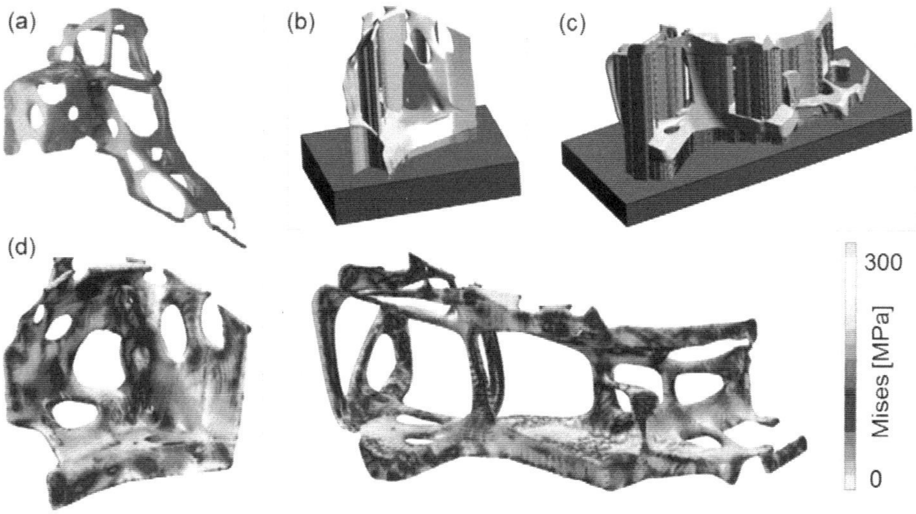

Figure 8. Topology optimized shock tower; (a) real part, (b) build setup of part 1, (c) build setup of part 2 and (d) the simulation results of the two parts using the inherent strain method.

7. Conclusions

Simulation of 3D metal printing and the post treatment steps has a growing interest in research and industry due to the wide application of the process. A general thermomechanical approach has been implemented to model the moving heat source on mesoscale and a layer-by-layer approach on macro scale. The inherent strain method has been implemented as-well to test the possibility to obtain sufficiently accurate results for this simplified approach.

Results have been compared to measurements for different parts. For the considered examples it was possible to get good agreement for both the residual stresses before machining and deformations after machining. The integrated modeling of printing and heat treatment made it possible to predict and analyze the stress relaxation at elevated temperature.

Finally, it was possible to simulate the stress level in large structural automotive parts in reasonable simulation times to make it realistic to check different designs and process conditions.

Acknowledgements

CustoMat_3D was sponsored by German Federal Ministry of Education and Research (BMBF) with No. 03XP0101A „CustoMat_3D“ – Maßgeschneiderte LAM-Aluminiumwerkstoffe für hoch-funktionale variantenreiche Strukturbauteile in der Automobilindustrie.

References

[1] Bruna-Rosso C, Demir A G and Previtali B 2018 *Mater. Des.* **156** 143–53
[2] Criales L E, Arısoy Y M and Özel T 2016 *Int. J. Adv. Manuf. Technol.* **86** 2653–66
[3] Sih S S and Barlow J W 2004 *Part. Sci. Technol.* **22** 427–40
[4] Ma N, Nakacho K, Ohta T, Ogawa N, Maekawa A, Huang H and Murakawa H 2016 *Proc. of OMAE 2016: June 19-24 2016, Busan, South Korea* (ASME) OMAE2016-54184
[5] Zhang Y, Guillemot G, Bernacki M and Bellet M 2018 *Comput. Methods Appl. Mech. Eng.* **331** 514–35
[6] Zhang W, Tong M and Harrison N M 2019 *Addit. Manuf.* **28** 610–20
[7] Strantza M, Ganeriwala R K, Clausen B, Phan T Q, Levine L E, Pagan D, King W E, Hodge N E and Brown D W 2018 *Mater. Lett.* **231** 221–4

Study of thermoelectric magnetohydrodynamic convection on solute redistribution during laser additive manufacturing

A Kao[1], T Gan[1], C Tonry[1], I Krastins[1,2] and K Pericleous[1]

[1] Centre for Numerical Modelling and Process Analysis, University of Greenwich, Old Royal Naval College, Park Row, London, SE109LS, United Kingdom
[2] Institute of Physics, University of Latvia, Miera iela 32, Salaspils, LV-2169, Latvia

E-mail: a.kao@gre.ac.uk

Abstract. Melt pools formed in laser additive manufacturing (AM) are subject to large thermal gradients, resulting in the formation of thermoelectric currents due to the Seebeck effect. When in the presence of an external magnetic field, a Lorentz force is formed which drives fluid flow in the melt pool. This Thermoelectric Magnetohydrodynamics (TEMHD) phenomenon, can have a significant impact on the melt pool morphology and can alter the microstructural evolution of the solidification process. By coupling steady-state mesoscopic melt pool calculations to a microscopic solidification model, predictions of the resulting microstructure for multiple deposited layers have been obtained. The results indicate that the magnetic field can have a transformative effect on the microstructure and solute redistribution. This study highlights the theoretical potential for using magnetic fields as an additional control system to tailor AM microstructures.

1. Introduction

Additive Manufacturing (AM) is the process of creating complicated 3d objects by joining materials together, typically in a layer-by-layer approach. For the AM processing of metals, a high energy source, such as a laser, is used to melt metallic powder layers successively onto each other. This allows them to join as they solidify. A common AM process is selective laser melting (SLM) [1], which successively melts a thin layer of powder on the print bed. These techniques create a liquid melt pool that travels with the energy source and solidifies in its wake. From a modelling point of view, laser AM can be considered to be a fast welding process. In the melt pool, flow, heat transfer and phase change need to be considered. Consequently, heat loss by radiation [2] and evaporation [3], Marangoni convection [4], spattering [5] and denudation [6] need to be included along with other complex physics. As the melt pool solidifies a microstructure is formed; this is then modified or remelted as the scan is repeated for other layers. This repeated remelting of layers leads to a number of defects that affect the integrity of components. This paper explores a novel control mechanism that has the potential to eliminate some of these defects.

There are two frequently used methods for modelling the microstructure of AM. The Phase Field method (PF) [7-8] can capture the dendritic scale but requires a very fine computational mesh and so only small sections can be modelled. The second is the Cellular Automata method (CA), which is generally used to predict the grain scale and can capture larger sections of the build [9-11].

AM is becoming widespread in industry with diverse applications from biomedical implants [12], energy [13] to aeronautical engine parts [14]. Process related defects [15] can be detrimental to the structural integrity of the final components. They include the formation of pores [16], cracks [17] and large columnar grains [1]. Additional control mechanisms are therefore desirable to suppress these defects. One relatively unexplored technique is the use of external magnetic fields to exploit the Thermoelectric Magnetohydrodynamic (TEMHD) phenomenon.

Thermoelectric (TE) currents are an inherent part of many solidification processes as they are generated by the combination of large thermal gradients and changes in the Seebeck coefficient. When a magnetic field is applied, these TE currents interact and generate a Lorentz force that drives flow, transporting heat and mass. Significant changes to the solidification process due to TEMHD have been observed experimentally and predicted theoretically [18-22]. The solidification conditions of AM mean that there are high thermal gradients and moderate solidification velocities compared to high undercooled growth and so TEMHD effects should be quite significant. To the authors' knowledge there is a lack of systematic investigation into the effects of TEMHD in AM. Sparse information includes a recently published experimental paper [23], which demonstrates that a 0.12 T magnetic field can lead to beneficial effects on the printed material. There have been a greater number of papers concerning welding. Chen *et al.* [24] conducted welding experiments in a 0.415 T static magnetic field, showing a reduction in width within the transitional region in the microstructure. Another study [25] showed that with a 0.3 T magnetic field the "humping" welding defect could be supressed.

Though the research may currently be sparse it has demonstrated the potential benefits of the application of a magnetic field to AM. The effect of magnetic field direction has been studied more systematically in a recently published paper by the authors [26]. This contribution studies in greater detail how redistribution of solute in the melt pool changes the distribution in the final solidified structure, as a direct consequence of TEMHD-dominated flow modifying the melt pool hydrodynamics by the introduction of the magnetic field.

2. Methodology

The governing equations for each of the physical phenomena present as well as the employed numerical methods are described in this section. A more detailed description is given in [26]. Thermoelectric currents, $\mathbf{J}$, are calculated from a modified Ohm's law [27] given by

$$\mathbf{J} = \sigma(-\nabla\psi + \mathbf{u} \times \mathbf{B} - S\nabla T), \tag{1}$$

where σ, ψ, $\mathbf{u}$, $\mathbf{B}$, S and T are the electrical conductivity, electric potential, fluid velocity, magnetic field strength, Seebeck coefficient and temperature, respectively, and current conservation $\nabla \cdot \mathbf{J} = 0$. The resulting currents interact with the magnetic field generating a Lorentz force, $\mathbf{F} = \mathbf{J} \times \mathbf{B}$. In the melt, this force drives the incompressible fluid flow governed by the Navier-Stokes equations,

$$\rho\left(\frac{\partial \mathbf{u}}{\partial t} + \mathbf{u} \cdot \nabla\mathbf{u}\right) = -\nabla p + \mu\nabla^2\mathbf{u} + \mathbf{J} \times \mathbf{B}, \quad \nabla \cdot \mathbf{u} = 0, \tag{2}$$

where t, ρ, p and μ are time, density, pressure and dynamic viscosity. Heat transport is governed by

$$\frac{\partial T}{\partial t} = \nabla \cdot \alpha\nabla T - \mathbf{u} \cdot \nabla T, \tag{3}$$

where α is the thermal diffusivity. Latent heat of fusion, L_f, is related to temperature through the volumetric enthalpy, H, which is defined as the sum of latent and sensible heats

$$H = fL_f + c_pT, \tag{4}$$

where c_p and f are the specific heat capacity and liquid fraction. The Navier-Stokes equations (2) are solved using the lattice Boltzmann method (LBM) [26, 28, 29] yielding a solution for $\mathbf{u}$. The change of phase and release of latent heat (4) are calculated using an enthalpy-based method. Solutions for the electric potential (1), heat transport (3) and enthalpy (4) use finite difference-based schemes [19, 30].

Equations (1-4) are solved throughout the entire computational domain. Additional boundary conditions represent surface phenomena, including the Marangoni force and heat sources/sinks: the heating laser, heat loss due to evaporation and radiation. The other domain boundaries represent far field conditions, namely the ambient temperature. The domain is electrically insulated. Free slip condition for velocity is set for the top surface of the melt pool.

In this work, microstructure evolution is predicted using a Cellular Automata method (CA) based on the μMatIC code [31-33]. For alloy solidification the concentration of solute in the solid, C_s, and liquid, C_l, are governed by $C_s = kC_l$, where k is the partition coefficient of solute. The concept of solute trapping is introduced to account for both the physical non-equilibrium effect, but also a numerical effect where the computational resolution is much larger than the inter-dendritic spacing. A cell is assumed to be a volume average of both dendrites and the high concentration eutectic structures. The Continuous Growth Model [34] accounts for this by proposing a change to k given by

$$k = \frac{k_e + V/V_D}{1 + V/V_D}, \tag{5}$$

where V is the growth velocity and V_D the 'diffusion' velocity. In this case, $V = u_s$ is the scan velocity and $V_D \approx D_l/L$, where L is some characteristic length typically the interface thickness, but to account for numerical trapping it is much larger, and close to the cell size, giving a value of k = 0.95. At the solid-liquid interface, the solid fraction, ϕ_s, rate of change is given by

$$C_l(1-k)\frac{\partial \phi_s}{\partial t} = -\nabla \cdot (D_e \nabla C_l) + \left[1-(1-k)\phi_s\right]\frac{\partial C_l}{\partial t}, \tag{6}$$

where $D_e = (1-\phi_S)D_l + \phi_s D_s$ is the equivalent diffusion coefficient and $T^i = m_l(C_l - C_0)$ is the equilibrium interface temperature with m_l and C_0 being the liquidus slope and reference concentration. The equivalent concentration $C_e = (1-\phi_S)C_l + \phi_s C_s$ enters the solute transport equation given by

$$\frac{\partial C_e}{\partial t} + \mathbf{u} \cdot \nabla C_l = \nabla \cdot (D_e \nabla C_l) \tag{7}$$

with zero mass flux prescribed at the boundaries.

The multiple layers of the AM process are captured by the microstructure model. A moving mesh approach is employed in such a way that the solidified cells exit at the back of the domain and are reintroduced at the front but lowered by the powder layer depth. A new powder layer is introduced above this. A schematic of the AM process is given in figure 1.

3. Results

The numerical model represents the SLM process of a 70 % efficient 100 W laser, scanning at 0.1 m/s with a beam diameter of 100 μm. In this work two cases are shown, a baseline case with no magnetic field, $|\mathbf{B}| = 0$, and one with a transverse magnetic field. The computational domain is a cuboid with 800 μm in the scan direction, x, 300 μm in the build direction, z, and 600 μm in the magnetic field orientation, y. The mesh comprises 1 μm length cubic cells. Initially the domain is set to the powder phase with a particle size ≈ 10 μm and each particle assigned a random crystallographic orientation. Material properties of Al10Si are used for each the liquid, solid and powder phases [26].

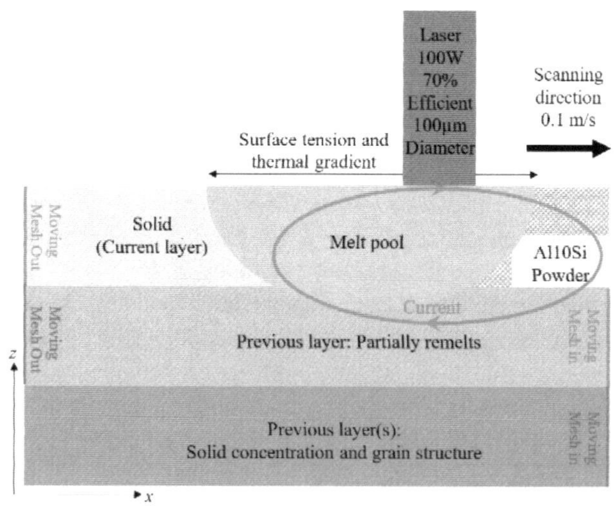

Figure 1. Schematic of the AM process, highlighting the TE current circulation and the moving mesh.

The strong thermal gradients and large variations in interface temperature due to latent heat inherent to AM and welding processes can lead to substantial thermoelectric currents [25]. Electric current flows downstream in $+x$ from the laser spot to the front of the melt pool then circulates through the solid/powder before re-entering in $+z$ upstream of the melt pool. For clarity upstream is defined as the

IOP Conf. Series: Materials Science and Engineering **861** (2020) 012009 doi:10.1088/1757-899X/861/1/012009

region where the laser has passed and is solidifying and downstream as the melting region at the front of the melt pool. In general, the current in $+y$ is weak compared to the other two components. The structure of these currents is shown in figure 1. When a transverse magnetic field is applied, the Lorentz force has two components, the first is a force in $-z$, downward from the surface of the melt pool and the second is in $+x$ toward the front of the melt pool.

With $|\mathbf{B}| = 0$, fluid flow is governed by Marangoni flow with fluid flowing radially from the laser spot, circulating down the sides of the melt pool and back up to the laser spot. However, for the magnetic field orientation considered, the Lorentz force acts against the returning flow along the centre and if the Lorentz force is dominant it can reverse flow in this region. The force acting in $+x$, stretches the upstream vortex and suppresses the downstream vortex until a single vortex encompasses the melt pool in the x-z plane. In the x-z plane fluid flows down the centre of the melt pool and up the sides.

In the magnetic field case, the melt pool shape is significantly altered compared to the $|\mathbf{B}| = 0$ case. TEMHD introduces a keyhole-like mechanism where hot fluid from the laser spot flows toward the front of the melt pool and downward, deepening the melt pool. However, not only does the change in fluid flow modify heat transfer but it also has a significant effect on solute transport. Schematics of the contrasting behaviour between typical Marangoni flow with no magnetic field and this TEMHD dominated case are given in figures 2 and 3 for no magnetic field and magnetic field respectively.

Using the mesoscopic thermal and velocity profiles as inputs into the microscopic model, a one-way coupling allows for qualitative predictions of solute redistribution. Figure 2 shows cross sections and the grain structures for the $|\mathbf{B}| = 0$ case. The grain structures clearly show that the grains preferentially follow the thermal gradient. Inspecting the dendritic structure, the results indicate that these grains have orientations that align with the thermal gradient. From a thermodynamic perspective the thermal gradient, which also represents the direction of heat flux will favour those with orientations aligned. Therefore, while the powder particles are each assigned random orientations, an initial competition occurs on a short length scale with favourable orientations winning and going to form long needle-like dendrites due to the small primary spacing. However, as the thermal gradient is varying spatially and changes relatively per layer, a secondary competition occurs with larger grains competing, giving rise to very long dendrites from epitaxial growth.

Another consistent feature observed between the numerical model and experimental observations is the formation of a low concentration melt pool boundary, indicated by the white bands in the figures. These form at the very initial stages of solidification on the sides and bottom of the melt pool. With a relatively homogeneous melt pool, solidification initiates without a solute boundary layer and partitioning allows more solute to be ejected, hence a lower concentration in the solid. At the back of the melt pool the solute boundary layer will already have formed and so this low concentration melt pool boundary region forms a shape analogous to the outer surface horizontal cylindrical segment.

At the front of the melt pool melting of the powder doesn't have much effect on the solute distribution. Powders are formed through atomisation processes with finer microstructures than AM and so at the length scales considered here, the powder can be assumed to be homogeneous and of bulk composition. However, when melting a previous layer, solute variations are released back into the melt, an effect that can be seen at the front of the melt pool in the x-z cross section. These variations are captured by the downstream fluid circulation and they are pushed forward allowing time for remixing. However, upstream of the melt pool the fluid circulation passing down the back wall, where the solute boundary is persistent, transports solute back into the melt pool. This can be seen by the plume of solute ejected back into the melt in figure 2(d). However, Marangoni flow also circulates in the y-z plane along the side walls bringing high concentration solute to the bottom centre of the melt pool. This build-up of solute at the base of the melt pool gets captured into the solid forming a thin and long high concentration region along the centre line. This can be seen in the x-z plane just above the low concentration melt pool boundary and also in the y-z plane in the centre. This effect is generally not captured in the x-z plane of experimental results, however we believe this may be simply a case of the difficulty of sectioning perfectly along the centre of the scan axis and also that other investigators may not be looking for this feature. This solute region is around 10-20 microns wide and so any angled section would only appear as a small region of high concentration. However, experimental sections in the *y-z* plane do consistently show that there is a higher concentration region at the base next to the melt boundary. A good example of this can be seen in the work of Singh *et al.* [35]. For sections in this plane, a small angle deviation will still capture this feature, an analogy is to consider the sectioning of a very long cylinder horizontally vs vertically. By offsetting the *x-z* plane by 20 microns from the centre axis, this solute enriched region vanishes while all the other key microstructure features are still clearly visible. This focused high concentration region is also captured by the upstream Marangoni flow circulation pushing the solute to the top of the melt pool and back. This gives a small increase in the concentration along the centre of the solidified region. However, this effect is somewhat subtle as due to the relatively long time the plume has to circulate through the melt pool, considerable mixing occurs and so the concentration approaching the interface is only slightly elevated.

With the application of the magnetic field the hydrodynamics and melt pool morphology are drastically changed, which consequently has a significant effect on solute redistribution and the microstructure. Figure 3 shows the concentration and grain structure for the case with an applied magnetic field. The grain growth follows the modified thermal gradient, and with a much deeper melt pool the grains become more tilted into the scan direction. There are still epitaxial grains along the centre of the melt pool and they are larger compared to the $|\mathbf{B}| = 0$ case in the *x-z* plane, however, they occupy a smaller transverse *y-z* section. As hot fluid is pushed to the bottom of the melt pool and circulates back up the side walls it cools allowing the side grains to overgrow those from the bottom. This can be seen by the longer side grains in figure 3(a) compared to those in figure 2(a).

Due to the modified hydrodynamics, solute transport is also quite different to the B=0 case. At the front of the melt pool remelting of previous layers brings solute back into the liquid and is swept down into the melt pool and towards the sidewalls. The solute then remixes in the melt pool without a noticeable effect on solidification. However, where solidification begins on the side walls, flow is now going up, transporting high solute concentration to the top sides of the melt pool before circulating back down the centre. Consequently, these high solute regions are captured into the solidification structure near the top of the melt pool, appearing as dark bands in the *x-y* cross section that looks down on the melt pool. The high concentration region at the bottom centre of the melt pool is no longer present due to flow up the side walls transporting concentration from this region away. Redistribution of this high concentration region away from the melt pool boundary may have beneficial effect on the mechanical properties. Experimental studies have shown that the melt pool boundary is a likely site of failure [1], but by reducing the discontinuity in composition, hence material properties this could be mitigated.

In both cases shown here, the laser scans in the same place with each layer. In practice, building macroscopic components will utilise more complex scanning strategies such as hatching, where neighbouring passes of the layers will interact from the sides. Changes to thermal diffusivity and electrical conductivity as the sides are now solid rather than powder, may modify the current density and consequently TEMHD. This may impact the melt pool dynamics, but also remelting and the subsequent regrowth from neighbouring passes may introduce additional microstructure growth features not captured by this idealised scanning strategy.

4. Conclusions

Melt pool dynamics and morphology were predicted using a mesoscopic enthalpy-based method for a Marangoni dominated flow with no magnetic field and a TEMHD dominated flow from a transverse magnetic field. One-way coupling of the thermal and velocity profiles to a microstructural model allowed for predictions of the resulting grain and dendritic structures. Without a magnetic field, features typical of AM processing were well captured and mechanisms of observed solute distribution in the solid were explained. With the application of the transverse magnetic field, changes in melt pool dynamics led to a keyhole-like effect creating a deep melt pool. The TEMHD flow significantly modified the solute distribution with high concentration regions swept to the top of the current layer. The effects of epitaxial growth were analysed over multiple layers in both cases demonstrating the potential of the model used. However, this is an idealised case where scanning occurs in the same location with each pass, further investigation is necessary to understand how scanning strategy combined with magnetic field strategy can be used to tailor the microstructure in a favourable way. Two- way coupling may also reveal more complex interactions between solute redistribution and phase change.

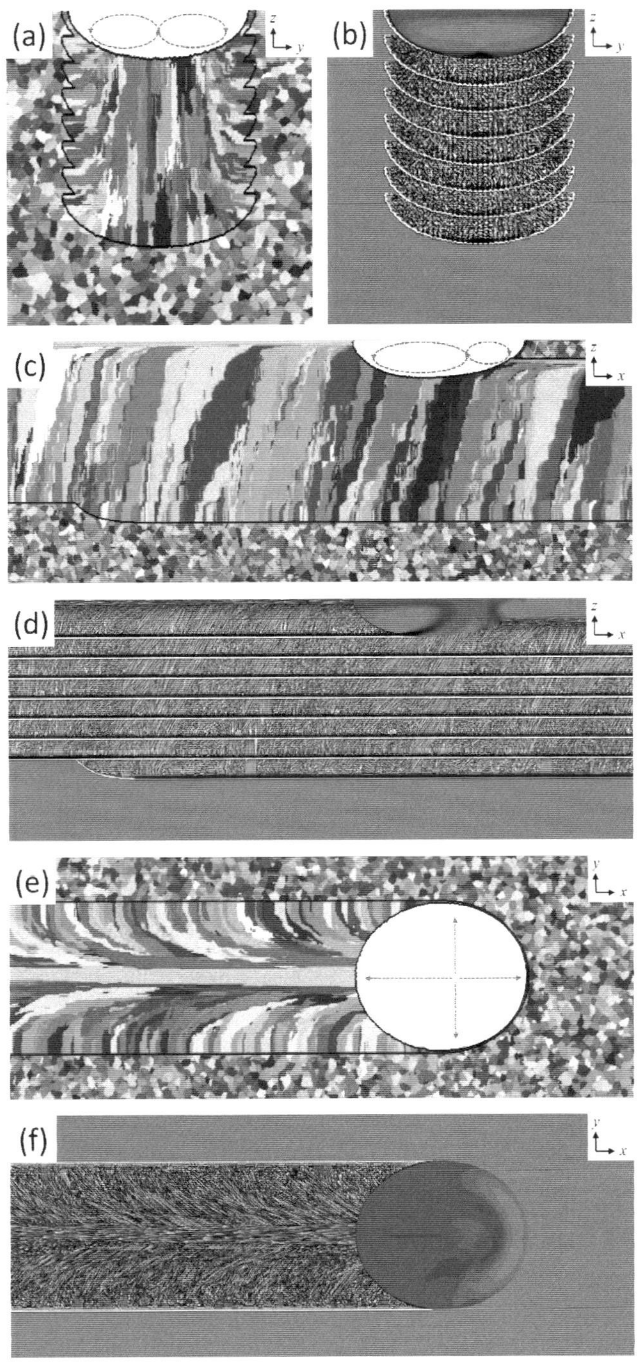

Figure 2. No magnetic field grain structures (a,c,e) and solute concentrations (b,d,f), with darker regions solute enriched, in the x-z plane (a,b), the y-z plane (c,d) and the x-y plane (e,f). Red arrows are schematics of fluid flow. Black lines in (a,c,e) are boundaries between liquid, solid and powder.

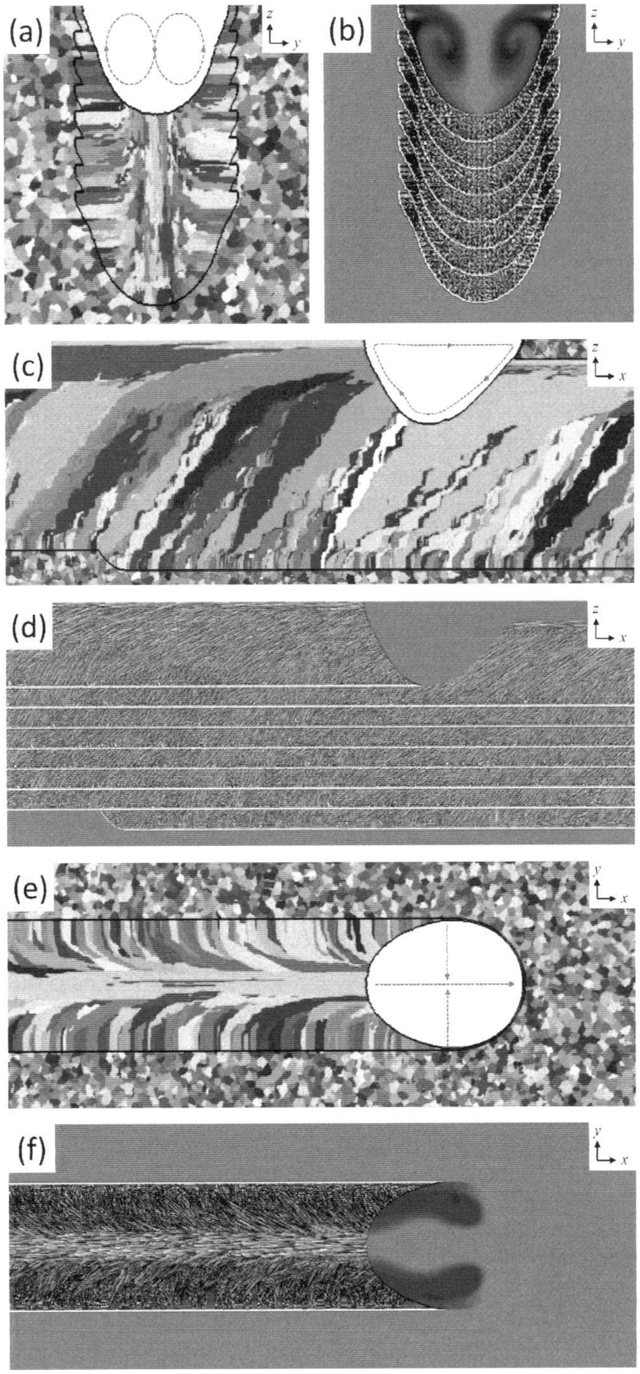

Figure 3. Magnetic field grain structures (a,c,e) and solute concentrations (b,d,f), with darker regions solute enriched, in the *x-z* plane (a,b), the *y-z* plane (c,d) and the *x-y* plane (e,f). Red arrows are schematics of fluid flow. Black lines in (a,c,e) are boundaries between liquid, solid and powder.

Acknowledgments
The EPSRC financial contribution in project 'TECalloy' (EP/K011413/1) led the foundations for the TEMHD work. T Gan and I Krastins, gratefully acknowledge the financial assistance offered by the University of Greenwich VC scholarship scheme for their PhD studies.

IOP Conf. Series: Materials Science and Engineering **861** (2020) 012009 doi:10.1088/1757-899X/861/1/012009

References

[1] Shifeng W, Shuai L, Qingsong W, Yan C, Sheng Z and Yusheng S 2014 *J. Mater. Process. Technol.* **214** 2660–7
[2] Gusarov A, Yadroitsev I, Bertrand P and Smurov I 2009 *J. Heat Transfer* **131** 072101
[3] Geiger M, Leitz K H, Koch H and Otto A 2009 *Prod. Eng.* **3** 127–36
[4] Yuan P and Gu D 2015 *J. Phys. D: Appl. Phys.* **48** 035303
[5] Khairallah S A, Anderson A T, Rubenchik A and King W E 2016 *Acta Mater.* **108** 36–45
[6] Matthews M J, Guss G, Khairallah S A, Rubenchik A M, Depond P J and King W E 2016 *Acta Mater.* **114** 33–42
[7] Wang X and Chou K 2018 *Int. J. Adv. Manuf. Tech.* **100** 2147–62
[8] Radhakrishnan B, Gorti S B, Turner J A, Acharya R, Sharon J A, Staroselsky A and El-Wardany T 2018 *Metals* **9** 14
[9] Lian Y, Gan Z, Yu C, Kats D, Liu W K and Wagner G J 2019 *Mat. Des.* **169** 107672
[10] Koepf J A, Gotterbarm M R, Markl M and Körner C 2018 *Acta Mater.* **152** 119–26
[11] Rai A, Markl M and Körner C 2016 *Comput. Mat. Sci.* **124** 37–48
[12] Singh R, Lee P D, Dashwood R J and Lindley T C 2010. *Mater. Technol.* **25** 127–36
[13] Ibrahim K A, Wu B and Brandon N P 2016 *Mat. Des.* **106** 51–9
[14] Murr L E 2016 *J. Mat. Sci. Tech.* **32** 987–95
[15] Zhang B, Li Y and Bai Q 2017 *Chinese J. Mech. Eng.* **30** 515–27
[16] Leung C L A, Marussi S, Atwood R C, Towrie M, Withers P J, Lee P D 2018 *Nat. Comm.* **9** 1355
[17] Li R, Shi Y, Wang Z, Wang L, Liu J and Jiang W 2010 *Appl. Surf. Sci.* **256** 4350–6
[18] Fautrelle Y, Wang J, Salloum-Abou-Jaoude G, Abou-Khalil L, Reinhart G, Li X, Ren Z M and Nguyen-Thi H 2018 *JOM* **70** 764–71
[19] Kao A, Cai B, Lee P and Pericleous K 2017 *J. Cryst. Growth* **457** 270–4
[20] Kaldre I, Fautrelle Y, Etay J, Bojarevics A and Buligins L 2013 *J. Alloys Compd.* **571** 50–55
[21] Gao J, Han M, Kao A, Pericleous K, Alexandrov D V and Galenko P K 2016 *Acta Mater.* **103** 184–91
[22] Kao A, Gao J and Pericleous K 2018 *Philos. Trans. Royal Soc. A* **376** 20170206
[23] Du D, Haley J C, Dong A, Fautrelle Y, Shu D, Zhu G, Li X, Sun B and Lavernia E J 2019 *Mat. Des.* **181** 107923
[24] Chen J, Wei Y, Zhan X and Pan P 2017 *Int. J. Adv. Manuf. Tech.* **92** 1677–86
[25] Kern M, Berger P and Hugel H 2000 *Weld. J. Res. Suppl.* 72–8
[26] Kao A, Gan T, Tonry C, Krastins I and Pericleous K 2020 *Philos. Trans. Royal Soc. A* In press.
[27] Shercliff J A 1979 *J. Fluid Mech.* **91** 231–51
[28] Kao A, Krastins I, Alexandrakis M, Shevchenko N, Eckert S, Pericleous K 2018 *JOM* **71** 48–58.
[29] Kao A, Shevchenko N, Alexandrakis M, Krastins I, Eckert S and Pericleous K 2019 *Philos. Trans. Royal Soc. A* **377** 20180206
[30] Voller V R, Cross M and Markatos N C 1987 *Int. J. Numer. Methods Eng.* **24** 271–84
[31] Wang W, Lee P and McLean M 2003 *Acta Mater.* **51** 2971–87
[32] Yuan L and Lee P D 2010 *Modell. Simul. Mater. Sci. Eng.* **18** 055008
[33] Yuan L and Lee P D 2012 *Acta Mater.* **60** 4917–26
[34] Aziz M J and Kaplan T 1988 *Acta Metall.* **36** 2335–47
[35] Singh A, Ramakrishnan A and Dinda G 2017 *J. Weld. Join.* **35** 67–73

IOP Conf. Series: Materials Science and Engineering **861** (2020) 012010 doi:10.1088/1757-899X/861/1/012010

Capturing Marangoni flow via synchrotron imaging of selective laser melting

S J Clark[1,2], C L A Leung[1,2], Y Chen[1,2], L Sinclair[1,2], S Marussi[1,2] and P D Lee[1,2]

[1] Mechanical Engineering, University College London, WC1E 7JE, UK
[2] Research Complex at Harwell, Harwell Campus, Didcot, OX11 0FA, UK

E-mail: samuel.clark@ucl.ac.uk

Abstract. Marangoni flow has a substantial influence on the quality of components fabricated via laser powder bed fusion (LPBF). However, Marangoni flow in melt pools is rarely quantified due to the opacity of liquid metals and the necessity for *in situ* evaluation. Here we report the findings of high-temporal-resolution synchrotron x-ray radiography experiments tracking the flow in the melt-pool. Dense, highly attenuating tungsten carbide particles are seeded within an elemental powder blend of aluminium and copper of varying composition. Due to the extremely high temporal resolution of greater than 50 kfps at the 31-ID-B beamline at the Advanced Photon Source, USA, we can track the position of tracer particles from frame to frame. This data provides valuable process guidance for optimising mixing and informs the development and validation of multiphysics models.

1. Introduction

Marangoni flow has a substantial influence on the quality of laser additive manufactured components. It has previously been shown that unstable melt flow can be responsible for increased porosity and surface imperfections in laser powder bed fusion (LPBF) [1]. The effects of melt-pool flow have been studied in related process such welding where the magnitude of the velocity and the direction of the flow has been shown to have a potent influence on the weld pool geometry [2–7]. The understanding of the melt-pool flow nature is thus critical for microstructural and process control. However, the nature of the Marangoni flow is rarely evaluated due to the opacity of liquid metals and the necessity for *in situ* evaluation.

Previous studies have deployed the use of highly attenuating tungsten tracers in x-ray experiments to track the melt-pool flow using high-speed synchrotron x-ray radiography [8] and using an innovative lab-source based stereographical apparatus [9] in arc welding and high power laser welding respectively. Leung *et al.* [10] traced the motion of gas porosities during the laser additive LPBF of invar powders during a build simulating an overhang condition where the deposit rests on powder rather than a base plate or previous layers of the build. Recently, Guo *et al.* [11] have reported the findings of their high-speed synchrotron radiographic studies of melt-pool flow tracking in AlSi10Mg and Al-6061. Guo *et al.* [11], revealed a sharp contrast between the full field flow patterns under both conduction-mode and keyhole-mode melting processing parameters. In conduction-mode, the melt pool was shown to exhibit a simple flow analogous to the glow platters typically observed in arc welding [8]. However, the flow pattern in depression-mode was observed to be complicated, and it was suggested that this was due to the coaction of driving forces. Marangoni flow dominates behind the laser whereas at and close to the laser-matter interaction site vaporization dominates the fluid flow along the keyhole walls combined with pressure and buoyancy effects. It is suggested that in order for an *in situ* alloying to be viable means of synthesizing novel alloys in LPBF [12, 13]. For this to be feasible it is suggested that the Marangoni flow in the melt-pool must be sufficient in order that the alloying elements in the melt-pool be homogeneously blended. In this work, we explore the potential for *in situ* alloying strategies via visualising the Marangoni flow in the melt-pool. We report the findings of our synchrotron x-ray radiography flow tracking experiments performed independently and without knowledge of the experiments by Guo *et al.* [11]. The reported flow pattern in keyhole-mode is schematically shown in figure 1.

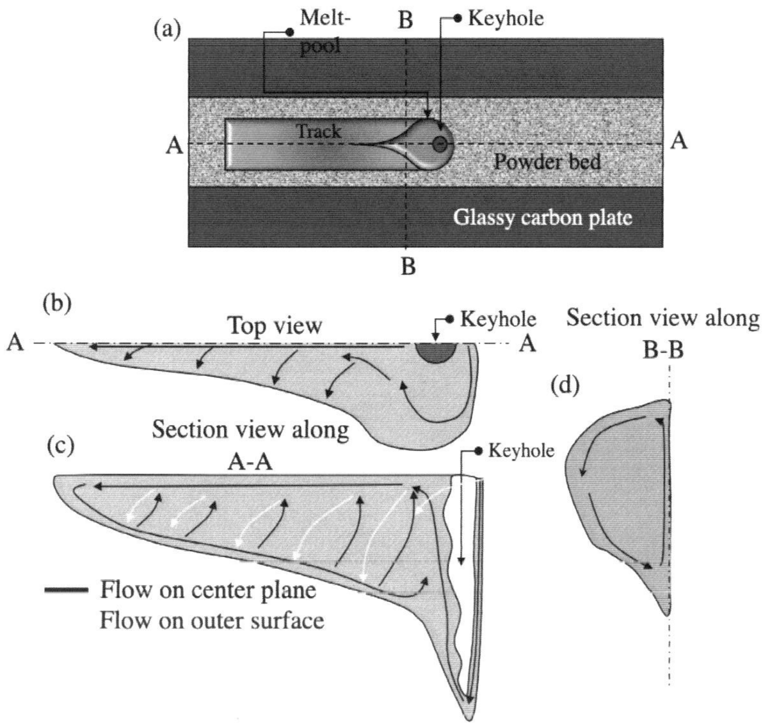

Figure 1. (a) Schematic top view of the melt-pool and track, (b) top view of the melt-pool annotated with the flow pattern, (c) section view along A-A annotating the flow observed on the central plane and that of the outer surface, (d) section view along B-B annotated with the in-plane flow pattern.

2. Materials and methods

The *in situ* and *operando* process replicator ISOPR is an evolution of the original design used in previous studies [10, 14, 15]. The apparatus consists of a miniaturised LPBF system with the functionality to automatically lower the substrate and spread a layer powder. The multi-layer capability facilitates the faithful replication of the fabrication of multilayer parts associated with industrial LPBF machines. The laser system consists of a continuous wave 520 W Yb-doped fibre laser, with a wavelength of 1070±10 nm (IPG YLR-500-AC, USA). The laser optics consisted of a collimator followed by a beam expander, and finally, an X–Y galvanometer scanner (intelliSCAN$_{de}$ 30, SCANLAB GmbH, Germany) coupled with an f-theta lens. The optical system was tuned to produce a spot size of $\approx 42\ \mu m$, which was determined through measuring the averaged throat of the keyhole observed in the x-ray radiographs, at the powder bed surface.

The unique ISOPR environmental chamber is shown in figure 2. It contains a miniature powder bed which is formed from a substrate of 99% pure aluminium (Goodfellow, UK) with dimensions of 46 x 17 x 0.8 mm sandwiched between two glassy carbon plates of 1 mm in thickness (HTW, Germany). The substrate is remotely lowered by ≈60 μm to create a cavity of 46 x 0.06 x 0.8 mm. A schematic view of the sample assembly is shown in figure 1(a). Gas atomised (GA) Al-6061 series powder (Goodfellow, UK) with a particle size range of 15 – 45 μm was blended by manually tipping the powders back and forth with 50 wt.% of tungsten powder (Alfa Aesar, USA) with a maximum particle size of 12 μm. The blended powders were deposited evenly within the cavity using a gravity-fed powder hopper subsystem. Prior to operation, the environmental chamber is evacuated and back-filled to a modest positive pressure of <10 kPa with technical grade argon gas at a flow rate of 4 L min^{-1}, this flow of argon was then maintained throughout the experiment.

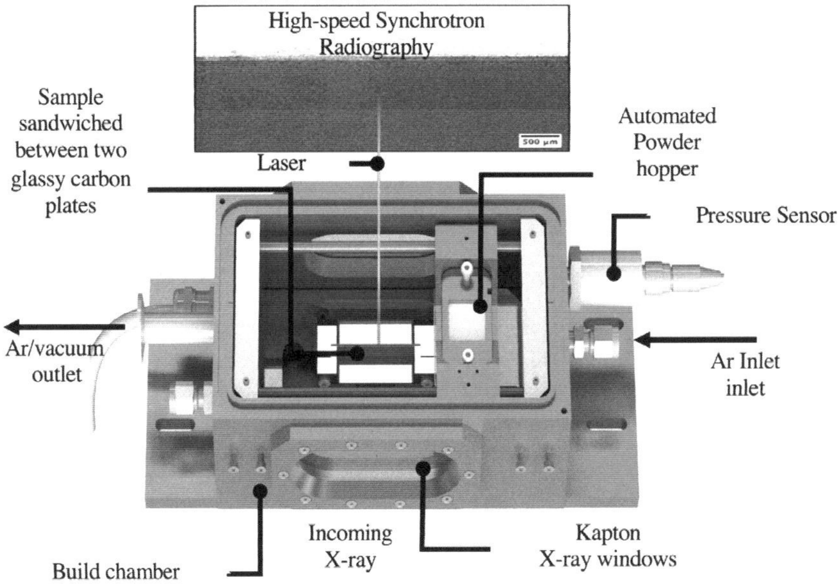

Figure 2. Schematic diagram of the ISOPR LPBF process replicator

In situ X-ray radiography experiments were undertaken at the 32-ID-B imaging beamline at the Advanced Photon Source (APS), in Chicago, USA. Recently, this beamline has been used for several high-speed radiographic studies of powder bed fusion [11, 16–18]. The beamline was configured to achieve a polychromatic pink beam with the harmonic energy at 24 keV and an energy bandwidth of 5-7%. The ISOPR process replicator is positioned directly in the path of the beam such that it is selectively attenuated by the sample assembly contained within. The attenuated x-ray beam is then converted to optically visible light using a 100 μm thick LuAg:Ce scintillator and recorded with a FASTCAM SA-Z (Photron, USA) with 10x magnification, at 50 kfps and a pixel size of 1.87 μm. The resultant field of view was 512 pixels (0.96 mm) in width by 680 pixels (1.27 mm) in height. In this experiment, the laser traverses the powder bed at a velocity of 756.7 mm s^{-1} (calibrated by tracking the position of the keyhole) over the powder bed with an output power of 500 W.

3. Results and discussion

3.1. Melt-pool reconstruction and quantification

To quantify the dimensions of the melt-pool, the stack of 1024 radiographs were processed via a local-temporal background subtraction implemented in Matlab (MathWorks inc., USA). Each image is background-subtracted through looping progressively through the stack dividing each image by its 50 nearest neighbours. A median image of the resultant sub-stack of 50 images is computed. The effect of this process is to remove any features in each image which are on average common to 50 nearest neighbours. Due to the large size of the melt-pool and the limited field-of-view of the imaging setup in no single acquired raw image is the entire melt-pool visible. Figure 3 shows a reconstructed image of the melt-pool, it is generated by extracting the velocity of the laser beam (in this case 756.7 mm s^{-1}) by tracking the bottom of the keyhole and then translating each image such that the laser beam appears to be stationary. Computing a median image of the translated stack yields a single image containing the entire melt-pool. It is shown that with the processing parameters outlined above the melt pool is of 2100 µm measured from the rear wall of the keyhole to the final point to solidify furthermore, the keyhole is shown to be 492 µm in depth measured from the top surface of the 99% pure aluminium substrate plate. Ahead of the keyhole a thin film 28 µm thick is melted. The projected area of the melt-pool behind the keyhole is 0.38 mm^2.

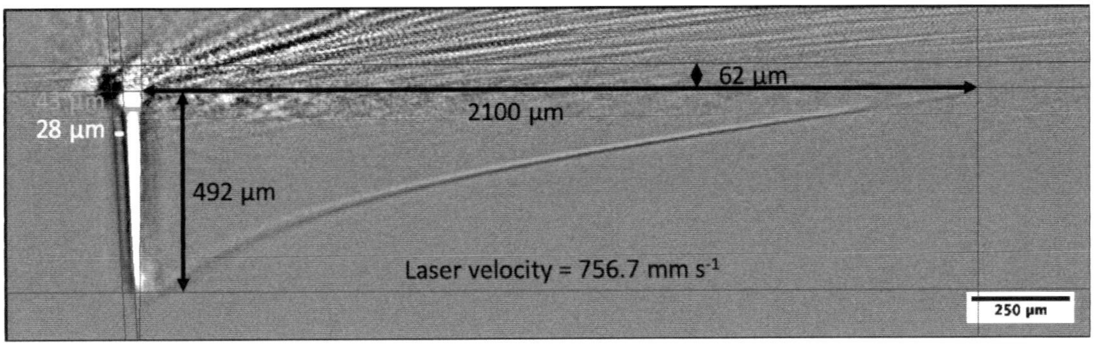

Figure 3. Reconstructed melt-pool

3.2. Elucidation of the flow pattern and projected tracer speed

Figure 4(a)-(e) shows a time series of high-magnification images of the region around the keyhole the passage of a single tungsten tracer particle has been highlighted in yellow. The average speed of highlighted particle was measured to be 968.8 mm s^{-1} in the moving reference frame of the laser. In our experiments, tungsten tracer particles were not observed to be transported via the thin melt ahead of the front keyhole wall. It is suggested that the experimentally observed high-velocity vapour jet emanating from keyhole [19–21] denudes any powder directly in the path of the laser beam via entrainment. It appears that powder particles typically bypass the keyhole joining the melt-pool from the outer surface. Figure 4(f) shows a composite image generated by extracting the pixelwise minimum intensity from a stack of 250 images in which a portion of the melt-pool is visible. Comparing the traces shown in figure 4(f) with the schematic flow lines shown in figure 1 reveals a general agreement between the traces observed in this work and the flow direction suggested by Guo *et al.* [11] on the outer surface of the melt-pool. However, it is also evident from figure 4(f) that the tungsten particles entering the melt-pool on its outer surface typically only have a sufficient residence time to be conveyed to bottom of the melt-pool. This finding has important connotations for the potential use of elemental powder blends for the *in situ* creation of novel alloys as it appears to be unlikely that sufficient mixing will occur in order to achieve a homogeneous alloy composition.

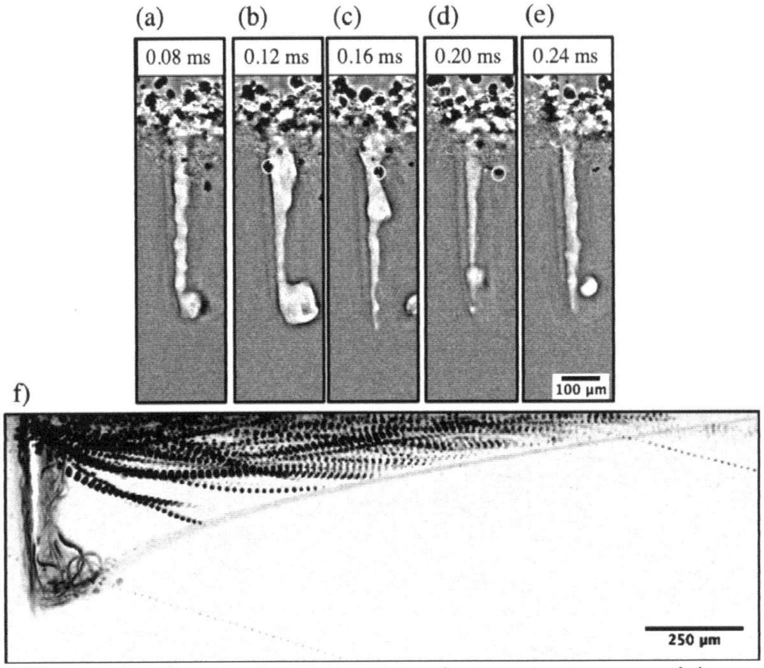

Figure 4. (a)-(e) Time series showing the progression of a tungsten tracer particle around the side of the keyhole. (f) The trajectories of tungsten tracers through the entire reconstructed melt-pool.

4. Conclusions

The nature of the melt-pool flow has been captured in the LPBF of Al-6061 using fine tungsten flow tracing particles using synchrotron x-ray radiography. It was observed in the processing parameters adopted that the flow tracing particles did not enter the melt-pool via the thin film of melt ahead of the keyhole rather they bypassed the keyhole being conveyed around its outer periphery. It is concluded that the powder particles ahead of the melt-pool in the direct path of the laser are complexly denuded via entrainment in the high-velocity vapor jet emanating from the keyhole. It is observed that the tungsten particles entering the melt-pool on its outer surface typically only have a sufficient residence time to be conveyed to bottom of the melt-pool. This highlights one of the challenges of developing processes exploiting elemental powder blends for the *in situ* creation of novel alloys. It is suggested that insufficient mixing will occur in order to achieve a homogeneous alloy composition. This experiment provides valuable data and process guidance for the optimization of LPBF and can be used to inform the development and validation of multiphysics models. Specifically, it is suggested the visualization and evaluation of the melt-pool depth, length, keyhole dimensions and melt-pool flow directions and magnitude enable the modelling community to validate complex multiphysics models.

Acknowledgements

This research is supported under MAPP: EPSRC Future Manufacturing Hub in Manufacture using Advanced Powder Processes (EP/P006566/1) and a Royal Academy of Engineering Chair in Emerging Technology. We also acknowledge the use of facilities and support provided by the Research Complex at Harwell and thank the Advanced Photon Source for providing the beam-time (213874) and staff at the 32-ID beamline for their assistance.

References

[1] Qiu C, Panwisawas C, Ward M, Basoalto H C, Brooks J W and Attallah M M 2015 *Acta Mater.* **96** 72–9
[2] Woods R A and Milner D R 1971 *Weld. J.* **50** 163–73

[3] Heiple C R and Roper J R 1982 *Weld. J.* **61** 97-102
[4] Kou S and Sun D K 1985 F *Metall. Trans. A* **16** 203–13
[5] Lee P D, North T and Perrin A R 1988 *Model. Control Cast. Weld. Process. IV* ed A F Giamei and A G J pp 131–40
[6] Lee P D, Quested P N and McLean M 1998 *Philos. T. R. Soc. A Math. Phys. Eng. Sci.* **356** 1027–43
[7] Mills K C, Keene B J, Brooks R F and Shirali A 1998 *Philos. T. R. Soc. A Math. Phys. Eng. Sci.* **356** 911–25
[8] Aucott L *et al.* 2018 *Nat. Commun.* **9** 1–7
[9] Nakamura H, Kawahito Y, Nishimoto K and Katayama S 2015 *J. Laser Appl.* **27** 032012
[10] Leung C L A, Marussi S, Atwood R C, Towrie M, Withers P J and Lee P D 2018 *Nat. Commun.* **9** 1–9
[11] Guo Q, Zhao C, Qu M, Xiong L, Hojjatzadeh S M H, Escano L I, Parab N D, Fezzaa K, Sun T and Chen L 2020 *Addit. Manuf.* **31** 100939
[12] Kang N, Coddet P, Dembinski L, Liao H and Coddet C 2017 *J. Alloys Compd.* **691** 316–22
[13] Hanemann T, Carter L N, Habschied M, Adkins N J E, Attallah M M and Heilmaier M 2019 *J. Alloys Compd.* **795** 8–18
[14] Leung C L A, Marussi S, Towrie M, del Val Garcia J, Atwood R C, Bodey A J, Jones J R, Withers P J and Lee P D 2018 *Addit. Manuf.* **24** 647–57
[15] Leung C L A, Marussi S, Towrie M, Atwood R C, Withers P J and Lee P D 2019 *Acta Mater.* **166** 294–305
[16] Zhao C, Fezzaa K, Cunningham R W, Wen H, De Carlo F, Chen L, Rollett A D and Sun T 2017 *Sci. Rep.* **7** 1–11
[17] Guo Q, Zhao C, Escano L I, Young Z, Xiong L, Fezzaa K, Everhart W, Brown B, Sun T and Chen L 2018 T *Acta Mater.* **151** 169–80
[18] Zhao C, Guo Q, Li X, Parab N, Fezzaa K, Tan W, Chen L and Sun T 2019 *Phys. Rev. X* **9** 021052
[19] Matthews M J, Guss G, Khairallah S A, Rubenchik A M, Depond P J and King W E 2016 *Acta Mater.* **114** 33–42
[20] Matthews M J, Guss G, Khairallah S A, Rubenchik A M, Depond P J and King W E 2017 *Addit. Manuf. Handb. Prod. Dev. Def. Ind.* **114** 677–93
[21] Bidare P, Bitharas I, Ward R M, Attallah M M and Moore A J 2018 *Acta Mater.* **142** 107–20

IOP Conf. Series: Materials Science and Engineering **861** (2020) 012011 doi:10.1088/1757-899X/861/1/012011

Effect of evaporated gas flow on porosity and microstructure of IN718 parts produced by LPBF-processes

J Jakumeit[1], C Huang[1], R Laqua[1], J Zielinski[2] and J H Schleifenbaum[2]

[1] Access e.V., Intzestr. 5, 52072 Aachen, Germany
[2] RWTH Aachen University - Digital Additive Production (DAP), Campus Boulevard 73, 52074 Aachen, Germany

E-mail: j.jakumeit@access-technology.de

Abstract. Laser Powder Bed Fusion (LPBF) is the most widely used metal additive manufacturing (AM) process. A laser beam melts metal powder and partly the underlying solid metal to add metal layer by layer in selected regions. Temperatures up to the boiling temperature are reached resulting in a violent gas stream which affects solidification and can lead to porosity. Due to the small size of the melting zone, a detailed experimental analysis of the process is very difficult and simulation is an important tool to understand the manufacturing process and the influence of process parameter on the quality of components. In this work a three-phase melting and solidification simulation methodology has been used to investigate the melting, evaporation and solidification in a LPBF process. The methodology, originally developed to model casting processes, was extended by a model to calculate the heating of the metal powder by the laser beam and a metal evaporation model to include the effect of the evaporated metal gas on the total gas flow. The methodology was applied to analyse the melting, evaporation and solidification of single track experiments with IN718 powder. The simulation results were validated by comparing the calculated melt pool depth with the experimental findings. High laser power combined with slow laser speed leads to high energy densities at the melt pool, which result in a deep keyhole and increased porosity. The simulations revealed that evaporation of melt leads to a violent gas flow with velocities up to 100 m/s.

1. Introduction

Additive Manufacturing (AM), also known as 3D-printing, is a revolutionary fabrication technology, particularly for the metal forming industry. Aeronautic, aerospace, bio-engineering sectors, among many others, aspire to take advantage of the freedom at the design stage, without the restrictions imposed by the specific fabrication technologies such as in casting or forging. Moreover, the natural integration between the optimization strategies and the manufacturing technology make AM very appealing for weight saving, for enhancing performance and for the on-demand (personalized) production (e.g. bio-engineering applications).

In the laser-based additive manufacturing of metallic materials, defects inherent in the process such as connection defects and residual porosity play a major role, because these defects limit the mechanical strength of additively manufactured components. The understanding of the influence of the powder characteristics on the melting behaviour in the process and the prognosis of the resulting defect distribution are necessary steps to derive from this defect distribution a reliable life-time-prediction for additive-produced materials.

Laser Powder Bed Fusion (LPBF) is the most widely used powder bed process for the additive fabrication of metallic components. The powder is applied in a layer thickness of about 30 - 50 microns and locally melted by a laser beam according to a CAD model using a laser beam. Subsequently, the building platform is lowered by the layer thickness and a new layer of powder is added. The previous layer is partially or completely remelted, whereby the layers are joined together [1]. The range of construction materials studied includes alloys based on iron [2], nickel [3] and aluminium [4]. The additive manufacturing of metallic alloys has not been used on a large scale so far, because the properties of the components cannot be reliably predicted. They depend, for example, on the equipment and parameters used as well as on the specific properties of the powders used [5].

Khairallah *et al.* [6] showed in a 2014 study for the first time a comprehensive transient 3D modelling of the LPBF process on this spatial and time scale (mesoscale model). The simulation approach includes beam tracing for laser power input into the powder, realistic powder size distribution (also bimodal distributions), Marangoni convection, energy loss through the substrate, the powder bed and radiation. For the treatment of the evaporation, the energy loss due to evaporation and the recoil pressure on the melt front were taken into account.

A similar level of modelling is shown in the work by Guertler *et al.* [7]. The investigation of the influence of particle inhomogeneity on the surface quality of the components was carried out with the software package OpenFOAM. Both approaches are single-phase models that model only the melt phase but not the gas phase. A recent, very comprehensive work by Panwisawas *et al.* [8] gives an overview of the multiscale modelling of the LPBF process. A simulation approach based on the Lattice-Boltzmann method was developed by Körner *et al.* for particle-resolved simulations of the LPBF process [9]. A single-phase model is used which neglects the gas phase. This model was extended by the consideration of metal vapour [10] and a coupling to a cellular automaton for the prediction of the resulting microstructures [11].

The influence of powder particle size distribution on porosity was investigated by Guertler *et al.* [7] for various aluminium alloys both experimentally and by simulation. One result of these investigations is that a powder bed with many smaller particles can better balance in-homogeneities in the distribution and leads to a more stable process. Panwisawas *et al.* [8] use the multiphase approach to compare simulated porosity with experimental values. If the simulation was performed with a disordered distribution of powder particles, a good match between simulation and experiment could be found.
The following chapter gives a short description of the methodology used in this work. Section 3 describes the single track LPFB experiments and presents first simulation results, followed by the validation of the methodology and investigations on the melt pool dynamics in section 4. Section 5 focusses on the resulting porosity and the work is summarised in section 6.

2. Theory

Simulation of the particle resolved LPBF-process is based on a finite-volume method using control volumes (CVs) of arbitrary polyhedral shape. The transport equations for mass, momentum, energy, and phase in integral form are applied to each CV, whereby the surface and volume integrals are approximated using the midpoint rule. Linear equation systems for each variable are solved using algebraic multi-grid iterative solvers. Mass conservation, pressure and velocity conservation are coupled via the SIMPLE algorithm. The transient term is discretized based on an implicit Euler-segregated concept. Details of discretization and the solution method are available in [12, 13]. For solidification modelling, the volume fraction of solidified liquid is determined using a tabulated fraction solid vs. temperature curve ($fs(T)$). Latent heat L is released in proportion to the change in fraction solid ($L*d(fs(T))/dT$). Details of the methodology are presented in [13].

For correct calculation of air entrainment, the method used to track the motion of the free surface must provide a sharp interface. The volume of fluid (VoF) approach, in combination with a high resolution interface capturing (HRIC) scheme, is used to tackle the problem: the entire fluid domain is considered to be filled by a fluid, the properties of which vary according to the distribution of volume fractions of melt C_m, solid C_s and gas C_g ($Cm + Cs + Cg = 1$). The transport of melt, solid and gas is computed by solving transport equations for their volume fractions with a source term for the phase change from melt to solid. To achieve sharpness of the interface, an HRIC scheme [14] is used, which typically resolves the interface within one cell.

The normal force due to surface tension is treated using the continuum surface force (CSF) model proposed by [15], which defines a volumetric source in the momentum equation. Flow resistance in the mushy zone is calculated assuming that the mushy zone acts like a porous medium. The fluid velocity resistance in the mushy zone can be approximated as a pressure drop [16]. Permeability can be deduced from the Kozeny-Carman equation as $K = ((1-fs)^3\lambda_2^2)/(180fs^2)$, where λ_2 is the secondary dendrite-arm spacing (SDAS).

For the evaporation of melt, a simple version of the model developed by Klassen *et al.* [10, 17] was adapted to LBPF (from EBF) and extended to incorporate the creation of a gas phase by evaporation. The evaporation model uses the following assumptions:

- The evaporated metal gas is treated as ideal gas.
- The density of the evaporated metal gas is low, so that absorption of laser light by the gas can be neglected.
- Alloy is treated as pure material having effective properties. The different partial vapour pressures of the alloying elements are not taken into account.
- LPBF processes at atmospheric pressure using medium laser intensities are modelled. The vapour flow remains subsonic and for the Mach number at the outer boundary of the Knudsen layer is smaller than 1. Therefore, using the maximum value of 0.82 for the evaporation coefficient Φ is a plausible approximation (see also [19]).
- Since the movement of gas bubbles in the melt pool is of main interest to analyse porosity Marangoni convection is included.

Based on these assumptions the net mass transport from the evaporating metal surface can be calculated by: $j_{net} = \Phi \cdot j_+$, $j_+ = P_s \sqrt{m_A / 2\pi k_B T_s}$, where $j+$ is the evaporating flux and Φ the evaporation coefficient, which gives the fraction of particles that can escape from the surface and are not scattered back to the surface or compensated by condensation of particles. For $MaKn(Ts) < 1$ Φ approaches the maximum value of 0.82 (see [17, 18]). This value is used throughout this work. In the evaporating flux $j+$ calculation m_A is the atomic mass, k_B is Boltzmann's constant and P_s and T_s are the pressure and temperature at the vapour side of the phase interface, respectively. P_s can be interpreted as the saturation vapour pressure at T_s at the melt surface and calculated based on the Clausius-Clapeyron equation in its approximated form and the temperature dependent latent heat of vaporization $L_{vap}(Ts) = L_{vap,0} * [1-(Ts/Tcrit)^2]^{1/2}$ (see [17] for details):

$$P_s = P_{atm} \cdot exp\left[-\frac{L_{vap,0} \cdot m_A}{k_B}\left[\frac{1}{T_s}\sqrt{1-\left(\frac{T_s}{T_{crit}}\right)^2} - \frac{1}{T_{boil}}\sqrt{1-\left(\frac{T_{boil}}{T_{crit}}\right)^2} - \frac{1}{T_{crit}}\left(sin^{-1}\left(\frac{T_s}{T_{crit}}\right) - sin^{-1}\left(\frac{T_{boil}}{T_{crit}}\right)\right)\right]\right], \quad (1)$$

where $L_{vap,0}$ is the latent heat of vaporization at absolute zero temperature, $P_{atm} = 1$ bar and T_{boil} and T_{crit} are the boiling temperature at standard atmosphere (1 bar) and the critical temperature, respectively. The evaporation model is linked to the continuum equations describing the LPBF process by source terms to the momentum and energy equation.

A crucial point for the comparison of experiments with simulation results is the model used to calculate the amount of laser power, absorbed by the powder. Such a model must treat the amount of powder irradiated by the laser and the absorption and reflection by the powder. While the laser power can be measured, the reflectivity of the powder and the amount of powder irradiated is difficult to measure and fluctuates during the process. A simple and often used approach is a volumetric heat source, which adds a defined amount of heat in each cell of the irradiated area. The amount of heat can be calculated using analytical functions describing the absorption of the powder bed and solidified melt underneath. (see for example [17, 19]) A limitation of this approach is, that energy is not only deposit at the surface of the metal and shading of areas by overlying metal areas and irradiation by reflected light cannot be modelled. All these effects can be correctly accounted by using ray tracing algorithms to calculate the irradiation of the powder and solidified metal by the laser in a physical correct way. Such an approach was firstly applied to LPBF-process simulation by Khairallah *et al.* [6] and used by Bayat *et al.* for the investigation of keyhole burning [20]. While this approach gives a physical description of the radiation effects, its precision is limited by the quality of calibration of the model parameter like reflectivity of hot melt surfaces, which are difficult to obtain by measurement.

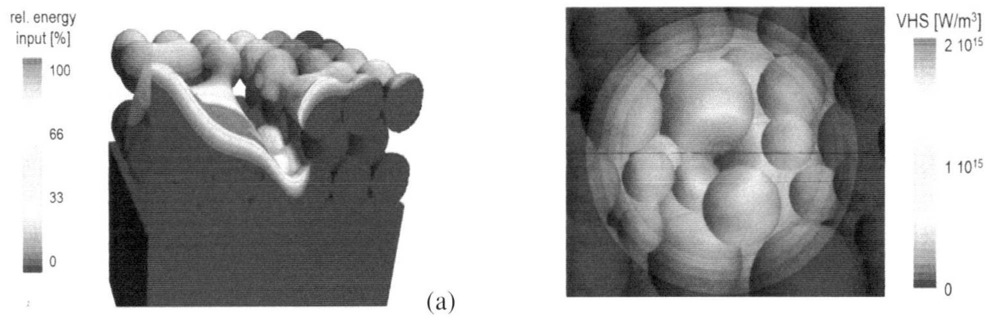

Figure 1. (a) Surfaces irradiated by laser light are marked including shading effects and (b) the absorbed Gaussian distributed laser energy is applied using a volumetric heat source.

Here, a new two stage approach is followed which is a compromise between the simple volumetric heat approach and the complex ray tracing method:

First, using a participating media radiation model, the melt surface hit by the laser beam is marked. From a light source above the metal surface light is radiated towards the powder and substrate. Metal surfaces hit by the light are marked for energy deposition (see figure 1(a)) and shading of metal surfaces is correctly modelled. Since the light of the participating media radiation model is not monochromatic and not restricted to the laser beam, a direct calculation of the heat transfer to the melt surface is not possible.

Therefore, in a second step, heat is deposited in the marked metal cells at the surface of the metal using a volumetric heat source. The Gaussian shape of the laser intensity is transferred to a radial Gaussian profile of the volume heat source term from the centre of the laser beam. Figure 1(b) shows the spatial heat distribution on the powder surface. In the simulation the amount of energy absorbed by the metal is directly accessible and, therefore, used as process parameter. The volumetric heat source is adjusted to add the total energy absorbed from the laser light to the marked metal volume. For comparison with experiments the laser power is calculated from the amount of energy absorbed by the metal by an absorption coefficient, which in reality depends on the metal surface size and form, the amount of reflected light absorption and the shading of laser light by evaporated metal gas and is difficult to be determined experimentally. This novel approach allows to use the laser absorption coefficient as fitting parameter between experimental findings and simulation result (see section 4).

Table 1. Material and powder parameters

$T_{liquidus}$	1335 °C
$T_{solidus}$	1250 °C
$T_{evaporation}$	2227 °C
T_{crit}	7395 °C
Latent heat evaporation	6 10^7 J/kg
Evaporation coef.	0.82
Mean particle diameter	20 μm

3. Simulation of single track LPBF-processes

The methodology was applied to analyse the melting, evaporation and solidification of single track experiments with IN718 powder (Ni+, 18.1 Fe, 18.6 Cr, 5.2 Nb, 3.02 Mo, 1.01 T, 0.54Al, 0.11Mn, 0.16 Si [wt%]) in a DAP laboratory LPBF-machine. Material and powder parameter are given in table 1.

The IN718 powder was melted using a laser beam diameter of 80µm (Gaussian distribution with standard deviation of 20 µm assumed), laser power between 142 W and 285 W and a laser speed between 0.6 m/s and 1.1 m/s.

The powder bed was created using the DEM-model of STAR-CCM+ with a raining model. Particles with a Gaussian distribution with mean value 20 µm fall on the IN718 substrate until a 60 mm layer was achieved. The DEM-created powder bed was then transferred to a polyhedral mesh (see figure 2). The size of the computation domain is 1000 µm x 500 µm x 500 µm (length x width x height) and the mesh has around 2.1 million cells.

Figure 3 shows the temperature distribution on the melt surface (left) and the pressure and velocity (right) after 0.3, 0.4 and 0.5 ms for a process with a laser power of 250 W and a lase speed of 1100 m/s.

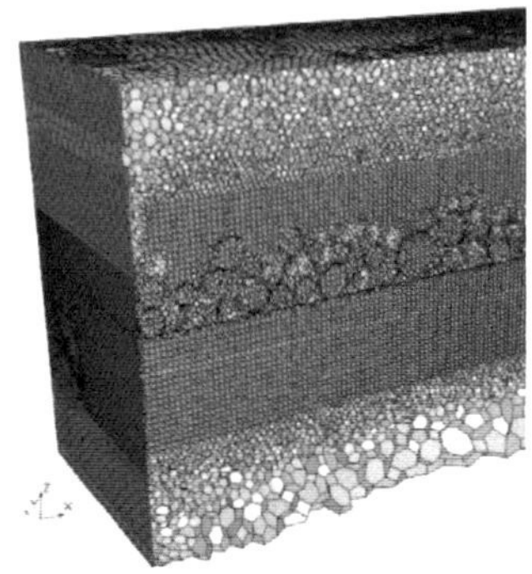

Figure 2. Polyhedral mesh with resolved particles.

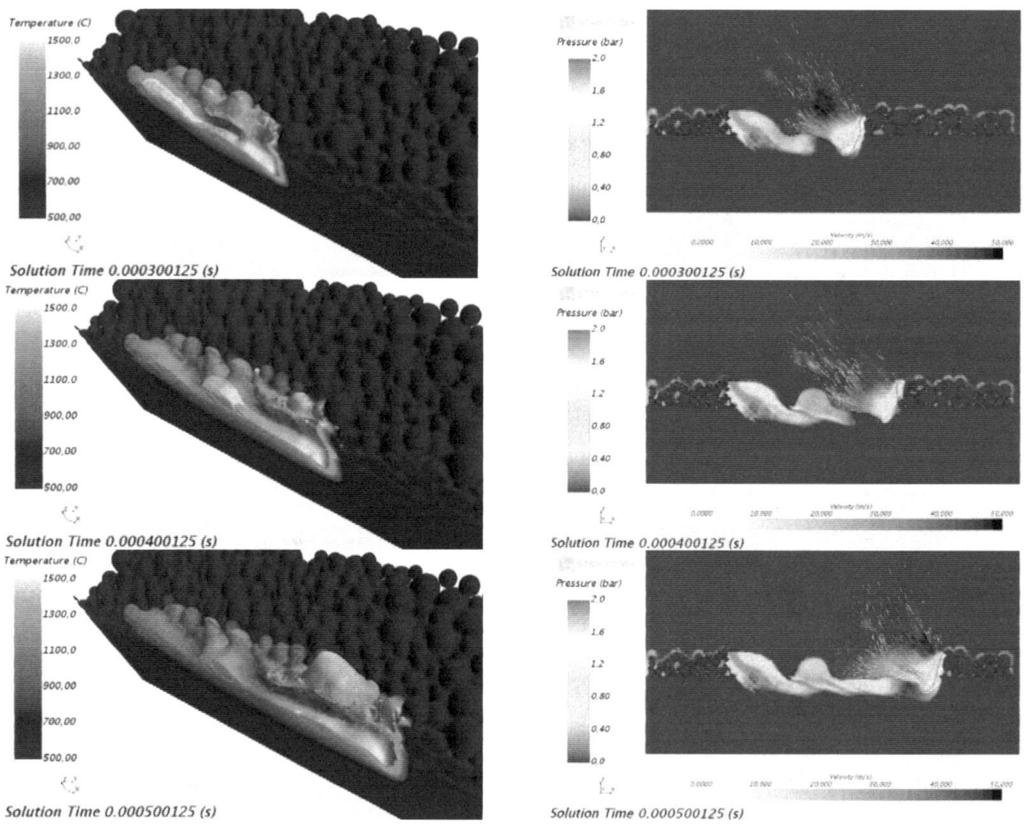

(a) (b)

Figure 3. (a) Temperature distribution on the melt surface and (b) the pressure and velocity field and velocity distribution after 0.3, 0.4 and 0.5 ms process time for a process with a laser power of 250 W and a laser speed of 1100 m/s.

4. Validation

The simulation was validated by comparing the calculated melt pool depth with the experimental findings for the single track experiments without (0 µm) and 60 µm powder layer. The powder layer height of the experiments is manually applied and verified with an infinite focus microscope. Table 2 gives for 5 combinations of laser power and laser speed the melt pool depth found in the experiments and related simulations. For the simulations the desired total volumetric heat source (VHS) were calculated assuming a constant absorption coefficient of substrate and powder of 30 %, a value often reported for metal surfaces [19]. The simulation gives a fair agreement with the experimental findings. The general tendency is reproduced by the simulation, but the slope of the melt pool depth vs line energy is slightly smaller (exp-0µm: 0.51, exp-60µm: 0.54, sim-0µm: 0.32, sim-60µm: 0.39 [1/J]).

Table 2. Experimental and simulation parameters and results

Experiment					Simulation		
Laser power [W]	Laser speed [mm/s]	Line energy [J/mm³]	Melt pool depth (no powder) [µm]	Melt pool depth (60 µm powder) [µm]	Total VHS [W]	Melt pool depth (no powder) [µm] (diff.)	Melt pool depth (60 µm powder) [µm] (diff.)
142	960	0.15	48	73	42.6	67 (28%)	88 (17%)
250	1100	0.23	85	120	75	100 (15%)	120 (0 %)
285	960	0.30	130	172	85.5	126 (3 %)	160 (7.5%)
250	660	0.38	150	176	75	145 (3.4 %)	
285	600	0.47	212	252	85.5		214 (18%)

To understand the found difference in the slope experimentalists at ILT suggest to use a power dependent absorption coefficient instead of the constant value of 30 %. Such a dependency is supported by the simulation result that the melt volume directly heated by the laser (marked cells) depends linearly on the line energy (see figure 4). In addition, Bayat *et al.* reported a linear dependence of the absorption coefficient on the laser power for Ti6Al4V powder in case of keyhole burning [20]. Using a linear dependency of the absorption coefficient vs line energy with slope of 0.3 mm/J, an almost perfect fit between simulation and experimental melt pool depths could be achieved, especially for the case with powder layer (see figure 5).

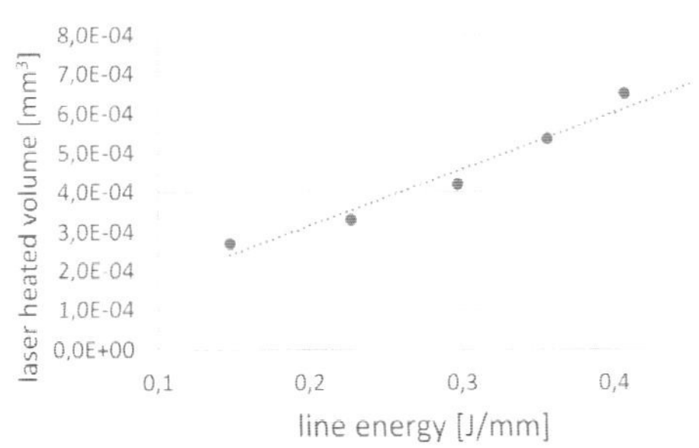

Figure 4. Melt volume directly heated by the laser as function of line energy.

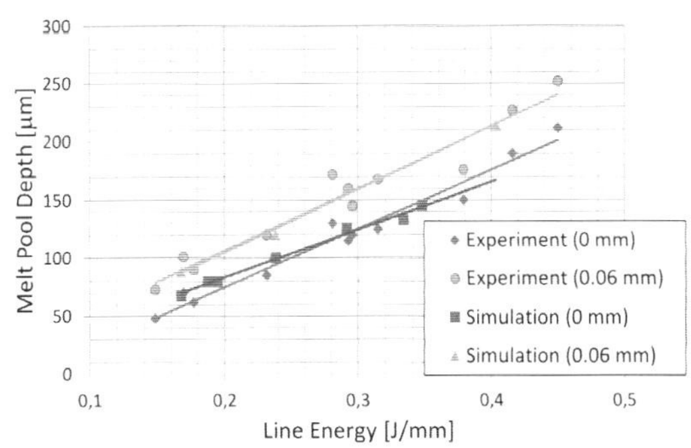

Figure 5. Comparison of experimental and simulation results for melt pool depth for single track experiments without powder and with a 60 µm powder layer.

5. Porosity

The porosity or density (=1-porosity) of the produced IN718 layer can be calculated by volume integral of the remaining gas in the melted and solidified region. Figure 6 shows the line energy dependence of the density. An almost linear increase of the density can be found with increasing line energy until keyhole burning leads to increasing porosity at high line energies. A similar behaviour was reported by Peng *et al.* [21] for 316L steel and by Bayat *et al.* [20] for Ti6Al4V.

The correct handling of the evaporated metal as compressible gas makes it possible to investigate the interplay between the evaporated gas and the gas stream across the powder layer. The standard setting is no gas flow across the powder layer (v = 0 m/s) and the top of the simulation box was handled as wall. Table 3 gives the influence of an open top of the simulation box (gas can escape freely) and a gas stream of 1 m/s across the powder layer in scan direction and against scan direction. The gas stream leads to a slightly reduced porosity, but the overall effect is small.

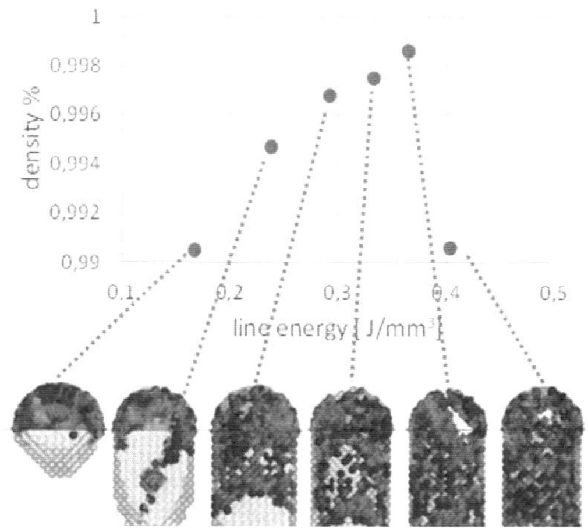

Figure 6. Energy dependence of the layer density.

Table 3. Influence of gas flow above the powder layer on the porosity

	Closed top no gas flow	Open top no gas flow	Closed top 1 m/s in laser dir.	Closed top 1 m/s against laser
Melt pool depth [μm]	120	112	125	122
Porosity [%]	0.57	0.52	0.46	0.44

6. Conclusion

Metal additive manufacturing by LPBF processes was analyzed using a three-phase melting and solidification simulation methodology originally developed to model casting processes. The methodology was extended by a model to calculate the heating of the metal powder by a laser beam, which combines a participating media radiation model with a volumetric heat source. The laser beam heats the metal powder to temperatures, at which metal evaporation leads to a violent gas flow, modelled by a newly implemented evaporation model. The methodology was applied to analyze the melting, evaporation and solidification of single track experiments with IN718 powder.

Keyhole burning at higher line energies lead to an increase of the light absorbing metal surface and, thereby, to a linear increase of the absorbability of the metal powder. Taking this into account an excellent agreement between the predicted melt pool depth and experimental findings could be achieved. In good agreement with published results for different alloys an increase of the part density with increasing line energy could be observed until significant keyhole burning leads to increasing porosity at high line energies. Only a small effect of a cooling gas flow across the powder layer was found, which can lead to slightly reduced porosity.

Acknowledgments

The work was supported by BMWi (Bundesministerium für Wirtschaft und Energie), INNOKOM project Fast Solid, funding contract 49VF170031. The authors acknowledge the support by Siemens PLM, agreement 60068580, for providing STAR-CCM+ licenses.

IOP Conf. Series: Materials Science and Engineering **861** (2020) 012011 doi:10.1088/1757-899X/861/1/012011

References

[1] Sun S, Brandt M and Easton M 2017 *Powder bed fusion processes.* In *Laser Additive Manufacturing* ed M Brandt (Elsevier) pp 55-77

[2] Hanzl P, Zetek M, Bakša T and Kroupa T 2015 The Influence of Processing Parameters on the Mechanical Properties of SLM Parts. *Procedia Engineer.* **100** 1405–13

[3] Karapatan NP, Egger G, Gygax P E and Glardon R 1999 *Optimization of Powder Layer Density in Selective Laser Sintering.* In *Proc. International Solid Freeform Fabrication Symposium*

[4] Mower T M and Long M 2016 *Mat. Sci. Eng. A* **651** 198–213

[5] Yadollahi A and Shamsaei N 2017 *Int. J. Fatigue* **98** 14–31

[6] Khairallah S A and Anderson A 2014 *J. Mater. Process. Tech.* **214** 2627–36

[7] Guertler F J, Karg M, Dobler M, Kohl S, Tzivilsky I and Schmidt M 2014 In *Proc. 25th Annual International Solid Freeform Symposium,* ed Bourell D (Austin, USA) pp 1099-117

[8] Panwisawas C, Sovani Y, Anderson M J, Turner R, Palumbo N. M, Saunders B C, Choquet I, Brooks J W and Basoalto H C 2016 In *Proc. 13th Intenational Symposium of Superalloys,* ed. Hardy M *et al.* (Hoboken, USA) pp 1021-30

[9] Bauereiß A, Scharowsky T and Körner C 2014 *J. Mater. Process. Tech.* **214** 2522–28

[10] Klassen A, Forster V E, Juechter V and Körner C 2017 *J. Mater. Process. Tech.* **247** 280–8

[11] Rai A, Markl M and Körner C 2016 *Comp. Mater. Sci.* **124** 37–48

[12] Jana S, Kattlitz O, Hediger F, Jakumeit J and Aguilar J 2012 *IOP Conf. Ser.: Mater. Sci. Eng.* **33** 012007

[13] Teskeredzic A, Demirdzic I, and Muzaferija S 2002 *Numer. Heat Tr. B* **42** 43

[14] Muzaferija S and Peric M 1999 *Computation of free surface flows using interface-tracking and interface-capturing methods.* In *Nonlinear Water Wave Interaction* ed Mahrenholtz O *et al.* (Southampton: WIP Press) pp 59-100

[15] Brackbill J U, Kothe D B and Zemach C 1992 *J. Comput. Phys.* **100** 335

[16] Sabau S and Viswanath S 2002 *Met. Mat. Trans. B* **33** 131

[17] Klassen A, Scharowsky T and Koerner C 2014 *J. Phys. D: Appl. Phys.* **47** 12

[18] Anisimov S I 1968 *Sov. Phys JETP* **27** 182

[19] Gusarov A V, Yadroitsev I, Bertrand P and Smurov I 2009 *J. Heat Transf.* **131** 072101

[20] Bayata M, Thanki A, Mohanty S, Witvrouw A, Yang S, Thorborg J, Tiedje N S and Hattel J H 2019 *Addit. Manuf.* **30** 100835

[21] Peng T and Chen C 2018 *Int. J. Pr. Eng. Man. - GT* **5** 55–62

IOP Conf. Series: Materials Science and Engineering **861** (2020) 012012 doi:10.1088/1757-899X/861/1/012012

Melt pool morphology in directed energy deposition additive manufacturing process

Y Chen[1,2], S Clark[1,2], A C L Leung[1,2], L Sinclair[1,2], S Marussi[1,2], R Atwood[1,2], T Connoley[3], M Jones[4,5], G Baxter[4] and P D Lee[1,2]

[1] Department of Mechanical Engineering, University College London, WC1E 7JE, UK
[2] Research Complex at Harwell, Harwell Campus, Didcot, OX11 0FA, UK
[3] Diamond Light Source Ltd, Harwell Campus, Didcot, OX11 0DE, UK
[4] University of Sheffield, Sir Robert Hadfield Building, Sheffield, S1 3JD, UK
[5] Rolls-Royce plc, PO Box 31, Derby, DE24 8BJ, UK

E-mail: yunhui.chen@ucl.ac.uk, peter.lee@ucl.ac.uk

Abstract. Directed Energy Deposition Additive Manufacturing (DED-AM) is one of the principal AM techniques being explored for both the repair of high value components in the aerospace industry as well as freeform fabrication of large metallic components. However, the lack of fundamental understanding of the underlying process-structure-property relationships hinders the utilisation of DED-AM for the production or repair of safety-critical components. This study uses in situ and operando synchrotron X-ray imaging to provide an improved fundamental understanding of laser-matter interactions and their influence on the melt pool geometry. Coupled with process modelling, these unique observations illustrate how process parameters can influence the DED-AM melt pool geometry. The calibrated simulation can be used for guidance in an industrial additive manufacturing process for microstructure and quality control.

1. Introduction

Directed Energy Deposition Additive Manufacturing (DED-AM) [1] is one of the most promising subsets of Laser Additive Manufacturing (LAM) and is considered a key technique for both the repair [2] of existing components [3] as well as freeform fabrication [4]. However, the lack of fundamental understanding of the underlying process-structure-property relationships hinders the utilisation of DED-AM for safety-critical components such as turbine aerofoil repair [2]. The ability to predict and control microstructure during the laser deposition process holds the key to realising a high-quality product. The thermal gradient at the solidification interface is a crucial parameter for controlling the final microstructure [5] and is directly related to the process parameters (e.g. laser power, spot size, traverse speed, etc). The reliance upon ex situ characterisation of the final microstructure [1, 6, 7] and mechanical properties [8–11] of DED-AM deposits is an inefficient and costly means of process optimisation. Furthermore, it is unable to elucidate the mitigations to the aforementioned restraints on DED-AM utilisation.

In situ and operando high-speed X-ray imaging enables researchers to characterise the time-transient phenomena in LPBF (Laser Powder Bed Fusion) [12–16] and the piezo-driven DED-AM process. High-brilliance X-rays can penetrate through sufficient depths of dense, optically opaque metallic samples [17] with high spatial and temporal resolution [18]. Using synchrotron X-rays and image analysis, we can estimate the transient melt pool geometry, shape, and volume. These values determine the geometry and quality of the deposited track. The melt pool shape also directly relates to the microstructural and the mechanical performance of the build [19] as the melt pool geometry is a result of the thermal gradient and cooling rate introduced by the laser beam. The purpose of this study is to provide observations that can be used as standards for stainless steel and to aid the development and validation of advanced thermo-physical models.

Rosenthal's analytical solutions [20] are well known for describing a moving point heat source traversing an infinite substrate. It has been widely used to calculate thermal process maps for controlling microstructure in the laser-based fabrication process [21]. A carefully calibrated Rosenthal solution for thin walls has been developed and applied to predict the microstructure, and thus the quality and performance of the deposit. The aim of this work is to study SS316 deposited using the DED-AM process and to capture the melt pool geometries in situ and operando using synchrotron X-ray radiography to elucidate the physical phenomena during DED-AM. These observations are then used to calibrate the Rosenthal solution, enabling the prediction of melt pool geometry at various combinations of process parameters not imaged. The simplified model of Rosenthal solution in this study could provide a fast way of process prediction in the industry. The results present in this work enable an enhanced understanding of DED-AM processes and facilitate improved manufacturing practice.

2. Materials and Methods

2.1. Synchrotron X-ray imaging of DED-AM process

The Blown Powder Additive Manufacturing Process Replicator (BAMPR) is developed to faithfully replicate a commercial DED-AM system. The instrument is designed to be integrated into synchrotron beamlines, e.g. I12: the Joint Engineering, Environmental, and Processing (JEEP) at Diamond Light Source, Harwell [22], as illustrated schematically in figure 1. The system is encased within a Class I laser enclosure and comprises an inert environment chamber, a high precision 3-axis platform (Aerotech, US), a coaxial DED nozzle and a laser system (SPI Lasers Ltd, UK). An industrial powder feeder (Oerlikon Metco TWIN-10-C) delivers powder to the system in a stream of argon gas. The laser is a 1070 nm Ytterbium-doped fibre laser (continuous-wave (CW) mode) with controllable laser power (P) of 200 W. The laser is coupled with tuneable optics (Optogama, Lithuania) to facilitate a controllable focused spot size of between 200 and 700 μm. The laser is positioned to be concentric with the powder delivery stream blown from the nozzle and normal to the substrate plate in a carrier of 6 litres min-1 Argon gas. In this work, the build platform is able to translate through 25 mm in width, 50 mm in length, and 50 mm in height. A gas atomised SS316 powder is used for this study and its powder size distribution ranges from 45 to 90 μm with a d50 of 70 μm. The substrate plate is positioned inside the environmental build chamber which has Kapton windows. Kapton is used due to its low X-ray absorption in the energy range of beamline I12: JEEP. The chamber is constantly flushed with a combined flowing argon atmosphere (powder + carrier gas and shielding gas) of 17 l min-1. The speed of the sample stages in both cases is controlled to be between 1 to 5 mm s-1 to enable a continuous track to be formed, as indicated by preliminary laboratory trials. A monochromatic beam is used for all trials with an X-ray photon energy of 53 keV. The synchrotron X-rays are attenuated by the powder and the deposit and then are converted to visible light by a scintillator, lens-coupled to a high-speed CMOS camera (MIRO 310M, Vision Research Inc.), recording at 200 - 5000 fps. The optical configuration provided an imaging resolution of approximately 6.67 μm per pixel at 200 fps. The resulting radiographs (or movies) reveal the time-resolved multi-layer melt track morphology evolution of a SS316 track during DED-AM, as sequenced in figure 1(c).

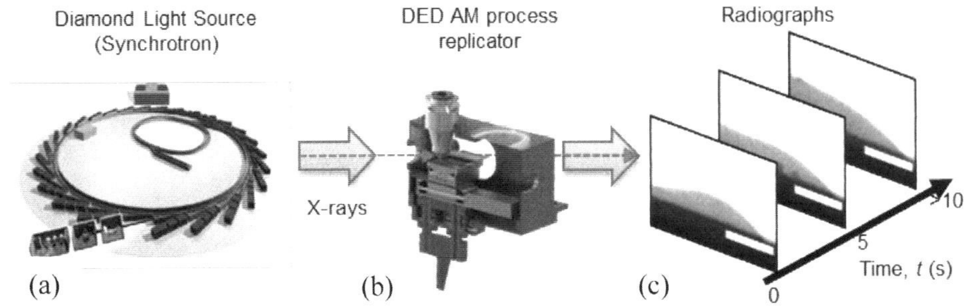

Figure 1. In situ and operando synchrotron X-ray imaging of DED-AM process. (a) Diamond Light Source (DLS) synchrotron facilities. Image credit to Diamond Light Source website. (b) BAMPR – DED-AM process replicator. (c) Radiographs acquired using in situ synchrotron X-ray technique. Scale bar = 500 μm.

2.2. Thin-walled Rosenthal solution for laser heat conduction

The thin-walled Rosenthal solution [20] provides a quasi-steady-state solution to describe a moving point heat source moving within a thin-wall geometry and is expressed in dimensionless form as [23]

$$\bar{T} = e^{-\bar{x}_0} K_0\left(\sqrt{\bar{x}_0^2 + z_0^2}\right) \qquad (1)$$

where K_0 is the modified Bessel function of the second kind, order zero. The dimensionless variables in equation (1) are defined in terms of the absorbed laser power αQ and velocity V

$$\bar{T} = \frac{T - T_0}{\alpha Q/\pi k b},\ \bar{x}_0 = \frac{x_0}{2k/\rho c V},\ \text{and}\ \bar{z}_0 = \frac{z_0}{2k/\rho c V} \qquad (2)$$

In the above normalisations, T is the temperature at a location (x_0, z_0) relative to the moving point source and T_0 is the initial temperature of the wall. The thermo-physical properties ρ, c, and k are the density, specific heat and thermal conductivity of the substrate material, respectively. They are assumed to be temperature-independent. b is the thickness of the substrate.

3. Results and Discussion

3.1. X-ray imaging of melt pool geometry in DED-AM process

Figure 2(a) shows a radiograph taken during the second layer build of the DED-AM process. The melt tracks were deposited with the substrate plate traversing bi-directionally by alternating the build direction through 180°. The laser beam is shown to partially consolidate the previous track and incorporate. New powder particles into a melt pool, forming a melt track and subsequently building the 3D component.

The flow dynamics in the molten pool are a complex interplay of several phenomena, incorporating laser-induced buoyancy force and Marangoni flow, the quenching effect of powder particles fed into the melt pool and the traverse of the substrate [19]. As a result, the molten pool presents an elliptical-cap shape during the build process, as shown in figure 2(b). The depth and length of the melt pool are defined as the radius and the height of the elliptical cap respectively. A relatively smooth surface finish is achieved, although some un-melted powder particles are partially embedded into the melt track when it is semi-solidified.

IOP Conf. Series: Materials Science and Engineering **861** (2020) 012012 doi:10.1088/1757-899X/861/1/012012

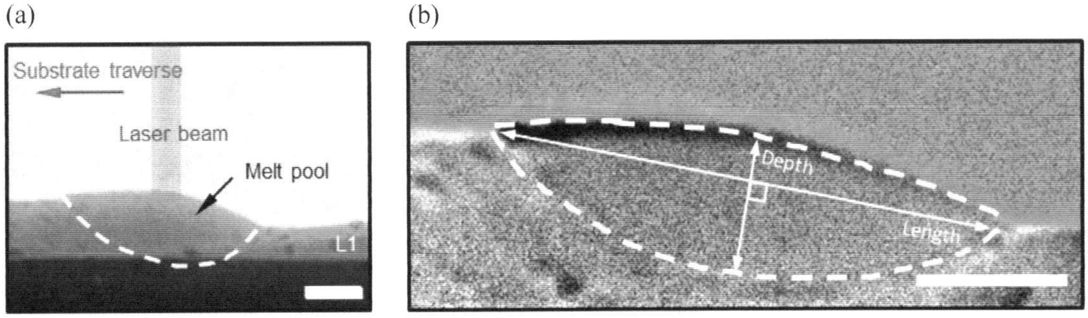

Figure 2. *In situ* and *operando* X-ray imaging of the DED-AM process. (a) A typical radiograph acquired during DED of a multi-layer build of SS316 with parameters: $P = 200$ W, $v = 1.67$ mm s^{-1} captured at 200 fps. Scale bar = 500 μm. (b) A schematic of the melt pool geometry measurement and the corresponding X-ray image of the melt pool. Measurements see figure 4. Scale bar = 500 μm.

3.2. Thin-walled Rosenthal solution for melt pool geometry prediction

The volume and shape of the melt pool is determined by the thermal gradient and cooling rate. This determines the geometry and quality of the deposit, which in turn directly influences the microstructure and mechanical properties of the build [20]. The upper abutted spherical cap profile and the height of the deposited track are strongly dependent on the powder mass flow rate and thus also influence the thermal gradient in the melt pool [25]. The thin-walled Rosenthal solution was adopted to simulate the thermal gradient when the laser is moving within the bounds of a thin substrate plate, neglecting the effect of powder entering the melt pool. The melt pool is assumed to be the boundary where the temperature of the substrate is equal to the material's melting temperature ($T = T_m$) see isothermal line in figure 3. The process parameters for SS316 used for the simulation are listed in table 1. The thickness of the substrate plate (b) is calibrated to be 0.5 mm.

Figure 3 shows the predicted profile of the melt pool in terms of depth and length variations across laser power and traverse speed measured using X-ray imaging and simulated via Rosenthal solution. With a decrease in the effective laser power density (either the decrease of laser power or the increase of traverse speed), the melt pool geometry decreases accordingly [24]. The increase of traverse speed also decreases the amount of powder deposited into the melt pool, which further decreases the size of the melt pool above the level of the previous layer. Increasing the traverse speed increases the surface area (per unit time) exposed to laser radiation, therefore, an increase in the melt pool length.

Table 1. The parameters used for thin-walled Rosenthal solution.

Parameters	SS316
Specific heat (c)	490 J/kg·K
Thermal conductivity (k)	13 W/m·K
Melting termperature (C)	1371 C
Mass density (ρ)	8000 kg/m^3

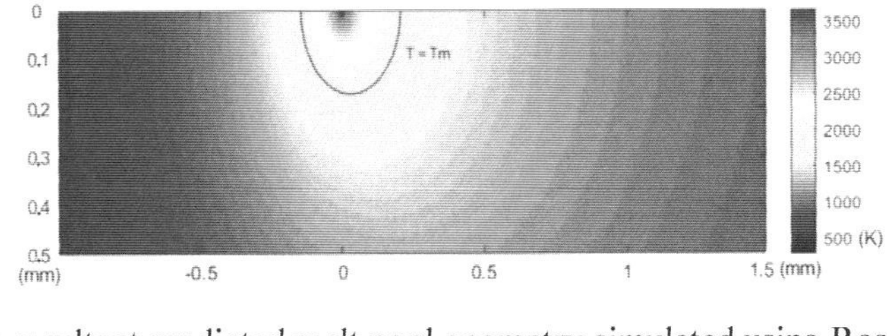

Figure 3. The resultant predicted melt pool geometry simulated using Rosenthal solution.

After the thin wall thickness calibration, the simulated melt pool geometries follow the similar pattern as the experimental results, as shown in figure 4. The relationship between the experimental results and simulation is shown to be complex. However, the deviation is smaller at high thermal gradient conditions, i.e. higher laser power and lower traverse speed.

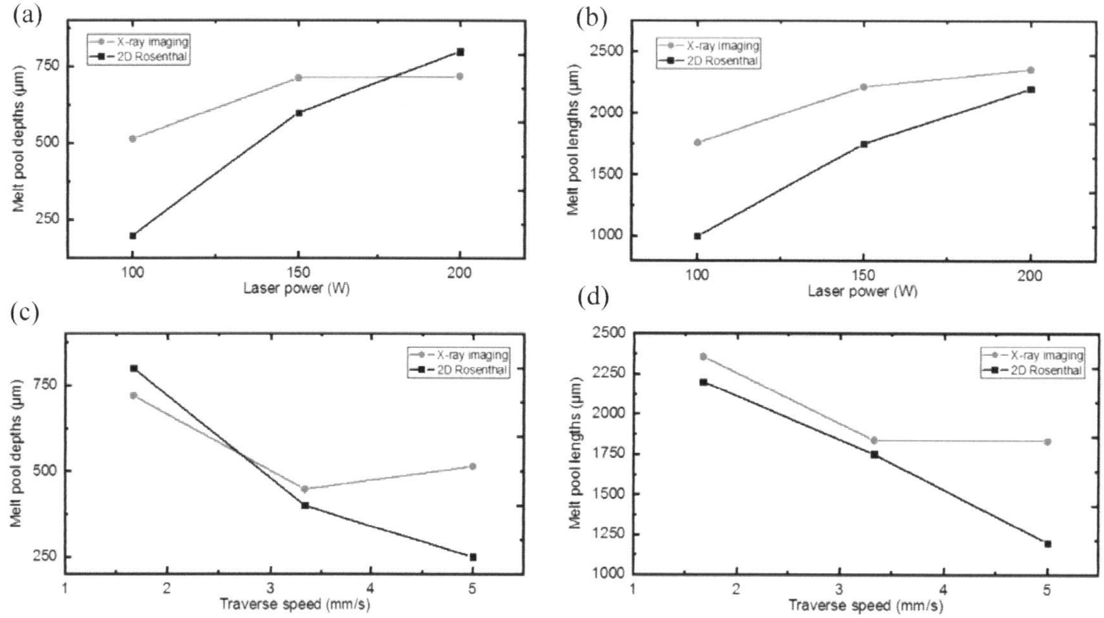

Figure 4. Melt pool depths and lengths variations across laser power and traverse speed of DED-AM of SS316. (a) Melt pool depth across laser power. (b) Melt pool length across laser power. (c) Melt pool depth across traverse speed. (d) Melt pool depth across traverse speed.

4. Conclusions

In summary, in situ and operando synchrotron X-rays and the thin-walled Rosenthal solution were used to understand the melt pool geometry in Directed Energy Deposition Additive Manufacturing (DED-AM) processing of SS316. An accurate characterisation of the melt pool morphology across building parameters was presented using radiographs. The analysis of the radiographs reveals the relationship between laser power, traverse speed, and the geometry of the melt pool. The results were used to calibrate the thin wall Rosenthal solution for melt pool geometry prediction. It is found that Rosenthal's solution can simulate the melt pool geometry but the requirement for the wall thickness to be known a priori is a significant limitation.

Acknowledgments

This research is supported under MAPP: EPSRC Future Manufacturing Hub in Manufacture using Advanced Powder Processes (EP/P006566/1) and a Royal Academy of Engineering Chair in Emerging Technology. This research is also supported by Rolls-Royce plc. via the Horizon 2020 Clean Sky 2 WP5.8.1 and Strategic Partnership (WP2C) programmes. We also acknowledge the use of facilities and support provided by the Research Complex at Harwell and thank the Diamond Light Source for providing the beam-time (MT20096-1) and staff at the I12 beamline for their technical assistance.

References

[1] Kumara C, Segerstark A, Hanning F, Dixit N, Joshi S, Moverare J and Nylén P 2019 *Addit. Manuf.* **25** 357–64
[2] Leino M, Pekkarinen J and Soukka R 2016 *Phys. Procedia* **83** 752–60
[3] Panesar A, Ashcroft I, Brackett D, Wildman R and Hague R 2017 *Addit. Manuf.* **16** 98–106
[4] Mazumder J 2016 *Laser-aided direct metal deposition of metals and alloys* (Elsevier Ltd)

[5] Karagadde S *et al.* 2015 *Nat. Commun.* **6**
[6] Yan F, Xiong W and Faierson E J 2017 *Materials* **10** 1260
[7] Kistler N A, Corbin D J, Nassar A R, Reutzel E W and Beese A M 2019 *J. Mater. Process. Tech.* **264** 172–81
[8] Leuders S, Thöne M, Riemer A, Niendorf T, Tröster T, Richard H and Maier H J 2013 *Int. J. Fatigue* **48** 300–7
[9] Shamsaei N and Simsiriwong J 2017 *Procedia Struct. Integr.* **7** 3–10
[10] Alafaghani A, Qattawi A and Castañón M A G 2018 *Int. J. Adv. Manuf. Technol.* **99** 2491–507
[11] Arifin A, Sulong A B, Muhamad N, Syarif J and Ramli M I 2014 *Mater. Des.* **55** 165–75
[12] Zhao C, Fezzaa K, Cunningham R W, Wen H, De Carlo F, Chen L, Rollett A D and Sun T 2017 *Sci. Rep.* **7** 3602
[13] Leung C L A, Marussi S, Atwood R C, Towrie M, Withers P J and Lee P D 2018 *Nat. Commun.* **9** 1–9
[14] Calta N P *et al.* 2018 *Rev. Sci. Instruments* **89** 055101
[15] Martin A A *et al.* 2019 *Materials Today Adv.* **1** 100002
[16] Álvarez-Murga M *et al.* 2017 *J. Synchrotron Radiat.* **24** 240–7
[17] Nommeots-Nomm A, Ligorio C, Bodey A J, Cai B, Jones J R, Lee P D and Poologasundarampillai G 2019 *Materials Today Adv.* **2** 100011
[18] Kareh K M, Lee P D, Atwood R C, Connolley T and Gourlay C M 2014 *Nat. Commun.* **5** 1–7
[19] Hofmeister W and Griffith M 2001 *JOM* **53** 30–4
[20] D. Rosenthal 1946 *Trans. ASME* **68** 849–866
[21] Bontha S and Klingbeil N W 2003 *SFF Proc.* 219–26
[22] Drakopoulos M *et al.* 2015 *J. Synchrotron Radiat.* **22** 828–38
[23] Vasinonta A, Beuth J L and Ong R 2001 *Proc. 12th Solid Free. Fabr. Symp.* 432–40
[24] Matys J, Dominiak M and Flieger R 2015 *J. Clin. Diagn. Res.* **9(12)** ZL01-ZL2

IOP Conf. Series: Materials Science and Engineering **861** (2020) 012013 doi:10.1088/1757-899X/861/1/012013

Part-scale mechanical modelling of LPBF including microstructural evolution effects

D De Baere, M Bayat, S Mohanty and J H Hattel

Department of Mechanical Engineering, Technical University of Denmark, Produktionstorvet, Building 425, 2800 Kgs. Lyngby, Denmark

E-mail: ddbae@mek.dtu.dk

Abstract. Due to its nature as a layer-by-layer production technique, the stresses and subsequent deformation from laser-based powder bed fusion are different from the ones observed in other manufacturing techniques. Additionally, because of the cyclic heating and cooling, the material undergoes significant microstructural changes during the process. Especially for the material Ti-6Al-4V, this microstructural change is pronounced, since the microstructure changes from β to α+β during the LPBF process. In this work, two models are coupled together in a novel way. First, a reduced-fidelity part-scale thermo-mechanical model will predict the stresses and deformation. This model uses both the meta-layer concept, and flash heating as methods for decreasing the computational cost. Secondly, a Johnson-Mehl-Avrami-Kolmogorov-based non-isothermal microstructural model is implemented as a state variable to estimate the change in phase fraction of the constituting phases. The results show that the part is made up mostly of a mix of α and α' phases, and that the microstructural change only leads to a small change in the residual stress after the LPBF process.

1. Introduction

Metal additive manufacturing (AM), and particularly laser-based powder bed fusion (LPBF) has become an industrially relevant process. However, due to the unique thermal history, certain aspects of the LPBF process are underexplored. In this paper, a part-scale thermo-mechanical model is two-way coupled with a microstructural model, in order to evaluate the effect of the change in microstructure on the change in the stress state.

In LPBF, a part is produced in a layer-by-layer manner. The LPBF machine deposits a thin layer of metal powder on top of a solid build plate. A laser rapidly melts a cross section of the desired part into the powder layer. The machine then lowers the build plate one step, a new layer is deposited and the process repeats. This cycle continues until the part is produced. Due to this layer-by-layer method of production, relatively high stresses and deformations are present in the final part, affecting the geometrical precision of the final part [1].

Thermo-mechanical modelling of LPBF is a research field that receives much attention in literature, which is illustrated with a non-exhaustive selection of recent works. Chen *et al.* [2] developed a model to predict the effect of the scanning pattern on the stresses in a thin plate-like specimen. Zhang *et al.* [3] performed part-scale simulations to investigate the effect of the mesh on the final part. Ganeriwala *et al.* [4] show, using validation with real samples, the strength of numerical models to predict the stresses in LPBF parts. Finally, Gouge *et al.* [5] performed part-scale models for specific beam-like specimens, showing close correlation between the computed and measured stress profiles. Although some of these models are capable of simulating the LPBF process on a macro scale, none of these to address the effect of the microstructure on the residual stress level.

In addition to the novel mechanical strain field, the rapid heating and cooling leads to non-equilibrium phases occurring in the part. In this work, the material of choice is Ti-6Al-4V. This metal is commonly used for metal AM, has good corrosion and wear resistance, and excellent mechanical properties, at least in part due to its two-phase nature. Both the α and β phase are equilibrium phases in Ti-6Al-4V, as can be seen in the schematic isopleth in figure 1(b). However, due to the rapid heating and cooling they might not be the resultant phases after the LPBF process. The works by Vrancken *et al.* [6] and Yang *et al.* [7] indicates experimentally that the microstructure after the LPBF process is made up of so-called α', a hexagonal closed packed martensitic non-equilibrium phase. This transformation is schematically represented in figure 1(c). As usual for non-equilibrium phases, heating up the part will result in a transformation from α' into the equilibrium fraction of α and β [8].

Modelling of microstructural transformations in Ti-6Al-4V typically starts with the so-called Johnson-Mehl-Avrami-Kolmogorov (JMAK) equation [9]. This equation describes homogeneous nucleation and growth, which accurately predicts the different phase fractions during an isothermal uniform heat treatment. Modifications to this model have been proposed by Murgau *et al.* [10], who modified the model for non-isothermal heat treatments. They divide the heat treatment into small steps, in which the temperature is assumed constant. Salsi *et al.* [11] adapted this model for the metal AM process, to include the effect of rapid cooling, which is a typical consequence of the laser heating. Both of these models include the decomposition of the α' phase into the equilibrium phases.

Each of these microstructural models calculates the overall microstructural composition of the part. However, to couple the microstructural and the mechanical models. Tan *et al.* [12] formulated such a localised model, which allows them to investigate the effect of the transformation from α to β, and from β to α', the former during heating up and the latter while cooling down to room temperature. What they fail to include is the lagging effect that follows from the kinetics of the phase transformation.

The model presented in this paper aims to improve upon the model presented by Tan *et al.* [12], by using a layer-by-layer part-scale model, which includes flash heating, and couple it with the model proposed by Murgau *et al.* [10]. The geometry is a bridge-type part, similar to the one used by Ganeriwala *et al.*[4]. The geometry is shown in figure 1(a).

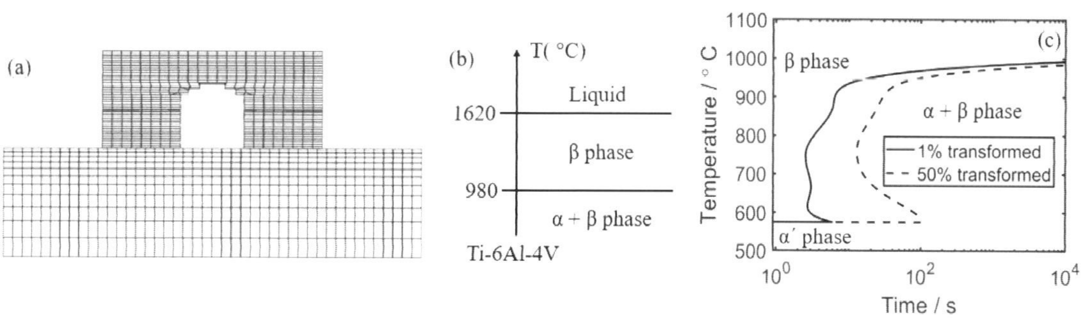

Figure 1. Geometry and material properties of Ti-6Al-4V. (a) The bridge geometry. (b) A schematic isopleth of Ti-6Al-4V, indicating the different equilibrium phases. (c) is a schematic TTT diagram, based on [11] cooling from the β phase.

2. Numerical model

The model that is used to mimic the LPBF process is set up in the commercial suite Abaqus FEA. As the name implies, the discretisation is handled using the finite element discretisation. This implies that the values for the different variables are calculated in the nodes and interpolated between the nodes.

2.1. Governing equations

The model consists of three main components: a thermal part, a mechanical part and a microstructural part. The governing equation for calculating the temperature, *T*, in the model is the traditional transient heat conduction equation:

$$\rho C_p \frac{\partial T}{\partial t} = \left(kT_{,i}\right)_{,i} - \rho \Delta H_{sl} \frac{\partial f_{liq}}{\partial t} + Q_V \tag{1}$$

In equation (1), ρ is the temperature dependent density, C_p is the specific heat capacity, k the thermal conductivity, ΔH_{sl} the latent heat of fusion and f_{liq} the fraction liquid. Q_V is the term representing the heat from the laser. In this work, the flash heating (FH) [13] strategy is used. In FH, a region in the domain, called a meta-layer is heated up at once. Each meta-layer represents a number of real layers. The actual value of the volumetric heat source can be found from the following equation [14]:

$$Q_V = \frac{\alpha P}{H v_{laser} \Delta t_{contact} \delta} \tag{2}$$

This represents the total energy applied to a single real layer. This total energy can be calculated from the effective laser power, represented by the laser power *P* multiplied by the absorptivity of the powder α. This effective laser power is divided by the time each layer is exposed to it, indicated by the hatch spacing (*H*), layer thickness (δ) and the laser velocity (v_{laser}). The volumetric heat source can be calculated by including the length of time each ($\Delta t_{contact}$) meta-layer is exposed to the laser energy.

To obtain the temperature field, each of the aforementioned meta-layers are activated separately, starting from the base plate, until the entire part is activated. The resulting temperature field will have a reduced thermal gradient inside of each meta-layer, but the model gains speed due to a reduction in computational cost.

The basis of the mechanical model is the equation describing static equilibrium:

$$\sigma_{ij,j} = 0 \tag{3}$$

where σ_{ij} is the stress tensor. This is related to the elastic strain via the generalised Hooke's law:

$$\sigma_{ij} = \frac{E}{1+v}\left[\frac{1}{2}\left(\delta_{ik}\delta_{jl} + \delta_{il}\delta_{jk}\right) + \frac{v}{1-2v}\delta_{ij}\delta_{kl}\right]\varepsilon_{kl}^{el} \tag{4}$$

E and ν are the Young's modulus and Poisson coefficient of the material respectively. δ_{ij} is the Kronecker delta, which is one for same, and zero for different indices, and ε_{kl}^{el} is the elastic strain tensor.

An important part of the mechanical model are the plastic and thermal components of the total strain. Finally, because the microstructure is also included in the presented simulations, there is an additional transformation strain component, which plays a role during the simulation. The increments of the elastic, plastic, thermal and transformation strain components are summed to form the total strain increment:

$$\dot{\varepsilon}_{ij}^{\mathrm{total}} = \dot{\varepsilon}_{ij}^{\mathrm{el}} + \dot{\varepsilon}_{ij}^{\mathrm{pl}} + \dot{\varepsilon}_{ij}^{\mathrm{th}} + \dot{\varepsilon}_{ij}^{\mathrm{tr}} \tag{5}$$

Using standard J2 flow theory, the plastic strain increment is given by:

$$\dot{\varepsilon}_{ij}^{pl} = \frac{9}{4\sigma_e^2}\left[\frac{E-E_t}{EE_t}\right]\dot{\sigma}_{kl} s_{ij} s_{kl} \tag{6}$$

The thermal strain increment couples the thermal calculation to the mechanical simulation:

$$\dot{\varepsilon}_{ij}^{\mathrm{th}} = \gamma \delta_{ij} \Delta \dot{T} \tag{7}$$

where γ is the thermal expansion coefficient. Parameters for the mechanical model can be found in [14].

2.2. Microstructural model

As mentioned previously, the temperature, which the part is subjected to during the LPBF process, will affect the composition of the microstructure. In Ti-6Al-4V, there are two major transformations, which will affect the mechanical properties. After the powder has melted, it starts to cool down. Immediately after the part has passed the melting temperature, the entire computational domain is made up of the β phase. This phase is only stable at high temperatures. When the part cools down further, this β phase decomposes into the equilibrium phase fractions of α and β. This work uses the Castro model [15] to compute these equilibrium fractions:

$$\begin{aligned} X_\beta^{eq} &= 0.925\exp\big(-0.0085(980-T)\big)+0.075 \quad &\text{for } T<T_\beta \\ X_\beta^{eq} &= 1 &\text{for } T>T_\beta \end{aligned} \tag{8}$$

This experimental relation was obtained by cooling slowly from the β phase field, and quenching the bar of the titanium alloy. Experimental observations by Gil Mur *et al.* [8] show that the martensitic phase will decompose into α and β when heated above 400 °C, a process attributed to nucleation and growth of both of these phases.

The JMAK equation for isothermal heat treatments is given by:

$$X_2 = 1 - \exp\left(k_{12}t^{\,n_{12}}\right) \tag{9}$$

where X_2 is the fraction of phase 2, and k_{12} and n_{12} the parameters governing the kinetics of the phase transformation from phase 1 to phase 2.

Although theoretically sufficient, in order to apply equation (9) in the FE framework, which is used for calculating the temperature and stresses during the LPBF process, it needs to be modified. Cahn [16] derived an additivity rule, to take into account partial phase transformations in a small time interval Δt. Leblond and Devaux [17] modified the traditional JMAK equation to include the effect of equilibrium phase fractions, and finally, Murgau *et al.* [10] combined these effects for modelling Ti-6Al-4V. In this work, it is assumed that the calculated phase fractions represent the volume around the FE integration point, and the nodal temperature is representative for the temperature in this finite domain.

After reaching the liquidus temperature of 1620 °C, the entire microstructure is initialised as β. If the cooling rate exceeds 20 °C/s, and the temperature is below the martensitic start temperature (575 °C) the martensite fraction is updated (from the nth numerical iteration to the $n+1^{\text{th}}$):

$$X_{\alpha'}^{n+1} = X_\beta^{n} \tag{10}$$

Otherwise, the ordinary α phase forms according to the modified JMAK equations:

$$X_\alpha^{n+1} = \left(1-\exp\left(-k_{\beta\alpha}\tau^{n_{\beta\alpha}}\right)\right)\left(X_{beta}^{n}+X_{alpha}^{n}\right)X_{\alpha,eq}^{n+1} \tag{11}$$

with τ being the modified time step:

$$\tau = \Delta t + \left[-\ln\left(1-\frac{X_\alpha^n/X_{\alpha,eq}^{n+1}}{X_\beta^n+X_\alpha^n}\right)\frac{1}{k_{\beta\alpha}}\right]^{\frac{1}{n_{\beta\alpha}}} \tag{12}$$

where Δt is the time step size.

Since the temperature will both increase and decrease during the LPBF process, it is possible that too much of the initial β phase transforms into α according to equations (11) and (12). Therefore, decompositions of the newly formed α phase is necessary. To ensure a smooth evolution of the phase fractions, the decrease of the α phase is [10]:

$$X_\alpha^{n+1} = 1 - X_{\beta,eq}^{n+1} f_{Diss}\sqrt{\Delta\tau} \tag{13}$$

$$\tau = \left(\frac{X_\beta^n}{2.2\times10^{-31}T^{9.89}X_{\beta,eq}^{n+1}}\right)^2 \tag{14}$$

Finally, the dissolution of the martensitic α' phase is discussed. It is known that, when a part is heat treated above 350 °C, but below the β transus temperature, the martensitic phase will decompose into the equilibrium phase fractions of α and β. This transformation is included in the microstructural model as follows:

$$X_{n+1,\alpha'} = \left(1 - \exp\left(-k_{\alpha'}t^{\,n_{\alpha'}}\right)\right)\left(X_{n,\beta} + X_{n,\alpha'}\right) \tag{15}$$

Each of these transformation laws (equations (10), (11) and (13)) are accompanied by complimentary statements ensuring that the sum of the phase fractions is at any point equal to one. The values of the JMAK kinetic parameters are shown in figure 2.

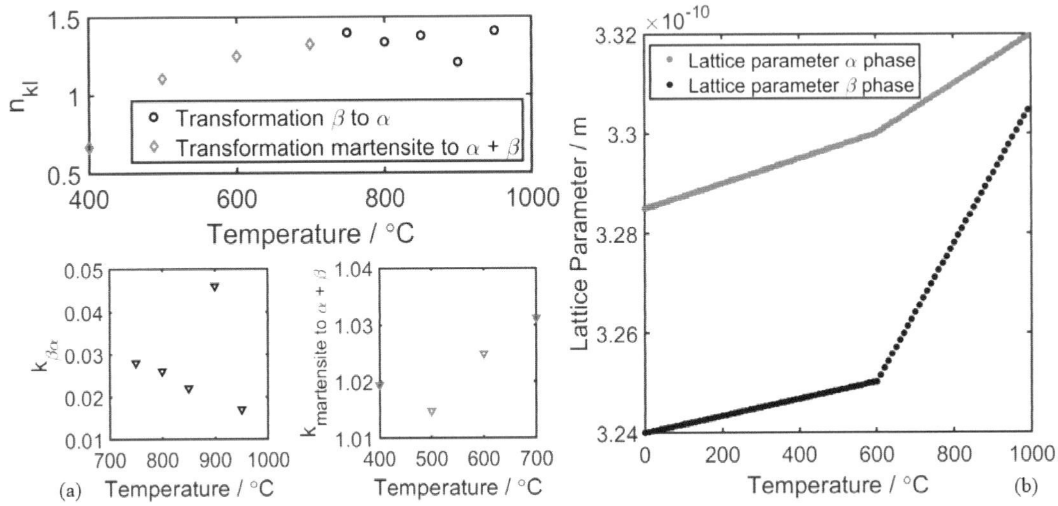

Figure 2. Parameter for the microstructural model. (a) shows the kinetic parameters for phase transformation [8], (b) the lattice parameter of the α and β phases [18].

2.3. Coupling of the microstructural and mechanical models

The microstructural model is clearly coupled to the temperature calculations via the various transformation equations and parameters, which are temperature dependent. Coupling the microstructural model back to the mechanical model is less obvious. This work chooses to implement and additional strain component, namely ε_{ij}^{tr} in equation (5). The transformation strain illustrates the effect of the change in the volume of the different constituting phases. For any phase transformation, the transformation strain can be denoted as [12]:

$$\dot{\varepsilon}_{ij}^{tr} = \delta_{ij} \sum_{kl} \varepsilon_{kl}^{\Delta V} \dot{f}_{kl} \tag{16}$$

where $\varepsilon_{kl}^{\Delta V}$ is the volumetric strain and $\dot{f}_{kl}$ the change in the phase fractions (from k to l). $\varepsilon_{kl}^{\Delta V}$ is calculated from the unit cell volumes of the different phases:

$$\varepsilon_{kl}^{\Delta V} = \frac{\sqrt[3]{V_l} - \sqrt[3]{V_k}}{\sqrt[3]{V_k}} \tag{17}$$

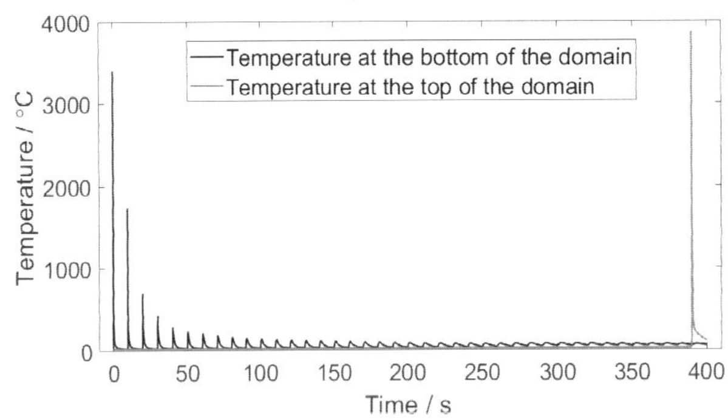

Figure 3. Temperature in the first layer of the simulation and in the final layer.

The different possible transformations are chosen to be α' to α+β, β to α and β to α'. The temperature dependent transformation strain for each of these phases is given in figure 2(b) [18]. Ab initio computational studies reveal that, for the transformation from martensite to the equilibrium fractions, and expansion of 0.8 percent is expected [19].

2.4. Modelling methodology

The model presented developed in this work uses the flash heating method for mimicking the LPBF process. In each simulation step, one meta-layer, consisting of multiple layers, is activated. The entire meta-layer is subjected to the volumetric heat source, denoted in equation (2). The layer is allowed to cool down, during which the microstructural model continuously updates the phase fractions, based on the temperature in each node and the size of the time steps. Together with the thermal strain, the transformation strain contributes to the stress calculation.

The part is nine millimetre high and divided in 40 meta-layers. The height of each meta-layer is 2.25e-4 millimetre, corresponding to approximately five real layers. To simulate the entire domain of the part within a reasonable amount of time, the mesh is chosen relatively coarse, while at least one layer of elements is present for each meta-layer.

3. Results and discussion

Because FH heats up an entire meta-layer at once, no mayor thermal gradients can be observed inside of each meta-layer. The strongest gradients will exist in the build direction of the part, leading to a stronger variation in stresses and microstructure in this direction. Ssince this model is focussed on part-scale modelling of the process, and this requires reductions in the computational cost, this is an acceptable sacrifice to make, along with the previously chosen coarser mesh.

Figure 3 shows the evolution of the temperature in the first meta-layer of the part, and in the final layer, both on the side of one of the pillars of the bridge. The temperature evolution shows the expected shape, with a single peak in temperature for each deposited layer, when the flash-heating heat flux inputs energy into the domain. The temperature probe at the bottom of the bridge also reveals a significant heating effect of the next layer on the already deposited one, which greatly exceeds the martensite start temperature. When looking at the temperature at the top of the domain, it can be observed that the temperature peak is significantly higher than the one near the build plate. This is most likely due to the heat-sink effect of the build plate being able to conduct more heat away from the part than the part itself.

Figure 4 shows the evolution of the microstructure in the bridge. Figure 4(a) shows the phase fractions on the right of the bridge in the last layer, at the top of the bridge. Figure 4(b) shows the same, but in the first layer, which is located directly on top of the build plate. Observing figure 4(b) reveals that there are three instances where the β fraction jumps one. This corresponds with the first three peaks in temperature at approximately the same points in time in figure 2. The difference in exact position stems from the actual location of the nodal points (where the temperature is calculated) and the integration points (where the phase fractions are defined in Abaqus FEA). However, the temperature only exceeds the β transus temperature for the first two. Moreover, the temperature for these first two peaks is above the melting temperature, indicating that this β fraction actually represents molten material. Since the nodal point is located at the bottom of the domain, the temperature will be lower than at the integration point, which is located slightly higher. Due to the large gradients in this problem, the β-transus temperature is most likely just passed at the phase fraction probe. The first two peaks in the β phase fraction only remain for a very small amount of time, immediately afterwards, the temperature decreases enough that only the equilibrium β fraction remains, while the rest of the domain is comprised of a mix of α and α'. The simulations suggest that, during the initial peaks, a significant amount of α forms, which converts to α' when cooling down further. The third α peak reveals a different behaviour. First of all, β converts significantly slower, due to the high equilibrium fraction at an elevated temperature. This β transforms into α, which is illustrated in the increase of the volume fraction of the latter after 28.7 seconds. Afterwards, the α phase fraction stabilises at 0.448. Since the temperature does not exceed the β-transus temperature at later stages, no large changes in the phase fractions can be found. The amount of α decreases when looking at later layers, but the simulation still shows a significant lack of decomposition of α' into $\alpha + \beta$. Rather, the amount of α at the end of the simulation, which is about

44.8 % near the build plate, but around 7 % when halfway up the bridge, is retained α in a martensite matric.

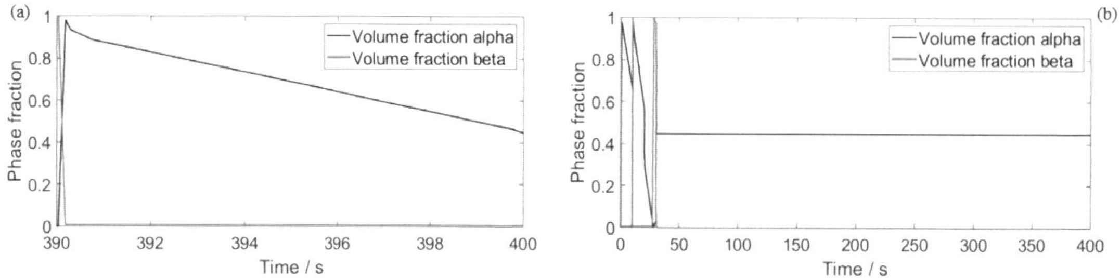

Figure 4. Phase fractions of α and β in the first and last layer. (a) shows the evolution of the β phase fraction in the last layer at the top of the bridge, (b) at the bottom of the bridge, in the first layer.

The phase fraction of β near the top of the domain only undergoes a major change once, due to the deposition of this top layer. On the other hand, understanding of the evolution of the α phase fraction is less straight forward. Figure 4(a) shows that the α phase fraction decreases as time progresses, indicating that the α phase fraction transforms into α'. A possible explanation for this is the order in which the phase fractions are calculated during the simulation: it is possible that the β-phase fraction forms in a time step, if the temperature is high enough to have an appreciable increase of the equilibrium β phase fraction. However, during the same time step, this β fraction will transform into α', due to the high thermal gradient, resulting in this apparent decomposition of α into α'.

Vrancken *et al.* [6] find from light optical microscopy images that the microstructure is entirely martensitic. However, they do not comment on the variation of the microstructure throughout the domain. Salsi *et al.* [11] calculate the α' volume fraction based on X-ray diffraction experiments and find that the volume fraction of a with a similar energy density as the one in this work, varies throughout the part. They find that the martensite volume fraction is around 60 % in the centre of the part. Figure 4(a) and (b) illustrate that the presented model results in final phase fractions which are in between these two values, but are closer to the ones from Salsi *et al.* [11].

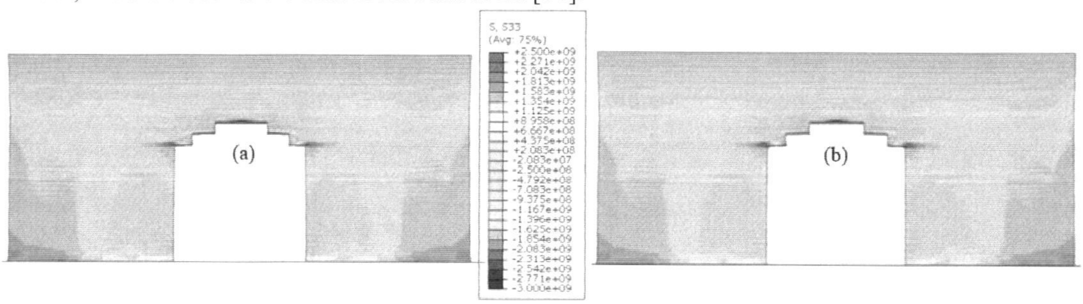

Figure 5. Stress in the bridge. (a) shows the S33 when the microstructural change is simulated, (b) the same when the microstructural change is not included.

Finally, the stress in the bridge is displayed in figure 5, for two cases, one where the microstructural change is accounted for, and one where this was not the case. The second simulation represents the currently most common simulations. Figure 5 shows that there is a minor effect of the phase transformation. As expected from the work by Tan *et al.* [12], the shighest observed stress will increase in the part, however, due to the similarity in the lattice parameter between the α and α', this effect seems rather small. The main differences in the stress contours can be found when looking at the extremities of the bridge-pillars, where the region with the highest stress is slightly larger when the microstructural change is included.The origin of this limited change in stress levels is most likely due to the previously mentioned limited difference between the α and α' lattice parameter: in this work, the change from martensite to the equlibrium phases only results in a 0.8 percent increase of the lattice parameter, based on the study by Mei *et al.* [19]. Additionally, the microstructural model does not take into account the local grain shape, and the possibility of non-homogeneous grain transformation, which can cause local deviations in the stress profile.

Finally, this paper weighs of the relevance of using the microstructural model. In the present model, with the FH method and JMAK-based microstrucutral model, a relatively fast, part-scale simulaitons are possible, however, the effect of the microstructural change is very small when looking at the final stress profile. Therefore, the additional computational cost is hard to justify.

4. Conclusion

In this work, an LPBF process simulation is expanded to include the microstructural evolution that can take place during the AM of a part in Ti-6Al-4V. The microstructural simulation is based on the well-established JMAK equation, and includes three phases, namely β, α and α'. The conclusions are the following:

- The temperature causes a cyclic transformation from β to α during deposition of a new meta-layer, and the first two layers deposited on top of it. After the heat source is more than three meta-layers away, the microstructure does not change significantly, and a α/α' mixture is predicted by the simulation
- The stress is about five percent higher when including the microstructural change, showing that including the effect of the microstructure on the mechanical stress will improve the simulation of the LPBF process.
- Due to the computational cost and the complexity of the models, including the effect of microstructure can currently only be justified when the exact value of the residual stress and post-process deformation is required.

Acknowledgements
This project has received funding from the European Union's Horizon 2020 research and innovation programme under the Marie Skłodowska-Curie grant agreement No 721383.

References

[1] Kruth J P, Froyen L, Van Vaerenbergh J, Mercelis P, Rombouts M and Lauwers B 2004 *J. Mater. Process. Tech.* **149** 616–22
[2] Chen C, Yin J, Zhu H, Xiao Z, Zhang L and Zeng X 2019 *Int. J. Mach. Tools Manuf.* **145** 103433
[3] Zhang W, Tong M and Harrison N M 2019 *Addit. Manuf.* **28** 610–20
[4] Ganeriwala R K, Strantza M, King W E, Clausen B, Phan T Q, Levine L E, Brown DW and Hodge N E 2019 *Addit. Manuf.* **27** 489–502
[5] Gouge M, Denlinger E, Irwin J, Li C and Michaleris P 2019 *Addit. Manuf.* **29** 100771
[6] Vrancken B, Thijs L, Kruth J P and Van Humbeeck J 2012 *J. Alloys Compd.* **541** 177–85
[7] Yang J, Yu H, Yang H, Li F, Wang Z and Zeng X 2018 *J. Alloys Compd.* **748** 281–90
[8] Gil Mur F X, Rodríguez D and Planell J 1996 *J. Alloys Compd.* **234** 287–9
[9] Avrami M 1939 *J. Chem. Phys.* **7** 1103–12
[10] Murgau C C, Pederson R and Lindgren L E 2012 *Model. Simul. Mater. Sci. Eng.* **20** 055006
[11] Salsi E, Chiumenti M and Cervera M 2018 *Metals* **8** 633
[12] Tan P, Shen F, Li B and Zhou K 2019 *Mater. Des.* **168** 107642
[13] Zaeh M F, Branner G 2010 *Prod. Eng.* **4** 35–45
[14] Bayat M, Klingaa C, De Baere D, Thorborg J, Skat Tiedje N and Hattel JH 2020 *Preprint*
[15] Castro R and Séraphin L 1966 *Mémoires Sci. la Rev. métallurgie.* **58** 1025–58
[16] Cahn J W 1956 *Acta Metall.* **4** 572–5
[17] Leblond J B and Devaux J 1984 *Acta Metall.* **32** 137–46
[18] Elmer J W, Palmer T A, Babu S S and Specht E D 2005 *Mater. Sci. Eng. A* **391** 104–13
[19] Mei W, Sun J and Wen Y 2017 *J. Mater. Res.* **32** 3183–90

IOP Conf. Series: Materials Science and Engineering **861** (2020) 012014 doi:10.1088/1757-899X/861/1/012014

Columnar-equiaxed transition in continuous casting based on a micro-macro solidification model with long-range solutal mixing

M Uddagiri[1], J Spee[2], S Hubig[1] and I Steinbach[1]

[1] Interdisciplinary Centre for Advanced Materials Simulation (ICAMS), Ruhr-University of Bochum, Universitaetsstrasse 150, 44780 Bochum, Germany
[2] thyssenkrupp Steel Europe AG, Kaiser-Wilhelm-Strasse 100, 47166 Duisburg, Germany

E-mail: murali.uddagiri@rub.de

Abstract. A novel model is proposed to describe the columnar to equiaxed transition (CET) in continuous casting. The model bases on Rappaz and Thevoz's solute diffusion model for equiaxed dendritic growth, combined with a 1-Dimensional solidification model normal to the slab surface. The unique feature of the proposed model is the combination with a mixing term between interdendritic and extradendritic melt, representing long-range solutal mixing by convection. The model can also be applied to predict equiaxed to columnar transition (ECT), i.e. the chill zone thickness. The model consists of modules such as nucleation, growth kinetics, solute and heat balance, and a solute mixing module. Nucleation is considered with a fixed nucleation undercooling. The growth kinetics of the dendrites are treated according to the LGK model. A finite difference scheme is employed for solving 1-Dimensional heat transfer equations and finally, volume averaged solute balance equations are solved in a staggered scheme. Mixing of inter- and extradendritic liquid is, as a first step, treated ideally fast. When applied to Fe-C binary system with the thermo-physical properties obtained from literature and CALPHAD simulations, the model successfully predicts recalescence, phase fraction evolution, and concentration profiles in different phases. Realistic boundary conditions of the continuous casting process are obtained from macroscopic FEM simulation.

1. Introduction

Continuous casting technology is the most economical way of steel making and currently it is responsible for more than 95 % of world's steel production [1]. A typical solidification structure of a steel grade contains both columnar and equiaxed dendrites. At microstructural scale, the decisive factor that defines the homogeneity of a steel grade is the amount of equiaxed dendritic region [2]. Hence, one of the primary objectives of continuous casting is to reduce the amount of columnar dendritic region and optimize the equiaxed crystal region. In this regard, numerous studies have been made to predict and control columnar to equiaxed transition which determines the amount of equiaxed dendritic region. However, due to numerous mechanisms that are active at various time and length scales, computational modelling of CET has still remained as a challenging task [3]. With the present model, we aim to propose a novel CET mechanism by considering long-range solutal mixing which arises due to convection.

1.1. CET Models

Over the past decades, computer modelling has played a significant role in understanding CET under various solidification conditions. Although CET and the factors influencing the transformation are qualitatively well understood, simulation models have not reached a stage to make quantitative predictions. This is mainly due to the complexities in the interaction mechanisms [3]. At present, numerous models are available in the literature which offer both direct and indirect predictions of CET and related microstructural evolution features. However, the transformation mechanisms employed in these models differ a lot from each other thereby making it difficult for the beginners who want to understand the phenomena of CET [4].

Witzke *et al.* [5] modelled solutal and temperature distributions ahead of a columnar front and suggested that CET requires two conditions. First, the liquid metal ahead of the columnar front should reach sufficient degree of solutal undercooling and second, the volume of the undercooled liquid zone should be high enough. According to them, these two are the necessary conditions for nucleation of new equiaxed crystals which hinder the growth of columnar front. Further, Lipton *et al.* [6] have also suggested similar mechanism where they proposed that CET is mainly governed by the undercooling of the columnar interface and width of the undercooled boundary layer at the columnar interface.

In 1986, based on the extended volume fraction of equiaxed grains, Hunt proposed a widely accepted criterion for CET in 1D solidification [7]. He argued that CET happens if the volume fraction of grains in the undercooled melt ahead of the columnar front is higher than 0.49. However, the value of 0.49 is selected purely on geometrical considerations and lacks a strong physical explanation.

Wang and Beckermann [8, 9] proposed a novel technique to predict CET under the conditions of diffusion driven dendritic growth solidification. The model equations are derived by following multiphase approach in which each control volume is assumed to be consisting of three phases, solid, interdendritic liquid and extradendritic liquid. However, they have also employed mechanical blocking criterion to predict CET.

Further, to eliminate the usage of mechanical blocking criterion, Martorano *et al.* [10] proposed a new mechanism, where they took solutal interactions between equiaxed grains into account. They have implemented their model for different nucleation densities and found that CET does not always occur at 0.49 grain volume fraction. For the case of higher nucleation density, CET occurred at much lower volume fraction of grains whereas for the case of lower nucleation density, CET is delayed.

Owing to these differences in proposed mechanisms and complexities associated with implementing them at the realistic macro scale, most of the CET models at industrial scale are still relying on the fundamental Hunt's model, even though there was no strong experimental validation supporting it.

1.2. Objectives of the current model

The objectives of the present model are multifold. First, to develop a 1-D numerical simulation model which can predict both CET and chill zone thickness. Second, to propose a new CET mechanism which bases on tracking the solute undercooling ahead of the columnar front. Third, to include the effects of long range solutal mixing in the solidification model. And finally, to enable direct coupling of temperature data obtained from industrial scale macro simulation models to the microscopic solute distribution equations so that the model can make effective predictions based on the actual process conditions.

2. Volume averaged solidification model in 1-D

The system under investigation is shown in figure 1. A slab of uniform thickness produced by continuous casting technique is considered. Due to symmetry of the slab geometry and applied boundary conditions, a quarter of the slab is generally sufficient for macroscopic modelling (see figure 1(a)). Further, due to uniform cooling conditions on the surface of the strand, a 1-D line from surface to the centre is taken as the system for microstructural investigation (see the red line in figure 1(a)). The outer surface is subjected to convective cooling and adiabatic heat transfer conditions are applied at the centre of the slab. The arrow (blue colour) represents the direction of pulling of the slab from mould to the secondary rollers.

The solute distribution is treated according to Rappaz and Thevoz's solute diffusion model [11, 12]. As shown in figure 1(b), each grain, irrespective of columnar or equiaxed dendritic microstructure, consists of 3 phases: solid dendrites (phase s), interdendritic liquid (phase d) and extradendritic liquid (phase l). The solid dendrites along with the interdendritic liquid trapped within the dendritic arms is termed as 'grain' (phase g). The two liquid phases, 'd' and 'l' are separated by a dendritic envelope due to differences in their interfacial length scales. Hence, dendrite envelope is defined as an interface that separates interdendritic and extradendritic liquid phases. Further, key assumptions of the model include, local equilibrium at the solid-liquid interface, finite diffusion of solute from interdendritic to extra dendritic phase and no diffusion of the solute in the solid phase. Also, peritectic reaction is not modelled and the initial solidification of delta phase and subsequent gamma phase are treated as uniform solid phase.

In each control volume, first the heat diffusion equation is solved to get the nodal temperatures. Based on the nodal temperatures, solute concentrations are obtained from linear phase diagram. Nucleation is assumed to happen instantaneously at a nucleation undercooling temperature and the growth kinetics of the envelope is treated according to LGK model [13]. Finally, based on the equations adopted from Wang and Beckermann's model [8, 9], the solute balance is performed in each phase. The novel feature of the present model is consideration of effects of macrosegregation, which is treated with a solute mixing module explained in detailed in the corresponding section.

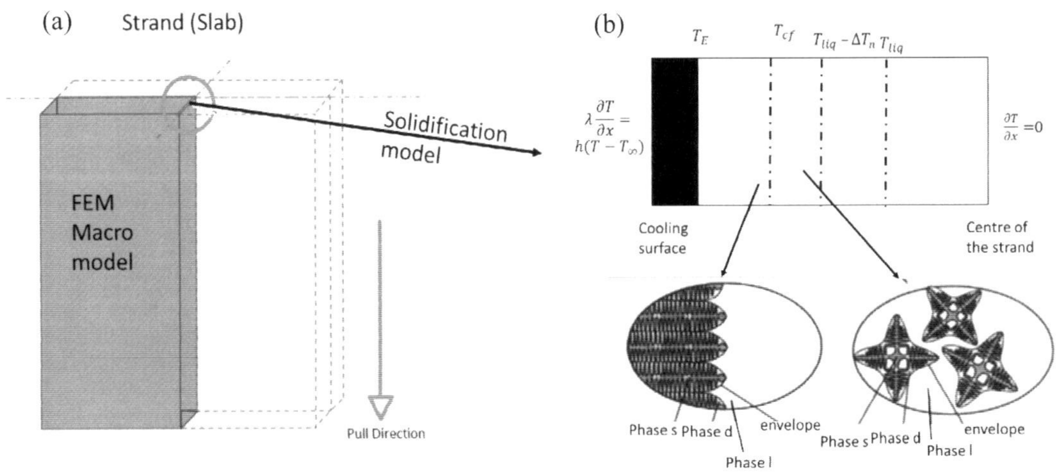

Figure 1: (a) Schematic representation of quarter of a strand (slab), red colour line is highlighting the 1D-system under investigation. (b) Solidification system with boundary conditions. Convective boundary conditions are applied on left boundary and adiabatic heat transfer is applied on the right-hand boundary. T_E, T_{cf}, T_L are temperatures corresponding to eutectic, columnar front and liquidus point. ΔT_n is nucleation undercooling. (figure 1(b) is based on figure 1 in [8, 9])

2.1. Model governing equations

Based on the volume conservation approach, Wang and Beckermann developed the governing equations for heat and mass balance in all the existing phases. Since the detailed explanation for all the equations are given in the original paper [8, 9], they are only summarized here.

The temperature distribution is calculated according to the following heat conduction equation with latent heat and appropriate boundary conditions.

$$\rho C_P \frac{dT}{dt} - L \frac{df_s}{dt} = \lambda \nabla^2(T). \quad (1)$$

Solute balance in the solid phase (phase 's') neglecting the back diffusion, is calculated from,

$$\frac{d}{dt}\left(f_s C_{s_{avg}}\right) = k C_l^* \frac{df_s}{dt}. \tag{2}$$

Solute Balance in interdendritic liquid (phase 'd') with infinite mixing is calculated from,

$$C_l^*(k-1)\frac{df_s}{dt} + f_d \frac{dC_l^*}{dt} = -\frac{S_e D_{liq}}{\delta}(C_l^* - C_{liq}^t);\ C_l^* = \frac{T - T_{iron}}{m_{liq}}. \tag{3}$$

Solute balance in extradendritic liquid (phase 'l') with finite diffusion, is calculated from,

$$\frac{d}{dt}(f_l C_l) = C_l^* \frac{df_l}{dt} + \frac{S_e}{\delta} D_{liq}(C_l^* - C_{liq}^t). \tag{4}$$

Further, the interrelationship between the three phases in each control volume is given as follows.

$$f_s + f_d + f_l = 1;\ f_g = f_s + f_d;\ f_l = 1 - f_g;\ \frac{df_g}{dt} = -\frac{df_l}{dt} = S_e V_{tip}, \tag{5}$$

where ρ is density, C_p is heat capacity, L is latent heat, λ is heat conductivity (taken as a constant), f_s, f_d, f_g, f_l are volume fractions of solid, interdendritic liquid, grain and extradendritic liquid respectively, $C_{s_{avg}}$, C_l^*, C_l are the average solute concentrations in solid, interdendritic liquid and extradendritic liquid of each control volume. C_{liq}^t is the nominal composition of solute in the remaining liquid metal which is obtained from the solute mixing module. Further, k is the equilibrium partition coefficient, m_{liq} is the slope of the liquidus line, D_{liq} is the diffusion coefficient in the liquid phase, δ is the diffusion length , S_e is the envelope surface area concentration and V_{tip} is the dendrite tip velocity.

With an assumption of spherical dendrite, S_e and R_g are calculated from,

$$S_e = \frac{3R_g^2}{R_{avg}^3};\ R_g = (f_g{}^{1/3})R_{avg}, \tag{6}$$

where, R_g is the radius of the grain or envelope and R_{avg} is the average spacing between dendrite envelopes.

As stated earlier, LGK kinetics is employed to calculate V_{tip}. Hence, V_{tip} is given by,

$$V_{tip} = \frac{D_{liq} m_{liq}(k-1)C_l^*}{\pi^2 \tau}\left[0.4567\left(\frac{\Omega}{1-\Omega}\right)^{1.195}\right]^2; \qquad \Omega = \frac{C_l^* - C_{liq}^t}{C_l^*(1-k)}, \tag{7}$$

where τ is the Gibbs-Thomson coefficient and Ω is a dimensionless quantity known as the super-saturation.

Note that the equation used to calculate supersaturation Ω is different from the previous CET models. In their original model, Rappaz and Thevoz [11, 12] used initial alloy concentration C_0 to calculate supersaturation ahead of the dendritic front. Wang and Beckermann [8, 9] also used C_0, at least to calculate supersaturation in columnar dendritic region. Hence, they had to use mechanical blocking criterion to predict CET. Later, Martorano *et al.* [10] modified the equation by replacing C_0 with C_l for both columnar and equiaxed growth. However, the effects of long-range solute mixing have not been considered in any of the existing analytical models. Hence, in the present model, a solute mixing module is developed to incorporate such effects and resulting macrosegregation.

2.2. Solute enrichment due to melt convection

Due to the complexities associated with fluid interactions, most of the existing CET models have neglected the effects of the turbulence and macrosegregation. Thus, the predictions for CET and other microstructural quantities obtained from such models are independent of melt convection effects. However, fluid flow in continuous casting process, especially in the mold part, has a very strong influence on the solidification microstructure of the slab. To bridge this gap, the present model aims to demonstrate a way to include the effects of a macrosegregation through a simple and easy to implement approach.

The key idea behind the proposed solute enrichment module is that, the nominal solute concentration i.e. the concentration of the solute available at the start of the solidification increases as the solidification proceeds from surface to the centre of the slab. Due to solute transport at the macro scale resulting from the turbulence inside the melt, the liquid metal away from the outer surface solidifies with higher nominal solute concentrations. In the present model, it is realized by allowing infinite mixing of the amount of solute available in extradendritic liquid phase of each control volume. This results in uniform solute concentration in the far away liquid, C_{liq}^t.

Having known the fraction of liquid and solute concentration in extradendritic liquid in each control volume, solute enrichment due to mixing is calculated from, $C_{liq}^t = \frac{\sum_{i=0}^{N} f_l C_l}{\sum_{i=0}^{N} f_l}$, which is of course an extreme, limiting case. Here, N is the total number of control volumes or nodes in the system, f_l is the fraction of liquid in each control volume and C_l is the solute concentration in extradendritic liquid in each control volume. Further, C_{liq}^t is calculated as long as the solidification front reaches the centre of the slab model.

2.3. *The CET mechanism*

A typical microstructure of a steel slab contains three dendritic regions, chill zone near to the cooling surface (outer equiaxed) followed by columnar dendritic, and globulatic region close to the centre (inner equiaxed). Also, we know that nucleation and growth of equiaxed grains require the melt to be undercooled. From this information, we can conclude that, during the solidification, the local undercooling in outer and inner equiaxed regions must be higher than nucleation undercooling and that of the columnar region should be below the nucleation undercooling. Figure 3 schematically explains the same. The blue curve which represents the amount of solute undercooling ahead of the solidification front is seen to decreasing from outer surface and again increasing towards the centre of the slab.

According to Kurz *et al.* [2], the amount of undercooling ahead of the solidification is resulting from

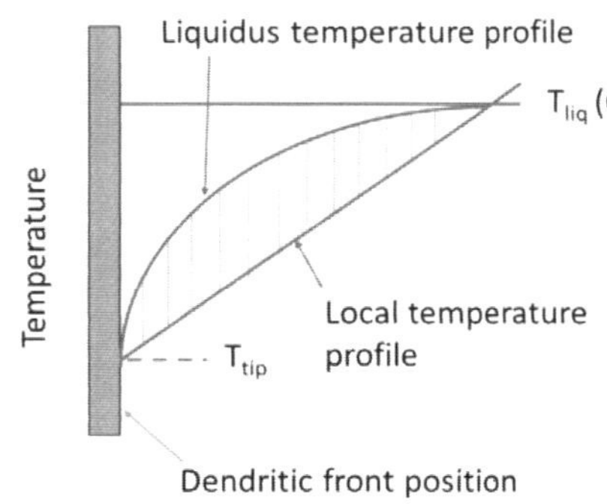

Figure 2: The equilibrium liquidus temperature and local temperature profile ahead of a dendritic front, after [2]

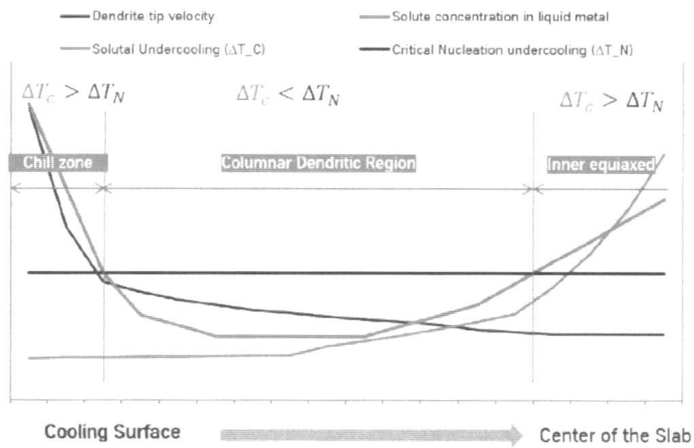

Figure 3: Schematic representation of proposed CET mechanism. ΔT_c and ΔT_N are solutal and nucleation undercoolings respectively.

solute redistribution due to diffusion and it has quadratic relationship with the front velocity. Further, at low velocity regimes, the undercooling is driven by temperature gradients in the liquid metal (refer to figure 2). However, at the realistic scale, the temperature gradients in the liquid metal are almost negligible hence it alone cannot explain the rise of undercooling towards the centre of the slab. Hence, in the present model, we propose that the rise in local undercooling towards the centre of the slab is not only due to temperature gradient in the liquid metal but also due to solute enrichment that results from both diffusion and macroscopic solute mixing.

By assuming ideal solute mixing, we calculate average solute concentration in the whole liquid phase i.e. C_{liq}^t and this concentration is used to calculate supersaturation and local undercooling. As shown in figure 3, by comparing the local undercooling with nucleation undercooling the model predicts both chill zone thickness and CET.

2.4. *Model implementation*

A numerical solution is obtained by discretizing the heat and mass balance equations and solving them with a finite difference method. The heat balance equation is solved with implicit finite difference scheme. Cellular automaton based front tracking algorithm is employed to track the solidification front. Once a control volume is activated for solidification, C_l^* is obtained from the liquidus line of a binary equilibrium phase diagram. Then the quantities such as V_{tip},f_g, f_l, and C_l are obtained by iteratively solving the corresponding equations. Finally, equations 1 and 3 are solved simultaneously to obtain both T and f_s. Knowing volume fractions and solute concentrations of each phase in every control volume, all the remaining quantities are calculated.

3. Results

The volume averaged solidification model coupled with long-range solute mixing is implemented to predict CET, chill zone thickness (ECT) and other microstructural evolution features of Fe-0.15 *wt %* C alloy. Thermo-physical properties of the binary alloy are as follows: liquidus temperature, T_{liq}=1791 *K*, eutectic Temperature, T_{eut}=1400 *K*, density of the liquid metal, ρ=7440 kg/m^3, specific heat, C_p= 720 *J/kg/K*, heat conductivity, λ=30 *W/m-K*, melting point of pure iron T_{pure}=1809 *K*, slope of the liquidus line m_{liq}=-90 *K/wt %*, equilibrium partition coefficient *K*=0.36, diffusion constant of the liquid D_{liq}=2.0x10^{-8} m^2/s, gibbs Thomson coefficient τ= 1.35x10^{-7} *Km*, and latent heat *L*= 2.71x10^5 J/m^3.

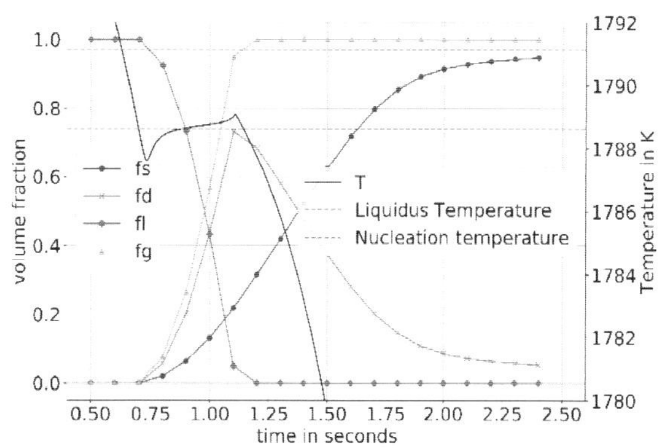

Figure 4. Evolution of phase fractions in a control volume, f_s, f_d, f_g and f_l represents volume fraction of solid, interdendritic liquid, grain and extradendritic liquid respectively.

A slab of half thickness 100 mm is taken as the 1-Dimensional linear system and is divided into number of control volumes. The initial temperature of the liquid metal was uniform with a superheat of 10 K. Adiabatic boundary conditions are applied at the centre of the slab (right boundary) and constant cooling boundary conditions are applied at the cooling surface (left boundary) by $q = h(T - T_\infty)$ where the heat transfer coefficient was h = 2000 W/m^2-*K* and ambient temperature $T_\infty = 300\,K$. Fitting parameters such as nucleation density and nucleation undercooling are taken as follows: nucleation density n= $2.39\,x\,10^8\,m^{-3}$ and nucleation undercooling ΔT_n=*2.5 K*. Fragmentation of dendrite tips due to solutal mixing is considered indirectly by choosing high volume of nucleation grain density.

First, the results obtained for the evolution of all the involved phases in a control volume are presented in figure 4. The associated cooling curve (T), liquidus temperature and nucleation temperatures are also shown in the same figure. As it can be seen in the figure, at the start of solidification, the control volume is completely in liquid state i.e. $f_l = 1$ and $f_s = f_d = f_g = 0$. As the temperature decreases due to cooling, dendritic growth proceeds and the solid phase starts to form out of the liquid. Hence f_g and f_s increase and f_l decreases. The interdendritic liquid phase, f_d, raises initially with f_g until the latter reaches its full volume fraction i.e. until f_g=1. Once the grain growth stops, the solid phase continues to grow from the remaining interdendritic liquid. Further, the model can also calculate the solidification time and metallurgical length of the slab which are of industrial interest. For the chosen cooling rate and process parameters, the model predicts that a slab of thickness 200mm takes about 700 seconds for complete solidification.

Finally, figure 5 presents the results obtained for dendrite tip velocity, solute undercooling and nominal solute concentration observed in each control volume. From the figure, it can be seen that, when the solute mixing is active, the undercooling is decreasing from the cooling surface and again increasing towards the centre of the slab, whereas in the absence of solute mixing, the undercooling does not rise towards the centre. This change in slope undercooling in the first case, is attributed to the solute enrichment, resulting from macroscopic solute mixing. As solute undercooling is a function of both dendrite tip velocity and nominal solute concentration, it continued to decrease until the drop in the tip velocity is significant. Once the drop in the tip velocity has become relatively insignificant, solute undercooling continued to rise due to increase in solute concentration. By comparing the solute undercooling observed in each control volume with the pre-defined nucleation undercooling, one can easily identify the three dendritic regions, i.e. chill zone, columnar dendritic and inner equiaxed regions. In both chill zone and inner equiaxed regions, the solute undercooling is higher than nucleation undercooling, whereas in columnar dendritic region, the nucleation undercooling is higher than solute undercooling. For the chosen process parameters and nucleation undercooling of 2.5 K, the model predicts a chill zone thickness of 6 mm and CET at around 75 mm from the surface. The model is further validated for different nucleation density and nucleation undercooling values and the results are found to be satisfactory.

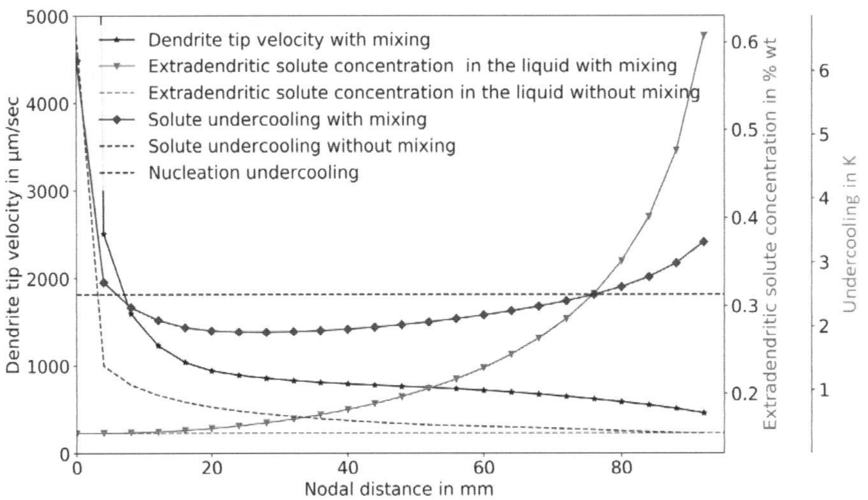

Figure 5. Maximum dendrite tip velocity and solute undercooling observed in each control volume. Nominal solute concentration of the control volumes and nucleation undercooling is also shown.

3.1. *Results obtained for temperature data generated from the FE simulation model*

The model setup enables direct coupling of temperature data obtained from macroscopic FE simulation models. To validate the functioning and applicability of the current model for the same purpose, it is implemented with temperature data obtained from a 3-Dimensional FE simulation model developed by thyssenkrupp steel Europe AG. The macroscopic FE simulation model calculates the temperature distribution in the strand shell with realistic process parameters and boundary conditions.

The temperature data is generated for a low carbon steel slab of 140 mm half thickness and of 0.15 wt% C carbon composition. The thermo-physical parameters are adopted accordingly. To calculate the chill zone thickness and CET, nodal temperatures along the half thickness of the slab, as represented by a straight line in figure 1(a) are considered. Owing to the massive size of the actual slab casting, the nodal spacing and time steps used in the FE simulation are bigger than those suitable for microstructural simulation. Hence linear interpolation is used to obtain the temperatures as per the grid spacing and time step requirement of the present numerical model.

For a nucleation density of 2.39 x 10^8 m^{-3} and nucleation undercooling of 3 K, the model is able to qualitatively predict solute concentration profiles in both solid and liquid phases and predicts a chill zone thickness of 3 mm and CET at 75-80 mm from the cooling surface. These results are compared with an etched section of the corresponding steel slabs and the location of CET is found to be in reasonable agreement.

4. Discussion

The model predictions for CET are comparatively in good agreement with the literature and experimental observations available for similar grades [14]. However, the results obtained for chill zone thickness are not satisfactory. Since the chill zone thickness is strongly influenced by the initial cooling rate and available nucleation density, the observed discrepancy is likely to be a mere artefact of the chosen boundary conditions. Further, the mixing algorithm used in the present model is based on ideal mixing which is the extreme case of mixing. In reality, the fluid flow during the solidification of continuous casting of steels is driven by numerous factors ranging from electromagnetic forces, slab geometry, gravity, and turbulences. To be able to make quantitative predictions, the model needs improvements in two key areas. First, the mixing algorithm should be improved by taking reference from fluid simulations and second, the model should be extended to multi component alloy system.

5. Conclusion

Taking macroscopic solute mixing into consideration, a numerical model is proposed to predict both chill zone thickness and CET. The numerical model is developed based on the solute diffusion equations derived from volume averaged technique and the effects of macrosegregation are included through ideal solute mixing. When applied to Fe-C binary system, the model predictions obtained for temperature distribution, dendrite tip velocity, solute concentrations, and solute undercooling were in accordance with the fundamental understanding of solidification in steel. Additionally, the model is also tested for temperature data obtained from an industrial macroscopic FE simulation model. The results obtained have proved that the model can effectively predict CET for a given binary composition and process parameters. However, in order to make quantitative predictions and to compare the model results to actual casting microstructure, further improvement of solute mixing module is required.

Acknowledgement

This work was financially supported by thyssenkrupp Steel Europe AG.

References

[1] Stahl, https://en.stahl-online.de/index.php/topics/technology/steelmaking/3/, accessed: 2019 Dec
[2] Kurz W, Bezencon C and Gaumann M 2001 *Sci. Technol. Adv. Mat.* **2**(1) 185
[3] Mirihanage W U, Dai H, Dong H and Browne D J 2013 *Adv. Eng. Mater.* **15** 216-29
[4] Spittle J A 2006 *Int. Mater. Rev.* **51** 247-69
[5] Witzke S, Riquet J P and Durand F 1981 *Acta Metall.* **29** 365
[6] Lipton J, Kurz W and Heinemann W 1983 *Concast Techn. News* **22** 4
[7] Hunt J D 1984 *Mater. Sci. Eng.* **65**(1) 75–83
[8] Wang C Y and Beckermann C 1993 *Metall. Mater. Trans.* A **24**(12) 2787–802
[9] Wang C Y and Beckermann C 1994 *Metall. Mater. Trans.* A **25**(5) 1081–93
[10] Martorano M A, Beckermann C and Gandin C 2003 *Metall. Mater. Trans. A* **34**A 1657
[11] Rappaz M and Thevoz P 1987 *Acta Metall.* **35**(7) 1487–97
[12] Rappaz M and Thevoz P 1987 *Acta Metall.* **35**(12) 2929–33
[13] Lipton J, Glicksman M E and Kurz W 1984 *Mater. Sci. Eng.* **65**(1) 57–63
[14] Straffelini G, Lutterotti L, Tonolli M and Lestani M 2011 *ISIJ Int.* **51**(9) 1448-53

IOP Conf. Series: Materials Science and Engineering **861** (2020) 012015 doi:10.1088/1757-899X/861/1/012015

Modelling viscoplastic behavior of solidifying shell under applied electromagnetic breaking during continuous casting

A Vakhrushev[1], A Kharicha[1], M Wu[2], A Ludwig[2], G Nitzl[3], Y Tang[3], G Hackl[3], J Watzinger[4] and C M G Rodrigues[2]

[1] Christian-Doppler Lab for Metallurgical Applications of Magnetohydrodynamics, University of Leoben, Franz-Josef-Str. 18, A-8700 Leoben, Austria
[2] Chair of Simulation and Modeling of Metallurgical Processes, University of Leoben, Franz-Josef-Str. 18, A-8700 Leoben, Austria
[3] RHI Magnesita, Austria
[4] Primetals Technologies, Austria

E-mail: alexander.vakhrushev@unileoben.ac.at

Abstract. Since several decades the continuous casting (CC) process became one of the dominant technologies for the metal production. The quality optimization and the production rate growth are the primary targets. Nowadays the numerical modelling is a valuable tool to assist in these aims. Moreover, it efficiently competes with the physical experiment and the industrial trials. One of the key issues with the high casting speeds especially for the thin slab products are the strong turbulent flow of the fresh melt being feed from the submerged entry nozzle (SEN) and the non-uniformity of the solidifying shell thickness. Thereby the electromagnetic braking (EMBr) is typically applied to damp the hot jets and to evenly redistribute the superheat. In the previous work it was shown by the authors that the presence of the highly conductive solid shell plays crucial role for the melt flow under the applied magnetic field. Excluding this interaction does not allow predicting the EMBr effects on the melt flow correctly during the solidification. Recently a viscoplastic deformation model was implemented to model a withdrawal of the solidified shell in a funnel type CC mold for a full 3D engineering geometry. An extended model was used to predict the macro-segregation during twin-roll casting of an Al-alloy using 2D assumption due to the large width / thickness ratio of the casted sheet. In the current study an effort is done to combine the viscoplastic deformation model of the solidified shell and the magnetohydrodynamics effects of the EMBr. The flow field alternation as well as the thickness of the solidified shell during CC are presented and analyzed with and without the magnetic field been applied.

1. Introduction

The thin slab casting (TSC) of steel is an effective continuous casting (CC) technology, featuring high casting speeds, reduced macrosegregation, applicability in the direct rolling concept and significant energy savings. However, the development of a highly turbulent flow in a narrow funnel-shape mold causes instability of the meniscus, the local remelting and thinning of the shell and enhanced risk of the slag and non-metallic inclusions entrapment. The electromagnetic brake (EMBr) is one of the widely applied flow control techniques [1]. Along with that, the numerical simulation became a valuable tool to investigate and to optimize this technology for the foundry field, whereas the immediate observation of the EMBr effects is mostly impossible at the real casting mill.

The mechanical properties of the alloys drastically vary in the two phase region [2]. Recently a viscoplastic deformation model was implemented by the current work authors with an aim to reflect this phenomenon during a withdrawal of the solidified shell in a funnel type CC mold [3]. Next, an extended two-phase approach was used to predict the macro-segregation during twin-roll casting of an Al-alloy sheets [4].

In current work the solidification model considering viscoplastic Norton-Hoff stress in the solidified shell is for the first time combined with the magnetohydrodynamics (MHD) Lorentz force. It is applied to simulate a thin slab casting process including a submerged entry nozzle (SEN) region, the mold and the strand part. The coupling is done using mixture volume averaging combining previously developed models [3, 5] with the MHD model based on the electric potential method [6]. The alternation of the liquid melt flow and the solidified shell thickness under the applied EMBr is simulated and discussed in the presented study.

2. Numerical model

2.1. General equations

Current work uses a constant density assumption for the solidifying melt. Thereby the incompressible mixture volume averaging gives the following continuity and momentum equations:

$$\nabla \bullet \mathbf{u} = 0, \tag{1}$$

$$\rho\left[\frac{\partial \mathbf{u}}{\partial t} + \nabla \bullet (\mathbf{u} \otimes \mathbf{u})\right] = \nabla \bullet [\mathrm{dev}(\boldsymbol{\Sigma}_{\mathrm{tot}}) - p\mathbf{I}] + \mathbf{F}_{\mathrm{L}}. \tag{2}$$

The deviatoric part of the total stress $\boldsymbol{\Sigma}_{\mathrm{tot}}$ combines the viscoplastic $\boldsymbol{\Sigma}^{\mathrm{vp}}$ and the Newtonian stresses based on the solid f_s and liquid f_ℓ phase fractions:

$$\mathrm{dev}(\boldsymbol{\Sigma}_{\mathrm{tot}}) = f_{\mathrm{s}} \cdot \boldsymbol{\Sigma}^{\mathrm{vp}} + f_\ell \cdot 2\mu_\ell \cdot \mathrm{dev}(\dot{\boldsymbol{\varepsilon}}). \tag{3}$$

No turbulence modelling is used in the current study, thus the liquid viscosity μ_ℓ is assumed to be constant. The Lorentz force $\mathbf{F}_{\mathrm{L}}$ in equation (2) represents magnetohydrodynamics effects from the applied magnetic field.

2.2. Viscoplastic behavior of the solidified shell

The detailed implementation of the mixture-based deformation model was previously described by the authors in reference [3] based on the hot steel viscoplastic behavior assumption from the book of Rappaz and coauthors [7]. Excluding the compressibility of the mush [8, 9], the solidifying shell is considered as an incompressible 'creeping solid'. The Norton-Hoff viscoplastic stress model combines the strain rate tensor and the equivalent strain rate as

$$\boldsymbol{\Sigma}^{\mathrm{vp}} = 2K\left(\sqrt{3}\,\dot{\varepsilon}_{\mathrm{eq}}\right)^{m-1}\mathrm{dev}(\dot{\boldsymbol{\varepsilon}}) \tag{4}$$

with the viscoplastic consistency K and strain rate sensitivity m being the model parameters [3, 7]. Full coupling between the deformation model (4) with the momentum equation (2) is done in the iterative manner by employing the improved both-side diffusion method to promote the stress-velocity coupling to overcome the stiffness of the stress tensor term [10].

2.3. Solidification model

To consider the solidification and the formation of the shell during the continuous casting the energy equation for the temperature T is taken as

$$\rho C_{\mathrm{p}}\left[\frac{\partial T}{\partial t} + \nabla \bullet (\mathbf{u}\,T)\right] = \nabla \bullet \lambda \nabla T + S_{\mathrm{e}}, \tag{5}$$

where the latent heat advection term is

$$S_e = \rho L \left[\frac{\partial f_s}{\partial t} + \nabla \bullet (f_s \cdot \mathbf{u})\right]. \tag{6}$$

with the specific heat C_p and the latent heat of fusion L.

2.4. Modelling magnetohydrodynamics effects

The Lorentz force $\mathbf{F}_L$ in momentum equation (2) is a cross product of the induced current density $\mathbf{j}$ and the externally applied magnetic field $\mathbf{B}_0$:

$$\mathbf{F}_L = \mathbf{j} \times \mathbf{B}_0. \tag{7}$$

The electric potential method is applied for the Maxwell's equations closure [11]. By introducing the electric potential φ, the e-current $\mathbf{j}$ can be calculated from the Ohm's law:

$$\mathbf{j} = \sigma(-\nabla\varphi + \mathbf{u} \times \mathbf{B}_0), \tag{8}$$

where σ is the electrical conductivity. The e-potential φ is calculated from the corresponding Poisson equation:

$$\nabla \bullet (\sigma\, \nabla\varphi) = \nabla \bullet \big(\sigma\, (\mathbf{u} \times \mathbf{B}_0)\big), \tag{9}$$

Electric conductivity σ is considered to vary ~1.4 times between liquid melt and solidified shell. The combined MHD / solidification solver is developed in the open-source CFD package OpenFOAM® [12]. According to the numerical model described in equations (1)–(9) a corresponding solver including fluid flow, viscoplastic stresses in the shell and action of the Lorentz force was developed. The MHD model is previously verified with the experimental measurements [13]. Initial studies for the influence of the highly conductive solid were presented in Liu *et al.* [14].

3. Simulation results

The combined model considering solidification, viscoplastic behavior of the shell and the effect of the EMBr was applied for the simulation of the thin slab casting process. The schematics of the simulated domain including SEN, primary (mold) and secondary (slab) cooling zones is shown in figure 1(a). The heat flux distribution along the casting direction is plotted in figure 1(b). The growth of the solidified shell is represented by solid fraction / temperature curve in figure 1(c). The magnetic field distribution of the 3-pole EMBr applied across the mold at 560 mm below meniscus with the peak value of 188 mT is in figure 1(d). The EMBr is aligned to act in the area of the side and central jets coming out of the 4 port SEN. All boundaries excluding domain's outlet and inlet are electrically insulated.

The 0.06C 0.1Ni 0.13Mn 0.1Si 0.08Cu 0.035Al 0.015P 0.012S alloy properties, the applied electromagnetic brake settings and casting conditions are listed in table 1. Viscoplastic model parameters correspond to the Model IA in Kozlowski *et al.* [15] and are previously analyzed and presented by the authors in [3].

The numerical mesh is originally refined near the mold walls and the slab water cooled surface to correctly resolve the temperature gradient. After the solidified shell develops an additional refinement is done at the region of the mushy zone based on the time averaged results. All in all, the mesh size reaches 10 million cells. The time integration step is adjusted dynamically to keep Courant number below 1.

The results of the thin slab CC simulation are shown in figures 2-4. The comparison between the no EMBr case and with the applied magnetic field is done. Firstly, the flow field alternation is presented in figure 2. For the no EMBr case a developed double-roll pattern is observed with the strong side and central jets (see figure 2(b)). When the EMBr is applied, the strong braking of the flow occurs. The jets become more stable, no impinging on the narrow walls is observed as can be seen in figure 2(c). At the lower part of the mold and in the slab region the melt flow becomes more uniform. The sub-meniscus velocities are reduced as well.

Table 1. Alloy properties, EMBr settings and casting conditions.

Properties	Symbols	Units	Quantities
Alloy density	ρ	kg m^{-3}	6998.49
Specific heat	C_P	J kg^{-1} K^{-1}	838.2
Thermal conductivity	λ	W m^{-1} K^{-1}	35
Liquid dynamic viscosity	μ_ℓ	Pa s	0.0054
Latent heat of fusion	L	J kg^{-1}	243000
Viscoplastic consistency	K	Pa s^m	65.16
Strain rate sensitivity	m	–	0.138
Electrical conductivity:	σ	S m^{-1}	
liquid melt			770000
solidified shell			1075000
Casting velocity	$\mathbf{u}_{\text{pull}}$	m min^{-1}	4.3
Casting (inlet) temperature	T_{cast}	K (°C)	1825 (1552)
Liquidus temperature	T_{liqudus}	K (°C)	1798 (1525)
Solidus temperature	T_{solidus}	K (°C)	1755 (1482)
EMBr magnetic field	$\mathbf{B}_0$	mT	188

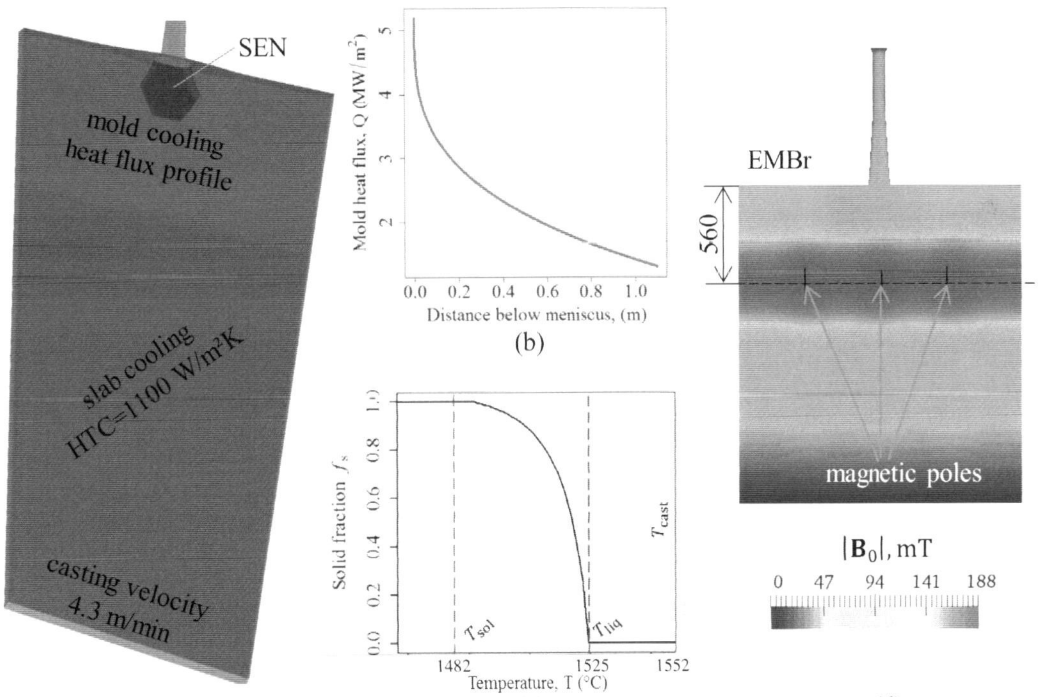

Figure 1. Simulation setup: (a) numerical domain layout with the cooling and casting conditions; (b) mold heat flux profile; (c) solid fraction vs. temperature curve; (d) applied EMBr magnetic field.

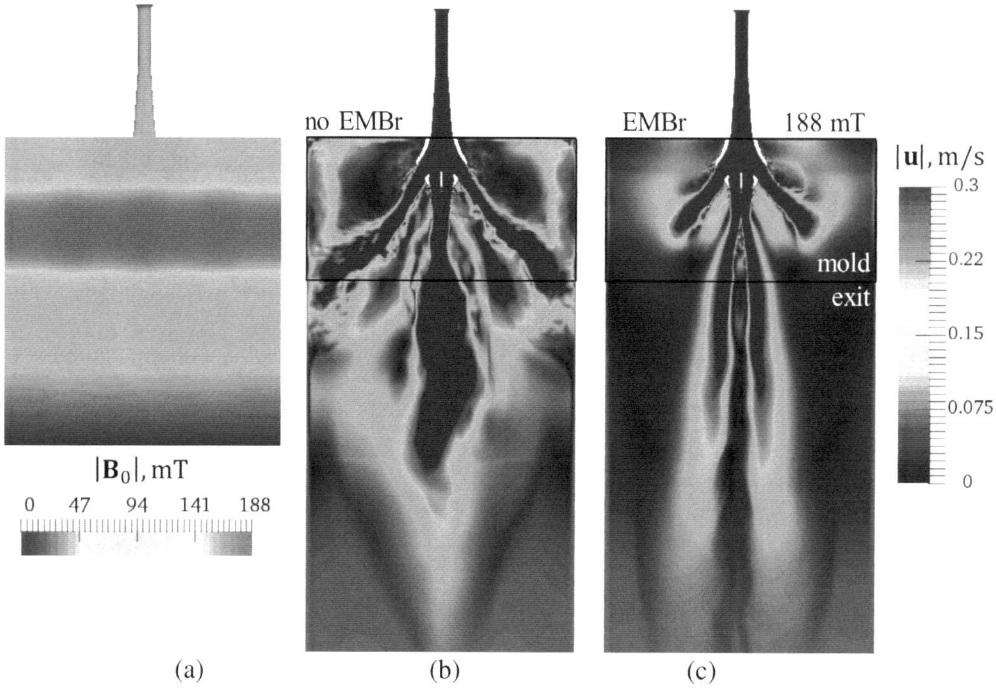

Figure 2. The simulated (instantaneous) flow field for the thin slab casting: (a) applied magnetic field distribution; (b) velocity magnitude distribution in the mid-plane without and (c) with applied EMBr.

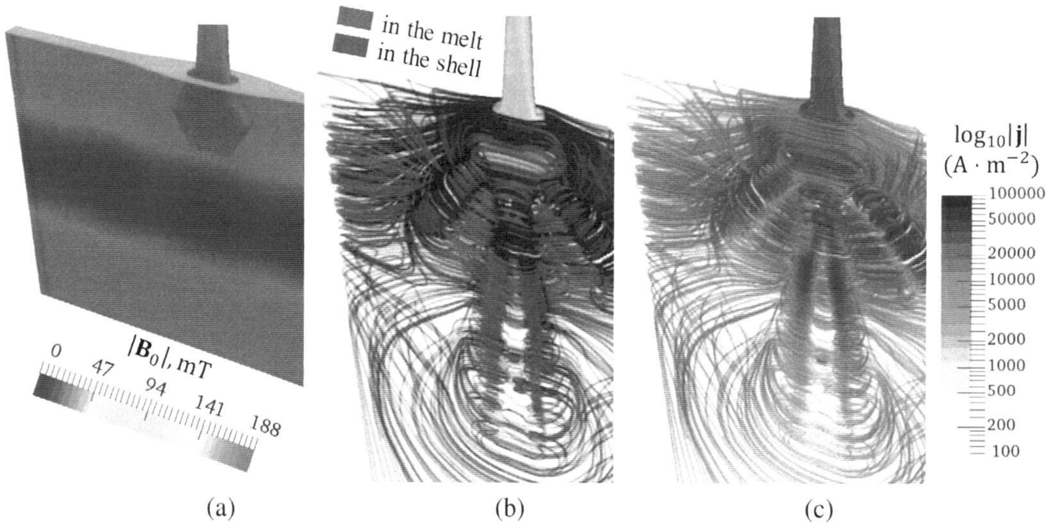

Figure 3. The induced current density lines: (a) applied magnetic field distribution; (b) e-current lines located in the liquid bulk (blue) and conducted through the solid shell (red); (c) e-current lines with the corresponding induced current density magnitude.

The strong braking action of the Lorentz force occur due to the fact, that the induced current density lines tend to close in the form of loops through the more conductive solid shell, which is indicated in figure 3(b) by corresponding colors. Thereby the e-current lines are "attached" to the shell and do not allow to develop the flow instabilities. At the same time, it is observed, that the most e-current is generated in the jets regions and is concentrated in the upper part of the domain.

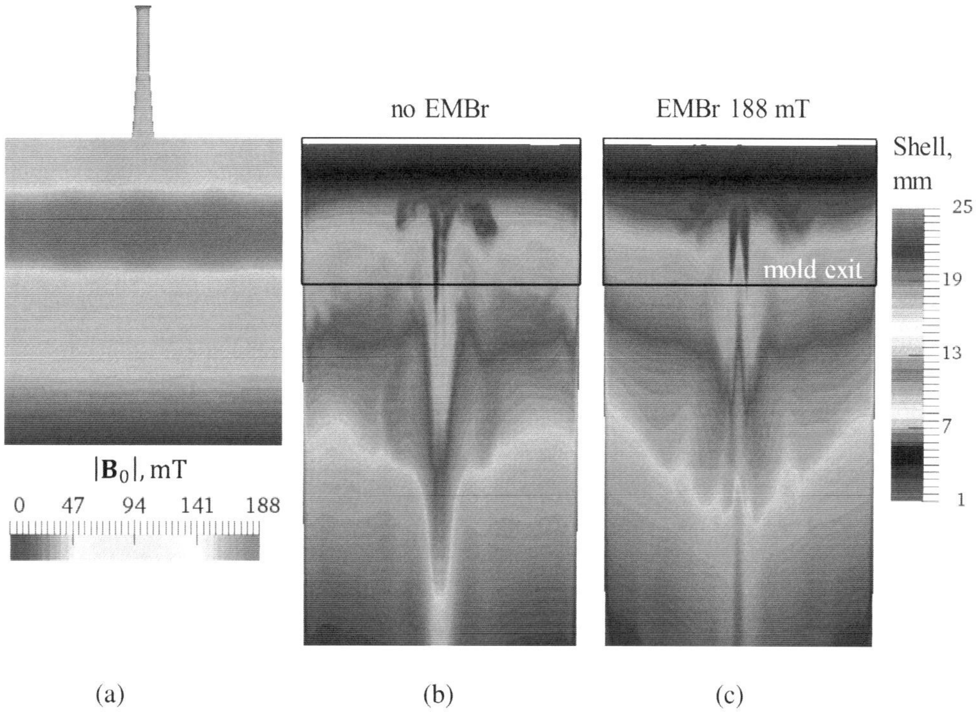

Figure 4. The simulated (instantaneous) solid shell for the thin slab casting: (a) applied magnetic field distribution; (b) shell thickness along a wide face without and (c) with applied EMBr. Results are shown for the 100% solid fraction iso-surface.

The corresponding changes of the hot melt flow consequently lead to the improved solid shell growth. The shell thickness is displayed in figure 4 representing the iso-surface of 100% solid fraction. For the initial configuration with no EMBr (figure 4(b)) the strong remelting of the shell is observed in the region of the side jets and at the center of the mold's wide side. The shell thickness is strongly non-uniform in the longitudinal and especially in the transversal direction. By applying the EMBr due to the flow pattern improvement and because of the braking effect, the shell thickness becomes significantly even. However further increase of the magnetic field value will lead to the formation of the opposite meniscus roll, thereby the EMBr should be combined with the advanced SEN designs. This topic as well as the interaction of the applied magnetic field with the viscoplastic stresses in the shell are the subjects for the future studies.

4. Conclusions

In the presented study the solidification model including a viscoplastic stress in the two-phase region was newly combined with the magnetohydrodynamics Lorentz force. Full coupling is done in the iterative manner including: modelling of the liquid melt flow; motion of the solidifying shell under a withdrawal according to the Norton-Hoff relation for the deviatoric stress tensor; the electromagnetic braking due to the induced electric current density under the applied magnetic field.

The effect of the applied EMBr on the flow pattern and the solidifying shell thickness for the real 3D geometry of the thin slab caster was investigated. The reduction of the hot melt impingement on the solidification front was observed, leading to more uniform solid shell profile. The developed methodology can be applied for the wide range of the EMBr settings and different mold / SEN designs to investigate the meniscus profile and fluctuations along with the submeniscus velocities, which are important reasons for applying EMBr technique.

Acknowledgments
The authors acknowledge the financial support by the Austrian Federal Ministry of Economy, Family and Youth and the National Foundation for Research, Technology and Development within the framework of the Christian Doppler Laboratory for Metallurgical Applications of Magnetohydrodynamics.

References
[1] Thomas B G and Cho S M 2018 *IOP Conf. Ser.: Mater. Sci. Eng.* **424** 012027
[2] Flemings M C 1991 *Metall. Trans. A* **22A** 957–81
[3] Vakhrushev A, Kharicha A, Wu M, Ludwig A, Nitzl G, Tang Y, Hackl G, Watzinger J and Rodrigues C M G 2019 *IOP Conf. Ser.: Mater. Sci. Eng.* **529** 012081
[4] Rodrigues C M G, Ludwig A, Kharicha A, Wu M and Vakhrushev A 2019 *Metall. Mater. Trans. B* **50** 1334–50
[5] Vakhrushev A, Wu M, Ludwig A, Tang Y, Hackl G and Nitzl G 2014 *Metall. Mater. Trans. B* **45B** 1024–37
[6] Vakhrushev A, Kharicha A, Liu Z, Wu M, Ludwig A, Nitzl G, Tang Y, Hackl G and Watzinger J 2019 *STEELSIM Conf. Proceed.* **PR-503-064** 615
[7] Rappaz M, Bellet M and Deville M 2003 *Numerical Modeling in Materials Science and Engineering* (New York: Springer) chapter 6 p 309
[8] Fachinotti V D, Le Corre S, Triolet N, Bobadilla M and Bellet M 2006 *Int. J. Numer. Meth. Engng.* **67** 1341–84
[9] Bellet M 2007 *AIP Conf. Proc.* **908:1** 1369–74
[10] Fernandes C, Araujo M S B, Ferrás L L and Nóbrega J M 2017 *J. Non-Newtonian Fluid Mech.* **249** 63–78
[11] Davidson P 2001 *An Introduction to Magnetohydrodynamics* (Cambridge: Cambridge University Press) chapter 9 p 304
[12] Weller H G, Tabor G, Jasak H and Fureby C 1998 *Comput. Phys.* **12(6)** 620–31
[13] Thomas B G, Singh R, Vanka S P, Timmel K, Eckert S and Gerbeth G 2015 *J. Manuf. Sci. Prod.* **15(1)** 93–104
[14] Liu Z, Vakhrushev A, Wu M, Karimi-Sibaki E, Kharicha A, Ludwig A and Li B 2018 *Metals* **8** 609
[15] Kozlowski P F, Thomas B G, Azzi J A and Wnag H 1992 *Metall. Mater. Trans B* **23A** 903–18

Simulation of longitudinal surface defect in steel continuous casting

M L S Zappulla and B G Thomas

Mechanical Engineering Department, Colorado School of Mines, 1500 Illinois St, Golden, Colorado, USA, 80401

E-mail: BGThomas@mines.edu

Abstract. Cracks and depressions in steel continuous casting often start during initial solidification in the mold and evolve later in the process. To understand the mechanism(s) of formation of these defects, a finite-element model of transient thermal-mechanical behavior during steel solidification has been developed and applied to simulate longitudinal depressions. The thermal model features a detailed analysis of heat transfer across the interfacial gap, which is fully coupled with a thermal-mechanical elastic-viscoplastic finite-element model that calculates the evolving size and shape of surface depressions. The results are compared with analysis of a sample with a longitudinal surface depression and crack from a commercial continuously cast slab. In this scenario, a thermal event caused a local drop in heat transfer, followed by frictional tension on the domain due to slab shrinkage, led to shell necking and the formation of a localized U-shaped depression. Finally, a new visualization methodology has been developed to better reveal defect formation in real time, such as longitudinal depressions, which persist as they move down the mold at the casting speed.

1. Introduction

Quality during continuous casting is greatly affected by heat transfer across the interfacial gap between the copper mold walls and the steel strand, which governs initial solidification of the steel shell and the formation of surface cracks and other defects [1–3]. Previous work has shown that cracks correlate with high and/or variable heat flux in the mold, especially near the meniscus region [4]. Thus, quality monitoring systems routinely include the measured mold heat flux signal, based on heat up of the cooling water as it passes through the mold as well [5, 6] as well as other methods [6–8].

Combining advanced computational modeling with careful plant measurements and forensic analysis of defects is an important methodology to identify how the defects form. Online visualization of real-time measurements is a useful tool to identify defect formation, hopefully in time to take corrective action, and is better if augmented by calculations based on the defect formation mechanism. This paper presents benchmark measurements obtained during continuous casting when longitudinal depressions and cracks formed, and compares the measurements with corresponding predictions of a thermal-stress model.

2. Benchmark plant measurements

On June 18, 2017, a campaign of low-carbon steel was cast at ~1.4 m/min at slab caster 21 at Tata Steel Europe in Ijmuiden, Netherlands. For this campaign, the loose broad face mold copper was instrumented with a 38 x 70 grid of fiber Bragg grating thermocouples that recorded temperature histories at more than 2000 locations [9]. The instrumented region extended across the entire mold width from above the meniscus (90mm below top of mold) to 285 mm below the meniscus. Although such molds are complex, and expensive to make and maintain, this new methodology enables faster response time, and smaller spacing between sensors than conventional thermocouples, which can be presented in real time, and recorded with high resolution [10,11]. Figure 1 shows severe longitudinal depressions and cracks in a ~1300 mm-wide slab Z9407-101, which is a relatively rare occurrence. Of particular interest is the depression at ~265 mm from the North narrow face, (-394 mm from the strand centerline), which is about 250 mm long.

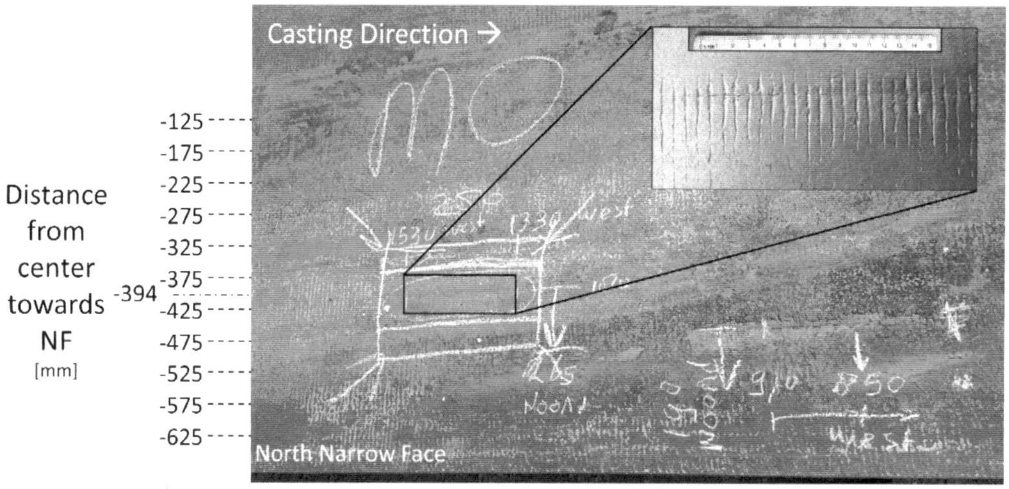

Figure 1. Top surface appearance of slab Z9407-101 with severe longitudinal defects off the North NF and inset image of a slab sample containing a depression that was subjected to detailed metallurgical analysis.

Figure 2 shows the instantaneous temperature profiles measured across the mold subsurface at several times during the casting campaign, which correspond to the ~25 s time period that the 250 mm-long longitudinal depression of interest was in the instrumented portion of the mold. The resolution of the data is indicated by the straight-line segments plotted between measured points, which are sufficiently dense to capture the real profile behavior under most conditions. Figure 2(a) shows typical steady-state temperatures across and down the mold after conditions returned to normal. As expected, temperatures in the mold generally decrease with distance below the meniscus, and steeply decrease towards the narrow faces, (strand corner), indicating the cast width of the strand.

Most importantly, temperatures recorded near to the depression location are lower. Figure 2(b) shows temperatures before, during, and after the residence time of the defect in the mold. It reveals a region of lower temperature that persisted across the mold width during the transient time that the depression developed, which is the general thermal signature of this type of defect. This temperature dip is partly due to the thermal event that initiated the defect and partly due to the consequent drop in heat transfer across the locally larger interfacial gap between the strand and mold wall, which thus experiences a larger thermal resistance. These temperature variations also fluctuate in time.

The center of the depression at -394 mm in figure 1, shown as a dashed vertical line in figure 2(b), passed between two columns of FBG sensors at -375 and -425 mm from the broad face center. A time average of the temperature history over the 30s duration of the depression is shown as a red line in figure 3(b), taken at the FBG sensor column at -375 mm from the wideface centerline, which was closest (~19 mm away) to the depression center (solid vertical line in figure 2(b)). Instantaneous time-averaged far field temperatures were taken at +475mm (opposite side of the slab) and are shown in black in figure 3(a). This temperature profile quantifies the significant drop in temperature from above the meniscus to the end of instrumentation, where it approaches back to the far field value.

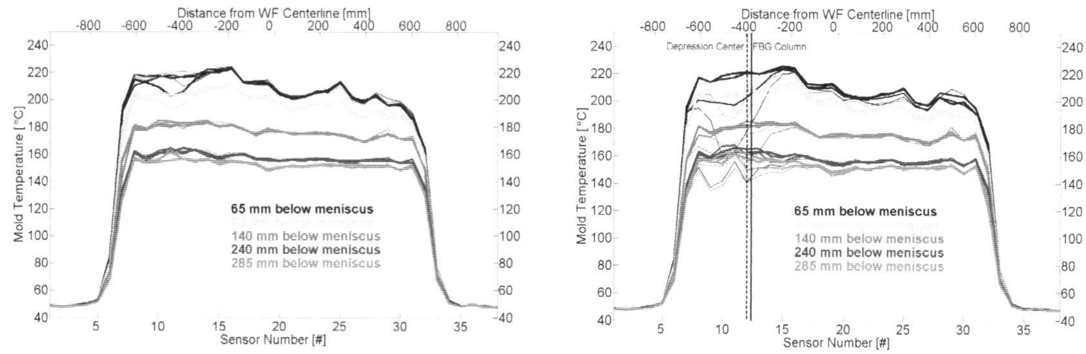

(a) (b)

Figure 2. Measured local temperature profiles across the mold width at specific times and locations down the mold. (a) Uniform casting conditions after the region containing the defect and (b) Before, during, and immediately after formation of the defect region with the depression.

3. Mold data mapping methodology

A new methodology has been developed to better enable operators to interpret continuous online measurement of dense mold temperature data to detect formation of longitudinal defects [12]. Figure 3(b) shows a 3D surface plot of temperatures down the mold for 70 s surrounding the residence time of the surface depression. Lines of constant time run parallel to the axis of distance down the mold. Temperatures are taken from the same vertical line of FBG sensors ~19 mm from the depression center. In this figure, movement in the casting direction at 25 mm/s of two points down the strand in the depression is visualized by two translucent vertical planes, spaced 11s apart. The oldest data is sliced by the right plane, where the shell starts at the meniscus (back right) at 22 s and exits at 37 s.

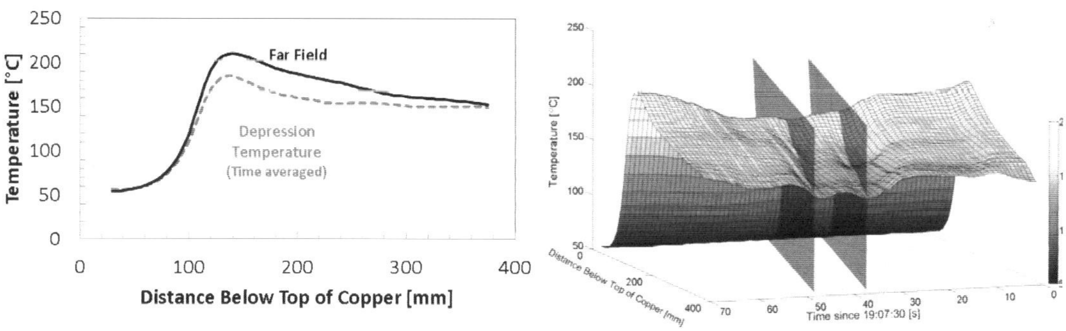

(a) (b)

Figure 3. (a) Temperature profiles down the mold for the far field and the FBG sensor closest to the depression center (19 mm away) averaged over the instrumentation residence time of the depression (b) 3D plot of temperature evolution along the same single vertical line near the depression.

In figure 3(b), there are clearly visible valleys in the temperature plot, delineated by the vertical planes. These temperature valleys represent the time duration of this depression. The angled nature of the valleys matches close to the casting speed planes, which indicates that the depression started at the meniscus and then translated down the strand at the casting speed. The steep gradient near the meniscus on the right plane indicates that the depression formed rapidly; temperatures stayed depressed for ~11 s until the left translucent plane, and then gradually returned to far field conditions. Slight changes in the shape of the valley indicate that the depression shape changed with time, growing rapidly when it first formed, stagnating for a short distance, continuing to grow, and finally receding and disappearing.

3.1. Mapping mold data to slab data

The translating nature of defect formation, shown in the previous section, suggests a better methodology to visualize the formation of defects in the mold. With the high sampling frequency and spatial resolution of FBG molds, it is possible to track detailed time histories that correspond with points on the slab surface moving down the mold at the casting speed, by matching position down the strand according to the time it was adjacent to a given distance down the mold. Then, the instantaneous drop in temperature at a given measured point in the mold surface is found by subtracting T_{Standard}, which is a 2D array of standard temperatures at the instrumented points of the mold, calculated previously by averaging data when no defects occurred, according to the given steel grade and casting conditions:

$$\Delta \mathrm{T(t)} = \min(0, T_{\text{Mold}}(x_{Slab}, y_{Slab} - \int_{t_0}^{t} V_c(t_i)dt_i, t) - T_{\text{Standard}}(x_{Slab}, y_{Slab} - \int_{t_0}^{t} V_c(t_i)dt_i)) \quad (1)$$

While a location of interest on the slab surface, y_{Slab}, is within the instrumented region of the mold, $t_{Enter} < t < t_{Exit}$, the criterion, T_{IDrop}, evolves with time by integrating the history of the temperature drop at that moving location:

$$T_{IDrop}(x_{slab}, y_{slab}, \mathrm{t}) = \int_{t_{Enter}}^{t} \Delta T(t_i)dt_i \quad (2)$$

where t_{Enter} is the time when the location, y_{Slab}, enters the instrumented mold region, $y_{Slab} = y_{Moldtop}$.

Large values of T_{IDrop} at the end of the instrumented region indicate regions of the slab surface which have experienced a mold temperature history that is cooler than standard conditions. This indicates lower heat flux across the interfacial gap and a hotter shell temperature. This is often due to surface depressions, which often lead to cracks, owing to a locally hotter shell. These regions can be tracked, subjected to extra inspection, and perhaps downgraded. This methodology and criterion can be applied to systems with arbitrary spacing of the measurement points, provided there is enough spatial resolution to achieve reasonable accuracy. A contour map of this criterion indicates regions on the strand surface that have experienced prolonged, periods of temperature drops relative to the standard profile, likely due to a surface depression. Details of this methodology are provided elsewhere [12].

3.2. Defect visualization

A console display setup for post-mortem (forensic) defect analysis using this new defect indicator is suggested in figure 4. Figure 4(a) shows the instantaneous contour map in the instrumented region displayed by the sensors at the current time, using a blackbody color map without interpolation.

Figure 4(b) shows a top-down photograph (adjusted to remove perspective) of the final surface appearance of the region of the cast slab corresponding to the temperatures in figure 4(a) at the time it resided in the instrumented region in the mold. The translucent gray rectangle at the top indicates that portion of the image which was above the meniscus level at that time.

Figure 4(c) shows a contour plot of the instantaneous deviations between the standard and current temperatures for each location in the array. Values are always negative or zero, because depression defects always tend to lower mold temperature. Increased negative values indicate deeper depressions.

Figure 4(d) shows an example of the output results from the new methodology, obtained by accumulating the results in figure 4(c) for 11 s (245 mm past the bottom of the instrumented region), according to the new defect criterion function. The time period spent within the instrumented region of the mold is represented by the space between the thick black horizontal lines, labeled top and bottom of instrumentation. Once a surface location passes out of the instrumented region, the contours are fixed, and the calculation for that surface point is completed. Again, increasingly negative numbers indicate higher likelihood of surface quality problems. The longitudinal depression at the left of figure 4(b) is clearly visualized in figure 4(d), but is hardly noticeable in the raw temperature data. This illustrates the value of this new defect identification methodology, which is described in detail elsewhere [12].

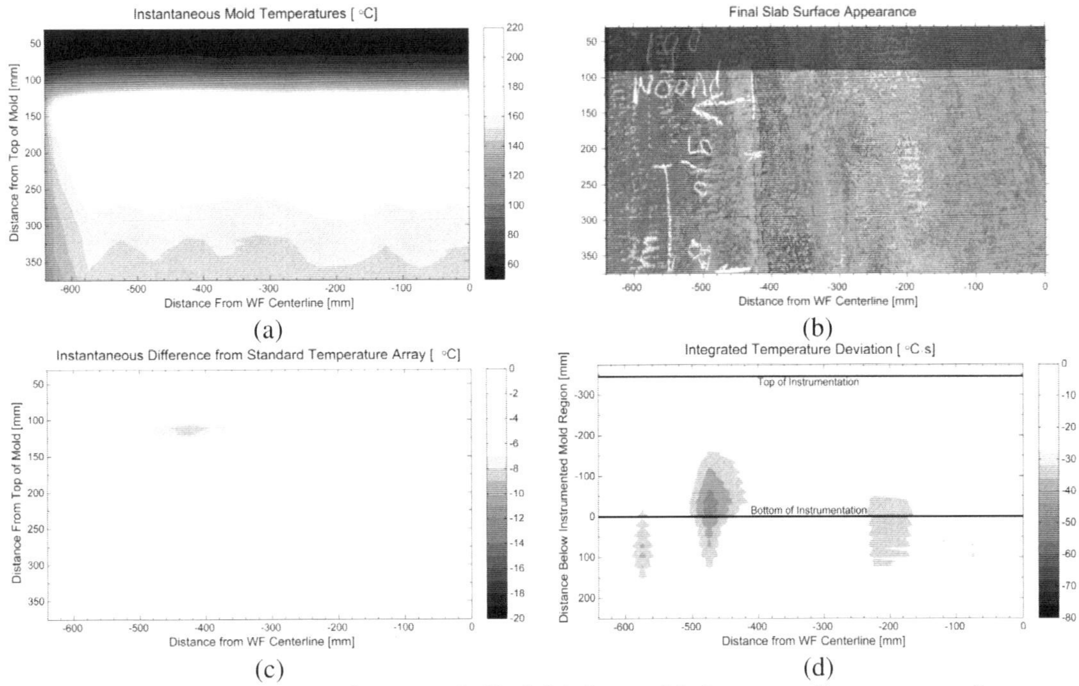

Figure 4. Contours that map defects over half of slab from wide face centerline to narrow face: (a) Instantaneous mold temperature, (b) Final slab surface appearance, (c) Instantaneous difference from standard temperature (d) Defect criterion based on integrated temperature drop, T_{IdropF}.

4. Thermal-mechanical model formulation

The thermal-mechanical model in this work focuses on a two-dimensional (2-D) slice through the solidifying steel shell in a state of generalized plane strain in both x and z directions [3], as it moves down the mold at the casting speed in a Lagrangian reference frame. The model solves the transient heat conduction and momentum equations to predict the temperature, displacement, strain, and stress histories of the shell. The governing equations are solved using the finite-element method in ABAQUS/Standard (implicit). The transient analysis uses coupled-temperature displacement elements and step-wise coupling between the thermal and mechanical calculations. The 20 mm thick x 40 mm wide simulation domain, shown in figure 5(a), represents a portion of the solidifying shell somewhere along the broad face away from the corner.

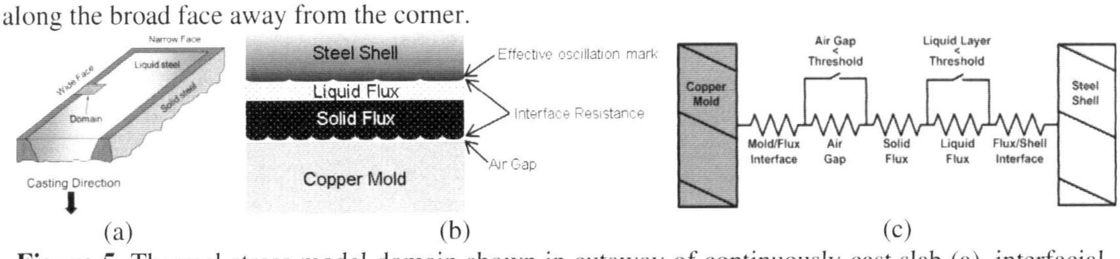

Figure 5. Thermal-stress model domain shown in cutaway of continuously cast slab (a), interfacial gap layers in this model (b), and numerical implementation as a thermal resistor model (c).

The model is applied here to the case with FBG measurements shown in the previous section. The initial temperature is set to the reported tundish temperature, 1554°C in the steel, and 100°C in the mold. Coupling between the solidifying shell and the copper mold was accomplished using a GAPCON subroutine. Given the mold hot face and steel shell surface temperatures, and the total gap thickness calculated by the mechanical model at each time step, the thermal model calculates appropriate local solid and liquid layer thicknesses using: flux solidification temperature 1045°C, and resistances (mm^2

K/mW) of the different layers comprising the gap: the mold/slag interface, 0.0004, air gap, $D_{Air}/0.06$, solid slag layer, $D_{Sol}/3.0$, liquid slag layer, $D_{Liq}/1.0$, and slag/shell interface, 0.251 exp(-0.0054T_{ls}). D_{Air}, D_{Sol}, and D_{Liq} are thicknesses of the air gap, solid slag, and liquid slag layers respectively, and T_{ls} is the temperature of the liquid slag/steel shell interface. The slag/shell interface resistance is non-linear and numerically constrained between 1,000 and 20,000 W/K.

To calculate accurate mold temperatures, a Reduced Order Model (ROM) of the mold was applied [13]. Modeled as a single row of elements along the width, the ROM mold has a calibrated thickness of 28.35 mm to account for water slots in the transverse direction. A constant thermal conductivity of 315 W/m·K and variable convection boundary condition is applied to the mold cold face.

The displacement and stress state of the model is calculated by solving the static stress equilibrium equations including frictional contact with μ=0.15. Details of this model, including the elastic-visco-plastic constitutive behavior with different temperature-, and composition-dependent properties in each of the steel phases: liquid, δ-ferrite, and γ-austenite can be found elsewhere [1–3].

5. Thermal model calibration and validation

First, the continuous-casting heat transfer model, CON1D [15] was calibrated to match both the heat flux measured from heat up of the mold cooling water, FBG temperature measurements at the far field, and two rows of embedded thermocouples, as shown in figure 6(a). For these far-field conditions, CON1D suggests a flux layer thickness profile down the gap, as shown in figure 6(b).

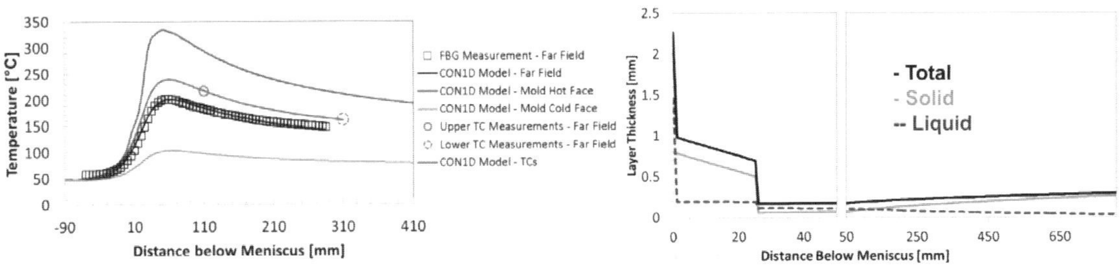

(a) (b)

Figure 6. (a) Temperature results from CON1D with plant measurements for FBG and thermocouples and (b) con1d output results for flux layer composition within the mold.

Of particular interest is that figure 6(b) indicates the presence of a significantly thicker layer of flux extending to 26 mm below the farfield meniscus; Slag thickness then decreases to ~0.2 mm and gradually increases with distance down the mold. The extended slag rim, of ~1 mm thickness, retards heat transfer in the meniscus region, and delays initial formation of the shell below the farfield meniscus. It is needed to enable CON1D to match the plant measurements. This provides strong evidence that such extended slag rims during casting are real. This is further supported by anecdotal evidence from this particular facility where several extended slag rims with similar thickness profiles were found after casting.

After calibration of CON1D with measurements, the GAPCON model in Abaqus was adjusted to match CON1D. Due to the 2D nature of the model, axial heat conduction is neglected. Instead, a negative volumetric heat generation was applied, decreasing from -25 [W/m^3] to zero over the first 10s. Figure 7(a) shows the match with the farfield CON1D heat flux profile, and figure 7(b) shows a comparison with the corresponding mold temperatures, including the FBG data. Note that the peak heat flux appears over 20 mm below the farfield meniscus, owing to the extended meniscus slag rim discussed previously.

Results from the Abaqus model are also compared with the FBG sensor data 19 mm away from the depression center. Figure 7(a) shows that the heat flux profile in the depression center is substantially lower than that of the far field. This is due to the extra thermal resistance created in that region, in addition to the consequent physical opening of the depression, which was assumed to create an air gap. Figure 7(b) shows a substantially higher peak temperature predicted by the Abaqus model near the meniscus, which is a modeling artifact due to missing copper above the meniscus that removes heat axially and leads to the smoother and lower temperatures of the FBG and CON1D profiles in this plot.

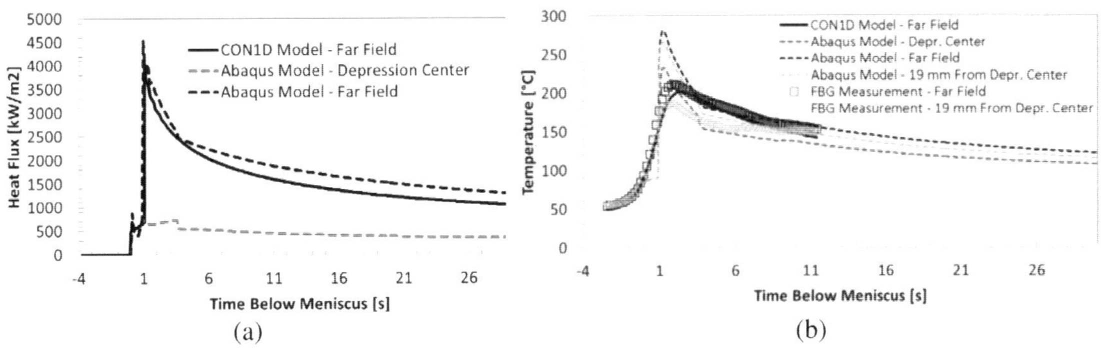

Figure 7. (a) Calculated heat flux profiles down mold comparing Abaqus and CON1D models; (b) Corresponding mold temperature profiles also comparing with FBG measurements.

6. Thermal-mechanical model results

Depression shape results of the thermal-mechanical model are compared with a digital microscope scan and measurements of the sample in figure 8. The ~4mm deep crack is indicated by the vertical dashed line. The inset compares a micrograph cross section through the depression (right) with the predicted shape (left). The model slightly overpredicts the depression depth, likely due to overprediction of the temperature gradient near the meniscus (figure 7(b)), which is due to incomplete incorporation of the axial heat flow near the meniscus into the Abaqus slice model.

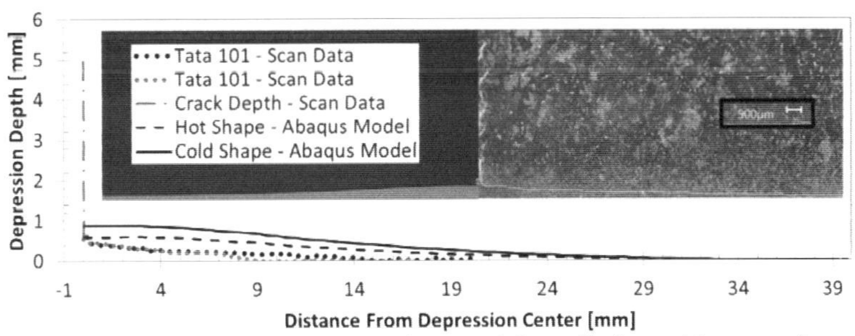

Figure 8. Slab surface depression shape comparing model prediction with scan of sample with defect

Figure 9 shows the results of a parametric study examining the effect of the thermal disruption start location (a); as well as the final deformed shapes, both at mold exit, and after cooling to near ambient (b). The thermal disruption was likely caused by a transient event, such as severe meniscus level fluctuations from turbulent flow. It is observed that delaying the start of the thermal disruption to a lower distance below the meniscus causes the depth of the resultant depression to decrease. It has also been found that the depression stops growing once the surface temperature reaches ~700°C, as shown in figure 9(b), which is substantially deeper than at mold exit. Furthermore, the depression is always much deeper when the disruption starts closer to the meniscus. This finding is consistent with the accepted understanding that longitudinal depressions and cracks tend to initiate close to the meniscus.

7. Conclusions

New models and visualization methods have been developed to enable new insights into the continuous casting solidification process, specifically revealing how temperature differences are related to longitudinal defect formation. A new computational thermal-stress model has been developed to investigate longitudinal surface defects, as well as to predict the corresponding mold temperatures. Results show that one possible scenario for the specific longitudinal crack and depression defect studied in this case was a ~20 mm wide local thermal disruption, combined with a crack and some applied tension which led to a necking style depression and a U shape defect with a crack at the base.

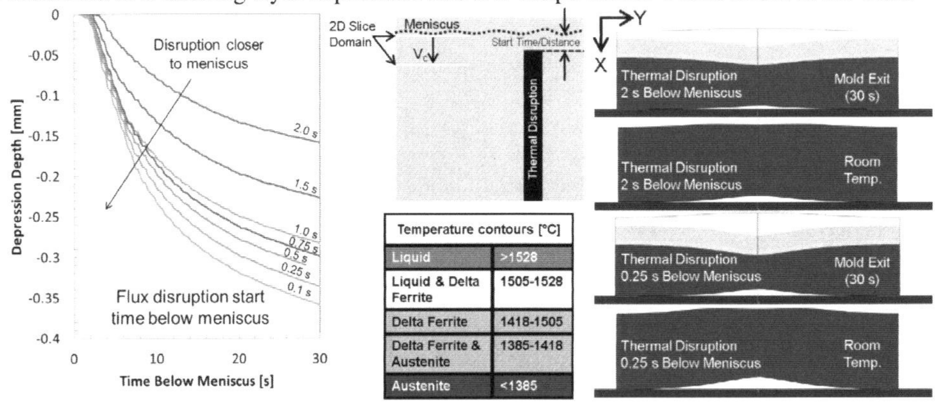

(a) (b)

Figure 9. Effect of thermal disruption initiation time on depression evolution: (a) Depth history for varying disruption initiation times, (b) Deformed shape (scaled 5x) of depression at mold exit and after cooling to near ambient, for late disruption (top) and early disruption (bottom).

Acknowledgements

The authors wish to thank the member companies of the Continuous Casting Center at the Colorado School of Mines for support and plant data. Particular thanks are extended to: Begoña Santillana, Mathias Mul, Ton Spierings, Stephen Carless, Gert Abbel, Dirk Van Der Plas, Arnoud Kamperman, of Tata Steel Europe, and Martin Seden of ABB.

References

[1] Zappulla M L S and Thomas B G 2018 *THERMEC' 2018* (Paris, France)
[2] Zappulla M L S and Thomas B G 2017 *The Materials Society Annual Meeting* **F6** (San Diego, CA) pp 501–10
[3] Zappulla M L S, Hibbeler L C and Thomas B G 2017 *Metall. Mater. Trans. A* **48** 3777–93
[4] Kawamoto M, Tsukaguchi Y, Nishida N, Kanazawa T and Hiraki S 1997 *ISIJ Int.* **37** 134–9
[5] Kumar S, Meech J A, Samarasekera I V, Brimacombe J K and Rakocevic V 1999 *Ironmak. Steelmak.* **26** 269–84
[6] Zhang Y, Vaculik V and Dudzic M 2003 Integrated Monitoring Solution to Start-Up and Run-Time Operations for Continuous Casting *IFAC Proc.* **36** 561–6
[7] Emling W H 2003 *The Making, Shaping and Treating of Steel: Casting Volume* ed A W Cramb (Pittsburgh, PA: AISE Steel Foundation)
[8] Tirian G O, Filip I and Proştean G 2014 *Neurocomputing* **125** 236–45
[9] Spierings T, Kamperman A, Hengeveld H, Kromhout J and Dekker E 2017 *AISTech 2017 Proc.* 1655–64
[10] Badri A B and Cramb A W 2002 *Steelmaking Conference Proceedings* (Warrendale, PA: Iron and Steel Society) pp 65–76
[11] Thomas B G and Okelman M 2011 *Sensors, Sampling, Simul. Process Control* 127–34
[12] Zappulla M L S, Cho S M and Thomas B G 2019 *Steel Res. Int.* **90**
[13] Hibbeler L C, Chin See M M, Iwasaki J, Swartz K E, O'Malley R J and Thomas B G 2016 *Appl. Math. Model.* **40** 8530–51

IOP Conf. Series: Materials Science and Engineering **861** (2020) 012017 doi:10.1088/1757-899X/861/1/012017

Integrated numerical simulation of the industrial continuous casting process

J Fainberg, E Hepp, W Schäfer and C Krafthöfer

MAGMA GmbH, Kackertstraße 11, 52072 Aachen, Germany

E-mail: e.hepp@magmasoft.de

Abstract. The production of a continuously cast strand involves a large number of coupled phenomena involving flow, heat transfer, solidification and thermo-mechanics. In this paper, different approaches are described for the numerical modelling of industrial continuous casting processes at different levels of detail, dependent on the questions to be investigated. In the simplest case to obtain a rough estimate of the process conditions, the solidification of the strand is considered with a prescribed convective flux profile in the strand body based on the withdrawal speed. In the most complicated case, the process phase of filling of the empty mould is first calculated, followed by the phase of the withdrawal of the strand including consideration of the flow in both strand and tundish.

1. Introduction

Modelling continuous casting is as complex as the process itself. In order to achieve a robust model for industrial application with acceptable calculation times, the handling of the changing of boundary conditions at the inlet and free surface of the metal are important. As an example, one can simplify the handling of the melt free surface in the tundish, launder and boundary conditions for the fluid flow. This can be used to significantly speed up the fluid flow computations, due to the long duration of the withdrawal phase. Although it is possible to continue to consider the free surface in the withdrawal phase, for the copper and aluminium applications shown in this paper experience has shown that this is not necessarily required.

The handling of the momentum equations in the moving strand and the remaining liquid pool during withdrawal by using two different coordinate systems will be discussed. A special enhancement is applied in the predictor step of the PISO method for the incompressible flow in order to account for the simultaneous treatment of fluid flow in two different coordinate systems.

1.1. Solidification simulation with prescribed convective flux

The first and easiest approach is a thermal simulation with a prescribed convective flux as a block profile in the strand body, see figure 1(a). In that case, the enthalpy equation is solved and no velocities in the liquid regions have to be calculated. The convective flux is unidirectional with a value given by the casting speed. The casting speed can be defined time dependent. Typical results can show the temperature distributions and the corresponding liquid pool depth. Additionally, the thermal criteria like thermal gradient, cooling rate, solidification time and microporosity are calculated. The flow effects on these criteria and on the temperature distribution are very simplified, but it can nonetheless be useful for a first rough estimation of the process conditions for primary and secondary cooling, in order to verify the defined boundary conditions and heat transfer coefficients between applied materials.

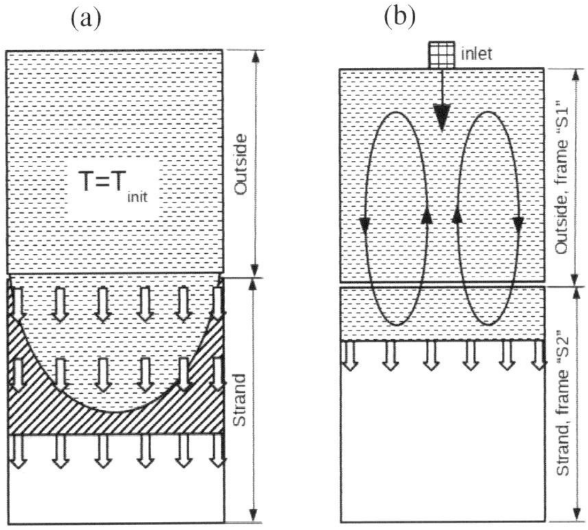

Figure 1. Two different approaches to continuous casting process modelling. (a): most simplified without any fluid flow calculation, the block velocity profile is prescribed in the strand region for the convective heat flux. The temperature above the strand is fixed. (b): The withdrawal again starts with a filled volume over the strand, but the fluid flow is computed everywhere and the temperatures over the strand are not fixed. No melt free surface movement is considered as the volume over the strand is completely filled from the beginning.

1.2. Fluid flow without filling of a tundish

A more realistic representation of the entire process has to allow for the assessment of the flow conditions during start-up and subsequent strand withdrawal. The volume above the strand is taken as completely pre-filled at the beginning of the withdrawal process, see figure 1(b). The melt free surface is considered as fixed and coincident with the top of the domain/mould. A Dirichlet boundary condition for pressure is applied at the inlet while the outlet boundary condition is set at the bottom strand edge. Using this approach, the computation of the withdrawal starts immediately. The transient heat transfer during the filling of the mould is not considered, instead a homogeneous temperature is prescribed at the beginning of the calculation. In comparison to the first approach, it is now possible to calculate the forced and buoyancy driven melt flow with conjugate heat transfer, the latent heat release and solidification. The convective heat transfer will cause changes in the temperature distribution compared to the approach using a prescribed convective flux as in figure 1(a).

1.3. Fully integrated approach including filling of the launder and tundish

The most complex option considers a coupled thermal and flow simulation consisting of at least two subsequent process phases. The first phase is a classical mould filling with an empty mould at the beginning. The process is switched to the second phase of strand withdrawal, when a predefined control condition such as filling time or volume percent filled or the melt level in the mould is achieved. The modelling of the melt free surface and the control of the inlet melt flux can be also switched to a simplified approach (figure 2(a)) either at the start of withdrawal or during withdrawal when an analogous second switch condition is reached. In the first case, the computation of the melt free surface (figure 2(b)) is omitted. These algorithms are described in detail in the next chapter.

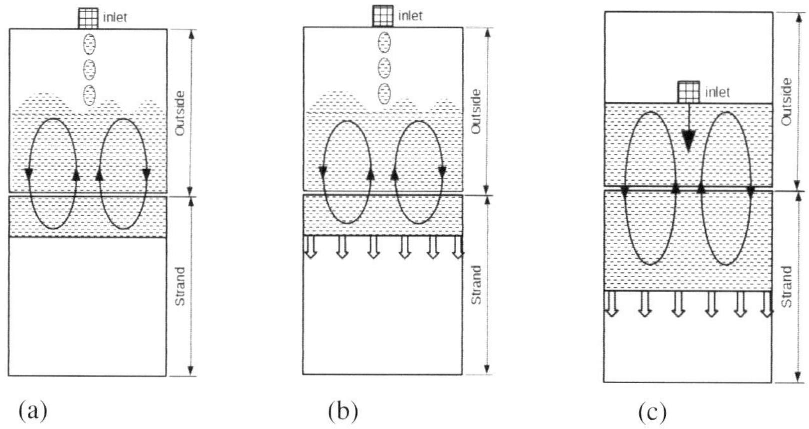

Figure 2. Phases of the integrated process model with fluid flow. (a) Filling of the empty volume over the strand and a part of the strand. (b) Withdrawal simultaneously with cavity filling. (c) The consideration of the melt free surface is terminated at the very beginning of withdrawal or in the withdrawal phase, because a nearly stationary state of the free surface may be expected. The melt free surface is smoothed, the inlet is transferred onto the location of the actual melt surface. The inlet flux is set equal to the mass flux given by the withdrawal velocity.

2. Modelling the continuous casting process

2.1. *Fluid flow approach*
The fluid flow in the continuous casting model is handled using the Finite Volume method [1] on a Cartesian grid. A staggered variable arrangement is used. The pressure is stored in the middle of the grid cell, while Cartesian velocity components are stored for each cell in the middle of its three faces in +X, +Y and +Z directions. The governing equations are assembled and solved in parallel using the MPI distributed memory approach with dynamic load balancing by means of grid partitioning into load balance units (LBU's) [2]. The momentum and mass continuity equations are solved by the PISO method for incompressible flow [3]. The melt free surface is treated using the volume of fluid method (VOF) [4]. In this respect, the continuous casting model follows the concept of the commercial MAGMASOFT® software package for filling in shape casting processes [5]. A special treatment is needed to account for the melt flow in the moving strand volume during the withdrawal and for the additional heat sources there.

2.2. *Solidified cells*
The velocity of the solidified melt is usually set to zero in filling for shape casting processes. The continuous casting process is more complex. We continue to treat the solidified melt as stationary everywhere before the withdrawal starts. In the withdrawal phase, the solidified fraction can move freely as equiaxed crystals driven by the melt flow and their buoyancy, or can stick somewhere on the outside strand wall, or can be moved with the prescribed withdrawal velocity as a rigid body in case of columnar crystals. This causes a different relative velocity between the solid and liquid fractions. Here, the equiaxed crystals are not modelled directly as a substantive phase within the single-phase fluid flow model. The effective viscosity of the solidifying melt is modelled as an exponential function of the solid fraction.

A grid cell is considered as "solidified" if some critical fraction solid is reached. The solidified strand cells are moved with a prescribed withdrawal velocity. This means that the convective flux and the momentum diffusion between liquid and solid fractions have to be accounted for. This is done using two different reference coordinate frames, see figure 3(a).

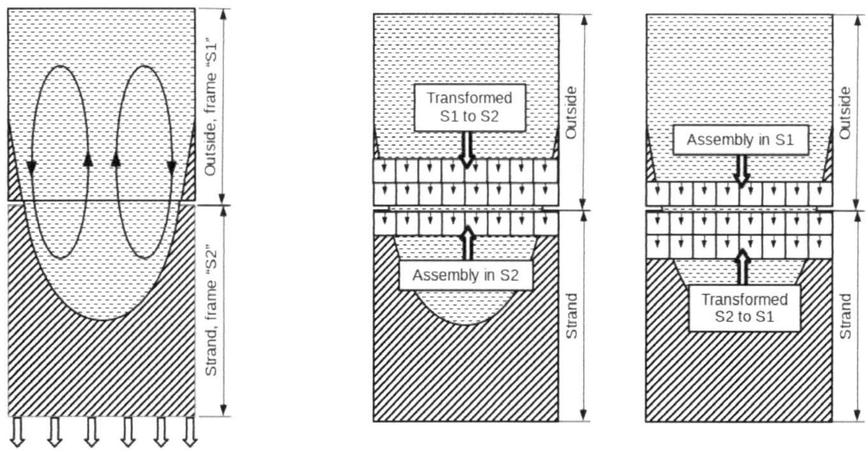

(a) (b) (c)

Figure 3. (a) Logical subdivision into regions of strand and of the melt above it. The liquid and solid are dashed horizontally and inclined, respectively. Momentum equations are not solved in the solid. The used coordinate frames for the momentum and mass continuity equations are stationary S_1 outside the strand and S_2 moving with the withdrawal velocity in the strand. The grid cells at the S_1/S_2 interface and their velocity components are shown for the momentum predictor steps. (b) The velocity transformation into the moving reference frame S_2 precedes the predictor step for the interface cells in the strand. (c) The same for the cells over the strand region, the predictor step for them is preceded by the velocity transformation in the upper strand cells.

2.3. Momentum equations and mass continuity in withdrawal

Velocities at the beginning of the time step are given in the stationary coordinate frame S_1. Before the momentum predictor starts, velocities in the strand grid cells are transformed into the moving S_2 coordinates i.e. $V \rightarrow (V - V_{strand})$. Solidified cells have zero velocity in S_2, hence the relative velocity liquid-solid coincides with the transformed liquid velocity there and no changes are needed in the standard momentum diffusion formulation.

This procedure requires a special treatment of the S_1/S_2 interface. First, new momentum values are computed everywhere in the PISO predictor step. Next, corrections are done for the horizontal grid layers located directly at the S_1/S_2 interface. The predictor step is repeated for the selected top grid layer of the strand, see figure 3(b). The old velocity values of two bottom grid layers over the strand are transformed into the moving reference frame S_2. Next, the predictor step for the bottom grid layer over the strand is repeated. Again, old velocities in two top strand grid layers are transformed into S_1. Each new velocity value is now computed for the stencil velocities in the same coordinate frame. The same holds formally for the horizontal continuous casting setup, only the setup in figure 3 has to be rotated into the horizontal direction.

Velocities in the strand material are transformed back into the stationary S_1 at the end of each time step. They are used in the transport equations for conjugate heat transfer, in calculating criteria functions and, when desired, for the transport of chemical species and particle movement simulation.

A further implication of the two reference frame scheme is a need to adjust the pressure correction equation for the cells located at the interface over the strand, as the velocity at their bottom face is stored for the adjacent strand cell in the moving S_2 coordinate frame. The correction is a negative mass sink term equal to the product of the bottom face area with the withdrawal velocity.

The solidified material in the mould above the strand can move with the withdrawal velocity for those cells with a solidified connection to the strand, as the differentiation between cells in both domains is only a logical one. This is in any case recommended for the continuous casting of steel with a thin and extended solidified skin, but this can cause unrealistic results for other alloys especially when the solidification reaches to the melt free surface at the top of the mould and a downward velocity is prescribed there.

Finally, the heat source term due to advection of the latent heat in the pulled strand is specific for continuous casting and is incorporated in terms of specific latent heat, withdrawal velocity and gradient of the solid fraction [6].

2.4. Melt free surface and inlet flux control in withdrawal
The free surface can be optionally tracked during the whole withdrawal phase, see figure 2(b). Then, the melt level in the mould varies dependent on the process conditions. An increased computational time results due to the adjustment of the CFL-based times step for large velocities in the free-falling stream under the inlet. If small deviations of the melt level from its equilibrium position during withdrawal are not of primary interest, calculation of the melt free surface movement can be omitted immediately or after some transient time as indicated in figure 2(c). This enables achieving the same low computational times as for a prefilled mould. For most industrial applications analysed thus far, the consideration of the dynamics of the melt free surface was not essential.

The filling state of all the regions in the casting setup is analysed automatically before the switch. This is not a trivial task, because the regions may be distributed arbitrarily and furthermore many various continuous casting setups exist. The applied procedure is to calculate the percent of filled volume for each horizontal grid level individually. The obtained layer wise distribution is analysed for each particular region. One looks for the transition between a value higher than a threshold of 40% not filled and a value smaller than the threshold in the next lower layer starting from the top. The first found position becomes the position of the interface planarization in the considered region.

Finally, the inlet position is transferred downwards to the shifted planarized melt interface. The inlet flux is set balanced with the mass flux of the pulled strand, which is required in order to assure the stationary condition of the melt free surface. There are many other details such as filling of "holes" somewhere in the nozzle in order to assure a realistic configuration with mass flux continuity. The applied method works well in most cases.

2.5. Temperature trace with movable thermocouples
The thermal history at any point attached to the strand material can be traced by thermocouples moving together with it, being stationary in S_2. Their start position may be defined inside of the prefilled strand or above it. In the latter case, the movable thermocouple will reach the strand after some delay. The host MPI grid partition as well as the grid cell has to be updated for retrieving the thermocouple values for every timestep during strand withdrawal.

2.6. Thermal quality criteria
The continuous casting model considers fluid flow and solidification phenomena simultaneously. The computation of thermal criteria is typically calculated for the solidification only, hence it has to be adjusted for the convective transport due to the strand motion. The computed 3D fields such as solidification time, liquidus to solidus time, thermal gradient, cooling rate, Niyama criteria and microporosity are advected one grid layer downward every time the strand is lengthened by the next grid layer. A special treatment is applied for the cooling rate criterion. The time derivative is composed of the convective component due to the strand motion and the substantive one due to the local temperature change. A 3D criterion "sump depth" shows the distance from the given location in the liquid pool to the strand top.

3. Application examples

The above described models are implemented in the commercial software MAGMA CC. It is a fully integrated solution for the virtual design and optimization of continuous casting processes for aluminium and copper. The software considers the flow, heat transfer, solidification and stress generation in the inflowing metal, the solidifying strand and the mould, and can be used for vertical and horizontal casting processes. As a typical application example for aluminium, the simulation of the start-up phase is shown figure 4. The flow and heat balance in the launder is analysed coupled with the withdrawal process. The start can be controlled by the melt level in the mould or in the launder system. Sometimes the withdrawal starts already before a final meniscus height is reached. Experiments were performed together with the company TRIMET Aluminium SE, Germany on a R&D full-scale casting facility to validate the flow and solidification conditions. Figure 4 shows such a comparison between simulation and the real casting process.

(a) (b)

Figure 4. The right hand picture (b) shows an experimental setup of a casting trial using a hot top system for the production of a round shaped aluminium billet (with courtesy of TRIMET Aluminium SE, Essen). Measurements of temperatures with the help of thermocouples were used in this experimental setup to calibrate the simulation and to get the needed heat transfer coefficients. The left hand picture (a) shows the corresponding simulated temperature result.

The simulation allows a combined consideration of forced and natural convection in the liquid metal and in the mushy zone, due to both the inflow and the temperature gradients in the metal. Figure 5 illustrates the application for the two common direct chill casting processes. There is on the one hand the classical process with a floating nozzle and on the other hand the hot top system shown in figure 4 above.

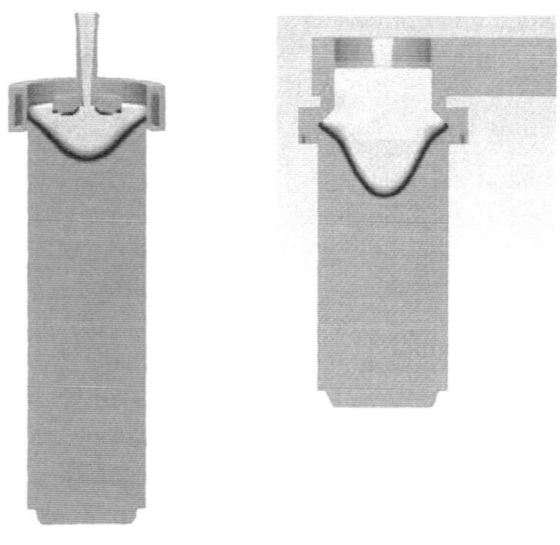

(a) (b)

Figure 5. For DC casting of aluminium, the two most common processes are a hot top system or a floating system with a nozzle. The picture (a) shows the temperature distribution for the floating system and the corresponding sump profile. The right picture (b) shows the sump profile in the hot top system.

IOP Conf. Series: Materials Science and Engineering **861** (2020) 012017 doi:10.1088/1757-899X/861/1/012017

In figure 6 an application for the vertical continuous casting of a copper alloy is visualized. The cast material is a Cu-Zn alloy. In the case of copper continuous casting, the mould is much longer compared to aluminium processes. The secondary cooling looks also different due to the different spray cooling zones. The results in figure 6 show a comparison between air cooling and water cooling in the secondary cooling. It can be clearly identified that the effect of spray water leads to a faster solidification. But, the effect of the temperature gradient is also a reason for residual stress build up and consequently for crack formation. Especially copper alloys without lead have a tendency for the formation of hot tears. With simulation and a virtual design of experiments, a good strategy for an optimized cooling can be achieved.

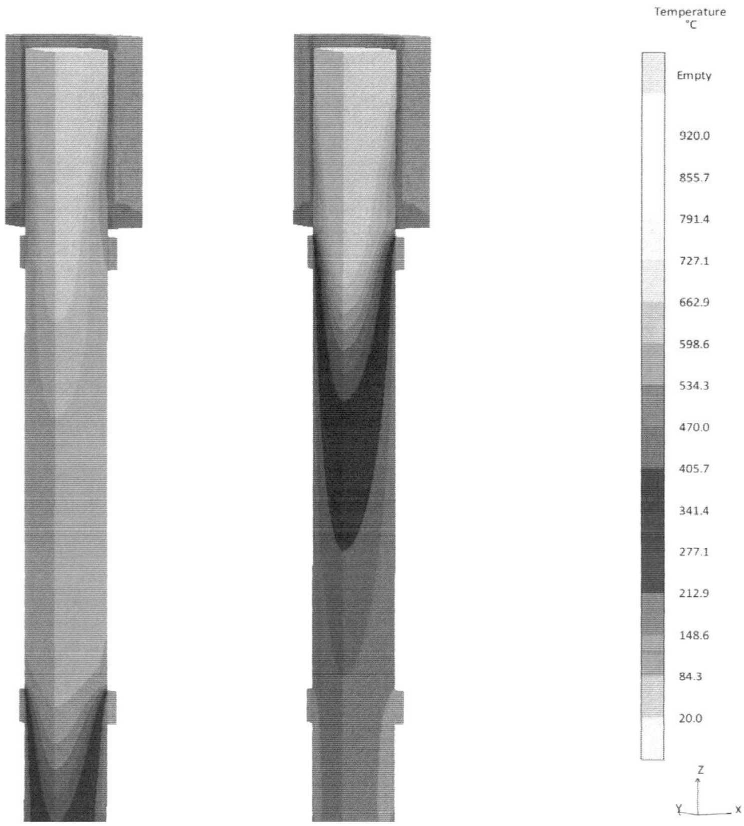

(a) (b)

Figure 6. On the left side (a) the temperature distribution is shown for a secondary cooling totally without water. This can be necessary especially for alloys with a high tendency to crack. The right picture (b) shows, for the same temperature range between RT and casting temperature, the resulting distribution if water spray cooling is used.

Another example for the use of simulation in continuous casting of copper is shown in figure 7. Here a rectangular shaped format is cast and in the centre a t-shape nozzle is used for pouring. With the obtained results, the sensitivity of a correct position of the nozzle in the mould should be demonstrated. In the left picture the nozzle is not positioned very carefully. This will directly lead to a different velocity field. Inclusions or other flow related defects are the consequence. The optimal design and position of nozzles are a typical application for such a virtual experimentation.

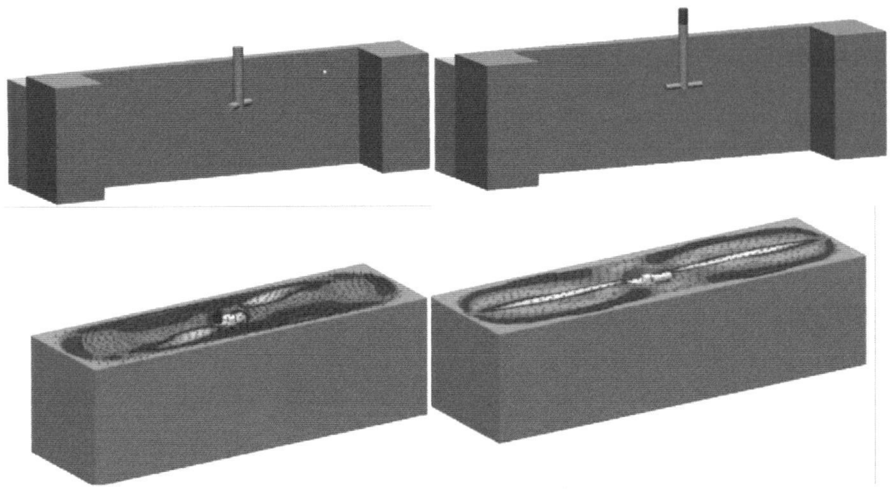

(a) (b)

Figure 7. The picture (b) is a process using a t-shape nozzle in a correct way. The nozzle was carefully positioned in the centre, which leads to a velocity distribution to the outsides of the mould. In the picture (a) the nozzle was shifted slightly, which causes turbulence and can be a reason for inclusions and other flow related defects.

4. Summary

This paper describes a modelling approach to simulate the flow and solidification in continuous casting processes. A strong focus is hereby the starting phase. Different possibilities are available. At the end it is always a question of accuracy in competition with computational time. The description of the melt free surface and also the convective flow during solidification has to be considered to get a reliable result.

Furthermore, examples for the application of the models for copper and aluminium continuous casting processes show the benefits for the user. This can at the end help to find robust process conditions to achieve a good product quality. Special thermal and mechanical quality criteria can be used for such virtual experiments.

The next development step for an improved modelling of heat transfer will be to consider an air gap formation between the strand and the mould, including reduced heat transfer due to thermal contraction. For that reason, a full thermomechanical coupling is necessary.

References

[1] Ferziger J and Peric M 2002 *Computational method for fluid dynamics* (Berlin: Springer-Verlag)

[2] Schäfer W 2002 *Formfüllungs- und Erstarrungssimulation von Gußteilen auf Rechnerarchitekturen mit verteiltem Speicher* (RWTH Aachen, Germany)

[3] Jasak H 1996 *Error analysis and estimation for the finite volume method with application to fluid flows* (PhD thesis, Imperial College, London, UK)

[4] Hirt C W and Nichols B D 1981 *Volume of fluid (VOF) method for the dynamics of free boundaries J. Comp. Phys.* **39(1)** 201-25

[5] Lipinski D M 1996 *Mold filling simulation for casting processes* (PhD thesis, RWTH Aachen, Germany)

[6] Hardin R A, Liu K, Kapoor A and Beckermann C 2003 A transient simulation and dynamic spray cooling control model for continuous steel casting *Metall. Mater. Trans. B* **34** 297-306

IOP Conf. Series: Materials Science and Engineering **861** (2020) 012018 doi:10.1088/1757-899X/861/1/012018

A meshless thermomechanical travelling-slice model of continuous casting of steel

B Mavrič[1,2], T Dobravec[1], R Vertnik[2,3] and B Šarler[1,2]

[1] Institute of Metals and Technology, Lepi pot 11, 1000 Ljubljana, Slovenia
[2] Faculty of Mechanical Engineering, University of Ljubljana, Aškerčeva cesta 6, 1000 Ljubljana, Slovenia
[3] Štore Steel, Železarska cesta 3, 3220 Štore, Slovenia

E-mail: bozidar.sarler@fs.uni-lj.si

Abstract. The travelling slice approximation of continuous casting process is widely used in industrial practice to model the heat transfer and solidification. Short computational times that are possible due to the reduction from three to two dimensions compensate for lower accuracy introduced by the approximation. In the same spirit, the travelling slice approximation is also used for the thermomechanical model in the present paper. The model includes contributions from metallostatic pressure, thermal contraction and viscoplastic deformation to describe the stress in the material slice. It incorporates simple models to predict the hot tearing and cracking. The model is solved with a novel strong-form meshless method. It allows flexible adaptive node distribution to accurately describe the behaviour in the solidifying region. In this paper we focus on the two-way coupling of the heat transfer and deformation of the strand to model the air-gap formation in the initial stage of the casting process and to study the emergent behaviour that is the result of this coupling. We find that the inclusion of the air-gap model significantly changes the stress distribution in the corners of the strand because of the reheating caused by the reduced heat-transfer.

1. Introduction

The travelling slice models of continuous casting are widely used in casting practice by both plant operators and caster designers to model the heat transfer during the casting process and to estimate the influence of casting parameters on the position of the solid-liquid interface [1,2]. The models are also used in optimization of the caster operation, based on productivity, quality, safety, maintenance and resources requirements and can serve also for caster on-line control [3]. The travelling slice model follows a slice of the strand through the caster. Such simplification is possible because of the slow changes in temperature field along the casting direction as compared to the variation in the orthogonal direction. This results in the model where the only mechanism for transfer of heat along the casting direction is the movement of the material slice. Although the travelling slice model introduces significant simplification by reducing the transport mechanisms to only diffusion/forces in perpendicular direction of the casting and to only advection in the direction of the casting. This provides much shorter computational times that offset the reduced accuracy of the physical description. The resulting models have thus achieved widespread use in industry.

The same idea is in the present paper applied also to the mechanical equilibrium in the shell. In the areas where most of the loading is introduced, either through thermal contraction of the material or through the action of the metallostatic pressure on the inside surface of the shell, the problem can be reduced to finding the stress equilibrium of a two-dimensional slice. Such conditions are encountered in the mould and several very informative papers have been produced on the topic [4–6], with special focus on the air-gap formation in the mould.

In contrast with the established way to solve the stress equilibrium problem with the finite element method, the present paper uses an original strong-form meshless method to discretize the governing equations. This method has been in the past used to solve the problems in thermoelasticity [7, 8] and plasticity [9] and has been applied to develop a model of hot-tearing during DC casting of aluminium [10] and a model of hot-rolling of steel [11]. In the considered problems it matches the performance of the finite element method and outperforms other meshless methods, such as the meshless local Petrov-Galerkin method [8]. The meshless nature of the method simplifies the adaptive node refinement that follows the position of the mushy zone by removing the need to remesh the computational domain. Similar strong-form meshless methods have been successfully applied to several other aspects of casting process and solidification. The thermal and the mechanical models are introduced, and the inter-model coupling is discussed first, followed by a short description of the numerical method. The paper is concluded by the presentation of the results and the discussion of the emergent behaviour of the solidified shell.

2. Model formulation

2.1. Heat transfer

The heat transfer model tackles the solidification problem in mixture enthalpy formulation. The relation between the enthalpy and the temperature, which guides the solidification, is obtained from JMatPro® [12] and is specific to the steel composition under consideration. The heat removal is described via specified heat flux at the surface. The calculation of the heat flux is performed with the resistor model of Li and Thomas [4], which relates the heat flux to the difference in temperature between the shell wall and the cooling water. Their relation considers the presence of the air-gap, the amount of radiative heat transfer and the conductivity of the mould. The heat transfer model also includes the influence of spray cooling and heat transfer to support rolls in the later stages of simulation. The details of the model can be found in [2].

2.2. Mechanical equilibrium

The mechanical model is stated in small-deformation approximation and considers additive decomposition of the strain into elastic, thermal and viscoplastic parts. The thermal part is determined by

$$\varepsilon^t = I\int_{T_{ref}}^{T} \alpha(T)d\mathrm{T} \ , \tag{1}$$

where $\alpha(T)$ is the temperature-dependent coefficient of thermal expansion obtained from [12]. The viscoplastic part is given by the Garafalo law

$$\dot{\varepsilon}^{vp} = A_0 \exp(-q/T)\left(\frac{\sigma_e}{\sigma_0}\right)^n \tag{2}$$

The parameters A_0, q, n and σ_0 are determined from the least squares fit of the high-temperature flow-stress data obtained from JMatPro.

The influence of the metallostatic pressure is introduced through an additional body force term given by $\mathbf{f}_b = p_m \nabla f_s$, where p_m is the metallostatic pressure at the centre of the slice and f_s is the liquid fraction. The resulting plane strain equilibrium equation for the displacement field $\mathbf{u}$ is

$$\begin{aligned} -p_m \nabla f_s = {} & G\nabla^2\mathbf{u} + (G+\lambda)\nabla\nabla\cdot\mathbf{u} + \nabla\lambda\nabla\cdot\mathbf{u} + \nabla G\left(\nabla\mathbf{u} + (\nabla\mathbf{u})^T\right), \\ & -\nabla\cdot\left(G\varepsilon^{vp}\right) - \nabla\cdot\left((3\lambda+2G)\varepsilon^t\right). \end{aligned} \tag{3}$$

where λ and G are the Lamé parameters.

The influence of the mushy zone is considered in several ways. The influence of metallostatic pressure is introduced by a two-phase analysis of stress in the mushy zone. The changing material properties in the mushy zone are introduced through the temperature dependence, obtained from [12]. All the terms are updated during the simulation with the information from the thermal model.

The boundary conditions for the solid mechanics model are either free-slip, on the parts of the boundary where the shell is determined to be in contact with the mould, or traction-free, on the parts where there exists an air gap between the surface of the shell and the position of the mould.

2.3. Coupling of the models

A two-way coupling is provided between the thermal and the mechanical models. The coupling of the mechanical model is performed explicitly through the temperature field, which determines the thermal deformations, and implicitly, through the fraction of solid, which determines the term describing the influence of the metallostatic pressure. The material parameters related to the elasticity of the mechanical model are temperature dependent.

The coupling of the thermal model with the solid mechanics model is performed through the boundary conditions for the heat transfer to the mould via the calculated air-gap heat flux.

3. The solution procedure

The two components of the model are discretized separately, using the same family of strong-form meshless methods, based on local interpolation with radial basis functions. The method is described in more detail in [2] for the thermal model and in [7] for the solid mechanics model. The basic idea of the method is to first introduce a node arrangement that describes the interior and the boundary of the computational domain. A subdomain consisting of neighbouring nodes of each node in the node arrangement is then selected to construct a local interpolation with radial basis functions, which is then used to calculate the derivatives that are present in the governing equations. Such approach is very flexible with respect to the node positions, dimensionality flexible and highly accurate [13].

In this paper, the interpolation with fifth-order polyharmonic splines, augmented with a second order polynomial, is used over local subdomains, consisting of 13 nodes.

The discretization in time is explicit for the thermal model and implicit for the solid mechanics model. The two models function in lock-step. First, the thermal field is updated and the updated thermal field is used to recalculate the material properties and the thermal loading of the mechanical model. Then the stress equilibrium is calculated with the updated data. After the deformations of the surface are known, they are used to update the heat transfer to the mould. The coupling procedure has no iterations within the time step to converge on either the air-gap size or the temperature field, since the time step $\Delta t = 2 \cdot 10^{-3}$ s is limited by the stability of the heat transfer, and is small enough to facilitate a frequent information transfer between the models.

The node arrangement, on which the solid mechanics calculations are performed, is non-uniform and is periodically updated to remain refined near the solidification front. The node arrangement at the beginning of the simulation and at the exit from the mould is shown in figures 1(a) and 1(b), respectively. The node spacing varies from 0.5 mm in the mushy zone to 1 mm in the solid shell and 5 mm in the liquid part of the computational domain. The node arrangement of the thermal model is regular square grid with spacing of 1.84 mm and remains unchanged throughout the simulation.

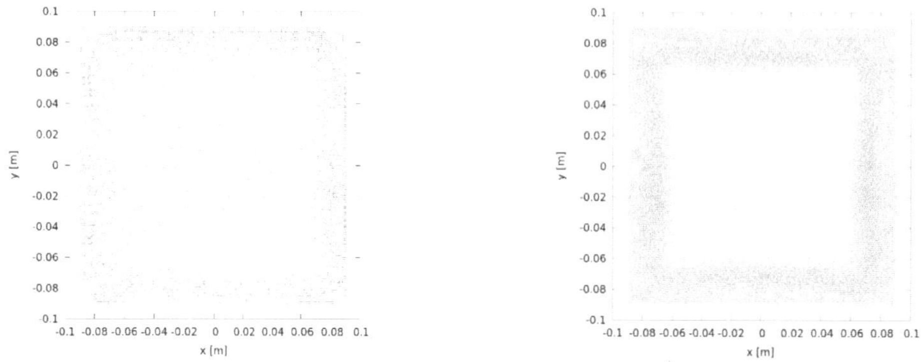

(a) (b)

Figure 1. Illustration of the node arrangement at (a) the start of the simulation, and (b) at the exit from the mould.

4. Results

The results presented in this paper are given for continuous casting of a square 184 mm billet of high-strength steel grade 51CrV4. The casting temperature is 1529 °C and the withdrawal rate is 1.44 m/min. The mould geometry is simplified to have a square shape with a taper rate of $3 \cdot 10^{-4}$, resulting in the dimensions of the mould, changing from 184.3 mm at the top (at vertical position 0 m) to 184 mm at the bottom (at vertical position -1 m). The presentation of the results is split into a set that discusses the influences on the surface of the billet and a set that discusses the cross-section of the billet.

The first set of the results, given in figures 2, 3 and 4, is used to compare the conditions on the surface for the simulations that consider the air-gap and the simulations without the air-gap. The variables of interest are plotted on unwrapped surface of the billet. A cut is performed along a corner and the billet surface is then unwrapped, displaying periodic structure of the plotted values caused by the very high symmetry being present in the solution. In figure 2, the calculated air-gap size is shown. There are two observations that can be made about the results. Firstly, we see that the gap, calculated without the coupling, is larger by approximately a factor of two. Secondly, the evolution of the gap is different. In the uncoupled case, the gap starts to form and reaches certain steady state value that remains essentially the same throughout the mould. The results of the coupled model show a contrasting behaviour. The air gap is formed and increases in the first 100 mm of the simulation and is then slightly reduced in the rest of the simulation.

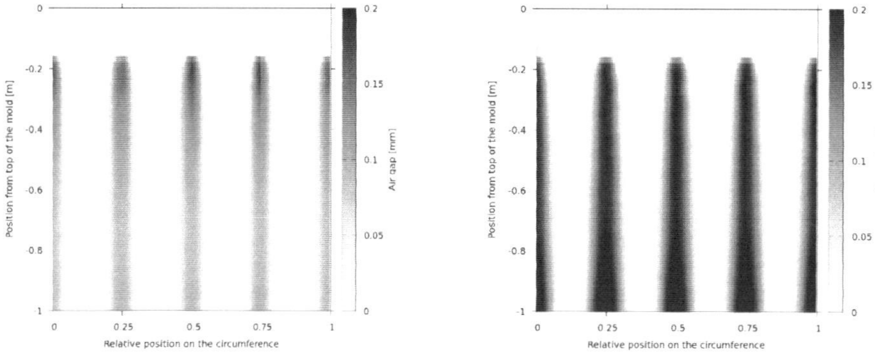

(a) (b)

Figure 2. The size of the air gap in the mould. (a) The results of the model for the case with air-gap dependent heat transfer model and (b) without the air-gap dependence.

The temperature field on the surface is shown in figure 3. Again, the effect of the inclusion of the coupling is evident. In the case without the coupling, illustrated in the right plot, we can see that the corners of the billet get significantly colder than the centre of the billet and remains so throughout the simulation. The case that includes the coupling, shows a more dynamic behaviour of the corner temperature. At the beginning, before the air gap is formed, the corners are undercooled, but they reheat, after the air-gap is formed. They remain warmer than the centre of the face at the exit of the mould.

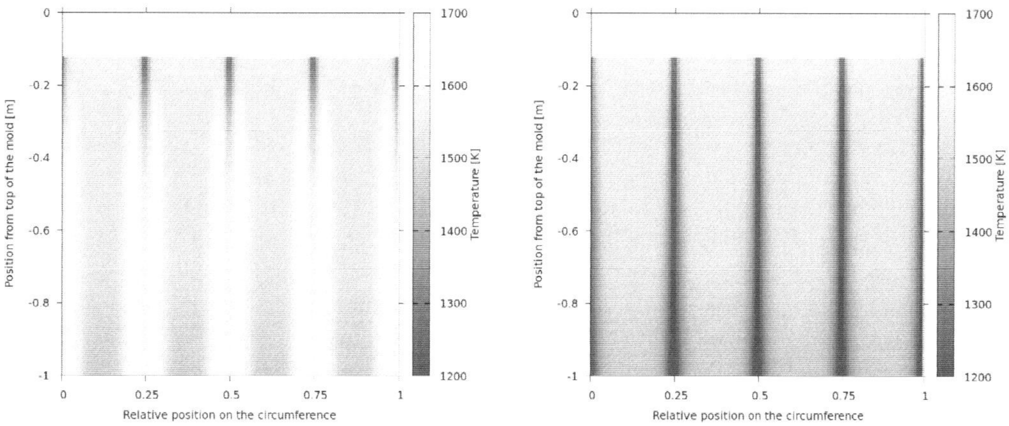

(a) (b)

Figure 3. Temperature at the surface. (a) The results of the model for the case with air-gap dependent heat transfer model and (b) results without air-gap dependence.

The plots in figure 4 show the heat flux on the surface of the billet. In the plot on the right, the heat transfer only depends on the surface temperature, since the effect of the air-gap is neglected. The plot on the left shows the behaviour of the heat flux that includes the effect of the air-gap. The heat flux is reduced by a factor of 10 in the corners when the air-gap is formed, which results in re-heating of the corners of the billet.

Overall, the results are qualitatively the same as the ones presented in relevant literature [4–6] and they match the observations and experience of the industrial partner. Further quantitative evaluations will follow in the future.

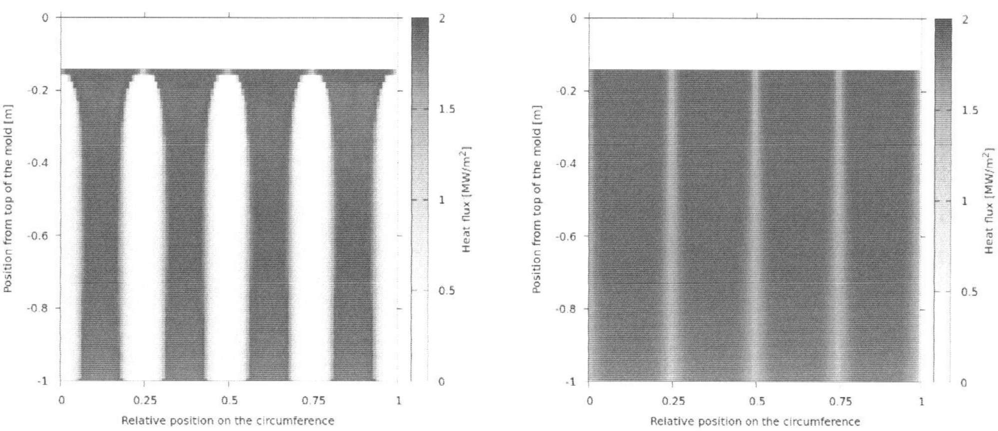

(a) (b)

Figure 4. Heat flux at the surface. (a) The results of the model for the case with the air-gap dependent heat transfer model and (b) results without air-gap dependence.

4.1. The interior of the billet

Figures 5 and 6 show the temperature and the equivalent stress over the slice, respectively. These plots are arranged in two rows; the top row displays the state of the coupled model and the bottom row the state of the uncoupled model. The plots in each column are given at specific position of the slice. The first column gives plots at z = -0.17 m, the second at z = -0.41 m, and the third at the exit from the mould at z = -1 m. In the plots, the shape of the slice is deformed by the displacement vector, scaled by a factor of 20, to give an idea about the shape of the gap at the specific positions in the mould.

The plots in figure 5 show the temperature field in the slice. We notice the fact that the re-heating of the corners in the case of coupled model causes slice to have a very gentle temperature profile in the corners, while the sharp temperature profile remains present at the parts of the surface that are in contact with the mould.

This change in the temperature field is reflected on the stress state that is illustrated by the equivalent stress in figure 5. It is noticed that when the air-gap forms in the mould, the increase in the temperature in the corners reduces the thermal strains in the corners, which are the most important contribution in the mechanical loading of the slice. This significantly reduces the equivalent stress that is present in the corners. Large equivalent stress remains present in the parts that remain in the contact with the mould.

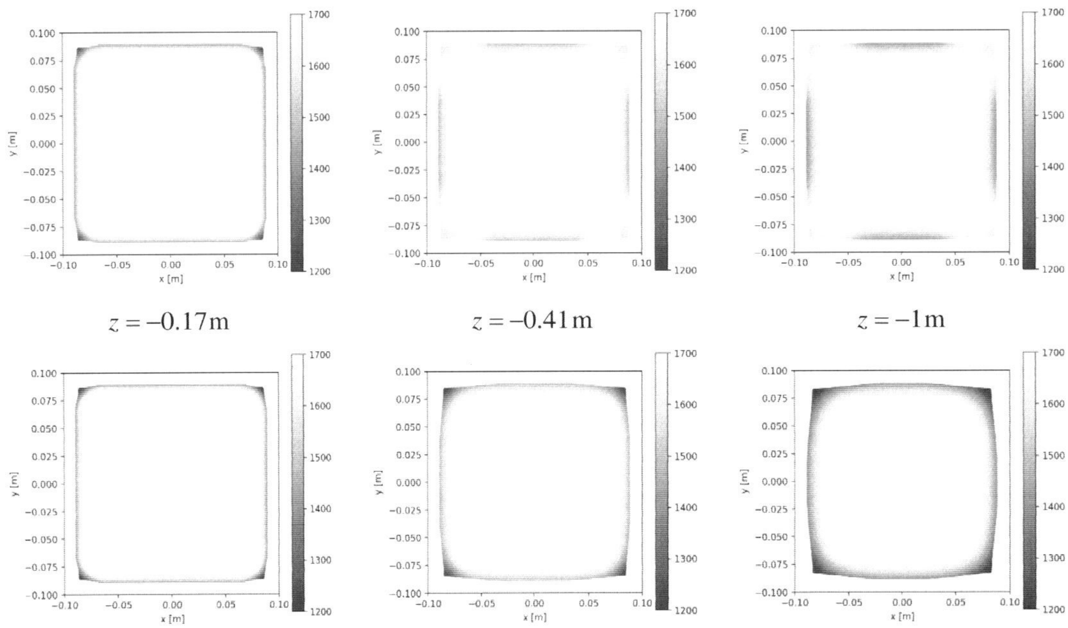

Figure 5. Temperature field (in degrees Kelvin) of the cross-section of the billet. The top row shows the results of the coupled and the bottom row the results of the uncoupled model.

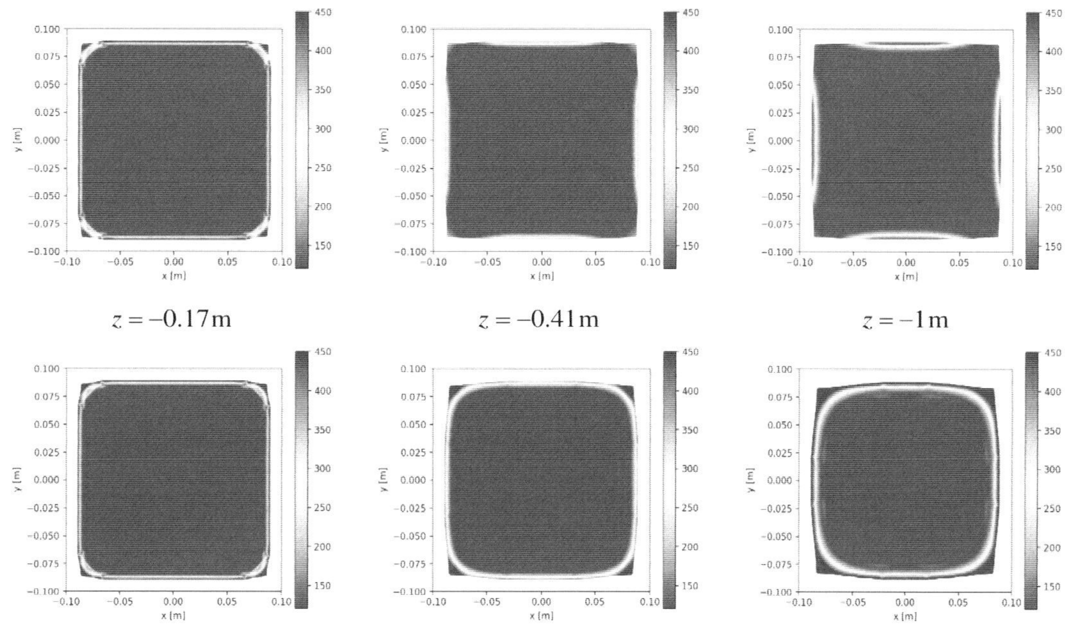

Figure 6. The equivalent stress in MPa. The top row shows the results of the coupled and the bottom row the results of the uncoupled model.

5. Conclusions

Implementation of a meshless mechanical travelling slice model for continuous casting of steel is presented. The model uses results of thermal travelling slice to calculate the thermal strains, which are the most important driving term for the mechanical model. The mechanical model is used to calculate the air gap between the mould and the solid shell. This information is used to influence the heat flux between the shell surface and the mould. A comparison of the results that incorporate this two-way coupling and the results that do not use it, shows that the coupling has a stabilising effect. The air gap of the coupled model is smaller than the air gap of the uncoupled model, moreover the equivalent stress in the corner is reduced while the temperature of the corners of the billet is increased. This illustrates that the coupling between the solid mechanics and the thermal conditions in the strand is very important and should be considered in the models. In the future, we will compare the predicted temperature of the shell at the exit of the mould with the measurements to validate the predicted behaviour and to fine-tune the material parameters, present in the resistor model of the heat transfer.

Acknowledgements

This work was supported by the Slovenian Research Agency (J2-1718 Advanced meshless modelling and simulation of multiphase systems; L2-9246 Multiphysics and multiscale modelling for competitive continuous casting; P2-0162 Transient two-phase flows).

References

[1] Vušanović I, Vertnik R and Šarler B 2012 A simple slice model for prediction of macrosegregation in continuously cast billets *IOP Conf. Ser.: Mater. Sci. Eng.* **27** 012056

[2] Šarler B, Vertnik R, Lorbiecka A Z, Vušanović I and Senčič B 2012 A multiscale slice model for continuous casting of steel *IOP Conf. Ser.: Mat. Sci. Eng.* **33** 012021

[3] Šarler B, Vertnik R, Šaletić S, Manojlović G and Cesar J 2005 Application of continuous casting simulation at Štore Steel *Berg Huettenmaenn. Monatsh.* **150** 300–6

[4] Li C and Thomas B G 2004 Thermomechanical finite-element model of shell behavior in continuous casting of steel *Metall. Mater. Trans. B* **35** 1151–72
[5] Koric S, Hibbeler L C, Liu R and Thomas B G 2010 Multiphysics model of metal solidification on the continuum level *Numer. Heat Tr. B-Fund.* **58** 371–92
[6] Cai Z Z and Zhu M Y 2014 Simulation of air gap formation in slab continuous casting mould *Ironmak. Steelmak.* **41** 435–46
[7] Mavrič B and Šarler B 2014 A collocation meshless method for linear thermoelasticity in 2D. In *Proc. of ThermaComp2014 Third International Conference on Computational Methods for Thermal Problems* (Bled) pp 279–282
[8] Mavrič B and Šarler B 2017 Application of the RBF collocation method to transient coupled thermoelasticity *Int. J. Numer. Meth. H.* **27** 1064–77
[9] Hanoglu U and Šarler B 2018 Multi-pass hot-rolling simulation using a meshless method *Comput. Struct.* **194** 1–14
[10] Mavrič B and Šarler B 2015 A RBF-based local collocation method for modelling thermomechanical phenomena during DC casting of aluminium billets. *Coupled problems 2015 : proceedings of the VI International Conference on Coupled Problems in Science and Engineering* Coupled problems 2015 (Venice: International Center for Numerical Methods in Engineering (CIMNE)) pp 160–8
[11] Hanoglu U and Šarler B 2019 Hot Rolling Simulation System for Steel Based on Advanced Meshless Solution *Metals* **9** 788
[12] Saunders N, Guo U K Z, Li X, Miodownik A P and Schillé J-P 2003 Using JMatPro to model materials properties and behavior *JOM* **55** 60–5
[13] Bayona V, Flyer N, Fornberg B and Barnett G A 2017 On the role of polynomials in RBF-FD approximations: II. Numerical solution of elliptic PDEs *J. Comput. Phys.* **332** 257–73

IOP Conf. Series: Materials Science and Engineering **861** (2020) 012019 doi:10.1088/1757-899X/861/1/012019

Performance of turbulence models for flow prediction in a mould of continuous steel caster

J Gregorc[1], A Kunavar[1] and B Šarler[1,2]

[1] Laboratory for Fluid Dynamics and Thermodynamics, Faculty of Mechanical Engineering, Aškerčeva 6, SI 1000 Ljubljana, Slovenia
[2] Laboratory for Simulation of Materials and Processes, Institute of Metals and Technology, Lepi pot 11, SI 1000 Ljubljana, Slovenia

E-mail: bozidar.sarler@fs.uni-lj.si

Abstract. The flow of molten steel in a mould during continuous casting process is nowadays most often studied using numerical approach. With such an approach, appropriate numerical tools have to be selected to achieve physically sound results. Within this work we focus on performance of different turbulence models for the case of casting of steel billets with single SEN port oriented in vertical, downward direction. Only fluid dynamics aspect of the casting process is considered, while all other phenomena are disregarded at this stage. Simulations were performed with commercial CFD package ANSYS Fluent with k-ε, k-ω and SAS turbulent models. The geometric model, material properties and process conditions are closely matched to those from our experimental water model studies, which is typically not done. Hence, the difference between the experiments and the numerical simulation originates only due to turbulence model performance. The experimental PIV data from the water model experimental system is used to benchmark the performance of various turbulence models. The models are compared based on lateral dispersion of the liquid jet exiting submerged entry nozzle (SEN) and formation of recirculation zones around the SEN. The results show that while all three models are capable of capturing general flow features, only the SAS model predicts finer turbulent structures and reasonably captures the flow behaviour in temporal domain.

1. Introduction

Steel has been a source of survival and development throughout the human history and continues to be one of the most important materials at our disposal. Steel production continuous to increase even in the first decade of the 21st century. Steel can be recycled and reused in new products, making it one of the earliest sustainable materials in modern economy. With time the steel production underwent huge improvements in sense of productivity, product quality and lately also energy consumption as well as other elements of sustainability [1].

The advances in computer capabilities and understanding of physics in the recent decades gave unique ability to study, predict and control processes happening in environment and/or conditions where measurements and trial & error type of approach is nor possible nor safe. As a consequence, the numerical simulations of continuous casting of steel casting are extremely important [2–6].

The meshless local radial basis function collocation method (LRBFMC) for solution of incompressible turbulent flow, based on k-epsilon model, was applied for numerical solution of solidification in continuous casting of steel in [5]. The method proved to be efficient and was upgraded also to account for electromagnetic stirring [7, 8].

In the past, RANS approach to turbulence modelling was typically used [9, 10]. With increased computational power, however the so called "scale resolved simulations", like LES, are becoming more and more affordable [11, 12], however, the computational effort for such a simulation is still huge. The aim of our current work is to assess the capabilities of two RANS turbulence models and a SAS (scale adaptive simulation) model that belongs to so called "RANS-LES hybrid" class of models. It is still capable to resolve at least some of the scales while requiring less computational effort. ANSYS Fluent code, without any alterations to turbulence models, was used to predict the flow structures in the geometric model of water experiment for continuous casting. Majority of the published papers, both experimental and numerical, focus on slab casting using SENs with multiple side ports of different shapes and sizes [6, 9, 10, 13]. In our case, however, casting of steel billets using SEN with single vertical downward port is considered. The results are compared with the measurements conducted on experimental rig without any geometric scaling and by using the same material properties for both, numerical simulation and experimental investigation.

Based on the assessment of the fit of the simulations with our own experimental data an informed decision is made in the present paper on which turbulence model to implement into own LRBFMC code in order to improve its prediction capabilities.

2. Numerical setup and procedure

The goal of this numerical study is to assess the performance of three different turbulence models; the first belonging to the k-ε family, the second to the k-ω, and the third is a SAS model.

2.1. Governing equations

Flow in the computational domain is assumed to be turbulent, incompressible, two-phase (gas and liquid with a free surface in-between), isothermal and unsteady. For such a case mass conservation reduces to the conservation of volume fraction (given by equation (1)) and momentum conservation reduces to the equation (2).

$$\frac{\partial \alpha_L}{\partial t} + \boldsymbol{\nabla} \cdot (\alpha_L \mathbf{V}) = 0$$
$$\alpha_L + \alpha_G = 1 \tag{1}$$

$$\frac{\partial}{\partial t}(\rho \mathbf{V}) + \boldsymbol{\nabla} \cdot (\rho \mathbf{V}\mathbf{V}) = -\boldsymbol{\nabla} p + \boldsymbol{\nabla} \cdot \bar{\bar{\boldsymbol{\tau}}} + \rho \tag{2}$$

The free surface at the top of the domain is described by VOF model using an algebraic approach for interface tracking. The surface tension effects were not considered. Material properties, e.g. density, viscosity for two phase system were computed for each cell based on the volume fraction by using equation (3).

$$\rho = \alpha_G \rho_G + (1 - \alpha_G)\rho_L$$
$$\mu = \alpha_G \mu_G + (1 - \alpha_G)\mu_L \tag{3}$$

2.1.1. Turbulence modelling. Three different RANS/URANS turbulence models were considered in this study: The Model 1 (M1): Realizable k- ε model [14] with Enhanced wall functions (Rk-ε EWT), the Model 2 (M2): SST (Shear Stress Transport) k-ω model [15] and the Model 3 (M3): SAS model [16]. All belong to two "so-called" equation classes of turbulence models where two additional transport equations are solved, for turbulence kinetic energy (k) and its dissipation rate (ε) / specific dissipation rate (ω). The general form of transport equation for k (equation (4)) is similar for all three models. G_k and G_b are production terms for k due to mean velocity gradients and buoyancy, respectively and Y_k is the destruction term for k. The G_b term is equal to 0 in case of M2 and M3 models.

$$\frac{\partial}{\partial \mathrm{t}}(\rho k) + \frac{\partial}{\partial \mathrm{x_j}}(\rho k u_j) = \frac{\partial}{\partial \mathrm{x_j}}\left(\left(\mu + \frac{\mu_t}{\sigma_k}\right)\frac{\partial \mathrm{k}}{\partial \mathrm{x_j}}\right) + G_k + G_b - Y_k \tag{4}$$

Transport equations for turbulent kinetic energy dissipation rate ε (equation (5)), in case of M1 model and turbulent kinetic energy specific dissipation rate ω in case of M2 and M3 models, are presented in equations (5), (6) and (7). The difference between these equations are that $G_{b,\varepsilon}$ term (dissipation production due to buoyancy) is only present in ε equation and D_ω (cross-diffusion modification) term is only present in both ω based models to allow for the blending of the standard k-ω and the standard k-ε approach.

$$\frac{\partial}{\partial \mathrm{t}}(\rho\epsilon) + \frac{\partial}{\partial \mathrm{x_j}}(\rho\epsilon u_j) = \frac{\partial}{\partial \mathrm{x_j}}\left(\left(\mu + \frac{\mu_t}{\sigma_\epsilon}\right)\frac{\partial \epsilon}{\partial \mathrm{x_j}}\right) + G_\epsilon + G_{b,\epsilon} - Y_\epsilon \tag{5}$$

$$\frac{\partial}{\partial \mathrm{t}}(\rho\omega) + \frac{\partial}{\partial \mathrm{x_j}}(\rho\omega u_j) = \frac{\partial}{\partial \mathrm{x_j}}\left(\left(\mu + \frac{\mu_t}{\sigma_\omega}\right)\frac{\partial \omega}{\partial \mathrm{x_j}}\right) + G_\omega - Y_\omega + D_\omega \tag{6}$$

$$\frac{\partial}{\partial \mathrm{t}}(\rho\omega) + \frac{\partial}{\partial \mathrm{x_j}}(\rho\omega u_j) = \frac{\partial}{\partial \mathrm{x_j}}\left(\left(\mu + \frac{\mu_t}{\sigma_\omega}\right)\frac{\partial \omega}{\partial \mathrm{x_j}}\right) + G_\omega - Y_\omega + D_\omega + Q_{SAS} \tag{7}$$

The equations for M2 and M3 models are essentially the same, except for one term, named Q_{SAS}. This term introduces von Karman length scale into the system of equations; allowing the model to adjust its turbulence length scale to the already resolved scales. This effectively means that the model is capable to resolve the additional turbulence structures in "LES-like" manner, down to the grid spacing, making this model a kind of a hybrid between the RANS and the LES models. For a detailed information on the source term definitions and the implementation of the models used, is the reader referred to [14–17].

2.2. Discretization

The geometry model is a 1:1 copy of the industrial caster used by our industrial partner (figure 1). At the same time, it is identical to the geometry of experimental test section of the water model. To reduce the total number of the computational cells, the upper part (100 mm) of the mould above the free surface, was cut, while still maintaining enough space to avoid numerical artefacts due to the moving interface boundary between the gas and the liquid.

Block structured, hexahedral approach was used to build a computational mesh on geometric model. O-grid type meshing was adopted to ensure high quality cells between the cylindrical SEN and the hexagonal mould (see figure 2). Special attention was devoted to area at the bottom of the SEN (as can be seen from figure 2(b) and figure 2(c)) where the inlet boundary is created, assuring smooth transition of the mesh in all three directions. The mesh details on the top and the bottom of the domain are given in figure 2(a) and figure 2(d), respectively. All together 43 blocks were needed to construct the mesh with 2,287,800 hexahedral cells. The mesh was built in accordance with the guidelines for RANS turbulence modelling without any wall functions, ensuring $y^+\sim 1$ on the walls.

The mesh independence was performed on three different meshes with constant refinement factor in accordance with the best CFD practice. A simplified Richardson's extrapolation reveals that the relative discretization error for the mesh with 2,287,800 cells is below 5%. Therefore, this mesh is considered in all the calculations that follow. Further refinement would also increase already extensive computational efforts. The simulations are run on 24 core server (AMD Opteron8439SE 2.8 GHz) with computation time ranging from 3-5 days.

2.3. Boundary conditions and solution procedure

There were three different boundary patches defined in the discrete model: inlet, at the bottom of SEN, outlet at the top of the domain and outlet at the bottom of the domain. The inlet was defined as velocity inlet, where velocity was prescribed as vector components in form of a profile obtained by separate simulation of fluid in the SEN. Pressure outlet was used for both, the domain top and the bottom. At the top, gauge pressure was set to 0 Pa, on the bottom the value of gauge pressure was such that the free surface was kept at the correct position. The materials were water and air. Their properties are listed in table 1.

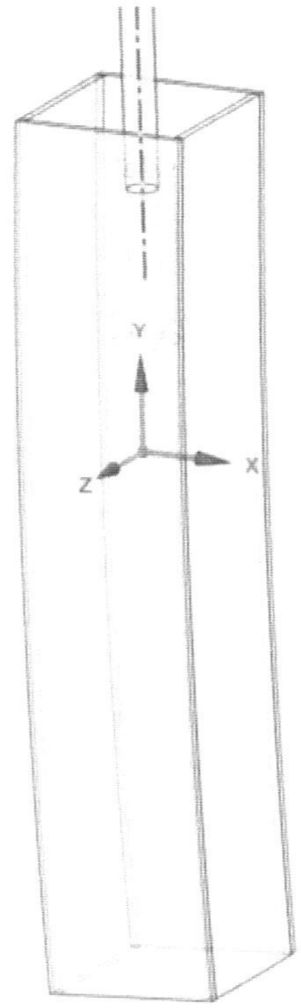

Figure 1. The basis for the geometric model.

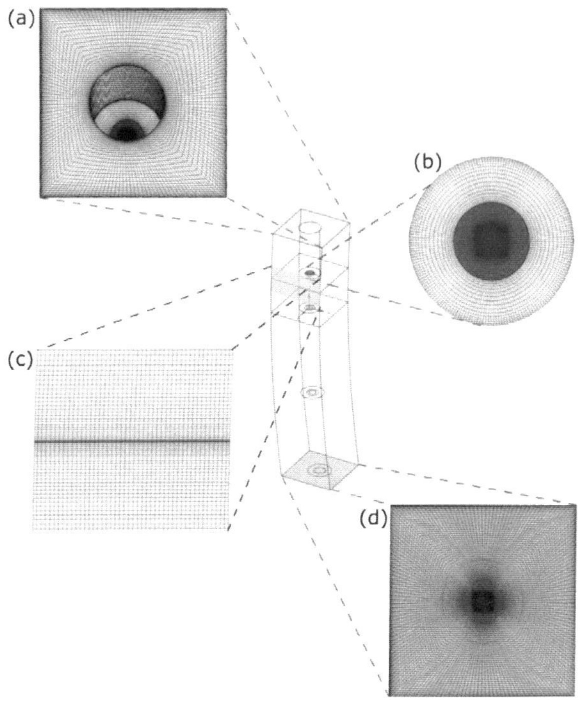

Figure 2. Overview of the mesh on the critical locations.

The temporal discretization is done using the second order implicit formulation. SIMPLE algorithm is used for pressure-velocity coupling. Spatial discretization is based on using Least Squares Cell Based, PRESTO! and compressive scheme for gradients, pressure and volume fraction, respectively. In case of momentum equation and turbulence quantities, second order upwind scheme is used.

Equations are solved by using ANSYS Fluent 18.2 which incorporates the finite volume framework. The time steps are between 10^{-3} and 5×10^{-3} s. Minimal simulated time is determined by expression $t_{min} = 3L_{domain,Y}\,(V_{inlet\text{-}mean})^{-1}$, where $L_{domain,Y}$ is the domain length in vertical direction and $V_{inlet\text{-}mean}$ is the mean velocity at the inlet. The convergence is checked at every time step by using the absolute convergence criteria with the values of globally scaled residuals set to 10^{-4}.

3. Experimental setup and methods

Without any doubt, the experimental data on flow behaviour in the mould are of key importance as they provide an insight into flow behaviour (essential in the design phase of numerical models) and can be used to validate the developed numerical tools.

Experimental test section considered for this work is based on industrial caster for steel billets used by Štore Steel company in Slovenia. The test section is made of two parts; the mould and SEN. The water model of the mould is made from 8 mm thick acrylic. It is slightly curved in vertical direction and has a square cross section with the side 190 mm (figure 3(a)). The SEN has inner and outer diameter of 35 mm and 65mm, respectively (figure 3(b)). The liquid exits in vertical direction.

Experimental rig is schematically shown in figure 4. The system operates as follows: water exits from tank (3) into the pump (4), that is set to the desired flowrate controlled with flowmeter (5). At the top, water enters the SEN and then the mould. Two valves are placed in series at the exit from the mould. The ball valve (7) is used during starting, stopping and filling procedures, membrane valve (8), on the other hand, is used for precise positioning of the moving surface at the top of the mould.

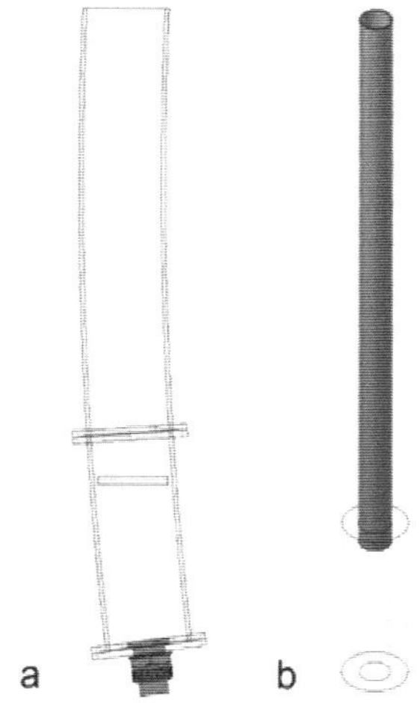

Figure 3. Experimental test section: 3D model of (a) the mould and (b) SEN.

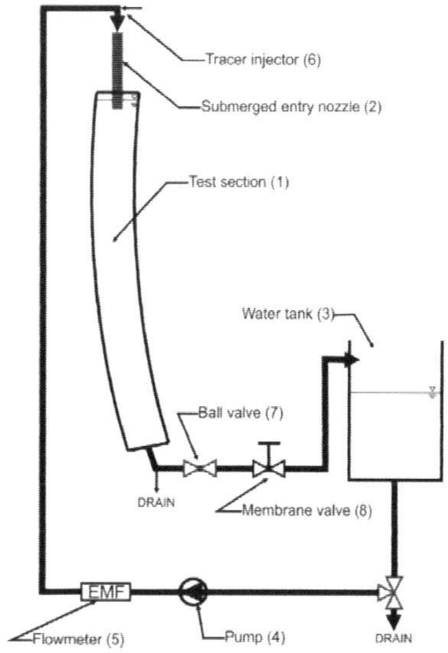

Figure 4. A schematic representation of the test rig.

3.1. Experimental conditions and process similarity

Experiments are performed at isothermal ambient conditions; hence, all thermal effects are disregarded. Care was taken when performing study of flow similarity between the industrial conditions using molten steel and laboratory tests, conducted with water to ensure that results would be relevant. Investigation was performed at industrially relevant casting speed 1.4 m/min. Based on true (industry) casting speed and mould cross section, the volume flow rate of the material passing through the mould was estimated. It is worth noting that the material shrinkage due to solidification cannot be considered. With known geometry of the mould, volume flow rate and material properties (see table 1) at typical casting temperatures, Reynolds number for molten steel was estimated. Using this value and material properties for water (see table 1) at ambient conditions, the necessary flowrate of water was calculated. The mean velocity at the end of SEN was estimated to be 1.02 m/s.

Table 1. Material properties for similarity analysis.

Material	Density (kgm^{-3})	Dynamic viscosity ($kgm^{-1}s^{-1}$)
Water (liquid)	$0.9982\times10^{+3}$	1.003×10^{-3}
Air (gas)	$1.2250\times10^{+3}$	1.820×10^{-5}
Steel (liquid)	$6.8839\times10^{+3}$	5.318×10^{-3}

3.2. PIV measurements

Particle Image Velocimetry (PIV) measurements are performed by using 2D PIV system from Dantec Dynamics. For seeding 10 μm rhodamine particles (encapsulated rhodamine) is used. They are illuminated by dual pulse laser capable of delivering 65 mJ per pulse at wavelength 532 nm. The Δt between the pulses is set to 1.7 ms. The particle reflections are recorded with 4 Mpix high sensitivity, double exposure camera. The system operates at 15 Hz. The duration of the data acquisition is 10 s, providing a total of 150 velocity fields per measurement. The vector fields and data post processing are structured in DynamicStudio software.

4. Results

The first task is to estimate the discrepancy between the numerical prediction and the measurements. This is done by comparing the contours of the velocity magnitude in *x-y* (figure 5) and *y-z* (figure 6) planes, both passing through SEN axis. The results presented are instantaneous, taken from 214th second of simulations and one randomly selected time in measured time series. As seen in both figures, it is clear that all three models are capable of capturing basic shape of the main jet exiting the SEN. Velocity field is more or less steady in case of M1 and M2 models. The shapes of the velocity magnitude contours are symmetrical to SEN axis. For M3, on the other hand, the flow doesn't really become steady, but rather resembles true "turbulence like" behaviour that seems to be far more realistic and closer to what was experimentally observed.

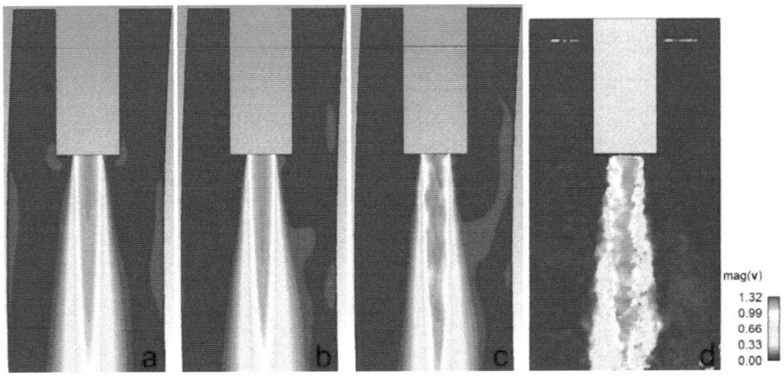

Figure 5. Velocity magnitude contours in x-y plane at $z = 0$ (middle of the SEN) for M1 (a), M2 (b), M3 (c) and experiment (d).

There are certain regions in experimental contour plots, that show high local velocity that seems to be unphysical/unexpected; e.g. region around the interface or bottom corners. These artefacts are products of light scattering from the free surface at the top or result of poor laser light intensity at the limits of the observed area (view window).

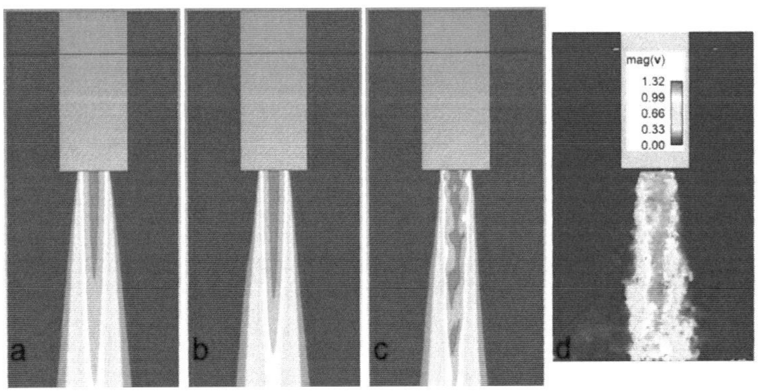

Figure 6. Velocity magnitude contours in y-z plane at $x = 0$ (middle of the SEN) for (a) M1, (b) M2, (c) M3 and (d) experiment.

The plots of y velocity components over the mould in x direction at $z = 0$ were made to further investigate the match between the numerical and the experimental results. The profiles presented in figure 7 are taken from two vertical positions: y =- 50 mm (50 mm below SEN bottom, part a) and y = -150 mm (150 mm below SEN, part b) using the same temporal settings as in figure 5 and figure 6. The overall agreement between the results is acceptable, considering the fact that the experimental data and M3 data is changing from time step to time step, hence it is practically impossible to align the results in the temporal domain. On both presented locations are the maximal velocity values for all three models in range with the experimental data. The same can be concluded for widening of the jet cone. In the region of high velocity, the experimental results show significant velocity fluctuations. It is clear, that from the three turbulence models being tested, only M3 is capable of predicting at least some degree of these fluctuations.

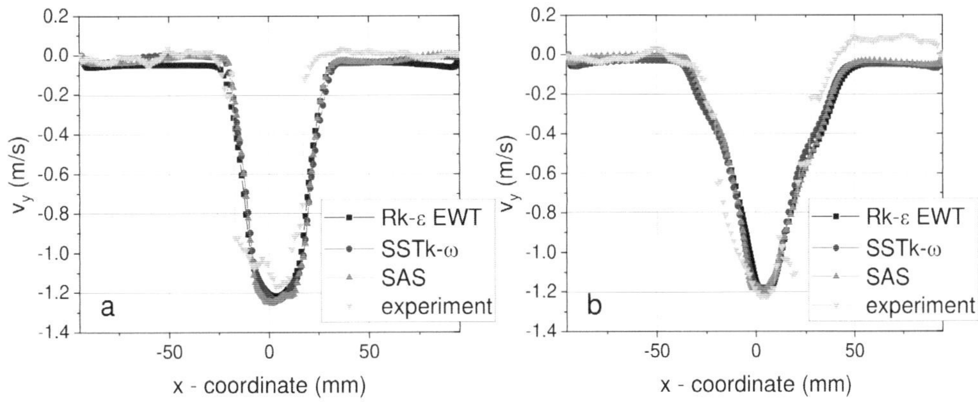

Figure 7. Profiles of v_y in x direction at $z = 0$ and (a) $y = -50$ mm and (b) $y = -150$ mm.

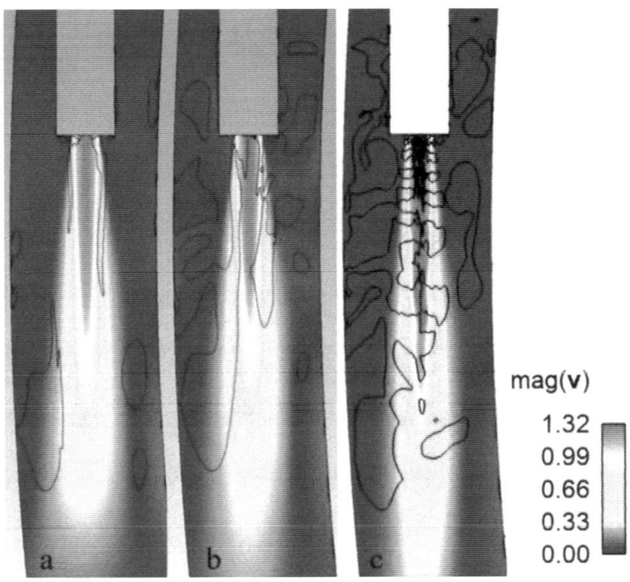

Figure 8. Turbulent structures represented as iso-values of $\lambda_2 = -0.15\ s^{-2}$ along with the velocity magnitude contours on x-y plane at $z = 0$ m for (a) M1, (b) M2 and (c) M3.

The results presented here conclusively show that M3 model can describe jet unsteadiness and its behaviour better than the rest two models used in this study. The fact that the general characteristics of the jet, like maximum velocity or velocity profile in the jet are predicted reasonably well even with simpler models, makes it difficult to "pick a winner". This is especially true if one also considers computational time and resources needed to perform M3 simulation compared to RANS. However, there are scenarios where RANS approach simply isn't enough. The study of the inclusions path (to estimate where the inclusions would accumulate) for example, would require to resolve at least some part of the turbulence structures (at least a few different length scales) as they would most certainly affect where inclusion would go. M3 as a hybrid RANS-LES model, capable of doing that, as can be seen from figure 8, where turbulent structures are represented iso-values of λ_2 parameter.

Within this study, the flow in the mould was assumed to be pure hydrodynamic problem. In case of flow in the caster at industrial conditions one also has to consider other physical phenomena of solidification that are coupled with the turbulence. The assessment of the performance of these models, when coupled also with the energy equation, EM field, and possibly solidification would be beneficial for proper prediction of the transport phenomena in continuous casting of steel.

5. Conclusions

Within this paper a performance of three different turbulence models is tested for the case of water flow in a mould of continuous steel caster for billets. With similarity analysis boundary values were estimated to assure hydrodynamic similarity between the flow in the industrial caster and the water model. The simulations were performed by using CFD package ANSYS Fluent 18.2. Results with M1, M2, and M3 models are presented and compared to PIV experimental data from the water model. The key findings are:

- All three models are capable of predicting general flow features in a reasonably good agreement with the experimental data.
- With M1 and M2 models the simulations reach steady state (with minor fluctuations) even if simulations are running in a transient manner.
- With M1 and M2 models is the shape of the velocity contours of the primary jet symmetrical to SEN axis.
- In case of M3 model, the steady state solution is not found, the flow around SEN axis is not symmetrical and the velocity fluctuations within the primary jet never fade out, which agrees with experimental observations
- Model M3 is computationally the most expensive of the three, followed by M2 and M1 that are quite comparable.
- For anticipated further study of the interaction between the inclusions and the fluid flow is the M3 model to pick from the three.
- Future work needs to be done to test the performance of this models also when turbulence is coupled with other physical phenomena.

Acknowledgments

The authors gratefully acknowledge support by Slovenian Research Agency (grants P2-0162 and L2-9246), and Štore-Steel company (www.store-steel.si).

References

[1] Honeycombe R and Bhadeshia H K D H 1995 *Steels : microstructure and properties* (London: Edward Arnold)
[2] Šarler B, Vertnik R and Maček M 2017 Simulation of continuous casting of steel with EM stirring *9th ECCC European continuous casting conference - ECCC 2017* (Vienna) pp 903–12
[3] Šarler B, Vertnik R, Lorbiecka A Z, Vušanović I and Senčič B 2014 *BHM Berg- und Hüttenmännische Monatshefte* **159** 99–107
[4] Šarler B, Vertnik R, Lorbiecka A Z, Vušanović I and Senčič B 2012 *IOP Conf. Ser.: Mat. Sci. and Eng.* **33** 012021
[5] Vertnik R and Šarler B 2009 *Int. J. Cast Met. Res.* **22** 311–15
[6] Thomas B G 2018 *Res. Int.* **89** 1700312
[7] Vertnik R, Mramor K and Šarler B 2019 *Eng. Anal. Bound. Elem.* **104** 347–63
[8] Košnik N, Vertnik R and Šarler B 2014 *Mater. Sci. Forum* **790–791** 390–5
[9] Hershey D E, Thomas B G and Najjar F M 1993 *Int. J. Numer. Methods Fluids* **17** 23–47
[10] Thomas B G and Najjar F M 1991 *Appl. Math. Model.* **15** 226–43
[11] Yuan Q, Thomas B G and Vanka S P 2003 *ISSTech 2003 (Conf. Proc.)* **1** 141–156
[12] Liu Z Q, Qi F S, Li B K and Jiang M F 2011 *Adv. Mater. Res.* **402** 432–435
[13] Cho S-M, Thomas B and Kim S-H 2018 *Metall. Mater. Trans. B* **50** 52–76
[14] Shih T-H, Liou W W, Shabbir A, Yang Z and Zhu J 1995 *Comput. Fluids* **24** 227–38
[15] Menter F R 1994 *AIAA J.* **32** 1598–605
[16] Menter F R and Egorov Y 2010 *Flow, Turbul. Combust.* **85** 113–38
[17] ANSYS 2015 *User Guide for ANSYS 18.2* (Cotonsburg, PA: ANSYS Inc.)

Modelling of localised shrinkage pressure in a directionally solidifying domain

B Nenchev[1] and S Gill[1]

[1] Department of Engineering, University of Leicester, Leicester, LE1 7RH, UK

E-mail: spg3@leicester.ac.uk

Abstract. A novel approach to modelling shrinkage during solidification has been proposed. It combines Phase-field (PF) simulation with fluid mechanics for a solidifying medium. A mass balance method, based on a viscous creeping flow in two phases (liquid, solid and interface) with an applied penalisation coefficient is proposed. The role of this coefficient is to continually separate the fluid from the solid domain in correspondence to any change due to a phase transformation process. The advantage of the method is that the medium is modelled with no explicit tracking of the interface which significantly improves computational cost and the ease of implementation. This is achieved via the application of a monolithic 'single-field' formulation to the system of equations. The method has the potential to locally calculate shrinkage pressure by taking into account multiple parameters, including thermal gradient, permeability, viscosity and morphology.

1. Introduction

During solidification, a metal melt undergoes a transition from liquid to solid under the influence of temperature and/or composition gradient. This transition generally occurs by the nucleation and growth of dendrites. Heat flow and crystallographic anisotropy dictate the most energetically favourable directions of dendritic growth and are the primary cause for the formation of complex dendritic networks. Deep within those networks, between the solid and liquid phases, an interface exists in which a drastic variation in behaviour and thermo-physical parameters such as density, viscosity and motion are present. In particular, the density variation across the interface causes shrinkage to occur due to mass conservation between both phases. This effect can generate situations where liquid pressure is substantial enough to cause deformation of the already solidified melt.

The deformation of the dendritic networks strongly depends on the coherency state and the inflow and outflow of liquid. The two most common defects formed due to shrinkage during solidification, porosity and hot cracking, are both contributed to the hydrostatic depression of the semi-solid (mushy) zones [1]. This depression is caused by the insufficient liquid feeding in the shrinking mushy zones. Due to the complexity of the mechanisms involved in porosity and hot-tearing formation, most of the developed models are semi-empirical or based on the solidification interval, most popular of which are Niyama criterion [2, 3] and RDG criterion [4]. In those models, solidification is generally approximated via Scheil equation and mass flow is taken into consideration via Darcy's law with an empirical determination of the permeability term [5, 6]. However, permeability, which is determined from the tortuosity of the dendritic networks and the solidification interval, is very sensitive to the solidification conditions. These semi-empirical models are also highly sensitive to the coherence value of the array that can be in the range from 0.9 to 0.99 volume fraction of solid [7]. There is a large ongoing research on multiphysics solidification models but most of them mainly focus on natural and forced convection flows during solidification [8, 9], with shrinkage effects ignored. The current understanding of the effects of fluid flow and pressure deep within the mushy zone is still not completely understood, especially when predictive methods are attempted to be developed.

The aim of the current work is to demonstrate a method for fluid-pressure calculation specially aimed for domains undergoing phase transition. The formulation coupled with a micro-solidification model such as Phase-field (PF), allows for local calculation of flow and pressure within the dendritic arrays. First, a simple PF model capable of dendritic solidification is built and then, a methodology for fluid dynamics is described. Shrinkage is implemented within a growing array via coupling the PF field variable and the Navier-Stokes equation for local mass conservation.

2. Framework

2.1. Phase-field

In phase-field simulations, the sharp interface is replaced by a diffuse interface with a finite thickness, where the phase-field variable φ changes smoothly from $\varphi = 0$ in liquid to $\varphi = 0$ in solid. Such an interface enabled us to simulate the interface migration without tracking the interface. The significance of this method for modelling dendritic growth was brought to light by Kobayashi [10] and it still remains one of the most powerful and versatile methods for simulating growing structures. Kobayashi's model dealt with a very thin steep interface layer in a dimensionless relationship between the field variable and the physical parameters, latent heat and anisotropy.

The computations presented in this work are based on his model in order to model two impinging solidifying fronts. Kobayashi's formulation offers the complexity of dendritic arrays but short computational times. Pure material solidification with incorporated interface anisotropy is simulated by the following evolution equation:

$$\tau\frac{\partial\phi}{\partial t} = \nabla.(\varepsilon^2\,\nabla\phi) - \frac{\partial}{\partial x}\left(\varepsilon\varepsilon'\frac{\partial\phi}{\partial y}\right) + \frac{\partial}{\partial y}\left(\varepsilon\varepsilon'\frac{\partial\phi}{\partial x}\right) + \phi(1-\phi)(\phi-0.5+m) \tag{1}$$

where m is the driving force for solidification, ε-anisotropy and $\varepsilon' = \partial\varepsilon/\partial\theta$.

The thermodynamic driving force for interface motion is a function of temperature T and for pure metal it can be obtained by the expression: $m(T) = (b/\pi)\tan^{-1}[\gamma(T_m-T)]$. To specify anisotropy, the relationship $\varepsilon = \varepsilon_0[1+\delta\cos(j(\theta-\theta_0))]$ can be used, where θ is the angle between the x-axis and the interface normal defined as $\theta = \tan^{-1}((\partial\phi/\partial y)/(\partial\phi/\partial x))$.

The dimensionless equation describing latent heat release during dendrite growth of a pure material is:

$$\frac{\partial T}{\partial t} = c\,\nabla^2 T + K\frac{\partial\phi}{\partial t} \tag{2}$$

where K is a dimensionless latent heat, proportional to the undercooling and c is diffusion constant. All simulation parameters and operating conditions are listed in table 1 [10].

Table 1. Non-dimensionalised parameters used for solidification model

Simulation parameters	Value
τ	0.0003
ε_0	0.01
b	0.9
γ	10
δ	0.05
j	4
θ_0	0
K	1.4
c	1.5

2.2. Fluid flow

Fluid flow formulations generally require explicit tracking of the interface or complicated meshing techniques such as level-set [11], IBM [12] or ALE [13, 14]. In the current work, however, a monolithic solution for prescribing interface boundaries is developed by borrowing an idea from the area of topology optimisation problems [15-17]. The unique idea is to model the entire domain as fluid but enforce a location based damping term in order to simulate a rigid body surface, as shown in figure 1. Numerically, this causes the velocity of the fluid to approach zero at the interface, implicitly enforcing 'no-slip' condition. This term can be included into the steady-state Navier-Stokes equation as follows:

$$\rho(v.\nabla)v = -\nabla.pI + \nabla.[\mu(\nabla v + (\nabla v)^T)] - \alpha v \ in \ \Omega_f \tag{3}$$

$$\nabla.v = \beta\frac{\partial\phi}{\partial t} \tag{4}$$

where μ is viscosity, ρ – density. The term $'\alpha v'$ is a 'volume' or 'friction' force and the coefficient α can be treated as a penalisation constant:

$$\begin{cases} \alpha \approx \infty & in \ \Omega_s \\ \alpha = 0 & in \ \Omega_f \\ \alpha = \dfrac{1}{k} & in \ \Omega_I \end{cases}$$

In this work, the interface is considered as a part of the liquid domain with regards to the coefficient α. However, a more in-depth research can be performed to investigate the relationship between k and μ. Research already done on this area by Guest *et al.* [17], showed that k is inversely proportional to μ, similarly to Darcy's equation. However, no investigation has been performed on solidifying structures. The shrinkage of the moving interface is expressed in the mass conservation equation via the change of phase field variable φ with time, representing the change in solid fraction in the domain. β is shrinkage coefficient and it is a function of the solid and liquid density ratios. Since only a localised interdendritic flow is modelled, convection is assumed negligible.

As shown in figure 1, the penalisation constant determines the phases in the Navier-Stokes equation. There is no need to decouple the equation into solid and liquid parts, since a solution is obtained only for the fluid regions, whereas the solid remains rigid.

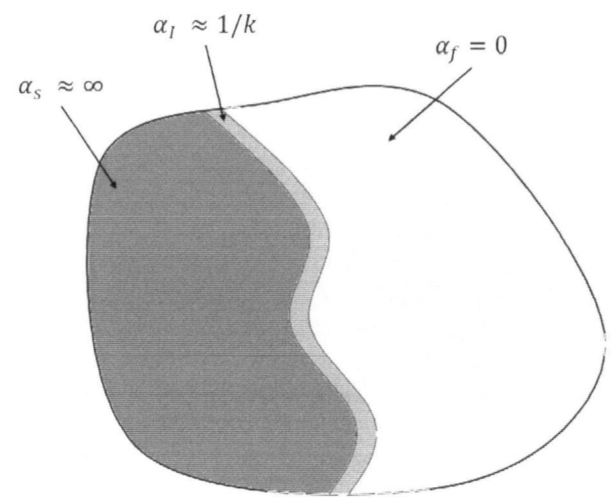

Figure 1. Topology domains with shown inverse permeability coefficient for solid, fluid and interface denoted as α_s, α_f and α_I, respectively

2.3. Configuration

The model is set to simulate two solidifying fronts growing opposite to each. A square domain of size $L_0 = 8\times8$ was created, with cell size $h_0 = 0.08$, and a time step of $t_0 = 0.002$ was set. As the entire formulation is non-dimensionalised, the time step and the grid size are artificial. The domain and the main boundary conditions are shown in figure 2. Symmetry boundaries are applied to the phase field equation. Heat is extracted from the left and right side of the domain with heat flux of $q_0 = -2$, where the top and bottom boundaries are set to be symmetric. For the Navier-Stokes' equation, fluid is allowed to freely leave or enter at the top and bottom and 'no-slip' condition is imposed on the rest of the boundaries. Two seeds with radial size less than 3 times the grid size are positioned as shown in the diagram below. Phase-field, heat equation and Navier-Stokes formulation are solved via finite element analysis (FEA) with a linear Lagrange discretisation on COMSOL Multiphysics® software. The total simulation time was *11 min 26 s* on a university PC (Intel® Core™ i5-8500 CPU @3.00GHz, 16GB RAM).

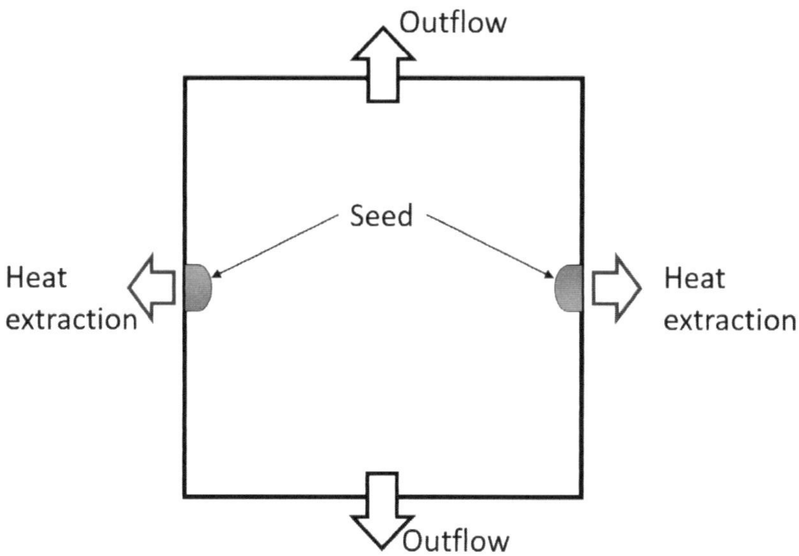

Figure 2. Schematic of computational domain and boundary conditions

The fluid flow is simulated on the same grid as the PF simulation, hence no interpolation or multi-level approaches are needed. The coefficient α is set as a function of the field variable ϕ where $\alpha(\phi>0.99)=10^4$ is considered solid and $\alpha(\phi=0)=0$ is liquid. The magnitude of α determines how impervious a material is. If α_s is too small, the solid domain allows fluid to enter while if α_s is too large, a numerical singularity would ensue. An optimum parameter exists such that solid permeability is optimised. This parameter has been investigate in literature for different physical processes and was found to be dependent on the mesh size as $\alpha=1/(0.05hx^2)$, for square grid [17-19]. It was found that this value does indeed provide a good first approximation but also that a range of values exists spanning across several magnitudes. Any of these optional values for α, cause an insignificant difference in the solution for pressure.

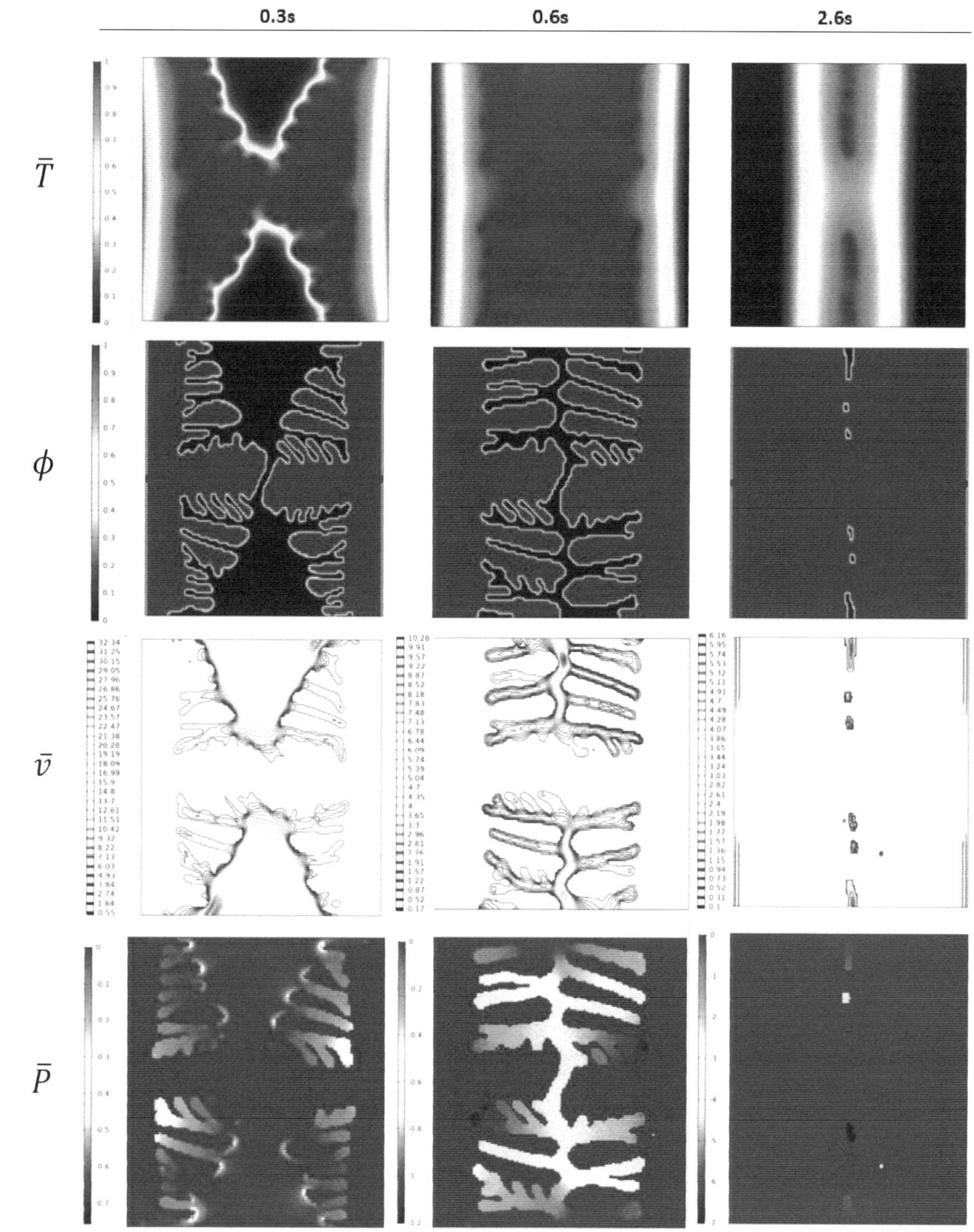

Figure 3. Colour maps of the four main processes taking place: heat transfer, solidification, velocity and pressure at three different times: 0.3 s, 0.6 s and 2.6 s. Each time step represents a different stage of growth, where 0.3 s corresponds to free fronts, 0.6 s – impinging fronts and 2.6 s – final stages. It is important to note that the pressure at the liquid pockets (0.6 s) is 5-6 orders of magnitude lower than that of the other cases.

IOP Conf. Series: Materials Science and Engineering **861** (2020) 012020 doi:10.1088/1757-899X/861/1/012020

3. Results

The four main processes heat transfer, phase-field, fluid flow and pressure are plotted in colour map format in figure 3. Three time steps are chosen, representing the three distinct stages from the process of impinging dendrite fronts. Up to 0.3 s both solidifying fronts have grown freely in the domain. High solidification rates and high velocity can be observed. It is important to note that at this stage, low pressure zones occur both at the tips of the dendrites and deep into the network. This can be contributed to the high solidification rates which give rise to high shrinkage at the dendrite tips. However, these pressure zones are easily accessible by fluid, hence pressure differences are relatively low. At 0.6 s, the two fronts are impinged. At this stage, heat has not been dissipated yet and the fronts have little space to grow, leading to reduced solidification rates. Fluid velocity, caused by shrinkage, also decreases and mostly occurs at the zones close to the boundaries. Pressures deep into the dendritic networks have begun decreasing. The pressure depression in these zones occurs due to the lack of feeding caused by their remoteness from the open boundaries and complexity of the channels. Results are also plotted at 2.6 s in order to represent the final stages of solidification, where the domain is 98% solidified. The channel no longer exists and only a few liquid pockets remain in the now solidified melt. There is barely any flow from the open boundaries as they are cut off by a solid phase. Important to observe is that the pressure depression in the liquid pockets is 5-6 orders of magnitude lower that in the other two cases. Such low pressures can potentially lead to pore formation and hot cracking.

Solid fraction and normalised fluid pressure in the entire domain, both plotted with time, are shown in figure 4. The three stages explained above can clearly be observed. Up to 0.3 s, the domain solidifies rapidly until it reaches 78% solid fraction. Barely any hydrostatic depression occurs due to the small pressure differences between tips and network zones. At 0.5 s however, when most of the dendrite arms impinge, a sudden drop in pressure can be observed. After this a steady-state occurs, where the domain slowly solidifies and pressure slowly recovers. At 2.6 s or 98% solid fraction, a sudden drop in pressure is observed. This, as already explained, is due to the cut-off channels and the formation of liquid pockets inside the solid.

The above described processes have been discussed in literature and proven experimentally, however, it is promising to see that our simulation results can clearly divide them into separate solidification stages. The advantage of the method, once applied to real alloy system, is that the exact time and solid fraction for each stage can be obtained. Furthermore, the exact positions for pore formation can investigated and identified with regards to depreciation pressure within dendritic arrays.

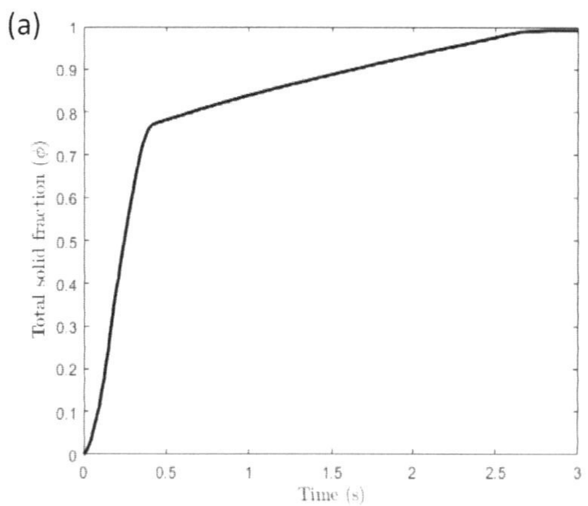

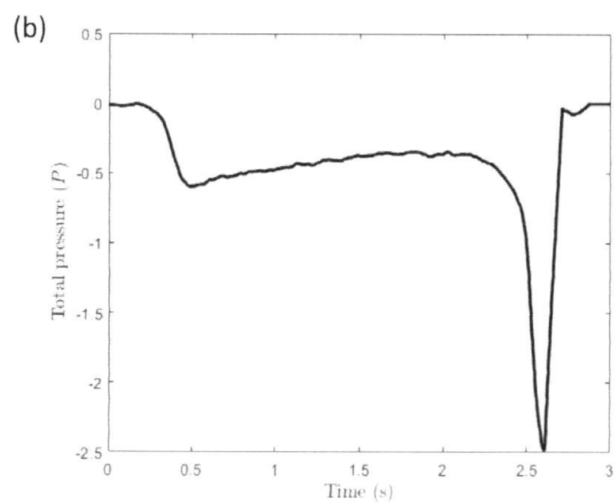

Figure 4. (a) Solid fraction in the entire domain (ϕ) with time, (b) pressure ($\bar{P}$) in the entire liquid domain with time.

4. Conclusions

A new framework for modelling fluid flow during phase transition process was developed and successfully implemented into modelling local shrinkage during solidification. A penalisation term was successfully applied into separating both domains, solid and liquid, without any re-meshing or front-tracking techniques. The new framework provided reasonable results, although only descriptive due to the non-dimensionalisation of all processes and variables. It was shown that 5-6 orders of magnitude larger drop in pressure occurs once liquid pockets form. The effect is contributed to the lack liquid feeding in those regions. The method is computationally efficient and has the potential to model pore and hot cracking formation during the late stages of solidification.

Acknowledgements

We gratefully acknowledge the financial support from the Centre for Doctoral Training in Innovative Metal Processing (IMPaCT) funded by the UK Engineering and Physical Sciences Research Council (EPRSC). COMSOL Multiphysics® research Licence No.7078652.

References

[1] Campbell J 2011 *Complete Casting Handbook* (Oxford: Butterworth-Heinemann) pp 465-97
[2] Carlson K D and Beckermann C 2009 *Metall. Mater. Trans. A* **40** 163-75
[3] Niyama E, Uchida T, Morikawa M and Saito S 1982 *Int. Cast. Met. J.* **7** 52-63
[4] Rappaz M, Drezet J-M and Gremaud M 1999 *Metall. Mater. Trans. A* **30** 449-55
[5] Melo M L N M, Rizzo E M S and Santos R G 2004 *Mat. Sci. Eng. A* **374** 351-61
[6] Coniglio N and Cross C E 2013 *Int. Mater. Rev.* **58** 375-97
[7] Clyne T and Davies G 1979 *in "Solidification and Casting of Metals"* (London: The Metals Society) pp 275–78
[8] Zhu M-F, Dai T, Lee S-Y and Hong C-P 2008 *Comput. Math. Appl.* **55** 1620-8
[9] Sun QY, Ren Y and Liu D-R 2019 *Results Phys.* **12** 1547-58
[10] Kobayashi R 1993 *Physica D* **63** 410-23
[11] Roshchenko A, Minev P D and Finlay W H 2015 *J. Fluid Struct.* **58** 109-26
[12] Sun X and Sakai M 2016 *Chem. Eng. Sci.* **139** 221-40
[13] Bordère S and Caltagirone J P 2014 *J. Fluid Struct.* **51** 344-53
[14] Bordère S and Caltagirone J P 2016 *Comput. Struct.* **164** 38-52
[15] Yoon G H 2010 *Int. J. Numer. Meth. Eng.* **82** 591-616
[16] Yoon G H 2014 *Comput. Meth. Appl. Mech. Eng.* **278** 499-523
[17] Guest J K and Prévost J H 2006 *Int. J. Numer. Meth. Eng.* **66** 461-84
[18] Bruns T E 2007 *Int. J. Heat Mass Tran.* **50** 2859-73
[19] Duan X-B, Ma Y-C and Zhang R 2008 *Appl. Math. Comput.* **202** 200-9

IOP Conf. Series: Materials Science and Engineering **861** (2020) 012021 doi:10.1088/1757-899X/861/1/012021

Multiphase-field modelling of hydrogen pore evolution during alloy solidification

A Zhang[1], Z Guo[1], Q Wang[2] and S Xiong[1, 3]

[1] School of Materials Science and Engineering, Tsinghua University, Beijing 100084, China
[2] Materials Technology, GM Global Product Group, Pontiac, MI 48340-2920, USA
[3] Key Laboratory for Advanced Materials Processing Technology, Ministry of Education, Tsinghua University, Beijing 100084, China

E-mail: smxiong@tsinghua.edu.cn

Abstract. Gas porosity is recognized as one of the most common defects that can severely deteriorate the mechanical properties of solidified materials. Simulating the gas porosity in castings is challenging due to intricate multiphase interaction. In this work, a multiphase-field model is developed to simulate the evolution of both solid phase and hydrogen pore during solidification in the Al-Cu alloy. The driving forces for the solid-liquid and liquid-gas transitions are the undercooling and pressure difference respectively. Results show that the presence of hydrogen pore substantially changes the dendrite network. The Cu element accumulates near the solid-liquid interface due to solute partition. The pore causes a local hydrogen reservoir and Cu vacuum region, which elevates the average hydrogen content and retards the dendrite growth by blocking Cu diffusion.

1. Introduction

Gas porosity in the interdendritic region of castings deteriorates mechanical properties of solidified materials, such as becoming crack source and impairing fatigue strength of engineering parts. The size, distribution, and morphology of the gas porosity change the eventual microstructure morphology by influencing both heat and mass transfer during solidification. Revealing the solid-liquid-gas interaction, together with the transport of solute and gas components, is important but challenging in metallurgy and applied physics [1].

Extensive experimental studies, including in-situ observation [2] and X-ray tomography [3], broaden our insight on the solid-liquid-gas interaction and subsequent porosity morphology. Nevertheless, the real-time precise measurements of the time-dependent quantities such as solute concentration and porosity distribution are still inaccessible for quantitative characterization. On the other hand, large amounts of modelling work are performed to predict the formation of porosity in castings. Lee *et al.* [4] reviewed the published models in the last century and classified those models into five groups by comparing their benefits, limitations and effectiveness. Those models focus more on the mean-field evaluation of porosity volume fraction and pore size, and they ignore the morphological evolution of solid network and the distribution of both solute and gas components.

As a powerful numerical tool to model microstructure evolution, the phase-field method (PFM) can recover Gibbs-Thomson effect and diffusion effect with rigor physics. An order parameter named the phase-field variable is introduced to characterize different physical states, which is beneficial to handle the evolution of complex pattern by avoiding explicit interface tracking. However, there is no mature PFM to reproduce the complex solid-liquid-gas interaction during solidification to our best knowledge. Felberbaum and Jacot [5] proposed a preliminary PFM to describe pore shape within static interdendritic liquid channels, and later their research team [6] further modified this model by using the multiphase-field concept to satisfy the mechanical and chemical equilibrium conditions at the triple junction. However, the solid growth is not considered and the dendrite array is artificially predefined. Carré *et al.* [7] developed another PFM to investigate the evolution of hydrogen pore in pure aluminum. This model is a modified version of the multiphase PFM developed by Eiken *et al.* [8], and it can handle large volume change of hydrogen pore. But the solute diffusion which is unavoidable in multicomponent system is not considered, and the solute redistribution caused by the presence of gas porosity remains unknown.

In this work, we develop a new multiphase-field model to investigate the complex solid-liquid-gas interaction. This model accounts for the growth of solid phase and the distribution of both solute and gas components, which allows for a direct description of micropores in a more realistic dendrite solid network. The hydrogen pore is modeled as a typical case of the gas porosity. The objective of the present work is to present the multiphase-field approach for a numerical description of the combined evolution of solid phase, hydrogen pore, solute concentration and gas content during alloy solidification.

2. Mathematical model

2.1. *Multiphase-field model*

A solid-liquid-gas multiphase-field model is developed to numerically describe the evolution of both solid phase and hydrogen pore during alloy solidification. The problem of interest represents a small section of a dendrite network. Three order parameters named the phase-field variables, ϕ_s, ϕ_l, and ϕ_g, are introduced to denote different phases, which correspond to local volume fractions and maintain at 1 at any location. ϕ_i ($i = s, l, g$) changes from 0 to 1 smoothly over an assumed diffuse interface and becomes 1 in the bulk i phase. Based on thermodynamic consistency, the evolution equations of ϕ_i are derived as

$$\tau_s \frac{\partial \phi_s}{\partial t} = K_s \left(\frac{\sigma_s}{h_s} \nabla^2 \phi_s - 2\phi_s (1-\phi_s)(1-2\phi_s) - \Theta_s + 15\phi_s^2 \phi_l^2 \frac{\Delta G_{ls}}{h_{ls}} - \frac{1}{3}\Lambda \right) \quad (1)$$

$$\tau_l \frac{\partial \phi_l}{\partial t} = K_l \left(\frac{\sigma_l}{h_l} \nabla^2 \phi_l - 2\phi_l (1-\phi_l)(1-2\phi_l) - \Theta_l - 15\phi_s^2 \phi_l^2 \frac{\Delta G_{ls}}{h_{ls}} - 15\phi_g^2 \phi_l^2 \frac{\Delta G_{lg}}{h_{lg}} - \frac{1}{3}\Lambda \right) \quad (2)$$

$$\tau_g \frac{\partial \phi_g}{\partial t} = K_g \left(\frac{\sigma_g}{h_g} \nabla^2 \phi_g - 2\phi_g (1-\phi_g)(1-2\phi_g) - \Theta_g + 15\phi_g^2 \phi_l^2 \frac{\Delta G_{lg}}{h_{lg}} - \frac{1}{3}\Lambda \right) \quad (3)$$

with

$$\Lambda = \sum_i \left[\frac{\sigma_i}{h_i} \nabla^2 \phi_i - 2\phi_i (1-\phi_i)(1-2\phi_i) - \Theta_i \right] \quad (4)$$

$$\Theta_i = \frac{\partial \sum_j \left[a_j \phi_i^2 \phi_k^2 (2\phi_i \phi_k + 3\phi_j) \right]}{\partial \phi_i} + \frac{\partial \sum_j b_j \phi_j^2 \phi_k^2 \phi_i^2}{\partial \phi_i} \quad (5)$$

where the introduction of Λ is to satisfy the constraint $\Sigma_i \phi_i = 1$, Θ_i corresponds to the triple-well potential with a tunable height of the saddle point on each binary interface [9], and a_j and b_j are two constants. τ_i, K_i, σ_i, and h_i are the relaxation time, interface mobility, interface energy, and potential energy coefficient which are dependent on the phase interfaces, respectively. It is assumed that $\tau_s = \tau_l = \tau_g = \tau_0$, $h_s = h_l = h_g$, where τ_0 is the time scaling. K_i can be used to adjust the phase-change rate, and a larger K_i corresponds to a faster interface advancing velocity. $\Delta G_{lg}/h = p_g - p_l$ is the driving force for the liquid-gas phase change, where p_i is the pressure in the i phase. $\Delta G_{ls}/h$ is the driving force for the solid-liquid phase change and is expressed as $-8\lambda(\theta + MC_\infty U)/15$ by mapping the multiphase-field equations to the solid-liquid phase-field model developed by Karma *et al.* [10, 11], where λ, θ, M, C_∞, and U are the coupling constant, dimensionless undercooling, dimensionless liquidus slope, far-field solute concentration, and dimensionless solute concentration, respectively, all of which have the same definitions as those in [11].

It is noted that the driving force for the solid-gas phase change is zero from equations (1)-(3), i.e. the interaction between the solid and gas phases is ignored. The gas phase only interacts with the liquid phase but liquid interacts with both solid and gas.

The evolution of the solute concentration and hydrogen content is governed by

$$\frac{\partial C^X}{\partial t} = \nabla \cdot \left(D^X \nabla C^X \right) + R\left(X \right) \tag{6}$$

where X denotes Cu for the Al-Cu alloy and hydrogen (H) for the gas pore, and D^X is the diffusivity of X and is determined by the volumetric averaging approach, i.e., $D^X = \phi_s D_s^X + \phi_l D_l^X + \phi_g D_g^H$. $R(X)$ denotes the partition of X at the phase interfaces, i.e.,

$$R\left(Cu \right) = n_s n_l C^{Cu} \left(1 - k^{Cu} \right) \frac{d\phi_s}{dt} \tag{7}$$

$$R\left(H \right) = n_s n_l C^H \left(1 - k^H \right) \frac{d\phi_s}{dt} \tag{8}$$

where n_i ($i = s, l, g$) is a step function, i.e., $n_i = 1$ if $\phi_i > 0$ and $n_i = 0$ otherwise. k^{Cu} and k^H are the equilibrium partition coefficients of Cu and H at the solid-liquid interface, respectively.

The region inside the pore is considered to be a local vacuum region of Cu element, i.e., Cu concentration inside the pore is zero [12]. The hydrogen content inside the pore is assumed uniform and is connected with the pressure by the perfect gas law. By combining the perfect gas law and the Sievert law, the gas content at $\phi_g > 0$ is obtained as

$$C^H = \frac{2p_g}{RT}\phi_g + k^H S_l \sqrt{\frac{p_g}{p_0}}\phi_s + S_l \sqrt{\frac{p_g}{p_0}}\phi_l \tag{9}$$

where the unit of C^H is atomic moles m^{-3}, R, T, p_0 and S_l are the universal gas constant, melt temperature, atmospheric pressure, and Sievert constant for the liquid with the unit atomic moles m^{-3}.

2.2. *Mapping to the two-phase solidification*

By setting ϕ_i ($i = s, l, g$) = 0, the multiphase-field equation is restored to the governing equation of the two-phase system. Here, taking $\phi_g = 0$ for instance, introduce $p = 1 - 2\phi_l$ and rewrite equations (1)-(2) in a unified form:

$$\tau_0 \frac{\partial p}{\partial t} = K_{sl} \left(\frac{\sigma_{sl}}{h_{sl}} \nabla^2 p + p - p^3 - \left(1 - p^2\right)^2 \lambda \left(\theta + MC_\infty U \right) \right) \tag{10}$$

where $K_{sl} = 1$, $\sigma_{sl}/h_{sl} = W^2$, and W is the interface thickness. The time and length are rescaled by $t' = t/\tau_0$ and $x' = x/W$. Equation (10) is the same as the thermosolutal phase-field equation of the binary alloy proposed by Ramirez *et al.* [11], and it is restored to the phase-field equation of a pure melt developed by Karma *et al.* [10], if the solute concentration U is neglected. Based on the definition of the antitrapping current [11] and the dimensionless expressions of θ and U,

$$\theta = \frac{T - T_M - mC_\infty}{\Delta T_0}, \quad U = \frac{\dfrac{2C/C_\infty}{1 + k - \left(1 - k\right)p} - 1}{1 - k} \tag{11}$$

the solute field equation (6) combining with equation (7) is restored to

$$\left(\frac{1+k}{2}-\frac{1-k}{2}p\right)\frac{\partial U}{\partial t}=\nabla\cdot\left(D\frac{1-p}{2}\nabla U+\frac{1}{2\sqrt{2}}\left[1+(1-k)U\right]\frac{\partial p}{\partial t}\frac{\nabla p}{|\nabla p|}\right)+\frac{1}{2}\left[1+(1-k)U\right]\frac{\partial p}{\partial t} \tag{12}$$

Accordingly, the multiphase-field model is restored to the two-phase solidification phase-field model. After mapping operation, the modeling parameters can be further determined by an asymptotic analysis of the thin-interface limit [11].

3. Results and discussion

Without considering the nucleation process, both the solid seed and a pre-existing hydrogen pore are initialized in the domain. The thermo-physical parameters are according to the Al - 4 wt% Cu alloy [13], and the other modeling parameters are listed in table 1. It is noted that the common symbol of hydrogen content in the casting industry is [H] and the corresponding unit is cc/100g. A unit conversion, $C^H = \rho_L[\mathrm{H}]/2241$, needs to be performed, where ρ_L is the alloy liquid density [14]. Since the computational domain is a small window centered on one pore and is much smaller than the space related to one pore, the initial hydrogen content needs to be large enough to offset that [6]. Once the initial pore radius r_g is given, the pressure inside the pore is determined by the Laplace pressure condition:

$$p_g = p_l + \gamma_{lg}/r_g \tag{13}$$

where γ_{lg} is the liquid-gas interface energy.

The temperature is assumed isothermal, and zero-Neumann boundary conditions are set at all sides for all variables including the three phase-field variables, Cu concentration, hydrogen content, temperature, and pressure.

Table 1. Parameters used in the phase-field simulations [6].

Parameter	Value
C^H_∞, far-field hydrogen content in liquid, cc/100g	1.0
D^H, diffusivity in liquid, m²/s	1×10^{-6}
S_l, Sievert constant in liquid, mol/m³	0.69
p_l, pressure in liquid, atm	0.9
p_0, atmosphere pressure, atm	1.0
γ_{sl}, solid-liquid interface energy, J/m²	0.15
γ_{lg}, liquid-gas interface energy, J/m²	0.868
ρ_L, alloy liquid density, kg/m³	2.61×10^3
θ, dimensionless undercooling	-0.25
R, universal gas constant, J/(mol K)	8.314

3.1. Interaction between single dendrite and hydrogen pore

The seed with radius of 1.6 is initialized at the domain center, and the pore with radius of 16 is set at $(0.5X, 0.2Y)$, where $X = Y = 512$ is the domain size in the unit of W. Figure 1 shows the evolution of both solid phase and hydrogen pore during solidification. The two rows correspond to the cloud map of the phase-field contour and Cu concentration. At the early stage, both the solid phase and the hydrogen pore evolve freely without interaction. The face-centered-cubic Al-Cu dendrite exhibits four-fold symmetry and the pore maintains the circle shape. Because of the solute partition, the Cu element

IOP Conf. Series: Materials Science and Engineering **861** (2020) 012021 doi:10.1088/1757-899X/861/1/012021

accumulates at the solid-liquid interface, especially at the dendrite root. The pore expands by absorbing the hydrogen dissolved in liquid until reaching the pressure equilibrium. When the downward dendrite tip approaches the pore, the dendrite wraps the pore by growing along the pore boundary. The pore expands by pressing the liquid aside, which causes local enrichment of Cu element around the pore and decrease of the dendrite growth velocity.

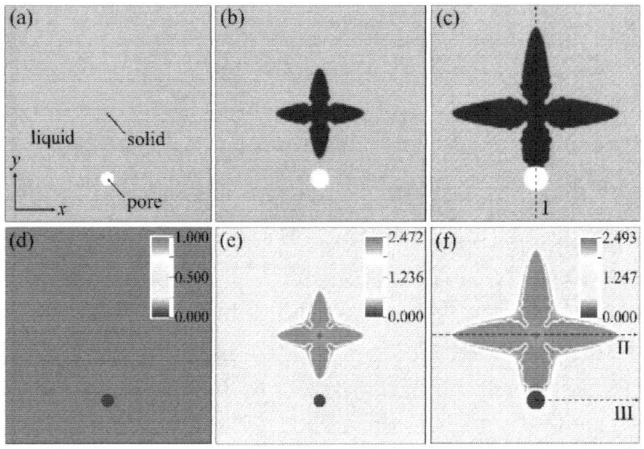

Figure 1. Evolution of the dendrite growth in the presence of hydrogen pore. Each row corresponds to the cloud map of the phase-field contour and Cu concentration, respectively.

Figure 2 shows the change of the three phase-field variables along the direction designated by line I in figure 1(c). The phase-field variables change smoothly at the phase interfaces, and the sum maintains at one at any location. Taking the liquid-gas interface in figure 2(b) for instance, $\phi_l = \phi_g = 0.5$ and $\phi_s = 0$ at $x/X = 0.142$. Due to the diffuse nature in the phase-field model, a triple junction is observed at $x/X = 0.252$, where $\phi_s = \phi_g = 0.38$ and $\phi_l = 0.24$. The solid-liquid-gas triple junction is like the triple point during eutectic solidification.

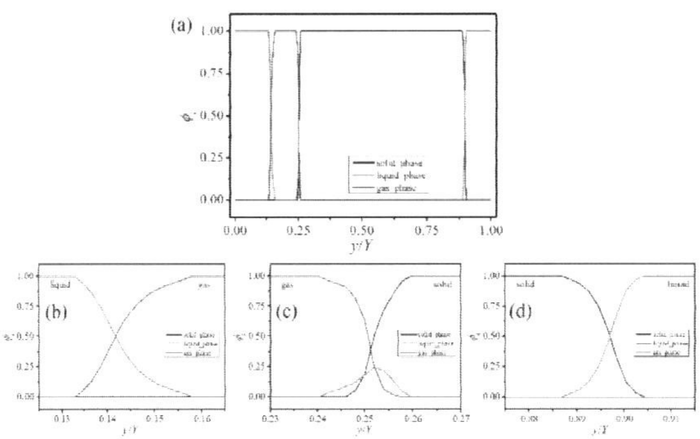

Figure 2. Three phase-field variables versus the distance along the direction designated by line I in figure 1(c). (b), (c) and (d) are the local enlarged images of (a).

The pore is a local Cu vacuum region which acts like an obstacle blocking Cu diffusion. When the diffusion of the rejected Cu element from the solidified dendrite is hindered, a local Cu enrichment region forms between the downward dendrite tip and pore boundary. According to the phase diagram, the enriched solute lowers the undercooling and thus slows down the downward solid growth. Such asymmetric growth behaviour is analogous to that under convection [15].

Figure 3 shows the distribution of Cu and H along the directions designated by lines II and III in figure 1(f). Along line II, the distribution is symmetric about the domain centre, and the concentration of both Cu and H obtains the extreme at the solid-liquid interface. After reaching the equilibrium at the solid-liquid interface, the Cu concentration inside the solid phase keeps unchanged. Along line III, the H concentration increases with the distance away from the pore, while that of Cu decreases. The H concentration near the pore boundary is lower than the average value in the liquid. The presence of hydrogen pore causes a local hydrogen reservoir. Thus the H diffuses towards the liquid-gas interface and into the pore, which increases the gas content inside the pore and promotes the pore expansion. The liquid-gas phase change, i.e., pore expansion or movement of the liquid-gas interface, is driven by the pressure difference, which is similar to the work performed by Meidani and Jacot [6].

The pore expands by pressing the liquid aside, and the Cu element accumulates around the pore. As the distance from the pore increases, the Cu concentration decreases until approaching the far-field value. It is noted that the liquid flow induced by the expansion of the pore and resultant convection transport are ignored, which deviates from the practical scenario and makes the concentration around the pore larger than the real value.

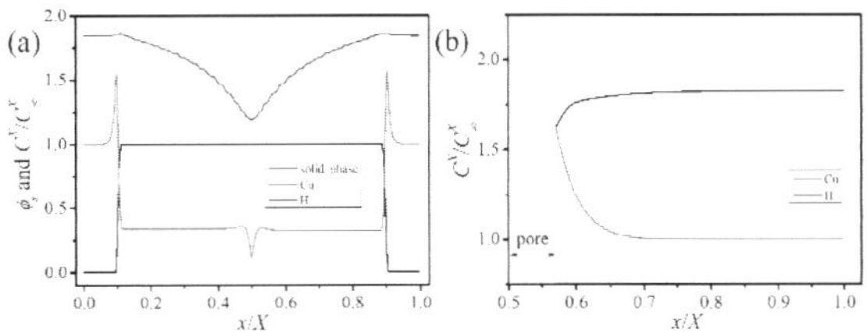

Figure 3. Distribution of Cu and hydrogen along the centrelines designated by lines II and III in figure 1(f).

3.2. Interaction between multidendrites and hydrogen pore

Ten solid seeds with radius of 1.6 are initialized randomly in the domain, and one pore with radius of 16 is set at (0.5X, 0.2Y), where $X = Y = 1024$ is the domain size in the unit of W, as shown in figure 4(a). The parameter configuration is the same as that in section 3.1. The multidendrites exhibit four-fold symmetry and grow competitively by impinging on neighbouring dendrite arms. The presence of the pore changes the complex dendritic morphology by inhibiting the solid growth.

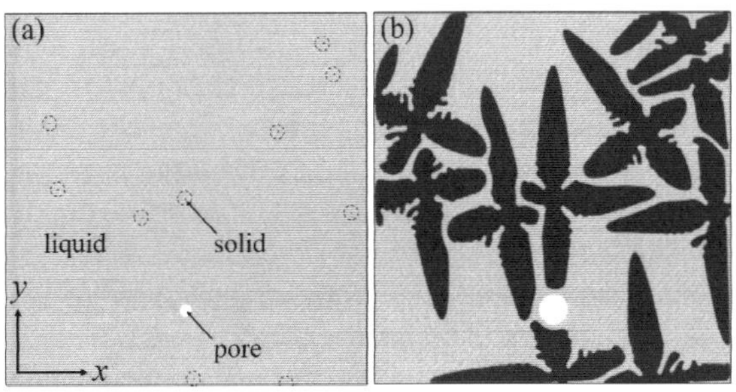

Figure 4. Evolution of the multidendrite case in the presence of pore. Ten seeds are circled by the dashed lines in (a).

Figure 5 shows the comparison of Cu concentration without and with the hydrogen pore. A local high-concentration Cu region exhibits near the pore boundary, but in other location, the distribution of Cu is not disturbed. The enriched Cu has two sources, i.e. solute partition at the solid-liquid interface and solute pileup due to pressed liquid by pore expansion. The distribution of hydrogen is more uniform than that of Cu, because the hydrogen diffusivity is about 300 times larger than the Cu diffusivity. The high-concentration H region is retained in the upper-right part due to larger interface area. The pore absorbs the neighbouring hydrogen, which promotes the rejection of hydrogen from the solidified dendrites and thus elevates the average hydrogen content inside the domain, i.e., from 1.77 to 1.95 mol/m^3.

The pore usually nucleates at high solid fraction and high hydrogen content. As the temperature decreases, the pore exhibits complex geometrical shape due to numerous contacts with neighbouring solid. The enriched Cu element slows down the solid growth, and the remaining liquid surrounding the pore makes the pore undergo negligible shape change.

The redistributed Cu element changes the relative density of melt, which can cause liquid convection and pore movement. The liquid convection is not considered in the present work and thus the pore centre stays still. In some in-situ observations reported on the directional solidification of organic alloys [2, 16], the bubble is stuck between the two glass covers, and the solid phase can engulf or entrap the pore as the solid-liquid interface advances. Those phenomena occur at a decreasing temperature, and the dendrites can have enough driving force to compensate the effect of the solute enrichment.

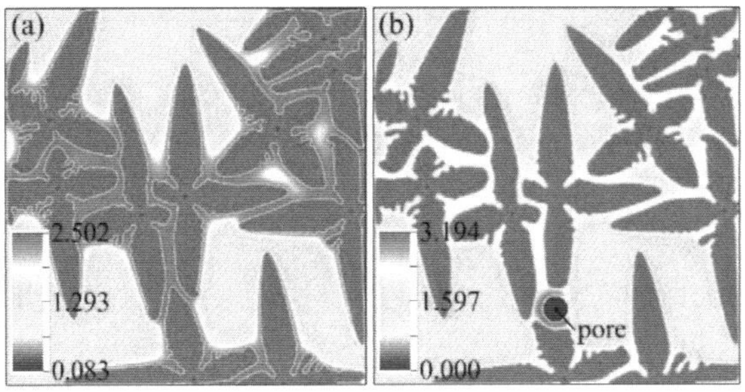

Figure 5. Comparison of Cu concentration without (a) and with (b) the hydrogen pore.

4. Conclusion and remark

A multiphase-field model is developed to simulate dendrite growth in the presence of the hydrogen pore during solidification. The present model incorporates the growth of solid phase, the expansion of hydrogen pore, and the redistribution of both solute and hydrogen at the phase interfaces. This model can be accurately mapped to the two-phase models, and the driving forces for the solid-liquid and liquid-gas phase changes are the undercooling and pressure difference. The presence of hydrogen pore substantially changes the dendrite symmetry by blocking solute diffusion and inhibiting solid growth. The pore acts like a local Cu vacuum region, which causes solute enrichment around the pore, but as a hydrogen reservoir, it elevates the average content of hydrogen by absorbing the neighboring hydrogen and promoting the rejection of hydrogen from the solidified dendrites. The solute-rich liquid surrounding the pore lowers the local undercooling and solidifies later, making the pore grow circularly and undergo less change.

The hydrogen diffusion is considered to be one of the most important factors governing the growth dynamics of the pore. To precisely characterize such growth dynamics, the liquid convection induced by the expansion of the pore and hence the hydrogen transport by convection need to be exploited quantitatively. Besides, the large pore deformation under convection poses a challenge on the conservative characteristic of the model in characterizing the complex topology interface. More sophisticated numerical models which can incorporate all these physical phenomena will be considered in future development.

Acknowledgments
This work is financially supported by the Joint Funds of the National Natural Science Foundation of China (grant number U1537202) and the Tsinghua-General Motors International Collaboration Project (grant number 20153000354). The authors would also like to thank the National Laboratory for Information Science and Technology in Tsinghua University for access to its supercomputing facilities.

References
[1] Dantzig J A and Rappaz M 2009 *Solidification* (Lausanne: EPFL Press)
[2] Xing H, Wang J Y, Chen C L, Jin K X and Shen Z F 2010 Morphological evolution of the interface microstructure in the presence of bubbles during directional solidification *Scripta Mater.* **63** 1228
[3] Lashkari O, Yao L, Cockcroft S and Maijer D 2009 X-Ray Microtomographic Characterization of Porosity in Aluminum Alloy A356 *Metall. Mater. Trans. A* **40** 991
[4] Lee P D, Chirazi A and See D 2001 Modeling microporosity in aluminum - silicon alloys: a review *J. Light Metals* **1** 15
[5] Felberbaum M and Jacot A 2009 A phase field model for a geometrical description of micropores constrained by a solid network. In: *Modeling of Casting, Welding, and Advanced Solidification Processes - XII,* ed S L Cockcroft and D M Maijer: TMS (The Minerals, Metals & Materials Society)) pp 369
[6] Meidani H and Jacot A 2011 Phase-field simulation of micropores constrained by the dendritic network during solidification *Acta Mater.* **59** 3032
[7] Carré A, Böettger B and Apel M 2010 Phase-field modelling of gas porosity formation during the solidification of aluminium *Int. J. Mater. Res.* **101** 510
[8] Eiken J, Boettger B and Steinbach I 2006 Multiphase-field approach for multicomponent alloys with extrapolation scheme for numerical application *Phys. Rev. E* **73** 066122
[9] Folch R and Plapp M 2005 Quantitative phase-field modeling of two-phase growth *Phys. Rev. E* **72** 011602
[10] Karma A and Rappel W 1998 Quantitative phase-field modeling of dendritic growth in two and three dimensions *Phys. Rev. E* **57** 4323
[11] Ramirez J C, Beckermann C, Karma A and Diepers H J 2004 Phase-field modeling of binary alloy solidification with coupled heat and solute diffusion *Phys. Rev. E* **69** 051607
[12] Zhang A, Du J, Zhang X, Guo Z, Wang Q and Xiong S 2020 Phase-field modeling of microstructure evolution in the presence of bubble during solidification *Metall. Mater. Trans. A* **51** 1023
[13] Zhang A, Du J, Meng S, Liu F, Guo Z, Wang Q and Xiong S 2020 Three-dimensional thermosolutal simulation of dendritic and eutectic growth *Comp. Mater. Sci.* **171** 109274
[14] Zhu J D, Cockcroft S L and Maijer D M 2006 Modeling of microporosity formation in A356 aluminum alloy casting *Metall. Mater. Trans. A* **37** 1075
[15] Zhang A, Meng S, Guo Z, Du J, Wang Q and Xiong S 2019 Dendritic growth under natural and forced convection in Al-Cu alloys: From equiaxed to columnar dendrites and from 2D to 3D phase- field simulations *Metall. Mater. Trans. B* **50** 1514
[16] Han S H and Trivedi R 1994 Primary spacing selection in directionally solidified alloys *Acta Metall. Mater.* **42** 25

IOP Conf. Series: Materials Science and Engineering **861** (2020) 012022 doi:10.1088/1757-899X/861/1/012022

Advanced modeling of shrinkage porosity in castings

P Esser[1], C Schankies[1], V Khalajzadeh[2] and C Beckermann[2]

[1] MAGMA Gießereitechnologie GmbH, Kackertstr. 11, 52072 Aachen, Germany
[2] Dept. of Mechanical Engineering, University of Iowa, Iowa City, IA 52242, USA

E-mail: p.esser@magmasoft.de, becker@engineering.uiowa.edu

Abstract. Porosity due to solidification shrinkage is a troublesome defect in all types of metal castings. It limits the performance of cast components by adversely affecting the material's strength, fatigue and creep properties. By reliably predicting porosity in casting process simulation, it can be minimized or eliminated. Here, a newly developed model for predicting porosity is presented. The model is based on the recent discovery that all shrinkage porosity nucleates and grows in regions of a casting where the solid fraction is the lowest. It calculates the feeding flows and pressure distribution in the liquid while accounting for the liquid density variation during cooling and solidification. It predicts the location, extent and amount of all types of shrinkage porosity in a casting, including riser pipes and large internal holes, surface sinks, and distributed micro-shrinkage. The model is numerically implemented in a standard casting simulation code. Comparisons to measurements in specially made steel castings demonstrate the capability of the model to accurately predict various types of shrinkage porosity.

1. Introduction

In metal casting, shrinkage porosity is a widespread problem. It forms during solidification, when liquid melt can no longer feed the loss of volume during contraction. This leads to defects (porosities) in the casting, which affect the quality and mechanical properties. In general, there are three types of porosity in a casting: surface sinks, internal macro-porosity and (internal) micro-porosity (see figure 1). We will present a mathematical model that can predict these types of defects adequately:

- It accurately predicts location, amount and appearance of porosity.
- It predicts locations of porosity nucleation and growth.
- It predicts the stages of surface and internal porosity formation.

This model is integrated into commercial simulation software.

2. Computational model

In order to model the formation of porosities, we use the following physical concepts:

- Porosity nucleates where solid fraction is the lowest.
- Feeding flows are calculated.
- Surface and internal porosity are calculated.

A schematic diagram of a solidifying casting volume is shown in figure 2. Multiple regions of porosity can form simultaneously. The boundary conditions at the casting surface and on the porosity containing regions are indicated in figure 2(a) and figure 2(c), respectively. The model allows for a representative elementary volume to be composed of three phases: solid (s), liquid (ℓ) and porosity (p).

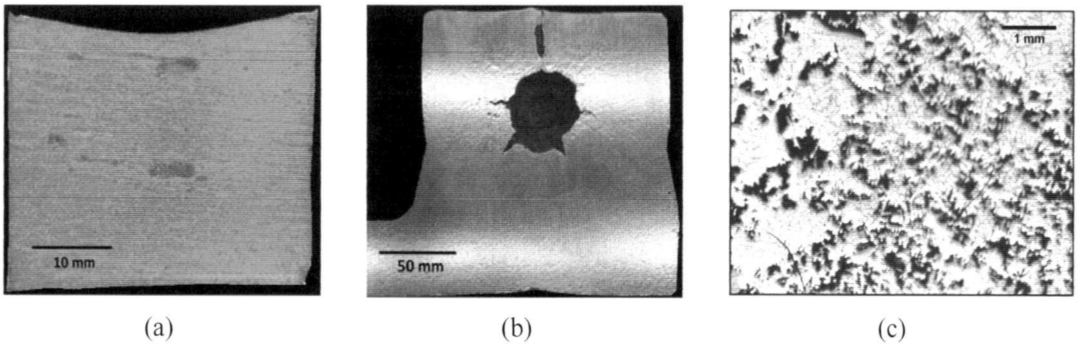

Figure 1. Various types of shrinkage porosity in castings: (a) a sink on the top surface of a cast aluminum block, (b) internal macro-shrinkage hole in a steel casting, and (c) distributed internal micro-shrinkage porosity.

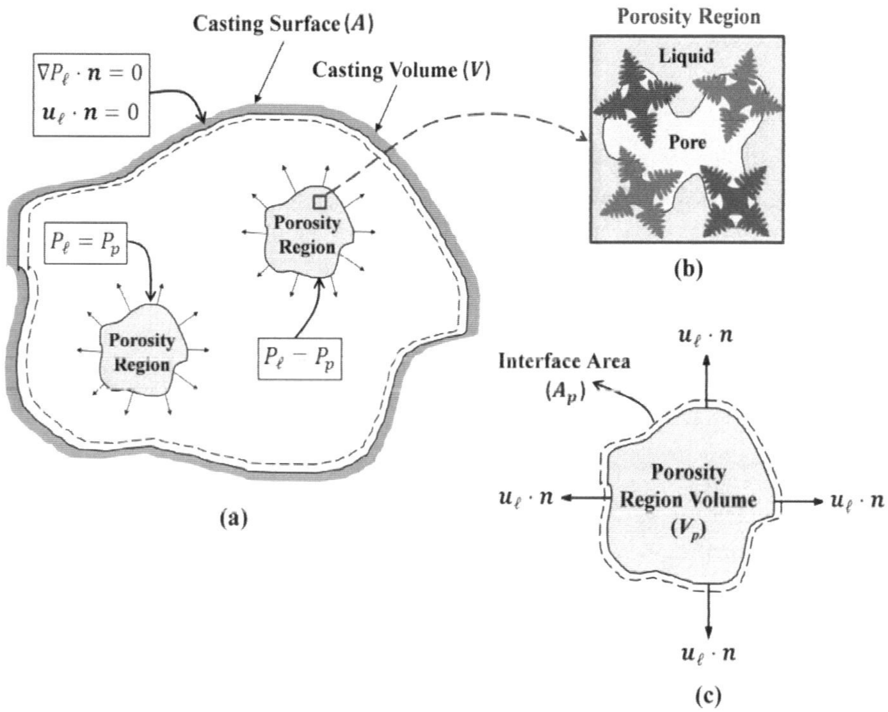

Figure 2. Schematic diagram of a solidifying casting with porosity: (a) casting volume with multiple porosity containing regions, (b) representative elementary volume containing solid, liquid and porosity, and (c) single porosity containing region

Using "g" to denote the volume fraction of a given phase, and the phases indicated by subscripts, the volume fractions must satisfy $g_\ell + g_s + g_p = 1$, as shown schematically in figure 2(b).

Key assumptions in the model are:

- Solid and porosity phases are stationary; they cannot move during solidification; however, solid may be replaced by porosity if the solid fraction is below a certain coherency limit.
- Flow of the liquid melt is driven by mixture density changes only; buoyancy-driven flow during solidification is neglected.

- Density is temperature-dependent above the solidus temperature and assumed constant when the metal is solid; the density of the gases inside of shrinkage pores is neglected.

The governing equations for the feeding flow and the procedures used for predicting the evolution of the pore volume fraction are given in the following sub-sections. The model assumes that the results from a standard thermal simulation of the casting are available. This includes the evolution of the temperature and solid fraction fields throughout the solidification process. The heat transfer is assumed to be unaffected by the porosity.

2.1. Continuity and momentum equations

Mass is conserved according to a mixture continuity equation

$$\frac{\partial \bar{\rho}}{\partial t} + \nabla \cdot (\rho_\ell \boldsymbol{u}_\ell) = 0 \tag{1}$$

where t is time, ρ_ℓ is the liquid density, and $\boldsymbol{u}_\ell$ is the superficial liquid velocity vector. The three-phase mixture density $\bar{\rho}$ is defined as $\bar{\rho} = \rho_s g_s + \rho_\ell g_\ell + \rho_p g_p$. The feeding flow in the mushy zone is assumed to be a creeping flow in a porous medium of variable permeability. For such flows the momentum equation is given by Darcy's law,

$$\nabla P_{\ell,t} = -\frac{\mu_\ell \boldsymbol{u}_\ell}{K} + \rho_\ell \boldsymbol{g} \tag{2}$$

where $P_{\ell,t}$ is the total liquid pressure, μ_ℓ is the dynamic viscosity of the melt, K is the permeability of the mush, and $\boldsymbol{g}$ is the gravity vector. Equation (2) is also used in the pure liquid regions, by setting the permeability to a sufficiently large value, $K_{\max}$. The total liquid pressure ($P_{\ell,}$) is the sum of the dynamic liquid pressure (P_ℓ) and the hydrostatic pressure (P_h), i.e., $P_{\ell,} = P_\ell + P_h$. Applying the gradient operator to this equation gives $\nabla P_{\ell,} = \nabla P_\ell + \nabla P_h$, where the hydrostatic pressure is given by $\nabla P_h = \rho_\ell \boldsymbol{g}$. Combining this and equation (2), the gravity term cancels and equation (2) simplifies to

$$\nabla P_\ell = -\frac{\mu_\ell \boldsymbol{u}_\ell}{K} \tag{3}$$

The permeability of the mush is a function of solid fraction (g_s) according to the Kozeny–Carman equation

$$K = K_0 \frac{(1 - g_s)^3}{g_s^2} \tag{4}$$

where K_0 is an adjustable constant permeability coefficient. To avoid division by zero in equation (3), a suitable minimum allowable value for the permeability, $K_{\min}$, is introduced. To ensure well-posedness of equation (4), suitable boundary conditions are needed. Here it is assumed that P_ℓ within regions having porosity is equal to the pore pressure: $P_\ell|_{\text{Porosity}} = P_p$, implying that surface tension effects are neglected. The value of the pore pressure P_p depends on the location in the casting. At an atmosphere-metal interface the pore pressure is assumed to be atmospheric: $P_p = P_{\text{atm}}$, at a mold-metal interface it is equal to some pre-defined mold atmosphere pressure: $P_p = P_{\text{mold}}$ and for internal porosity, the pores are assumed to be vacuum (negligible density), $P_p = 0$. By forcing the dynamic pressure to a uniform value, the liquid velocities vanish inside of porosity forming regions. Assuming for simplicity that ρ_ℓ in equations (1) and (2) is constant, equations (1) and (3) can be combined to form an equation for determining the pressure P_ℓ throughout the casting during solidification

$$\nabla \cdot \left(\frac{K}{\mu_\ell} \nabla P_\ell\right) = \frac{1}{\rho_\ell} \frac{\partial \bar{\rho}}{\partial t} \tag{5}$$

As shown in figure 3, the boundary condition for the above equation is given by a zero-mass flux condition at the casting surface, which can be written as $(\boldsymbol{u}_\ell \cdot \boldsymbol{n})_{\text{surface}} = (\nabla P_\ell \cdot \boldsymbol{n})_{\text{surface}}$, where $\boldsymbol{n}$ is the normal vector on the surface. Using this boundary condition and forcing the pressure in the porosity containing regions to $P_\ell|_{\text{Porosity}} = P_p$, the pressure distribution P_ℓ in the casting can be determined by solving equation (5).

2.2. Pore nucleation model

The presented pore nucleation model is based on the one developed by Khalajzadeh *et al.* [1]. Surface sinks and internal porosity do not evolve at the same time. Surface sinks develop first and internal porosity forms in a second stage. The transition between the two stages is modeled by introducing the critical solid fraction $g_{s,\text{sur}}$. When the solid fraction at the surface is below this critical value, a surface sink can form. Once the solid fraction everywhere on the surface is above $g_{s,\text{sur}}$ the surface becomes rigid and internal porosity can nucleate. The pore nucleation model is based on the inequality form of the Young-Laplace equation, which is given in the present study by

$$P_p - P_{\ell,t} \geq P_\sigma \tag{6}$$

where P_σ is a capillary pressure due to surface tension. As found by Khalajzadeh *et al.* [1], the capillary pressure P_σ in equation (6) can be taken as a linear function of the solid fraction as

$$P_\sigma = P_{\sigma,0}\, g_s, \tag{7}$$

where $P_{\sigma,0}$ is a reference capillary pressure at a solid fraction of unity. As in Khalajzadeh *et al.* [1], the Young-Laplace inequality is implemented in the model through a so-called Π parameter, which is defined as

$$\Pi = P_{\ell,t} - (P_p - P_\sigma) \tag{8}$$

Porosity is thus nucleated where the Π parameter is the lowest. To make the nucleation calculations numerically robust, porosity is nucleated where the Π value falls within a small dimensionless interval ε_{nuc} according to:

$$\left|\frac{\Pi - \Pi_{\text{min}}}{\Pi_{\text{max}} - \Pi_{\text{min}}}\right| < \varepsilon_{\text{nuc}} \tag{9}$$

where Π_{min} is the instantaneous minimum Π value in the computational domain and Π_{max} is the maximum possible Π value. Physically, the adjustable parameter ε_{nuc} controls the initial size of a newly nucleated porosity region. Other details of the pore nucleation model can be found in [1].

2.3. Pore growth model

As shown in figure 3(a), multiple regions of porosity can nucleate and grow in a casting simultaneously. Each region is a grouping of connected computational cells having $g_p > 0$. The volume and surface area of each porosity forming region are denoted by V_p and A_p, respectively, as shown in figure 3(c). The average pore growth rate $(d\bar{g}_p/dt)$ for each porosity containing region is calculated by integrating the mass conservation equation, equation (1), over V_p. Using the divergence theorem for the second term in equation (1) yields

$$\int_{V_p} \left(\frac{\partial \bar{\rho}}{\partial t}\right) dV + \int_{A_p} (\rho_\ell \boldsymbol{u}_\ell) \cdot \boldsymbol{n}\, dA = 0 \tag{10}$$

The density $\bar{\rho}$ should be distinguished from the solid-liquid mixture density in the absence of porosity, which is given by, $\bar{\rho}^{SL} = \rho_s, g_s^{SL} + \rho_\ell g_\ell^{SL}$, where $g_s^{SL} + g_\ell^{SL} = 1$. The density $\bar{\rho}^{SL}$ is assumed to be known as a function of temperature for a given metal alloy. Assuming that $g_s^{SL} = g_s$ and $\rho_p \ll \rho_s, \rho_\ell$ yields the following equation for the three-phase density

$$\bar{\rho} = \bar{\rho}^{SL} - \rho_\ell g_p \tag{11}$$

Assuming that ρ_ℓ in equation (11) is constant and that the pore growth rate is uniform within each V_p, substitution of equation (11) into equation (10) results in

$$\frac{d\bar{g}_p}{dt} = \frac{1}{\rho_\ell |V_p|} \int_{V_p} \left(\frac{\partial \bar{\rho}^{SL}}{\partial t} \right) dV + \frac{1}{|V_p|} \int_{A_p} \boldsymbol{u}_\ell \cdot \boldsymbol{n} \, dA \tag{12}$$

Equation (12) allows for the calculation of the average pore growth rate in each porosity forming region from the knowledge of the average solid-liquid mixture density variation within V_p and the feeding flow velocities normal to the boundary A_p. The average porosity increase from equation (12) is then evenly distributed over all active computational cells within each V_p to determine the local time rate of change in the pore fraction according to [1].

The pore volume fraction is typically allowed to increase until a computational cell becomes empty of liquid, i.e., when $g_s + g_p = 1$. This condition, however, would often result in the local pore fraction never being able to approach unity. A pore fraction of unity occurs, for example, when a surface sink or internal hole is free of solid. For the pore volume fraction to reach unity, the previously grown solid at that location must have moved or been pushed away by the growing porosity, which is frequently referred to as mass feeding. As in Khalajzadeh *et al.* [1], mass feeding is modeled by introducing an adjustable parameter called the coherency solid fraction $g_{s,\mathrm{coh}}$. The maximum pore volume fraction, $g_{p,\max}$, is then given by

$$g_{p,\max} = \begin{cases} 1, & g_s < g_{s,\mathrm{coh}} \\ 1 - g_s, & g_s \geq g_{s,\mathrm{coh}} \end{cases} \tag{13}$$

The two critical solid fractions $g_{s,\mathrm{sur}}$ and $g_{s,\mathrm{coh}}$ take typically the same value [1]. When a computation cell reaches the maximum pore fraction, it is designated as inactive. Once a pore forming region has no more active cells, but solidification is not yet complete, porosity can spread beyond the originally nucleated volume. As in Khalajzadeh *et al.* [1], such spreading of pore forming regions is modeled using the same Π parameter introduced in the previous subsection. Computational cells that neighbor a cell with porosity already present are activated when their Π parameter falls within the interval

$$\left| \frac{\Pi - \Pi_{\min}}{\Pi_{\max} - \Pi_{\min}} \right| < \varepsilon_{\mathrm{layer}} \tag{14}$$

The small dimensionless number $\varepsilon_{\mathrm{layer}}$ is an adjustable model parameter that controls the rate of spreading of porosity regions. It is emphasized here that the use of the Π parameter in both the pore nucleation and growth models was developed by some of the authors through careful comparisons with real-time video radiography experiments [1].

3. Experiments and simulation

To validate the model, castings were made from high manganese (Mn) steel in chemically bonded olivine sand [2]. The geometry is shown in figure 3. It consists of a gating system, a riser, two steps of different thickness, a thinner plate section, and a thick end block. The thermal simulation was calibrated with the help of thermocouples [3].

The experimentally observed porosity is shown in figure 4, while the porosity predicted by the present model, using the parameters listed in table 1, is shown in figure 5. We see in general good agreement of the model predictions with the porosity observed in the casting.

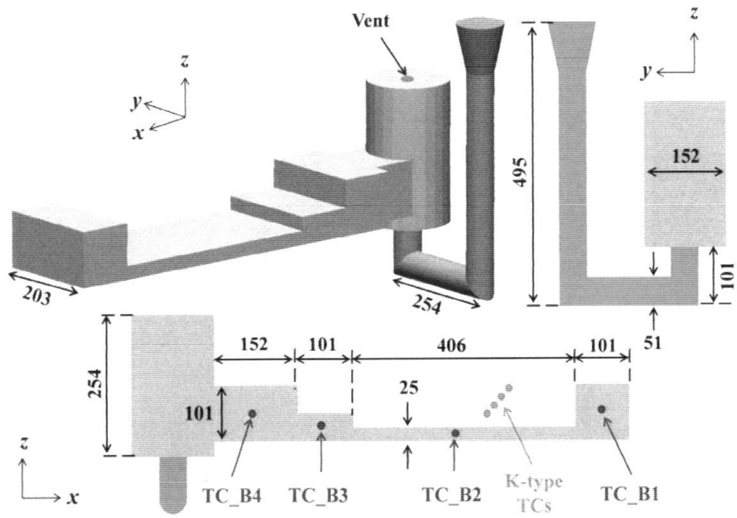

Figure 3. Mn-steel casting geometry used in the experiments; the dimensions are in millimeter (mm).

Figure 4. Porosity distribution observed in the Mn-steel casting: vertical slices through the casting at mid-width, with the red areas showing microporosity that is visualized using dye penetration testing.

Figure 5. Porosity predictions obtained by the present model for the Mn-steel casting shown in figure 4.

IOP Conf. Series: Materials Science and Engineering **861** (2020) 012022 doi:10.1088/1757-899X/861/1/012022

Table 1. Material properties and model parameter used in the simulation

Parameter	Value
K_0 (m^3)	1.7e-9
$K_{\min}$ (m^3)	1.0e-16
$K_{\max}$ (m^3)	1.0e-6
$g_{s,\mathrm{sur}}$ (-)	0.625
$g_{s,\mathrm{coh}}$ (-)	0.625
$\varepsilon_{\mathrm{layer}}$ (-)	5.0e-2
$\varepsilon_{\mathrm{nuc}}$ (-)	5.0e-3
P_{atm} (bar)	1.101325
$P_{\sigma,0}$ (bar)	P_{atm}
P_{mold} (bar)	0.4 P_{atm}
ρ_s (kg/m^3)	7124
ρ_ℓ (kg/m^3)	6894
μ_ℓ (Pa.s)	8.57e-3

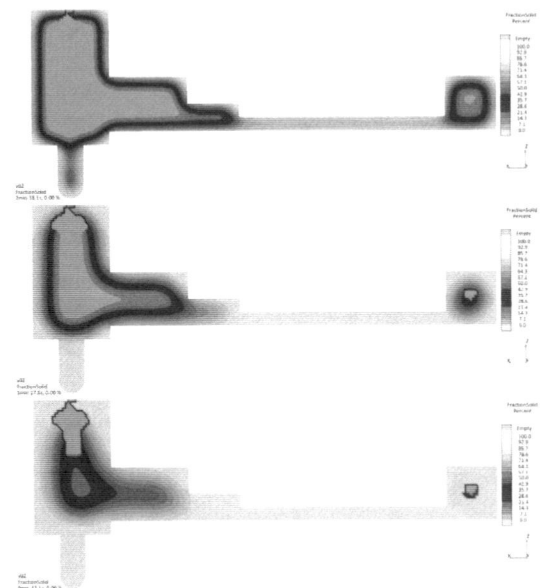

Figure 6. Fraction solid (50%, 75%, 90% solidified).

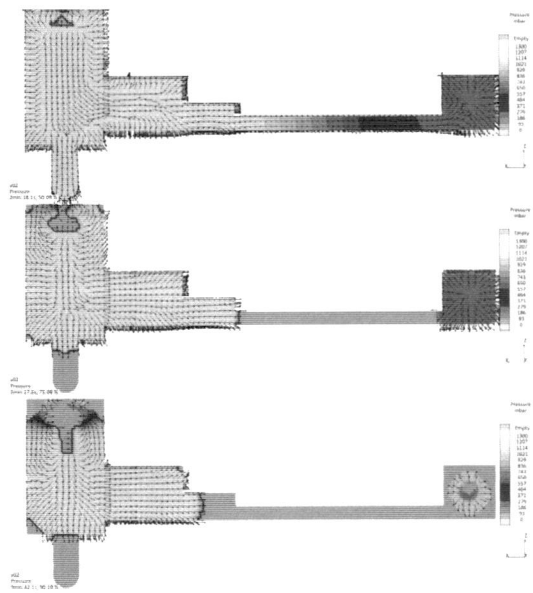

Figure 7. Total pressure and feeding velocity vectors (50%, 75%, 90% solidified).

Other results of the simulation are shown in figure 6 and figure 7. These show the evolution of the fraction solid g_s from the thermal simulation (figure 6) and the total pressure P_ℓ and feeding velocity vectors $\boldsymbol{u}_\ell$ from the present model (figure 7). The results are shown at the states of 50%, 75%, 90% solidified cells in all casting materials. Until almost 50%, all shrinkage of the metal during cooling and solidification is fed by the gating system. The total pressure distribution is mostly hydrostatic, because the dynamic pressure variations are still very small. The top of the riser is connected to the atmosphere through a small vent that induces a shrinkage pipe at that location once the pressure falls below atmospheric pressure. At 50% solidified cells, the permeability in the thin plate section is low enough that the liquid flow feeding the end block from the riser creates a large pressure drop. Eventually, the pressure in the end block approaches vacuum and porosity nucleates in the thermal center of the end block where the solid fraction is lowest. At 75%, the thin plate section between the riser and the end block is fully solidified and the pressure inside the block is forced by the internal porosity in the center of the block. Small areas of microporosity can be observed in the thin plate section. The rest of the casting continues to be fed by the downward growing hole at the top of the riser. At 90%, the riser pipe acquires a more conical shape and extends over the top one-third of the riser. Eventually, microporosity forms next to and below the open riser pipe and in the larger step (figure 5).

4. Conclusions

A model for the prediction of shrinkage porosity in metal castings is developed. It is based on the solution of the basic equations for the feeding flow and pressure distribution during solidification and an extended version of the pore nucleation and growth model of Khalajzadeh *et al.* [1]. The model uses the temperature and solid fraction results from a thermal simulation and predicts all kinds of shrinkage porosity, including surface sinks and riser pipes, open shrinkage holes internal to the casting, and micro porosity dispersed between dendrite arms. Good agreement is obtained with observations made in an experimental Mn-steel sand casting. The shape of the riser pipe, the open shrinkage hole in the thermal center of the end block, and the microporosity in the thin plate section and around the larger holes are all predicted correctly. No surface sinks were observed or predicted to form. Results from another test casting with surface sinks, including quantitative comparisons between measured and predicted pore fraction values, can be found in Khalajzadeh *et al.* [1,2].

References

[1] Khalajzadeh V, Goettsch D and Beckermann C 2019 *Metall. Mater. Trans. A* **50** 757-71
[2] Khalajzadeh V 2018 *Modeling of shrinkage porosity defect formation during alloy solidification* PhD thesis (University of Iowa, Iowa City, IA)
[3] MAGMASOFT® MAGMA GmbH Aachen Germany

IOP Conf. Series: Materials Science and Engineering **861** (2020) 012023 doi:10.1088/1757-899X/861/1/012023

Three-dimensional numerical simulation of solidification microporosity and microstructure of aluminum alloys

C Gu and A A Luo

The Ohio State University, Columbus, USA

E-mail: luo.445@osu.edu

Abstract. A three-dimensional model was developed to study microporosity evolution with grain growth during solidification by cellular automaton method. The simulations are able to show the evolution and morphology of both grain structure and porosity. The effects of shrinkage pressure and initial hydrogen content on porosity evolution and final porosity percentage are modeled and related to the location-specific microstructure of the solidified products. A wedge die casting, providing samples of different cooling rates, was used to validate the simulation results. X-ray micro-tomography was performed to obtain the relevant porosity information. The simulated porosity size, percentage and distribution considering both shrinkage pressure and hydrogen content were found to agree well with those from experimental measurements.

1. Introduction

Aluminum alloys have been widely used in industrial applications due to the good mechanical properties of high strength-to-weight ratio, good corrosion resistance, and so on. However, porosity exists in castings is known to reduce mechanical properties of casting products, in particular ultimate strength and fatigue resistance. Porosity observed in aluminum castings can be mainly divided into shrinkage porosity and gas porosity [1, 2]. Experiments have been used to investigate porosity of casting products under different processing parameters, but it is still hard to observe the porosity nucleation and evolution during solidification process and predict porosity in castings. Modeling and simulation of porosity formation [3, 4] can be used on the mechanisms of porosity formation for casting design and process optimization.

However, most of the relevant researches are only able to provide the predicted porosity level in solidification products, and there is little study on the prediction of porosity information such as distribution and size with graphical grain and porosity morphology [1]. The prediction of porosity coupled with grain of alloys is an essential part of microstructure to mechanical property predictions in integrated computational materials engineering (ICME) [5]. The understanding of the mechanism of morphological changes of grain and porosity is helpful for the porosity control in casting products, which also provides a critical link in ICME for the casting industry.

In this study, a numerical model was proposed to predict microporosity evolution during the solidification process by cellular automaton method. Both the effects of shrinkage pressure and the hydrogen content are considered in the model, which means the predicted porosity is developed by both the solidification shrinkage and the dissolved hydrogen. The evolution of porosity with multi-grains was simulated, and the porosity morphology, porosity distribution and size were obtained. The simulated results were validated with wedge die casting experiments.

2. Model description

2.1. Grain growth model

The continuous nucleation distribution [6] is used [7] for the nucleation of grains during solidification. Following nucleation, solute distribution including solute partitioning and diffusion can be calculated by:

$$C_i^{S*} = k_i C_i^{L*} \tag{1}$$

$$\frac{\partial C_i^E}{\partial t} = \nabla \cdot \sum_j D_{ij} \nabla C_i^E + (C_i^{L*} - C_i^{S*}) \frac{\partial f_s}{\partial t} \tag{2}$$

where C_i^{S*} and C_i^{L*} are the interface solute concentrations of the solute element i in solid and liquid respectively, k_i is the solute partition coefficient, and D is the diffusion coefficient. The interface solute concentration can be calculated by [7,8]:

$$\sum_i^{n-1} \frac{\partial T}{\partial C_i^L} (C_i^0 - C_i^{L*}) = T_0 - T^* - \Gamma\kappa \tag{3}$$

where $T_0 \approx T_P - \sum_i^{n-1} \frac{\partial T}{\partial C_i^L} C_i^0$ is the liquidus temperature, T^* is the local temperature, Γ is the Gibbs–Thomson coefficient, and κ is the interface curvature. To trace the interface, the solid fraction increment can be calculated based on the lever rule [7, 8].

2.2. Porosity evolution model

In this study, pore nucleation was considered by using the stochastic porosity nucleation model [9]. Based on Sievert's law, the hydrogen saturation can be calculated by [10]:

$$S_H^L = \sqrt{\frac{P_G}{P_0}} 10^{\left(-\frac{2760}{T} + 2.796 - 0.0119 C_{Si}^L + 0.017 C_{Mg}^L\right)} \text{ (mL/100g Al)} \tag{4}$$

where $P_G = P_0 + 2\gamma/r_P - P_S$ is the pore internal pressure, P_0 is the standard atmospheric pressure, γ is the surface tension, r_P is the pore radius, P_s is the shrinkage pressure.

During solidification, the hydrogen content in the liquid is increased. Meanwhile, the supersaturated hydrogen in the liquid is absorbed by adjacent nucleated pores, resulting in the growth of the pore. The pore volume increment can be calculated by considering all the hydrogen absorbed:

$$\Delta V = \frac{P_0}{P_G} V_G = \frac{P_0}{P_G} \sum_A (1 - f_S - f_G) V_{cell} \rho (C_H^L - S_H^L) \tag{5}$$

where V_{cell} is the cell volume, and ρ is the liquid density. According to the equations above, pore growth can be calculated considering the effects of both shrinkage pressure and hydrogen content. Thus, the evolution of microporosity and microstructure during solidification is obtained. It should be mentioned that the eutectic solidification and the effect of fluid flow are ignored in this paper.

3. Model verification

A V-shaped wedge casting experiment was developed to provide validation for simulations. Al alloy, A356, was used in the experiment. Figure 1 shows the schematic geometry and the cast sample of the proposed wedge die casting. With different wall thickness, the casting experiment is able to obtain different cooling rates from top to bottom. At first, the Al ingots were melted to about 700 °C, and the melt were kept at the temperature for 30 min. During pouring, K-type thermocouples were used to simultaneously measure the corresponding temperature profiles at the three locations A, B, and C. Since instant cooling rates were used in the simulations, the instant cooling rates were calculated at the liquidus temperature from the experiment cooling curves, which were 64.8, 10.7, and 2.5 K/s at point A, B, and C, respectively.

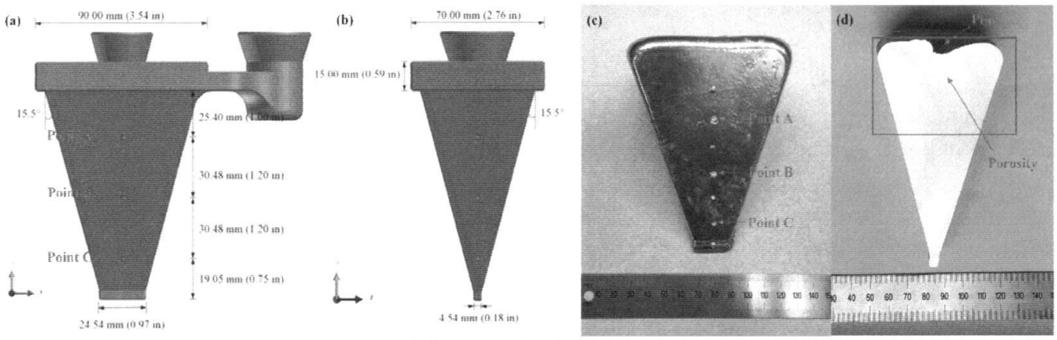

Figure 1. Schematic geometry (a) and (b), sample (c), and cross-section (d) of wedge die casting.

Reduced pressure test (RPT) [11] was performed to measure the equivalently initial hydrogen which was obtained as C_H^0 = 0.3 mL/100g Al. In addition, X-ray micro computed tomography (XMCT) was used to characterize the porosity information, including porosity morphology, percentage, and location of each pore. The samples were scanned at 90 kV, 80 μA and a resolution of 6.318 μm per voxel using HeliScan^{TM} microCT.

4. Results and discussion

4.1. Grain structure and porosity simulation

Based on the proposed 3D CA model, the evolution of grain as well as porosity of Al-7Si-0.3Mg ternary alloy can be simulated as shown in figure 2. In the simulation, the calculation domain was set as a 1000 × 1000 × 100 μm^3, and the cell size was 5 μm. A constant cooling rate of 10 K/s was adopted. The initial hydrogen concentration was 0.5 mL/100g Al, and the shrinkage pressure was 0. From the simulation result, the structures of grain and porosity are connected closely with each other. During the solidification, since the pores are encompassed by solidified grains, it is difficult for the pores to run away from the liquid and eventually result in porosity. From solute distributions of Si and Mg, the solute concentration is increased as temperature decreases due to solute partitioning. The distributions of Si and Mg are not influenced by the pores. Si and Mg concentrations in the remaining liquid are increased to their eutectic concentrations when the temperature decreases to the eutectic temperature. However, the hydrogen distribution shows much difference from the solute distributions. The hydrogen concentration shows to increase first due to hydrogen partitioning as well. As solidification proceeds, temperature continues decreasing and hydrogen content in the liquid continues increasing. When the hydrogen concentration is larger than the hydrogen saturation, the pore nucleation will happen as the nucleation criterion is fulfilled. The nucleated pores then start to grow by absorbing the adjacent saturated hydrogen, decreasing the local hydrogen concentration. Thus, hydrogen concentration decreases in the latter stages of solidification. It should be noted that with the change of temperature and local Si and Mg concentrations, local hydrogen saturation will also change, leading to the pore growth. It can be seen in the simulations that the nucleated pore continues absorbing the surrounded hydrogen. Since the diffusivity of hydrogen is much larger than that of solute, there is no obvious hydrogen gradient in the remaining liquid.

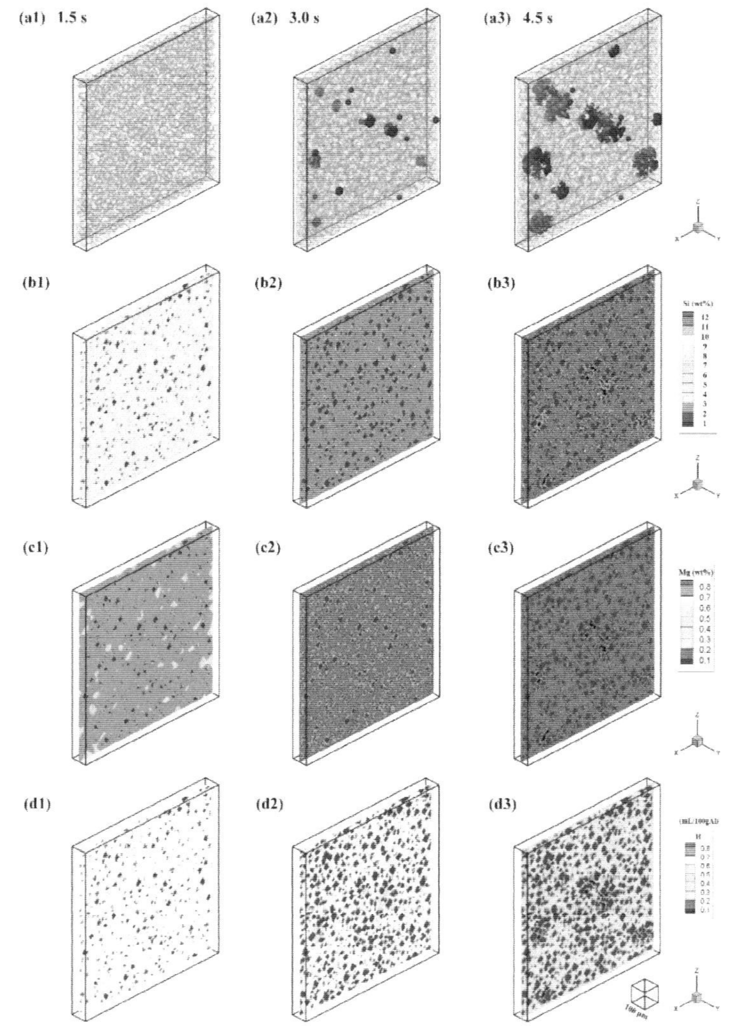

Figure 2. Simulation results of (a) grain and porosity, (b) Si distribution, (c) Mg distribution, and (d) hydrogen distribution at different times: (a1, b1, c1, d1) 1.5 s; (a2, b2, c2, d2) 3.0 s; (a3, b3, c3, d3) 4.5 s.

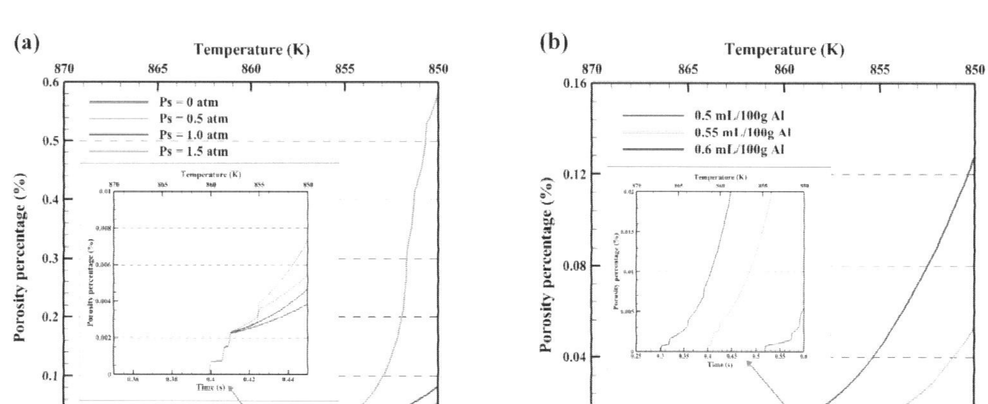

Figure 3. Porosity percentage as a function of temperature and time with (a) shrinkage pressure of 0, 0.5, 1.0, and 1.5 atm, and (b) initial hydrogen concentrations of 0.5, 0.55, and 0.6 mL/100g Al.

Figure 3 plots the porosity percentage vs. temperature under the effect of shrinkage pressures and initial hydrogen concentrations. The simulations were performed on Al-7Si binary alloy and in a calculation domain of 500 × 500 × 500 μm^3, and the cooling rates were 50 K/s. It should be noted that the porosity percentage increases with increasing shrinkage pressure, while the porosity nucleation is not influenced by the increase of shrinkage pressure in this study. It is because the porosity formation during eutectic at the end of solidification is not considered in this paper. Meanwhile, increasing initial hydrogen concentration will increase the porosity percentage. It shows that pores start to nucleate at a higher temperature with a higher initial hydrogen concentration, which is due to that higher initial hydrogen concentration will lead to more hydrogen enrichment and thus higher levels hydrogen will satisfy the nucleation criterion earlier, and the nucleation of porosity will occur more rapidly and at higher temperatures. When the initial hydrogen concentration is at a lower value, it will take more time for the local hydrogen concentration to be large enough and thus the nucleation occurs at lower temperature.

4.2. Experimental validation

As a validation with the experiments, 3D CA simulations with same condition were performed. The whole calculation domain was 1 mm × 1 mm × 1 mm which is the same with the microCT measurements as shown in figure 4. In the real casting, the volume change during solidification, $\Delta V_{c,w}$. can be calculated by:

$$\Delta V_{c,w} = \left(1 - \frac{\rho_l}{\rho_s}\right) V_{0,w} f_{s,w} \tag{6}$$

where ρ_l and ρ_s are the liquid density and solid density, $V_{0,w}$ is the casting volume, and $f_{s,w}$ is the solid fraction. The shrinkage pressure, P_s, is calculated by:

$$P_0 V_{0,w}(1 - f_{s,w}) = (P_0 - P_s)\left[V_{0,w}(1 - f_{s,w}) + \Delta V_{c,w}\right] \tag{7}$$

$$P_s = P_0 \left(1 - \frac{1 - f_{s,w}}{1 - \frac{\rho_l}{\rho_s} f_{s,w}}\right) \tag{8}$$

As shown in microCT measurements in figure 4(a), (d) and (g), the pores show different morphologies due to the constraint of the surrounded dendrites. With decreasing cooling rate, the pore number decreases while the pores appear to have much larger sizes. Figure 4(b), (e), and (h) show the simulated porosity with transparent dendrites at the end of solidification. In the simulated results, as cooling rate decreasing, the porosity size increases and the pore number decreased as expected.

In order to clearly observe the morphology of dendrites and pores, 2-D optical microstructure with different cooling rates was measured and compared with the 2-D simulated results as shown in figure 5. The simulations and optical microstructure clearly show the porosity and dendrite morphologies. The simulated dendrite morphologies match well with the experiments as shown in figure 5(c) and (d). It should be noted that at the end of solidification, the remaining liquid will transform to eutectic as shown in both simulation and experiment. As the cooling rate decreased, the pores appear to be larger, while the morphologies of the pores are not shown to be round shape. It is because that 2-D optical microstructure is only able to provide a slice of complete 3D information [12, 13]. Even so, the 2-D simulated results of both the dendrite and porosity morphology agree well with the experiments.

Figure 6 shows the relationship between porosity percentage and cooling rate. Increasing cooling rate decreases the porosity percentage. By coupling both hydrogen and shrinkage, the simulation results show well agreement with the wedge casting experiments. It should be pointed out that the simulated results considering only hydrogen are lower than the experimental results compared to that considering both hydrogen and shrinkage. Porosity morphologies in figures 4 and 5 also point out this difference. It is because that micro-shrinkage increases the final percentage of porosity. It can be seen that the evolution of microporosity and microstructure can be predicted by using present model. The validation of the model provides an excellent opportunity for the prediction of location-specific microstructure which is helpful and essential for the casting design optimization. This microporosity and microstructure

IOP Conf. Series: Materials Science and Engineering **861** (2020) 012023 doi:10.1088/1757-899X/861/1/012023

model shows to be a critical link between process parameters and mechanical properties in ICME researches of casting products.

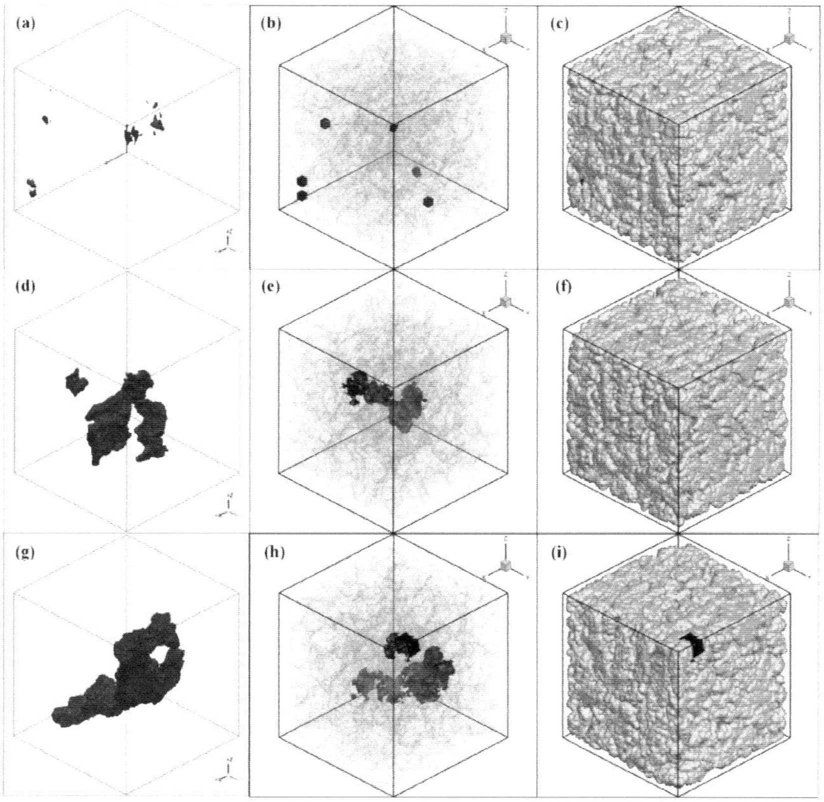

Figure 4. Porosity morphology of microCT measurements as a comparison with CA simulations: microCT (a), (d), and (g); simulated porosity (b), (e), and (h); simulated dendrite (c), (f), and (i). The cooling rates: (a)-(c) 64.8 K/s; (d)-(f) 10.7 K/s; and (g)-(i) 2.5 K/s.

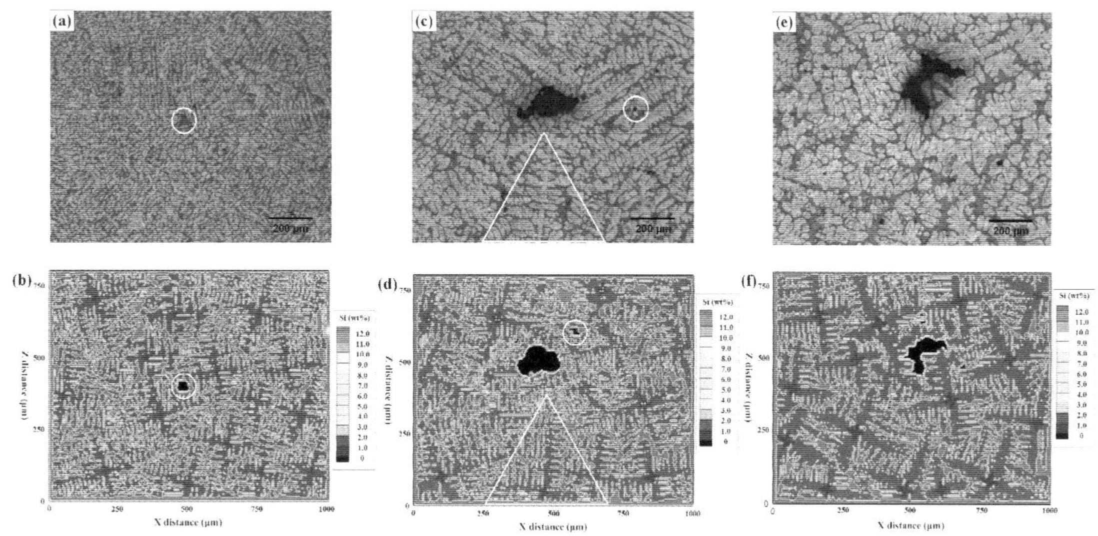

Figure 5. Optical microstructure at cooling rate of: (a) 64.8 K/s, (c) 10.7 K/s, and (e) 2.5 K/s; and simulated results: (b) 64.8 K/s, (d) 10.7 K/s, and (f) 2.5 K/s.

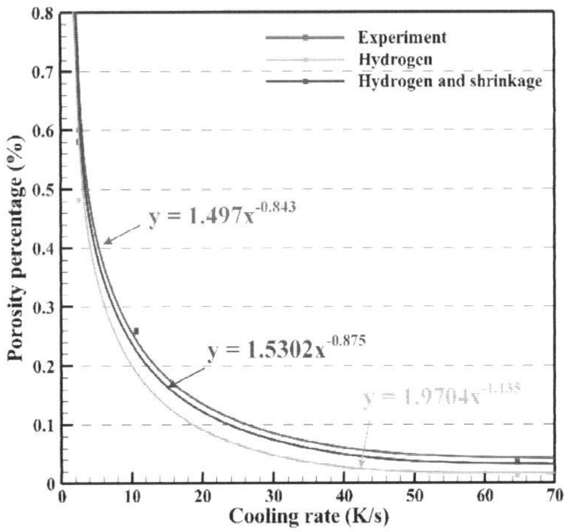

Figure 6. Porosity percentage as a function of cooling rate.

5. Conclusions

In this study, a numerical model was developed for the prediction of evolution of microporosity and microstructure during solidification of Al-7Si-0.3Mg ternary and Al-7Si binary alloys. By adopting the present model, the morphology and distribution of microporosity and microstructure can be simulated, and porosity percentage can be calculated. The effect of initial conditions were investigated. With initial hydrogen concentration increases, porosity percentage increases. In this study with high hydrogen content and pore nucleates before shrinkage occurs, with increasing shrinkage pressure the porosity percentage increases. The simulated porosity percentage and the pore distribution considering both shrinkage pressure and hydrogen content show well agreement with the experiments. The proposed 3D CA model provides an essential link between process parameters and location-specific mechanical properties for further ICME researches.

References

[1] Lee P D, Chirazi A and See D 2001 *J. Light Met.* **1** 15–30
[2] Kaufman J G and Rooy E L 2004 *Aluminum Alloy Castings: Properties, Processes, and Applications* (ASM International)
[3] Gao Z, Jie W, Liu Y and Luo H 2017 *Acta Mater.* **127** 277–86
[4] K D Carlson, Z Lin, C Beckermann, G Mazurkevich and M C Schneider 2006 *Modeling of Casting, Welding and Advanced Solidification Processes – XI* 627-34
[5] Luo A A 2015 *Calphad Comput. Coupling Phase Diagrams Thermochem.* **50** 6–22
[6] Rappaz M and Gandin C 1993 *Acta Metall. Mater.* **41** 345–60
[7] Gu C, Lu Y, Cinkilic E, Miao J, Klarner A, Yan X and Luo A A 2019 *Comput. Mater. Sci.* **161** 64–75
[8] Gu C, Ridgeway C D and Luo A A 2019 *Metall. Mater. Trans. B.* **50** 123–35
[9] Zhu M, Li Z, An D, Zhang Q and Dai T 2014 *ISIJ Int.* **54** 384–91
[10] Atwood R C, Sridhar S, Zhang W and Lee P D 2000 *Acta Mater.* **48** 405–17
[11] La-Orchan W 1994 *Quantification of the reduced pressure test* (McGill University)
[12] Anson J P and Gruzleski J E 1999 *Mater. Charact.* **43** 319–35
[13] Sigworth G K and Wang C 1993 *Metall. Trans. B.* **24** 349–64

IOP Conf. Series: Materials Science and Engineering **861** (2020) 012024 doi:10.1088/1757-899X/861/1/012024

Modeling mixed columnar-equiaxed solidification of Sn-10wt%Pb alloy under forced convection driven by travelling magnetic stirring

Z Zhang, M Wu, H Zhang, E Karimi-Sibaki, A Ludwig and A Kharicha

Chair of Simulation and Modelling of Metallurgical Processes, Department of Metallurgy, Montanuniversitaet Leoben, A-8700 Leoben, Austria.

E-mail: menghuai.wu@unileoben.ac.at

Abstract: A mixed columnar-equiaxed solidification model was used to investigate the solidification benchmark experiments, as performed at SIMAP Laboratory. The first experiment case of the benchmark made under natural convection condition was successfully simulated previously. The current work is to simulate the second experiment case, i.e. the casting under travelling magnetic stirring (TMS) which is applied in the direction of natural convection to enhance the natural convection. Crystal fragmentation of columnar dendrites is assumed as the sole origin of equiaxed crystals. Through the analysis of the simulation results, deep understanding to the experimentally reported phenomena was achieved. In comparison to the first experiment case of pure natural convection condition, the TMS-enhanced convection provides a favorable condition for columnar-to-equiaxed transition (CET) by homogenizing the temperature distribution in the bulk liquid region, enhancing the fragmentation of the columnar secondary arms and the transport of the fragments to bulk liquid region. Contradicting with the previous knowledge, the TMS-enhanced convection in this experiment case does not enhance the global macrosegregation significantly. However, it strengthens the channel segregates.

1. Introduction

Series of solidification experiments were performed to investigate the formation of as-cast structure and macro-/meso-segregation in rectangular castings (100 × 60 × 10 mm^3) of Sn-10wt%Pb alloy under different convection conditions: 1) natural convection; 2) travelling magnetic stirring (TMS) applied in the same direction of natural convection to enhance the convection; 3) TMS in the reverse direction of natural convection; 4) TMS in alternative directions [1]. These four experiment cases were precisely controlled/monitored: two heat exchangers to control the heating/cooling rate on both lateral walls of the sample ingot; 50 thermal couples to monitor the temperature field during cooling/solidification; metallographic analysis for the as-cast structure and macrosegregation, etc. Previously, the first experiment case (under natural convection) was successfully simulated by the authors [2]. The as-cast structure and macrosegregation shown a satisfactory agreement with the experiment. The current study will simulate the second experiment, i.e. the case under the TMS which is used to enhance the natural convection. The TMS device is placed in the bottom of the sample [3-7]. The inducted Lorentz force acts in the melt region near the bottom surface of the sample and decreases exponentially in the direction away from the bottom surface.

A three phase mixed columnar-equiaxed solidification model, as developed and further extended by the authors [8-10], was used to investigate the effect of TMS on solidification process, especially on the formation of as-solidified structure and macrosegregation. The primary goal of this work is to further validate the numerical model for the ability to 'reproduce' the experiment under the TMS.

2. Model description

The mixed columnar-equiaxed solidification model [8] was introduced previously. Main features are listed as follows:

- Three phases (liquid, equiaxed and columnar) are considered. Their volume fractions, f_ℓ, f_e and f_c, sum up to 1.0. The liquid and equiaxed are treated as the moving phases, and their velocities are obtained by solving the corresponding Navier-Stokes equations. The columnar dendrites, growing from the mold wall, are treated as stationary.
- Ideal morphologies are assumed for the two solid phases: globular morphology for equiaxed and step-wise cylinders for columnar with constant primary dendritic arm spacing λ_1. Considering the drag force and other hydrodynamic interactions, a grain envelope is assigned to equiaxed dendritic crystal. The solid dendrite and inter-dendritic melt are included in this envelope. In this study, the volume ratio of solid dendrite to equiaxed grain envelope f_{si} is set as 0.5.
- Fragmentation of columnar dendrites is considered as the sole source for equiaxed grains. The mass transfer rate by fragmentation from columnar phase to equiaxed phase is calculated with the formula: $M_{ce} = -\gamma \cdot \vec{u}_\ell \cdot \nabla c_\ell \cdot \rho_e$. The coefficient γ is introduced to bridge the unknown correlation between the fragmentation-induced mass transfer rate and other factors contributing to the fragmentation [10].
- The columnar-to-equiaxed transition (CET) occurs when the volume fraction of equiaxed grain envelope reaches to the criterion ($f_e^{env} = f_e/f_{si} = 0.49$) at the columnar tip front.
- All the volume-averaged concentrations (c_ℓ, c_e, c_c) are calculated and the macrosegregation is represented by a segregation index: $c^{index} = 100(c_{mix} - c_0)/c_0$, where the c_{mix} and c_0 are mixture concentration and initial concentration, where $c_{mix} = (c_\ell \cdot f_\ell \cdot \rho_\ell + c_e \cdot f_e \cdot \rho_e + c_c \cdot f_c \cdot \rho_c)/(f_\ell \cdot \rho_\ell + f_e \cdot \rho_e + f_c \cdot \rho_c)$. Thermodynamic equilibrium is assumed at the liquid-solid interface. Thus, the concentration difference between equilibrium (c_ℓ^*) and melt (c_ℓ) works for the driving force for the solidification.
- Electromagnetic forces acting on the melt and equiaxed grains are: $\vec{F} = \vec{B} \times \vec{J}$. It is calculated with the ANSYS Maxwell software.

3. Benchmark configuration

The experiment setup was described by Hachani and co-authors [1, 11, 12]. The geometry of sample ingot (Sn-10 wt% Pb) is a quasi-two dimensional rectangular ($100 \times 60 \times 10$ mm^3). Two heat exchangers set at the two lateral walls were used to control the solidification unidirectionally (figure 1). The temperature difference between these two exchangers was 40 K, and the cooling rate was 0.03 K/s. Because the heat contact resistance exists between the heater and the two lateral walls, the temperature difference between these two side walls is much smaller, approximately 20 K [1, 13]. On each of two heat exchangers, 9 thermocouples were arranged to measure the heat flux extracted from the ingot. Except for these two narrow walls, all others were maintained adiabatic approximately. An array of 50 thermocouples fixed on the wide surface of the crucible by laser-welding was used to record the transient temperature field. Thermodynamic, thermal physical and other modeling parameters [2, 14] used in this study can be found in table 1.

A linear motor which generates the TMS was placed 5 mm underneath the bottom wall of the sample. The calculated electromagnetic force with the ANSYS Maxwell software is shown in figure 2. The inducted Lorentz force acts in the same direction of the thermo-solutal buoyancy to strengthen the melt convection and the motion of equiaxed phase along the sample bottom.

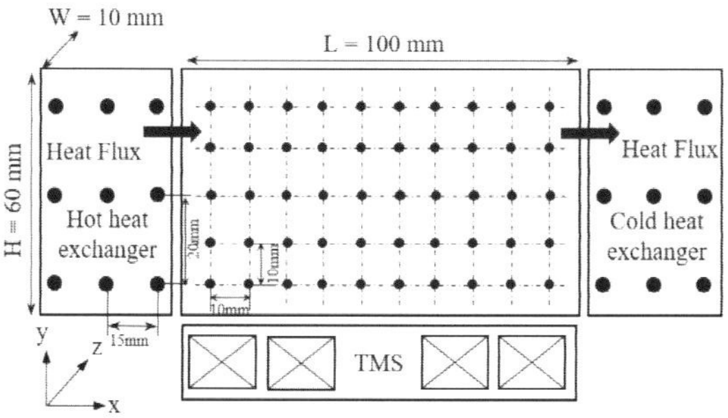

Figure 1. Sketch of the experiment setup: the location of the thermocouples (black dots), two heat exchangers, the travelling magnetic stirring.

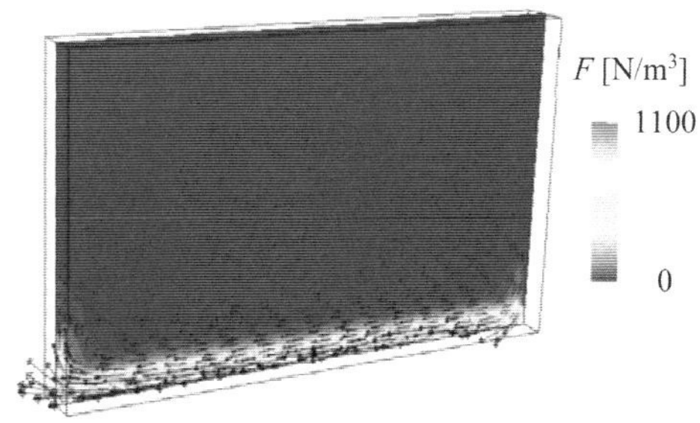

Figure 2. Distribution of electromagnetic force as shown in vectors and color scale.

Table 1. Thermo-physical properties of the alloy and other modeling parameters.

Property/parameters	Symbol	Units	Value
Nominal concentration of alloy (Pb)	c_0	wt%	10
Liquidus temperature	T_{liq}	K	492.46
Melting temperature of pure Sn	T_{m}	K	505.15
Partition coefficient	k_{p}	-	0.0656
Liquidus slope	m	K/(wt%)	-1.2826
Eutectic temperature	T_{e}	K	456.57
Eutectic concentration	C_{e}	wt%	38.1
Reference density	ρ_{ref}	kg/m^3	7000
Solid density	ρ_{s}	kg/m^3	7430
Thermal expansion coefficient	β_{T}	K^{-1}	6.0×10^{-5}
Solutal expansion coefficient	β_{C}	(wt%)$^{-1}$	-5.3×10^{-3}
Primary dendritic arm spacing	λ_1	m	2.25×10^{-4}
Secondary dendritic arm spacing	λ_2	m	6.5×10^{-5}
Diffusion coefficient in liquid	D_ℓ	m^2 s^{-1}	4.5×10^{-9}
Diffusion coefficient in solid	D_{s}	m^2 s^{-1}	1×10^{-12}
Latent heat	L	J/kg	6.1×10^4
Specific heat	$c_{\text{pl}}, c_{\text{ps}},$	J (kg K)$^{-1}$	260
Thermal conductivity	$k_\ell, k_{\text{e}}, k_{\text{c}}$	W(m K)$^{-1}$	55
Viscosity	μ_ℓ	Pa s	1.0×10^{-3}
Gibbs-Thomson coefficient	Γ	m K	6.5×10^{-8}
Fragmentation coefficient	γ	-	0.5

4. Results

4.1 Solidification process

The calculated solidification sequence is shown in figure 3. A full 3D calculation was made, but here only the result on the middle vertical plane (z = 0.005 m) is presented. The starting point (t = 0 s) refers to the moment when the lowest temperature in the sample reaches the liquidus temperature.

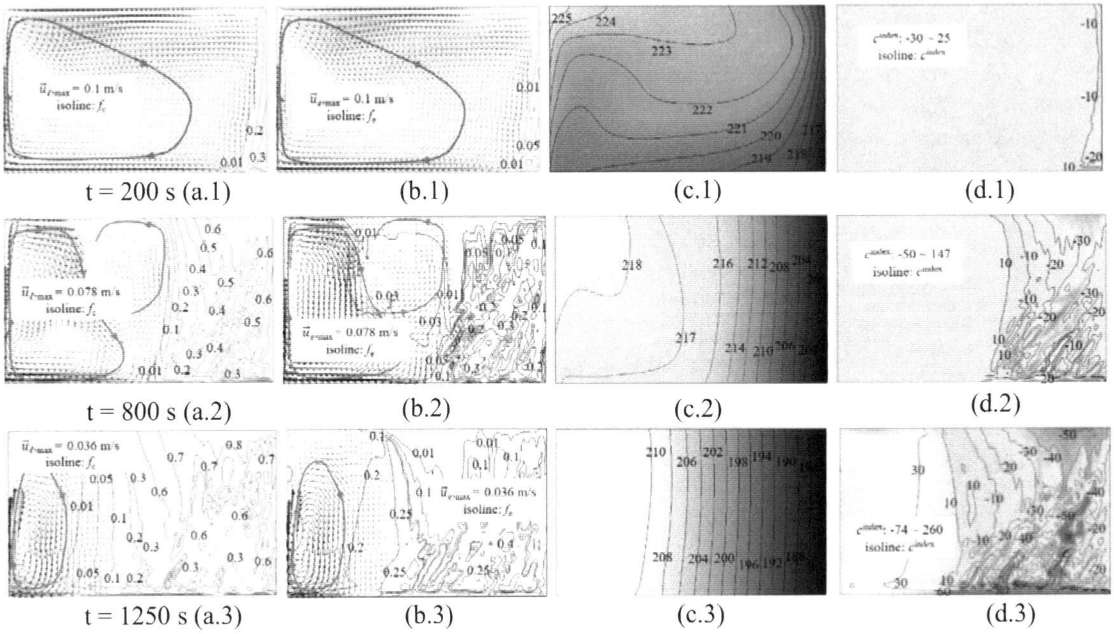

Figure 3. Solidification sequence of the casting sample (on the plane z = 0.005 m): (a.1 - a.3) liquid velocity overlaid with isolines of volume fraction of columnar phase, (b.1 - b.3) equiaxed velocity overlaid with isolines of volume fraction of equiaxed phase, (c.1 - c.3) temperature (°C) in grey scale and isolines, (d.1 - d.3) contour of the macrosegregation with the same c^{index} color scale (-80 ~ 80) and isolines, the real distribution range of c^{index} are labeled in the figures.

At the early stage of solidification (t = 200 s), due to the effect of thermo-solutal convection as enhanced by the TMS, a clockwise vortex biased towards the lower-left corner forms (figure 3(a.1)). The maximum velocities of liquid and equiaxed are equal to 0.1 m/s. As the superheat is extracted from the right side wall and the clockwise convection gathers the cold melt to the lower-right corner, the columnar dendrites initialize and grow from this region. Meanwhile, the equiaxed grains are produced by fragmentation mechanism in front of the columnar-liquid mushy zone (figure 3(b.1)). These heavier grains tend to sediment and also gather at the lower-right corner. Negative segregation starts to build up from the right-bottom corner; while the bulk region remains almost the initial concentration (figure 3(d.1)). The form of the isotherms can clearly show the effect of convection (figure 3(c.1)). The vortex caused by the TMS-enhanced convection makes a visible deformation of the isotherms.

At t = 800 s, the columnar dendrites continue to grow left-wards. Owing to the fact that the liquid density increases with the solute (Pb) enrichment and the increase in the number of equiaxed grains in the bulk region, the velocities of liquid and equiaxed decrease significantly. The maximum values of them are ca. 0.078 m/s (figure 3(a.2) - (b.2)). The TMS is still the dominant force to drive the melt flow and the motion of the equiaxed phase. To this moment, a new small vortex forms in the middle upper region. The newly formed equiaxed grains sediment and are brought into the left region. Additionally, the intensity of the melt flow is strong enough to bring some small grains upwards along the left wall. The equiaxed grains as located in the upper part grow and settle down again with the decrease of the temperature, further interact with the melt flow. The temperature field is also affected by this flow pattern (figure 3(c.2)). Channel segregates initialize and form in the columnar region. Negative segregation region extends leftwards. Correspondingly, positive segregation develops in the bulk melt region (figure 3(d.2)).

At t = 1250 s, the growth of the columnar primary dendrites tips is blocked by the accumulation of the equiaxed phase (figure 3(b.3)). Thus, the CET occurs ($f_e^{env} = f_e/f_{si} = 0.25/0.5 = 0.5 > 0.49$). The vortex is confined to the left part of rest liquid region, and the maximum velocities of liquid and equiaxed are gradually reduced to ca. 0.036 m/s. The isotherms are almost vertical in the solid region, while the temperature gradient almost disappears in the rest liquid region (figure 3(c.3)). More segregation channels form in the columnar region. Solute enriched melt gathers to promote more serious positive segregation in the rest liquid region (figure 3(d.3)).

4.2 As-cast structure
The simulated as-solidified structure, in comparison with the metallography of the experimental result is shown in figure 4. The global mixed columnar-equiaxed structure is well predicted by the simulation, i.e. the columnar structure occupies the right part, while the equiaxed region is mostly distributed in the left part (figure 4(c)). The isoline ($f_e^{env} = 0.49$) indicates the CET. The rest eutectic forms in the left-bottom corner, consistent with the experimental result as well (figure 4 - Zone1). However, some details of the as-cast structure can still not be modelled properly, e.g. the area of columnar phase in the upper region is overestimated by the simulation.

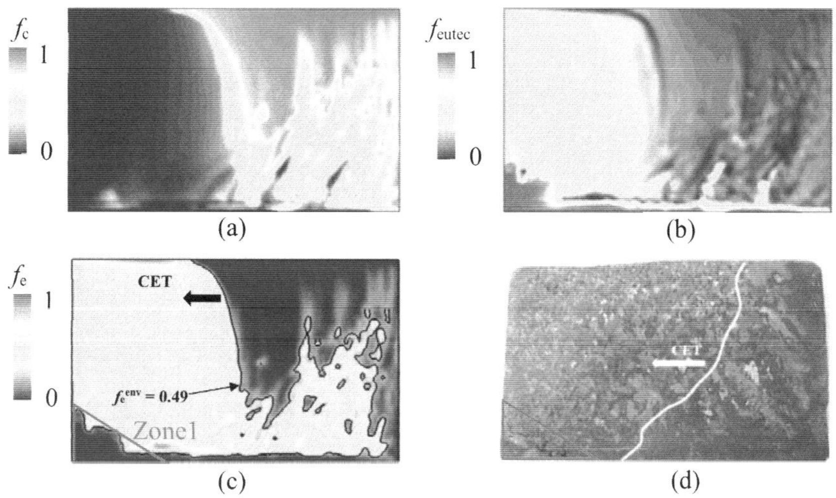

Figure 4. As solidified structure: (a) volume fraction of columnar, (b) volume fraction of the rest eutectic, (c) volume fraction of equiaxed phase overlaid with the isoline of $f_e^{env} = 0.49$, (d) metallographic analysis of the as-cast structure in the laboratory experiment.

4.3 Macrosegregation
The simulated macrosegregation map is compared with the X-radiography of the as-cast casting sample, as shown in figure 5. The X-radiography gives the macrosegregation information of the casting sample through the entire sample thickness, while the simulation result shows the c^{index} distribution on the vertical middle plane. This comparison shows a qualitative agreement between the simulation and the experiment. In the bright region of the lower-left corner, as shown by both simulation and experimental results, there is strong positive segregation; some segregation channels are found in the region of lower-right corner; negative segregation mostly occurs in the upper part.

Quantitative simulation-experiment comparison of the macrosegregation profiles along five horizontal lines (x-direction) were made in figure 6. Fifty measurement points were extracted from the as-cast sample. The positions of these points correspond to the thermocouple positions as shown in figure 1. The chemical method coupled with inductive coupled plasma (ICP) technique were used to get the mean Pb concentration of each point. With this measurement method the channel segregates cannot be analyzed. In order to make a meaningful simulation-experiment comparison, the simulation results of c^{index} were also averaged along the sample thickness (z-direction). The figure 6 (a)-(e) correspond to the c^{index} profiles at y = 0.05, 0.04, 0.03, 0.02, 0.01 m (distance from the sample bottom). The experimental values and the simulation curves show very good agreement regarding to the segregation tendency. Generally, the simulation has overestimated the positive segregation in the left-side of the casting sample.

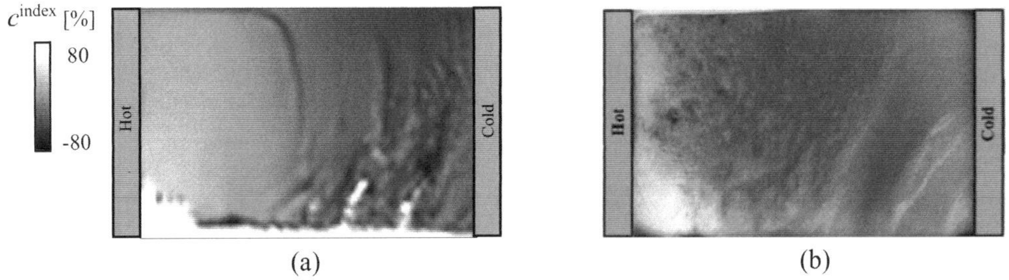

Figure 5. (a) Simulated macrosegregation distribution (c^{index}) on the vertical middle plane (z = 0.005 m), (b) X-radiography of the as-cast casting sample. The simulation result is shown in grey scale, with bright for positive segregation region and with dark for the negative segregation region.

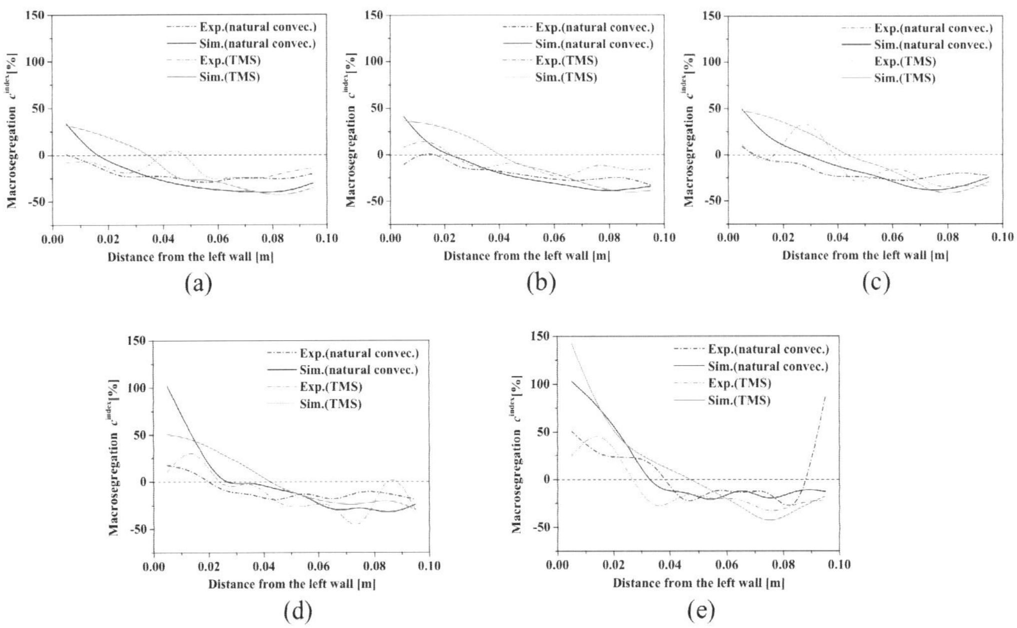

Figure 6. Comparison of the Pb-concentration between experimental analysis (black dotted line) and simulation (black solid line) along the horizontal lines: (a) - (e) correspond to the c^{index} profiles at y = 0.05, 0.04, 0.03, 0.02, 0.01 m (distance from the sample bottom). As reference (or comparison), the experimental analysis (red dotted line) and simulation (red solid line) for the experiment case of pure natural convection without applying TMS were also presented here.

5. Discussion

In this special experiment case, the TMS is applied in the bottom region of the casting sample and the inducted Lorentz force is used to intensify the thermo-solutal convection. In order to understand the effect of this TMS on the solidification process and the formation of as-cast structure and macrosegregation, the calculation for the case of pure natural convection is also made, and its result is compared with the calculation of the current case with TMS. The simulation settings of these two cases are identical except for the TMS. Figure 7 shows the calculated temperature distribution at time t = 900 s. For the case without TMS the temperature range in the sample is distributed between 221 °C (T_{max} = 494 K) and 193 °C (T_{min} = 466 K); while for the case with TMS the temperature range is between 218 °C (T_{max} = 491 K) and 199 °C (T_{min} = 472 K). It means the TMS-enhanced convection can homogenize the temperature field in the bulk liquid region effectively. This condition would favorite the growth of equiaxed structure. As-cast structure and macrosegregation between the two experiment cases are compared in figure 8. As shown in figure 8(b.1), a wider equiaxed zone forms for the case with TMS. The positive segregation zone, which corresponds to the last-to-solidified region, is more confined in the lower-left corner by the TMS-enhanced convection. Meanwhile, this forced convection seems to strengthen the channel segregates in the lower-right columnar region (figure 8(b.2)). However, contradicting to the general knowledge that stronger flow should result in more intensive macrosegregation, the TMS-enhanced convection in the current experimental condition does not strengthen the global macrosegregation significantly. This finding was also supported by the laboratory experiment as shown in figure 6, i.e. the experiment case with TMS did not show obvious stronger macrosegregation than that of the experiment case of pure thermo-solutal convection.

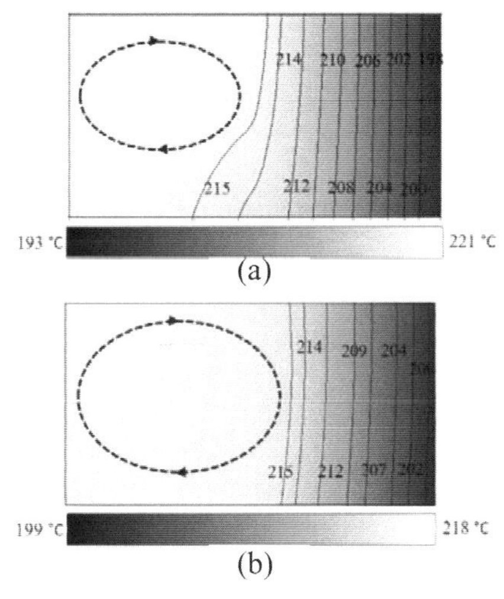

Figure 7. Temperature distribution of two different experiment cases (t = 900 s): (a) natural convection, (b) natural convection and TMS.

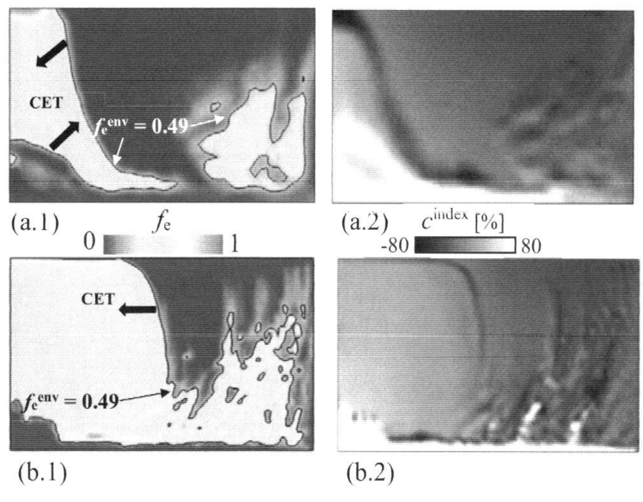

Figure 8. As-solidified structure (volume fraction of equiaxed overlaid with CET line) and macrosegregation (c^{index}) of two different experiment cases: (a.1-a.2) natural convection, (b.1-b.2) natural convection and TMS.

Although the distribution tendency of macrosegregation is well predicted by the numerical model, the quantitative difference between the simulation and experiment is still significant. As shown in figure 6, the positive segregation is overestimated. One main reason may be the ignorance of the remelting of the equiaxed crystals. During the late period of solidification, equiaxed crystals are brought by the flow from columnar tip front region into the bulk melt (left) region. Part of the crystals should have been re-melted there, and dilute the positive segregation to some extent in the lower-left corner region, but this remelting phenomenon is not considered. This model imperfection may be also responsible for some discrepancy between the simulated as-solidified structure and experimental as-cast structure (figure 4). The second reason is the calculation of the travelling magnetic force (TMF). The design of the electromagnetic stirring device (necessary for the ANSYS Maxwell settings) has to be estimated due to lack of sufficient information. Additionally, most thermo-physical properties are not treated as temperature dependent, except for the buoyancy term in the momentum conservation equation where the density of the melt is considered as temperature and concentration dependent.

6. Conclusion

A mixed columnar-equiaxed solidification model was used to simulate the solidification of Sn-10wt%Pb alloy under forced convection condition driven by travelling magnetic stirring (TMS). The TMS is applied in the bottom region of the solidification sample, and the inducted Lorentz force acts to enhance the natural convection. The simulation result was compared with the experiment as performed by Hachani *et al.* [1]. The following conclusions were drawn.

(1) The simulation results agree qualitatively with the experimental results, regarding to the as-solidified mixed columnar-equiaxed structure and macrosegregation pattern. On this agreement, one can safely use the modeling results to analyze the complex solidification sequence, hence to achieve better understanding to the formation mechanisms of the as-cast structure and macrosegregation.

(2) In comparison with the experiment case of pure natural convection condition, the TMS-enhanced convection can create a favorable condition for CET by homogenizing the temperature field in the bulk liquid region, enhancing the fragmentation of the columnar secondary arms and the transport of these fragments to bulk liquid region. These fragments act as the origin of the equiaxed grains. Therefore, the TMS-enhanced convection leads to a wider equiaxed zone.

(3) In the current experiment case, the TMS-enhanced convection does not enhance the global macrosegregation significantly, but it strengthens the channel segregates.

Acknowledgments

The authors acknowledge the financial support from the Austrian Research Promotion Agency (FFG) through the project Bridge I (No. 868070).

7. References

[1] Hachani L, Zaidat K and Fautrelle Y 2015 *Int. J. Heat Mass Transf.* **85** 438-54
[2] Zheng Y, Wu M, Karimi-Sibaki E, Kharicha A and Ludwig A 2018 *Int. J. Heat Mass Transf.* **122** 939-53
[3] Avnaim MH, Mikhailovich B, Azulay A and Levy A 2018 *Int. J. Heat Fluid Flow* **69** 23-32
[4] Wang X, Fautrelle Y, Etay J and Moreau R 2009 *Metall. Mater. Trans. B* **40** 82
[5] Wang X, Moreau R, Etay J and Fautrelle Y 2009 *Metall. Mater. Trans. B* **40** 104-13
[6] Zaidat K, Sari I, Boumaaza A, Abdelhakem A, Hachani L and Fautrelle Y 2018 *IOP. Conf. Ser. Mater. Sci. Eng.* **424** 012052
[7] Yesilyurt S, Motakef S, Grugel R and Mazuruk K 2004 *J. Cryst. Growth* **263** 80-9
[8] Wu M and Ludwig A 2006 *Metall. Mater. Trans. A* **37** 1613-31
[9] Wu M, Ludwig A and Kharicha A 2019 *Metals* **9** 229
[10] Zheng Y, Wu M, Kharicha A and Ludwig A 2017 *Model. Simul. Mat. Sci. Eng.* **26** 015004
[11] Hachani L, Saadi B, Wang X, Nouri A, Zaidat K, Belgacem-Bouzida A, Ayouni-Derouiche L, Raimondi G and Fautrelle Y 2012 *Int. J. Heat Mass Transf.* **55** 1986-96
[12] Wang X and Fautrelle Y 2009 *Int. J. Heat Mass Transf.* **52** 5624-33
[13] Boussaa R, Hachani L, Budenkova O, Botton V, Henry D, Zaidat K, Hadid HB and Fautrelle Y 2016 *Int. J. Heat Mass Transf.* **100** 680-90
[14] Bellet M *et al.* 2009 *Int. J. Therm. Sci.* **48** 2013-6

IOP Conf. Series: Materials Science and Engineering **861** (2020) 012025 doi:10.1088/1757-899X/861/1/012025

In situ observation of directional solidification in Ga-In alloy under a transverse DC magnetic field

S He[1,2], N Shevchenko[1] and S Eckert[1]

[1] Helmholtz-Zentrum Dresden-Rossendorf, 01328 Dresden, Germany
[2] State Key Laboratory of Advanced Special Steel & School of Materials Science and Engineering, Shanghai University, 200072 Shanghai, P. R. China

E-mail: n.shevchenko@hzdr.de

Abstract. The directional solidification of a Ga-25wt%In alloy under the effect of a transverse DC magnetic field is investigated by X-ray radiography. The magnetic field pointing parallel to the X-ray beam is generated by two ring-shaped permanent magnets. The magnetic field reaches values up to ~0.19 T in the field of view. The dendritic growth and the flow patterns of Ga-rich plumes migrating along the solidification front are captured and analysed. It shows that the local fluctuations of solute concentration are partially damped by the magnetic field. At the temperature gradient of 1 K/mm, the growth velocities of solidification front and plumes are not affected. In the case of higher temperature gradient (~2 K/mm), the magnetic field causes an increase of the plume velocity in the horizontal direction and a decrease in the vertical direction while the velocity of the solidification front remains constant. Additionally, it is found that the magnetic field damps the fluctuations of tip velocity and refines the primary arm spacing. Above phenomena are discussed based on the thermoelectric magnetic and electromagnetic braking effects. The in situ experimental data are important for verification and improvement of the existing numerical simulations of solidification under the magnetic field.

1. Introduction

Extensive experimental and theoretical investigations have been performed to reveal the influence of magnetic fields on metals solidification, such as phase migration [1], alignment and orientation of crystal/phase [2], microstructure transition [3], solute redistribution [4] and forced/damped flow in liquid [5]. Particularly, the application of X-ray radiography has attracted considerable attention since it can provide the visualization of alloys solidification [6, 7]. Thus, in situ and real-time observations of microstructure evolution under a magnetic field are widely reported. For instance, many studies have observed the motion of dendrite fragments and equiaxed grains for Al-Cu alloys under a steady magnetic field by means of synchrotron radiation providing the experimental evidence of the thermoelectric magnetic force (TEMF) acting on the solid [8-10]. Meanwhile, natural and forced convection during alloys solidification under various magnetic fields have been preliminarily studied. Wang *et al.* [11] found that an initially tilted solid-liquid interface was changed into a nearly flat shape together with the motion of Cu-rich melt during directional solidification of Al-Cu alloy under a transverse magnetic field, which uncovered the existence of thermoelectric magnetic convection (TEMC). Cao *et al.* [12] observed the local movement of Sn-rich plumes showing the forced melt flow induced by applying a traveling magnetic field. For Ga-In alloys, Shevchenko *et al.* [13, 14] studied the microstructure and the pattern of Ga-rich plume under the impact of an electromagnetic driven convection. These works extended the understanding of alloys solidification under the magnetic field.

In recent years, a large number of interesting phenomena have been found during the directional solidification under the transverse magnetic field, such as deformation of interfacial morphology, enrichment of solute and modified crystal orientation [15-17]. The effect of the magnetic field on the melt convection has been proposed to be responsible for these phenomena in most publications. Nevertheless, the interrelation between convection and dendritic growth is still an open question. In the present work, directional solidification of Ga-25wt%In alloy under a 0.19 T transverse DC magnetic field is experimentally studied using a microfocus X-ray tube. The dendritic growth and melt convection are in situ observed and quantitatively analysed. Further, possible reasons for the variations of plume motion and dendritic growth under the transverse magnetic field are discussed.

2. Experimental procedures

The schematic diagram of experimental apparatus is shown in figure 1(a). The microfocus X-ray tube (XS225D, Phoenix) produces a horizontally aligned divergent beam. In this work, a voltage of 63 kV and an electric current of 50 μA are chosen. The X-ray image intensifier with CCD camera captures the attenuated X-ray beam passing through the sample and then converts the beam into images with a scan rate of 50 half frames per second and the images are integrated over a period of 1 s to reduce the noise level of single image. More details of the X-ray radiography setup can be found in reference [18]. The transverse DC magnetic field is generated by two ring-shaped permanent magnets (Nd-Fe-B) with an internal intensity of 1.17 T. The inner diameter of the magnet is 10 mm, the outer diameter is 20 mm, the height is 6 mm. An iron core is designed to constrain the magnetic flux to the sample volume. The magnets are fixed on the iron core and the sample is located in between, as shown in figure 1(b). Based on this configuration, both the direction and the distribution of the magnetic field within the field of view (⌀10mm) are numerically simulated using the commercial finite element software *COMSOL Multiphysics*. The simulation shows that the direction of the magnetic field is approximately perpendicular to the sample surface, as displayed in figure 1(c). Figure 1(d) compares the magnetic field intensities obtained from the simulation and the measurements by a Gauss meter. The good agreement gives rise to the assumption that the distribution of magnetic field is almost uniform within the field of view and the magnetic field intensity B in the sample is approximately equal to 0.19 T.

The Ga–25wt%In alloy prepared with high-purity Gallium (99.99 %) and Indium (99.99 %) is selected because its low melting temperature ($T_{Liquidus}$=25.7 °C) enables the experiments to be implemented efficiently and flexibly. The alloy is melted and fully mixed in a heating furnace and then filled into a 30×30 mm^2 Hele-Shaw quartz cell with a gap of 150 μm for subsequent directional solidification experiments. The heating/cooling system contains two sets of independently controlled Peltier units which are contacted with the top and bottom edges of the cell, respectively. Miniature K-type thermocouples are used to monitor the temperature. The heating/cooling process and temperature gradients setting are controlled by adjusting the temperatures at both ends. During the experiments, the cell is firstly heated to 55 °C at a constant rate of 0.1 K/s, the temperature gradients G are set to 1 K/mm and 2 K/mm. Then, the cell is solidified at a cooling rate of 0.01 K/s. The entire process is recorded by the microfocus X-ray system. To ensure the statistics of the data, each experiment has been repeated three times. After experiments, the image processing procedure follows the algorithm proposed by Boden *et al.* [18], from which the velocities of solidification front, plume, tip growth and primary dendritic arm spacing can be obtained.

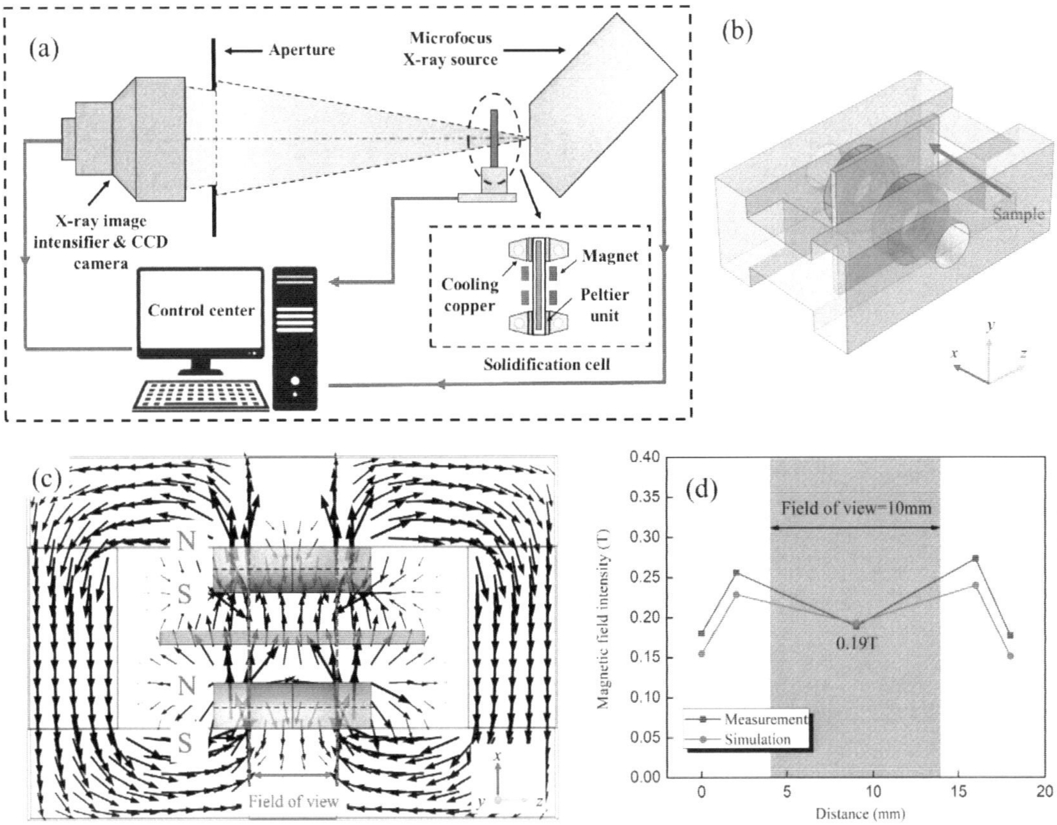

Figure 1. (a) Schematic sketch of the experimental apparatus; (b) configuration of the magnet device; (c) simulated distribution of the magnetic flux; (d) comparison of measured and simulated magnetic field intensities along the y-direction.

3. Results and discussion

The Ga-25wt%In alloy is directionally solidified from the bottom upwards applying two temperature gradients with and without a transverse magnetic field of 0.19 T. During the directional solidification, the primary Indium dendrites grow while Ga is rejected at the solid-liquid interface into the melt. Since the density of Ga is much lower than that of the initial melt, Ga-rich melt is driven to move upward by buoyancy force and thus solute plumes form. The motion of plumes uncovers the convection pattern in the melt. The morphologies of solidification microstructure and Ga-rich plumes for directionally solidified Ga-25wt%In alloy under various conditions are characterized using X-ray radiography, as shown in figure 2. The colour level represents the solute composition. Red colour refers to the Ga-rich area while blue colour indicates the Indium dendrites. From figure 2, it is obvious that typical columnar dendrites and upward-moving Ga-rich plumes appear for all parameter sets considered here. Moreover, the pattern of Ga-rich plume also exhibits some discrepancies comparing the different experimental conditions. For the experiments at the temperature gradient of 1 K/mm, the effect of the magnetic field on the plumes is imperceptible, most of the plumes show large-scale dimensions and extend particularly in vertical direction, as shown in figure 2(a)-(b). In the case of a temperature gradient of 2 K/mm, the typical patterns of natural convection with applied magnetic field appear much smaller and are limited to a smaller region in front of the solid-liquid interface, as shown in figure 2(c)-(d). It suggests that the transverse magnetic field modifies the pattern of natural convection during directional solidification.

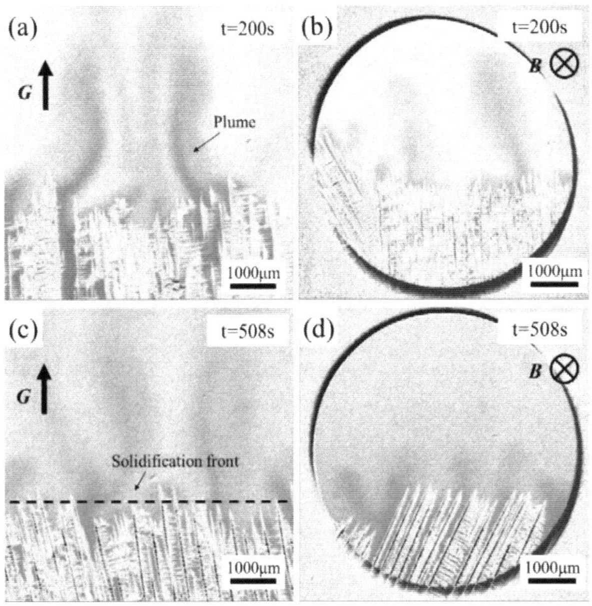

Figure 2. The morphologies of dendritic structure and Ga-rich plumes of directionally solidified Ga-25wt%In alloy at different temperature gradients with and without the transverse magnetic field: (a) G=1 K/mm, B=0 T; (b) G=1 K/mm, B=0.19 T; (c) G=2 K/mm, B=0 T; (d) G=2 K/mm, B=0.19 T.

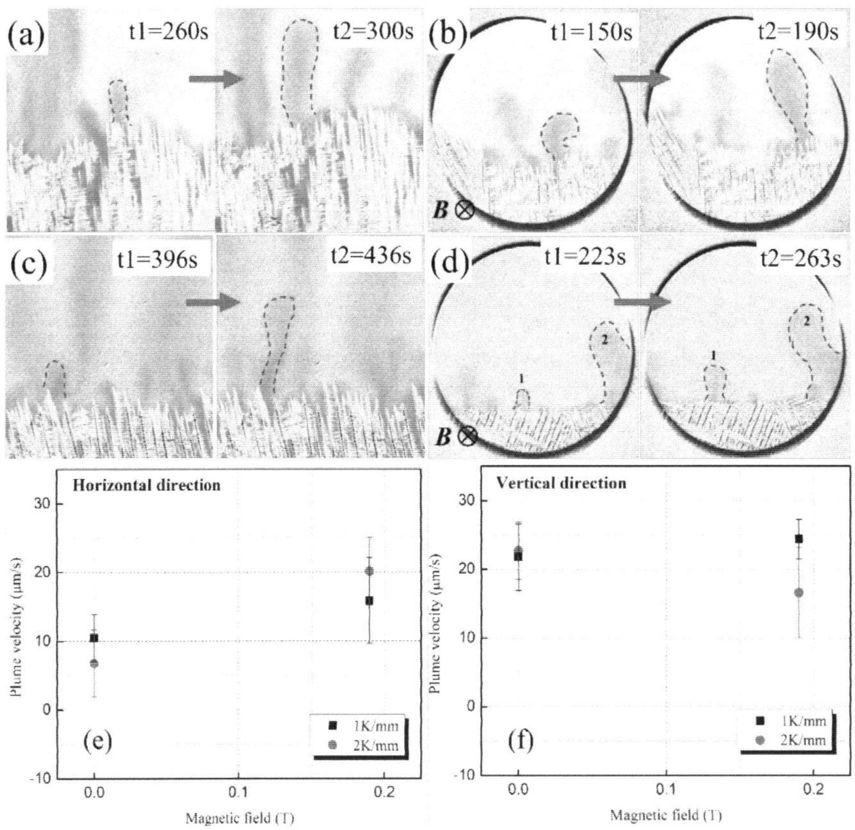

Figure 3. Evolution of Ga-rich plumes during the directional solidification of Ga-25wt%In alloy: (a) G=1 K/mm, B=0 T; (b) G=1 K/mm, B=0.19 T; (c) G=2 K/mm, B=0 T; (d) G=2 K/mm, B=0.19 T, and (e) the horizontal velocities of plumes at different conditions, (f) the vertical velocities of plumes at different conditions.

Figure 3 (a)-(d) show typical X-ray images at successive times during the directional solidification of the Ga-25wt%In alloy under various conditions, in which mobile plumes can be observed. Generally, the movement of plume consists of horizontal and vertical components. Thus, the plume velocities in both of directions are quantitatively analysed, as indicated in figure 3(e)-(f). It is found that the plume velocities in both directions are not influenced by the transverse magnetic field at the temperature gradient of 1 K/mm. In contrast, the application of the transverse magnetic field promotes the horizontal velocity and reduces the vertical velocity when the temperature gradient is 2 K/mm.

Above results demonstrate that the melt convection is changed under the action of the transverse magnetic field. It results in an essential influence on the migration and redistribution of Ga-rich melt in front of the solid-liquid interface. It is well-known that the magnetic field causes an electromagnetic braking (EMB) effect on the melt convection, which has been widely used to suppress the thermal fluctuation and solute segregation during the crystal growth of alloys and semiconductors [19]. In this work, the directions of the plume motion and the magnetic field are perpendicular. As a result, an eddy current will be generated in the conductive melt according to Faraday's law. The interaction between the eddy current and the magnetic field produces a Lorentz force whose direction is opposite to the plume's motion. This Lorentz force will decelerate the movement of the plumes. Thus, it can be deduced that the EMB effect should be responsible for the decrease in vertical velocity, while the increase in horizontal velocity must be attributed to another reason.

As mentioned in previous publications [5,11,20], the existence of the thermoelectric magnetic (TEM) effect near dendritic tip under the magnetic field has been confirmed. It originates from the interaction between the magnetic field and the thermoelectric current (TEC) induced by a temperature gradient and a difference in thermoelectric power S between the liquid and solid phases at the interface. During the directional solidification, the TEM effect generates a forced convection (TEMC) in the melt. The TEMC possibly modifies the movement of the plumes since many experimental and theoretical studies have found that the TEMC induced by the transverse magnetic field can promote the solute transportation in radial direction and finally enhance segregation during directional solidification [11,21,22]. Here, due to $S_{liquid}<S_{solid}$ [23], it is reasonable to infer that the TEC flows from the melt into the dendritic tip driving a TEMC from right to left (figure 4). This effect explains the migration of Ga-rich plumes along the interface and the enhancement of the horizontal velocity of the plumes. In the more general case of larger samples, the kinetics of plumes will also be influenced by the flow field in neighbouring regions. However, such effects cannot be investigated in this study owing to the geometrical limitations of the solidification cell.

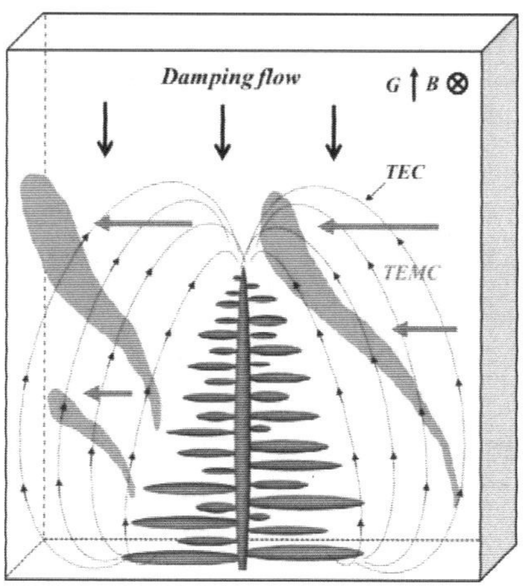

Figure 4. Schematic illustration of flow effects near the dendritic tip under the transverse magnetic field.

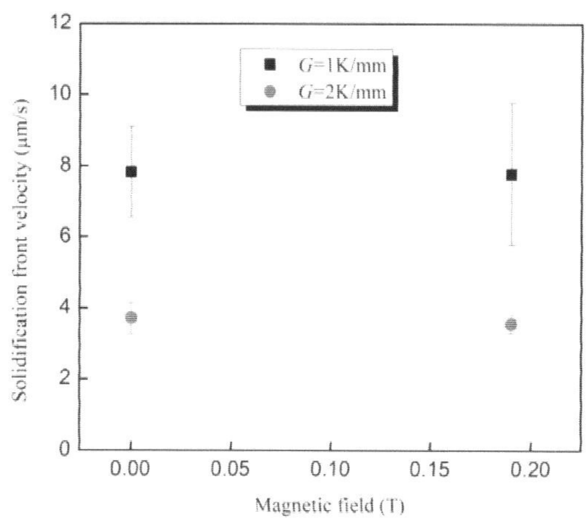

Figure 5. The solidification front velocities of Ga-25wt%In alloy directionally solidified at various temperature gradients with and without transverse magnetic field.

In a next step, the behaviour of dendritic growth under the transverse magnetic field is investigated. The mean velocities of the solidification front in different conditions are determined, as shown in figure 5. The results indicate that the solidification front advances faster at the lower temperature gradient. It is known that the relationship between the temperature gradient (G), cooling rate (R) and solidification front velocity (V_{Front}) can be written as $V_{Front}=R/G$. Thus, for a given cooling rate a decrease in G will lead to an increase in V_{Front}. In addition, the application of the transverse magnetic field has no significant impact on the solidification front velocity at either of the temperature gradients. Figure 6 shows the variations of the tip growth velocity with solidification time for selected single dendrites. It reveals that the dendritic tip growth velocity always fluctuates with time, while the amplitude of the fluctuation is considerably reduced by the transverse magnetic field. This means, the transverse magnetic field stabilizes the dendritic growth. From the X-ray images, it is obvious that the concentration pattern near the tip becomes uniform by applying the magnetic field. The reduction of local gradients of the solute concentration leads to a smooth tip velocity curve.

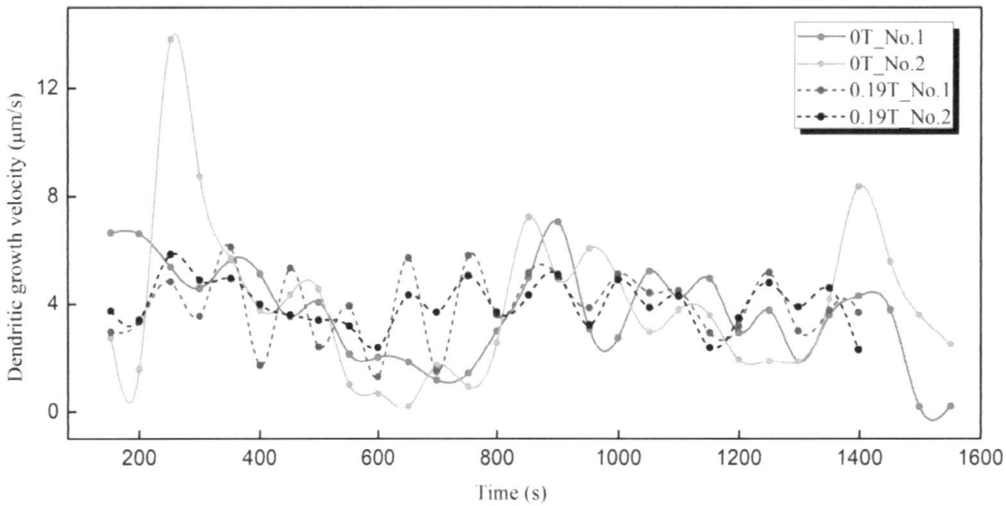

Figure 6. The relationship between tip growth velocity and solidification time at the temperature gradient of 2 K/mm with and without the transverse magnetic field.

Further, the final primary dendritic arm spacings of Ga-25wt%In alloy directionally solidified at temperature gradient of 2 K/mm with and without the transverse magnetic field are measured, which are 196.44±2.87 μm and 231.55±6.12 μm respectively. The statistic results show that the transverse magnetic field causes the reduction of primary dendritic arm spacing by 15.2 %. The refinement of primary dendritic arm spacing under a magnetic field has been also found in previous publications [24-26]. Based on Kurz's calculation [27], the primary dendritic arm spacing λ_1 can be written as:

$$\lambda_1=4.3(\Delta T_0-\Delta T^*)^{1/2}\left(\frac{D\Gamma}{\Delta T_0 k}\right)^{1/4}V^{-1/4}G^{-1/2} \quad (1)$$

where D is the liquid diffusion coefficient, Γ is the Gibbs Thompson parameter, ΔT_0 is the difference between liquidus and eutectic temperature, ΔT^* is the tip undercooling, k is the equilibrium partition coefficient, V is the growth rate of interface. Although the liquid flow is not considered in Kurz's calculation, this model still can reflect the general variation of the primary dendritic arm spacing during solidification [28]. In this work, V is found to keep constant (figure 5). G is adjusted by the Peltier elements. The values of Γ, ΔT_0 and k are not affected by the transverse magnetic field [29,30]. Therefore, the modification of primary dendritic arm spacing may be attributed to the changes of the liquid diffusion coefficient and/or the tip undercooling by the transverse magnetic field. On one hand, the magnetic field accelerates the melt flow in front of the solid-liquid interface and enhances the liquid diffusivity. However, an increase in D would lead to an enlargement of λ_1, which is inconsistent with the observations. On the other hand, it can be observed that the Ga-rich plumes migrate along the interface under the magnetic field which causes a depletion of Ga concentration near the tip resulting in a higher liquidus temperature and an increase in tip undercooling. According to equation (1), it can be concluded that the increasing ΔT^* will reduce λ_1. In addition, due to the low resolution of the images, it is not possible to obtain accurate results of the secondary dendritic arm spacing (SDAS). Thus, the effect of magnetic field on the SDAS is not discussed here.

4. Conclusion

The directional solidification of Ga-25wt%In alloy with and without a 0.19 T transverse magnetic field is experimentally investigated by means of X-ray radiography. In situ observations of Ga-rich plumes show the convection pattern in front of the solid-liquid interface. For G=1 K/mm, the plumes are not affected by the transverse magnetic field. For G=2 K/mm, the plumes are shifted from right to left and their fluctuation are suppressed to some extent compared with those developing without magnetic field.

The magnetic field increases the horizontal velocity of plumes and decreases the vertical velocity. The TEM and EMB effects are deduced to be responsible for the modifications respectively. Further, it is found that the transverse magnetic field refines the primary dendritic spacing and stabilizes the tip growth velocity at G=2 K/mm, which should be attributed to the variation of Ga concentration in front of the interface.

Acknowledgments

In this work, S He was funded by the Natural Science Foundation of China (Nos. 51690162, 51701112, 51604171), National Science and Technology Major Project (No. 2017-VII-0008-0102) and China Scholarship Council (No. 201806890052).

References

[1] Li M and Tamura T 2019 *Philos. Mag.* **99** 1-15
[2] Li L, Bi Y, Zhang H and Cui J 2019 *Mater. Trans.* **60** 939-43
[3] Hou Y, Ren Z, Zhang Z and Ren X 2019 *ISIJ Int.* **59** 60-8
[4] He S, Li C, Guo R, Xuan W, Ren Z, Li X and Zhong Y 2018 *ISIJ Int.* **58** 899-904
[5] Shercliff J A 1979 *J. Fluid Mech.* **91** 231-51
[6] Neumann-Heyme H, Shevchenko N, Lei Z, Eckert K, Keplinger O, Grenzer J, Beckermann C and Eckert S 2018 *Acta Mater.* **146** 176-86
[7] Zeng G, Callaghan M D, McDonald S D, Yasuda H and Nogitac K 2019 *J. Alloys Compd.* **797** 804-10
[8] Wang J, *et al.* 2012 *Appl. Phys. Lett.* **101** 251904
[9] Wang J, Lin X, Fautrelle Y, Nguyen-Thi H and Ren Z 2018 *Metall. Mater. Trans. B* **49** 861-5
[10] Abou-Khalil L, Wang J, Salloum-Abou-Jaoude G, Garrido M, Li X, Ren Z, Reinhart G, Nguyen-Thi H and Fautrelle Y 2019 *Int. J. Therm. Sci.* **145** 106047
[11] Wang J, Fautrelle Y, Ren Z M, Nguyen-Thi H, Salloum-Abou-Jaoude G, Reinhart G, Mangelinck-Noël N, Li X and Kaldre I 2014 *Appl. Phys. Lett.* **104** 121916
[12] Cao F, Yang F, Kang H, Zou C, Xiao T, Huang W and Wang T 2016 *J. Cryst. Growth* **450** 91-5
[13] Shevchenko N, Roshchupkina O, Sokolova O and Eckert S 2015 *J. Cryst. Growth* **417** 1-8
[14] Roshchupkina O, Shevchenko N and Eckert S 2015 *IOP Conf. Ser.: Mater. Sci. Eng.* **84** 012080
[15] Li H, Fautrelle Y, Hou L, Du D, Zhang Y, Ren Z, Lu X, Moreau R and Li X 2016 *J. Cryst. Growth* **436** 68-75
[16] Lu Z, Fautrelle Y, Ren Z and Li X 2018 *Sci. Rep.* **8** 10641
[17] Shen Z, *et al.* 2018 *Metall. Mater. Trans. A* **49** 3373-82
[18] Boden S, Eckert S, Willers B and Gerbeth G 2008 *Metall. Mater. Trans. A* **39** 613-23
[19] Yasuda H, Toh T, Iwai K and Morita K 2007 *ISIJ Int.* **47** 619-26
[20] Lehmann P, Moreau R, Camel D and Bolcato R 1998 *Acta Mater.* **46** 4067-79

[21] Du D, Lu Z, Gagnoud A, Fautrelle Y, Ren Z, Lu X, Moreau R and Li X 2014 *J. Cryst. Growth* **402** 319-24
[22] Lu Z, Zhang Y, Ren Z, Fautrelle Y and Li X 2017 *ISIJ Int.* **57** 84-90
[23] Plevachuka Y, Sklyarchuk V, Shevchenko N and Eckert S 2015 *Int. J. Mater. Res.* **106** 66-71
[24] Kaldre I, Fautrelle Y, Etay J, Bojarevics A and Buligins L 2013 *J. Alloys Compd.* **571** 50-5
[25] Kao A, Cai B, Lee P D and Pericleous K 2017 *J. Cryst. Growth* **457** 270-4
[26] Shen Z, Zhou B, Zhong Y, Zheng T, Ren W, Lei Z and Ren Z 2019 *J. Cryst. Growth* **517** 54-8
[27] Kurz W and Fisher D J 1981 *Acta Metall.* **29** 11-20
[28] Spinelli J E, Rosa D M, Ferreira I L and Garcia A 2004 *Mater. Sci. Eng. A* **383** 271-82
[29] Aoki Y, Hayashi S and Komatsu H 1992 *J. Cryst. Growth* **123** 313-6
[30] Sun Z, Guo X, Guo M, Li C, Vleugels J, Ren Z, Van der Biest O and Blanpain B 2012 *J. Phys. Chem. C* **116** 17676-81

IOP Conf. Series: Materials Science and Engineering **861** (2020) 012026 doi:10.1088/1757-899X/861/1/012026

Comparison of two-phase and three-phase macroscopic models of equiaxed grain growth in solidification of binary alloy with electromagnetic stirring

T Wang[1,2,4], S Semenov[3], E Wang[1,4], Y Delannoy[2], Y Fautrelle[2] and O Budenkova[2]

[1] Key Laboratory of Electromagnetic Processing of Materials (Ministry of Education), Northeastern University, P. R. China
[2] Univ. Grenoble Alpes, CNRS, Grenoble INP, SIMAP, F-38000 Grenoble, France
[3] On leave from Univ. Grenoble Alpes, CNRS, Grenoble INP, SIMAP, F-38000 Grenoble, France
[4] School of Metallurgy, Northeastern University, Shenyang 110004, P. R. China

E-mail: olga.budenkova@grenoble-inp.fr

Abstract. Simulations of equiaxed solidification using two-phase and three-phase models are performed for the experimental benchmark AFRODITE with electromagnetic stirring. A three-phase model presented by authors elsewhere accounts for solid phase, inter- and extradendritic liquid phases. With respect to that model, the two-phase approach can be considered as reduced or simplified, yet, this implies also less number of assumptions regarding closure relations. In simulations, as expected, final segregation obtained with two–phase model is stronger, yet, it is qualitatively similar to the segregation pattern obtained with three-phase model.

1. Introduction

In industrial casting production, a large equiaxed grain region is often expected because it is supposed to provide better homogeneity than the structure consisting of columnar grains [1]. Both industrial and laboratory experiments indicate that forced convection may promote formation of equiaxed grains. Yet, it is also known that interaction between convection and solidifying structure may lead to appearance of channels, freckles and zones of macrosegregation. Post-mortem analysis of sample provides information only about the final state of solidification and can be rather puzzling since evolution of various processes in time remains unclear. As far as in-situ experiments are concerned, they often deal with rather weak convective flow because of small sizes of samples, therefore, these observations could not be easily transposed to large-scale solidification processes. Consequently, numerical modelling is needed to understand how emerging columnar or equiaxed solid structure affects convection and solute transport during solidification process. Numerical simulation of solidification with equiaxed grains is more challenging than with columnar ones because such phenomena as nucleation, grains' drag by liquid phase, and their packing due to interaction with walls and other grains have to be considered. Currently, volume averaged method makes simulation of equiaxed solidification under forced convection at macroscale possible [2]. However, despite the variety of existing equiaxed solidification models [3–7], reliability of results obtained with their use remain uncertain mainly because of two reasons. First, these models rely on various physical assumptions used to close system of equations and contain lots of empirical parameters to which numerical results are quite sensitive [8–13]. Second, there is a lack of experimental data which could help to validate model of solidification in a purely equiaxed regime in presence of convection. Consequently, it is important to perform numerical studies of a particular solidification problem using different parameters or closure relations or even models based on different approaches. For example, in the simulation of 2.45 ton ingot solidification by Li *et al.* [9], it was found that the presence of equiaxed grains cause severe negative segregation at the bottom, however, if maximum nuclei density is larger than some critical value, the segregation will be relieved. With a two-phase equiaxed model, Krane *et al.* [13] simulated a benchmark solidification case using different packing limit fraction. They concluded that for a two-phase model the packing limit fraction had to be very likely smaller than 0.637, which is the value traditionally used in three-phase model. Wang *et al.* [8] demonstrated crucial effect of solute diffusion length on the calculation results of macrosegregation in Hebditch-Hunt case using three-phase model. Wu *et al.* [3] found that in solidification of Al–4.7 wt.% Cu alloy two-phase globular equiaxed model predicts heavier segregation than three-phase dendritic one. Plotkowski *et al.* [11] compared three different grain attachment models. Cited studies helped to understand role of various parameters and assumptions in equiaxed solidification modelling, yet, none of them used experimental results to validate their models.

Present works deals with comparison of three-phase and two-phase modelling of equiaxed solidification of a binary Sn-10wt%Pb alloy under the action of electromagnetic convection. The modelling is based on data issued from the experiment, referred hereafter as AFRODITE, which is briefly described below while details and results of the experiment can be found elsewhere [14-15]. Application of a three-phase model to this experimental case using two-dimensional description was also reported previously [16]. In the present study we are interested if a two-phase model which can be considered reduced (or simplified) compared to three-phase provides similar results regarding grain growth, their motion and segregation pattern.

2. Description for solidification case

The experiment under consideration presents solidification of a binary Sn–10wt%Pb alloy in a rectangular cavity with inner size of 100×60×10 mm (figure 1). In the experiment a constant difference of 40 K is imposed for temperatures at the heaters adjacent to the lateral walls of the cavity, and cooling rate of CR = 0.03 K/s is applied to solidify the sample [17], other walls are regarded as adiabatic. Thermal resistance between heaters and the solidifying volume makes the temperature differences across the latter smaller and equal to nearly 15 degrees [15] that is taken as initial condition in calculations. A travelling magnetic field created with a linear motor placed horizontally along the cavity and below gave rise to Lorentz force acting mainly along the bottom and rapidly decreasing in the vertical direction. An analytical expression for the resulting Lorentz force presented in [18] was used in numerical simulations. Since Lorentz force counteracts the buoyancy, convection in the cavity is less intense compared to the case without the stirring [19], and there is no need to introduce additional damping force to account for front and back walls.

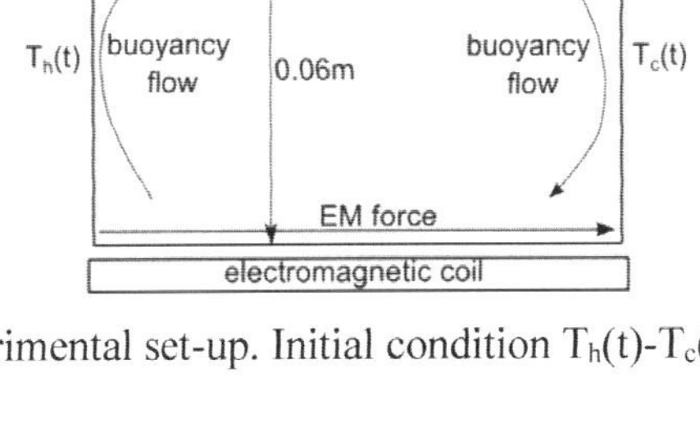

Figure 1. Scheme of the experimental set-up. Initial condition $T_h(t)$-$T_c(t)$=15^{0}C is used in simulations.

However, simulations of the melt flow performed with laminar approach, both for 2D and 3D configurations, could not be converged because of the interaction of two unstable vortices existing in the fluid. The vortex appeared due to the electromagnetic stirring cannot overcome the thermal convection near the hotter boundary. Consequently, for further simulations a realizable k-ε turbulent model is used while calculated turbulent viscosity is rather low that indicated weak turbulence.

3. Model description

The shape of equiaxed grain depends on cooling condition and grain size. Generally, equiaxed grains prefer to grow as globular ones when they are still small [7]. In most conditions, transition from globular shape to dendritic one happens due to trigging of faster velocity of dendrite tip compared to diffusion-controlled velocity [3]. In present work, similar to other ones [5, 20], we assume dendritic shape of grains for three-phase model and globular one for two-phase mode (figure 2). In three-phase model, the liquid phase is divided into inter- and extradendritic parts. The interdendritic liquid in the model is united with solid "skeleton" thus giving grain phase surrounded by an imaginary envelope. Details of three-phase model can be found elsewhere [8, 16], below only a two-phase model is shortly presented with its governing equations given in table 1. Major differences of two models are described at the end of this section and summarized in table 2.

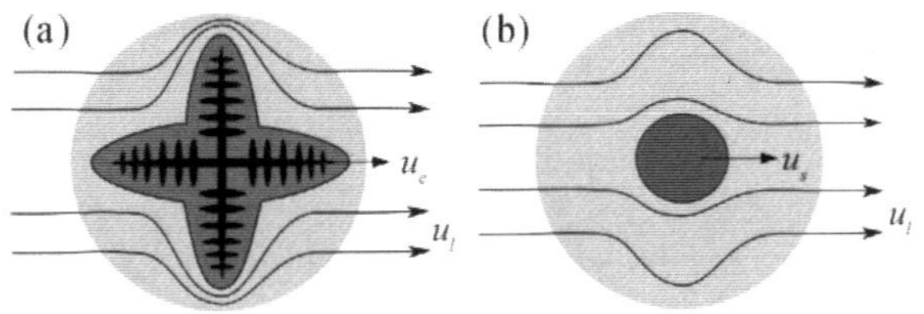

Figure 2. Schematic figure of an equiaxed grain with dendritic shape (a) and globular solid grain (b).

In two-phase model, the volume fraction of solid phase f_s and liquid phase f_l satisfy the constraint $f_s+f_l=1$. Both phases have their proper velocities, temperatures T_s and T_l and concentrations of the solute c_s and c_l. Initial state of the model is supposed to be pure melt of a nominal concentration c_0, whose temperature is above the liquidus temperature. With condition of thermal equilibrium between solid and liquid phases, an infinitely fast heat transfer between the phases is assumed which is provided in the model with a large value of heat transfer coefficient (table 1). Once the constitutional undercooling somewhere in the volume is larger than critical value, the nucleation happens with the nucleation rate N_Φ giving rise to grains number density (concentration of grains) n. It is supposed that nucleation can happen throughout the process if liquid fraction exists, conditions of constitutional undercooling are satisfied, and total number of grains is below of maximal one defined with n_{max}. Although assumption regarding importance of dendrites' fragmentation as a source of nuclei can be found elsewhere [21], this phenomenon is not taken into account in the models presented here to limit uncertainty of the latter. The transport equation is solved for n thus defining the local number of grains which is related to the size of grains. It is supposed that n is transported with the velocity of the solid phase whose small amount appears simultaneously with nucleation and whose growth occurs according to local undercooling, i.e. to the distribution of the concentration in surrounding liquid. In the vicinity of grain, the latter is defined by convective transport but also by diffusion of the solute rejected by the solid phase according to the phase diagram. Similar to [7] and based on our experience [8] we suppose that the solute diffusion length in the liquid l_l depends on intensity of liquid flow around the grain. The back diffusion in solid is neglected as well as shrinkage phenomenon, i.e. all densities in the conservation equations are constant and equal to the reference density: $\rho_l=\rho_s=\rho_{ref}$. Similar to Založnik and Combeau [5], to model the sedimentation (here the floating) phenomenon, in the buoyancy term, a constant difference between the solid phase density and a reference density is introduced. The Boussinesq approximation accounts for solutal and thermal convection in the liquid phase. To calculate the drag force in two-phase model we use Happel model [22] for a low solid fraction and when grains are packed the Kozeny-Carman model

[23] is applied. Note that in three-phase model the approach is similar but the drag force is calculated with grain fraction since it is applied to grains. Regarding grains packing, we choose to block those grains whose fraction is above a critical one and which are situated either near the wall or near already packed neighbouring grain. This allows us to avoid unphysical situation with blocking of grains brought to the cavity centre by forced convection.

Table 1. Conservation equations, source terms, and auxiliary expressions

1. Conservation equations

Mass:

$$\frac{\partial(f_l\rho_l)}{\partial t}+\nabla(f_l\rho_l\vec{u}_l)=M_{sl}-M_\Phi\ ;\qquad \frac{\partial(f_s\rho_s)}{\partial t}+\nabla(f_s\rho_s\vec{u}_s)=M_{ls}+M_\Phi$$

Momentum:

$$\frac{\partial(f_l\rho_l\vec{u}_l)}{\partial t}+\nabla(f_l\rho_l\vec{u}_l\vec{u}_l)=-f_l\nabla p+\nabla\left[\mu_l f_l\left(\nabla\vec{u}_l+(\nabla\vec{u}_l)^{\mathrm{T}}\right)\right]+\vec{u}_l M_{sl}+\vec{F}_{Bl}+\vec{U}_{sl}^D+f_l\vec{F}_{ems}$$

$$\frac{\partial(f_s\rho_s\vec{u}_s)}{\partial t}+\nabla(f_s\rho_s\vec{u}_s\vec{u}_s)=-f_s\nabla p+\nabla\left[\mu_s f_s\left(\nabla\vec{u}_s+(\nabla\vec{u}_s)^{\mathrm{T}}\right)\right]+\vec{u}_l M_{ls}+\vec{F}_{Bs}+\vec{U}_{ls}^D+f_s\vec{F}_{ems}$$

Energy:

$$\frac{\partial(f_l\rho_l h_l)}{\partial t}+\nabla(f_l\rho_l\vec{u}_l h_l)=\nabla(k_l f_l\nabla T)+LM_{ls}f_l+M_{sl}h_l+Q_{sl}$$

$$\frac{\partial(f_s\rho_s h_s)}{\partial t}+\nabla(f_s\rho_s\vec{u}_s h_s)=\nabla(k_s f_s\nabla T)+LM_{ls}f_e+M_{ls}h_l+Q_{ls}$$

where $h_l=\int_{T_{ref}}^{T_l}c_p^l dT+h_l^{ref}$, $h_s=\int_{T_{ref}}^{T_s}c_p^s dT+h_s^{ref}$

Solute:

$$\frac{\partial(f_l\rho_l c_l)}{\partial t}+\nabla(f_l\rho_l\vec{u}_l c_l)=\nabla(D_l f_l\nabla c_l)+J_{sl}\qquad \frac{\partial(f_s\rho_s c_s)}{\partial t}+\nabla(f_s\rho_s\vec{u}_s c_s)=J_{ls}$$

Grain number density:

$$\frac{\partial}{\partial t}n+\nabla(\vec{u}_s n)=N_\Phi$$

2. Source terms

Mass:

$$M_{ls}=-M_{sl}=\rho_l\cdot S_s\cdot v_{ls};\qquad M_\Phi=N_\Phi\rho_l\cdot\frac{1}{6}\pi d_0^3$$

Momentum:

$$\vec{F}_{Bl}=f_l\rho_l\vec{g}\left[\beta_T\left(T_{ref}-T_l\right)+\beta_c\left(c_{ref}-c_l\right)\right];\qquad \vec{F}_{Bs}=f_s(\rho_s^b-\rho_{ref})\vec{g}$$

$$\vec{U}_{ls}^D=-\vec{U}_{sl}^D=K_{ls}(\vec{u}_l-\vec{u}_s)$$

Energy:

$$Q_{ls}=-Q_{sl}=H^*(T_l-T_s),\ H^*=1\times10^9\ W\ m^{-3}K^{-1}$$

Solute:

$$J_{ls}=-J_{sl}=(M_{ls}+M_\Phi)\cdot c_s^*$$

Grain number density:

$$N_\Phi=\begin{cases}f_l(n_{max}-n)/\Delta t, & \text{if } (\Delta T>\Delta T_{nucl}) \text{ or } (n<1)\\ 0, & \text{else}\end{cases}$$

Auxiliary expressions

$$c_l^*=\frac{T_l-T_0}{m};\qquad c_s^*=k\cdot c_l^*;\qquad S_s=f_l\cdot(36\pi\cdot n)^{1/3}\cdot f_s^{2/3}$$

$$v_{ls}=\frac{D_l}{l_l}\cdot\frac{c_l^*-c_l}{c_l^*-c_s^*};\qquad K_{ls}=\begin{cases}4f_l^2\beta^2\dfrac{\mu_l}{d_s^2}, & \text{if } f_s<f_{s,p}\\ \dfrac{180\mu_l f_s^2}{f_l d_s^2}, & \text{if } f_s>f_{s,p}\end{cases};\qquad l_l=\frac{d_s}{2}\left(\frac{1}{1-f_s^{\frac{1}{3}}}+\frac{Sc^{\frac{1}{3}}Re^a}{3f_l}\right)^{-1}$$

$$\beta=\left\{\frac{9}{2}f_s\frac{2+4/3f_s^{5/3}}{2-3f_s^{1/3}+3f_s^{5/3}-2f_s^2}\right\}^{1/2}$$

The main difference between the three-phase and two-phase models is related to the phase transition mechanism that is governed by solute and heat transport, in kinematic interaction between the phases and in treatment of grain packing, each item is discussed below.

1. In two-phase model the phase transition occurs between two phases, solid and liquid, and is controlled by solute distribution in the surrounding liquid, rejection of the solute happens into liquid surrounding the grain. In three-phase model solidification can be regarded as strong coupling of two processes. The first one, notated in the table 2 as $l \to d$, consists of grain enlargement that is related to the tip growth and is seen as expansion of the grain envelope with the velocity v_{env}. This process is governed by the solute undercooling ahead of tips and results in transformation of liquid surrounding the grain into interdendritic liquid. The second stage (notated as $d \to s$) is actually transition of the interdendritic liquid to the solid skeleton, which is governed by the solute undercooling in the interdendritic liquid. Rejection of the solute happens at the interface between the solid and interdendritic liquid and further the solute is transported to the liquid surrounding the grain via molecular diffusion. Consequently, in three-phase model interdendritic phase plays role of "buffer" which defines difference in growth rate of grain and solid phase. Therefore, one can expect that solid fraction in the two-phase model will increase faster compared to the three-phase model. On the other hand, increase of grain fraction in three-phase model should be more intense than increase of the solid fraction in two-phase model.

Table 2. Phase transfer rate between phases in models

		Surface area concentration	Interface movement velocity	Phase transfer rate
Two phase model	$l \to s$	$S_s = f_l \cdot (36\pi \cdot n)^{\frac{1}{3}} \cdot f_s^{\frac{2}{3}}$	$v_{ls} = \frac{D_l}{l_l} \cdot \frac{c_l^* - c_l}{c_l^* - c_s^*}$	$M_{ls} = S_s \cdot v_{ls}$
Three phase model	$l \to d$	$S_e = f_l \cdot (36\pi \cdot n)^{\frac{1}{3}} \cdot f_e^{\frac{2}{3}}$	$v_{env} = \Phi_M \frac{D_l \cdot m_l \cdot (\kappa - 1) \cdot c_l^*}{\pi^2 \cdot \Gamma} \times \left[\frac{(c_l^* - c_l)}{c_l^*(1-k)}\right]^2$	$M_{ld} = S_e \cdot v_{env}$
	$d \to s$	$S_{ds} = f_e \cdot \rho_s \cdot \frac{2 \cdot f_d^e}{\lambda_2}$	$v_{ds} = \frac{D_l}{l_d} \cdot \frac{c_l^* - c_d}{c_l^* - c_s^*}$	$M_{ds} = S_{ds} \cdot v_{ds}$

Note: $\Phi_M = 0.683$ is growth shape factor assuming an octahedral grain envelope shape [10].

2. In two-phase model liquid and solid phase are moving independently and exchange momentum. In three-phase model we assume that the interdendritic liquid and solid phase share the same velocity field, and momentum interaction happens between grain and surrounding liquid. Consequently, with accounting of the first item, we assume stronger kinematic interaction between phases in three-phase model. Yet, further discussion depends on properties of material and solidification conditions: whether only natural convection is involved and what is the density relation between liquid and solid phases.

3. It is of high importance at which fraction grains are getting packed since it determines intensity of the convective transport between the packed grains and the latter, at large extent, defines final segregation pattern. In three-phase model the packing is applied to grains while solid fraction inside can be rather small, i.e. the region of interdendritic liquid can be large that assumes growth of grains with further enrichment of the interdendritic and extradendritic liquid. In two-phase model the packing is applied to the solid phase that means that to have similar situation with large amount of liquid between blocked solid the packing has to be performed earlier. It is widely accepted that the packing fraction limit for the grain phase is 63.7% [10, 12, 24], that corresponds to the packing fraction of randomly arranged monodisperse spheres. This value is kept in simulations with three-phase model. The packing limit for solid phase in two-phase models is generally accepted to be in range $f_{s,p}$=0.1-0.5 [13, 25, 26], large dispersion is related to the type of the solid grain, dendritic or globular. In present simulations for

a two-phase model a value of $f_{s,p}$=0.3 is defined, similar to the work [27]. It should be underlined that in two-phase model all liquid between grains can move while in three-phase it applies only to extradendritic liquid, that means that whatever is adjustment of the packing limit, behaviour of systems composed of three and phases will be different.

4. Results and discussion

To understand results presented below one should remember that solid phase is supposed to be lighter than liquid and this difference increases with enrichment of fluid by rejected solute. This means that growing grains float in the liquid instead of been sediment. Simulations start with overheated liquid in order to have the temperature and flow field installed before solidification is started. First nuclei appear nearly at 180 s of calculations at the upper corner of the colder boundary. The solid (in 2-phase model) and equiaxed (in 3-phase model) phase which starts to appear is brought to the centre of the cavity where it is remelted. Solidification along the whole colder wall begins at ~400 s of calculations. Then, the grains which nucleate closer to the bottom first travel along the cold wall that allows them to grow till the fraction at which they are packed while some of them can travel along the upper wall without been remelted. Since that time the solid grains in two-phase model and equiaxed grains in three-phase model start to fill the cavity. After 540 s of calculations temperature fields obtained with two-phase and three-phase models are different because of difference in convective flow (figure 3(a) and (b)). In two-phase model the packed grains completely occupy the upper boundary and damp their convection flow which otherwise would consist of interacting vortices due to natural and forced convection (figure 3(a) and (c)). The forced flow is concentrated in the lower part of the cavity where fraction of the solid phase is low. Liquid from the cold boundary is brought to the cavity centre by forced convection that creates a region of low temperature in the bottom half of the cavity. Also, forced convection defines transport of rejected solute, a part of which is captured in the centre of its vortex.

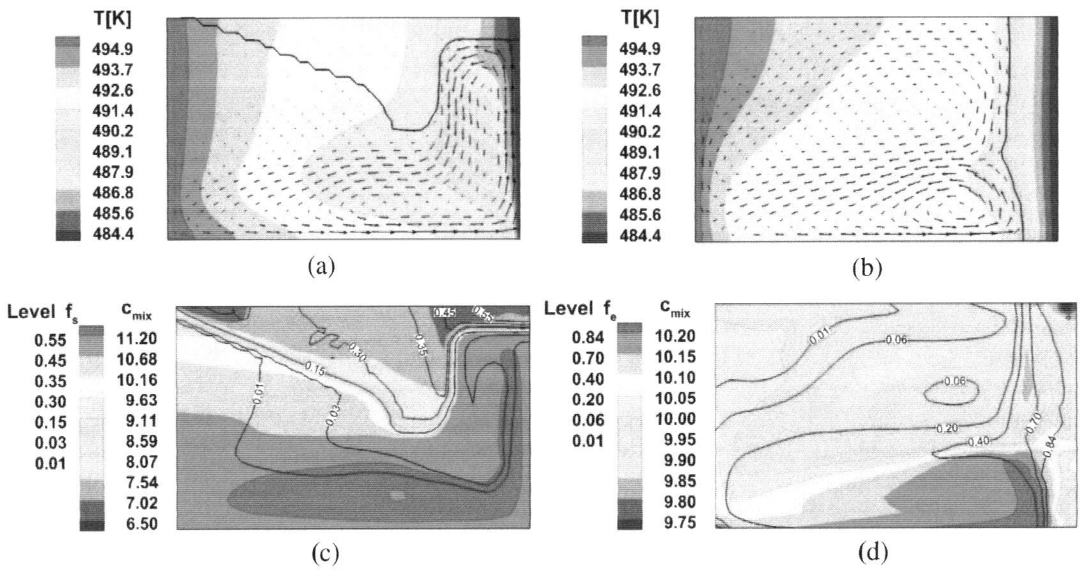

Figure 3. Results obtained at 540 s of simulations: temperature distribution in two-phase (a) and three-phase (b) model, the black line on these picture correspond to the edge of the zone of packed grains defined with $f_{s,p}$=0.3 in two-phase model and $f_{e,p}$=0.637 in three-phase model, velocity field in the liquid is superposed over temperature distribution; map of the average concentration and contours for solid fraction in two-phase model (c) and grain fraction for three-phase model (d).

In three-phase model two vortices with different direction still exist in the cavity at 540 s. The vortex generated by Lorentz force is located only at a lower half and a vortex related to thermal-solute effect occupies the upper half of the cavity (figure 3(b)). While equiaxed grains exist almost everywhere their growth mainly at the colder side, consequently, the packed zone is located there.

In final macrosegregation maps (figure 4), both models show negative segregation at the top and positive in the lower part of the cavity that can be explained as follows. At early stage of solidification grains are brought to the top by buoyancy and forced convection where they solidify taking less value of solute and rejecting the excess to the liquid which becomes heavier. Rejected solute is captured by the forced convection which redistributes it along the bottom but also drag slightly upward. At the same time heavier fluid tends to sediment throughout the process and mostly at the final stage of solidification, when the grains are packed yet liquid continues to move between them.

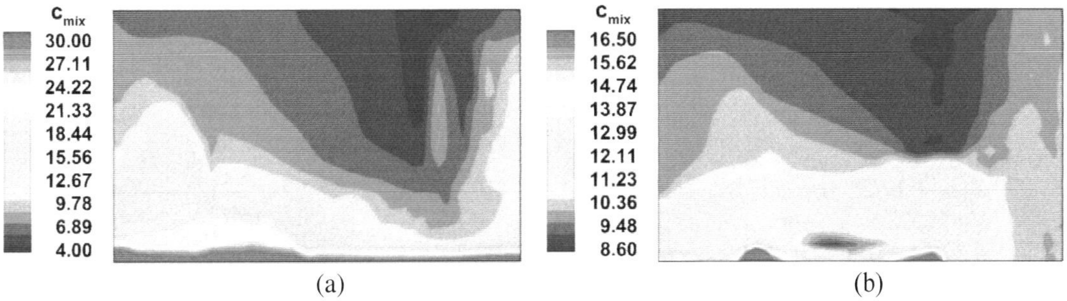

Figure 4. Final macrosegregation map obtained with two-phase (a) and three-phase (b) model.

Although the general character of the macrosegregation patterns is similar for two models, discrepancies are seen in details and some of them can be explained by different assumptions used in the models. In two-phase model both drag force and buoyancy are related to the solid fraction while in three-phase model the buoyancy is applied to the solid fraction and the drag force is related to the grain fraction. That means that at the beginning of solidification globular grains in two-phase model are subjected to stronger buoyancy force and weaker drag force hence grains in two phase model show stronger tendency to float up to the top of cavity and get packed there (figure 3(a)). At the end of solidification, the liquid flow between packed grains in two-phase model is more intense than in three-phase model, due to smaller value of packing fraction and since in three-phase model the interdendritic liquid is attached to the solid skeleton. Because of this in two-phase model solute rejected by grains descents and has time to be redistributed over the whole bottom. An interesting phenomenon which cannot be presented here because of a short format of the paper is the interaction of forced and buoyancy flow which is responsible for the shape of the zone of negative segregation at the upper part of the cavity.

It can be believed that through appropriate adjustment of parameters of both models it is possible to obtain results in the two-phase model closer to those of the three-phase model. To our opinion, results should become more similar if interdendritic fraction is kept as small as possible because, as we stated above, it serves as a buffer between the solid and the extradendritic liquid. In principle, this can be done by adjustment of solute transport through the interdendritic liquid via modification of two diffusion lengths and calculation of the surface concentration, i.e. using various form factors. The critical packing fraction should be modified accordingly.

5. Summary

A two-phase and three-phase equiaxed solidification models are applied to simulations of AFRODITE experiment on solidification of a binary Sn–10wt%Pb alloy under the forced convective flow driven by electromagnetic force. With chosen set of parameters both models provide qualitatively similar results which resemble final distribution of Pb in the sample obtained via X-Ray imaging (figure 12(c) in [15]). Analysis of results show that fraction at which grains are getting packed is one of the crucial parameters since it defines regions of action of forced and buoyancy convection in the case under consideration. Evolution of the convective flow is different in two models because it is subjected to the interaction with solid phase (in two-phase model) and grain phase (in three-phase). Different flow field leads to different temperature evolutions during the process. Surprisingly, despite these differences throughout the process, calculated final macrosegregation maps shows qualitatively similar results. In both cases negatively segregated layer at top is formed and positive segregation zone is observed at bottom similar to the experimental case. This indicates probably that final segregation is largely defined by residual flow through rigid solid network.

Acknowledgements
This work is a joint cooperation between the SIMAP laboratory of Grenoble INP (France) and the Key Laboratory of EPM of Northeastern University (P. R. China). The authors gratefully acknowledge financial support from the National Nature Science Foundation of China (Grant No. U1760206), the National Key R&D Program of China (Grant No. 2017YFE0107900), the Project of Introducing Talents of Discipline Innovation to Universities 2.0 (the 111 Project of China 2.0, No. BP0719037). The SIMAP laboratory acknowledges the financial support provided by the ESA-MAP MICAST project contract 14347/01/NL/SH.

References
[1] Wang F, Wang E, Zhang L, Jia P and Wang T 2017 *J. Manuf. Process.* **26** 364–71
[2] Wu M, Ludwig A and Kharicha A 2019 *Metals* **9** 229
[3] Wu M and Ludwig A 2009 *Acta Mater.* **57** 5621–31
[4] Ni J and Beckermann C 1993 *J. Mater. Process. Manuf. Sci.* **2** 217–31
[5] Založnik M and Combeau H 2010 *Comput. Mater. Sci.* **48** 1–10
[6] Wang C Y and Beckermann C 1996 *Metall. Mater. Trans. A* **27** 2754–64
[7] Appolaire B, Combeau H and Lesoult G 2008 *Mater. Sci. Eng. A* **487** 33–45
[8] Wang T, Semenov S, Wang E, Delannoy Y, Fautrelle Y and Budenkova O 2019 *Metall. Mater. Trans. B* **50** 3039-54
[9] Li J, Wu M, Ludwig A and Kharicha A 2014 *Int. J. Heat Mass Transf.* **72** 668–79
[10] Wu M and Ludwig A 2009 *Acta Mater.* **57** 5632–44
[11] Plotkowski A and Krane M J M 2017 *Metall. Mater. Trans. B* **48** 1636–51
[12] Leriche N, Combeau H, Gandin C A and Založnik M 2015 *IOP Conf. Ser.: Mater. Sci. Eng.* **84** 1–8
[13] Krane M J M 2004 *Appl. Math. Model.* **28** 95–107
[14] Wang X and Fautrelle Y 2009 *Int. J. Heat Mass Transf.* **52** 5624–33
[15] Hachani L, Zaidat K and Fautrelle Y 2015 *Int. J. Heat Mass Transf.* **85** 438–54
[16] Wang T, Wang E, Delannoy Y, Fautrelle Y and Budenkova O 2019 *IOP Conf. Ser.: Mat. Sci. Eng.* **529** 12030
[17] Boussaa R, Hachani L, Budenkova O, Botton V, Henry D, Zaidat K, Ben Hadid H and Fautrelle Y 2016 *Int. J. Heat Mass Transf.* **100** 680–90
[18] Wang X, Moreau R, Etay J and Fautrelle Y 2009 *Metall. Mater. Trans. B* **40** 104–13
[19] Botton V, Boussaa R, Debacque R, Hachani L, Zaidat K, Ben Hadid H, Fautrelle Y and Henry D 2013 *Int. J. Therm. Sci.* **71** 53–60
[20] Ludwig A and Wu M 2002 *Metall. Mater. Trans. A* **33** 3673–83
[21] Zheng Y, Wu M, Kharicha A and Ludwig A 2017 *Model. Simul. Mater. Sci. Eng.* **26** 15004
[22] Happel J 1958 *AIChE J.* **4** 197–201
[23] Bird R B, Stewart W E and Lightfoot E N 1960 *Transport Phenomena* (NewYork: John Wiley and Sons)
[24] Jiang D and Zhu M 2017 *Metall. Mater. Trans. B* **48** 444–55
[25] Plotkowski A and Krane M J M 2016 *Appl. Math. Model.* **40** 9212–27
[26] Olmedilla A, Založnik M, Messmer T, Rouat B and Combeau H 2019 *Phys. Rev. E* **99** 12907
[27] Založnik M, Kumar A and Combeau H 2010 *Comput. Mater. Sci.* **48** 11–21

IOP Conf. Series: Materials Science and Engineering **861** (2020) 012027 doi:10.1088/1757-899X/861/1/012027

Predicting the interfacial heat transfer coefficient of cast Mg-Al alloys using Beck's inverse analysis

M U Bilal and N Hort

Magnesium Innovation Centre - MagIC, Helmholtz-Zentrum Geesthacht, 21502 Geesthacht, Germany

E-mail: muhammad.bilal@hzg.de

Abstract. Apart from many governing parameters, the interfacial heat transfer coefficient (IHTC) has prime importance for the numerical simulation of casting as it quantifies the heat flux between casting and mold (or chill). Most Mg alloys are based on the Mg-Al system and casting is the commonly used production process. The experimental configuration makes it challenging to measure flux and surface temperatures required to evaluate the IHTC. In this study, the IHTC was predicted for a variety of Mg-Al compositions which were cast using a permanent cylindrical mold. Unidirectional heat flow was ensured in order to replicate the experimental conditions for solving the one-dimensional transient heat equation. The numerically determined mold and surface temperatures, using Beck's inverse methodology, were in good agreement with the experiments and analytical solution, respectively. Moreover, the heat transfer behavior across the interface depicted in the form of IHTC was analyzed, also various empirical and numerical aspects of the method are discussed.

1. Introduction

The IHTC is a key parameter for the simulation of casting. Experimentally, it is pretty challenging to quantify the heat transfer between the cast part and mold (or chill) since it is highly influenced by various factors, such as surface roughness, formation of shrinkage gap, alloy and mold temperature [1, 2] etc. In the last two decades, a number of numerical models have been introduced to predict the heat transfer in foundry processes. These predictions are mainly classified into two approaches. The first one takes into account the variation of interfacial gap as an alloy begins to solidify and correlates this gap size with the IHTC [3-5]. The other approach, which has been employed in this study, requires measuring temperature at interior points located at a finite distance from the interface followed by deriving the IHTC using the inverse method first proposed by Beck [6, 7].

Mathematically, Beck's formulation refers to the solution of inverse heat conduction problem (IHCP). It is an ill-posed problem since it does not satisfy the general conditions of existence, stability and uniqueness [7]. The IHTC requires knowledge of the casting and mold surface temperatures along with the heat flux between these surfaces. However, the physical configuration of the IHCP makes it extremely difficult to measure flux or surface temperatures directly either by using a sensor or thermocouple. To evaluate the unknowns, the known temperature history at an interior location serves as one of the boundary conditions of the numerical procedure, while the other boundary expressed in terms of the heat flux is estimated using the inverse analysis. In recent years, researchers have reported the IHTC of different alloy systems with varying process conditions and casting techniques using

Beck's inverse scheme. For example, Lau *et al.* studied the IHTC between cast iron and metallic mold and highlighted the characteristic heat transfer behavior [8]. Kumar *et al.* investigated the components of heat flux in bar and plate of an Al-Cu alloy and concluded that the heat flux did not show spatial distribution in case of bars as compared to plates [9]. Santos *et al.* determined the effects of superheat and mold thickness on the IHTC during the solidification of pure Al and Sn, Al-Cu and Sn-Pb alloys [10]. Meneghini *et al.* modelled the issue of metal head in the graving sand casting of an Al alloy and found that the metal head pressure increases the IHTC [11]. Dong *et al.* proposed an empirical expression of the IHTC during the investment casting of Ni based single crystal blades [12]. Zhang *et al.* included the release of latent heat in their model which resulted in an effective specific heat dependent IHTC [13]. All these and numerous other studies indicate the applicability of inverse analysis for a diverse range of problems, hence it is a dominant method for estimating the IHTC at metal-mold interface in shape castings.

In this work, we aim to present the IHTC of Mg-Al alloys. Mg and its alloys belong to the category of lightweight materials and have gained attention in recent times as being promising candidates for structural applications [14]. The alloys under consideration are: Mg-3Al, Mg-6Al and Mg-9Al. The objective is to highlight the heat transfer behavior and the resultant IHTC. For the sake of verification, the approximated mold and surface temperatures will be compared with the experimental and analytical results, respectively, also other numerical aspects of the method will be discussed. So far, the numerical prediction of the IHTC of Mg alloys has been a less addressed issue as compared to other non-ferrous alloys. Our effort in this regard will be a fruitful addition to the existing literature [15, 16].

2. Experimental procedure

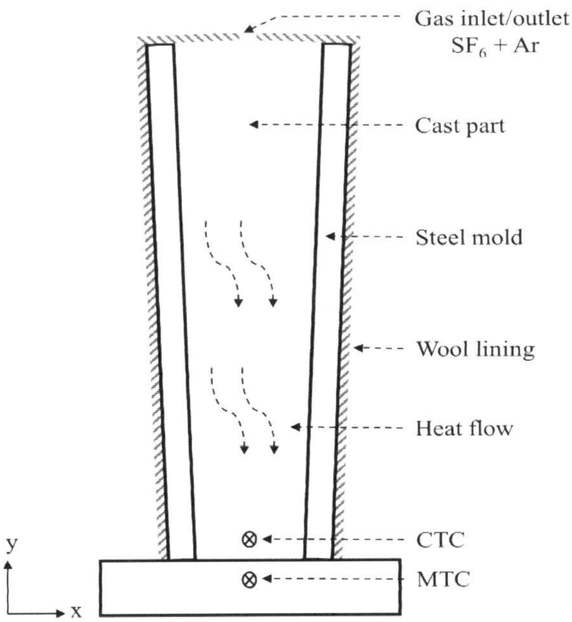

Figure 1. Schematic illustration of the casting setup.

A slightly tapered and cylindrical shaped permanent mold of steel was used to cast Mg-3Al, Mg-6Al and Mg-9Al. The common practice in IHTC studies is to restrict the heat flow in one direction. This experimental configuration can be easily replicated in the numerical procedure by employing one-dimensional heat equation, hence dimensional complexity is reduced. As depicted in figure 1, the adiabatic wool lining restricts the heat flow along the x-axis and in the positive y-direction, but allows flow in the negative y-direction. In this way, the heat is removed in a downward direction, while the alloy solidifies vertically upward. Furthermore, the mold is placed over a copper slab in order to maintain the directionality of heat flow towards the metal-mold interface. Since liquid Mg has an affinity for O_2, it is protected by a mixture of SF_6 + Ar to suppress oxidation [17]. Two K-type thermocouples are installed, i.e., casting thermocouple (CTC) and mold thermocouple (MTC), at approximately 5mm on the either side of the interface. Each thermocouple is attached to a portable data logger that records temperature values at an interval of 1s. The melt and mold are preheated to a temperature of 1023K (±5K). The steel mold is sprayed with the mold releasing agent for easy removal of cast part at the end of process. Before pouring, the melt is stirred and its surface is skimmed. The casting is cooled at room temperature and the recorded temperature profiles are then transferred to a computer program for further investigation.

3. Numerical implementation

The physics behind the IHTC problem can be reasonably estimated by the one-dimensional transient heat equation. Mathematically, its partial differential equation (PDE) is given as

$$\frac{1}{\alpha}\frac{\partial T(y,t)}{\partial t} = \frac{\partial^2 T(y,t)}{\partial^2 y} \qquad 0 \le y \le L, t > 0 \tag{1}$$

where T denotes the temperature in Kelvin, t is the time in seconds, y is the cartesian coordinate, L is the total length of domain in meter and α is the thermal diffusivity with units m^2s^{-1}. The volumetric heat source has not been explicitly taken into account, consequently, the effect of latent heat during solidification is not incorporated and the specific heat is assumed to be constant. This effect will be addressed in future studies. To evaluate this PDE, three conditions should be specified: one initial condition $T(y, 0)$ and two boundary conditions, i.e., $T(0, t)$ and $T(L, t)$. In a direct heat conduction problem, all three conditions are known and the transient temperature profiles at the interior points are determined. In the case of IHCP, one of the boundary conditions is unknown. Beck's inverse method effectively predicts the unknown boundary in terms of heat flux q and yields the casting T_c and the mold T_m surface temperatures required for the IHTC. The method minimizes the sum of squares between the measured S and the estimated temperature T as

$$F(q) = \sum_{i=1}^{l=mr} (S_{\eta+i} - T_{\eta+i})^2 \tag{2}$$

and the minimization is as follow

$$\frac{\partial F}{\partial q} = 0 \tag{3}$$

where r is the number of future temperatures and $m=\Delta\theta/\Delta t$. $\Delta\theta$ and Δt are the time steps for heat flux and temperature, respectively, and represent the same time if $mM=\eta$, where M is the index of last know q and η is the time index for t and T (or S). The unknown value of q is expressed as a vector of elements $q(q_1, q_2, \ldots \ldots \ldots, q_N)$ and each element represents a time step. A key presumption in this analysis is to set the same value of q for a given number of future temperatures, i.e., $q_{M+2}=q_{M+3}= \ldots \ldots \ldots =q_{M+r}=q_{M+1}$, where q_{M+1} is to be evaluated. Substituting equation (2) into (3) and using the Taylor series of $T_{\eta+i}$, we get

$$\nabla q_{M+1}^{l} = \frac{\sum_{i=1}^{l=mr}\left(S_{\eta+i} - T_{\eta+i}^{l-1}\right)\varphi_i^{l-1}}{\sum_{i=1}^{l=mr}\left(\varphi_i^{l-1}\right)^2} \tag{4}$$

where $q^l_{M+1}=q^{l-1}_{M+1}+\nabla q^l_{M+1}$ gives the correction for heat flux. φ in equation (4) is known as the sensitivity coefficient. It is given as

$$\varphi_i^{l-1} = \frac{T_{\eta+i}(q_{M+1}^{l-1}(1+\epsilon)) - T_{\eta+i}(q_{M+1}^{l-1})}{\epsilon q_{M+1}^{l-1}} \tag{5}$$

and calculates the sensitivity in the estimated temperature by altering the value of heat flux by a small factor, i.e., $\epsilon=10^{-2}$ - 10^{-5}. Starting with a guess value of q (for the first time step only), the analysis continues for l number of iterations until the following condition is satisfied

$$\frac{\nabla q_{M+1}^{l}}{q_{M+1}^{l-1}} < 0.005 \tag{6}$$

and the last iterated value of q is taken as an initial q for the next step and the estimation continues until the last time step. After finding all unknowns, the time dependent IHTC h can be written as

$$h = \frac{q}{T_c - T_m} \tag{7}$$

To discretize the numerical problem in this study, the finite difference method (FDM) with forward time marching has been employed. A detailed description of the FDM can be found in [18]. An important aspect, which is worthy to mention, is the stability of explicit FDM. For the sake of stability, the following must be satisfied

$$F_o = \alpha \frac{\Delta t}{\Delta x^2} \leq \frac{1}{2} \tag{8}$$

where F_o is the dimensionless Fourier number, Δt and Δx are the time and grid spacings, respectively. α is set to a constant value of 47×10^{-6} m^2s^{-1} for Mg-3Al, 32×10^{-6} m^2s^{-1} for Mg-6Al and 26×10^{-6} m^2s^{-1} for Mg-9Al [19]. Δt is 0.012s and Δx is adjusted, as indicated in figure 3(c), in such a way that F_o is less than or equals to 0.5.

4. Results and discussion

4.1. IHTC of Mg-Al Alloys

The time variation of IHTC has been presented in figure 2. This characteristic IHTC profile is reported in various studies [8, 20]. Before opening the discussion, it is worthy to mention that the freezing range increases with the increase in Al content, whereas the release of latent heat decreases [21]. In alloys with less Al, this excessive heat may alter the temperature, consequently affecting the IHTC, if measurements are made above the solidus temperature. The IHTC profile has been identified with three distinct stages.

In the first stage, all three alloys are fully liquid and have a perfect contact with the interface. This leads to a maximum heat transfer and the IHTC, in all three cases, reaches its peak value and marks the end of stage one.

In the following stage, a sharp decrease in the IHTC indicates that a previously developed perfect contact has been deteriorated. The underlying reason is the formation of air gaps [3-5, 8, 15, 16, 20]. During casting, nucleation happens heterogeneously and as soon as the melt approaches the liquidus temperature, an adequate fraction of solid appears on the interface. Subsequently, the solidified melt contracts and it moves away from the interface resulting in small gaps. Due to low thermal conductivity, heat transfer through gap/air is low as compared to liquid (or solid). Another noticeable aspect during the second stage is the spike in case of Mg-3Al. Theoretically, this composition will evolve the highest latent heat as compared to other two alloys within a short freezing range. Hence, the latent heat effect is only observed in Mg-3Al, and not in Mg-6Al and Mg-9Al due to its comparatively small magnitude.

The third stage starts with an increase in the IHTC in form of a sharp peak for Mg-3Al and Mg-6Al, while it is a smooth plateau for Mg-9Al. This local peak can be associated with the eutectic liquid. Mg alloys with more than 2% Al exhibit eutectic microstructure [22]. Hwang *et al.* [23] proposes that in the later stages of casting near the eutectic temperature, the exudation of liquid melt may decrease the size of air gaps resulting in an enhanced heat transfer for a small span of time and a local peak in the IHTC. However, this hypothesis does not take into account the nature and magnitude of this peak, also this is not the focus of our study. Afterwards, the remaining melt solidifies and the IHTC reaches a steady state which indicates that the size of air gaps is not changing anymore.

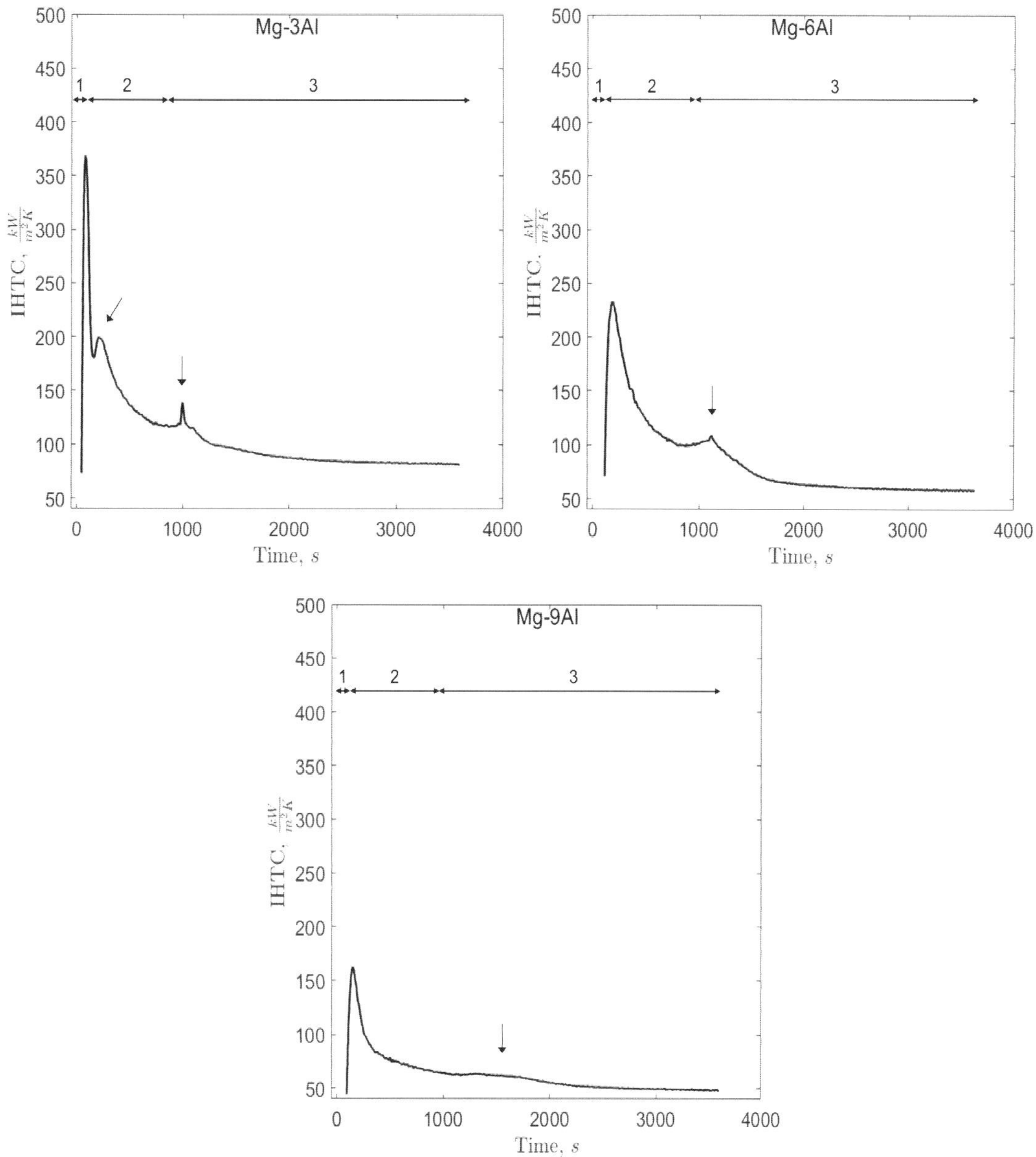

Figure 2. IHTC of Mg-Al alloys.

4.2. IHTC verification and FDM discretization error

The discussion in this section will focus on Mg-9Al, however, the observed results (or trends) are also similar in Mg-3Al and Mg-6Al alloys. To verify the applied numerical procedure, the deviation between the measured and predicted mold temperature should be within an acceptable range. Moreover, the predicted casting and mold surface temperatures, which cannot be determined experimentally, are compared with the analytical solution. Furthermore, if the stability condition of FDM, as in equation (8), is satisfied, the final approximation is reliable and satisfactory. Figure 3(a) shows that the mold temperature measured during experiment and the one predicted by the inverse analysis are quantitatively identical with a maximum absolute error of 0.32K (or °C). The observed deviation is less than 0.5K, which corroborates that the applied numerical procedure yields satisfactory estimates.

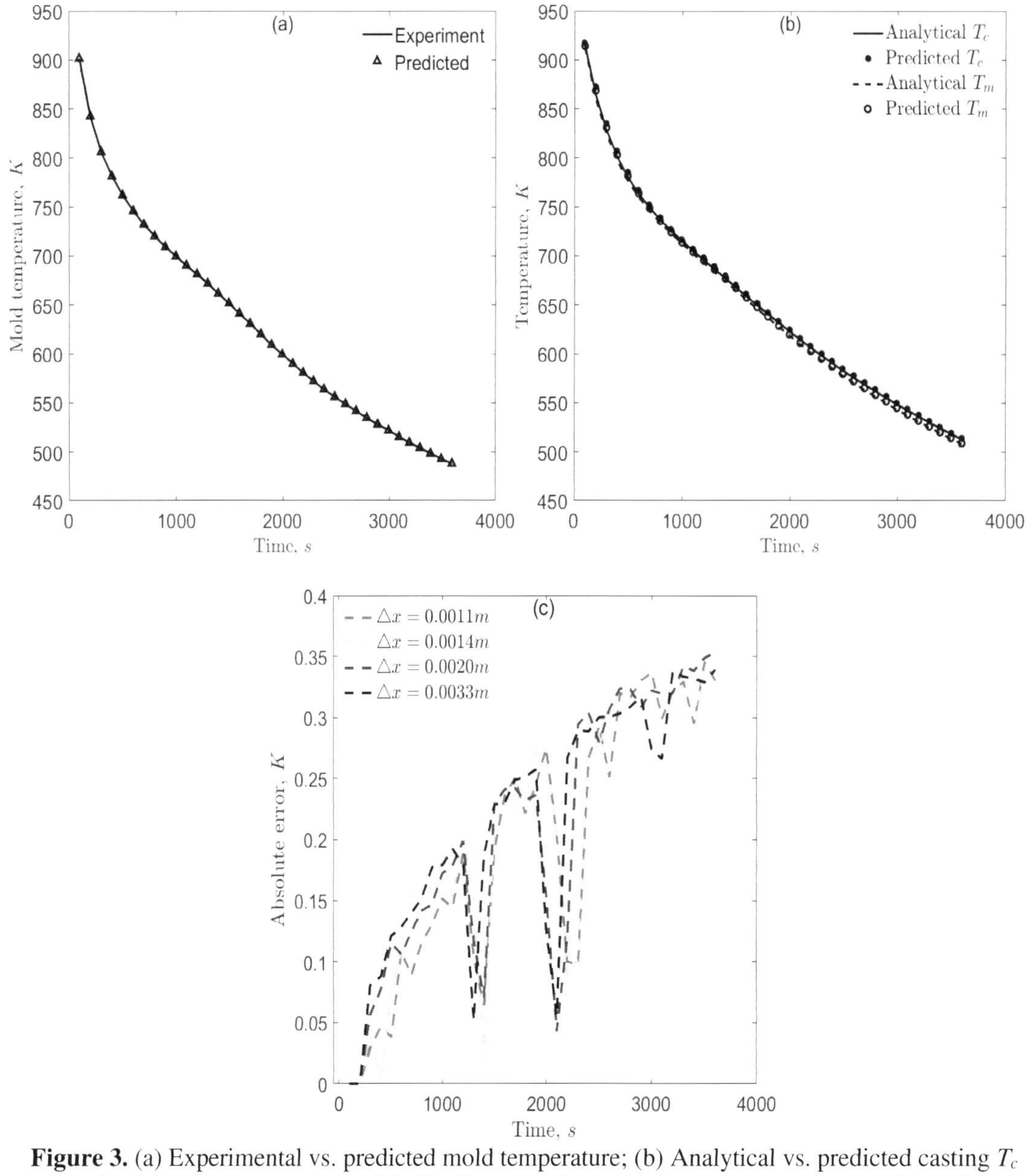

Figure 3. (a) Experimental vs. predicted mold temperature; (b) Analytical vs. predicted casting T_c and mold T_m surface temperatures; (c) Absolute error between the experimental and estimated mold temperature as a function of grid spacing.

On the other hand, T_c and T_m are obtained analytically. The analytical solution of the transient heat equation given in equation (1) with inhomogeneous boundary conditions, i.e., $T(y, 0)=f(y)$, $T(0, t)=a(t)$ and $T(L, t)=b(t)$, is given as

$$T(y,t) = a(t) + [b(t) - a(t)]\frac{y}{L} + \sum_{n=1}^{\infty} \tilde{\phi}_n e^{-\alpha\left(\frac{n\pi}{L}\right)^2 t} \sin\frac{n\pi y}{L} \tag{9}$$

where

$$\tilde{\phi}_n = \frac{2}{L}\int_0^L [f(y) - \left(1 - \frac{y}{L}\right) a(0) - \frac{y}{L} b(0)] \sin\frac{n\pi y}{L} dy \tag{10}$$

and $\tilde{\phi}_n$ corresponds to the Fourier coefficient of $f(y)$. Figure 3(b) represents the analytical solution obtained by employing equation (9) and the predicted surface temperatures. The numerically determined temperatures are fairly close to the analytical ones and the maximum absolute error is observed to be 17K for both T_c and T_m. This error is relatively large as compared to the one in the case of mold temperatures from the experiment and modelling. Nonetheless, the inverse analysis has produced satisfactory approximations, both qualitatively and quantitatively, and the absolute error up to 20K can be regarded as within tolerable limit. Additionally, the surface temperatures appear to be quite identical. The underlying reason is the size of domain taken for analysis. The domain size is around 10mm, while the casting and mold surfaces are only separated by a distance of 1-2mm. It is also noticeable that T_c has slightly higher values as it represents a hot body (or surface) while T_m refers to a cold surface with relatively smaller values.

Equation (8) establishes the dependency of Fourier number on the grid and time spacings, which means one has to limit both parameters to achieve stability. In our study, we fix Δt and vary Δx in order to observe the effect of coarse and fine discretization on the numerical results. Figure 3(c) depicts the increase in grid spacing amplifies the magnitude of fluctuations in the absolute error. Quantitatively, the error is within a reasonable range for all Δx, it is still recommended to have fine discretization in order to avoid large variations between the predicted and experimental values. However, the least possible value of Δx is constrained by equation (8).

5. Conclusion

The IHTC of Mg-Al has been predicted by the inverse analysis. The alloys are cast in a permanent cylindrical mold in a unidirectional heat transfer setup. The results illustrate that the first stage lasts only few seconds until the IHTC reaches its peak value. Afterwards, the second stage lasts few hundred seconds and marks the formation of air gaps as the IHTC decreases sharply. The last stage continues until the casting is fully solidified and the IHTC shows a steady value revealing no further change in the size of air gaps. There are local peaks in the IHTC profile which can be associated with the latent heat and/or eutectic liquid. Furthermore, the inverse analysis results of mold and surface temperatures are in good agreement with the experiments and analytical solution, respectively. Moreover, a coarse grid leads to large variations as compared to a fine grid and the stability condition of FDM limits the value of grid discretization.

Acknowledgements

The authors acknowledge the assistance of G Meister in the cast shop and the financial support from the Helmholtz Association.

References

[1] Griffiths W D 2000 *Metall. Mater. Trans. B: Process Metall. Mater. Process. Sci.* **31** 285–95
[2] Mehraram P 2010 *Measuring heat transfer during twin roll casting of metals* (University of Waterloo) 3–23
[3] Ho K and Pehlke R D 1983 *AFS Trans.* **91** 689–98
[4] Ho K and Pehlke R D 1984 *AFS Trans.* **92** 587–98
[5] Issac J, Reddy G P and Sharma G K 1985 *AFS Trans.* **93** 29–34
[6] Beck J V 1970 *Int. J. Heat Mass Transf.* **13** 703–16
[7] Beck J V, Blackwell B and Clair C R S 1985 *Inverse heat conduction. Ill-posed problems* (New York: J Wiley and Sons) 108-57
[8] Lau F, Lee W B, Xiong S M and Liu B C 1998 *J. Mater. Process. Technol.* **79** (1–3) 25–29
[9] Kumar T S P and Kamath H C 2004 *Metall. Mater. Trans. B* **35** (3) 575–85

[10] Santos C A, Quaresma J M V and Garcia A 2001 *J. Alloys Compd.* **319** (1–2) 174–86
[11] Meneghini A, Sangiorgi Cellini G and Tomesani L 2007 *Int. J. Cast Met. Res.* **20** (3) 159–62
[12] Dong Y, Bu K, Dou Y and Zhang D 2011 *J. Mater. Process. Technol.* **211** (12) 2123–31
[13] Zhang L, Reilly C, Li L, Cockcroft S and Yao L 2014 *Heat Mass Transf.* **50** (7) 945–55
[14] Kainer K U 2003 *Magnesium alloys and technologies* (Wiley-VCH) 1-22
[15] Chen L, Wang Y, Peng L, Fu P and Jiang H 2014 *Exp. Therm. Fluid Sci.* **54** 196–203
[16] Sun Z, Niu X and Hu H 2011 *TMS Magnesium Tech. (San Diego)* 137–42
[17] Hort N, Wiese B, Dieringa H and Kainer K U 2015 *Mater. Sci. Forum* **828–829** 78–81
[18] Smith G D 1985 *Numerical solution of partial differential equations: Finite difference methods* (Oxford university press) 11–77
[19] Pan H *et al.* 2014 *J. Mater. Sci.* **49** 3107–24
[20] Şahin H M, Kocatepe K, Kayikci R and Akar N 2006 *Energy Convers. Manag.* **47** (1) 19–34
[21] Khan S S, Hort N, Eiken J, Steinbach I and Schmauder S 2009 *Adv. Eng. Mater.* **11** (3) 162–68
[22] Wei-Chao Z, Shuang-Shou L, Bin T and Da-Ben Z 2006 *China foundry* **3** (4) 270–74
[23] Hwang J C, Chuang H T, Jong S H and Hwang W S 1994 *AFS Trans.* **102** 877–884

IOP Conf. Series: Materials Science and Engineering **861** (2020) 012028 doi:10.1088/1757-899X/861/1/012028

A meshless solution of a of lid-driven cavity containing a heterogeneous porous medium

V Hatić[1,2], Z Rek[2], K Mramor[1,2], B Mavrič[1,2] and B Šarler[1,2]

[1] Institute of Metals and Technology, Lepi pot 11, 1000 Ljubljana, Slovenia
[2] Faculty of Mechanical Engineering, University of Ljubljana, Aškerčeva cesta 6, 1000 Ljubljana, Slovenia

E-mail: bozidar.sarler@fs.uni-lj.si

Abstract. The main objective of the present paper is to define and evaluate a benchmark test for solving the Navier-Stokes equations in heterogeneous porous medium. The benchmark is of high relevance for proper numerical simulation of flow pattern in the liquid-solid phase-change related phenomena. The test case represents a lid-driven square cavity which consists of a solid, a liquid, and a mushy part. Three different variations of the problem are solved, in which the ratio between the liquid and the mushy part is modified. The Reynolds and the Darcy number of this case are equal to 400 and 2.5×10^{-10}, respectively. The Navier-Stokes equation, with an additional source term to take into account the Darcy effect, is solved with the local meshless diffuse approximate method. The method is structured by using the second-order polynomial basis vector and the Gaussian function to evaluate the weighted least squares approximation. The explicit Euler scheme is used to perform the time discretisation. The non-incremental pressure correction scheme is used to couple the pressure and velocity fields. Results are presented in terms of stream function, velocity magnitude, and mid-plane velocity profiles in the steady state. The accuracy of the numerical method is evaluated by a comparison with the finite volume solution obtained with a commercial solver and an investigation of the node arrangement effect.

1. Introduction

Solidification represents a key phenomenon in a spectra of classical casting and welding processes as well as in the contemporary technologies such as additive manufacturing [1]. The material structure and properties are direct results of this phenomenon. Understanding the solidification is therefore highly important when the defects, such as cracking, hot tearing, porosity, macrosegregation, surface roughness, etc., are tackled. Computer simulations are an extremely powerful and a very cost-effective tool used to perform numerical experiments and to investigate the related phenomena [2,3]. Simulations on the macroscopic scale are considered in this paper. One of the suitable frameworks for simulation of solidification is the volume-averaged formulation of the two-phase liquid-solid system [4, 5]. The conservation equations for the mass, momentum, heat, and species, along with the adjacent constitutive relations, form the phase change model [6]. Obtaining a numerical solution of this numerical model is not trivial, since it is highly non-linear and strongly coupled [7]. The numerical models are therefore prone to errors, which can be identified and reduced with verification tests. The test described in this paper is focused on the mass and momentum conservation part of the solidifying systems with a special attention to the proper fluid flow evaluation in the mushy zone. The main purpose is to investigate the fluid flow in the situation where the liquid phase fraction decreases from one to zero. This is a typical situation, encountered in solidification systems. In order to focus on proper evaluation of fluid flow in such a system, the model is decoupled from the energy and species conservation equations. The position of the mushy zone is thus constant and prescribed in advance. Such test enables the verification of the fluid flow in the mushy zone, which has a significant impact on the macrosegregation and other casting defects. This is the first benchmark of this type to the best of author's knowledge. Other benchmarks with porous media also exist and are available in the literature, but they mainly consider homogenous porous media and the Darcy's law.

Meshless numerical methods [8] represent one of the most rapidly developing fields of computational mechanics. They are praised for their flexibility in the choice of computational nodes. They do not require the mesh construction, which is beneficial in solidification problems. Most noticeably in cases where very fine node spacing is required in the mushy area and a sparse one in the solid phase, where time adaptive computational node arrangements are used to reduce the computational time, and where complex shaped geometries are considered (e.g. pressure die casting). Several successful applications of different types of meshless methods to multiphysics and multiscale phase-change related phenomena as well as other manufacturing phenomena have already been published in recent years [9–17]. The results presented in this paper are computed with the meshless diffuse approximation method (DAM) [18–20] and compared with the solution obtained with the reference finite volume model (FVM).

2. Model formulation

The volume averaged mass and momentum conservation equations for incompressible flow are solved at the macroscopic scale. The mushy zone is described as a porous medium with permeability K which is a function of position in the proposed benchmark

$$\nabla \cdot \mathbf{v} = 0,$$
$$\rho \frac{\partial \mathbf{v}}{\partial t} + \frac{\rho}{g_l}(\nabla \mathbf{v})\mathbf{v} = -g_l \nabla p + \mu \nabla^2 \mathbf{v} - \frac{\mu g_l}{K}\mathbf{v}, \tag{1}$$

where $\mathbf{v}$ is the volume averaged velocity, t is the time, ρ is the density, μ is the dynamic viscosity, p is the pressure, and g_l is the liquid fraction (= porosity). The volume averaged velocity is directly linked to the liquid velocity $\mathbf{v}_l$ with the liquid fraction: $\mathbf{v} = g_l\, \mathbf{v}_l$. The pressure-velocity coupling is implemented with the non-incremental pressure-correction scheme (see [13] for more details).

3. Numerical benchmark definition

3.1. Dimensionless numbers analysis

The dimensionless numbers that effect the Navier-Stokes equation with the Darcy source term are the Reynolds ($\mathrm{Re} = \rho v L / \mu$) and the Darcy number ($\mathrm{Da} = K_0 / L^2$). These two numbers are set to Re = 400 and Da = 2.5 10^{-10} in all cases considered in this paper. The value of these numbers is consistent with the conditions, encountered in direct-chill casting of aluminium alloys [21].

3.2. Geometry

The test resembles the laminar lid-driven cavity benchmark case [22]. Global computational domain is a square shaped cavity with unit sides ($H = L = 1$), which is partially filled with a heterogeneous porous media (see figure 1). The permeability is infinite at the top of the cavity ($y > S + M$) and starts to decrease linearly at the vertical position $y > S + M$, where S and M are the thickness of the solid and the porous medium, respectively. It decreases to 0 at the bottom of the cavity ($y = S$). The permeability is defined through the liquid fraction and is more thoroughly elaborated in the next section. Three different geometrical cases are simulated, where the porous media thickness M is varied (0.25, 0.50, and 0.75).

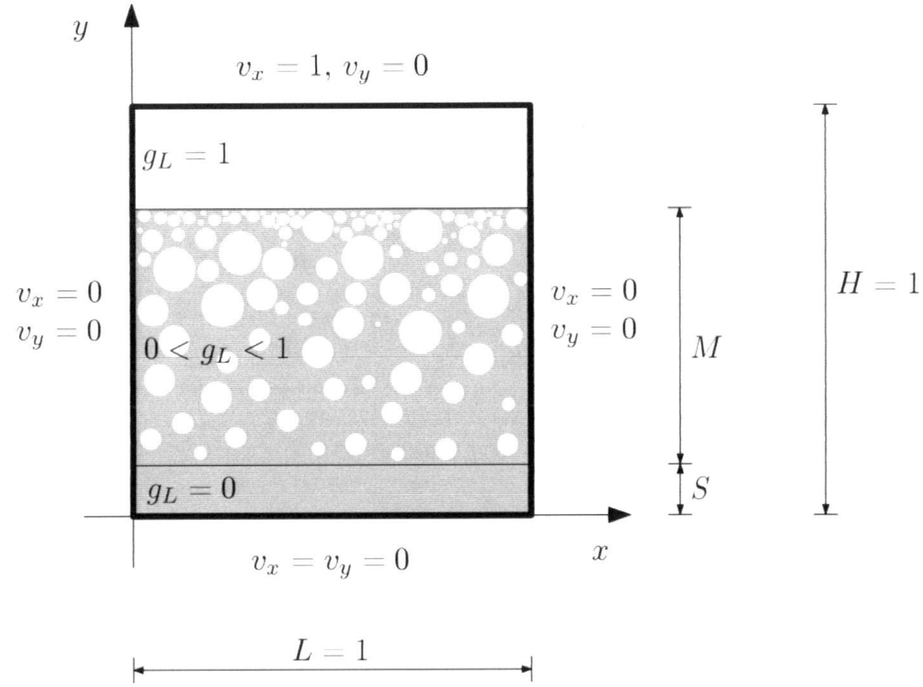

Figure 1. Geometry of the 2D porous cavity benchmark with the description of the boundary conditions.

3.3. Heterogeneous porous media

The permeability K in the porous media is modelled by using the Kozeny-Carman model

$$K = K_0 \frac{g_l^3 + \varepsilon}{(1 - g_l)^2}. \tag{2}$$

where K_0 is permeability constant, which is set to 2.5 10^{-10} m^2 in order to match the desired Darcy number. ε is a small value (e.g. 1 10^{-12}), used to avoid the division by zero in the momentum conservation equation in case $g_l = 0$. The liquid fraction g_l decreases linearly with the vertical position and is defined with the following relation

$$g_l(y) = \begin{cases} 1 & \text{if } y > M + S \\ M^{-1}(y - S) & \text{if } S \le y \le M + S, \\ 0 & \text{if } y < S \end{cases} \tag{3}$$

where

$$S = 0.1H. \tag{4}$$

3.4. Boundary conditions

The Dirichlet boundary conditions are used on all boundaries. The velocity in the x direction is set to $v_x = 1$ m/s on the top boundary and to $v_x = 0$ m/s on the bottom and the side boundaries. The velocity in y direction is set to $v_y = 0$ m/s on all boundaries.

3.5. Initial conditions

The initial values for velocity in the x and the y direction are set to 0 m/s and the initial pressure is set to 0 Pa in the whole domain.

4. The numerical procedure

The partial differential equations are solved with the meshless-diffuse-approximate method. The DAM uses the weighted least squares formulation to determine a locally smooth approximation from a discrete set of data:

$$\hat{f}(\mathbf{x}) = \mathbf{p}(\mathbf{x}, \mathbf{x}_0)\boldsymbol{\alpha} = \sum_{j=1}^{J} p_j(\mathbf{x}, \mathbf{x}_0)\alpha_j, \tag{5}$$

where $\boldsymbol{\alpha}$ is the vector of coefficients and $\mathbf{p(x,x_0)}$ is the basis vector. The second-order polynomials are used as the basis vector. The Gaussian function is used to weight the approximation error in the minimization expression:

$$\theta(\mathbf{x}, \mathbf{x}_0) = \exp\left(-c\frac{\|\mathbf{x}-\mathbf{x}_0\|^2}{h^2}\right), \tag{6}$$

where c and h are free-parameter and scaling factor, which are equal to 5 and to the Euclidian distance between $\mathbf{x}_0$ and the farthest position from $\mathbf{x}_0$ in the discrete set of data.

The method is localized by defining the smooth approximation for each computational node separately. This is performed by associating each node with a unique local neighbourhood, which is used for minimization. Either 9 or 14 nodes are included in the local subdomains for simulations. The stability of the advective term is attained with a shift of the Gaussian weight and evaluation location in the upwind direction. This approach is called the adaptive upwind weight function and it is a successful stabilization procedure [13]. Explicit Euler scheme is used for temporal discretization.

5. Time and spatial discretisation

Both structured and unstructured computational node arrangements are considered (see figure 2). Node arrangements with 50x50, 100x100, 150x150, 200x200, and 250x250 are used in order to observe the node spacing convergence. The time step is set to 1×10^{-4} s in all simulations. The criterion for the steady state is that the relative change in the x mid-plane minimum horizontal velocity between two time steps is less than 1×10^{-6}. The reference (FVM) steady state solution is obtained only for the structured node arrangement with 250x250 nodes. The FVM model is calculated with the steady-state, pressure-based solver. It is solved in 2D with a double precision. The pressure-velocity coupling is performed with SIMPLE approach. Second order upwind stabilisation is used for the convective term. Residual limit of the steady-state solver for the continuity and velocity components are set to 1×10^{-6}.

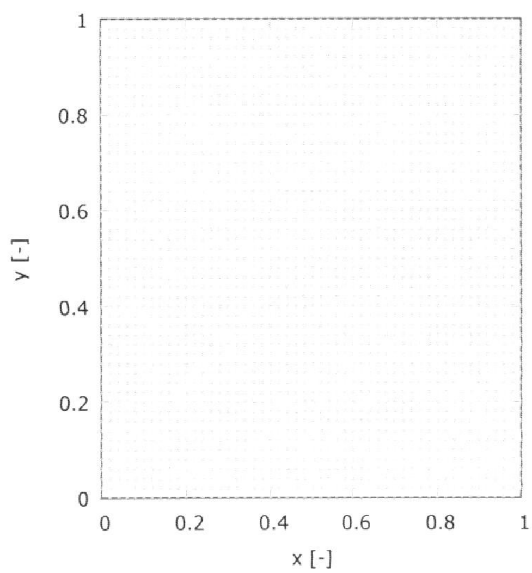

(a)

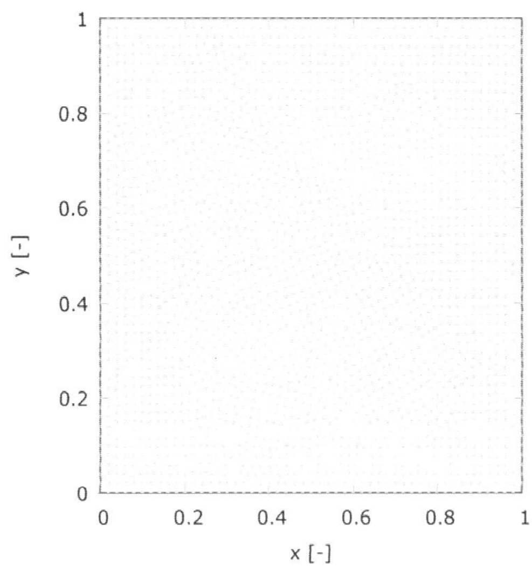

(b)

Figure 2. Comparison of (a) structured and (b) unstructured computational node arrangement (50x50).

IOP Conf. Series: Materials Science and Engineering **861** (2020) 012028 doi:10.1088/1757-899X/861/1/012028

6. Results

The final velocity field calculated with DAM is presented with the velocity magnitude contours and velocity direction vectors in figure 3. The results are additionally presented with figure 5, where the streamline contours and the mid-plane horizontal velocities obtained with DAM and FVM model are compared. The velocity magnitude decreases exponentially in the porous media. The horizontal velocity magnitude is therefore plotted with the logarithmic scale in the right part of figure 5.

The results show that the flow is restricted by the decreased permeability. The effect of porous media is very large even at very high liquid fraction, causing the main vortex to form almost exclusively in the liquid phase. The size of the main vortex is therefore controlled by parameter M. Recirculation is observed for $M = 0.25$ and 0.50, while only the main vortex is observed for 0.75.

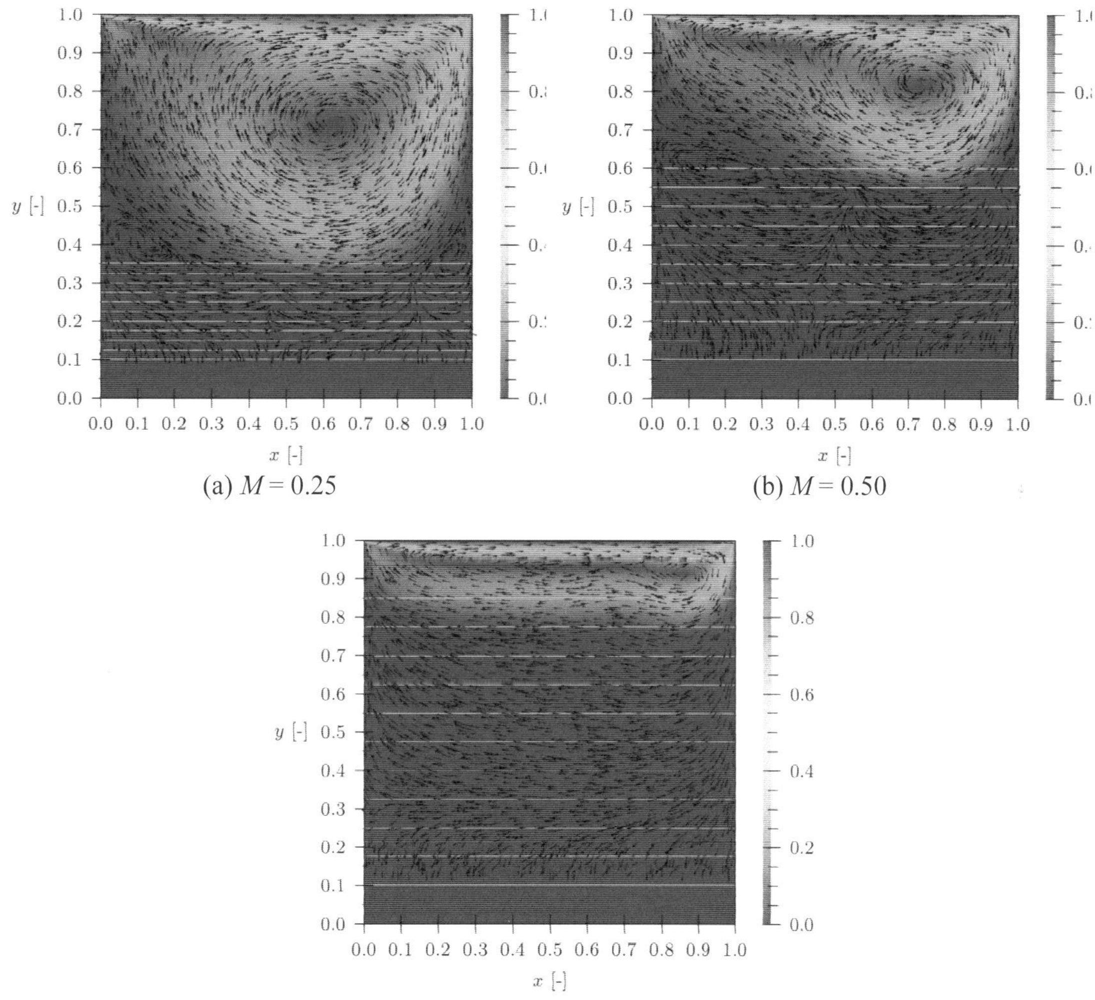

(a) $M = 0.25$ (b) $M = 0.50$

(c) $M = 0.75$

Figure 3. Direction vectors of the velocity field. The DAM results for 250x250 structured node arrangement is shown.

6.1. Convergence

The convergence of the node spacing is evaluated for all three cases in figure 4. The minimum horizontal component of the velocity vector along the vertical line through the geometric centre of the cavity is compared. This value is simply referred as the minimum velocity in the further text. The results show that the solution on the finest node spacing (h = 0.004) is independent of node distribution. This observation is valid for both structured and unstructured node arrangement.

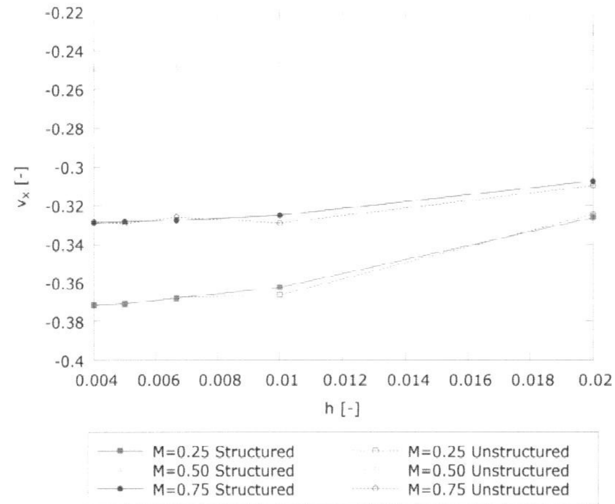

Figure 4. Node spacing convergence for the structured and unstructured node arrangement.

6.2. Verification

The solution obtained with the presented meshless model is compared with the solution obtained with the reference FVM model. Comparisons are presented in figure 4 on the finest node density. The results have a very good match, especially at the top, liquid part of the cavity, where the streamlines overlap completely. The relative difference in the minimum velocity is calculated for the 250x250 (structured) node arrangements

$$E = \frac{\mathrm{v}_{x,\,\min}^{\mathrm{DAM}} - \mathrm{v}_{x,\,\min}^{\mathrm{FVM}}}{\mathrm{v}_{x,\,\min}^{\mathrm{FVM}}}, \tag{7}$$

where superscripts DAM and FVM are used to denote which model is used to obtain the velocity value. The relative difference between the results is equal to -1.09×10^{-2}, -6.33×10^{-3}, and -7.43×10^{-3} for M = 0.25, 0.50, and 0.75, respectively.

The agreement between the results is good, especially in the upper part of the porous media, where permeability is relatively high ($g_l > 0.9$). The difference between the results increases when the absolute value of minimum velocity is below 1×10^{-6}. This is expected, because the FVM solution is calculated with a steady state solver, with the residual limit for the velocity components set to 1×10^{-6}. It is observed that the velocity in the non-permeable (solid) phase is not zero for the FVM solution, which is probably a numerical error. This is not the case in DAM solution, where an additional condition is added to the solver so that the momentum conservation equation is not solved in the solid phase. Therefore, the comparison of the results is not valid in the solid phase.

6.3. Comparison of computational node arrangements

The solution of the problem is obtained on both structured and unstructured node arrangement (see figure 2). The visual comparison shows a slight difference between the results (see figure 4 and figure 5). The difference is evaluated by calculating the relative difference in the minimum velocities for the finest node arrangement

$$E = \frac{\mathrm{v}_{x,\,\min}^{\mathrm{Unstructured}} - \mathrm{v}_{x,\,\min}^{\mathrm{Structured}}}{\mathrm{v}_{x,\,\min}^{\mathrm{Unstructured}}}, \tag{8}$$

where the superscripts are used to denote the type of node arrangement on which the solution is obtained. The relative difference between both node arrangements is equal to 4.35×10^{-4}, -1.71×10^{-3}, and -1.90×10^{-3} for M = 0.25, 0.50, and 0.75, respectively.

IOP Conf. Series: Materials Science and Engineering **861** (2020) 012028 doi:10.1088/1757-899X/861/1/012028

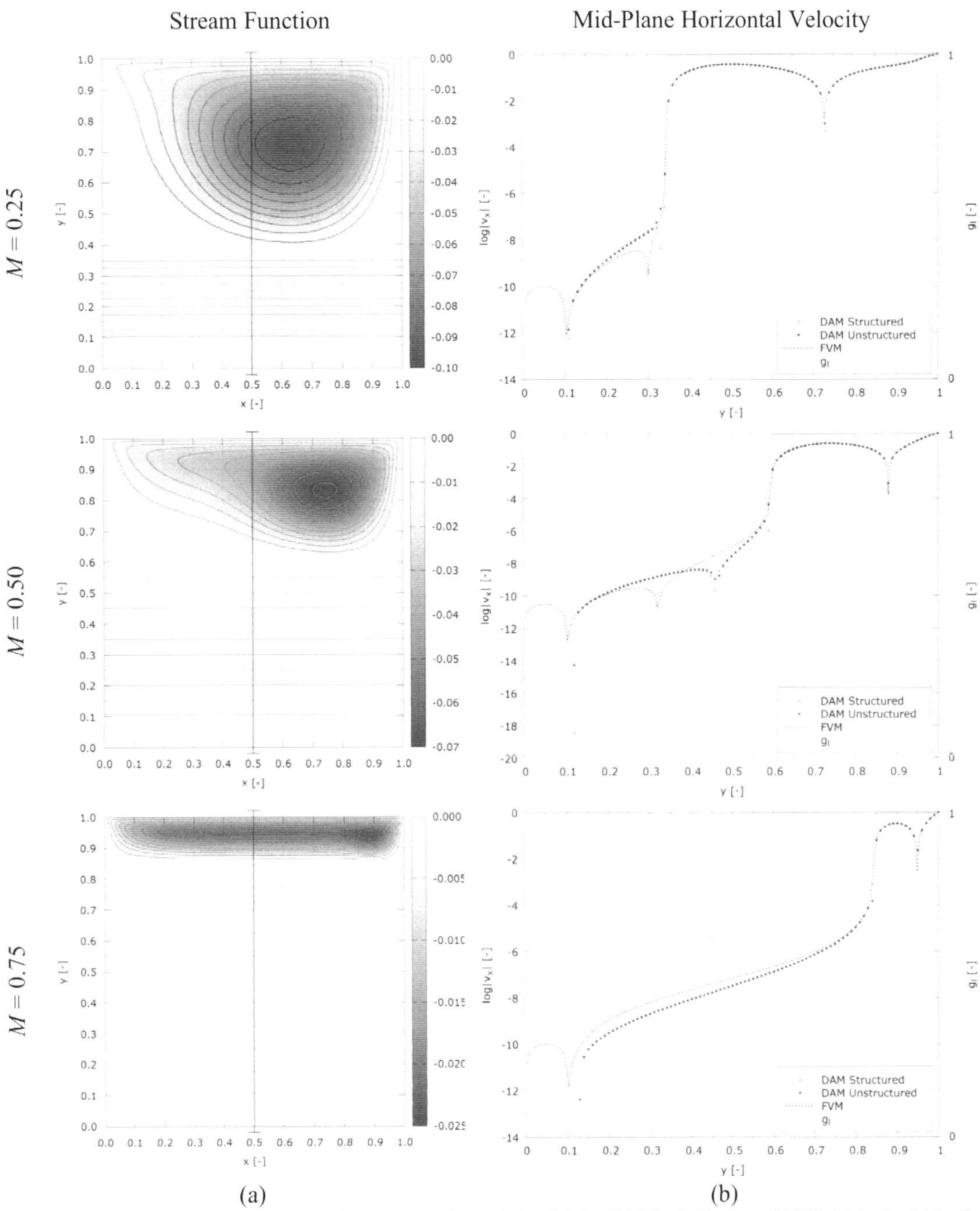

Figure 5. Comparison of DAM and FVM results. (a) DAM (solid black line) and FVM (dashed black line) stream function contours. The horizontal green lines are used to display the liquid fraction isolines. The vertical blue line is used to show the position of the vertical mid-plane cross section. (b) Horizontal component of the velocity magnitude in the logarithmic scale is compared at the vertical cross section. The dashed line is used to plot the FVM results. The red and the blue dots are used to plot the DAM results on the structured and unstructured computational node arrangements. The red line is used to show liquid fraction/porosity.

7. Conclusions

A new benchmark for fluid flow in heterogeneous porous medium is proposed. The test case is designed in such a manner that the conditions are similar to the ones during direct-chill casting of aluminium alloys. Three different cases are studied, where the thickness of the porous media is varied. The benchmark is solved by using the local meshless diffuse approximate method. The meshless solutions are obtained on the structured and the unstructured computational node arrangements with different node spacing. These results show that the solution is independent of the node distribution type. Furthermore, the solution is compared with the reference FVM model. Agreement with the reference results is very good, especially in the liquid and in the highly permeable part of the porous region. The results of this paper show that DAM is suitable for solving the Navier-Stokes equation in heterogeneous porous media. Furthermore, this means that the method is suitable for simulation of melt flow in solidification problems, including the industrial processes of direct-chill casting. The upgrade of the test will include the time dependent solution of the problem with variation of the mushy zone dynamics as well as the anisotropic behaviour of the permeability.

Acknowledgements

This work was supported by the Slovenian Research Agency (J2-1718 Advanced meshless modelling and simulation of multiphase systems; L2-9246 Multiphysics and multiscale numerical modelling for competitive continuous casting; P2-0162 Transient two-phase flows); European Structural and Investment Funds and Slovenian Ministry of Education, Science and Sport (20.00369 MARTINA; 20.03531 MARTIN).

References

[1] Dantzig J A and Rappaz M 2016 *Solidification* (Lausanne, Switzerland: EPFL Press)
[2] Rappaz M, Bellet M and Deville M 2003 *Numerical Modeling in Materials Science and Engineering* (Berlin Heidelberg: Springer-Verlag)
[3] Glicksman M E 2011 Principles of Solidification - An Introduction to Modern Casting and Crystal Growth Concepts (Springer)
[4] Ganesan S and Poirier D R 1990 *Metall. Trans. B* **21** 173–81
[5] Poirier D R, Nandapurkar P J and Ganesan S 1991 *Metall. Trans. B* **22** 889–900
[6] Ni J and Beckermann C 1991 *Metall Trans B* **22** 349–61
[7] Založnik M and Combeau H 2010 *Comp. Mater. Sci.* **48** 1–10
[8] Li H and Mulay S S 2013 *Meshless Methods and Their Numerical Properties* (Boca Raton, FL: CRC Press)
[9] Bertrand O *et al.* 1999 *Int. J. Therm. Sci.* **38** 5–26
[10] Šarler B and Vertnik R 2006 *Comput. Math. Appl.* **51** 1269–82
[11] Reuther K and Rettenmayr M 2012 *Acta Mater.* **60** 2128–34
[12] Kosec G and Šarler B 2014 *Eng. Anal. Bound. Elem.* **45** 36–44
[13] Hatić V, Mavrič B and Šarler B 2018 *Int. J. Numer. Method. H* **28** 361–80
[14] Hatić V, Cisternas Fernández M, Mavrič B, Založnik M, Combeau H and Šarler B 2019 *Int. J. Therm. Sci* **142** 121–33
[15] Hanoglu U and Šarler B 2019 *Metals* **9** 788
[16] Vertnik R, Mramor K and Šarler B 2019 *Eng. Anal. Bound. Elem.* **104** 347–63
[17] Dobravec T, Mavrič B and Šarler B 2020 *Comp. Mater. Sci.* **172** 109166
[18] Nayroles B, Touzot G and Villon P 1992 *Comput. Mech.* **10** 307–18
[19] Sadat H and Prax C 1996 *Int. J. Heat Mass Tran.* **39** 214–8
[20] Prax C, Sadat H and Dabboura E 2007 *Appl. Math Comput.* **186** 1040–53
[21] Šarler B *et al.* 2019 *Stroj Vestn-J Mech. E* **65** 658–70
[22] Ghia U, Ghia K N and Shin C T 1982 *J. Comput. Phys.* **48** 387–411

IOP Conf. Series: Materials Science and Engineering **861** (2020) 012029 doi:10.1088/1757-899X/861/1/012029

Permeability tensor for various columnar dendrite structures

Y Mitsuyama[1], T Takaki[2], S Sakane[1], Y Shibuta[3] and M Ohno[4]

[1] Graduate School of Science and Technology, Kyoto Institute of Technology, Matsugasaki, Sakyo–ku, Kyoto 606–8585, Japan
[2] Faculty of Mechanical Engineering, Kyoto Institute of Technology, Matsugasaki, Sakyo–ku, Kyoto 606–8585, Japan
[3] Division of Materials Science and Engineering, Faculty of Engineering, Hokkaido University, Kita 13 Nishi 8, Kita–ku, Sapporo 060–8628, Japan
[4] Department of Materials Engineering, The University of Tokyo, 7–3–1 Hongo, Bunkyo–ku, Tokyo 113–8656, Japan

E-mail: takaki@kit.ac.jp

Abstract. Permeability is a very important parameter determining fluid dynamics in macroscale casting simulations. For a columnar solidification structure, the permeability has anisotropy and becomes a tensor quantity. In our previous study, we proposed a permeability tensor and confirmed its validity. In this study, the applicability of the permeability tensor is shown through a series of permeability computations that are conducted for some columnar dendritic structures with different morphologies.

1. Introduction

In simulations of casting and macrosegregation prediction, we need to know values of a permeability expressing the ease of liquid flow in the mushy region, where solid and liquid coexist [1, 2]. The permeability is computed by Darcy's law:

$$\boldsymbol{U} = -\frac{1}{\mu}\boldsymbol{K}\nabla P \tag{1}$$

where $\boldsymbol{U}$, $\boldsymbol{K}$, ∇P, and μ are the average liquid flow velocity vector, permeability tensor, average pressure gradient vector, and viscosity, respectively. While the permeability tensor, $\boldsymbol{K}$, is isotropic for an equiaxed structure [3, 4], it is anisotropic for a columnar structure [5-9]. Permeabilities of flow normal and parallel to a columnar structure have been expressed as a function of primary arm spacing and secondary arm spacing [5, 7, 8]. However, the accuracy of those equations has not been clarified [9] and more efforts are needed to obtain a conclusive equation for expressing the permeability of a columnar structure. Also, no studies concerning the permeability tensor for a columnar structure have been performed to the best of our knowledge.

To predict permeability for various morphologies of solidification microstructure, numerical simulation is a powerful and promising approach [9-12]. In particular, the phase-field method can predict dendritic solidification microstructures with high accuracy. We have developed a permeability prediction method by bridging the phase–field and lattice Boltzmann methods [13]. In this method, large–scale simulation using a GPU–rich supercomputer enabled the permeability prediction for realistic columnar structures with multiple primary arms [14-16]. Through investigations of the permeability for columnar dendritic/cellular structures, it was concluded that dimensionless permeabilities for normal and parallel flows are well approximated by the Kozeny–Carman (KC) equation [17] with a coefficient $k_c = 9$ and 3, respectively [13, 18]. Here, the KC equation is expressed as

$$K = \frac{\left(1 - f_S\right)^3}{k_C S_S^2 f_S^2} \tag{2}$$

where f_s is the solid fraction, S_S is the specific interface area, and k_C is a coefficient. The specific interface area, S_S, is defined by $S_S = A_{SL}/V_S$, where A_{SL} is the solid–liquid interface area and V_S is the volume of solid. We also proposed a permeability tensor for a columnar structure. The validity of the proposed permeability tensor was confirmed by simulations of flow in various directions to the columnar structure [18]. However, it is important to test the validity of the proposed permeability for different columnar dendritic structures to understand its applicability.

In this study, we carry out testing of the validity and applicability of the permeability tensor by systematically computing permeabilities for some columnar dendritic structures of an Al–15wt%Cu binary alloy.

2. Permeability prediction method

The permeability prediction procedure used in this study consists of the following three steps [13]:

Step 1 A dendrite solidification structure is computed by the phase–field method. In the present study, the directional solidification simulations using frozen temperature approximation are conducted to obtain columnar dendritic structures for a binary alloy.

Step 2 Interdendritic liquid flow in a dendrite structure obtained in Step 1 is computed by the lattice Boltzmann method, where the phase–field variable computed in Step 1 is used to distinguish between the solid and liquid, and a pressure gradient is applied to a direction where the permeability is to be computed.

Step 3 Computing the average flow velocity in the pressure gradient direction; permeability in the pressure gradient direction is computed based on Darcy's law, equation (1).

The phase–field and lattice Boltzmann simulations in steps 1 and 2 were conducted in large–scale system. For the large–scale simulations, parallel computation was performed using multiple GPUs on a GPU–rich supercomputer (TSUBAME3.0) at Tokyo Institute of Technology [19-23].

2.1. *Phase–field equations (Step 1)*

A quantitative phase–field model for a dilute binary alloy [24] is applied to the directional solidification simulation. Time evolution equations of the temperature T (frozen temperature approximation), phase–field ϕ ($\phi = 1$ in a solid and $\phi = -1$ in a liquid), and nondimensional supersaturation u are expressed as

$$T\left(z\right) = T_0 + G\left(z - V_p t\right) \tag{3}$$

$$\tau\left(\nabla\phi\right)\left[1-\left(1-k\right)u'\right]\frac{\partial\phi}{\partial t} = \nabla\cdot\left[W\left(\nabla\phi\right)^2\nabla\phi\right] + \sum_{i=1}^{3}\frac{\partial}{\partial x_i}\left[W\left(\nabla\phi\right)\frac{\partial W\left(\nabla\phi\right)}{\partial\phi_{,x_i}}\left|\nabla\phi\right|^2\right] - \left(-\phi+\phi^3\right) - \lambda^*\left(1-\phi^2\right)^2\left(u+u'\right) \tag{4}$$

$$\frac{1}{2}\left[1+k-\left(1-k\right)\phi\right]\frac{\partial u}{\partial t} = \nabla\left[D_L q\left(\phi\right)\nabla u - j_{AT}\right] + \frac{1}{2}\left[1+\left(1-k\right)u\right]\frac{\partial\phi}{\partial t} - \nabla\cdot J \tag{5}$$

where T_0 is the reference temperature at $z = 0$ and $t = 0$, G is the temperature gradient, z is the spatial coordinate along the direction of G, V_p is the pulling velocity, and t is the time. In equation (4), $u' = (z_T - V_p t)/l_T$ denotes the additional supersaturation, where $l_T = |m|(1-k)c_0/(kG)$ is the thermal diffusion length, k is the partition coefficient, c_0 is the initial concentration in the liquid, m is the liquidus slope, and z_T is the coordinate in the G direction with an origin at $u = -1$. $\tau(\nabla\phi) = \tau_0 a_s(\nabla\phi)^2$ and $W(\nabla\phi) = W_0 a_s(\nabla\phi)$ are the phase–field relaxation time and interface thickness, respectively, with a crystalline anisotropy $a_s(\nabla\phi) = 1-3\varepsilon_4+4\varepsilon_4(\phi_{,x}^4 + \phi_{,y}^4 + \phi_{,z}^4)/|\nabla\phi|^4$, where ε_4 is the anisotropic strength. Here, $\phi_{,xi}$ is the spatial derivative of ϕ with respect to x_i. λ^* is a coupling constant associated with the thermodynamic driving force given by $\lambda^* = a_1 W_0/d_0$, with $a_1 = 0.88388$ and the chemical capillary length $d_0 = k\Gamma/(|m|(1-k)c_0)$, where Γ is the Gibbs–Thomson constant. The nondimensional supersaturation u is defined as $u = (C_L - C_L^e)/(C_L^e - C_S^e)$, where C_L is the concentration in the liquid, and C_L^e and C_S^e are the equilibrium concentrations in the liquid and in the solid, respectively. We followed the Kim–Kim–Suzuki model [25] and applied the relations $k = C_S^e/C_L^e = C_S/C_L$. The concentration C is given by $C = C_S(1+\phi)/2 + C_L(1-\phi)/2$. In the solute diffusion equation, equation (5), j_{AT} represents the anti–trapping current, given by $j_{AT} = -(1-kD_S/D_L)/(2\sqrt{2})W_0[1+(1-k)u](\partial\phi/\partial t)\nabla\phi/|\nabla\phi|$, in which D_S and D_L are the diffusion coefficients in the solid and in the liquid, respectively. J is the fluctuating current [26] and $q(\phi)$ is an interpolating function given by $q(\phi) = [kD_S + D_L + (kD_S - D_L)\phi]/(2D_L)$.

2.2. Lattice Boltzmann equations (Step 2)

The lattice Boltzmann equation is expressed by [13]

$$f_i(\boldsymbol{x}+\boldsymbol{c}_i\Delta t, t+\Delta t) = f_i(\boldsymbol{x},t) - \frac{1}{\tau_{LBM}}\left[f_i(\boldsymbol{x},t) - f_i^{eq}(\boldsymbol{x},t)\right] + w_i\rho\left[3\frac{\boldsymbol{c}_i - \boldsymbol{U}}{c^2} + 9\frac{(\boldsymbol{c}_i\cdot\boldsymbol{U})\boldsymbol{c}_i}{2c^4}\right]\cdot\boldsymbol{G}\Delta t - 3\rho w_i(\boldsymbol{c}_i\cdot\nabla P)\Delta t \tag{6}$$

where f_i is the particle distribution function, f_i^{eq} is the equilibrium distribution function, τ_{LBM} is the relaxation time, $\boldsymbol{c}_i$ is the discrete particle velocity, w_i is the weight function, $c = \Delta x/\Delta t$ is the lattice velocity, Δx is the grid size, and Δt is the time increment. The density ρ and flow velocity $\boldsymbol{U}$ are computed by $\rho = \sum_{i=0}^{Q-1} f_i$ and $\rho\boldsymbol{U} = \sum_{i=0}^{Q-1}\boldsymbol{c}_i f_i$ with the number of discrete velocities Q. In this study, $Q = 19$ because the D3Q19 model is employed. The kinetic viscosity ν can be computed by $\nu = \Delta t c^2(\tau_{LBM} - 0.5)/3$. To satisfy the no–slip condition at the solid–liquid interface, the dissipative drag force $\boldsymbol{G} = -2\rho\nu h/W_0^2((1+\phi)/2)^2\boldsymbol{U}$ [27, 28] is introduced to a diffuse interface, where ν is the kinetic viscosity and $h = 2.757$ [27]. The forced convection is caused by applying a constant pressure gradient ∇P [29].

2.3. Permeability computation (Step 3)

When computing the average flow velocity U_{xi} in the pressure gradient direction, the permeability in the x_i–direction is computed by $K_{xi} = -\mu U_{xi}/(\mathrm{d}P/\mathrm{d}x_i)$.

3. Permeability tensor

In our previous study [18], a permeability tensor for a columnar structure was proposed as

$$\boldsymbol{K} = \boldsymbol{R}^T\boldsymbol{K}_n\boldsymbol{R} \tag{7}$$

where $\boldsymbol{K}_n$ is the diagonal permeability tensor and $\boldsymbol{R}$ is the coordinate transformation matrix. Here, $\boldsymbol{K}_n$ and $\boldsymbol{R}$ are expressed as

$$\boldsymbol{K}_n = \begin{bmatrix} K_x & 0 & 0 \\ 0 & K_y & 0 \\ 0 & 0 & K_z \end{bmatrix}, \quad \boldsymbol{R} = \begin{bmatrix} \cos\theta & -\sin\theta & 0 \\ \sin\theta & \cos\theta & 0 \\ 0 & 0 & 1 \end{bmatrix} \begin{bmatrix} \cos\psi & 0 & \sin\psi \\ 0 & 1 & 0 \\ -\sin\psi & 0 & \cos\psi \end{bmatrix} \tag{8}$$

where K_x, K_y and K_z are the permeabilities of flow in the x–, y– and z–directions, respectively. It is convenient that the x–, y– and z–directions correspond to the preferred growth directions [100], [010] and [001] of a cubic material. Thus, K_x and K_y are the permeabilities for normal flow, and K_z is the permeability for parallel flow, as the temperature gradient, G, is set to the z–direction. After coordinate transformation, the permeability for the (θ, ψ) direction can be expressed by

$$K(\theta,\psi) = K_{11} = (\cos\theta\cos\psi)^2 \cdot K_x + (\sin\theta\cos\psi)^2 \cdot K_y + \sin^2\psi \cdot K_z \tag{9}$$

By comparing this equation to simulation results, the validity of the permeability tensor, equation (7), is confirmed.

4. Simulations and results

4.1. Columnar dendritic structures

We prepared columnar dendritic structures by performing phase–field simulations for directional solidification of a single–crystal Al–15wt%Cu binary alloy. The temperature gradient G and pulling velocity V_p in the z–direction were set as G = 50 K/mm and V_p = 20 μm/s, respectively. The material and simulation parameters are summarized in table 1. From a preliminary simulation in a large computational domain which allows growth of multiple primary arms, we obtained the average primary arm spacing λ_{ave} = 174.7 μm (= 447.3Δx) by a counting method assuming a regular hexagonal array. In our previous study, we confirmed that the permeability computed for a columnar structure with multiple primary arms becomes nearly the same as that of a columnar structure with periodical regular hexagonal array with the same primary arm spacing. This means that we can drastically reduce the computational cost for permeability computations. Thus, a minimum domain to express the regular hexagonal array was employed as shown in figure 1. In addition, to obtain columnar dendritic structures with different morphologies, four computational domains with different size were used as shown in figures 1(a–d). Here, the primary arm spacing, $\lambda = (L_x^2 + L_y^2)^{0.5}$, becomes (a) $\lambda = 0.68\lambda_{ave}$, (b) $0.93\lambda_{ave}$, (c) $1.22\lambda_{ave}$, and (d) $1.47\lambda_{ave}$ for each model. The number of grid points in all directions was set to a multiple of 8 for efficient computation, and the size in the z–direction covers the solid fraction $f_S = 0 \sim 0.55$. Figures 1(a–d) also show dendrite morphologies at a steady–state growth (~10^7th step). Many computational steps were required to achieve steady–state growth due to the low pulling velocity. Since a moving frame algorithm was employed, the top region with 1068Δx consists of only liquid. As shown in figure 1, we can see different dendritic morphologies depending on the domain size. For a larger computational domain, the secondary arms grow longer and competitive growth among secondary arms is observed.

Table 1. Material and simulation parameters.

Quality	Symbol	Value	Reference
Initial concentration	c_0	0.0697 at.frac.	
Pulling velocity	V_p	20 μm/s	
Temperature gradient	G	50 K/mm	

Diffusion coefficient in liquid	D_L	2.4×10^{-9} m²/s	[30]
Diffusion coefficient in solid	D_S	1.15×10^{-13} m²/s	[30]
Partition coefficient	k	0.14	[31]
Anisotropic strength	ε_4	0.012	[32]
Gibbs-Thomson constant	Γ	1.36×10^{-7} Km	[33]
Liquidus slope	m	–655 K/at.frac.	[34]
Reference temperature at z = 0	T_0	885.89 K	
Grid size	Δx	0.3906 μm	
Interface thickness	W_0	Δx/0.6 = 0.651 μm	
Time increment	Δt	9.083×10^{-6} s	

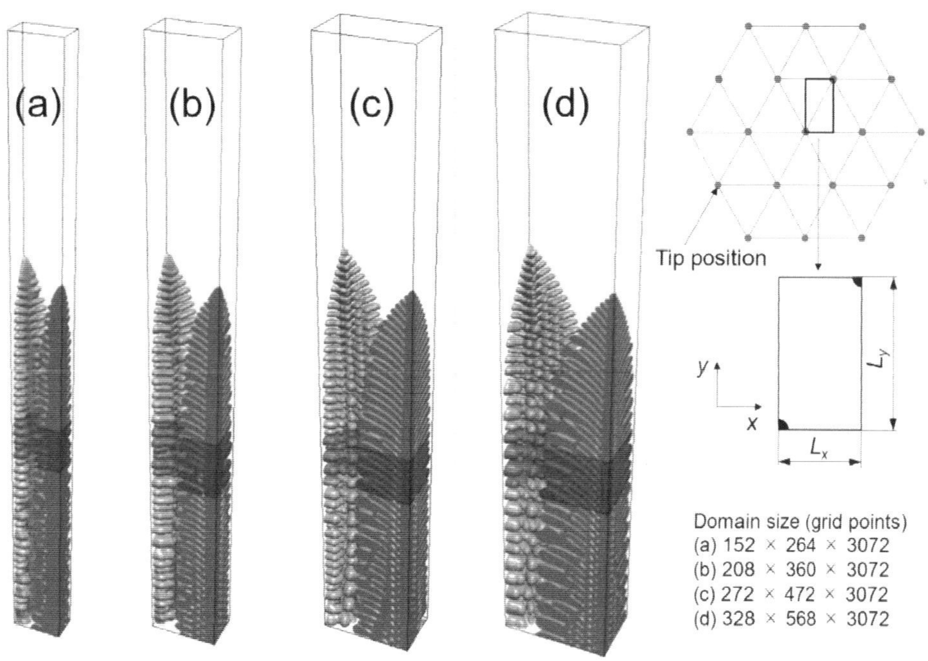

Figure 1. Columnar dendritic structures at a steady–state growth condition (~10^7th step) for four different domains with a periodic regular hexagonal array. The computational domain sizes are also indicated in this figure as the number of grid points.

4.2. Interdendritic liquid flow

To compute interdendritic liquid flow in the dendritic structures presented in figure 1, four models shown in figure 2 were developed. These models were constructed by mirror symmetry of a red domain shown in figure 1 in all axial directions, which enabled us to compute the liquid flow in arbitrary directions [18]. The size in the z–direction of the red region shown in figure 1 was determined as a length

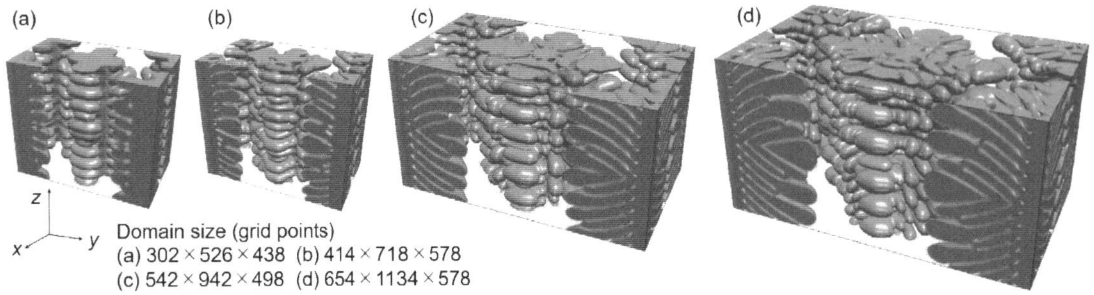

Figure 2. Models with $f_S = 0.4$ to compute interdendritic liquid flow in arbitrary directions: (a) Model A, (b) Model B, (c) Model C and (d) Model D. These models were constructed by mirror symmetry of the red region shown in figure 1 in all axial directions. Domain sizes are also indicated in this figure as the number of grid points.

of roughly five secondary arms. The solid fraction f_S, interface area density S_V, and specific interface area $S_S = S_V / f_S$ for the four models shown in figure 2 are indicated in table 2. The value of f_S for all models is a constant of 0.4. S_V and S_S increase as the domain size increases. This means that the complexity of solid morphology becomes greater with domain size.

Table 2. Solid fraction f_S, interface area density S_V, specific interface area S_S and permeabilities in all axial directions for four models shown in figure 2.

Model	f_S	S_V [μm^{-1}]	S_S [μm^{-1}]	K_x [μm^2]	K_y [μm^2]	K_z [μm^2]
A	0.398	0.0479	0.120	26.4	15.2	48.6
B	0.399	0.0571	0.143	33.7	21.1	58.0
C	0.399	0.0732	0.183	35.1	21.4	67.0
D	0.399	0.0729	0.183	40.6	21.8	76.7

4.3. Permeability

Computed permeabilities in the x–, y– and z–directions for models shown in figure 2 are indicated in table 2. For all models, the permeability is largest for K_z, followed by K_x and then K_y. Thus, the liquid can flow more easily in the z–direction, followed by the x– and y–directions shown in figure 2. Because the liquid flow path in the z–direction is established as being straight, it would be reasonable to conclude that the parallel direction to the columnar structure is the easiest flow direction. The same tendency has been observed in other studies [7, 9]. The relation of $K_x > K_y$ depends on the primary arm spacing in the x– and y–directions as shown in figure 2. As the primary array in real columnar structures is not regular, the permeabilities in the x– and y–directions should be the same for columnar structures with massive number of primary arms. We confirmed this point in our previous study for columnar structures with multiple primary arms [13, 15]. It is also observed all the permeabilities increase with increasing domain size, because the spacing among solids becomes larger for a larger domain size.

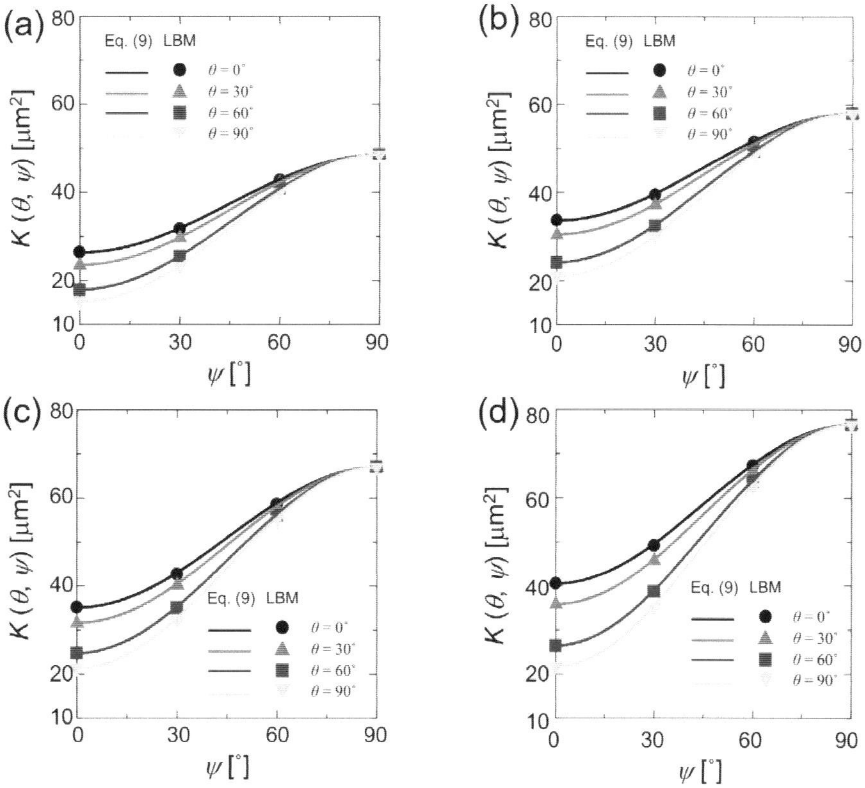

Figure 3. Permeabilities computed for different angles of θ and ψ for (a) Model A, (b) Model B, (c) Model C and (d) Model D shown in figure 2.

4.4. Permeability tensor

To confirm the validity and applicability of the permeability tensor defined by equation (7), we conducted 16 lattice Boltzmann simulations for each model shown in figure 2 by changing the angles of θ and ψ from 0° to 90° with an increment of 30°. The results are shown in figure 3. Here, the horizontal axis is the angle ψ, and the vertical axis is the permeability $K(\theta, \psi)$ in the (θ, ψ) direction. The marks are the simulation results, and the solid lines correspond to equation (9). Equation (9) agrees well with the simulation results in all cases. Thus, it can be confirmed that the permeability tensor, equation (7), can be applied to various columnar dendritic structures.

5. Conclusions

To express the applicability of the permeability tensor proposed in the previous study, permeabilities were computed for four columnar dendritic structures with different morphologies. Here, the columnar dendritic structures were obtained by conducting the phase–field simulations during directional solidification of a single–crystal of Al–15wt%Cu binary alloy, and the interdendritic liquid flow in the columnar dendritic structures were computed by the lattice Boltzmann simulations. As a result, for all dendritic structures, the computed permeabilities agreed well with the results of the permeability tensor equation. Thus, the wide applicability of the permeability tensor equation could be confirmed. In conclusion, it was shown that we need only two components K_n and K_p, where K_n is the permeability of normal flow and K_p is for parallel flow to the columnar structure. The K_n can be computed as an average value of K_x and K_y shown in equation (8), $K_n = (K_x + K_y)/2$.

Acknowledgements

This research was partly supported by KAKENHI, Grant–in–Aid for Scientific Research (A) 17H01237. This research also relied on computational resources of the HPCI system provided by Tokyo Institute of Technology through the HPCI System Research Project (Project ID: hp190067).

References

[1] Flemings M C 2000 *ISIJ Int.* **40** 833-41
[2] Beckermann C 2002 *Int. Mater.Rev.* **47** 243-62
[3] Bernard D, Nielsen Ø, Salvo L and Cloetens P 2005 *Mat. Sci. Eng. A* **392** 112-20
[4] Ehsan K and Daan M M 2012 *Modell. Simul. Mater. Sci. Eng.* **20** 035004
[5] Murakami K, Shiraishi A and Okamoto T 1984 *Acta Metall.* **32** 1423-28
[6] Nasser-Rafi R, Deshmukh R and Poirier D R 1985 *Metall. Trans. A* **16** 2263-71
[7] Poirier D R 1987 *Metall. Trans. B* **18** 245-55
[8] Heinrich J C and Poirier D R 2004 *Comptes Rendus Mécanique* **332** 429-45
[9] Natsume Y, Takahashi D, Kawashima K, Tanigawa E and Ohsasa K 2013 *ISIJ Int.* **53** 838-47
[10] Böttger B, Haberstroh C and Giesselmann N 2016 *JOM* **68** 27-36
[11] Ludwig A, Kharicha A, Hölzl C, Domitner J, Wu M and Pusztai T 2014 *Eng. Anal. Bound. Elem.* **45** 29-35
[12] Diepers H J, Beckermann C and Steinbach I 1999 *Acta Mater.* **47** 3663-78
[13] Takaki T, Sakane S, Ohno M, Shibuta Y and Aoki T 2019 *Acta Mater.* **164** 237-49
[14] Takaki T, Sakane S, Ohno M, Shibuta Y, Aoki T and Gandin C-A 2018 *Materialia* **1** 104-13
[15] Takaki T, Sakane S, Ohno M, Shibuta Y, Shimokawabe T and Aoki T 2016 *Acta Mater.* **118** 230-43
[16] Takaki T, Sakane S, Ohno M, Shibuta Y, Shimokawabe T and Aoki T 2016 *ISIJ Int.* **56** 1427-35
[17] Carman P C 1956 *Flow of gases through porous media* (New York: Academic Press)
[18] Mitsuyama Y, Takaki T, Sakane S, Shibuta Y and Ohno M 2020 *Acta Mater.* **188** 282-7
[19] Takaki T, Shimokawabe T, Ohno M, Yamanaka A and Aoki T 2013 *J. Cryst. Growth* **382** 21-5
[20] Takaki T 2014 Phase-field modeling and simulations of dendrite growth *ISIJ Int.* **54** 437-44
[21] Sakane S, Takaki T, Rojas R, Ohno M, Shibuta Y, Shimokawabe T and Aoki T 2017 *J. Cryst. Growth* **474** 154-9
[22] Shibuta Y, Ohno M and Takaki T 2015 *JOM* **67** 1793-804
[23] Shibuta Y, Ohno M and Takaki T 2018 *Adv. Theory Simul.* **1** 1800065
[24] Ohno M and Matsuura K 2009 *Phys. Rev. E* **79** 031603
[25] Kim S G, Kim W T and Suzuki T 1999 *Phys. Rev. E* **60** 7186-97
[26] Echebarria B, Karma A and Gurevich S 2010 *Phys. Rev. E* **81** 021608
[27] Beckermann C, Diepers H J, Steinbach I, Karma A and Tong X 1999 *J. Comput. Phys.* **154** 468-96
[28] Rojas R, Takaki T and Ohno M 2015 *J. Comput. Phys.* **298** 29-40
[29] Succi S 2001 *The lattice boltzmann equation: for fluid dynamics and beyond* (United Kingdom: Oxford University Press)
[30] Rebow M and Browne D J 2007 *Scripta Mater.* **56** 481-4
[31] Kurz W and Fisher D J 1989 *Fundamentals of Solidification* (Switzerland: Trans Tech Publications Ltd)
[32] Morris J R 2002 *Phys.Rev. B* **66** 144104
[33] Rocha O L, Siqueira C A and Garcia A 2003 *Metall. Mater. Trans. A* **34 A** 995-1006
[34] Murray J L 1985 The aluminium-copper system *Int. Metals Rev.* **30** 211-34

Modelling of an industrial die casting process for the production of aluminum automotive parts

J Ou[1], C Wei[1, 2], D Maijer[1], S Cockcroft[1], L A[2], Y Zhang[2], Z Chen[2] and Z Zhu[2]

[1] The Department of Materials Engineering, The University of British Columbia, 6350 Stores Road, Vancouver, BC, V6T 1Z4, Canada
[2] CITIC DICASTAL Co., Ltd, No 185, Long Hai Road, Economic and technological development zone, Qinhuangdao City, Hebei Province, China

E-mail: jun.ou@ubc.ca

Abstract. Die (permanent) casting is one of the major manufacturing processes for aluminum automotive parts thanks to its ability to provide a good balance between mechanical performance and production efficiency. In the die casting industry, computational modelling has been widely adopted to analyze, troubleshoot and optimize die design and process parameters. In this work, a computational modelling methodology including key process boundary conditions and material properties has been developed to simulate an industrial die casting process of an aluminum automotive part. It has been demonstrated that the model is accurate and robust with respect to temperature and defect prediction by comparison with a large amount of high-quality plant-trial data. The data was acquired from an in-house designed data acquisition system (DAQ) connected to an instrumented die and industrial production equipment in a production facility.

1. Introduction

In the auto industry, the trend continues towards increased usage of aluminum automotive parts produced by permanent die casting processes thanks to its lightweight, relatively low cost and good mechanical performance. Die casting processes have become very sophisticated to improve quality, however, various types of casting-related defects continue to challenge production engineers. For example, the large-scale (normally >1 mm) irregular shape porosity may be caused by large liquid encapsulation when directional solidification is not achieved. Small-scale (normally <1 mm) round-shape porosity may also form due to the reduction of hydrogen solubility associated with the liquid to solid phase transformation. Oxide film entrapment may occur during filling when the free surface of liquid aluminum is disturbed because of turbulence.

As a complement to experience-based design, numerical simulations have been adopted to provide insights and guidance to the design activities for dies and process operation in the casting industry. It has successfully helped to reduce the risk of critical defect formation. However, in industry, there is still a strong desire to improve model accuracy and robustness in order to expand their use. This requires a systematic evaluation and characterization of the relevant physics, e.g. material properties and key boundary/interface conditions associated with interface behavior and cooling. The transport phenomena and the relevant boundary conditions are often complicated and have been investigated by a numerous researchers. Examples include the studies carried out by Nishida *et al.* [1] and Wei *et al.* [2] for the investigation of the die/casting interface behavior and the associated heat transfer. Moayedinia *et al.* [3] and Hu *et al.* [4] conducted work to characterize the heat transfer of water cooling in die cooling channels. Drawing on this body of fundamental work, researchers have been able to develop computational models of casting processes with improved accuracy. For example, Zhang *et al.* [5] developed a 3D thermal model of the low pressure die casting (LPDC) of A356 aluminum alloys wheels using ABAQUS. It correlated well to measured data with respect to die temperature and shrinkage-based porosity.

For the industrial applications, assessing a computational model's accuracy is often limited by the availability of high-quality, widespread data from the production line which can be challenging to acquire due to technical barriers and cost. Another challenge is that models are often inaccurate because of insufficient understanding and description of the complex physical phenomena that occur during the casting processes. This is exacerbated when the process conditions and die structure are significantly different from past experience. This study seeks to apply the modelling methodology that was originally developed for an LPDC wheel production process to a novel Counter Pressure Casting process for an automotive control arm. The LPDC modelling methodology being applied here has been refined over the last 20 years and incorporates relevant material properties and boundary/interface conditions that have been characterized by measurements on the lab- and industrial-scale and/or augmented with data collected from literature. The methodology has been validated for a range of process conditions and die structures. The CPC process has similar operation principles and employs a permanent die like the LPDC, with unique features – refer to the *CPC Process Description* section later for more details. Quantitative temperature data has been acquired from an industrial CPC process to validate the model. This work will present an assessment of the accuracy and robustness of the modelling methodology as applied to the CPC process, and provide insights into the comparison of the CPC and LPDC processes from the modelling perspective. *Note: since this is part of an applied research project with an industrial partner, confidential process information must be protected and will not be presented in this manuscript.*

2. CPC process description

The CPC process is a relatively new casting process that can be used to produce aluminum automotive parts. There is less known about this process in the casting industry compared to the conventional LPDC process. In principle, CPC process operation is similar to the LPDC process. The major difference is that the die is placed in a controlled pressure chamber in the CPC process. This allows the pressure in the chamber to be increased during both the filling and solidification stages. For a detailed introduction to the CPC process, readers are referred to the manuscript titled "*Toward the Development of a Thermal-Stress Model of an Industrial Counter Pressure Casting Process*" submitted by *Chunying Wei* in the proceedings of this conference.

As the CPC process allows the filling and solidification stages to occur under a higher pressure over the LPDC process, it has been argued that it benefits the casting quality in several ways [6]: 1) low oxide film entrainment because of reduce free surface turbulence; 2) refined microstructure due to enhanced cooling rate caused by improved die/casting interfacial heat transfer; and 3) reduced number and/or size of shrinkage-based porosity from solidification at relatively high counter pressure. At present there is little practical evidence to support these assertions.

3. Model development

The model was developed using the Finite Element Method (FEM) based, commercial software package, ProCAST™. It is able to simulate various casting processes and is capable of describing the transport phenomena occurring in the different stages of casting, i.e. filling (fluid flow), solidification, die opening, part ejection and die closing. It should be noted that the solution of the fluid flow (momentum conservation via the Navier-Stoke Equations) is only activated in the filling stages to improve computational efficiency. Whereas, the energy conservation is always computed throughout the process to update temperature distributions in the casting and die.

In addition, it should be noted that the modelling methodology was originally developed for the LPDC process. There are a number of parameters that common to both processes, such as the need to turn cooling on/off with specific timing, die manipulation timings and pressure curves, that are implemented in a similar manner or are based on calculation. One significant except is the heat flux representing die cooling because of residual water in the cooling manifolds - introduced in detail below.

Please note that the cooling configuration in the *CPC process is different from that in the LPDC process in terms of the physical structure and layout.*

3.1. Geometry and mesh

Figure 1 shows the geometry and mesh of the control arm casting and die / insert sections included in the model. The top surface of the top die is obscured to hide the detailed design of cooling configuration due to confidentiality. The mesh was generated using the following approach:

- The casting and the die/casting interface were meshed with a relatively fine size – 4~6 mm;
- At the locations where strong heat transfer exists, i.e. in proximity to the die cooling locations, a mesh size of 2 mm was selected;
- The mesh size was increased gradually moving away from the die/casting interface; and
- A maximum mesh size of 10 mm was used for the balance of the die structure

After applying the above, the resulting mesh has 73,996 surface cells and 448,137 body cells.

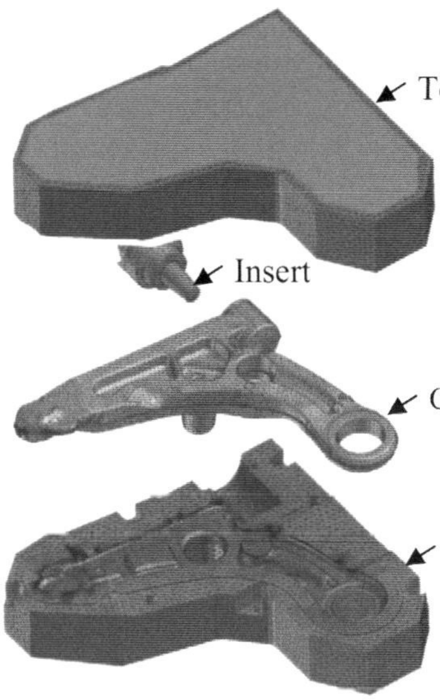

Figure 1. Geometry and mesh of the die/insert sections and the control arm (casting)

3.2. Part materials and thermo-physical properties

The materials used are listed in table 1. Due to space limitation, the thermo-physical properties have been omitted from this paper, but they are based on those found in the references [5] and [7].

Table 1. Parts and their materials

Part	Material
Die structure (top and bottom dies, and the insert)	H13
Casting	A356

3.3. Initial conditions

At the start of a casting cycle, the die does not contain liquid metal and only the initial temperature needs to be specified. The CPC process is a cyclic process where the temperature of the die after each casting cycle becomes the initial temperature condition for the start of the next cycle. If there are no process condition changes, eventually the process will reach cyclic steady-state where the temperature distribution at the start and end of each cycle is the same. The initial condition does not impact the final temperature distribution at cyclic steady-state. However, the number of cycles required to reach steady-state will vary depending on the initial conditions. In this work, the temperature field calculated from the previous run was employed as the initial condition. This approach significantly reduces the number of cycles required to reach steady state compared to using a uniform temperature for the initial conditions.

3.4. Boundary conditions

Two categories of boundary conditions, i.e. thermal and fluid flow boundary conditions, are required to solve the energy conservation equations for temperature and momentum conservation equations for fluid flow, respectively. As noted earlier, fluid flow was only solved for in the casting domain during the filling stage.

3.4.1. Thermal boundary conditions. Water cooling - water cooling is used to rapidly extract heat from the die, and therefore, it is one of the most important process parameters for the process control. The water cooling configurations are carefully designed with respect to both the locations and on- and off-timings of the cooling. Figure 2 shows the locations in the die where water cooling is used (labeled CC) and the on- and off-timings for each channel. *Note: only the relative water-on durations are provided.* The heat transfer phenomena occurring during water cooling in the die are complicated for both the water-on and -off stages. When water is flowing (i.e. water-on) to a cooling location, heat is removed by forced convection. When water begins to flow to a location, boiling will occur for a short period of time because the die temperature is relatively high (i.e. 300-500°C). When cooling water flow is turned off, heat will continue to be extracted from the die as the water remaining in the channel boils. Eventually, the residual water will evaporate leaving the cooling channel dry, which reduces heat transfer from the die.

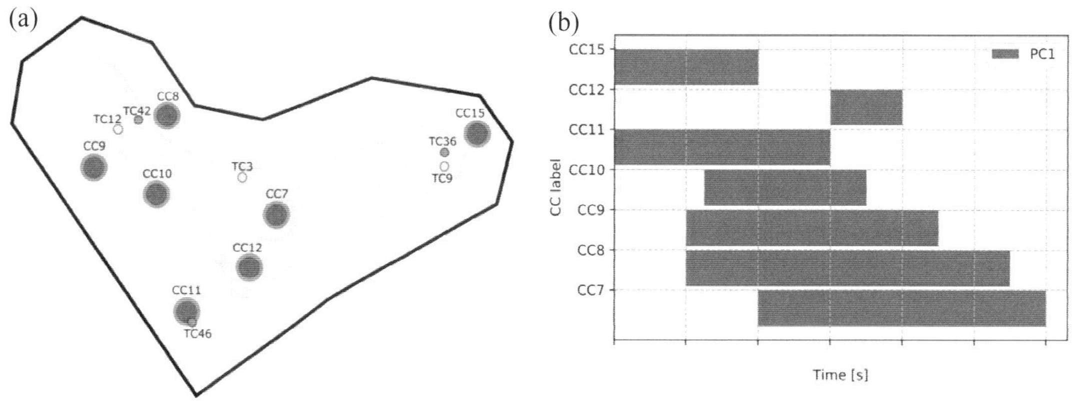

Figure 2. (a) Locations of TCs and cooling configurations and (b) cooling timings of the cooling configurations.

From the modeling perspective, the heat transfer for the water-on stage is described by a thermal boundary condition:

$$q_{cc} = h_{cc}(T_{surf} - T_{water}) \quad (1)$$

where q_{cc} is the heat flux, h_{cc} is the heat transfer coefficient (HTC), and T_{surf} and T_{water} are the surface temperature and water temperature, respectively. h_{cc} is determined using the method developed in reference [3], which is provides a temperature-dependent relationship incorporating the effects of boiling.

At the water-off stage, a constant heat flux (derived by a trial-and-error method) is applied to represent the heat removed by the residual water (including boiling). It is acknowledged that this method is a simplified description that may be improved using a calculated temperature-dependent HTC. However, this requires the application of two temperature-dependent HTCs on one boundary, corresponding to the two stages (time-periods), i.e. water-on and water-off. This requires the use of user-subroutines in ProCAST and was not available for this study.

Heat transfer to the environment - heat loss through the die outer surface to the ambient environment in the pressure chamber occurs via two mechanisms: air convection and radiation. In the model, the convective HTC is set to 20 W/(m^{2}K) and the environment temperature to 85°C (measured). For radiation, an emissivity of 0.8 is used with the same environment temperature.

Inlet temperature - the temperature of the incoming metal is set to 700°C at the entrance to the sprue based on the average temperature of liquid aluminum measured in the holding furnace.

3.4.2. Interface heat transfer. The heat transfer across the casting/die and die/die interfaces exhibits a variety behaviors including time, temperature and location dependence. For the die/die interfaces that maintain contact throughout the casting process, e.g. the top die/center pin interface, a constant interface HTC is applied. In contrast, some die/die interfaces are broken at specific times in a casting cycle to facilitate different operations. For example, when the die system opens, the top die/bottom die interface is broken and the exposed surfaces on both dies are open to the environment. In the model, the contact interface (boundary) condition is changed to an environment boundary condition (refer to the previous section) for the two surfaces.

Turning to the casting/die interface, this interface starts with a HTC selected to represent liquid metal in contact with the solid die during the filling stage. As the casting solidifies at the interface, the interface changes to a solid (casting)/solid (die) type of interface. For this type of contact, depending on the temperature field and geometrical structure of the die system, the solid/solid interface can experience increased pressure across the interface or the formation of a gap between the two surfaces due to thermal contraction/expansion. At the end of solidification, once the die is opened and the casting is ejected, the casting/die interface is broken and the environmental boundary condition is applied to all exposed surfaces. In the current version of the model, a temperature-dependent interface HTC has been adopted to describe the heat transfer associated with the aforementioned interface behavior. A typical interface HTC profile as a function of temperature is given in figure 3. The interface HTC is high when the liquid metal of the casting is in contact with the die and decreases linearly as solidification and thermal contraction occur. Enhanced interfacial HTC due to increased interface pressure has not been implemented in this model at this time.

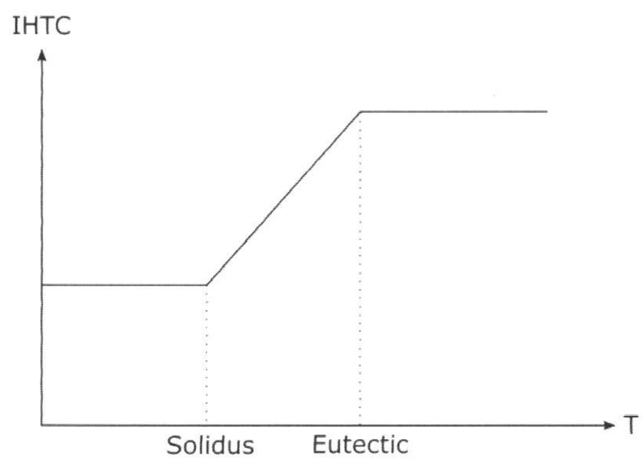

Figure 3. Example casting/die contact interface HTC as a function of temperature

3.4.3. Fluid flow boundary conditions. Pressure inlet - the fluid flow during filling in the CPC process is driven by the pressure difference, referred to as the counter pressure between the metal holding furnace and the pressure chamber. Currently, the model employs a pressure inlet boundary condition during the filling process. Note that the computational domain of the model does not include the transfer tube underneath the sprue, and therefore the pressure applied at the model inlet has been modified to offset the pressure needed to fill the transfer tube – equation 2.

$$P_{model} = P_{counter} - \rho_{A356} g h_{tube} \tag{2}$$

where P_{model} is the pressure applied at the model inlet, $P_{counter}$ is the counter pressure, ρ_{A356} is the density of aluminum alloy A356, g is the acceleration due to gravity and h_{tube} is the height of the transfer tube between the sprue and the liquid metal surface in the holding furnace.

It should be pointed out that from the modeling perspective, the only difference between the LPDC process and the CPC process is the pressure curve implemented. The LPDC process uses the furnace pressure as an input for the pressure inlet, while the CPC uses the counter pressure. In both processes, adequate venting is required during the filling stage to ensure that no back pressure develops.

3.5. Simulation settings

In ProCAST, the time-steps were dynamically updated to meet the solution convergence tolerance. The initial time-step size and maximum time-step size were defined for the filling and solidification stages, respectively. Tests were conducted to ensure that the default convergence tolerance was adequate for this model. Time-steps of 0.001s, 0.1s and 1s were used for the initial time-step size and the maximum time-steps for the filling and solidification stages, respectively.

4. Industrial production process assessment

An operational, industrial-scale CPC process was instrumented with 65 Type-K thermocouples (TCs) as follows: 31 in the top die, 13 in the bottom die, 3 in the casting component, 5 in the environment surrounding the die, 9 in the cooling channels and 4 in the bottom plate. Temperature data was collected from these thermocouples to provide a comprehensive and quantitative understanding of the temperatures in the process and to validate the model.

5. Results and discussion

5.1. Temperature

The temperatures were measured for 13 cycles over 1 hour. The near steady-state cyclic response from the process was identified and the temperature history from 6 TCs will be presented to define the characteristic behavior of the process. Figure 4 shows the measured temperatures for 3 TCs located in the top die (labeled TC3, TC9 and TC12), and 3 in the bottom die (labeled TC36, TC42 and TC46) – refer to figure 2 for their locations. The temperatures predicted by the model are also plotted in figure 4 for comparison.

Firstly, looking at the measured data, in general, the temperature histories exhibit a similar pattern. The temperatures are relatively low at the beginning of the cycle for a period of time before the liquid metal has filled the die – the system is building up pressure in the holding furnace and the pressure chamber. Once the liquid aluminum starts to fill into the die cavity, the die heats up, and consequently, the measured temperatures show a rapid rise as the heat diffuses to the TC locations (between ~40s to ~70s). The water-cooling is turned on with the sequence shown in figure 2(b). Within the die, once the rate of the heat removal, due to water-cooling and diffusion to cooler areas of the die, surpasses the heat supplied (from the casting), the temperature decreases – after ~75s. Another observation based on comparing the data from the top and bottom dies is that there are significant temperature gradients in the top die (e.g. TC12 and TC3 starts at ~290°C and 430°C, respectively). In contrast, the bottom die temperatures appear very close (within ~20°C of each other). This observation was attributed to the fact that the top die has 7 active water-cooling locations and the bottom die has none.

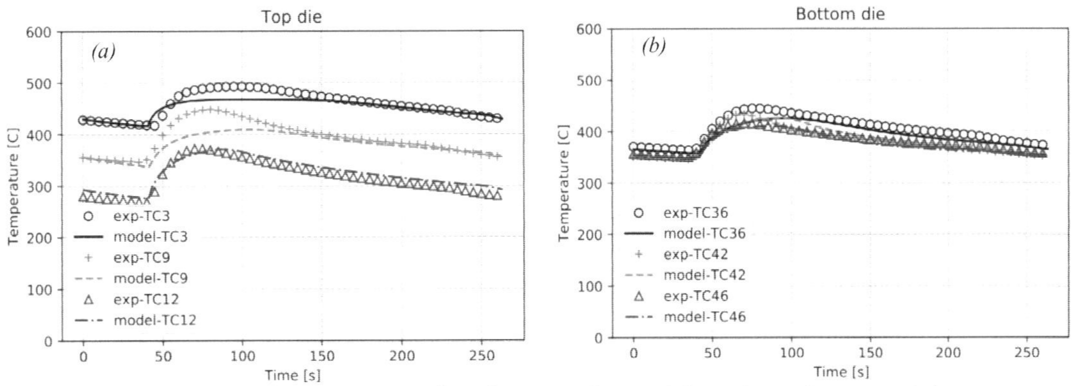

Figure 4. Temperature comparison between the model results and measured data

Turning to the comparison between the predicted and the measured temperatures, overall the model predictions correlate well with the measurements at all six locations. This indicates that the model has incorporated the key physics properly with respect to both heat transfer and fluid flow. However, it can be seen that at some locations, e.g. TC3 and TC9, the fit is relatively poor in the peak temperature region (~60s to ~120s). It is speculated that the simplified description of the complicated casting/die interface behavior may be reason for this discrepancy. Recent research has confirmed that the interface behavior during the casting process, particularly the casting/die interface, is very complicated due to the formation and evolution of an air gap or contact pressure along the interface caused by thermal contraction/expansion. A coupled temperature-fluid flow-displacement analysis is needed to properly describe this behavior because the deformation and heat transfer across the interface influence each other.

5.2. Solidification direction (shrinkage-based porosity)

The solidification direction predicted by the model has been used to assess likely locations of shrinkage-porosity formation. The desired direction is that the solidification occurs from the farthest end of the casting and works its way towards the inlet of liquid (sprue) – the so-called directional solidification. Failure to directional solidification leads to liquid encapsulation and may cause shrinkage-based porosity. However, a quantitative correlation between liquid encapsulation and shrinkage-based porosity has not been established, even though large and long-lasting liquid encapsulation is prone to form shrinkage-based porosity. Examination of the model results (figure 5) shows that, in general, directional solidification is achieved and only one small liquid encapsulation is observed existing for 4 s (figure 5(b)). Porosity in exemplar control arm castings was examined by Computed Tomography (CT) imaging (figure 6). CT results confirm that there are no locations of large-sized shrinkage porosity. Some small-sized porosity was found at the two locations highlighted in figure 6(b). The small number and size of this porosity and the fact that it occurs in non-critical locations permits these casting to pass inspection.

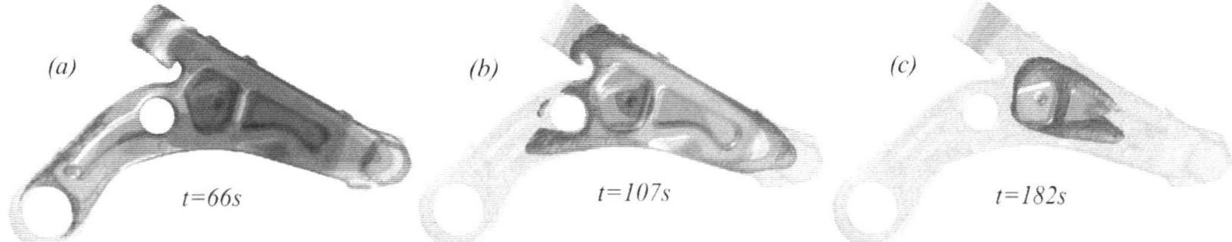

Figure 5. Solidification direction - represented by cut-off view at solid fraction = 0.7.

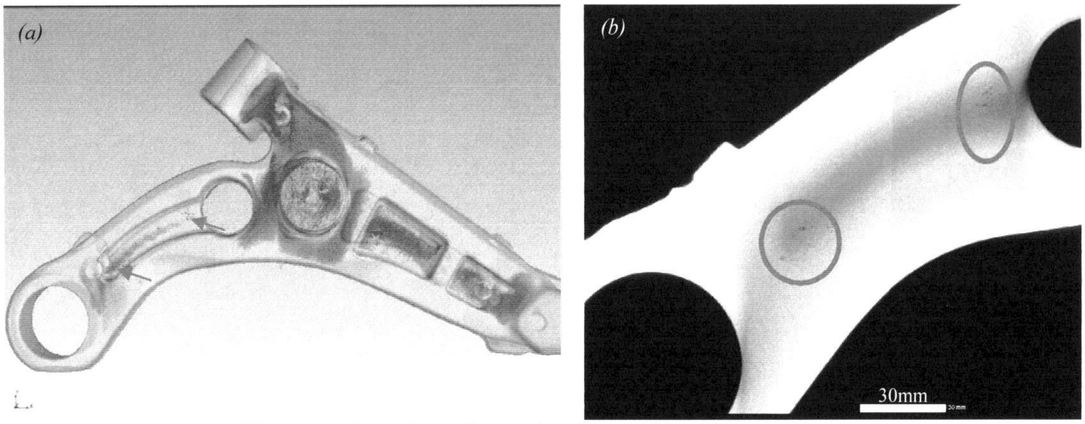

Figure 6. Location of porosity detected by CT scanning.

6. Summary

A numerical modelling methodology has been developed using ProCAST for the industrial CPC process of aluminum automotive parts. The model incorporates the key process boundary conditions and material properties. Validation against data measured on the production process has demonstrated the model's accuracy. Given that the modelling methodology was originally developed for the LPDC process and has also been validated by the measured data from a LPDC process for wheel production, it proves the modelling methodology is robust and applicable for both the LPDC and CPC processes.

Acknowledgement

The authors thank CITIC Dicastal Co. Ltd for the technical and financial support of this work.

References

[1] Nishida Y, Droste W and Engler S 1986 *Metall. Mater. Trans. B* **17** 833–44
[2] Wei X 2014 Thermal mechanical analysis of interfacial behavior in aluminum alloy wheel casting process (University of British Columbia)
[3] Moayedinia S 2014 Quantification of cooling channel heat transfer in low pressure die casting (University of British Columbia)
[4] Hu H, Chen F, Chen X, Chu Y L and Cheng P 2004 *J. Mater. Process. Technol.* **148** 57–67
[5] Zhang B, Maijer D M and Cockcroft S L 2007 *Mater. Sci. Eng. A* **464** 295–305
[6] Vijayaram T R 2012 *Metalworld* 24–6
[7] Mills K C 2002 Recommended values of thermophysical properties for selected commercial alloys (Woodhead Publishing)

IOP Conf. Series: Materials Science and Engineering **861** (2020) 012031 doi:10.1088/1757-899X/861/1/012031

Optimisation of the filling process in counter-gravity casting

M Papanikolaou, P Saxena, E Pagone, K Salonitis and M Jolly

Manufacturing Theme, Cranfield University, College Rd, Cranfield, Bedford MK43 0AL, UK

E-mail: m.papanikolaou@cranfield.ac.uk

Abstract. Metal casting is one of the most energy-intensive manufacturing processes with limited resource efficiency. To solve the problem of high energy consumption, a novel counter-gravity casting process has been earlier introduced. This process also referred to as CRIMSON (Constrained Rapid Induction Melting Single Shot Up-Casting) makes use of melting metal, just enough to fill one mould cavity at a time. The molten metal is subsequently pushed into the mould with the help of a piston, using a counter-gravity controlled method. Although CRIMSON has been proven to be a highly efficient process with the potential to produce high quality final cast products, there is still room for optimisation of the process. The objective of this investigation is to estimate the optimum ingate velocity in order to ensure smooth filling of the mould and eventually reduce turbulence and the likelihood of defects in the final cast product. For this purpose, a computational framework integrating a CFD solver and an optimisation algorithm has been developed. The obtained results suggest that the optimised ingate velocity can contribute towards the smooth filling of the mould and effectively contribute towards the reduction of entrained air and surface defect concentration in the final cast product.

1. Introduction

Sand casting is one of the oldest [1], most widely used and versatile metal forming processes. This process consists of two main stages which involve (a) pouring molten metal into a sand mould, (b) allowing for solidification and (c) removing the cast product by breaking the sand mould. However, producing good quality castings using this method is a not a trivial issue. The mechanical properties of the final product depend on the concentration of defects in the solidified material. One of the most important factors leading to defective castings is turbulence which is responsible for the inclusion of oxide bi-films into the melt [2]. Although turbulence cannot be eliminated, especially in the case of gravity casting, it can be significantly reduced via designing proper running systems and effectively controlling the metal velocity. John Campbell in [3] advocated some running and gating system designs leading to the minimisation of entrainment defects and superior quality castings. More specifically, the author emphasised the importance of placing filters across the runner in order to prevent backflow and promote uniform priming of the runner. Moreover, Campbell distinguished two gating system designs, namely, the Trident Gate and the Vortex Gate which can potentially lead to the elimination of entrainment defects. The potential of the aforementioned gating system designs to reduce entrainment defects was also verified via means of numerical simulations [4].

Counter gravity processes have been developed in order to address the issue of defects entrained in the casting. As indicated by their name, these methods eliminate turbulence arising during the fall of the melt down the sprue. The Constrained Rapid Induction Melting Single Shot Up-Casting method (CRIMSON) developed by Jolly *et al.* [5] is a counter-gravity sand casting method which has been proven to offer energy savings and the potential to minimise the concentration of double oxide films (DOF) and air entrainment in the casting [6]. In CRIMSON, molten metal is pushed into the mould with the help of a piston whose velocity is regulated via a computer-controlled method. Although CRIMSON is a counter-gravity casting method, a small amount of defects might still arise during filling or solidification [7]. Therefore, it of utmost importance to exploit the advantages of this method, such as the controlled filling rate, in order to produce enhanced quality cast products.

Since the early 1950s the optimisation of casting processes relied mostly on the intuition and experience of the foundry engineers. Later on, water models were used to visualise turbulence in order to geometrically optimise various components of the filling system [8–10]. In the early 1980s the development of numerical models capable of accurately reproducing a wide range of phenomena occurring during casting elucidated the process dynamics and allowed for the optimisation of the process. Nowadays, numerical modelling is widely accepted by foundry engineers as a means to improve quality and increase energy efficiency [11]. The foundations of the numerical modelling of casting processes were laid by Eyres *et al.* [12] back in 1946. Since then, numerical models have been developed for calculating the temperature profiles during solidification [13], identifying the factors leading to shrinkage porosity [14] and micro-porosity formation [15]. More contemporary investigations have been focused on the modification or optimisation of casting processes using numerical modelling. Sun [16] implemented the Taguchi method in order to optimise a set of performance characteristics including filling velocity, shrinkage porosity and product yield of a sand-casting process by identifying the optimal geometrical features of the gating system design. Moreover, in [7] an optimisation framework consisted of an optimiser and a CFD solver was implemented to estimate the optimum dimensions of the feeding system leading to the minimisation of defects.

As elaborated in the previous paragraphs, numerical modelling has been proven to be a powerful tool for the optimisation of casting processes. The objective of this investigation is to exploit the design advantages of the CRIMSON process in order to improve the quality of the casting expressed in terms of entrained air and surface defect concentration. More specifically, an optimisation framework has been developed in order to determine the optimum ingate velocity leading to minimum entrainment defects. The objective function was defined as the weighted sum of the inverse fill fraction, normalised surface defect concentration and normalised air entrainment. The obtained results indicate that, even in counter gravity casting, controlling the metal ingate velocity is crucial for delivering superior quality products.

2. Methodology

2.1. Simulation

The CRIMSON process makes use of a plunger which pushes the molten metal into the cavity of a sand mould. A detailed overview of the process can be found in [17]. In order to simplify the simulation process and reduce the computational cost, the motion of the plunger was replaced by a mass source located at the bottom of the mould as illustrated in figure 1. In order to further reduce the computational cost, only a quarter of the simulation domain was simulated by exploiting symmetry boundary conditions. The steps and requirements for the filling stage are listed below:

- The mass source depicted in figure 1 was activated. The source normal velocity was considered as the first optimisation input variable (v_1).
- When the liquid metal reached the level of Probe 1 the value of the source velocity was changed to v_2. This is the second optimisation input variable.
- When the molten metal reached the level of Probe 2 the simulation was terminated.
- The filling time should be less than 4 s; the simulation was terminated at t=4s even if the liquid metal did not reach the level of Probe 2.

The reason behind the positioning of Probe 1 at the bottom of the tensile bars was that turbulence is expected at this critical point due to the sudden change of the cross section. In other words, it is crucial to control the meniscus velocity right at the point when molten metal enters the tensile bars which are the final cast product. As it will be elaborated in the following subsection, the optimisation objective was to estimate the optimal set of velocity values [v_1, v_2] leading to the minimisation of the objective function.

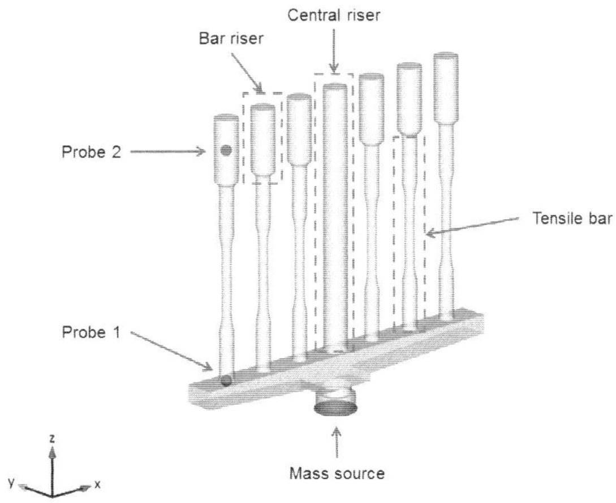

Figure 1. Simulation setup.

The simulated fluid was Aluminium LM-25 while the pouring temperature was set to 680 °C. A solid mesh with orthogonal hexahedra with a cell size equal to 2.5 mm was conformed to the open volume of the simulation domain with an overlap length equal to 1 cm. The cell size was selected so as to have at least 3 cells expanding across the thinner cross section of the mould, which is located at the middle of the tensile bars. Symmetry boundary conditions were applied to all directions with the exception of z_{max}, where a pressure boundary condition was applied. The developed CFD model accounted for surface tension [18] while the RNG k-ε method was used to model turbulence [19]. Surface defect concentration and air entrainment were modelled using the corresponding models [20, 21] of the Flow-3D software [22]. The material properties and simulation parameters used in this study are summarised in table 1 and table 2 respectively.

Table 1. LM-25 properties.

Density [kg/m^3]	2417
Viscosity [kg/(m·s)]	0.00138
Thermal Conductivity [W/(m·K)]	70.15
Surface Tension Coefficient [kg/s^2]	0.8
Latent Heat [J/kg]	3.58·10^5

Table 2. Simulation parameters.

Mould initial temperature [°C]	25
Pouring temperature [°C]	680
Air temperature [°C]	25
Cell size [mm]	2.5
Total number of cells	59,000

2.2. Optimisation

As mentioned in the previous sections the objective of this investigation is to exploit the advantages offered by the CRIMSON process design in order to investigate its potential to produce further improved results. One of the advantages offered by CRIMSON is that it is a computer-controlled counter-gravity casting method, i.e. the filling velocity can be controlled at all times. In this study the mass source (ingate) velocity values v_1 and v_2 have been used as the optimisation input variables, as discussed in the previous subsection. Linear scalarisation has been used in order to solve this multi-variable and multi-objective problem. The objective function was defined as the sum of weighted individual objective functions:

$$OF = w_1 OF_{ff} + w_2 OF_{ae} + w_3 OF_{sdc} = w_1 \frac{ff_0}{ff} + w_2 \frac{ae}{ae_0} + w_3 \frac{sdc}{sdc_0} \quad (1)$$

where ff (dimensionless), the fluid fraction ae (dimensionless) the entrained air volume fraction, sdc (kg/m^3) the surface defect concentration and w_i is a set of weights with $\sum_{i=1}^{3} w_i = 1$. The individual objective functions were normalised by their corresponding values obtained in the first iteration of the optimiser (ff_0, ae_0 and sdc_0); this was due to their different orders of magnitude between the different objectives. The values of ff, ae and sdc were calculated by averaging their corresponding cell values over the region of the tensile bars (which are the final cast product) at the last timestep of the filling simulation. Among the 3 individual optimisation objective functions the most important is OF_{ff}, i.e. the tensile bars should be necessarily completely filled. In order to fulfil this requirement, the highest weight was assigned to the specific individual objective (w_1=0.5). Moreover, the values of the initial guess [v_{10}, v_{20}] were selected so as to ensure that the tensile bars are completely filled (ff_0=1). Therefore, although very low ingate velocities led to zero air entrainment and surface defect concentration they were eventually rejected as solutions as they led to incomplete filling (ff<1).

The developed optimisation framework consisted of a commercial CFD software (Flow-3D™) and a python code responsible for the following tasks: (1) post-processing of the CFD results, (2) estimation of the objective function (*OF*), (3) feeding of the calculated *OF* value to the optimiser, (4) optimisation and (5) modification of the input file based on the new input variables proposed by the optimiser. The aforementioned process ran in a loop until the value of objective function converged to a minimum. The L-BFGS-B constrained multi-variable optimisation method [23] was used while both input variables (v_1 and v_2) were bounded between 0 and 0.5 m/s.

3. Results

3.1. Filling simulation

In the simulation case presented below the mass source was removed at the end of the filling process and an additional 10 s interval was allowed for equilibration. It has to be noted that the equilibration phase was neglected when performing the optimisation loop as described in the methodology section. This was because the values obtained for the average surface defect concentration and air entrainment volume fraction in the area of the tensile bars with or without the equilibration phase were not significantly different. Moreover, simulating the equilibration phase for each iteration would dramatically increase the computational cost.

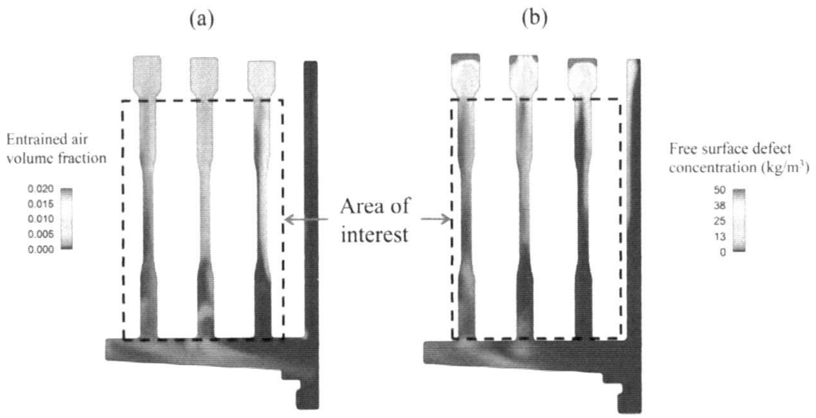

Figure 2. (a) Entrained air volume fraction and (b) free surface defect concentration distribution at the end of the equilibration process.

The entrained air volume fraction and surface defect concentration distribution at the end of the equilibration process are illustrated in figure 2. By comparing figure 2(a) and figure 2(b) it is evident that there are some significant similarities and discrepancies between the two distributions. Although both profiles are similar, surface defect concentration appears to sharply increase in the vicinity of the risers in contrast to the entrained air volume fraction which remains relatively constant. Similarly to entrained air, surface defects arise at the fluid free surface and subsequently diffuse into the bulk. However, in contrast to air entrainment, surface defect concentration depends on the exposure time of the free surface. As a result, its value continuously increases during the equilibration phase. However, due to the fact that the fluid velocity becomes almost negligible as equilibration proceeds defects cannot be diffused and are confined to the vicinity of the risers.

3.2. Optimisation

Linear scalarisation was used to minimise the objective function of equation (1). The fill fraction (*ff*), air entrainment (*ae*) and surface defect concentration (*sdc*) were evaluated at the last timestep of the filling simulation in the area of the tensile bars as illustrated in figure 2. It has to be noted that runs

IOP Conf. Series: Materials Science and Engineering **861** (2020) 012031 doi:10.1088/1757-899X/861/1/012031

performed with a higher resolution mesh did not affect the average values of air entrainment and surface defect concentration. Two sets of weight distributions (*wd)* have been tested. For both cases the weight assigned to the fill fraction objective function (OF_{ff}) was equal to 0.5 as this is the most important of all optimisation objectives. In the first case, reduced air entrainment was considered more important than surface defect concentration (w_2=0.4 and w_3=0.1) while in the second one the inverse scenario was examined (w_2=0.1 and w_3=0.4).

(a) (b)

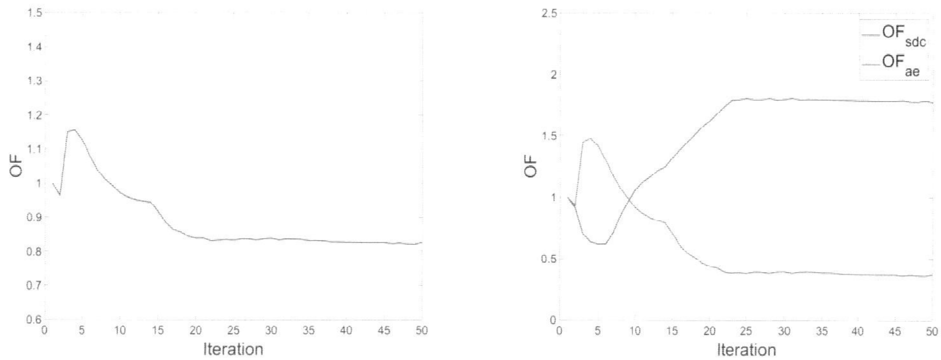

Figure 3. (a) Total and (b) individual objective functions versus number of iterations for wd_1.

As shown in figure 3(a) the optimiser has converged within 25 iterations while the objective function has been reduced by almost 18% at the end of the run. Since air entrainment has been assigned with a higher weight than surface defect concentration, the minimisation of OF_{ae} has been prioritised over OF_{sdc} and its value has been minimised as illustrated in figure 3(b). However, there is a cost for that: OF_{sdc} increased. This behaviour can be explained by observing figure 4 which illustrates the evolution of the velocity values selected until the final solution has been reached. It is clear that the reason for the penalty imposed on OF_{sdc} is due to the convergence to lower velocity values compared to the initial guess (v_{10}= v_{20}=0.21 m/s). Lower velocities lead to decreased turbulence but on the other hand increase the duration of the exposure of the moving free surface to air as well as the free surface defect concentration (OF_{sdc}) in the final casting. The solution obtained suggests using a lower ingate velocity during the first part of the filling process (while the fluid level is below the location of Probe 1) and a higher velocity for the rest of the filling process. This is because of the fact that the area of the free surface is much larger when being below Probe 1.

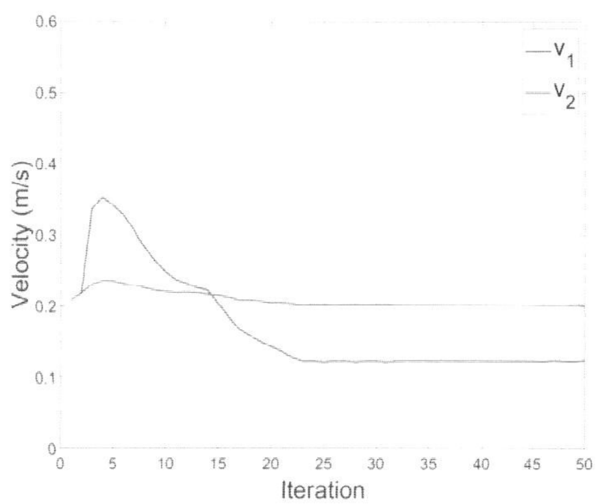

Figure 4. Input variables versus number of iterations (wd_1).

The trends observed above are inversed for the second weight distribution (wd_2) as expected. Although the total objective function has converged within the first 30 iterations as shown in figure 5(a), this time OF_{sdc} has been prioritised over OF_{ae} due to its higher weight. Similarly to wd_1, the penalty for this change is higher air entrainment, as shown in figure 5(b).

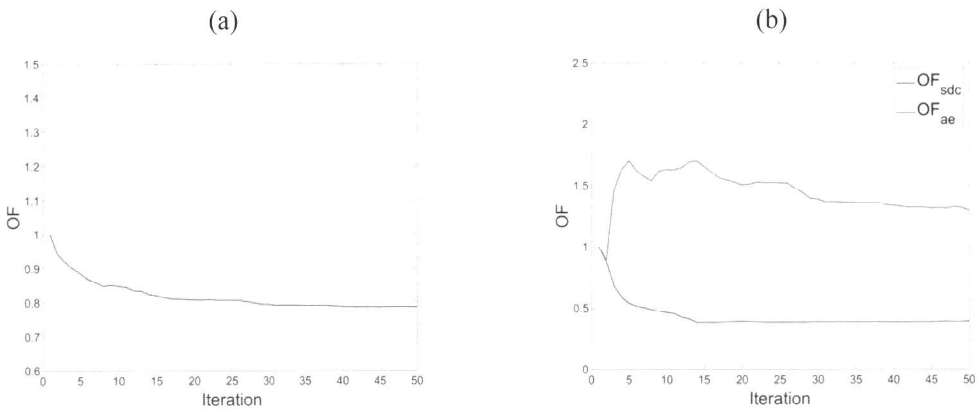

Figure 5. (a) Total and (b) individual objective functions versus number of iterations for wd_2

As shown in figure 6, in this case the obtained solution suggests using higher velocities than the ones of the initial guess in order to decrease the filling time and consequently the diffusion of surface defects to the bulk. More specifically, the proposed ingate velocity for the first filling stage v_1 is almost double the corresponding initial guess v_{10}. This is because the longer exposure of the free surface to the air the more the surface defect concentration. The proposed velocity values would be even higher if there wasn't the constraint imposed by the air entrainment criterion. This is evident by observing the curve for v_1 in figure 6. Between the 10th and 14th iteration the optimiser attempts to further increase the value of v_1. However, the value of OF_{ae} at the corresponding iteration increases sharply as shown in figure 5; this is the reason why v_1 is maintained to a lower and steady level. The same principles apply for v_2 which is maintained at lower levels due to the fact that the surface defect criterion is more dominant during the first filling stage to the increased free surface area.

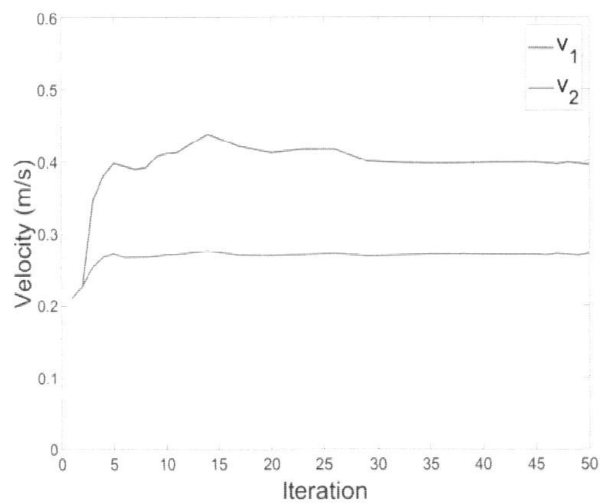

Figure 6. Input variables versus number of iterations (wd_2).

Table 3. Final solution for the 2 sets of weight distributions.

w_1 (OF_{ff})	w_2 (OF_{ae})	w_3 (OF_{sdc})	v_1 (m/s)	v_2 (m/s)
0.5	0.4	0.1	0.12	0.20
0.5	0.1	0.4	0.4	0.27

Table 3 summarises the effect of the weight distributions on the final solutions proposed by the optimiser. It is clear that low velocity values are selected when air entrainment is more significant as an optimisation criterion compared to surface defect concentration while the velocity values gradually increase as the importance of criteria is inversed. Moreover, it can be observed that v_1 has a larger variation than v_2. This is because the metal top surface area is larger during the first filling stage and it is more likely that surface turbulence, affecting both air entrainment and surface defect concentration, might arise at this stage compared to the second one. Thus, controlling the ingate velocity while the melt top surface is below the level of Probe 1 has a greater effect on the final result than controlling v_2.

In both cases examined in this investigation, the weight distribution between surface defect concentration and air entrainment is imbalanced. This results in two distinct solutions due to the prioritisation of one of two conflicting objectives each time. However, a good casting is defined based on the specifications set by the foundry engineers in reality. In the case where bubble damage or gas porosity have been identified as the major defects to be avoided, the air entrainment criterion should be prioritised. On the other hand, if the focus is laid on the elimination of potential sites for crack nucleation and propagation the surface defect concentration criterion should be assigned with a higher weight. If both of the aforementioned scenarios are likely and equally important both the two criteria should be assigned with equal weight (w_2=w_3=0.25); the developed framework allows the free allocation of weights to each criterion examined.

4. Conclusions and future work

In this investigation a computational framework for the optimisation of the Constrained Rapid Induction Melting Single Shot Up-Casting (CRIMSON) method has been developed. The proposed framework was consisted of an optimisation algorithm plugged to a commercial CFD software while the optimisation objective was the minimisation of entrained air and surface defects in the final cast product by controlling the ingate velocity during filling. Two scenarios with different weight distributions assigned to the individual optimisation objectives were examined. In the first scenario, air entrainment was considered more significant than surface defect concentration while in the second one the inverse case was examined. The L-BFGS-B constrained multi-variable optimisation method was used to solve this multi-variable and multi-objective optimisation problem. Linear scalarisation was also used for the estimation of the total objective function. The main conclusions drawn from this study are listed below:

- Very low ingate velocities lead to smooth filling but on the other hand increase surface defect concentration and the risk of incomplete filling and premature solidification.
- For all weight distributions examined the optimisation algorithm was capable of reducing the value of the total objective function by at least 15%. The design advantage of the CRIMSON process to control the ingate velocity can lead to superior quality castings.
- The optimisation algorithm was capable of converging to the minimum value of the objective function within the first 30 iterations.
- The velocity value during the first filling stage has more dominant effects on the quality of the final cast product than its corresponding value during the second stage. In a more general context, it is of utmost significance to control the filling velocity when the metal surface area is large. This is due to the fact that the majority of defects might arise at this point in time.

In the authors' opinion there is still room for optimisation of the CRIMSON process. CRIMSON offers the advantage of controlling the plunger velocity during the whole filling process with a high resolution, making the potential of defect free castings feasible. Therefore, future research should be directed towards using smooth ingate velocity profiles and avoid the use of steps in order to further reduce turbulence during filling.

Acknowledgements
This research was funded by the UK EPSRC project "Energy Resilient Manufacturing 2: Small is Beautiful Phase 2 (SIB2)" under grant EP/P012272/1.

References

[1] Weiss D 2018 Advances in the Sand Casting of Aluminium Alloys *Fundamentals of Aluminium Metallurgy* (Elsevier) 159–71
[2] Campbell J 2015 *Complete Casting Handbook: Metal Casting Processes, Techniques and Design* (Butterworth-Heinemann)
[3] Campbell J 2018 *Mini Casting Handbook.* (Aspect Design)
[4] Papanikolaou M, Pagone E, Jolly M and Salonitis K 2020 Numerical Simulation and Evaluation of Campbell Running and Gating Systems *Metals (Basel).* **10** 68
[5] Dai X, Jolly M R and Zeng B 2011 The capability enhancement of aluminium casting process by application of the novel CRIMSON method *Shape Cast. John Wiley Sons, Inc* 265–72
[6] Salonitis K, Jolly M R, Zeng B and Mehrabi H 2016 Improvements in energy consumption and environmental impact by novel single shot melting process for casting *J. Clean. Prod.* **137** 1532–42
[7] Papanikolaou M, Pagone E, Georgarakis K, Rogers K, Jolly M and Salonitis K 2018 Design optimisation of the feeding system of a novel counter-gravity casting process *Metals* **8** 817
[8] Swift R E, Jackson J H and Eastwood L W 1949 A study of principles of gating *AFS Trans.* **57** 76–88
[9] Grube K R and Kura J G 1955 Principles applicable to vertical gating *Trans AFS* **63** 35–48
[10] Hsu F Y and Lin H J 2009 A diffusing runner for gravity casting *Metall. Mater. Trans. B Process Metall. Mater. Process. Sci.* **40** 833–42
[11] Yu K-O 2001 Modeling for casting and solidification processing (CRC Press)
[12] Eyres N R, Hartree D R, Ingham J, Sarjant R J and Wagstaff J B 1946 The calculation of variable heat flow in solids *Philos. Trans. R. Soc. London. Ser. A, Math. Phys. Sci.* **240** 1–57
[13] Sulaiman S and Hamouda A M S 2004 Modelling and experimental investigation of solidification process in sand casting *J. Mater. Process. Technol.* **155–156** 1723–6
[14] Gunasegaram D R, Farnsworth D J and Nguyen T T 2009 Identification of critical factors affecting shrinkage porosity in permanent mold casting using numerical simulations based on design of experiments *J. Mater. Process. Technol.* **209** 1209–19
[15] Lee P D, Chirazi A and See D 2001 Modeling microporosity in aluminum-silicon alloys: A review *J. Light Met.* **1** 15–30
[16] Sun Z, Hu H and Chen X 2008 Numerical optimization of gating system parameters for a magnesium alloy casting with multiple performance characteristics *J. Mater. Process. Technol.* **199** 256–64
[17] Papanikolaou M, Pagone E, Salonitis K, Jolly M and Makatsoris C 2018 A Computational Framework Towards Energy Efficient Casting Processes *International Conference on Sustainable Design and Manufacturing* pp 263–76
[18] C.W. Hirt *Modeling Surface Tension*
[19] Prasad V.Tota 2009 Turbulent Flow Over a Backward-Facing Step Using the RNG k- ε Model *Flow Sci.* 1–15
[20] C.W. Hirt 2012 *Modeling Turbulent Entrainment of Air at a Free Surface* (Flow Science, Inc.: Santa Fe, NM, USA)
[21] Barkhudarov M R and Hirt C W 1998 Tracking defects *1st international Aluminium casting technology symposium*
[22] Anon 2019 *FLOW-3D® Version 12.0 [Computer software]* (Santa Fe, NM: Flow Science, Inc.)
[23] Zhu C, Byrd R H, Lu P and Nocedal J 1997 Algorithm 778: L-BFGS-B: Fortran subroutines for large-scale bound-constrained optimization *ACM Trans. Math. Softw.* **23** 550–60

IOP Conf. Series: Materials Science and Engineering **861** (2020) 012032 doi:10.1088/1757-899X/861/1/012032

Prediction of solidification structures in a 9.8 tonne steel ingot

B Gerin[1], H Combeau[1], M Založnik[1], I Poitrault[2] and M Cherif[2]

[1] Université de Lorraine, CNRS, IJL, F-54000 Nancy, France
[2] ArcelorMittal Industeel, 56 rue Clemenceau, F-71201 Le Creusot, France

E-mail: benjamin.gerin@univ-lorraine.fr

Abstract. The control of the carbon macrosegregation level in steel ingots is important for the structural integrity of the final component. Previous studies using the SOLID® multiscale modelling software have shown that in order to obtain predictive results for the macrosegregation and the grain structure (CET, grain morphology) in steel ingots, a model needs to account for fragmentation of columnar dendrites as a source of equiaxed grains. The goal of this study is to show that a numerical model that takes into account fragmentation can describe the formation of the structures and the macrosegregation during solidification of a large steel ingot. The present article describes how fragmentation is taken into account in the multiphase numerical model used in the simulations. The simulation results are compared to experimental data from a 9.8 tonne ingot cast in A5/6 steel by ArcelorMittal Industeel. The model correctly reproduces the major features of the experimentally observed structure and macrosegregation. We show that the structures are formed very early on during solidification, whereas macrosegregation develops much more gradually. Our results also underline the importance of liquid flow and grain movement in order to correctly predict the final structure and macrosegregation.

1. Introduction

Achieving homogeneous mechanical properties for large cast ingots is an industrial challenge because of the segregation of alloying elements during solidification. This is due to an enrichment of the liquid in alloying elements as solidification progresses. This can lead to large-scale inhomogeneity (macrosegregation) which will affect the local mechanical properties. Being able to predict and control macrosegregation is thus crucial for the structural integrity of the components.

Segregation is the result of the complex interaction of the solidification structures (columnar, equiaxed, grain morphology) and liquid flow. Additionally, the structures themselves can influence mechanical properties, so accurately predicting both the macrosegregation and the solidification structures is an important issue for steel ingot producers. The prediction of the structures requires taking into account the fragmentation of the columnar dendrites in the model [1]. The present study uses the SOLID® multiscale and multiphysics model to predict the structure and macrosegregation of a 1.44 m diameter, 9.8 tonne steel ingot by taking into account the fragmentation during solidification. The originality of this study lies in the prediction and description of the formation of the structures and the macrosegregation, which is not commonly found in literature for such large products.

Fragmentation of the columnar dendrites happens when secondary dendrite arms detach and are swept in the bulk liquid. If the local conditions are favourable, they will grow as equiaxed grains before sedimenting at the bottom of the ingot, leading to a columnar to equiaxed transition. Fragmentation is an important source of equiaxed grains, especially in castings with no inoculation. The detachment of secondary arms does not occur because of mechanical breakage [2] but rather due to local dissolution at the root of the secondary arms [3] or remelting of the arms due to temperature increase [4]. During normal coarsening of the columnar structure, secondary arm detachment is quite slow. However, certain situations massively increase fragmentation, such as unconstrained growth of the columnar front [5] or slowdown of the columnar tip velocity [6]. In addition, in steels, the peritectic transformation can also promote the increase in size of the equiaxed region [7].

2. Model used in the SOLID® software

The SOLID® software, developed on scientific models from IJL, supported by a French consortium of industrials in the framework of metallurgy and supported also by national and European funding is commercialized by SCC (see www.scconsultants.com) to be released in the industry. SOLID® uses a 2D finite volume method to discretise the set of partial differential equations of a multiphase model (liquid, columnar, equiaxed and inter-dendritic liquid for both solid phases). Each discretised volumetric element, or "cell" can contain both columnar and equiaxed solid, in addition to liquid. Cells containing both columnar and equiaxed grains are called mixed cells. Solidification begins with a columnar front starting on the mould walls which can grow and fragment. Fragmentation generates equiaxed grains in the undercooled region ahead of the columnar front. These equiaxed grains are not individually simulated but are instead represented by average grain size and grain density values in each cell. Fragmentation is implemented using the model developed by Nicolas Leriche in his PhD [8]. Fragmentation uses two parameters: the fragmentation flux density (number of generated grains per surface area of columnar front per second) and the thermal gradient activation threshold. The threshold means that the fragmentation occurs only when the temperature gradient at the columnar front is locally smaller than a given threshold value.

The model calculates the movement of the liquid phases (bulk liquid and inter-dendritic liquid inside the grain envelope) and of the equiaxed grains. There is a complex interaction between liquid flow and grain movement, each influencing the other. The grains can be swept up by the liquid flow and grow or remelt depending on local undercooling. Generally, small grains (< 1 mm) will follow the liquid movement, whereas larger grains have enough mass that the liquid flow cannot overcome their downward movement, and they sediment at the bottom of the bulk liquid. Grain movement in a cell is blocked when the liquid fraction is below 0.6. Additionally, there is no grain movement in cells containing columnar grains. The columnar front in a cell is blocked when the envelope fraction of equiaxed grains in that cell reaches 0.5, leading to a columnar to equiaxed transition (CET). Unlike the columnar structure, equiaxed cells cannot progress from one cell to the next. However, they can generate an equiaxed to columnar transition (ECT), producing new columnar cells. In the model, ECT is set to occur only if there are no equiaxed grains in the surrounding liquid. When ECT starts, a new columnar front is started in the adjacent liquid cells, which can then grow, fragment and can itself be blocked. Further information about the model can be found in previously published work [9].

3. Structures and macrosegregation experimentally observed in the 9.8 t ingot

The focus of the present article is a 9.8 t experimental research ingot cast in A5/6 steel by Industeel. It has a diameter of 1.44 m and a height of 0.74 m, which is much smaller than typical commercial Industeel ingots. This ingot was cast specifically for research purposes and was not destined to produce a component. Ingots of this shape (large diameter compared to height) are used to make flat component such as thick plates and disks [10]. The composition of the alloy, measured just before casting, is given in table 1.

Table 1. Chemical composition in wt% of the A5/6 steel used in the 9.8 t ingot.

C	S	P	Si	Mn	Ni	Cr	Mo	Cu	Sn	Al	Fe
0.184	0.026	0.008	0.245	0.59	0.17	0.23	0.035	0.250	0.025	0.018	Balance

The mould is made of cast iron and there is no hot top, however the liquid metal is covered by two layers of powder: an exothermal powder with an insulating powder on top. The total solidification time was less than 6 hours. After solidification, the ingot was cut in half along a diameter in order to analyse its macrostructure and segregation. Figure 1 shows the schematic representation of macrostructures such as dendrites and equiaxed grains, identified with a macro-etching. The different macrostructures were visually identified and manually colour-coded.

The stand-out features have been numbered in figure 1: the primary columnar region along the side and bottom walls (1); an equiaxed structure with fine grains (2) on the bottom and larger grains on top (3); a thin layer of equiaxed grains (4) separating the primary columnar with the central columnar region (5). Above the equiaxed region, a large 'step' in columnar thickness is visible (6). The top of the ingot is mostly columnar, with the centre region being a mix of columnar and equiaxed grains.

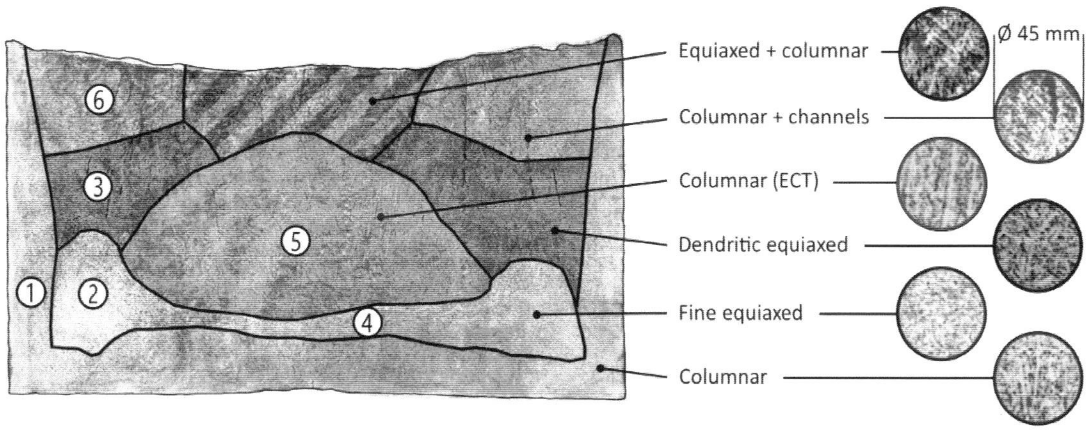

Figure 1. Experimentally observed structures of the 9.8 t ingot.

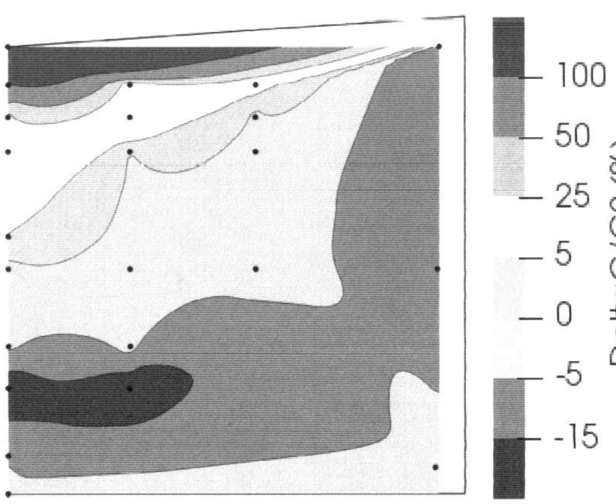

Figure 2. Experimental measurements of carbon concentration in a cross-section of the ingot, represented as percentage variation of nominal composition $\Delta C/C_0$. Contour lines were obtained by interpolating the experimental data. Only half of the ingot is represented, the centreline is on the left. Minimum value is –23% and maximum value is +290%.

Analysis was performed on the open face of the ingot cross-section to measure carbon segregation. The experimental measurements of carbon concentration are presented in figure 2. Each data point is located by a black dot and its value is given as a percentage variation of the nominal concentration $\Delta C/C_0$. The experimental data was interpolated in order to plot contour lines. The bottom part of the ingot is negatively segregated, reaching –23% of the nominal composition. The top centre region, which is the last to solidify, features a large positive segregation, with a maximum of +290% relative to the nominal composition.

4. Numerical simulation

The solidification of the 9.8 t ingot was simulated using the SOLID® software using an axial symmetry. In order to simplify the calculation, a binary Fe–C alloy is used, with 0.184 wt% carbon. The fragmentation parameters were determined by performing a sensitivity analysis on these parameters (not discussed in this article), and then choosing the values that gave the results closest to the experimental ingot. The chosen fragmentation parameters are: fragmentation flux density of 500 frag.m^{-2}.s^{-1} and activation of fragmentation at a thermal gradient below 26 K.m^{-1}. Both exothermal and insulating powders are present in the simulation. The exothermal reaction is simulated using an enthalpy of reaction, leading to an increase in temperature of the powder. The chosen enthalpy of reaction in this simulation leads to a powder temperature significantly lower than the liquidus of the alloy ($T_{exo} < T_{liquidus}$). Additionally, the formation of columnar grains is disabled on the top surface of the ingot, where the liquid metal is in contact with the exothermal powder. Filling is taken into account using a simplified line by line filling method. The simulation uses a mesh size of 1 cm, so the total number of cells is 21643 of which 5697 are for the ingot. With a time step of 0.1 s, the total computation time on one CPU core was 40 hours. Coarser mesh sizes give similar results, with a roughly equivalent columnar thickness.

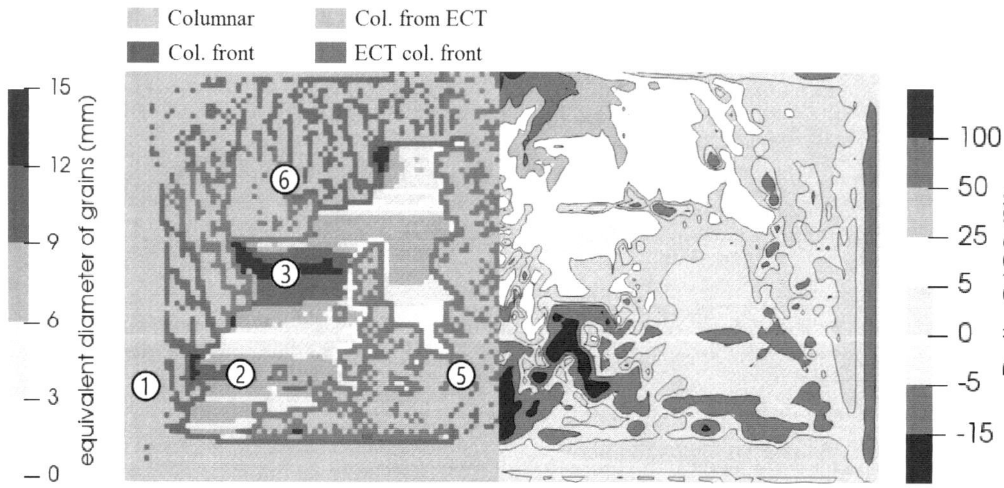

Figure 3. Simulation results for the 9.8 t ingot. The structure map is given on the left side, showing the columnar structure (either primary or from ECT) and the grain diameter for the equiaxed structure. On the right side, the carbon composition is given as a percentage of nominal composition.

The final structure map and carbon segregation pattern at the end of the simulation are given in figure 3. The same structure colour code was used as in figure 1: green and orange for columnar, blue for equiaxed. In addition, mixed columnar-equiaxed cells are distinguished by a darker columnar colour (dark green or dark orange).

The simulated macrosegregation is similar to the experimental results (figure 2), with a negative segregation reaching –34% in the bottom part of the ingot and a large positive segregation in the top part, with the maximum value (+129%) along the centre axis. The transition from negative to positive segregation along the axis happens at a similar point, roughly half the height of the ingot. The simulated structures also show similarities to the macrographic etching (figure 1), with almost all major features being reproduced. These have been numbered the same way as in figure 1: the primary columnar on the mould wall (1); the equiaxed region (2) of small grains (mostly smaller than 9 mm) with the transition to larger grains (3) (greater than 9 mm); the large central ECT region (5); and the 'step' in columnar thickness (6) above the equiaxed region. However, the layer of equiaxed grains separating the primary columnar from the center columnar (4) in figure 1, is missing. Additionally, the thickness of the primary columnar is overall greater in the simulation than in the experimental data.

IOP Conf. Series: Materials Science and Engineering **861** (2020) 012032 doi:10.1088/1757-899X/861/1/012032

(a) End of filling

(b) 900 s after the end of filling

(c) 3600 s after the end of filling

(d) 6300 s after the end of filling

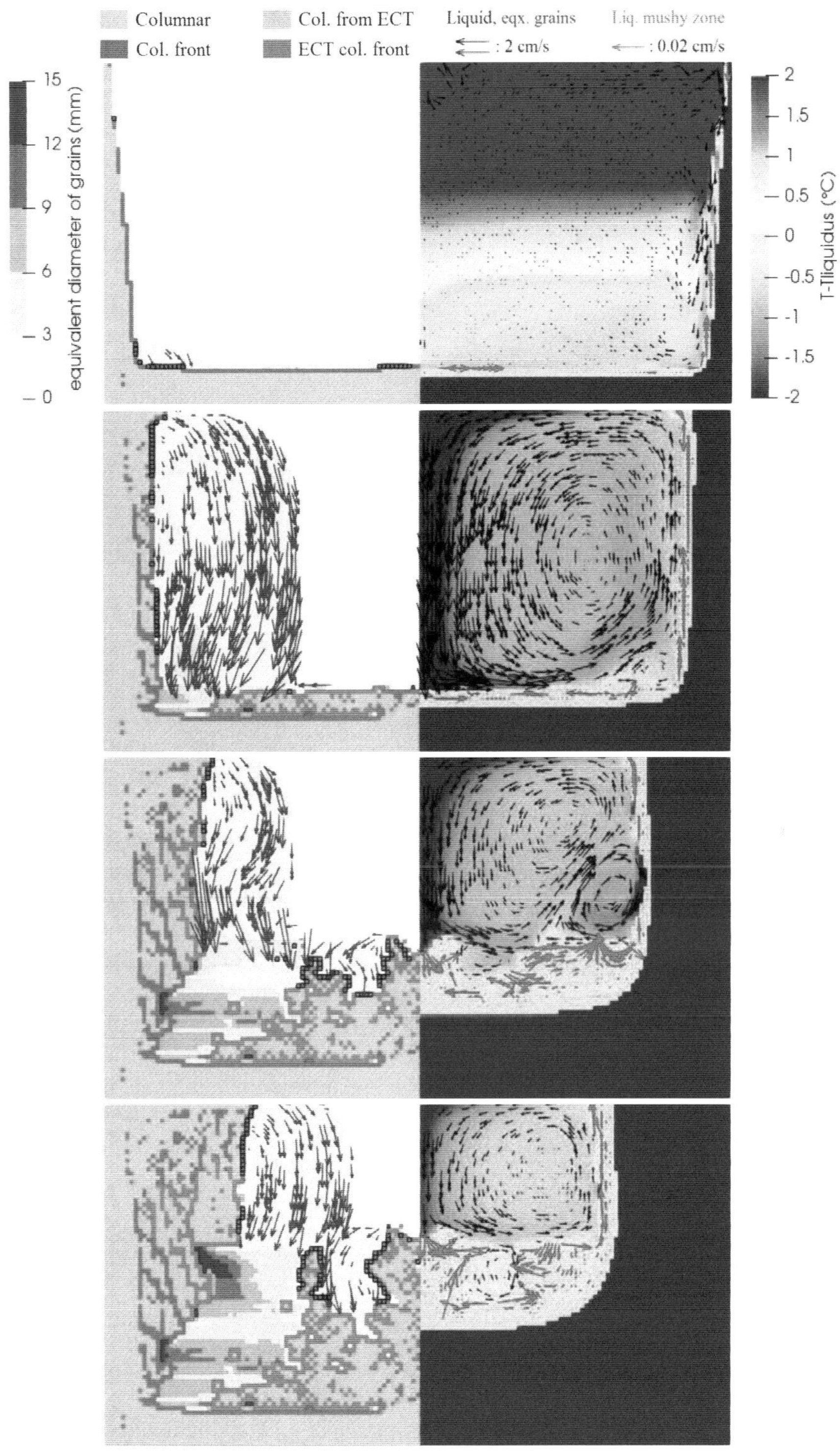

Figure 4. Simulation results for the 9.8 t ingot. The left side shows the columnar structure, with fragmenting cells outlined in black, the equiaxed grain diameter and grain movement (blue arrows). The right side shows $T - T_{liquidus}$ (positive: superheated; negative: undercooled), the liquid movement in black arrows, and the liquid movement in the mushy zone is in red arrows.

5. Discussion

The simulation shows the entire solidification sequence, explaining when and how the various structures were formed. The solidification sequence can be divided into the following stages, illustrated in figure 4:

- Solidification starts immediately with a columnar front initiating on the mould walls. Halfway during filling, a few cells start fragmenting at the bottom on the ingot, generating fragments in the liquid ahead of the columnar front.
- At the end of filling (figure 4(a)), the columnar front is already well developed, and the equiaxed grains produced by fragmentation have already blocked the columnar front on the bottom surface. The equiaxed structure starts forming at the bottom from the accumulation of sedimented grains. Because of the simplified filling model, there is little liquid convection during filling, and most of the liquid remains in superheat. After filling, convection is much more pronounced due to the heat exchange between the liquid and the exothermal powder (which is colder than the liquidus temperature of the alloy). The liquid loses its superheat 200 s after filling.
- 900 s after filling (figure 4(b)), liquid convection has stabilised into a large vortex with the liquid flowing downward in the middle of the ingot and upward along the columnar front. This movement is due to the liquid being simultaneously cooled at the top by the exothermal powder and enriched in carbon along the columnar front. Because the entire bulk liquid is undercooled, the equiaxed grains grow and sediment quickly. The liquid vortex prevents the sedimenting equiaxed grains (blue arrows) from reaching the centre of the ingot. The grains thus sediment near the mould wall, creating an equiaxed region on the outer edge of the ingot, and allowing the centre columnar region to grow unimpeded. The equiaxed and columnar regions grow upward roughly in parallel, creating the central structure seen in figure 1.
- Around 3600 s after filling, the small corner vortex in the liquid convection becomes much larger, as seen in figure 4(c). For the next 1000 s, the liquid convection has two counter-rotating vortices, after which the smaller corner vortex subsides. This smaller vortex changes the liquid flow along the vertical columnar front: instead of warm enriched liquid flowing upward, cooler less rich liquid is flowing downward. The falling equiaxed grains therefore grow much faster before sedimenting, and the sedimentation of larger grains leads to a rapid progression of the equiaxed region. This region of large equiaxed grains can be seen in figures 1 and 3.
- After the corner vortex disappears (around 4600 s after filling), the sedimenting grains are much smaller in size. This is because the liquid flowing up against the columnar front is enriched and less undercooled, but also because the grains have a shorter distance to grow before blocking. The sedimentation of smaller grains greatly slows down the upward progress of the equiaxed region. This gives time for the columnar front on the mould wall to grow horizontally. This is shown in figure 4(d): at 6300 s after filling, the columnar region on the side has grown horizontally to cover the region of large equiaxed grains. At that time, the ingot is 78% solidified, with the upper part still fully liquid and a small mushy zone in the equiaxed region (the red arrows in the figure shows the liquid convection in the mushy zone). At this stage, most of the macrostructures are already in place but the macrosegregation is still low: the carbon content in the bulk liquid is +15% relative to nominal composition.
- For the remainder of solidification ($t > 6300$ s), there are no major changes to the structure or liquid flow. Total solidification time is around 21300 s, with final solidification in the upper region at the centre of the ingot. The two columnar fronts slowly grow, competing with the equiaxed region to fill the remaining liquid volume. This results in the columnar and equiaxed structures that are visible in the top part of the ingot in figure 3. This final stage of the solidification is by far the longest (15000 s, 70% of total solidification time), during which the macrosegregation continues to evolve slowly due to liquid flow and grain sedimentation.

The simulation has a complex liquid flow that has a large influence on the sedimenting grains, and this ultimately affects the final structure and macrosegregation of the ingot. Liquid flow is driven by thermal convection but also by solutal convection (carbon enrichment reduces liquid density). In the simulation, thermal convection is largely influenced by the temperature of the exothermal powder that covers the top of the ingot. If the powder is sufficiently colder than the liquidus temperature of the alloy, cooling the liquid, then the liquid will flow downward along the central axis of the ingot. This is the case for the simulation presented in figure 3 and 4, with the resulting convection loop shown in figure 4. The convection generates an outward liquid flow along the bottom part of the ingot, chasing away the sedimenting grains and allowing for the large columnar region in the centre of the ingot. The convection also brings in fresh liquid into the growing columnar front, pushing enriched liquid outward and upward. This creates the region of negative segregation seen at the bottom of the ingot in figure 3. In the current model, this is achieved by using enthalpy of reaction values that give exothermal powder temperatures that are unrealistically low compared to the industrial process. Further work must be done to better understand the influence of the exothermal powder on liquid flow.

Conversely, if the powder is hotter than the liquid, then the top part of the bulk liquid stays hot and the cooling of the liquid happens from the side wall. This creates a convection loop in the opposite direction: liquid is flowing downward along the mould wall and upward along the centre axis. A convection loop in the opposite direction will bring in sedimenting grains and enriched liquid, instead of pushing them away. This drastically changes the macrosegregation and structure in the bottom part of the ingot. Figure 5(a). shows the result of a simulation with $T_{exo} > T_{liquidus}$ (other parameters being identical). The opposite convection loop in this simulation removes the central columnar region and leads to a much less pronounced negative segregation.

(a) Frag. parameters: 500 frag.m^{-2}.s^{-1} 26 K.m^{-1}

$T_{exo} > T_{liquidus}$

(b) Frag. parameters: 1500 frag.m^{-2}.s^{-1} 25 K.m^{-1}

$T_{exo} < T_{liquidus}$

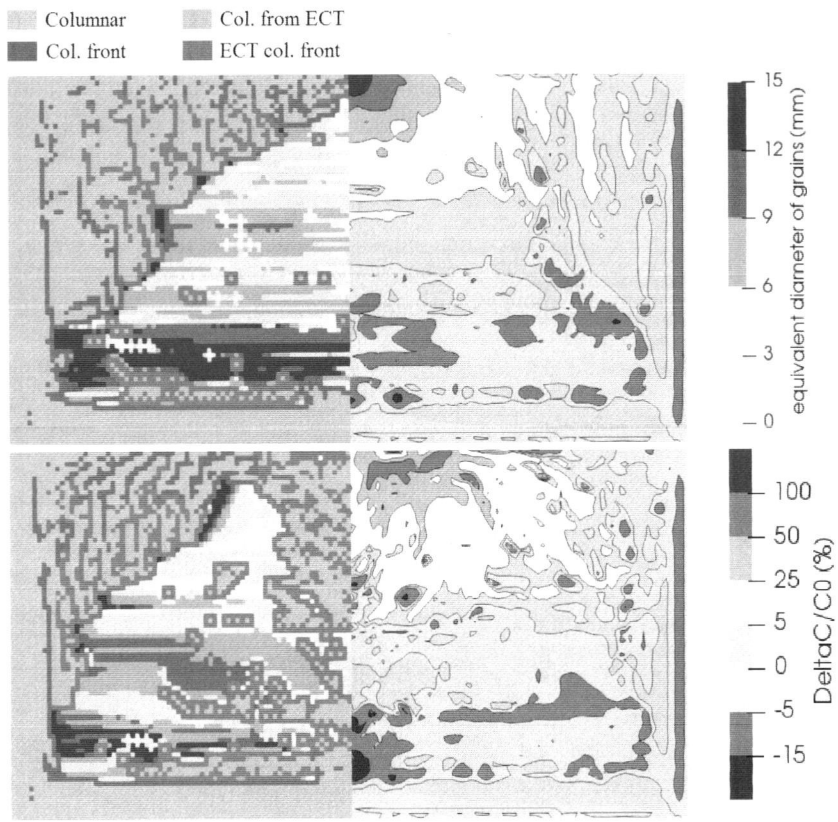

Figure 5. Simulation results for the 9.8 t ingot with different fragmentation parameters and exothermal powder temperatures.

These results show the large influence that liquid movement in the bottom part of the ingot has on structure and macrosegregation. An outward movement along the bottom will lead to a columnar region in the centre, with a pronounced negative segregation. Inward convection will instead lead to an equiaxed region and a reduced negative segregation. This effect is further illustrated in figure 5(b). In this simulation, the direction of liquid flow changes several times during solidification, and this creates alternating layers of equiaxed and columnar grains in the lower part of the ingot. Furthermore, these columnar regions have corresponding areas of negative segregation.

The absence of convection during filling in the simulations is probably the reason why the layer of equiaxed grains between the primary columnar and the central columnar is missing at the bottom of the ingot. The ingot being bottom poured, this will naturally lead to a large vortex with inward flow along the bottom, thus creating the missing equiaxed layer.

6. Conclusion

The simulation results show that by using only fragmentation as a source of equiaxed grains, it is possible to predict both the macrosegregation of carbon and the complex macrostructure of a steel ingot. The simulations show that liquid flow has a large influence on the development of the structures and the segregation. In addition, most of the structures are formed early on during solidification, 6300 s in our case, 30% of the total 21300 s. Conversely, the large positive segregation forms slowly over a much longer period

However, as the exothermal powder temperature used in the simulations is lower than what is used industrially, further work must be done to understand its effect and improve how it is modelled. Additionally, the lack of convection during filling is another area where the model will be improved.

Acknowledgements

The authors would like to thank the partners of this study for their support: Industeel, EDF, Framatome and Sciences Computers Consultants.

References

[1] Leriche N, Combeau H, Gandin Ch-A and Založnik M 2015 Modelling of columnar-to-equiaxed and equiaxed-to-xolumnar transitions in ingots using a multiphase model *IOP Conf. Series: Mat. Sci. & Eng.* **84** 012087

[2] Pilling J and Hellawell A 1996 Mechanical deformation of dendrites by fluid flow. *Met. & Mat. Trans. A* **27** 229-32

[3] Neumann-Heyme H, Shevchenko N, Lei Z, Eckert K, Keplinger O, Grenzer J, Beckermann C, and Eckert S 2018 Coarsening evolution of dendritic sidearms: from synchrotron experiments to quantitative modeling. *Acta Mat.* **146** 176-86

[4] Sato T, Kurz W, and Ikawa K 1987 Experiments on dendrite branch detachment in the succinonitrile-camphor alloy. *Trans. of the Japan Institute of Metals* **28(12)** 1012-21

[5] Gandin Ch-A 2000 From constrained to unconstrained growth during directional solidification. *Acta Mat.* **48(10)** 2483-501

[6] Liu S, Lu S-Z and Hellawell A 2002 Dendritic array growth in the systems NH4Cl-H2O and [CH2 CN]2-H2O: the detachment of dendrite side arms induced by deceleration. *J. of Crystal Growth* **234(4)** 740-50

[7] Tiwari S N and Beech J 1981 The role of the peritectic reaction in the formation of the equiaxes zone in iron base alloys. *Trans. ISIJ* **21** 554-8

[8] Leriche N 2015 *Etude de la Transition Colonnaire-Equiaxe dans les Lingots et en Coulée Continue d'Acier et Influence du Mouvement des Grains.* Phd Thesis, Université de Lorraine, Nancy

[9] Combeau H, Založnik M, Hans S, and Richy P E 2009 Prediction of macrosegregation in steel ingots: influence of the motion and the morphology of equiaxed grains. *Met. & Mat. Trans* B **40** 289-304

[10] Benhamou C, Poitrault I, Pisseloup J and Bocquet P 1985 Application of directional solidification ingot (LSD) in forging of PWR reactor vessel heads. *10th Int. Forging Conf.* 23-5

IOP Conf. Series: Materials Science and Engineering **861** (2020) 012033 doi:10.1088/1757-899X/861/1/012033

Modelling and simulation of ingot solidification with the open-source software *Code_Saturne*

C Demay[1], M Ferrand[1,2], S Belouah[1] and V Robin[1]

[1] EDF R&D, 6 quai Watier, 78400 Chatou, France
[2] CEREA Lab (joint laboratory Ecole des Ponts ParisTech – EDF R&D), 6-8 avenue Blaise Pascal, Cité Descartes, 77420 Champs-sur-Marne, France

E-mail: charles.demay@edf.fr

Abstract. In order to address the issue of macro-segregations in large industrial ingots, a solidification model has been recently developed in *Code_Saturne*, the general purpose open-source Computational Fluid tablDynamics (CFD) software developed and released by EDF R&D. This model is a liquid-solid mixture model inspired from the pioneer work of Bennon and Incropera, including mass, momentum, energy and species conservation equations. In particular, the energy conservation equation is formulated with the temperature variable, the thermosolutal convection is accounted for through the Boussinesq approximation and the interdendritic flow in the mushy zone is governed by the Darcy's law. Furthermore, a rigid solid phase is assumed and classical microsegregation models are used featuring lever-rule or Gulliver-Scheil assumptions. At the discrete level, a first-order upwind scheme is implemented with a SIMPLEC approach to solve the velocity-pressure coupling. The other couplings involving temperature and concentration fields are efficiently solved performing sub-loops with a PISO-like approach. The overall scheme is mainly implicit with additional treatments to ensure the equilibrium between the hydrostatic pressure gradient and the buoyant forces, and to deal properly with solid zones. The segregation predictions offered by the model are validated against both academic test case and industrial ingot configurations. A convergence study is also led focusing on time step and mesh size sensitivities.

1. Introduction

Ingot casting is a manufacturing process widely used in several industries including automotive, naval, aeronautic, petroleum or nuclear industry. The ingot results from a phase change where a liquid phase, generally an alloy, becomes solid. This phase change generally takes place over a range of temperatures, thus solid and liquid can coexist in equilibrium at various temperatures. The co-existence of both phases is macroscopically regarded as a mushy zone, where the solid phase forms a porous matrix through which the residual liquid flows. In the mushy zone, the liquid phase is enriched compared to the solid due to higher solubility in the liquid phase. This solute redistribution leads to segregations (chemical heterogeneities) at micro-scale, and may lead to macrosegregation, i.e. segregation at larger scale, when coupled with transport phenomena such as thermosolutal convection and solidification shrinkage, see [1] for some details. These macrosegregations are particularly unwanted by ingot producers as they may weaken the mechanical properties of the final piece. Thus, lots of efforts are led to finely understand their origin and control their expansion, with the help of both experimental and modelling studies.

IOP Conf. Series: Materials Science and Engineering **861** (2020) 012033 doi:10.1088/1757-899X/861/1/012033

The accurate modelling of macrosegregation phenomena is highly complex as it implies multi-scale and multi-physics problem. Several models are available in the literature, from purely thermal simulations to multi-phase models describing both liquid and solid phase dynamics, see [1] for a review. For instance, recent models include the prediction and description of the solidification structures and their link with macrosegregation, see [2] for instance. A classical in-between approach consists in assuming a motionless solid phase with a mushy zone governed by the Darcy's law. The thermosolutal convection is accounted for through the Boussinesq approximation and a strongly coupled systems of equations is obtained, including momentum, mass and energy conservation equations for the liquid-solid mixture.

In the following, a classical solidification model originally introduced by Benon & Incropera [3] is implemented in the open-source software *Code_Saturne* [4, 5] developed by EDF R&D. This software is used for industrial applications and research activities in several fields related to energy production and safety (thermal-hydraulics simulations of incidental situations), process engineering (plasma, electric arcs), gas and coal combustion, turbomachinery, aeraulics (ventilation, pollutant dispersion). It solves the Navier-Stokes or Reynolds averaged Navier-Stokes equations for 2D, 2D axisymmetric and 3D flows in steady or unsteady, laminar or turbulent, dilatable, incompressible, isothermal cases. The solver is based on a co-located finite volume approach and is highly parallelized. The model and the associated numerical scheme are described first in this paper. Then, the simulated segregation predictions are validated against both academic test case and industrial ingot configurations. A convergence study is also led focusing on time step and mesh size sensitivities.

2. Solidification model

The solidification model presented in this section is a mixture liquid-solid model based on the work of Bennon & Incropera [3]. It consists in averaging the local conservation equations of each phase on a Representative Elementary Volume (REV), typically the size of several dendrites, and summing the contributions of both phases to get the mixture dynamics. The volume fraction g_k of each phase in the REV is introduced ($k = l$ stands for the liquid while $k = s$ stands for the solid) such that $g_l + g_s = 1$, and a mixture variable φ is defined as $\varphi = g_l\varphi_l + g_s\varphi_s$. In addition, the following assumptions are considered:

i. The temperature is in equilibrium between both phases, i.e. the mixture temperature T verifies $T = T_l = T_s$.

ii. The density is constant and equal in both phases except for the buoyancy force where the Boussinesq approximation is applied; the mixture density ρ verifies $\rho = \rho_l = \rho_s$ and the buoyancy force $\boldsymbol{F}_\mathbf{b}$ writes:

$$\boldsymbol{F}_\mathbf{b} = \rho_\mathrm{b}(T,C)\boldsymbol{g} = \rho\left(1 - \beta_T(T - T_\mathrm{ref}) - \beta_C(C - C_\mathrm{ref})\right)\boldsymbol{g}. \qquad (1)$$

iii. The solid phase is motionless, i.e. the mixture velocity $\boldsymbol{u}$ verifies $\boldsymbol{u} = g_l\boldsymbol{u}_l$.

iv. The mushy zone is modelled as a porous medium where a drag force $\boldsymbol{F}_u$ derived from the Darcy's law applies:

$$\mathbf{F}_\mathbf{u} = -\frac{\mu_l}{\mathrm{K}(g_l)}\mathbf{u}, \qquad \mathrm{K}(g_l) = \frac{\lambda_2^2}{180}\frac{g_l^3}{(1-g_l)^2}. \qquad (2)$$

The resulting conservation equations governing the mixture dynamics are listed below:

- Mass and momentum conservation equation

$$\begin{cases} \mathrm{div}(\rho\boldsymbol{u}) = 0, \\ \partial_t(\rho\boldsymbol{u}) + \mathrm{div}(\rho\boldsymbol{u}\otimes\boldsymbol{u}) = -\nabla P + \mathrm{div}(\mu_l\nabla\boldsymbol{u}) + \rho_\mathrm{b}(T,C)\boldsymbol{g} - \dfrac{\mu_l}{\mathrm{K}(g_l)}\boldsymbol{u}. \end{cases} \qquad (3)$$

- Energy conservation (temperature formulation)

$$c_p(\partial_t(\rho T) + \mathrm{div}(T\rho \boldsymbol{u})) = \mathrm{div}(\lambda \nabla T) - \partial_t(\rho g_l \mathrm{L}). \tag{4}$$

- Species conservation

$$\partial_t(\rho C) + \mathrm{div}(\rho C_l \boldsymbol{u}) = \mathrm{div}(\rho \mathrm{D} \nabla C). \tag{5}$$

In order to close the model, the liquidus slope is linearized as:

$$T = T_m + \mathrm{m}_l C_l, \tag{6}$$

and a micro-segregation model is specified. In the following, the lever-rule model is chosen (perfect diffusion in the solid phase), which leads to $C_l = C/(g_l + g_s \mathrm{k_p})$ and to the following solidification path:

$$g_l(T,C) = 1 - \frac{1}{1 - \mathrm{k_p}} \frac{T - T_l(C)}{T - T_m}. \tag{7}$$

This model offers at least an interesting description of columnar solidification with a limited number of physical parameters to tune, compared to advanced solidification models [2, 6]. It may also be extended to multicomponent alloys adding for each component, a species conservation equation, its buoyancy force contribution and a micro-segregation model. The solidification is modified subsequently integrating all components in the thermodynamic equilibrium.

3. Numerical scheme

The solidification model under consideration displays strong coupling through source terms and includes an incompressible Navier-Stokes system to solve. In practice, it is proposed to solve first the scalar equations for T and C, deduce g_l and C_l through closure laws, and finally use a classical SIMPLEC algorithm [7] to handle the velocity-pressure coupling. The scalar fields included in the buoyancy force and the drag force are thus up to date when solving the momentum conservation equation. Sub-iterations are performed to ensure the convergence of the velocity field (10 sub-iterations are chosen in practice), following a PISO-like approach [7].

Denoting n the current time iteration, k the current sub-iteration, and Δt the time step, the time discretization scheme is of order 1 and writes in sequence:

i. Solve the scalar equations for T and C.

$$\begin{cases} \rho \dfrac{C^{n+1,k} - C^n}{\Delta t} + \mathrm{div}\left(C_l^{n+1,k-1} \boldsymbol{Q}^{n+1,k-1}\right) = \mathrm{div}\left(\rho \mathrm{D} \nabla C^{n+1,k}\right), \\ c_p \rho \dfrac{T^{n+1,k} - T^n}{\Delta t} + c_p \mathrm{div}\left(T^{n+1,k} \boldsymbol{Q}^{n+1,k-1}\right) = \mathrm{div}\left(\lambda \nabla T^{n+1,k}\right) - s_T^{n+1,k}, \end{cases} \tag{8}$$

$$\text{where } s_T^{n+1,k} = \partial_T(\rho g_l \mathrm{L})^{n+1,k-1} \frac{T^{n+1,k} - T^n}{\Delta t} + \partial_C(\rho g_l \mathrm{L})^{n+1,k-1} \frac{C^{n+1,k} - C^n}{\Delta t}.$$

ii. Update g_l and C_l following the phase diagram: use equations (6) and (7).

iii. Velocity prediction : $\left(\boldsymbol{u}^n \rightarrow \tilde{\boldsymbol{u}}^k\right)$

$$\rho \frac{\tilde{\boldsymbol{u}}^k - \boldsymbol{u}^n}{\Delta t} + \mathrm{div}(\tilde{\boldsymbol{u}}^k \otimes \boldsymbol{Q}^{n+1,k-1}) = -\nabla P^{n+1,k-1} + \mathrm{div}\left(\mu_l \nabla \tilde{\boldsymbol{u}}^k\right) + \rho_b^{n+1,k-1} \boldsymbol{g} - \frac{\mu_l}{K^{n+1,k}} \tilde{\boldsymbol{u}}^k. \tag{9}$$

iv. Pressure correction: ($\tilde{\boldsymbol{u}}^k \rightarrow \boldsymbol{u}^{n+1,k}$)

$$\begin{cases} \rho \dfrac{\boldsymbol{u}^{n+1,k} - \tilde{\boldsymbol{u}}^k}{\Delta t} = -\nabla q^k + \delta\rho_b^{n+1,k}, \\ \operatorname{div}\left(\boldsymbol{Q}^{n+1,k}\right) = 0. \end{cases} \tag{10}$$

with:

$$\begin{cases} \delta\rho_b^{n+1,k} = \rho_b^{n+1,k} - \rho_b^{n+1,k-1}, \\ P^{n+1,k} = P^{n+1,k-1} + q^k. \end{cases} \tag{11}$$

The steps i-iv are repeated until, $||\boldsymbol{u}^{n+1,k} - \boldsymbol{u}^{n+1,k-1}||_{L^2} \leq \epsilon||\boldsymbol{u}^n||_{L^2}$, with $\epsilon = 10^{-5}$. Note that the buoyancy force is also included in the pressure-correction step, which ensure its discrete equilibrium with the pressure gradient. Also, mass fluxes towards solidified cells are not allowed. The spatial discretization is a first-order upwind scheme and note that the variables $\boldsymbol{u}$, P, T and C are computed at the centre of cells (co-located approach) while the mass fluxes are computed at the faces. As already stated in the introduction of this paper, the *Code_Saturne* open-source software allows to solve this model in a parallel environment with 2D or 3D meshes composed of hexahedron. Numerical results on both academic and industrial configurations are presented in the next section.

4. Simulation results

In this section, two configurations are simulated. The first one is an academic configuration, corresponding to a binary Pb-18%Sn alloy [8], while the second one is an industrial configuration, corresponding to a 3.3 t steel ingot [6].

4.1. Hebditch and Hunt benchmark

This configuration is a numerical benchmark proposed in [8], based on the experiment of Hebditch & Hunt. As presented in figure 1, it corresponds to a rectangular cavity of 0.1 m wide and 0.06 m high. The top and bottom walls are assumed to be adiabatic, while the two vertical walls are cooled (Fourier conditions). For the flow, a non-slip condition is imposed on all the boundaries of the domain. Due to the symmetry of geometry and boundary conditions, and assuming the symmetry of the solution, only half of the domain was considered. At the initial time, the Pb-18%Sn alloy is assumed at rest (no velocity) and its temperature equal to the liquidus temperature corresponding to its nominal composition. The thermophysical properties are identical to the ones used in the literature [8].

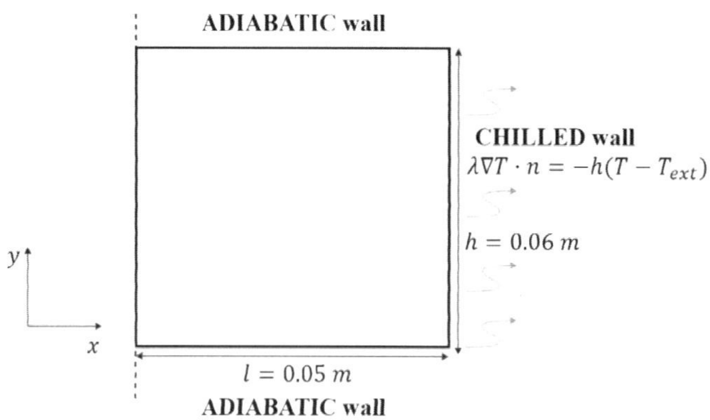

Figure 1. Geometry and boundary conditions for the Hebditch and Hunt benchmark [8].

The resulting segregation map of Sn is presented in figure 2, considering a time step of 10^{-3}s and decreasing the space step from dx = 1 mm (3000 cells) to dx = 0.25 mm (48000 cells). A comparison is

IOP Conf. Series: Materials Science and Engineering **861** (2020) 012033 doi:10.1088/1757-899X/861/1/012033

also proposed with the SOLID® software [2, 6, 9], which is based on a 2D finite volume solver and already validated for this configuration [8]. Results obtained by the 2 codes compare well on the finer mesh and the principal characteristics of macrosegregation can be identified: the last part to solidify, i.e. the upper left corner, is positively segregated, while the bottom right is negatively segregated. Segregated channels are also captured with better resolution when refining the mesh. As stated in [8], note that these channels are very sensitive to the mesh size and the numerical method such that no reference solution is currently available in the literature for this test case.

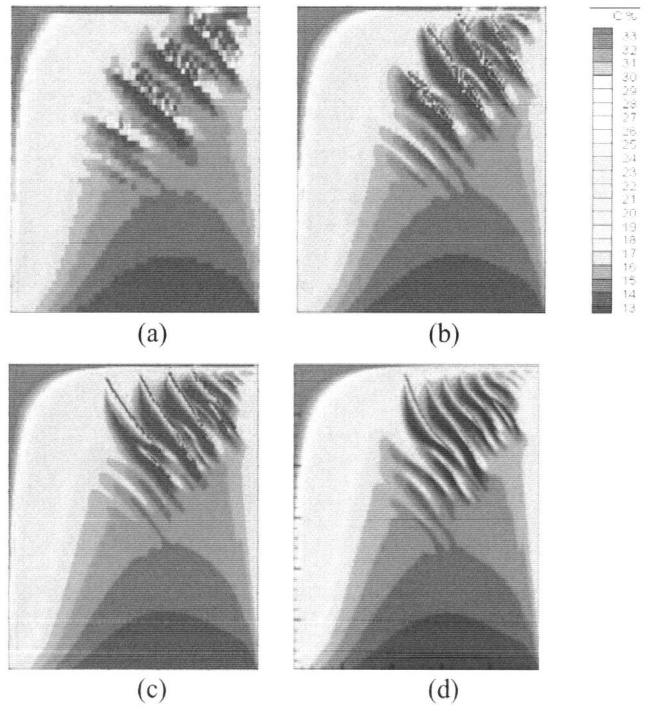

Figure 2. Sn segregation map at the end of solidification: mesh sensitivity with *Code_Saturne* and comparison with SOLID®. *Code_Saturne*: (a) dx = 1 mm, (b) dx = 0.5 mm, (c) dx = 0.25 mm, SOLID: (d) dx = 0.25 mm.

4.2. Aubert & Duval 3.3 t steel ingot

In this section, a 3.3 t steel ingot configuration described in figure 3 is simulated. The ingot is 2 m in height and 0.6 m in average diameter, and the steel is simplified as a binary Fe-C alloy with a nominal composition of C_0 = 0.36 wt%. Heat exchange are modelled within *Code_Saturne*, between the alloy, the mold and the exothermal powder. This configuration being widely studied in the literature for solidification solver validation, please report to [6] for its complete description. As presented on figure 3, the computational domain is 2D axisymmetric with hexahedron cells (7000 cells with dx = 1 cm). In the following, the numerical study focuses on time step sensitivity (figure 4) and mesh size sensitivity (figure 5) regarding segregation predictions.

Starting with the time step sensitivity analysis presented on figure 4, the expected segregation pattern is obtained, with a negative segregation cone at the bottom and a positively segregated hot top. Furthermore, some segregated channels are captured just below the hot top and could be referred to as A-segregation [1]. The solution differs notably from dt = 10^{-2} s to dt = 10^{-3} s while the smaller time step offers slight variations around the segregated channels. Thus, the intermediate time step dt = 10^{-2} s seems well adapted for this configuration, as for the previous benchmark.

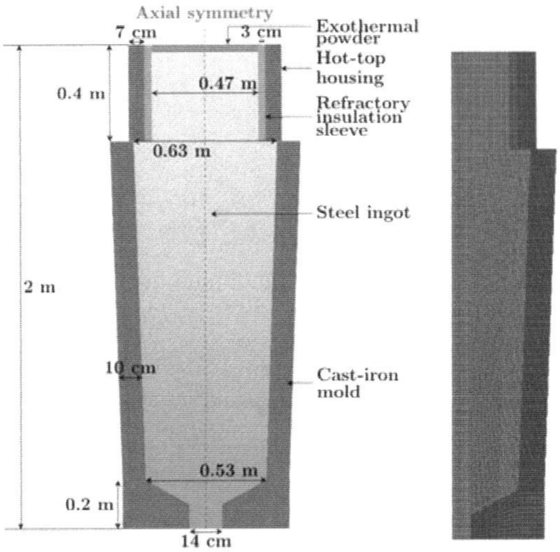

Figure 3. Geometry of the Aubert & Duval 3.3 t steel ingot [6] (left) and computational domain (right).

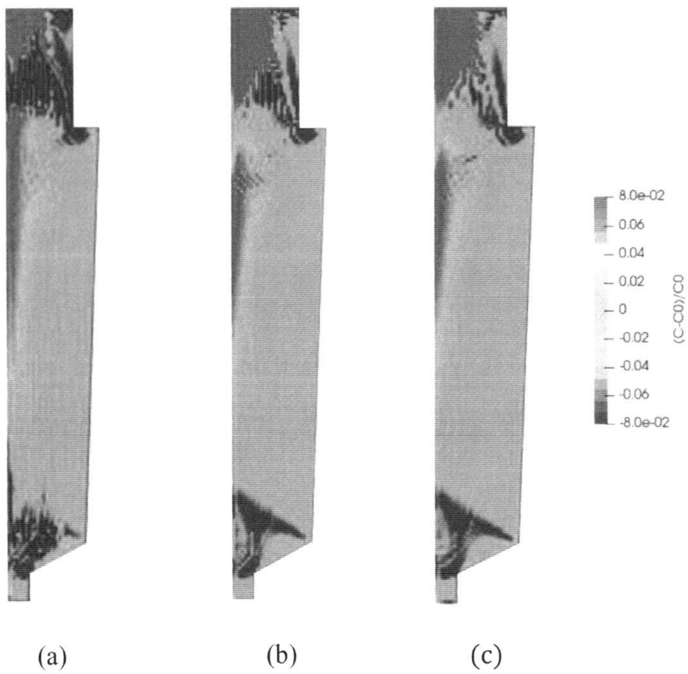

Figure 4. Carbon segregation map: time step sensitivity for a 3.3 t steel ingot (dx = 1 cm). (a) dt = 10^{-2} s, (b) dt = 10^{-3} s, (c) dt = 10^{-4} s.

Regarding the mesh sensitivity analysis on figure 5, ranging from 2000 cells (dx = 2 cm) to 30000 cells (dx = 0.5 cm), the main segregation patterns seem already well captured with the intermediate mesh composed by 7000 cells (dx = 1 cm). The finer mesh still yields different results when focusing on segregated channels. From these observations, it seems difficult to conclude about the physical and/or numerical origin of these channels; further numerical and experimental validation have to be conducted in that sense. At the end, the intermediate time step combined with the intermediate mesh seems to offer a reasonable compromise.

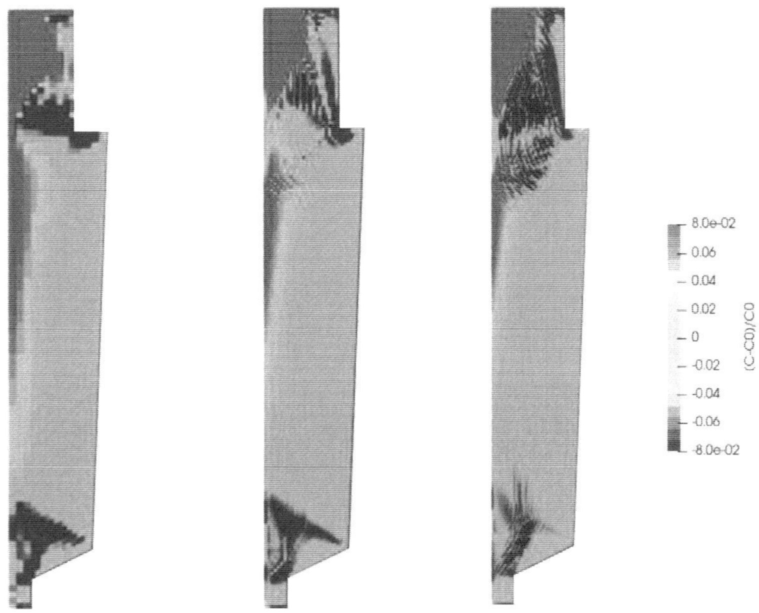

(a) (b) (c)

Figure 5. Carbon segregation map: mesh size sensitivity for a 3.3 t steel ingot (dt = 10^{-3} s). (a) dx = 2 cm, (b) dx = 1 cm, (c) dx = 0.5 cm.

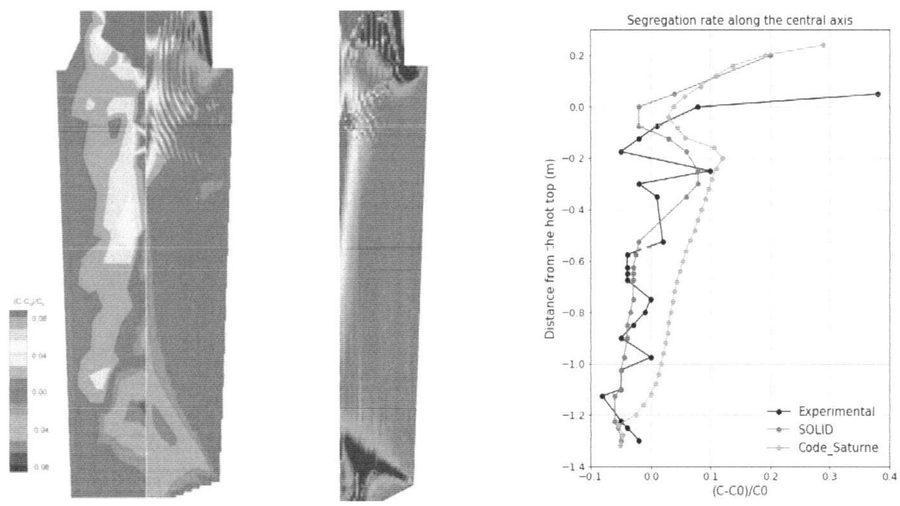

(a) (b) (c)

Figure 6. Comparison between experimental data, SOLID® and *Code_Saturne* results for a 3.3 t ingot. (a) Experimental (left) and SOLID results (right), (b) *Code_Saturne* results, (c) Axis segregation rate.

On figure 6, the segregation results are compared with the SOLID® and experimental results. Correct agreement is found between the 2 codes which are able to capture the main trends of the experimental data. However, note that the large positive segregation zone aside the central axis, which extends from the hot top to the negative segregation cone (in yellow on the experimental map), is not captured numerically. It might result from a non-modelled physical phenomena, such as the movement and sedimentation of equiaxed grains coming from nucleation or dendritic fragmentation, see [2] for instance.

5. Conclusion

This paper presents a solidification model with an associated numerical scheme implemented in the open-source software *Code_Saturne*. This model is a classical liquid-solid mixture model which accounts for thermosolutal convection and assumes a motionless solid phase. The numerical results are comparable with similar approaches available in the literature, offering mainly qualitative prediction of segregations. These results also highlight the difficulties to obtain converged results when focusing on segregated channels. Further development are conducted in that sense in *Code_Saturne* with the aim of improving the accuracy and the efficiency of numerical methods for solidification models.

Acknowledgements

This work is supported by EDF R&D and the authors would like to thank Marie-Ange Rasendra and Mathilde Cavelier for their work as intern.

Nomenclature

β_T	Thermal expansion coefficient (K^{-1})	k_p	Partition coefficient
β_C	Solutal expansion coefficient (K^{-1})	P	Pressure (Pa)
C	Bulk Concentration (wt. pct.)	Q	$\rho\mathbf{u}$ ($kg.m^{-2}.s^{-1}$)
C_k	Liquid or solid concentration	ρ	Density ($kg.m^{-3}$)
c_p	Specific heat per unit mass ($J.kg^{-1}.K^{-1}$)	ρ_b	Boussinesq density ($kg.m^{-3}$)
D	Diffusion coefficient ($m^2.s^{-1}$)	μ_l	Dynamic viscosity of the liquid (Pa.s)
Δt	Time step (s)	T	Temperature (K)
$\boldsymbol{g}$	Gravity acceleration ($m.s^{-2}$)	T_l	Liquidus temperature (K)
K	Permeability (m^2)	T_m	Melting temperature (K)
L	Latent heat ($J.kg^{-1}$)	$\boldsymbol{u}$	Velocity ($m.s^{-1}$)
λ	Thermal conductivity($W.m^{-1}.K^{-1}$)	g_k	Volume fraction of phase k
λ_2	Secondary Dendritic Arm Space (m)	l,s	Subscript referring to liquid and solid phase
m_l	Liquidus slope (K/wt.pct.)	ref	Subscript referring to reference value

References

[1] Pickering E J 2013 *ISIJ int.* **53** 935
[2] Gerin B, Combeau H, Založnik M, Poitrault I and Cherif M 2019 *Proc. IOP Conf. Ser. Mater. Sci. Eng.* **529** 012036
[3] Bennon W D and Incropera F P 1987 *Int. J. Heat Mass Trans.* **30** 2161
[4] https://www.code-saturne.org
[5] Archambeau F, Méchitoua N and Sakiz M 2004 *Int. J. Finite Vol.* **1** 1
[6] Combeau H, Založnik M, Hans S and Richy P E 2009 *Metall. mater. trans. B* **40** 289
[7] Versteeg H K and Malalasekera W 2007 *An introduction to computational fluid dynamics: the finite volume method* (Pearson education)
[8] Combeau H *et al.* 2012 *Proc. IOP Conf. Ser. Mater. Sci. Eng.* **33** 012086
[9] Ahmad N, Combeau H, Desbiolles J L, Jalanti T, Lesoult G, Rappaz J, Rappaz M and Stomp C *Metall. Mater. Trans. A* **29** 617

IOP Conf. Series: Materials Science and Engineering **861** (2020) 012034 doi:10.1088/1757-899X/861/1/012034

Multiscale solidification simulation of Sr-modified Al-Si-Mg alloy in die casting

B Zhou, G Laschet, J Eiken, H Behnken and M Apel

Access e.V., Intzestr. 5, 52072 Aachen, Germany

Email: b.zhou@access-technology.de

Abstract. Thermomechanical casting simulations incorporating solidification models are widely applied to improve the dimensional accuracy of casting components. The solidification path is commonly described based on Scheil-Gulliver assumptions which however fail to describe the effect of Strontium in Sr-modified Al-Si-Mg casting alloys. Strontium, even in small amounts, strongly affects the solidification morphology of the Al-Si eutectic together with its growth undercooling and hence the fraction solid–temperature relation during later stages of solidification. In order to address this problem, a dedicated micro-macro simulation approach is proposed here. The macroscopic thermomechanical casting simulation is linked with a spatially resolved multi-phase field simulation of the microstructure. Due to the fine fibrous morphology of the Si eutectic, a two-level homogenization scheme was applied to derive the effective mechanical properties of the Al alloy during its solidification. The proposed multiscale simulation and homogenization approach was applied to a permanent mould casting using an axisymmetric bowl cast from A356 for experimental validation.

1. Introduction

Al-Si based alloys are widely applied for light weight components due to their excellent castability, low weight, corrosion resistance and good strength. In commercial applications, additional chemical modifiers are added to the Al-Si alloys to modify the Al-Si eutectic in order to improve mechanical properties. Strontium (Sr) addition in range between 100ppm and 400ppm transforms the eutectic morphology from coarse plate-like structure to a fine coral-like fibrous networks [1, 2]. This finer and homogenous eutectic morphology strongly improves the ductility of the casting alloy and reduces the possibility of crack formation [3].

The mechanism of Sr-modification to the eutectic solidification has been well studied by several experimental works. Sr impacts both growth and nucleation of the eutectic Si phase. Lu and Hellawell [4] found that the absorbed Sr atoms at the interface hinder the growth of Si phase in twin-plane directions, but promote multidirectional twinning. Most recent studies rather focused on the impact of Sr on the nucleation of eutectic Si. In the commercial A356 alloy, phosphorus (P) normally exists as impurity. It reacts with Al in the liquid and produces a fine dispersion of aluminium phosphide (AlP) particles, which act as potent nuclei for eutectic Si. Dahle *et al.* [5] suggested that the potency of AlP particles to nucleate the Si phase is reduced by the competitive formation of Al_2Si_2Sr, which nucleates on AlP. In the numerical studies of Eiken *et al.* [6, 7], both aspects were considered and verified by 3D phase field (PF) simulations coupled with a thermodynamic database for the Al-Si-Sr-P system.

Macroscopic casting simulations incorporating solidification models are often applied to optimize process parameters and geometries of casting components. Commonly, Scheil-Gulliver approximations are employed in casting simulations to estimate the solidification path, i.e. the temperature dependent fraction of solid $f_s(T)$ and the latent heat release. However, the Scheil-Gulliver model fails to describe the effect of Sr on the nucleation and growth process of eutectic Si phase. This model leads to an inaccurate description of the solidification path, as it does not consider recalescence or growth undercooling of the eutectic.

In order to address this problem, a dedicated micro-macro coupled simulation approach is proposed here. In this approach, the thermomechanical finite element casting simulation is coupled with microstructure formation simulations based on a homoenthalphic method proposed by Böttger *et al.* [8]. The spatially resolved microstructure simulations were performed using a Calphad-based multicomponent multiphase-field (MMPF) model [9]. The dimensions of the representative volume element (RVE) has been selected based on the experimental observed average grain size. In the studied case, i.e. Sr-modified hypoeutectic Al-Si alloys, the interdendritic eutectic has a much finer microstructure, compared to primary fcc-Al dendrites. Therefore, the two-phase eutectic was not spatially resolved, but handled as an effective single phase in the MMPF simulation. The effect of Sr-modification was achieved by adjusting the nucleation model for eutectic Si phase, and by calibrating a growth parameter which characterizes the critical length scale of the eutectic microstructure. The MMPF simulation covers the physical complexity of the solidification process and provides an improved description of solidification path and latent heat release to the macroscopic casting simulation.

Moreover, to derive the effective mechanical properties of the mushy zone, a two-level homogenization scheme based on an asymptotic expansion of the involved periodic fields [10, 16] is adopted here. At the fine scale level, the fibrous eutectic morphology is homogenized based on a microstructure simulation on a smaller RVE that resolves the growth of the fine eutectic structure. Then, at the larger scale level, microstructures containing primary fcc-Al dendrite, effective eutectic and residual melt are homogenized with a mixed finite element discretization, which is suitable to model the melt incompressibility [10]. The proposed multiscale simulation approach was applied to a permanent mould casting using an axisymmetric bowl cast from A356 alloy for experimental validation.

2. Homoenthalpic approach for micro-macro coupling

In macroscopic casting simulations, the casting alloys are commonly treated as a liquid/solid coexisting system, which is characterized by the temperature dependent fraction solid $f_s(T)$. In practice, as it is more convenient to measure the specific heat at constant pressure than at constant volume, the state variable enthalpy is always used in the thermal analysis of solidification. Following Dantzig and Rappaz [11], the average form of the local energy balance in terms of specific enthalpy can be expressed as:

$$\frac{\partial\langle\rho h\rangle}{\partial t}+\nabla\cdot\langle\rho h\mathbf{v}\rangle=\dot{\mathrm{Q}}+\langle\rho r\rangle\ , \tag{1}$$

in which ρ and h denote the density and specific enthalpy, respectively. $\mathbf{v}$ represents the velocity vector, $\dot{\mathrm{Q}}$ is the heat flux, and $\langle\rho r\rangle$ is a heat source term. Here, $\langle\bullet\rangle$ represents the volume average of a quantity in the two-phase system. By introducing the specific heat capacity c_p and the latent heat release L_f during the solidification, the average enthalpy of the mixture can be expressed as

$$\langle\rho h\rangle=\int_{T_0}^{T}\langle\rho\, c_\mathrm{p}(\tilde{T})\rangle\ \mathrm{d}\tilde{T}+\langle\rho\rangle\ L_\mathrm{f}\left[1\text{-}f_\mathrm{s}(\mathrm{T})\right] \tag{2}$$

with its time derivative calculated as

$$\frac{\partial\langle\rho\, h\rangle}{\partial t}=\langle\rho\, c_\mathrm{p}\rangle\frac{\partial T}{\partial t}-\langle\rho\rangle L_\mathrm{f}\frac{\partial f_\mathrm{s}}{\partial T}\frac{\partial T}{\partial t}. \tag{3}$$

In absence of external heat sources, i.e. $\langle\rho r\rangle$=0, substituting equation (3) into equation (1) leads to

$$\left(\langle\rho\, c_\mathrm{p}\rangle+\langle\rho\rangle L_\mathrm{f}\frac{\partial f_\mathrm{s}}{\partial T}\right)\frac{\partial T}{\partial t}+\nabla\cdot\langle\rho\, h\,\mathbf{v}\rangle=\dot{\mathrm{Q}}\ . \tag{4}$$

In this coupling approach, the heat capacity, the latent heat release and the fraction solid are obtained by performing the Calphad-based MMPF microstructure simulation. Outcomes of the macroscopic process simulation, e.g. the local heat extraction rate $\dot{Q}$ and local temperature gradient, are applied as boundary conditions in microstructure simulations. In the MMPF simulation, the volume averaged enthalpy change in the current time step is calculated as

$$\mathrm{d}H^{t_n} = \frac{1}{V_c}\int_{V_c}\sum_{\alpha}\mathrm{d}\left(\phi_{\alpha}^{t_n}\, H_{\alpha}(T^{t_n})\right)\mathrm{d}v \;, \tag{5}$$

where ϕ_α and H_α are the phase fraction and the molar enthalpy of phase α, respectively. The increment of volume averaged latent heat release $\mathrm{d}\bar{L}_f^{t_n}$ is then calculated based on the average heat capacity from the previous time step, i.e.,

$$\mathrm{d}\bar{L}_f^{t_n} = \mathrm{d}\bar{H}^{t_n} - \bar{C}_p^{t_{n-1}}\,\mathrm{d}T^{t_{n-1}} \quad \text{with} \quad \bar{C}_p^{t_{n-1}} = \frac{1}{V_c}\int_{V_c}\sum_{\alpha}\phi_{\alpha}^{t_{n-1}}\, C_{p,\alpha}\mathrm{d}v \,. \tag{6}$$

The volume and phase-specific enthalpy H_α and heat capacity $C_{p,\alpha}$ are read from the thermodynamic database of ThermoCalc®. The updated fraction solid, latent heat release and heat capacity in dependency of temperature can then be applied for improving the macroscopic process simulation.

3. Macroscopic casting simulation model

The proposed approach was applied to simulate the solidification process of a die casting component. The macroscopic casting simulation was performed by using the finite element (FE) software Abaqus. The model was built according to the experimental set-up presented in [12]. As illustrated in figure 1(a), the model mainly consists of an inner steel core, a steel mould, a sand core on the top and the axisymmetric bowl shaped casting component. The bowl component has an outer diameter of 155 mm and a wall thickness that gradually varies from 15 mm up to 30 mm. The insulation on bottom and the sand core on top of the bowl assured a radial flow of the heat flux. The steel mould was built up with three identical 120° segments, which include cooling channels and thermo-elements. For this reason, one third of the entire set-up was modelled for FE simulations. According to the casting experiment, the melt, i.e. the bowl region had an initial temperature of 720 °C, while the mould temperature was set to 30 °C.

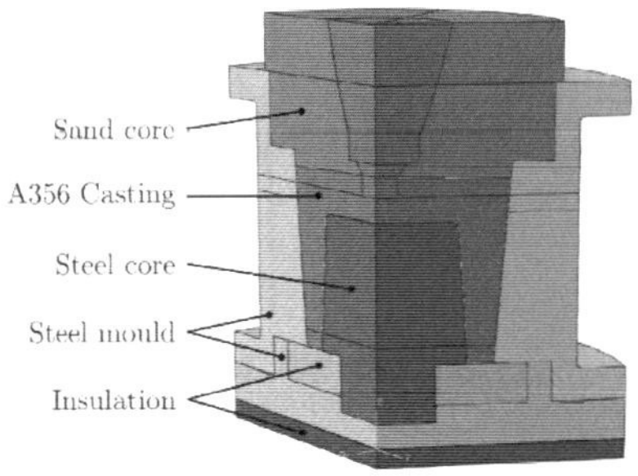

(a)

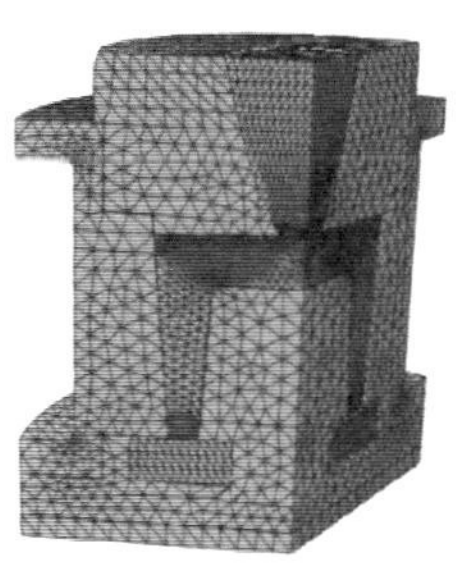

(b)

Figure 1. (a) Construction of the experimental set-up; (b) Finite element mesh with temperature-displacement coupled 4-node linear tetrahedral elements.

The model is meshed with temperature-displacement coupled 4-node linear tetrahedral elements, see figure 1(b). As in casting simulations the thermal and mechanical solutions strongly interact, a fully coupled thermal-stress analysis is performed. The A356 component was modelled as thermo-elastoplastic material, while the mould and core parts are thermo-elastic materials. In order to model the phase transformation, a user specific subroutine was written to specify the internal heat generation due to the latent heat release along the solidification path according to equation (4). As the bowl component undergoes a volume shrinkage during the solidification process, air gap could be formed between the casting component and mould. For further precision of the simulation, the contact condition between the bowl component and its surrounding mould was distinguished between fluid-solid and solid-solid surface contact. By modelling the solid-solid surface contact condition, the impact of the formed air gap on the heat transfer coefficient (HTC) was taken into account. In this case, the gap conduction model is written as a function of the local gap width and the local contact pressure [13].

In experiments, a commercial A356 with Sr-modification and grain refinement (TiB5) was applied for the casting. The grain refinement ensures equiaxial growth of fcc-Al dendrites over the entire casting component and a relatively homogenous grain size distribution. This behaviour allows the application of a uniform fraction solid $f_s(T)$ and enthalpy-temperature relations over the whole component in the casting simulation.

4. Microstructure simulation model

The phase formation sequence of unmodified A356 alloys during the solidification was firstly calculated via the thermodynamic computation based on Scheil-Gulliver model using the software ThermoCalc®. Here, the alloy system is defined by 92.7 wt% Al, 7 wt% Si and 0.3 wt% Mg. Further chemical elements normally contained in commercial A356 alloys, as Mn and Fe, are not considered. As shown in figure 2, the equilibrium eutectic temperature in the unmodified alloy under Scheil condition is 575.3 °C.

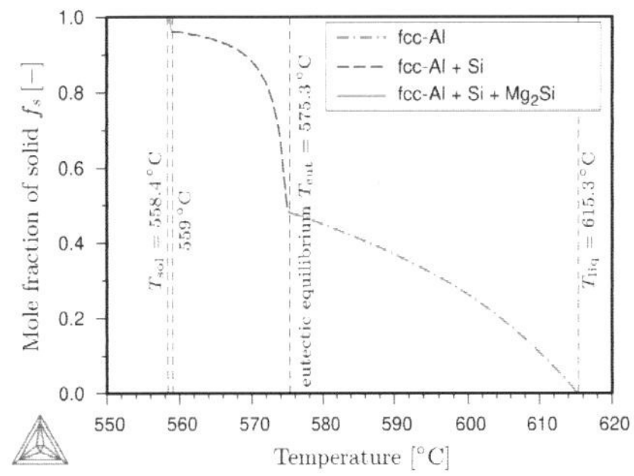

Figure 2. Scheil calculation for unmodified A356 alloy using ThermoCalc with database TCAL5.

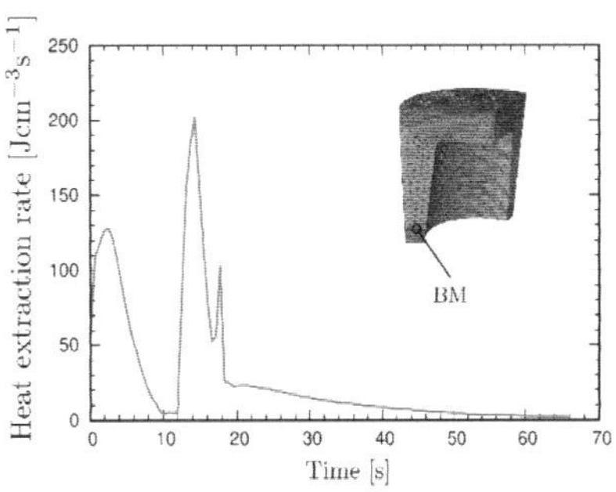

Figure 3. Heat extraction rate read from the casting simulation result on the bottom middle (BM) point of the bowl

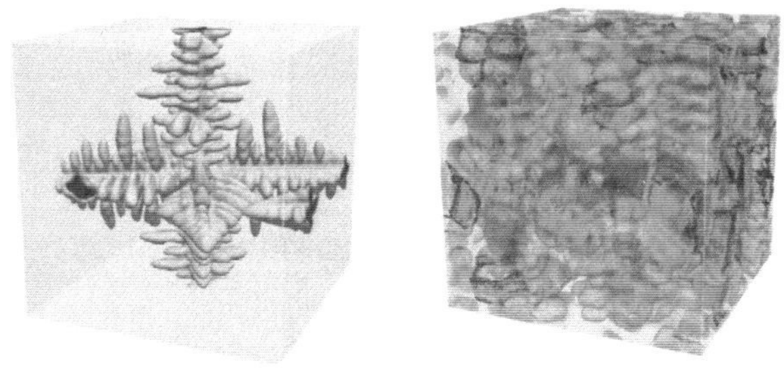

(a) (b)

Figure 4. Results of 3D microstructure simulation: (a) one single primary fcc-Al dendrite at t=0.8 s and T=599.4 °C; (b) primary fcc-Al and interdendritic eutectic (orange coloured) at t=10 s and T= 571.7 °C.

The microstructure simulations were performed using the software MICRESS®, which is based on the MMPF model specified in [9]. The Calphad-database are coupled via the TQ-interface. According to experimental observations, the simulation was performed on a 3D-domain with a size of 320 μm^3 and a grid resolution of 2 μm. The simulation started from 100% liquid at 615 °C. As boundary condition, heat was extracted from the RVE according to the casting simulation results (see figure 3). The eutectic was treated as an effective single phase. Iterative micro-macro simulations were performed to obtain consistent enthalpy-temperature relation. As no additional temperature gradient exists, an equiaxed single fcc-Al dendrite growths at the beginning of the solidification, see figure 4(a). At higher solid fractions, interdendritic eutectic was formed between primary fcc-Al dendrites and residual melt, leading thus to a rather complex microstructure, see figure 4(b).

4.1. The impact of Sr on the eutectic solidification

The impact of Sr on the eutectic solidification was considered by applying appropriate nucleation and growth parameters for the effective eutectic phase. Specifically, the effect of Sr on the growth was simulated by calibrating a parameter which characterize the critical length scale of the eutectic morphology, whereas its effect on the nucleation of eutectic Si was realized by adjusting the number of initial nuclei for the eutectic phase.

(a) (b) (c)

Figure 5. Distribution of eutectic Si nucleation sites (red coloured) at the beginning of eutectic solidification for different simulation scenarios (a) unmodified A356 alloy; (b) A356 with lower Sr content; (c) A356 with higher Sr content.

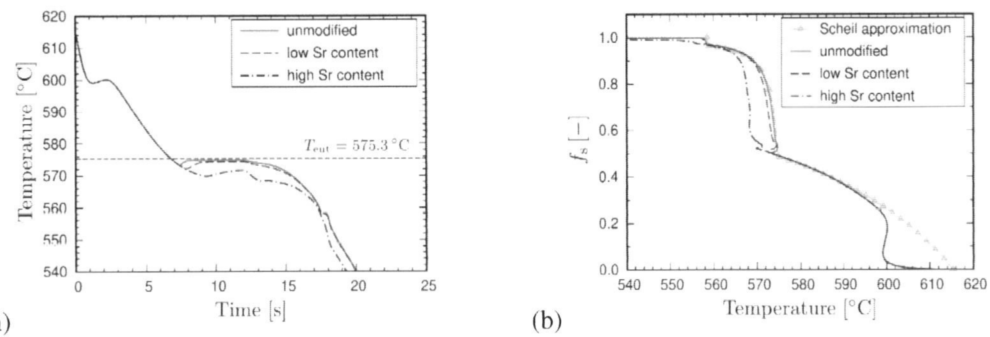

(a) (b)

Figure 6. Temperature evolutions (a) and fraction solid curve (b) of three different scenarios.

As shown in figure 5, three different scenarios have been simulated in 2D and compared. In scenario (a), i.e. unmodified A356 alloy, more potent nuclei for eutectic Si were distributed in the interdendritic region and lead to the growth of eutectic grains. In scenarios (b) and (c), modified A356 with different Sr contents were simulated. The higher the Sr content, the fewer Si nuclei appear in the simulation region. The effect of Sr-modification was fully embodied by the reduced eutectic growth temperature. As illustrated in figure 6(a), in the unmodified case the eutectic growth temperature is very close to the equilibrium temperature obtained by Scheil approximation, while by high Sr content it is nearly 5 °C lower. Additionally, recalescence has been observed in the primary solidification (fcc-Al). Consequently, the fraction solid curve differs to the Scheil approximation, see figure 6(b). The fraction solid curve obtained from the MMPF simulation was then transferred to the macroscopic casting simulation in order to take into account the effect of Sr-modification.

5. Coupled multiscale simulation results

The coupled multiscale simulation approach was performed in two iterative loops. It began with a FE casting simulation with the classical Scheil approximation. The obtained local heat extraction was then applied as boundary condition for the temperature calculation in the MMPF simulation based on the energy balance. In the second FE simulation, the new fraction solid curve and latent heat release obtained from the microstructure simulation was applied.

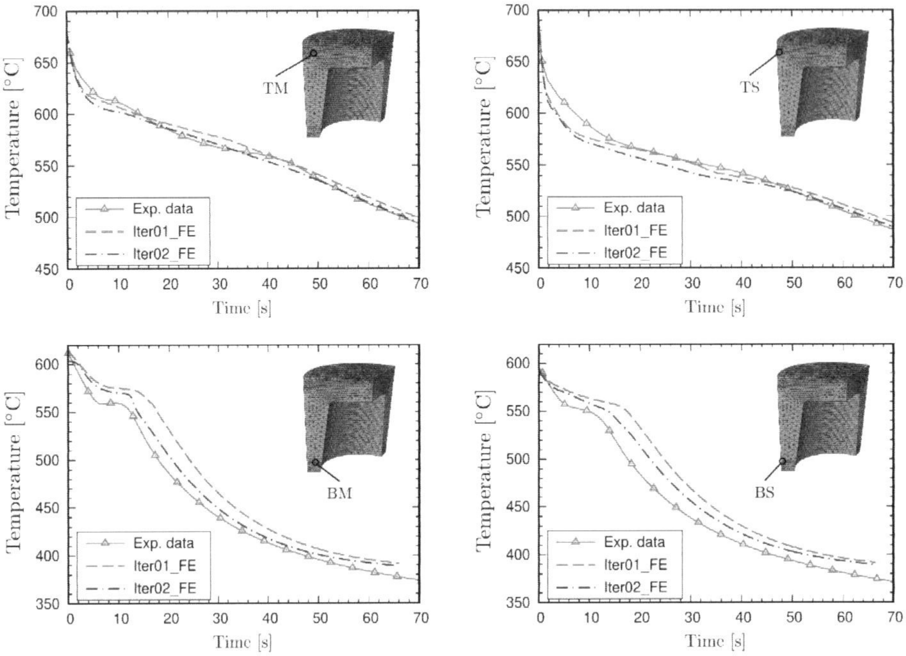

Figure 7. Temperature evolutions obtained by FE casting simulations on four experimentally observed locations, which are labelled as TM (top middle), TS (top surface), BM (bottom middle) and BS (bottom surface).

For validation, the simulated temperature evolutions in four different bowl locations have been compared with the experimental measurements in figure 7. It can be seen that, the FE simulation with the Scheil approximation (Iter01_FE) supplied good temperature predictions in locations TM and TS. However, on the bottom of the bowl (locations BM and BS) where the cooling rate is higher, the predicted temperature was significantly higher than in the experiment, especially at the later stage of solidification. By applying the fraction solid-temperature relationship and the latent heat release from the MMPF simulation in the second FE simulation (Iter02_FE), a better agreement between predicted and experimental temperatures was achieved. For further improvements of the FE simulation results, 3D MMPF simulations have to be performed for more accurate description of recalescence and the impact of Sr on the eutectic growth temperature.

6. Two-level homogenization of 3D A356 microstructures

As the experimental characterization of the mechanical behaviour of the mushy zone during Al alloy solidification is delicate, the locally performed 3D MMPF simulations were used to derive accurate effective elastic properties of the mush by applying a two-level homogenization scheme.

6.1. Homogenization of the coral-like structure of Si eutectic

As outlined in detail in [6, 7], the effect of Sr on the twinning of Si in eutectic was modelled on the finer scale implicitly by adopting an effective anisotropic surface mobility. From the converged 3D large scale simulation of the solidification of Sr modified A356, the temperature and concentration fields were locally extracted in the melt near primary dendrite arms. These local fields are applied as initial conditions on a fine discretized, small RVE at the lower scale. A single nucleation seed of Si was adopted in the 3D MMPF simulation of the eutectic formation. From this nucleus a complex, coral-like fibrous Si eutectic morphology grows around the primary fcc Al-dendrite during further solidification.

From this microstructure, four pure eutectic RVE's have been extracted (see figure 8, where two extracted RVE's are shown). Their Si content varies from 10.61% up to 15.35%, leading to a mean Si content of 12.90%.

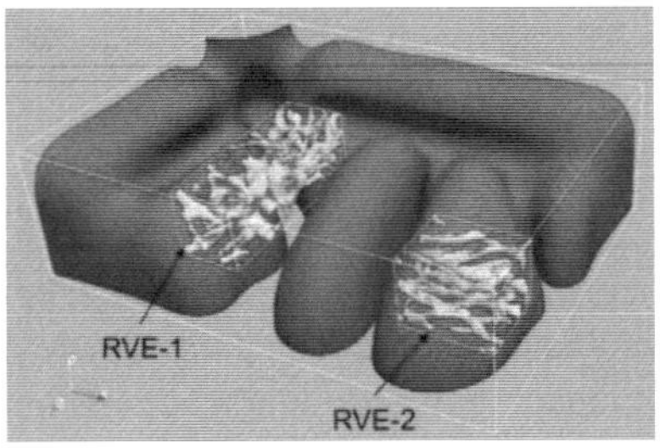

Figure 8. Extraction of RVE-1 and RVE-2 of pure eutectic near a fcc-Al dendrite.

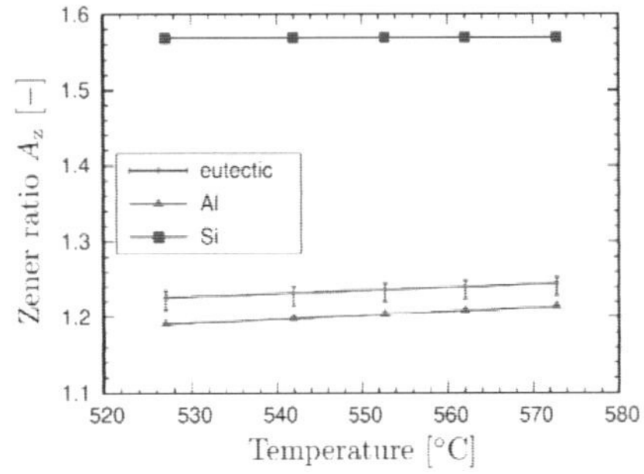

Figure 9. Temperature variation of the Zener ratio A_z and its scatter for effective eutectic, pure single crystal Al and Si.

The temperature dependent elastic properties of the single crystals Al and Si were taken from [14] and [15], respectively. The homogenization of the four RVE's was performed not only to determine the mean elastic constants of the eutectic but also to quantify the influence of the scatter of the Si content on their values. *The effective eutectic is cubic* and presents a noticeable effective Zener anisotropy coefficient A_Z, which increases slightly with the temperature, as shown in figure 9. Note that the Si content scatter affects mainly the effective shear modulus C_{44}; whereas the effective elastic constant C_{11} is nearly unaffected.

6.2. Homogenization of equiaxed dendritic 3D microstructures

At selected time steps of the 3D MMPF simulation of local A356 solidification, the actual microstructure was transferred to the homogenization tool HOMAT to calculate the effective mechanical properties. The implemented homogenization scheme is based on the asymptotic expansion method outlined in [16]. Effective thermal conductivities of the mushy zone have already been presented in [18].

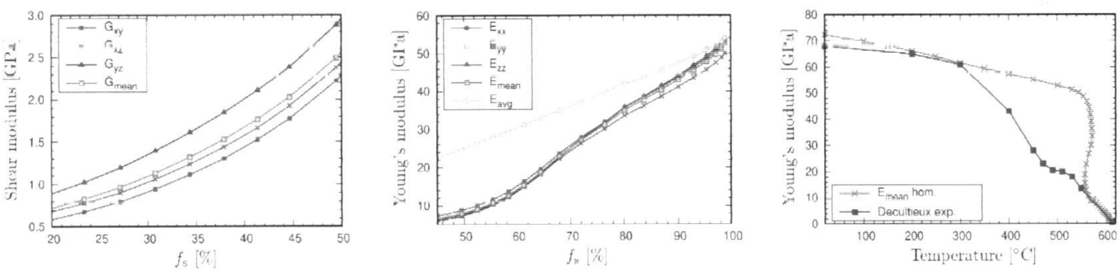

(a) (b) (c)

Figure 10. Variation of the effective shear modules (a) Young modules (b) with the solid fraction. Variation of the homogenized and experimental Young modules with the temperature (c).

The thermoelastic homogenization of the different microstructures provided amongst others the results outlined in figure 10. The variation of the effective shear modules with f_s for the primary solidified Al dendrites shows a pronounced orthotropy: $G_{yz}>G_{xz}>G_{xy}$. The in-plane shear modulus G_{xy} presents the lowest stiffness. This homogenization is performed for f_s above shear coherency ($f_s \approx 0.22$) [17], at which the much develops first shear resistant due to dendrites entanglement. Figure 10(b) displays the variation of the directional Young modules, their mean value and the volume averaged Young modulus with f_s above the estimated traction coherency $f_{s,tr} \approx 0.48$. Obviously, Si induces a stronger increase of the effective Young modules during the eutectic formation than during the pure Al dendrite formation. Due to the calculation of correction terms, the 1st order homogenization scheme [16] improves the volume averaging prediction mainly at the end of the primary solidification, where the solid-liquid interfaces are at the largest. Eventually, figure 10(c) shows the temperature evolutions of the mean effective Young's modulus and of an experimentally measured Young's modulus from uniaxial tensile tests [19] over the complete cooling path. Due to recalescence and undercooling effects, the $E(T)$ curve is not unique in the mushy region. Therefore, the more accurate $E(f_s)$ has to be derived in the mushy region. A large discrepancy between experimental and predicted Young's modulus curves is observed at higher temperatures ($T > 300°C$). It can be explained by the fact that in the experiment the viscous component, present always at higher temperatures, cannot be separated from the pure elastic one, as achieved in the homogenization, where only pure elastic Young's moduli were derived.

7. Conclusions

The present work confirmed that thermomechanical casting simulations based on fraction solid curves derived under Scheil-Gulliver assumptions are not able to provide a predictive simulation result in the case of Sr-modified Al-Si alloys. In the proposed multiscale simulation approach, the multicomponent multi-phase field based microstructure simulation effectively depict the effect of Sr-modification on the nucleation and growth of eutectic Si phase. The more realistic thermodynamic description of the solidification process provided by the MMPF simulations significantly improved the temperature prediction, especially in regions with high cooling rate. Effective mechanical material properties were derived based on a two-level homogenization approach. The results show that the complex fibrous morphology of the multiphase eutectic material exhibit an effective cubic behaviour. Moreover, it was found that the effective mechanical mushy zone properties have to be expressed as a function of fraction solid rather than as a function of temperature, as the temperature dependence is non-unique. An interesting task will be to study the impact of the homogenized Young's moduli on the local stress-strain state in future thermomechanical casting simulations. Virtual tests will be performed on the predicted microstructures in order to derive the effective viscoplastic flow behaviour of the A356 mush and to identify unknown parameters of the thermo-elasto-plastic constitutive laws, developed in ref. [17] for the coherent and non-coherent mushy zones.

Acknowledgments

The authors would like to acknowledge support from the German Research Foundation (DFG) in the framework of the Collaborative Research Centre SFB1120-236616214 "Precision Melt Engineering".

References

[1] Sigworth G K 2008 *Int. J. Met.* **2** 41
[2] Timpel M *et al.* 2012 *Acta Mater.* **60** 3920–28
[3] Lasko G, Apel M, Carré A, Weber U and Schmauder S 2012 *Adv. Eng. Mater.* **14** 236–247
[4] Lu S Z and Hellawell A 1987 *Metall. Trans. A* **18** 1721–33
[5] Dahle A K, Nogita K, McDonald S D, Dinnis C and Lu L 2005 *Mater. Sci. Eng. A* **413** 243–8
[6] Eiken J, Apel M, Liang S M and Schmid-Fetzer R 2015 *Acta Mater.* **98** 152–163
[7] Eiken J and Apel M 2015 *IOP Conf. Ser. Mater. Sci. Eng.* **84** 012084
[8] Böttger B, Eiken J and Apel M 2009 *J. Comput. Phys.* **228** 6784–95
[9] Eiken J, Böttger B and Steinbach I 2006 *Phys. Rev. E-Stat. Nonlinear, Soft Matter Phys.* **73** 1–9
[10] Laschet G, Nokhostin H, Koch C, Meunier M and Hopmann C 2020 *Mech. of Mat.* **140** 103225
[11] Dantzig J A and Rappaz M 2009 *Solidification* (EPFL Press)
[12] Wolff N, Pustal B, Vossel T, Laschet G and Bührig-Polaczek A 2017 *Materwiss. Werksttech.* **48** 1235–40
[13] Laschet G, Jakumeit J and Benke S 2004 *Zeitschrift für Met.* **95** 1087–96
[14] Sutton P M 1953 *Phys. Rev.* **91** 816–21
[15] Masolin A, Bouchard P O, Martini R and Bernacki M 2013 *J. Mater. Sci.* **48** 979–988
[16] Laschet G and Apel M 2010 *Steel Res. Int.* **81** 637–643
[17] Laschet G and Behnken H 2019 *IOP Conf. Ser. Mater. Sci. Engng.* **529** 012083.
[18] Laschet G, Vossel T, Wolff N, Apel M and Bührig-Polaczek A 2017 *SP17 Proc. 6th Decenn. Int. Conf. Solidif. Proceeding (Old Windsor)* ed Fan Z (Middlesex, UK: BCAST) pp 576–80
[19] Decultieux F 1996 *PhD thesis* (CEMEF, Paris Tech.)

IOP Conf. Series: Materials Science and Engineering **861** (2020) 012035 doi:10.1088/1757-899X/861/1/012035

Multi-scale simulation of hybrid light metal structures produced by high pressure die casting

J Jakumeit[1], H Behnken[1], R Laqua[1], J Eiken[1] and J Brachmann[2]

[1] Access e.V., Intzestr. 5, 52072 Aachen, Germany
[2] RWTH Aachen University – Foundry Institute, Intzestr. 5, 52072 Aachen, Germany

E-mail: j.jakumeit@access-technology.de

Abstract. In modern automotive structural components, metal sheets are combined with casting parts to enable a light and mechanical stable structure. In this study, the bonding between steel metal sheets and aluminium casting parts are performed as part of the high pressure die casting (HPDC) process. In this way additional bonding steps like welding can be omitted, which saves time and energy. The combination of sheet metal and cast is achieved by a structured surface (ribs) of the sheet, which is filled in the casting process with the liquid aluminium alloy and leads to a mechanical connection after solidification. Coupled flow and solidification simulation is used to analyse the hybrid casting process with a focus on the rapid solidification inside the ribs. A high-resolution (mesoscale) simulation of the interface between metal sheet and aluminium alloy during filling and solidification can capture the details of melt flow and rapid solidification. By coupling the mesoscale flow and solidification simulations to multiphase field simulations on the microscale, the resulting microstructure can be calculated. The solidification conditions inside the ribs as result of the mesoscale simulation can be validated by comparing the calculated microstructure with experimental results found in real castings, performed on a cold-chamber HPDC machine.

1. Introduction

High pressure die casting (HPDC) is an important manufacturing process for large, thin-walled structural parts, especially in the automotive industry. HPDC allows the fabrication of complex large aluminium or magnesium parts, which can be more efficient than combining several forged metal sheets by welding. An interesting new application is the combination of aluminium casting and metal sheets in one hybrid casting process step. In this way additional bonding steps like welding can be omitted, which saves time and energy. The combination of sheet metal and cast is achieved by a structured surface (ribs) of the sheet, which is filled during the casting process with the liquid aluminium alloy and leads to a mechanical connection after solidification. The mechanical connection is stronger than glued connections but weaker then welded ones. Important for the connection strength is the complete filling of the structural surface of the metal sheet.

Simulation of the HPDC filling process is challenging due to the very fast and chaotic filling of the cavity with a significant amount of air entrainment. Most commercial simulation programs use a one-phase approach to describe casting processes [1-4] with some special models to approximate the effect of remaining air. In this work a three-phase simulation of casting processes, including the air, melt and solid phases is used. Both, melt and air are considered as compressible fluids separated by a volume-of-fluid approach (VoF), including special treatment to keep the interface sharp. Reduced melt fluidity during the solidification process is handled by a porous media approach to describe the flow through the dendrite network. At a critical solid fraction value, the melt is stopped completely. This approach is implemented in the commercial casting simulation package STAR-CCM+ [5]. Correct treatment of the air and fast solidification which may result in misruns are seen to be important for the investigation of hybrid casting, since air entrainment and melt solidification before complete filling of the structured metal sheet surface can weaken the mechanical bond between metal sheet and casting.

Details of the simulation approach are given in the following section, followed by a description of the multiscale simulation approach. Section 4 investigates the filling of the ribs using different process configurations. A multiphase field simulation of the resulting microstructure is described in section 5, which allows to validate the solidification conditions given by the mesoscale simulation. All simulation results are summarized in section 6.

2. Theory

Simulation of the entire casting process is based on a finite-volume method using control volumes (CVs) of arbitrary polyhedral shape. The transport equations for mass, momentum, energy, and phase in integral form are applied to each CV, whereby the surface and volume integrals are approximated using the midpoint rule. Linear equation systems for each variable are solved using algebraic multi-grid iterative solvers. Mass conservation, pressure and velocity conservation are coupled via the SIMPLE algorithm. The transient term is discretized based on an implicit Euler-segregated concept. Details of discretization and the solution method are available in [6-8]. The conservation equation for total energy is solved for the solution variable temperature. For solidification modelling, the volume fraction of solidified liquid is determined using a tabulated fraction solid vs. temperature curve ($fs(T)$). Latent heat L is released in proportion to the change in fraction solid ($L*d(fs(T))/dT$). Details of the methodology are presented in [8].

For correct calculation of filling including air entrainment, the method used to track the motion of the free surface must provide a sharp interface. The volume of fluid (VoF) approach is used to tackle the problem: the entire fluid domain is considered to be filled by a fluid, the properties of which vary according to the distribution of volume fractions of melt C_m, solid C_s and gas C_g ($C_m + C_s + C_g = 1$). The transport of melt, solid and gas is computed by solving transport equations for their volume fractions with a source term for the phase change from melt to solid. To achieve sharpness of the interface, a high-resolution interface capturing scheme (HRIC) [8, 9] is used, which typically resolves the interface within one cell.

The normal force due to surface tension is treated using the continuum surface force (CSF) model proposed by [10], which defines a volumetric source in the momentum equation. Flow resistance in the mushy zone is calculated assuming that the mushy zone acts like a porous medium. The fluid velocity resistance in the mushy zone can be approximated as a pressure drop [11]. Permeability can be deduced from the Kozeny-Carman equation [12] as $K = ((1-fs)^3\lambda_2^2)/(180fs^2)$, where λ_2 is the secondary dendrite-arm spacing (SDAS).

The viscous porous resistance depends on the velocity of the melt and tends to zero as the velocity approaches zero. Hence, slight creeping of the melt is not hindered, especially if the melt is being pushed by a high force, as in the case of HPDC. To bring the melt to a complete stop, a flow-stop functionality was implemented. If the fraction solid is above a threshold value, the momentum equation is no longer solved for that particular cell. Flow stopped cells are of zero velocity (relative to the cell centre velocity), and fluxes through all adjacent faces are also zero. Pressure remains constant at the value obtained when the cell has been flagged to be stopped. Since the density of the melt depends on temperature, a mass compensation is needed to conserve the total melt mass. The flow-stop-mass-compensation functionality applies appropriate mass and energy sources to compensate for the unphysical mass change in stopped cells and maintains the total mass of each phase.

3. Multi scale simulation of hybrid HPDC processes

The multiscale simulation approach was applied to a hybrid test geometry casted at the foundry institute of the RWTH-Aachen. Figure 1 shows part with gating system, casting and metal sheet. Al-5wt%Mg-1.8wt%Si was filling into the cavity with a casting temperature of 680 °C, a maximal piston speed 1 m/s, die preheating temperature of 200 °C and an intensification pressure of 200 bar on a Buehler H630-SC HPDC machine. The connection between steel sheet and casting is achieved by a rib-structure on the surface of the metal sheet (see high-resolution cut on the left of figure 1). To analyse the filling of the ribs a high-resolution mesoscale simulation is used, which simulates just the ribs area with a fine mesh with cell sizes in the μm range.

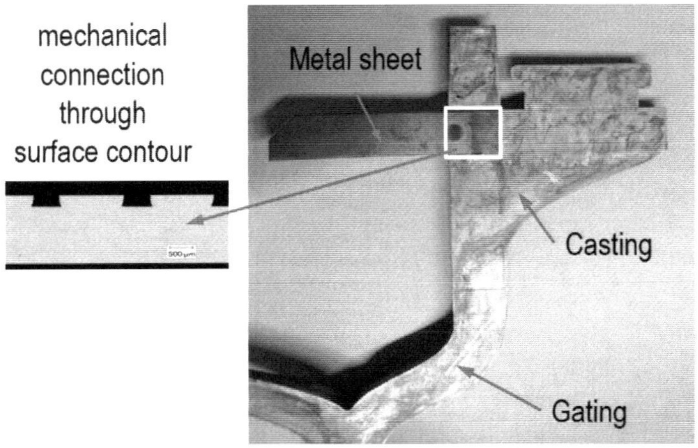

Figure 1. Hybrid HPDC casting with gating system, casting and metal sheet. The metal sheet is connected to the casting by the surface contour of the metal sheet (ribs), which is filled by the melt.

The initial and boundary conditions for the mesoscale simulation are calculated using a macroscale simulation of the whole casting process, calculating melt flow, cooling and solidification in a strongly coupled manner. Material data was taken from the STAR-CCM+ metallurgy material database. Figure 2 shows the temperature distribution on the melt during the critical phase of the hybrid process, when the melt fills the casting below the metal sheet. The melt cools down during filling but remains significantly above liquidus temperature. For the mesoscale simulation temperatures around 650°C and velocities between 2 m/s and 4 m/s can be extracted from the macroscale simulation as initial and inlet boundary conditions.

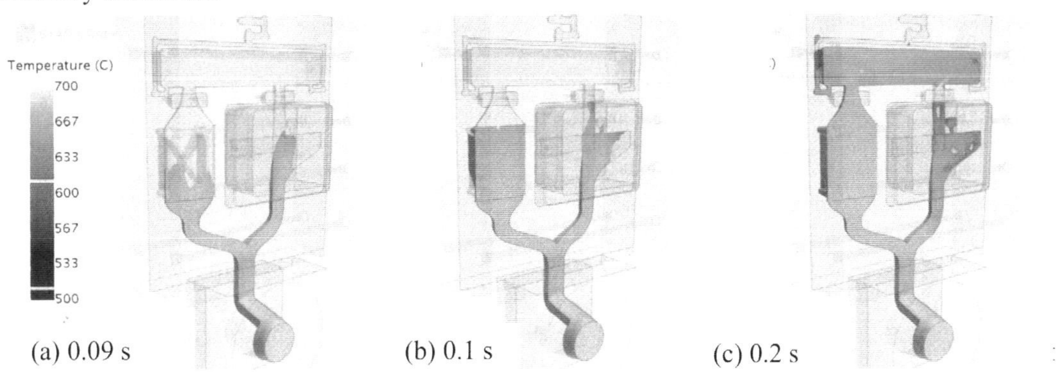

Figure 2. Simulation of the HPDC process coupling flow, temperature and solidification. After ~0.1 s (b) the critical region below the metal sheet is filled. After 0.2 s (c) the melt in the metal sheet area starts to solidify (green line = points at liquidus temperature).

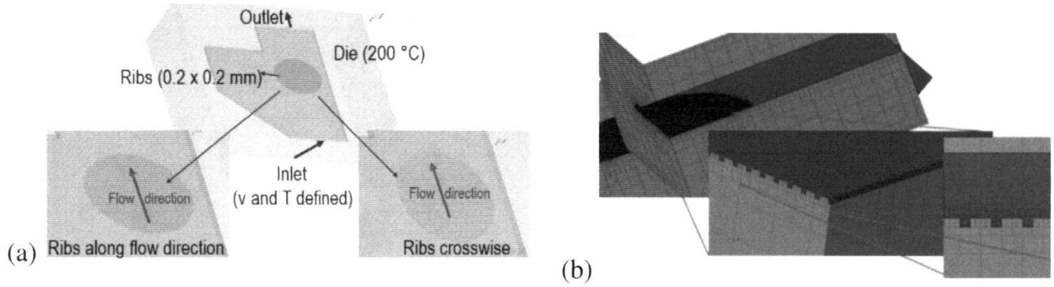

Figure 3. (a) Mesoscale geometry of the region below the metal sheet with 0.2 x 0.2 mm ribs and (b) the trimmed fine mesh. The round shape of the region with ribs gives possibility to use different orientations towards the melt flow direction.

The critical region below the metal sheet was schematically reconstructed for the mesoscale simulation with a round area of ribs with 200 µm width and depth for a systematic analysis of the ribs filling (see figure 3(a)). The round area was chosen to allow a rotation of the rib´s direction relative to the melt flow. Figure 3(b) shows the trimmed fine mesh used for the mesoscale simulation. The ribs are meshed with a cell width of 50 µm. This work focuses on the flow behaviour and not on the mechanical bonding. Therefore, the small inclination of the ribs for the mechanical bonding was neglected. It should have only a minor influence on the flowability, but would increase the numerical effort considerably.

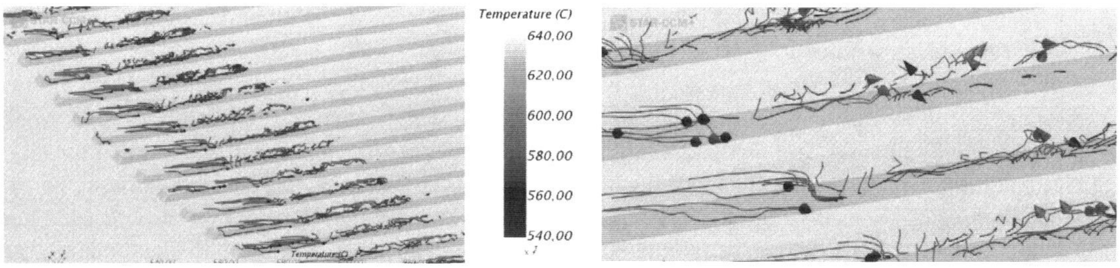

Figure 4. Streamlines showing the movement of the ribs and the temperature of the melt during the filling process after 25 ms.

4. Cavity and ribs filling

In figure 4 the filling of the ribs for a filling along the ribs direction is visualized using streamlines coloured with the melt temperature. The melt drops into the ribs and is quickly cooled down below solidus temperature by contact with the cold wall of the ribs. Higher layers of melt remain liquid and can stream out of the ribs again resulting in a chaotic flow pattern near the top of the ribs. The cooling curves given in figure 5 also show a fast solidification of melt deep inside the ribs. At several points inside (negative position, red) and above the rib (positive position, orange) the temperatures are recorded during filling of the ribs (the position of the probe points is shown in the small grey sketch on the right). The temperature of the melt drops below the eutectic temperature with a high cooling rate in the range of 5000 K/s and the melt remains solid during further filling. Melt above the ribs undergoes a much slower cooling with about 200 K/s. After 60 ms the cooling curves show a small jump to higher temperatures (green oval), indicating the starting pressure build up, which presses hot melt deeper into the ribs.

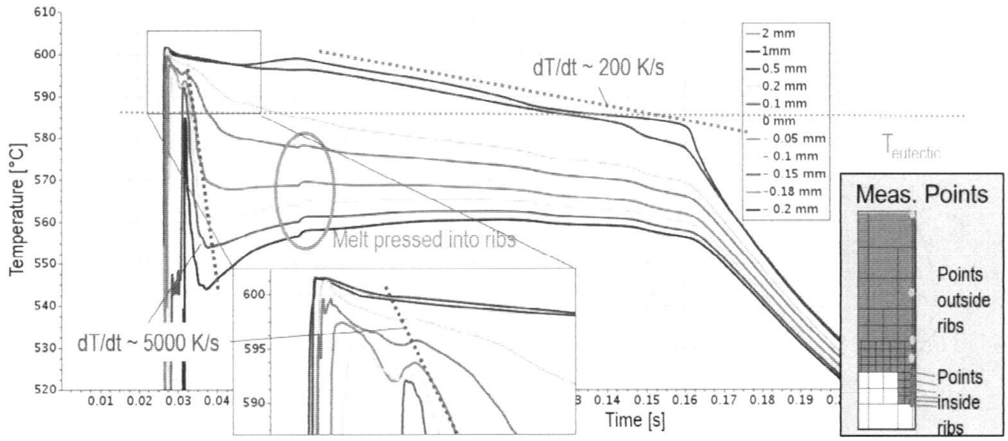

Figure 5. Cooling curves at different points inside (negative position) and above the ribs (positive position). The small sketch at the right shows the positions of the data recording points.

The effect of the pressure build-up can also be seen in the filling curves of the ribs in figure 6. The ribs are only completely filled after 60 ms, when pressure build-up in the filled cavity presses the melt into the ribs.

Figure 7 shows the influence of process and solidification parameter on the ribs filling. To compare different configurations with different total filling times, the time is normalized to the switching time, when the cavity is completely filled and the HPDC-process switches from filling to pressure build-up. The following configurations are compared:

- Advanced fraction solid curve from MICRESS® calculations (see below) for the solidification of the melt
- Fractions solid curves from Scheil approximation
- 90° rotation of the ribs from orientation along the melt flow to an orientation across the melt flow
- Filling of a 5 mm thick cavity above the ribs, compared to 2 mm in the standard configuration
- Filling with a velocity of 4 m/s

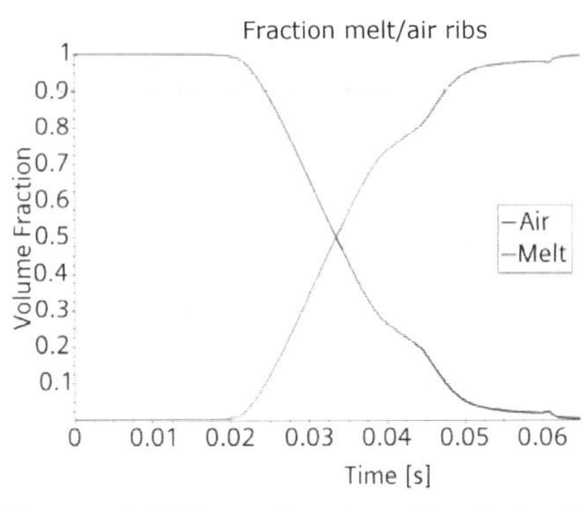

Figure 6. Volume fraction of melt (red) or air (blue) during filling.

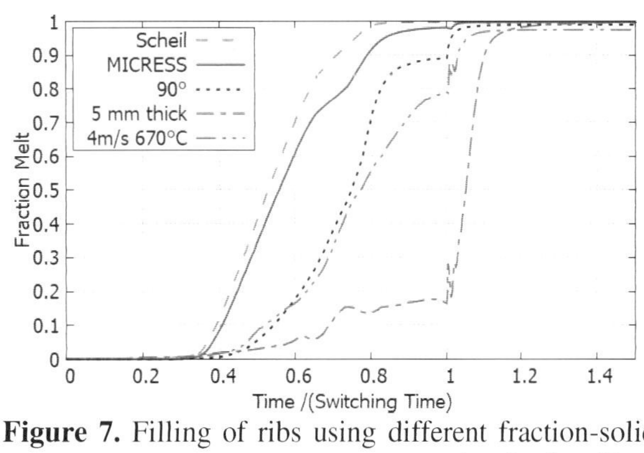

Figure 7. Filling of ribs using different fraction-solid curves, process parameters or geometries (broken lines use Scheil approximation).

The multiphase field simulation of the fast solidification (see section 5 below) gives a fraction solid curve with 20 K undercooling leading to a later start of the solidification compared to a Scheil approximation (see figure 8). The simulation using this fraction solid curve shows a slower filling than the calculation using Scheil approximation. The later start of the solidification is accompanied with a later increase of the viscosity. This hinders the small pressure build-up found when using Scheil approximation, which presses the melt earlier into the ribs. Consequently, undercooling of the melt, described by the fraction solid curve calculated using MICRESS®, results in a slower filling of the ribs.

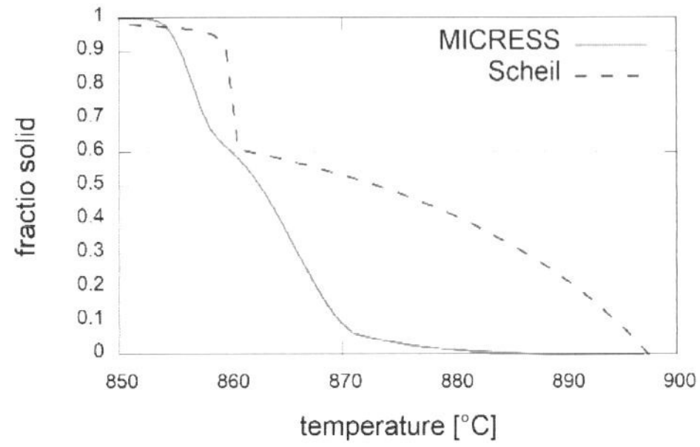

Figure 8. Fraction solid curves from Scheil approximation and phase-field simulation using MICRESS®.

Concerning the process settings the following effects on the ribs filling can be observed: The rotation of the ribs by 90° leads to a later filling of the ribs. Melt is more likely to flow across the ribs then into the ribs. The thickening of the cavity above the ribs to 5 mm hinders the pressure build-up above the ribs and significantly reduces the filling of the ribs. Thus, thin castings are more favourable for hybrid casting applications than thick parts. As to expect a higher melt velocity leads to a later filling of the ribs, because melt is more likely to flow across the ribs with a higher velocity.

Table 1 summarizes the effect the different configurations on the filling. The volume fraction of melt inside the ribs at switching time is compared. Clearly changing the process settings has a stronger influence than changes in the solidification parameter. This indicates that a Scheil assumption would be sufficient when different designs and process parameters are being evaluated.

5. Microstructure simulation

The solidification conditions obtained by the mesoscale simulation of the rib filling process where used as initial and boundary conditions for a phase-field simulation using MICRESS® [13, 14] phase-field software in the thin-interface limit with anti-trapping terms [15] and FD-correction [16]. All material data for the three phases (liquid, fcc-Al, Mg Si) was evaluated from the thermodynamic database TCAl5 via the TQ-interface [17]. Since the individual fcc(Al) and Mg_2Si lamellae are much too fine to be numerically resolved on this length scale, the eutectic was treated as an effective phase with thermodynamically consistent phase fractions and concentrations.

Table 1. Filling of ribs using different fraction solid curves or process parameter

Variants	Filling at switching time
fs - Scheil	99 %
fs - MICRESS	98 %
Ribs crosswise	89 %
5 mm thick	16 %
v = 4 m/s	79 %

The mesoscale simulation gives a heat extraction rate of $2x10^{11}$ W/m^3 for the cooling inside the ribs. A series of phase-field simulations were performed under systematic variation of cooling and nucleation conditions. Good agreement between experimental microstructure (a) and simulated (b) microstructure was obtained for a nucleation density of $5x10^{16}$ m^{-3}. Since direct temperature measurement of the fast cooling inside the ribs is very difficult if not impossible, comparing the microstructure is a valuable tool to validate the solidification conditions during filling, cooling and solidification of the metal in the ribs as result of the mesoscale simulation.

IOP Conf. Series: Materials Science and Engineering **861** (2020) 012035 doi:10.1088/1757-899X/861/1/012035

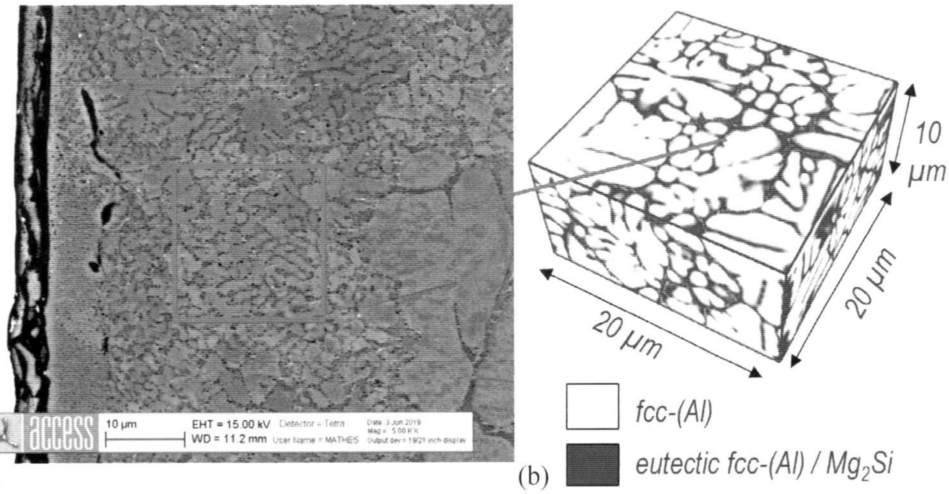

(a) (b)

Figure 9. A series of phase-field simulations were performed under systematic variation of cooling and nucleation conditions. Good agreement between (a) experimental microstructure and (b) simulated microstructure was obtained for an extraction rate of $2x10^{11}$ W/m^3 and a nucleation density of $5x10^{16}$ m^{-3}.

6. Conclusion

Three-phase coupled flow and solidification simulation is used to analyse the hybrid high pressure die casting process combining a metal sheet with the casting. The focus of the investigation lies on the rapid solidification inside the ribs on the metal surface, which are used to create a mechanical bond between metal sheet and casting part. A high-resolution (mesoscale) simulation of the interface between metal sheet and aluminium alloy during filling and solidification can capture the details of melt flow and rapid solidification, using velocities and temperatures of the macroscale process simulation as initial and boundary conditions. By coupling the mesoscale flow and solidification simulations to multiphase-field microstructure simulations on the microscale, the resulting microstructure can be calculated. The good agreement between calculated and experimental analysed microstructure demonstrate that the mesoscale simulations give the solidification conditions inside the ribs correctly.

The systematic investigation of different process configurations leads to the following findings:

- A pressure build-up above the ribs is important for a complete filling of the melt.
- Undercooling of the melt by 20 K (phase-field simulation result) leads to a slower filling of the ribs. The later start of the solidification is accompanied with a later increase of the viscosity and pressure build-up. The latter hinders a fast filling of the ribs.
- Orientation of the ribs perpendicular to the melt flow leads to a later filling of the ribs, since melt is less likely to flow into the ribs.
- Thickening of the cavity above the ribs hinders the pressure build-up above the ribs and reduces the filling of the ribs. This indicates that thin castings are more favourable for hybrid casting applications than thick parts.
- Higher melt velocities lead to a later filling of the ribs, since the melt flows at a higher speed over the ribs rather than into the ribs.

Acknowledgments

The work was supported by BMWi (Bundesministerium für Wirtschaft und Energie), INNOKOM project *Fast Solid*, funding contract 49VF170031 and joint project *SPEED* funding contract 03ET1516A. The authors acknowledge the support by Siemens PLM, agreement 60068580, for providing STAR-CCM+ licenses.

References

[1] The aluminum automotive manual 2002 (European Aluminium Association) https://www.european-aluminium.eu/media/1526/aam-manufacturing-1-casting-methods.pdf
[2] Beckermann C and Monroe A 2009 Porosity simulation for high pressure die castings *Die Casting Engineer* (The North American Die Casting Association) ISSN 0012-25JX
[3] Bodhayana M R and Ramesha N 2014 *Int. J. Res. Eng. Tech.* **03** (11) 1163
[4] Zhu J *et al.* 2009 *MCWASP XII* (Warrendale: The Minerals, Metals & Materials Society TMS) 361-8
[5] Jana S, Kättlitz O, Hediger F, Jakumeit J and Aguilar J 2012 *IOP Conf. Ser.: Mater. Sci. Eng.* **33** 012007
[6] Ferziger J H and Peric M 2002 *Computational Methods for Fluid Dynamics* (McGraw-Hill) ISBN 2-540-42074-6.
[7] Teskeredzic A, Demirdzic I and Muzaferija S 2002 *Num. Heat Trans. B* **42** 43
[8] Demirdzic I and Muzaferija S 1995 *Comput. Method Appl. Mech. Eng.* **125** 235-55
[9] Muzaferija S and Peric M 1999 *Computation of free surface flows using interface-tracking and interface-capturing methods.* In *Nonlinear Water Wave Interaction*, ed Mahrenholtz O *et al.* (Southampton: WIP Press) pp 59-100
[10] Brackbill J U, Kothe D B and Zemach C 1992 *J. Comput. Physics* **100** 335
[11] Sabau S and Viswanath S 2002 *Met. Mat. Trans. B* **33B** 131
[12] Poirier D R, Yeum K and Mapples A L 1987 *Metall. Mater. Trans. A* **18** 1979
[13] MICRESS®, multicomponent multi-phase-field software, www.micress.de
[14] Eiken J, Böttger B and Steinbach I 2006 *Phys. Rev.* E **73** 066122
[15] Carré A, Böttger B and Apel M 2013 *J. Cryst. Growth* **380** 5
[16] Eiken J 2012 *Mater. Sci. Eng.* **33** 012105
[17] Thermo-Calc Software: http://www.thermocalc.com

A verification of thermophysical properties of a porous ceramic investment casting mould using commercial computational fluid dynamics software

C A Jones[1], M R Jolly[1], A E W Jarfors[2], M Irwin[3], R Svenningsson[4], J Steggo[2] and J Eriksson[3]

[1] Sustainable Manufacturing System Centre, Cranfield University, Bedfordshire, MK43 0AL, UK
[2] Jönköping University, School of Engineering, Box 1026, 551 11 Jönköping, Sweden
[3] TPC Components AB, Brånstaleden 2, 734 32 Hallstahammar, Sweden
[4] RISE Research Institute of Sweden, Tullportsgatan 3, 553 22, Jönköping, Sweden

Email: c.a.jones@cranfield.ac.uk

Abstract. Defects in cast metals remain a common problem in many areas of the foundry industry, particularly in the investment casting of large area, thin-walled components for aerospace applications. During previous research, the thermophysical properties, density and porosity of a fibre reinforced ceramic investment casting mould were determined using several experimental techniques. Without verification, these experimental results remain nothing more than educated guesswork. The purpose of this study is to verify previous results and to more fully characterise the ceramic mould material with complementary measurements. A commercially available computational fluid dynamic (CFD) simulation package, Flow-3D®, was used in conjunction with a full-scale Ni-superalloy (IN718) casting to assess the accuracy of these results. By placing thermocouples strategically across the mould thickness, temperature profiles were determined and compared directly to predicted profiles extracted from the simulation by a custom-written Python script.

1. Introduction

This work continues from results obtained in a previous study [1] where the thermophysical properties of a fibre-reinforced, porous ceramic mould used in aerospace casting were experimentally determined. Techniques including laser flash analysis (LFA), differential scanning calorimetry (DSC) and gas pycnometry were used to obtain the thermal diffusivity, specific heat capacity and material density respectively. This paper reviews these properties in the light of improvements to the experimental methodology, re-presenting data along with a discussion of any differences relative to previously drawn conclusions. Fulfilment of the aim to verify these properties was achieved through comparison with results obtained from a CFD simulation package, Flow-3D®. Cast aerospace components have strict microstructural and residual stress limitations to ensure they are fit for service, the control of which requires accurate, substantiated simulations which are often widely overlooked by industrial entities. Simulation configurations are rarely transferable from one foundry situation to another making characterisation work an essential part of all operations.

IOP Conf. Series: Materials Science and Engineering **861** (2020) 012036 doi:10.1088/1757-899X/861/1/012036

2. Methodology

2.1. Experimental casting trials

A full-scale casting experiment was conducted at the TPC Components AB foundry in Sweden to provide data points for comparison with simulation results for property verification. A ten layer test mould was produced consisting of eight step structures organised into two groups of four divided into upper and lower groups; a computer generated version of which is shown in figure 1. This test mould had a structural composition comparable to that of the genuine aerospace moulds, the breakdown of which is detailed in table 1. It should be noted that the composition of sample 2 was altered between the sample preparation and the construction of the mould for experimental casting verification. This change was not radical and hence not anticipated to significantly impact the results, but should be noted as a variable.

Two parts in the upper grouping and two parts in the lower grouping (four in total) positioned directly opposite one another were coated with Superwool® Plus insulation. Six K-type thermocouples were positioned in the upper group of test parts in the mould, three in an insulated part and three in an uninsulated part. One S-type thermocouple was placed centrally in the cavity of the 8 mm section of the insulated component. The exposed wires of these thermocouples were sleeved with ceramic tubing for protection and were connected to a data-logger which in turn was placed inside an insulated container. In both parts, the K-type thermocouples were positioned with one on the second mould layer, one on the sixth layer and one on the eighth layer. All of the thermocouples were active and generated data throughout the experiment with the exception of the centrally positioned S-type thermocouple which was damaged as a result of manual handling. A Python script was produced to extract the data from the simulation output and compare it to the thermocouple results obtained in the casting trials.

Table 1. Compositional breakdown of the mould by layer

Sample Identity	Layer	Composition		
		Name	Details	
Sample 1	1	Cobalt Aluminate with 90 mesh Zircon sand	Chilches Microzir Flour 200M IC	$ZrSiO_2 > 63.5\%$ $TiO_2 < 0.2\%$ $Fe_2O_3 < 0.15\%$ $Al_2O_3 < 1.3\%$ $SiO_2 < 33\%$
			Cobalt Aluminate	$CoAl_2O_4$
			Primecote® Plus	Binder
Sample 2	2	Intermediate with 54 mesh Alumina sand	-	-
Sample 3	3	Backup with 30/80 mesh molochite sand	Ranco-Sil 140F	SiO_2
			Matrixol 30	Binder
			MXC Excel X2	Fibre Reinforcement
			30/80 Sand	-
Sample 4	4 5 6 7 8 9	Backup with 16/30 mesh molochite sand	Ranco-Sil 140F	SiO_2
			Matrixol 30	Binder
			MXC Excel X2	Fibre Reinforcement
			16/30 Sand	-

The thermocouples were secured onto the dried surface of the previous mould layer using ceramic slurry, with their position determined using a digital calliper to measure the mould thickness at the thermocouple site. These positions are recorded in figure 1 and the positions of the inner and middle thermocouples were matched to measurement nodes in the simulations. The thermocouples in the insulated and uninsulated have been assumed to be at the same positions as they were attached at the same time, following the same number of slurry dips.

The data logger was set to commence recording as the mould was placed into the preheat furnace. 1050°C was selected as the preheat temperature of the furnace, which the mould remained in overnight to allow time for a stable temperature to be reached. Immediately prior to casting, the mould was transferred to a vacuum chamber and the vacuum drawn to 3.8 x10^{-4} Torr (0.0507 Pa) at which point the Inconel 718 melt at 1468 °C was poured.

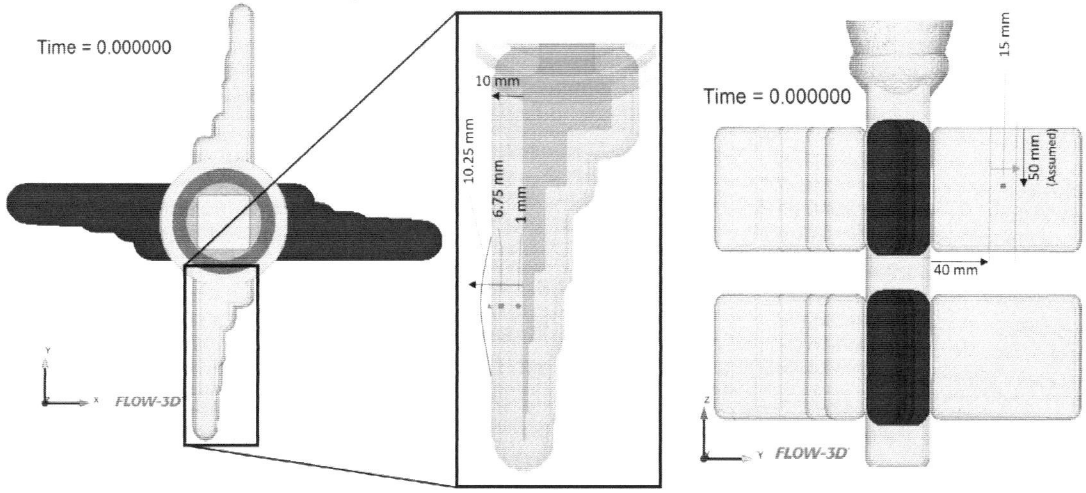

Figure 1. Computer generated representation of the casting geometry taken from Flow-3D® pre-processor. The locations of the three thermocouples are indicated in red with distances marked relative to the interface between mould and cavity. Blue indicates wool insulation.

2.2. Computer simulations

The computer simulation in Flow-3D® was configured to replicate these conditions where a mould geometry of a uniform 10 mm thickness, see figure 1, was placed within a time-varying pressure boundary condition to reflect the transitioning environment from atmospheric pressure to vacuum and back. Void pointers were used to define the difference in initial air conditions between the mould cavity, 1050°C, and the surrounding environment, 30°C, as it would have been the instant the mould was removed from the furnace. The complete geometry was meshed using two mesh regions; a coarser mesh of 5 mm cubes covering the complete environment and a finer mesh of 2.5 mm cubes surrounding the insulated and uninsulated parts of interest. This configuration of meshes ensures an improvement in simulation time when compared to a uniform fine mesh situation although when using multiple meshes, it must be ensured that the mesh planes intersect and there is no more than a factor of 2 between the sizes of overlapping mesh regions.

Due to the presence of the thermocouples in the real mould, a local thickening of the mould was seen, as indicated in figure 1, which was not reflected in the simulations because of the method by which the mould was formed. As a result, the outer thermocouple was not actually measurable from the simulation as it lay beyond the outer wall of the simulation mould. The heat transfer coefficients (HTC) between the fluid and mould was defined in a temperature varying manner in an attempt to account for the air gap that forms as the metal solidifies ranging from 3500 $Wm^{-2}K^{-1}$ at 2273K to 300 $Wm^{-2}K^{-1}$ at 273 K. The HTC between the mould and insulation was set at a constant value 1000$Wm^{-2}K^{-1}$ based on literature values [2]. The emissivity value for the surface of the mould was defined as 0.79 to reflect the fact the dominant material at the surface is Silica and 0.56 for the insulation, determined by proportionally taking the emissivity of the constituent components (Silica, MgO & CaO) relative their percentage abundance in the material. Based on the compositional makeup of the IN718 alloy, the TCFE7: Steels/Fe-Alloys v7.0 database in Thermo-Calc® software version 2019.1.36671-604 was used to determine the latent heat of fusion and solid fraction as a function of temperature to improve the accuracy of the values available in the Flow-3D® materials database.

2.3. Thermophysical properties

Differential scanning calorimetry (DSC) analysis was conducted on various layers of the mould to determine the effective specific heat capacity of the complete mould. When this assessment was conducted in [1] the samples considered were produced by taking scrapings which could result in the inclusion of material from different layers of the mould. In this work, samples of each mould layer – confirmed to contain no contaminants from other regions of the mould - have been investigated. The mould was divided into four samples, defined in table 1, each uniform in composition which together in the correct proportions form the complete mould. Sample masses of 20.4 mg, 20.0 mg, 20.5 mg and 20.7 mg for samples one, two, three and four respectively were selected for DSC analysis, values selected to allow calculations to be made relative to a 20 mg Sapphire reference sample. All samples were tested using the sample heating cycle of 25 °C to 1400 °C at a rate of 10 K per minute before returning to room temperature at the same rate, sampling data continuously.

The linear coefficient of thermal expansion (CTE) of the ceramic mould was measured using dilatometry. In this work, two measurements were taken along orthogonal axes: one in the direction parallel to the heat flux (radially through the mould) and the second in the direction perpendicular to the heat flux (in the plane of the layers). Two cylindrical samples of the complete ceramic mould with an approximate diameter of 6 mm and a length of 12.8 mm and 8.4 mm. Ideally, these samples would both be approximately 12 mm in length to allow comparison to the reference standard. However, as the mould was no thicker than 10 mm in the parallel direction an alternative reference was required in this case. The first sample was evaluated against a 12 mm polycrystalline aluminium oxide whereas the shorter second samples were evaluated against an 8 mm Alumina reference standard. It is important to note that the use of different reference standards does not impact the final results of CTE, it merely ensures the instrument was correctly calibrated to the sample length. Both experiments were run using the same heating cycle; 25 °C to 1400 °C increasing at a rate of 10 K per minute, continuously sampling data.

In order to determine the thermal conductivity of the ceramic mould, it was necessary to combine the heat capacity and the material density (ρ) with the thermal diffusivity (α), see equation (1), which was measured using Netzsch laser flash analysis (LFA). The CTE was supplied to the LFA software to allow compensation for changes in sample dimensions during the temperature cycle. Thin cylindrical wafers of the sample materials were prepared with an approximate diameter of 10 mm and a thickness

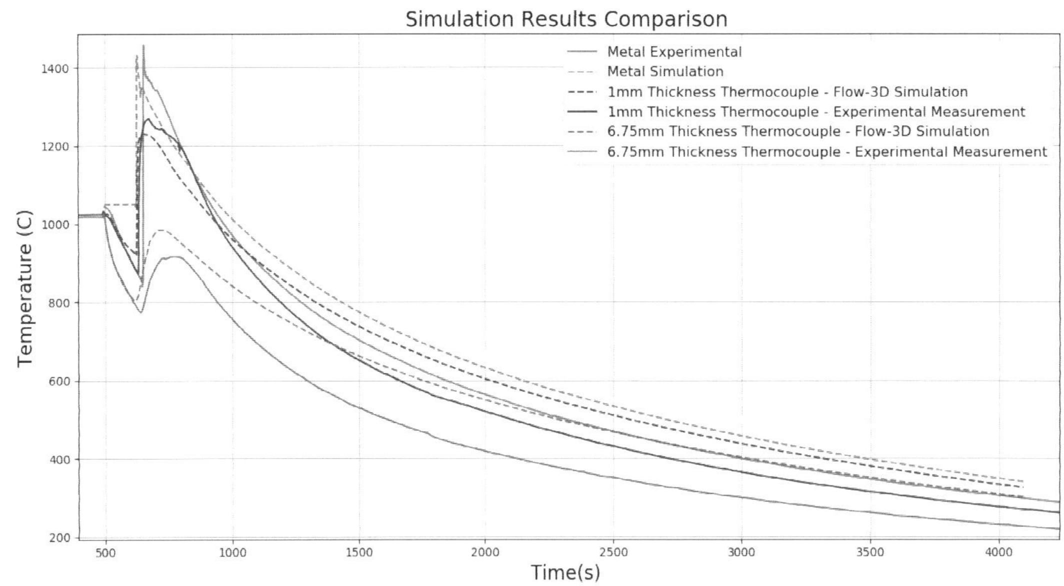

Figure 2. Experimental results obtained from the thermocouple contrasted against predictions from computer simulations

of 1.94 mm, 2.10 mm, 1.99 mm and 3.8 mm for samples 1, 2, 3 and 4 respectively. Difficulties were experienced during the determination of the thermal diffusivity in [1], largely as a result of the laser flash analysis setup where the suspected transparency of the mould to the laser wavelength (1064 nm) prevented the acquisition of data by obscuring the sample response, particularly at elevated temperatures. With the aim of obtaining improved data, the LFA analysis was repeated with the samples placed into a sealed crucible fitted with a Sapphire window in the lid and into the bottom to restrict the area upon which the laser was incident. The system was configured to increase the temperature from ambient to 1400 K at a rate of 50 K per minute with five measurements taken at each 200 K temperature increment.

$$\lambda = \rho \alpha C_P \tag{1}$$

3. Results

3.1. Casting trials and simulations

Figure 2 shows the comparison between casting experiment and simulation of the uninsulated part indicating a good agreement between the experimentally measured thermal gradients and those predicted by the simulation in both the metal and mould, hence validating the accuracy of the specific heat capacity, thermal expansion and thermal conductivity of the mould determined via the discussed methodologies. At this stage, only the uninsulated part has been considered in order to isolate the mould thermophysical properties in the comparison by removing the uncertainty associated with the properties of the insulation material. The properties of the insulation will require further specification before inclusion in the future. Spikes in the data profiles obtained by experiment were caused by exposed sections of the thermocouple wires making contact with one another and have been largely removed from the plot as outliers, with a linear interpolation conducted between the remaining data points to ensure continuity.

Initially in the mould, all thermocouples were constant at their maximum temperatures attained whilst in the preheat furnace which was approximately 1027.5 °C for the uninsulated at the inner thermocouple site and 1044.5 °C for the insulated at the inner thermocouple site. It was noted that despite the mould remaining in the preheat furnace overnight, none of the six thermocouples in the mould ever attained the targeted 1050 °C and there was a notable difference between those in insulated and those in uninsulated regions. Literature offers a possible justification for this observation. In [3], thermocouples were strategically positioned around a steam autoclave to investigate the temperature distribution which was found to be decidedly non-uniform. Of particular note was that the temperature recorded in the vicinity of the door was consistently lower than that recorded deeper in the autoclave. Although this was conducted in a steam autoclave, it can be conjectured that the same principle applies to the preheat furnace. Given that the mould was positioned near the door of the preheat furnace to accommodate the thermocouple wires emanating out along the shortest possible route, this would explain the lower mould temperature. Even though the S-type thermocouple was not serviceable for this case, an identical casting was conducted the previous day with approximately comparable environmental and preheat conditions with the thermocouple placed in the uninsulated part which did successfully record information about the melt temperature profile with time. By placing a measurement point in the cavity of the simulation it was possible to make a comparison; the exact positioning of the node is not so important in this case due to the very high conductivity of the metal. In the simulated metal case, the melt arrived at the measurement location slightly earlier than in the real case, most likely because of the currently poor understanding of the true mass flow rate of the metal into the mould cavity. Despite this, the profile was a reasonable match although the simulation does over predict the temperature of the metal following approximately 750 seconds.

From the moment the furnace door was opened the mould began cooling, a trend predicted well by the simulations, particularly at the 6 mm thermocouple position, until after approximately 2 minutes and 18 seconds where the temperature increases dramatically as melt fills the cavity. The agreement during this stage, particularly at the inner thermocouple's location (1 mm from the cavity) is very good. As would be expected, the gradient measured in response to the melt introduction shallows as one measures at positions further from the cavity. For example, the response of the innermost thermocouple is virtually instantaneous whereas the second displays a more gradual increase. This difference emerges as a result of the finite time required for thermal energy to diffuse through materials which increases with thickness. The cooling of the air within the mould cavity was not modelled in this instance resulting in simulated metal temperature appearing as constant with time prior to mould filling. As metal fills the cavity at approximately 2 minutes 18 seconds the temperature profile of the metal began to display the expected trend in agreement with the experimental measurements. The agreement between the simulation and experimental measurements begins to deviate during the final cooling as time progresses with the temperature consistently overpredicted by the simulation. This deviation could be attributed to the fact that the simulated mould was not an exact match to reality, in terms of wall thickness/surface roughness or to poorly captured heat transfer mechanisms. A less obvious source of variation may arise from the simulation mesh size of 2.5 mm in the area of interest, large compared to mould thickness, leading to a potential inprecision in the determination of the temperature profiles. Despite this deviation, the trend remains good with the gradients of the simulated and experimental results beginning to match as time progresses. This agreement suggests the heat transfer mechanism at this time was correctly captured by the simulation using the measured data. The initial discrepancy will require further investigation.

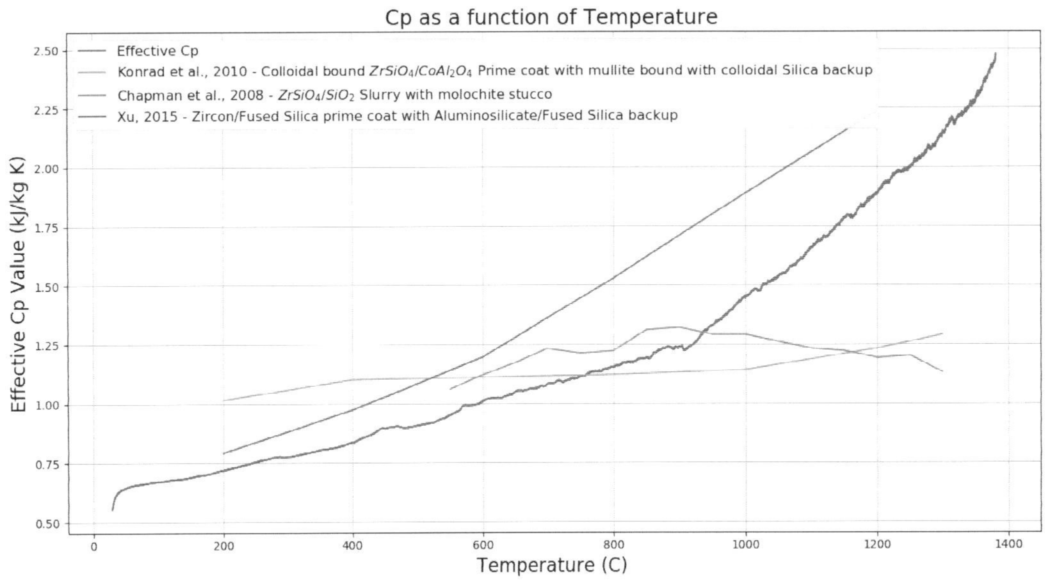

Figure 3. Plot of the effective specific heat capacity alongside other published data ([2, 4-5]).

3.2. Thermophysical properties

3.2.1. Specific heat capacity. The effective specific heat capacity of the ceramic mould shown in figure 3 was determined mathematically using equation 2, where A is a scaling coefficient, M is the sample mass and i is an integer, by combining the individual measured values of each layer based on their contribution to the total mass of the mould through a rule of mixtures approach [6]. The specific heat capacity measurements for samples 3 and 4 showed a significant increase in C_p with temperature, particularly at high temperature, which is not reflected to the same extent in samples 1 and 2 which show a very gradual increase with temperature. The scaling coefficients reflect that samples 1, 2, 3 and 4 contribute in mass percent 7.63%, 15.39%, 15.39% and 61.59% respectively, hence the high increase of sample 4 dominates. When comparing these results to the other published results, below 1000 °C the trend of Konrad *et al.* is in reasonable agreement with experimental data however, conversely exhibits a decrease at high temperatures. Chapman *et al.* on the other hand does report an increase in C_p at the higher temperatures although not to the same extent as that describes by the experimental data. The mould material investigated by Xu is arguably the most compositionally comparable material to that investigated in this work which, although consistently higher in value, has a strong similarity in trend with temperature. Data that can be found in [6] also made a consideration of the C_p from measuring two compositional elements of their mould; Zircon/Silica prime coat and a 200 mesh molochite. Their results indicated that within a range of 300-800 °C both the Zircon/Silica and molochite have a gradual increase in C_p with temperature from approximately 0.75 to 0.8 kJ kg^{-1} K^{-1} and 1 to 1.1kJ kg^{-1} K^{-1} respectively. These results are in good agreement with the measurement of the prime coat conducted in this work (blue line of figure 3) but less so for the molochite backups (red/green lines of figure 3).

$$C_p^{eff} = \sum_i \left(\frac{A_i M_i}{A_1 M_1 + A_2 M_2 + A_3 M_3 + A_4 M_4}\right) C_p^i \quad (2)$$

3.2.2. Coefficient of thermal expansion. It could well be expected that the inhomogeneous nature of the mould material would lead to an inconsistency in the thermal expansion properties depending on the axis one considers. The average results of assessing the thermal expansion in two orthogonal axes, parallel and perpendicular to the direction of heat flux, are shown in table 2. The CTE in the perpendicular direction is consistently lower than that in the parallel direction, although both compare well with similar materials from published sources [5]. It is interesting to note that when comparing the results to the Chapman *et al.* investigation of fused Silica, i.e. comparable to the experimental mould with the exception of Zircon, the results are notably higher indicating the presence of Zircon in a mould increases the CTE. The quoted uncertainty in measurements was determined with reference to previous literature [7], where the same Netzsch instrument reference samples were employed.

Table 2. Average values of the coefficient of thermal expansion

Direction	CTE
Perpendicular	3.53±0.06 x10^{-6} K^{-1}
Parallel	4.84±0.06 x10^{-6} K^{-1}

3.2.3. Thermal conductivity. Thermal diffusivity measurements were taken for each of the four samples with the intention of determining thermal conductivity by amalgamation with the individual thermal expansion, density and specific heat capacity measurements. Diffusivity measurements taken for sample 4 displayed a periodic, oscillatory behaviour that obscured the expected step response of the profile. It is most likely that as a result of the extended exposure time whilst attempting to determine a stable response the detector did in fact register the periodic temperature fluctuations of the furnace rather than the true response from the sample. As a consequence, the experimental values of thermal diffusivity from a previous investigation [1] were used in combination with the effective specific heat capacity determined in this work to calculate the thermal conductivity, as shown in figure 4. Plotted alongside this data is a set of data from other published sources [2,4], most interesting of which is that of Xu [2] which shows a remarkably similar trend in conductivity with temperature. Data obtained from Konrad *et al.* [4] shows a much lower variation in thermal conductivity with temperature compared with the experimental data obtained here.

4. Conclusion

The temperature profiles obtained from computer simulations have a good agreement with the experimentally determined temperature profiles, indicating that the measured thermophysical properties of the mould are reasonable. These thermophysical properties are comparable to similar materials presented in the literature. Published literature including [8] illustrating the difference in predictive capabilities of computer simulation packages available on the market, highlighting their sensitivity to the setup of elements including boundary conditions and mesh setting; factors that can have a significant impact on the results. Care should be taken when configuring simulations for different circumstances. It would be beneficial to investigate the variations in predictions in greater detail, although this is a topic for future research. With thermophysical data available for individual layers of the mould, a more complete model could be constructed by considering each layer individually in the simulation as opposed to combining them into an effective value of the parameter. Questions remain to be answered regarding the specific heat capacity profile of the mould; what is responsible for the increase in C_p observed in samples 3 and 4 at increasing temperatures? This will be investigated in detail in due course.

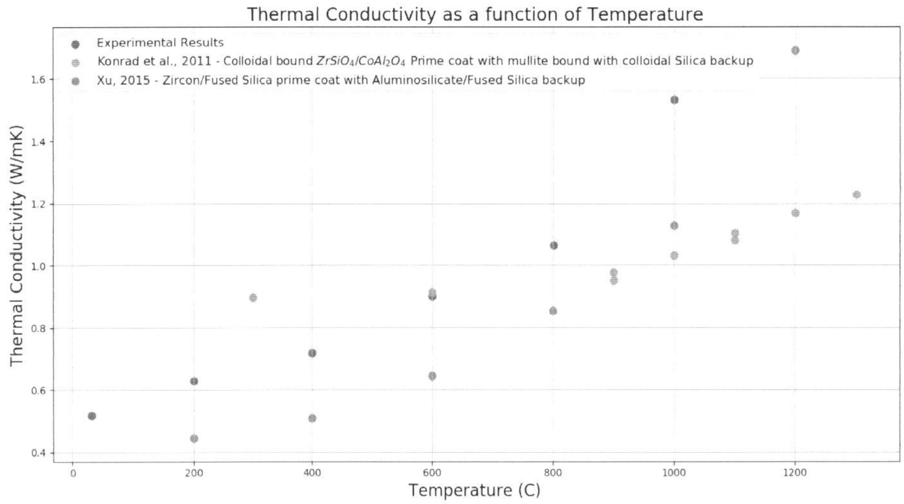

Figure 4. Thermal conductivity as a function of temperature alongside published results.

Acknowledgements
The authors would like to thank the Engineering and Physical Sciences Research Council (EPSRC) and TPC Components AB for providing their financial support towards this research. In addition,
The authors would like to specifically thank Daniel Kuivamäki (TPC Components AB) and personnel of the Flow 3D® support team for their technical support with the computer simulations.
The data used in this paper is described in the Cranfield Online Data Repository (CORD) at: http://doi.org/10.17862/cranfield.rd.9934280

References

[1] Jones C A, Jolly M R, Jarfors A E W and Irwin M 2020 *TMS 2020 149th Annual Meeting and Exhibition Supplemental Proceedings* (San Diego: Springer) pp 1095–106
[2] Xu M 2015 *Characterization of investment shell thermal properties* (Missouri University of Science and Technology)
[3] Jones S, Jolly M R, Gebelin J, Cendrowicz A and Lewis K 2001 *FOCAST 2nd Mini Conference* (Unpublished)
[4] Konrad C H, Brunner M, Kyrgyzbaev K, Volkl R and Glatzel U 2011 *J. Mater. Process. Technol.* 181–6
[5] Chapman L A, Morell R, Quested P N, Brooks R F, Brown P, Chen L-H, Olive S and Ford D 2008 *Properties of Alloys and Moulds Relevant to Investment Casting* (Teddington: National Physical Laboratory)
[6] Jones S 2000 *FOCAST 1st Mini Conference* (Unpublished)
[7] Matsushita T, Ghassemali E, Saro A, Elmquist L and Jarfors A 2015 *Metals* 5 1000–19
[8] Khan M A A and Sheikh A K 2018 *Int. J. Simul. Model.* 17 197–209

IOP Conf. Series: Materials Science and Engineering **861** (2020) 012037 doi:10.1088/1757-899X/861/1/012037

X-ray tomography, digital volume correlation and FE modelling: A synergistic combination to study the processing-structure-property relations in ductile iron

T Andriollo[1], Y Zhang[1], S Fæster[1], J Thorborg[1,2], N S Tiedje[1], V Kouznetsova[3] and J Hattel[1]

[1] Technical University of Denmark, Kgs. Lyngby, Denmark

[2] MAGMA GmbH, Aachen, Germany

[3] Eindhoven University of Technology, Eindhoven, The Netherlands

E-mail: titoan@mek.dtu.dk

Abstract. The mechanistic understanding of the processing-structure-property relations in ductile iron is still far from complete. One reason is that the impact on the mechanical properties of some of the microstructural features arising from the casting process can be hard or even impossible to investigate using experimental methods alone. The present work shows that a solution can be the synergistic combination of X-ray tomography, digital volume correlation (DVC) and finite element modelling, which are applied here to study the effect played by the Si micro-segregation and local residual stresses upon mechanical loading. First, miniaturized tensile and compact tension specimen are loaded incrementally while imaging with X-ray tomography. Then, the micro-scale displacement is reconstructed with DVC and used to prescribe the boundary conditions in high-fidelity 3D finite element models of the microstructure. Simulations are run considering or not the formation of the local residual stresses and build-up of micro-segregation during manufacturing. The numerical predictions are compared to the corresponding experimental data both at the macro-scale – applied load – and at the micro-scale – strain field reconstructed with DVC. This allows for a first assessment of the impact of the local residual stresses and Si micro-segregation on the mechanism of tensile deformation as well as of crack propagation of ductile iron.

1. Introduction

Since its commercial introduction in 1948, ductile iron, also known as nodular cast iron or spheroidal graphite iron (SGI), has constantly found new fields of application, ranging from the automotive sector to the wind power industry. Nowadays, this positive trend is challenged by the limitations in terms of reliability and range of applicability of the processing-microstructure-property relations available in the literature, which are rooted in the still incomplete understanding of the mechanisms controlling deformation and fracture at the microscale [1]. This is primarily a consequence of the complex material microstructure, composed of graphite nodules embedded in a matrix where multiple phases may coexist depending on chemistry and cooling conditions, which makes direct experimental investigations challenging.

In this respect, a breakthrough in the experimental characterization of SGI has been achieved in the last 10-15 years with the introduction of techniques based on X-ray computed tomography (CT), which have been applied to visualize the microstructure evolution in 3D during both manufacturing

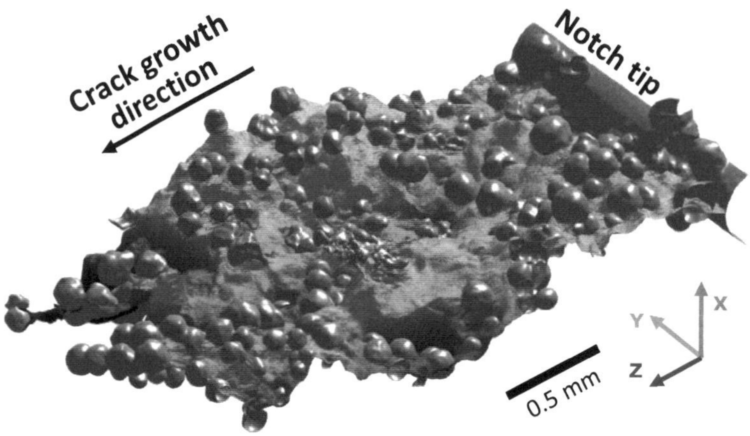

Figure 1. Fatigue crack path at the microstructural scale in a 1.5 mm thick compact tension specimen, reconstructed from CT via a dedicated segmentation procedure. Adapted from [2].

and room temperature mechanical loading [3, 4]. A great advantage of CT is that it can be combined with pattern tracking methods like digital volume correlation (DVC) and used to reconstruct the deformation field at the level of the microstructure during both monotonic [5] and cyclic loading [6]. This has opened up to the possibility to observe and quantify the interaction between the microstructural constituents with unprecedented level of detail. An example is given in figure 1, where a 3D reconstruction of a fatigue crack which grows by linking the nodules in a 1.5 mm thick compact tension specimen is depicted.

In spite of these powerful experimental techniques, some aspects of the deformation of SGI at the microscale remain challenging to investigate. Among these is the impact of those microstructural factors whose individual effects are hard or even impossible to isolate because of experimental difficulties in changing such factors without changing, at the same time, other important microstructural parameters. The Si micro-segregation and the local residual stresses are two examples of these. The former builds up during solidification due to the progressive decrease in Si content of the liquid combined with the Si rejection from the growing graphite, whereas the latter form during solid-state cooling due to the thermal contraction mismatch between the nodules and the matrix. Recent work pointed out that the local residual stresses can reach magnitudes of 100 – 150 MPa [7, 8]. At the same time, extensive Si micro-segregation [9] can be expected to produce localized variations of the matrix yield stress in the order of 50 – 100 MPa [10]. When these values are compared to the macroscopic yield stress of standard SGI grades, it is clear that both factors can potentially play a significant role in how the material behaves upon mechanical loading.

Recently, the authors introduced a framework based on the synergistic combination of CT, DVC and finite element (FE) modelling to predict the yield stress [11] and the fatigue crack growth direction [2] in SGI. In this work, the Si micro-segregation and local residual stresses are taken as test cases to demonstrate that the same framework can be used to study the impact of the microstructural factors that, for the reason stated above, are hard to investigate with experimental methods alone.

2. Methodology

The framework laid out in [2, 11] is summarized briefly in the following subsection. Next, the modifications needed to include the impact of the Si micro-segregation in the FE models are described. It is worth noting that the formation of the local residual stresses was already taken into account in the works [2, 11]. However, the associated effects were investigated in relation to the SGI yield stress only, not with respect to the crack propagation direction at the microscale.

2.1. Overall framework

The overall idea of the framework described in ref. [2, 11] is to perform a mechanical test on a miniaturized specimen while imaging in-situ with CT. Based on the CT data, the micro-scale displacement is reconstructed with DVC [12] and used to prescribe the boundary conditions in a 3D microstructure-resolved FE model of the region of interest. This approach offers two main advantages. First, the size of the model can be maintained small compared to that of the specimen, meaning that the computational complexity can be kept at a reasonable level. Second, the numerical predictions can be compared to the corresponding experimental data both at the macro-scale and at the micro-scale in a combined fashion.

The specific mechanical test considered in [11] is a uniaxial tensile test, conducted on a mainly ferritic SGIs with a small fraction of pearlite $\approx 5\%$ and highly spherical nodules. The left-hand side of figure 2 shows the specimen and the associated FE model. In the model, which corresponds to the region of the specimen between two cross-sections, the subdivision of the microstructure into matrix and nodules is taken into account based on the information from the CT. The matrix is assumed to obey an isotropic elastoplastic constitutive model with isotropic hardening, whose parameters are calibrated using experimental data of an alloy with equivalent metallurgical parameters. The nodules are assumed linear elastic and their highly inhomogeneous internal structure is modelled by dividing each nodule into an external shell and a core. While the former is assumed isotropic, different orientations of the elastic symmetry and anisotropic moduli are assigned to each mesh element in the core based on the element location with respect to the centroid of the nodule, in order to represent, approximately, the strong gradient in mechanical properties inside the nodules [13, 14]. The weak interface between the nodules and the matrix is modelled using the surface-based cohesive formulation available in ABAQUS to allow for interface debonding. The model is subjected to two simulation steps. First, to simulate the formation of the local residual stresses, the last part of the solid-state cooling is simulated by prescribing a uniform temperature change from 500 °C to 20 °C. Then, the displacement reconstructed with DVC is prescribed on the model boundaries to simulate the tensile testing. Further details can be found in [11].

Conversely, the mechanical test considered in [2] is a fatigue crack growth test. The right-hand side of figure 2 shows the compact specimen used in the test and the associated crack tip FE model. In this case, the region considered in the model is the volume surrounding the crack tip. Besides an internal

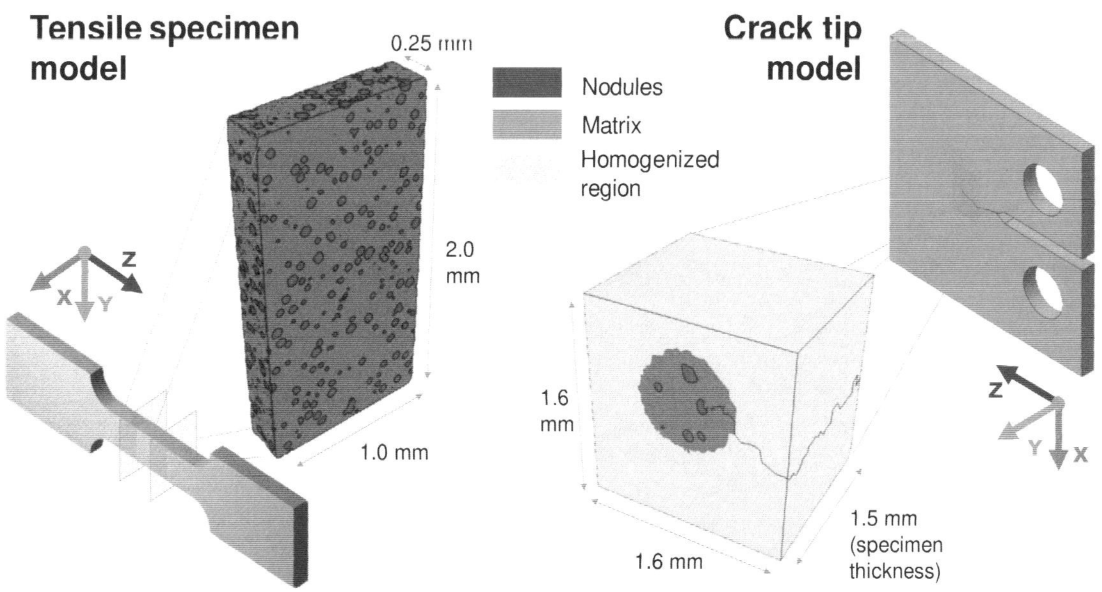

Figure 2. Miniaturized tensile and compact tension specimens and corresponding FE models.

microstructure-resolved region, the model features an additional external homogenized region, whose purpose is to transfer the load to the internal region in a more gradual fashion, as well as a stationary crack, which is inserted in the model based on the real crack geometry reconstructed with CT (shown in figure 1). The constitutive behaviour of the nodules and of the nodule-matrix interface is equivalent to that described in relation to the tensile specimen model. Conversely, an isotropic elastoplastic formulation with both isotropic and kinematic hardening is selected for the matrix, whose parameters are calibrated by constructing a representative volume element of the microstructure and fitting its macroscopic response – computed via first-order computational homogenization [15] – to cyclic experimental data for the considered SGI. The homogenized region obeys the same constitutive model of the matrix, although with a different set of parameters, chosen to represent the SGI effective behaviour. The solid-state cooling is simulated as in the tensile specimen model. Subsequently, five consecutive loading cycles with identical boundary conditions (prescribed boundary displacement at maximum and minimum load on the model boundaries) are simulated, in order to achieve stabilization of the micromechanical fields, i.e. hardening. Further details about this model can be found in [2].

2.2. Si micro-segregation

The micro-segregation of Si in the SGI microstructure is primarily a consequence of the positive Si partition coefficient in temperature range of the stable and metastable eutectic, which causes the Si content in the liquid to decrease progressively as solidification proceeds [16]. As the diffusivity of Si in the solid is limited, the room-temperature microstructure exhibits first-to-solidify regions characterized by a Si content larger than the average, and last-to-solidify regions where the Si content is significantly lower. On top of this, the Si rejection by the growing graphite plays a role also and tends to increase the Si concentration near the nodules.

The above arguments indicate that if a high-resolution 3D map of the Si content is not available, the Si segregation pattern can still be estimated based on the location of the first-to-solidify and last-to-solidify regions and on the spatial distribution of the graphite. The latter can be easily obtained with CT, whereas the former necessitates laborious procedures based on serial sectioning to be determined.

In the present work, since the 3D subdivision of the microstructure into first-to-solidify and last-to-solidify regions could not be determined, it was decided to estimate the Si segregation pattern based on the distribution of the graphite nodules only. Specifically, it was assumed that the Si concentration w at a given point in the matrix is a decreasing function of the distance r to the closest nodule, which can be computed by applying a distance transform algorithm to the segmented CT images. Despite this assumption might appear simplistic, it is remarked that since the nodules in the last-to-solidify regions are smaller and usually more distant from each other than those in the first-to-solidify regions, a lower Si concentration is estimated in the former regions than in the latter, in agreement with experimental findings.

In the matrix region of both the tensile specimen and crack tip models the Si concentration was prescribed according to the following function:

$$\begin{cases} w(r) = w_{max} - \dfrac{w_{max} - w_{min}}{r^*} r & \text{if } r \le r^* \\ w(r) = w_{min} & \text{if } r > r^* \end{cases} \tag{1}$$

where w_{max} is the concentration at the nodule-matrix interface and w_{min} the minimum concentration found in the last-to-solidify regions. The threshold r^* was introduced in order to avoid having an extremely low concentration in some remote areas (e.g. the corners of the model) and it was adjusted so that the volume satisfying $r > r^*$ was no more than 1 % of the total matrix volume. The values of w_{max} and w_{min} were set to 2.75 wt% and to 1.50 wt%, respectively, based on previous segregation studies [17,18] and on the requirement of having an average Si concentration w_{avg} equal to that of the SGI used for manufacturing the tensile and compact tension specimens (2.30 %). Figure 3 shows the Si concentration prescribed according to equation (1) over the central X-Y cross-section of the tensile specimen model.

In the FE models, the Si micro-segregation was assumed to affect the local yield stress of the matrix σ_y. Specifically, the following Voce-type relationship was assumed to apply:

$$\sigma_y = Q^0 + Q^\infty(1 - \exp(-b\varepsilon_{eq}^{pl})) \tag{2}$$

where ε^{pl}_{eq} is the equivalent plastic strain, b is a constant and the parameters Q^0 and Q^∞ depend linearly on the local Si concentration:

$$\begin{aligned} Q^0 &= Q^0_{avg} + k^0(w - w_{avg}) \\ Q^\infty &= Q^\infty_{avg} + k^\infty(w - w_{avg}) \end{aligned} \tag{3}$$

The proportionality factors k^0 and k^∞ were set to 57.9 MPa/wt% and 76.2 MPa/wt%, respectively [10].

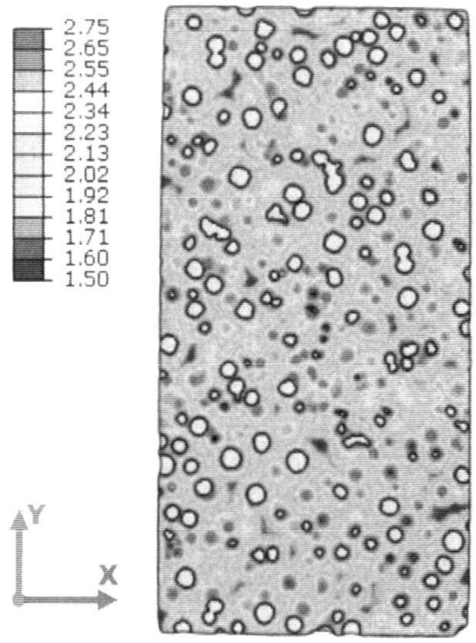

Figure 3. Si concentration over the central X-Y cross-section of the tensile specimen model. Values are in wt%. White areas represent the nodules.

3. Results

3.1. Monotonic tensile deformation

The model of the tensile specimen shown in figure 2 was used to simulate the deformation of the SGI microstructure upon tensile loading up to a macroscopic strain of 3.8 %. The nodal reaction forces on the Z+ boundary were recorded at regular intervals and used to compute the macroscopic tensile stress. The results are shown in figure 4 and relate to two simulations: one assumes uniform Si concentration whereas the other considers the Si segregation pattern described by equation (1). Hardly any changes in the predicted stress vs. strain curve can be detected when the Si segregation pattern is included in the FE model. That is, the non-uniform spatial distribution of the matrix properties does not seem to play a role. It is worth remarking that, as the average Si concentration is the same in both simulations, the average volumetric yield stress and strength of the matrix are also the same due to the linearity of equations (1) and (3). To a certain degree, this can explain why the macroscopic response of the SGI as predicted by the FE model is unaffected by the presence of Si segregation.

In order to investigate the deformation at the level of the microstructure, the predictions of the FE model were compared to the data from DVC. As discussed in [11], the comparison is not straightforward when a subvolume-based DVC algorithm is employed, due to the different locations at which the displacement vector is known, i.e. at the centre of each subvolume in DVC but at each node in the FE model. In the present work, as the FE mesh is much finer than the subvolume grid used in the DVC, it was decided to make the comparison at the level of the DVC grid. Accordingly, the displacement of each DVC subvolume, as predicted by the model, was considered equal to the mean displacement of all the FE nodes located inside that subvolume. Based on this "averaged" displacement, the equivalent strain predicted

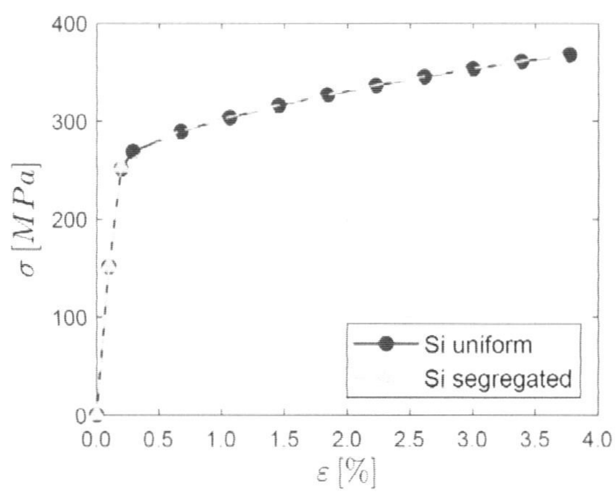

Figure 4. Tensile curves predicted by the FE model of the tensile specimen.

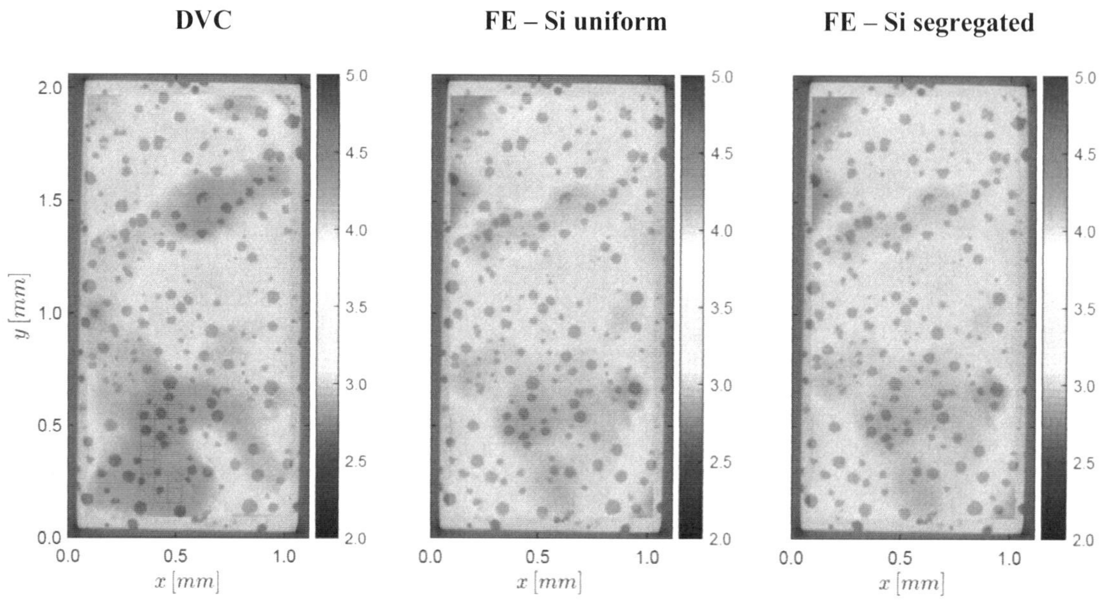

Figure 5. Contours of the equivalent strain (in %) over the central cross-section of the tensile specimen, superimposed onto the CT scan. The data relates to a macroscopic strain of 3.8 %.

by the model at the level of the microstructure was calculated and compared with the corresponding DVC data.

The comparison is shown in figure 5 for the central cross-section of the tensile specimen, where, as before, the predictions of two simulations which either account or not for the Si segregation are considered. Focusing on the DVC strain data, it can be observed that the deformation pattern seems to follow the underlying distribution of the nodules. The FE model that assumes uniform Si concentration can reproduce this pattern relatively well, even though some non-negligible differences exist (see e.g. the area near the bottom left corner of the cross-section, which is highly strained according to the DVC but not according to the FE simulation). Including the Si segregation in the FE model leads to negligible changes in the predicted strain values and, consequently, the agreement with the DVC data does not improve. In this respect, it is worth emphasizing that recent experimental studies demonstrate that the subdivision of the SGI matrix into first-to-solidify and last-to-solidify regions and the resulting spatial variation in mechanical properties do have an impact on how the material deforms at the level of the microstructure [19]. A possible explanation for why this effect is not seen in the present FE simulations is that, in addition to the Si segregation, there are other metallurgical factors that contribute to create a difference between the mechanical properties of the matrix in the former regions, which are not taken into account in the present model, e.g. porosities, grain size, pearlite fraction, etc.

3.2. Fatigue crack propagation

The model of the fatigue crack tip illustrated in figure 2 was used in [2] to show that the crack growth direction at the microstructural level correlates well with the direction perpendicular to the maximum principal stress predicted in the matrix. In the present work, the same model was used to investigate the extent to which the crack growth direction can be affected by the magnitude of the local residual stresses and by the Si segregation, which was neglected in [2]. To this end, four FE simulations were run, each corresponding to a different combination of the initial stress state (with/without residual stresses) and Si concentration pattern (uniform/segregated according to equation (1)). The output of each simulation was used to predict the crack growth angle in the X-Z plane (see figure 2) at 500 points along the crack front, based on the maximum principal stress direction. Subsequently, the difference $\Delta\theta_{ZX}$ with the propagation angle measured with CT upon reloading of the compact tension specimen was computed.

The results are presented in figure 6 in the form of box plots, where the horizontal band in each box indicates the median of the $\Delta\theta_{ZX}$ values along the crack front, the box edges represent the 25th and the 75th percentile and the whiskers extend to cover a length of 2.7 times the standard deviation. It can be seen that the four simulations give rise to almost identical box plots, which indicate, based on the position of the box edges, that the absolute value of $\Delta\theta_{ZX}$ is less than ≈ 10 degrees for 50 % of the points considered along the crack front. That is, there is good agreement between the predicted and the measured crack propagation angles independently of the simulation type. Consequently, the local residual stresses and the Si segregation do not appear to affect the crack propagation at the microstructural level significantly.

This result can be partially justified as follows. It is experimentally documented that the fatigue crack grows by linking the nodules, which tend to be approached by the crack along their radial direction. As the crack propagates perpendicularly to the maximum principal stress, it follows that the latter is roughly perpendicular to the nodule radius. Now, focusing on a single nodule, the local residual stresses around it can be considered spherically symmetric to a first approximation, and are such that any direction perpendicular to the nodule radius is a direction of maximum principal stress. Therefore, the local residual stresses and the actual stresses that drive the crack growth share the direction of the maximum principal stress. Therefore, a lower or a higher magnitude of the residual stresses cannot be expected to change significantly the resulting maximum principal stress direction and, therefore, the crack propagation direction. A similar reasoning might hold for the Si segregation pattern, given its approximate spherical symmetry near each nodule, see equation (1).

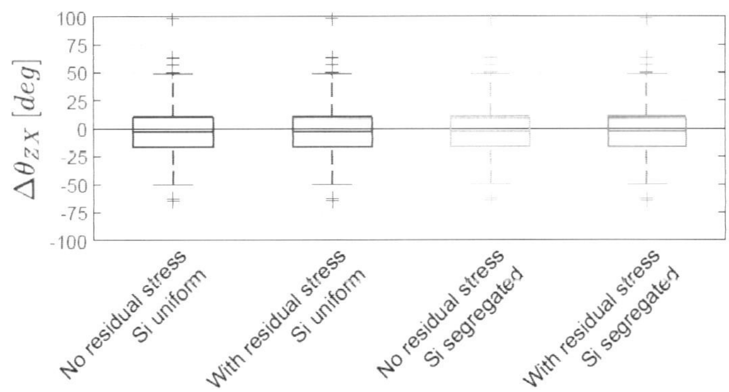

Figure 6. Statistical box plots of the angle $\Delta\theta_{ZX}$ between the predicted and the measured fatigue crack propagation directions at 500 points along the crack front.

4. Conclusions

The synergistic combination of CT, DVC and FE modelling allows developing computationally tractable microstructure-resolved thermo-mechanical models of SGI that can be validated at both the macro- and micro-scale. This opens up to the possibility of analysing the impact of some of those microstructural factors that are difficult to investigate with experimental methods alone. In this respect, it was shown here that the Si micro-segregation is not expected to affect significantly the material tensile response, either at the micro- or at the macro-scale, under the assumption that the Si concentration is a function of the distance to the closest graphite particle. The fatigue crack propagation direction in the matrix, computed based on the maximum principal stress direction, turned out to be unaffected by both the Si segregation and the local residual stresses forming during the solid-state cooling.

References

[1] Hütter G, Zybell L and Kuna M 2015 Micromechanisms of fracture in nodular cast iron: From experimental findings towards modeling strategies – A review *Eng. Fract. Mech.* **144** 118–41
[2] Andriollo T, Zhang Y, Faester S and Kouznetsova V 2020 Analysis of the correlation between micro-mechanical fields and fatigue crack propagation in nodular cast iron *Acta Mater.* **188** 302–314
[3] Verdu C, Adrien J and Buffière J Y 2008 Three-dimensional shape of the early stages of fatigue cracks nucleated in nodular cast iron *Mater. Sci. Eng. A* **483–484** 402–5
[4] Azeem M A, Bjerre M K, Atwood R C, Tiedje N and Lee P D 2018 Synchrotron quantification of graphite nodule evolution during the solidification of cast iron *Acta Mater.* **155** 393–401
[5] Buljac A, Helfen L, Hild F and Morgeneyer T F 2018 Effect of void arrangement on ductile damage mechanisms in nodular graphite cast iron: In situ 3D measurements *Eng. Fract. Mech.* **192** 242–61
[6] Limodin N, Réthoré J, Buffière J Y, Gravouil A, Hild F and Roux S 2009 Crack closure and stress intensity factor measurements in nodular graphite cast iron using three-dimensional correlation of laboratory X-ray microtomography images *Acta Mater.* **57** 4090–101
[7] Zhang Y B, Andriollo T, Fæster S, Barabash R, Xu R, Tiedje N, Thorborg J, Hattel J, Juul Jensen D and Hansen N 2019 Microstructure and residual elastic strain at graphite nodules in ductile cast iron analyzed by synchrotron X-ray microdiffraction *Acta Mater.* **167** 221–30
[8] Andriollo T, Hellström K, Sonne M R, Thorborg J, Tiedje N and Hattel J 2018 Uncovering the local inelastic interactions during manufacture of ductile cast iron: How the substructure of the graphite particles can induce residual stress concentrations in the matrix *J. Mech. Phys. Solids* **111** 333–57
[9] Freulon A, de Parseval P, Josse C, Bourdie J and Lacaze J 2016 Study of the Eutectoid Transformation in Nodular Cast Irons in Relation to Solidification Microsegregation *Metall. Mater. Trans. A Phys. Metall. Mater. Sci.* **47** 5362–71
[10] Abramowitz P and Moll R A 1970 Silicon-Carbon interaction and its effect on the notch toughness of mild steel *Metall. Trans.* **1** 1773–5
[11] Andriollo T, Zhang Y B, Fæster S, Thorborg J and Hattel J 2019 Impact of micro-scale residual stress on in-situ tensile testing of ductile cast iron: Digital volume correlation vs. model with fully resolved microstructure vs. periodic unit cell *J. Mech. Phys. Solids* **125** 714–35
[12] Thermo Scientific™ 2019 Avizo 9.6
[13] Bellini C, Di Cocco V, Favaro G, Iacoviello F and Sorrentino L 2019 Ductile cast irons: Microstructure influence on the fatigue initiation mechanisms *Fatigue Fract. Eng. Mater. Struct.* 1–11
[14] Andriollo T, Fæster S and Winther G 2018 Probing the structure and mechanical properties of the graphite nodules in ductile cast irons via nano-indentation *Mech. Mater.* **122** 85–95
[15] Matouš K, Geers M G D, Kouznetsova V G and Gillman A 2017 A review of predictive nonlinear theories for multiscale modeling of heterogeneous materials *J. Comput. Phys.* **330** 192–220
[16] Lacaze J 1999 Solidification of spheroidal graphite cast irons: III. Microsegregation related effects *Acta Mater.* **47** 3779–92
[17] Selig C and Lacaze A 2000 Study of microsegregation buildup during solidification of spheroidal graphite cast iron *Metall. Mater. Trans. B* **31** 827–36
[18] Boeri R and Weinberg F 1989 Microsegregation in Ductile Iron *AFS Trans.* **89** 179–84
[19] Xu C L, Andriollo T, Zhang Y B, Hernando J C, Hattel J and Tiedje N 2020 Micromechanical impact of solidification regions in ductile iron revealed via a 3D strain partitioning analysis method *Scr. Mater.* **178** 463–467.

IOP Conf. Series: Materials Science and Engineering **861** (2020) 012038 doi:10.1088/1757-899X/861/1/012038

Impact of local Si segregation on strain localization in ductile cast iron

C L Xu[1], T Andriollo[1], F Lin[2], Y B Zhang[1], J Hattel[1] and N Tiedje[1]

[1] Department of Mechanical Engineering, Technical University of Denmark, DK-2800, Kgs, Lyngby, Denmark
[2] School of Engineering, Jönköping University, SE-55111, Jönköping, Sweden

E-mail: chaoxu@mek.dtu.dk

Abstract. The distribution of Si content in tensile deformed ductile cast iron has been characterized using electron microscopy and correlated to the strain distribution determined based on 3D tomography data collected before and after tensile deformation and digital volume correlation analysis. The results show that the high plastic strain regions localize in bands consisting of large graphite nodules and deformed matrix with high Si content connecting the graphite nodules in the first-to-solidify regions. The bands are aligned about 45° with respect to the loading direction, which is close to the maximum shear direction.

1. Introduction

Ductile cast iron (DCI), due to its high performances and versatility at relatively low cost compared to steels with analogous performances, is widely used in different applications such as small and medium sized heavily loaded part for the automotive industry, e.g. gears and brakes, and very large industrial components with extreme demands for mechanical properties e.g. wind turbine hub [1]. DCI consists of very soft graphite nodules and steel matrix which have quite different mechanical properties. Analogous to other types of cast irons, is based on the Fe-C-Si system and contains some alloy elements intentionally added to enhance one or more useful properties. With increasing Si content, the thermodynamic activity of C in the liquid can be increased, promoting stable eutectics and inhibiting the formation of cementite. At the same time, Si can lower the carbon content of the eutectic promoting the formation of graphite nodules during solidification. Additionally, Si dissolves in the steel matrix, providing solid solution strengthening. However, Si segregates upon solidification leading to a heterogeneous chemical distribution in the steel matrix, which can lead to local strength difference, being higher at Si rich regions and lower at Si rare regions [2]. Lekakh *et al.* [3] has shown that such segregation plays an important role on the mechanical properties of ductile cast iron. Fernandino *et al.* [4] developed 2D multi-scale model for capturing how the damage proceeds in an inhomogeneous microstructure at microscale. He also reported that cracks preferentially initiated at the nodule-matrix interface and propagated across internodular ligaments avoiding the low-Si last-to-solidify (LTS) zones [5]. However, it is worth pointing out that both the materials and the mechanical behaviour are in 3D, which implies that the heterogeneities on the subsurface could affect the mechanical reactions on the sample surfaces. Therefore, a more accurate analysis of the correlation of local strain to inhomogeneous microstructure in 3D is desirable.

In a recent study, we have developed a new 3D method to correlate the local strain localization to the microstructural heterogeneities [6]. First we use 3D X-ray tomography to follow in situ the 3D microstructural evolution during tensile deformation, to determine the 3D strain distribution using a digital volume correlation (DVC) method [7]. Then the microstructure on a 2D section inside the 3D volume is characterized using optical microscopy for analysing the strain partitioning. Although the correlation is conducted on a 2D section, the analysis is representative for true 3D behaviour. In the present paper, we use the same method to study the strain partitioning with respect to the Si segregation, measured using energy dispersive X-ray spectroscopy (EDX) in a scanning electron microscope.

2. Experiments

A miniaturized dog-bone shaped tensile specimen with a gauge length of 10 mm and a cross-section of 1 x 2 mm^2 was extracted from a large DCI casting. The main alloy elements beside Fe were C (3.68 wt%), Si (2.3 wt%), Mn (0.22 wt%), Mg (0.11 wt%). It consists of almost spherical graphite nodules homogeneously embedded in a metal matrix being mainly ferrite with a small fraction of pearlite about 5%. The volume fraction and average size of the graphite nodules are 11.5% and 30 μm respectively. The specimen was mounted on a tensile test rig attached inside a Zeiss Xradia 520 Versa microscope and loaded in tension to a macroscopic strain of ≈ 10%. Before and after the tensile load, the specimen was scanned with X-ray computed tomography to monitor the microstructure evolution in 3D. Subsequently, DVC was employed to calculate the microscopic strain field attendant with the mechanical deformation with a spatial resolution of 60×60×60 μm^3. Based on this, the strain value calculated from the 3D volume was obtained via linear interpolation. For microstructural characterization of the tensile deformed specimen, a 2D section (≈120 μm from one lateral surface) of the 3D volume was ground and polished, followed by SEM and EDX characterization with dwell time of 1 s and 128 frames.

3. Microstructures characterized using SEM

The SEM images showing the microstructure after the tensile test on the polished surface are presented in figure 1, where graphite nodules and matrix grains can be distinguished. Within graphite nodules some internal structures, e.g. conical sector structure and steel particles are seen, which are typical for DCI [8]. The interfaces between the graphite nodules and surrounding matrix debond upon tensile deformation at the two opposite sides that are perpendicular to the loading direction. The spherical white region marked by an arrow in figure 1 is believed to be a graphite nodule that fell out during polishing. Only in a few cases, cracks are observed inside the graphite nodules.

This result in that the fracture mechanism is mainly nodule-matrix debonding differs from previous observations on the 2D sample surface [9, 10], suggesting that the results seen at the sample surface might not be representative of bulk behaviours. Iacoviello [11] and Di Cocco [10] have confirmed that the stress state around the nodules is linked to the role of graphite nodules and thus the damaging mechanisms. As the microscopic stress state on the specimen surface differs from that in the bulk, different damaging mechanisms can be active. Furthermore, Zhang [12] has demonstrated that the residual stress around the bulk nodules differs from that around the nodules exposed to the sample surface. It is worth pointing out that the characterized section is in the bulk while the sample was loaded and only exposed by removing a layer (≈120 μm) away, which might be the reason for more pure graphite-matrix debonding in the present study.

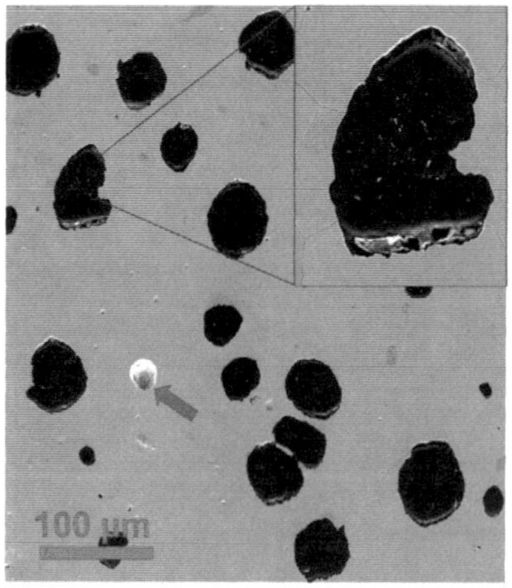

Figure 1. SEM image showing the debonding damage mechanism, and the inset presenting the nodule in high magnification

Additionally, the finding that only a few of cracks are observed in the graphite nodules is somewhat different from previous results [9, 13, 14]. A model with a 3D periodic cubic unit cell with a spherical nodule in the centre and a 2D square periodic unit cell with a circular central nodule subjected to plane-stress conditions were developed. These two models are supposed to represent the material behaviour in the bulk and on the specimen surface respectively. The model details can be found in [15]. The simulation results are shown in figure 2. The black curves represent the shape of undeformed nodules, while the blue and red solid curves are the shapes of deformed nodules in the bulk and on the sample surfaces respectively. R, X and Y are the radius of the undeformed nodules, length along and perpendicular to the loading direction. This figure is quite revealing and shows that the nodules on the surfaces become flatter than the nodules in the bulk, which means that the metal matrix compresses the nodules more and thus more cracks are likely to form in the nodules on the sample surfaces. Therefore, it is not particularly surprising to observe fewer cracks in the nodules in the present study, given the fact that the investigated section is in the bulk.

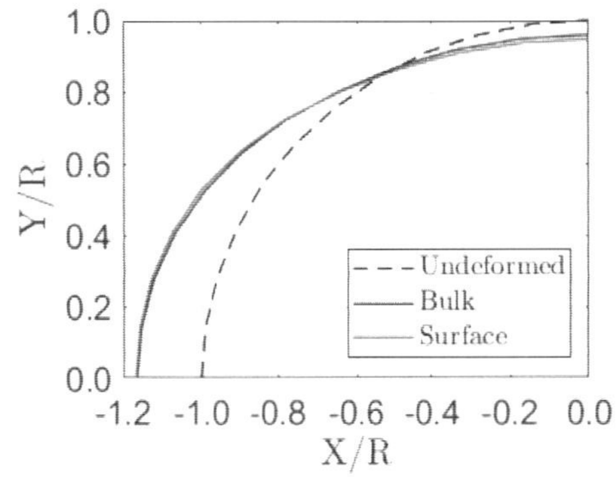

Figure 2. Shapes for undeformed nodules, and deformed nodules on the surface and in the bulk represented by the black broken line, the red solid line and the blue solid line respectively. R is the radius of the undeformed nodules, and X and Y are the length of nodules along and perpendicular to the loading direction

4. Si segregation

4.1. Si distribution

Si has a negative segregation during solidification of cast irons, which means that the Si preferentially concentrates in the austenite, leaving a lower Si content in the liquid melt during solidification. Consequently, the first-to-solidify (FTS) zones contain a higher Si content, while the LTS zones have the lowest Si content. Figure 3(a) shows the Si distribution map measured after the tensile test. In the map, the dark blue areas containing Si content lower than 0.5 wt% are graphite nodules and the yellow areas with Si content higher than 2.5 wt% are considered as the FTS zones. The areas with a relatively low Si concentration which are displayed in light blue surrounded by thin green areas are treated as the LTS zones.

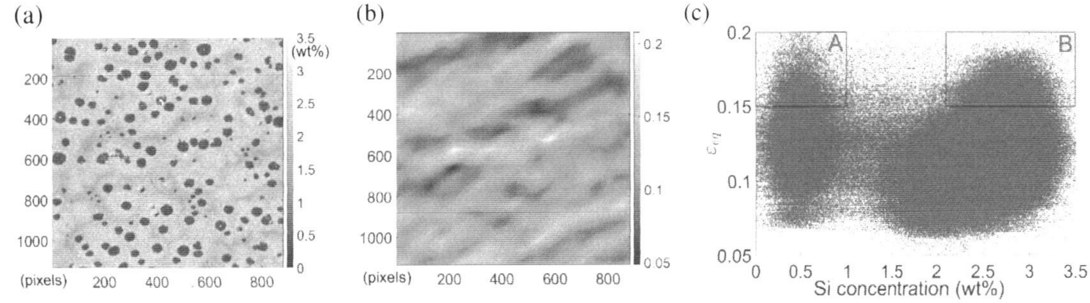

Figure 3. (a) Si concentration map, (b) strain distribution and (c) the correlation between them. Region A and B in (c) marked the high strained regions with low and high Si concentration respectively.

It can be noticed that a considerable amount of small nodules with size less than 25 μm are located mainly in the LTS zones, while almost all large nodules with size in the range 25 -75 μm are within the FTS zones. The large nodules nucleate at the beginning of the solidification and thus have long time to grow than the rest small nodules. It also indicates that the material solidifies in the very beginning with large nodules surrounded by austenite.

4.2. Strain distribution

It is well acknowledged that localized deformation generally develops within a broad range of ductile materials, which lead to intense deformation and final fracture. The von Mises equivalent strain obtained from DVC calculation is shown in figure 3(b). Yellow bands with strain values higher than 0.15 showing the strain localization are seen, which are oriented at approximately 45° with respect to the loading direction. Low-strain areas can be seen in areas between these bands. It, therefore, can be inferred that there are local softening regions in the tested materials, which allow the formation of localized strain bands.

4.3. Correlation of local strain to Si segregation

With the purpose of directly visualizing the role played by the local Si concentration on the mechanical behavior in microscale, the local strain values upon loading is plotted in figure 3(c), where the horizontal and vertical axes represent the Si content and strain values respectively. Noticeably, the Si content of voxels with strain values higher than 0.15 are mainly located in regions marked as A and B. By re-examining the Si distribution in figure 3(a), it can be seen that region A and B represent the strain of the graphite nodules and the FTS zones respectively. Furthermore, it is noticeable that the local strain in the LTS zones with Si concentration of 1.5 wt% to 2.1 wt% is lower than 0.15. This results clearly shows that the FTS zones tend to deform more plastically and thus with higher strain values compared with the LTS zones showing relatively lower strain values. Previous study shows that the matrix in FTS and LTS zones are mainly ferrite and pearlite respectively. The difference in local deformation behaviour of FTS and LTS zones thus might stem from two factors: the different mechanical properties between ferrite and pearlite or the effects of graphite nodules. On one hand, as well-acknowledged, the amount of carbon in ferrite is approximately 0.006% at room temperature since carbon atoms are larger than the lattice interstitial voids of ferrite by a factor of two. Ferrite is thus characterized by good ductility but relatively low hardness and strength, which is quite analogous to the mechanical properties of pure iron. Nevertheless, the structural components of pearlite are not only soft ferrite but also high-hardness cementite layers, which strengthen the pearlite. Consequently, the ferrite tends to be deformed more and easier than pearlite upon loading, and thus has higher strain values. On the other hand, the graphite nodules are soft and can lead to stress concentration around nodules.

However, it should be also noticed that most of voxels in FTS zones with high Si content shows low strain values. With the aim to clarify which voxels in the FTS zones and in the nodules are highly strained, the voxels in region A and B in figure 3(c) are extracted from the Si map, as shown in figure 4(a) and (b) respectively. As can be seen evidently, the voxels with high strain values both in the FTS zones and in the nodules are from the same bands, which are oriented at certain angles with respect to the loading direction.

Another striking observation to emerge from figure 3(c) is that although both nodules and FTS zones have some voxels with high strain values, the nodules have more voxels with strain values larger than 0.18. The finding indicates that the graphite nodules can be considered as a much softer phase than ferrite even though they cannot be treated as voids, which means the main load carrying area decreases. Therefore, it is expected that the matrix adjacent to the graphite nodules, ferrite, would have higher stresses and thus higher strain values compared with the pearlite mainly being located in the LTS zones. Furthermore, it can be clearly identified that the deformation bands are mainly located in areas of lines of graphite nodules, which is in good agreement with Fernandino's investigation that fracture occurs across internodular ligaments [5]. Much more ferrite and nodules along nodule lines as seen in figure 3(a) causing higher stress between nodules might make for the higher strains in bands.

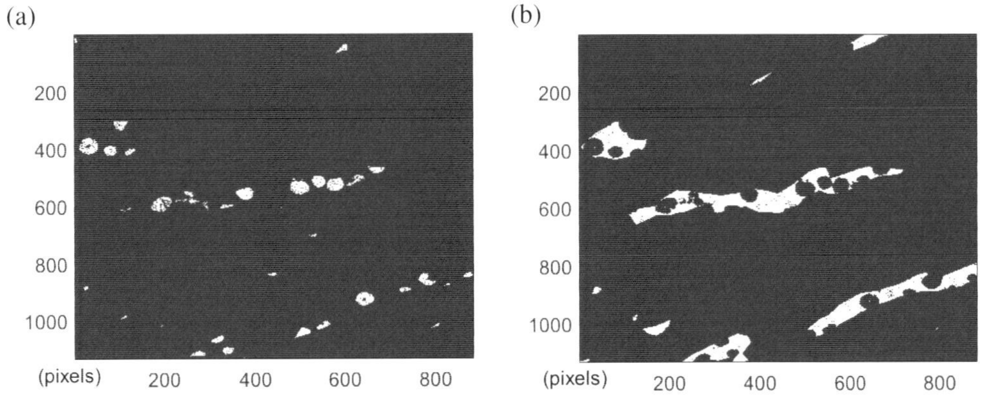

Figure 4. The location of voxels in (a) region A and (b) region B in figure 3(c).

5. Conclusions

The Si distribution and strain localization in tensile deformed DCI have been characterized using EDX, computed tomography and DVC. Possible correlation between them has been investigated. The results show that the high plastic strain regions localize in bands aligned about 45° with respect to the loading direction, which is close to the maximum shear direction. The bands mainly consist of large nodules and connected deformed matrix with high Si concentration from FTS regions.

References

[1] Tiedje N S 2010 *Mater. Sci. Technol.* **26** 505–14
[2] Steiner R 2001 *ASM Metals HandBook - Properties and Selections - Irons Steels and High and Performance* (Am. Soc. Met.) p 3470
[3] Lekakh S N, Richards V L and Medvedeva N 2012 *AFS Proc.* 12-031
[4] Fernandino D O, Cisilino A P, Toro S and Sanchez P J 2017 *Int. J. Fract.* **207** 1–26
[5] Fernandino D O and Boeri R E 2019 *Fatigue Fract. Eng. M.* **42(10)** 2220–31
[6] Xu C L, Andriollo T, Zhang Y B, Hernando J C, Hattel J and Tiedje N 2019 *Preprint*
[7] ThermoFisher Scientific 2018 Thermo Scientific Avizo Software
[8] Ghassemali E, Hernando J C, Stefanescu D M, Dioszegi A, Jarfors A E W, Dluhoš J and Petrenec M 2019 *Scr. Mater.* **161** 66–9
[9] Cocco V Di, Iacoviello F and Cavallini M 2010 *Eng. Fract. Mech.* **77(11)** 2016–23
[10] Cocco V Di, Iacoviello F, Rossi A and Iacoviello D 2014 *Theor. Appl. Fract. Mech.* **69** 26–33
[11] Iacoviello F, Di Cocco V and Cavallini M 2015 *Procedia Engineering* **109** 35–42
[12] Zhang Y B, Andriollo T, Fæster S, Liu W, Hattel J and Barabash R I 2016 iron *Acta Mater.* **121** 173–80
[13] Berdin C, Dong M J and Prioul C 2001 *Eng. Fract. Mech.* **68** 1107–17
[14] D'Agostino L, Di Cocco V, Fernandino D O and Iacoviello F 2017 *Procedia Struct. Integr.* **3** 201–7
[15] Andriollo T, Thorborg J and Hattel J 2015 *Proc. 8th Int. Conf. Comput. Plast. - Fundam. Appl. COMPLAS* 632–41

IOP Conf. Series: Materials Science and Engineering **861** (2020) 012039 doi:10.1088/1757-899X/861/1/012039

In situ analysis of cast irons mechanical behaviour using synchrotron x-ray tomography and 3DXRD

T Sjögren[1], S Hall[2], L Elmquist[3], E Dartfeldt[1], E Larsson[1], M Majkut[4], J Elfsberg[5], P Skoglund[5] and J Engqvist[2]

[1] RISE Research Institutes of Sweden, Borås, Sweden
[2] Division of Solid Mechanics, Lund University, Lund, Sweden
[3] Xylem Water Solutions Manufacturing AB, Emmaboda, Sweden
[4] ESRF European Synchrotron Radiation Facility, Grenoble, France
[5] Scania CV AB, Södertälje, Sweden

E-mail: torsten.sjogren@ri.se

Abstract. When subjecting cast irons to mechanical loading the deformation and damage mechanisms occur on a microstructural level and are dependent on the inherent microstructure. A deeper understanding of the relation between the different microstructural constituents and the macroscopic mechanical behaviour would be beneficial in material development efforts and for the ability to design and cast components with tailored properties. Traditionally, microscopy examinations on sectioned cast iron samples have been used when analysing the microstructure in cast irons. Since all microstructural heterogeneity is in three-dimensions (3D), methods that provide a three-dimensional characterisation are essential for a deeper understanding of, both the microstructural features as well as the deformation and damage of cast irons. Therefore, different cast iron grades have been studied using synchrotron X-ray tomography and 3D x-ray diffraction (3DXRD) at ESRF in Grenoble, France. The samples were stepwise loaded and unloaded in-situ at in the tomography/3DXRD set-up to study the deformation with regard to microstructural constituents and the microstructural evolution in 3D. Based on the 3D tomography image sequences, digital volume correlation (DVC) was used for full strain field analysis and for the analysis of damage and deformation mechanisms. In addition, 3DXRD data were analysed to provide details on the lattice parameters and lattice strain of individual ferrite grains. This work shows the possibilities of such synchrotron experiments for advanced study of the mechanical behaviour of cast iron.

1. Introduction

The development of highly sophisticated combustion engines is driven by the desire to improve efficiency and power density as well as to reduce emissions. Such developments will inevitably lead to an increase of the combustion pressure and temperature, which has the consequence of an increased loading on the materials. Today, spheroidal graphite iron (SGI) alloyed with silicon and molybdenum, SiMo-alloys, are commonly used in components such as manifolds in exhaust systems, where materials that can withstand the harsh conditions are needed. The mechanical behaviour of these materials has been studied by others [1-4]. However, the microstructural analysis used to correlate the mechanical behaviour to microstructural properties is performed on polished 2D surfaces taken from the position of final fracture of the specimen. For a detailed understanding of the microstructure and micromechanics of the material in-situ 3D analysis is needed. Also, a better understanding of the mechanical behaviour related to the 3D microstructure is essential to develop an understanding and models that can predict how the material will behave under load. In this work x-ray tomography, together with Digital Volume Correlation (DVC) and 3D x-ray diffraction (3DXRD) during in-situ loading (i.e., loading within the measurement set-up) has been used to identify and characterize the deformation mechanisms and corresponding phase interactions. The results will provide a better understanding of the influence of the different microstructural length scales on the macroscopic mechanical response and explain the observed crack propagation paths. In this paper an overview of possible analyses based on tomography and 3DXRD data are presented.

2. Experimental procedure

The experiment was performed at the European Synchrotron Radiation Facility (ESRF) in Grenoble, France, beamline ID11. For the experiment, tensile test samples were machined from 25 mm thick cast plates of SiMo material. A tensile test stage developed at the division of Solid Mechanics, Lund University, for the specific synchrotron measurement setup was used to apply and control tensile load on the specimen. Samples were monotonically loaded in steps, between which x-ray tomography and 3DXRD measurements were performed with the sample displacement held constant. The x-ray beam had a photon energy of 78.4 keV and a beam size of approximately 1500x1500 µm2. Figure 1 shows the sample geometry and the setup inside the beamline. Force and displacement data were acquired from which the macroscopic stress and strain were derived. However, the deformation measurement includes deformations taking place in all parts of the tensile stage, i.e. strains are not representative of the actual strain of the SiMo material. The 3DXRD data includes nine, 100 µm high slices over the sample height and covering the full gauge diameter width. Analysis of the 3DXRD data using the ImageD11 software (Wright, J., 2005, https://github.com/FABLE-3DXRD/ImageD11/), provides grain-by-grain unit cell parameters from which grain-strains are derived relative to "ideal" unit cell parameters.

Figure 1. Loading stage as mounted in the beamline hutch. The small pictures in the right lower corner show the mounted tensile test specimen and the specimen geometry (total length≈10 mm, gauge diameter≈1.4 mm).

2D and 3D microstructure investigations, using light optical microscopy (LOM) and scanning electron microscopy (SEM), enabled correlation and verification of the different phases observed in the tomography images. Samples for the microscopy analysis were produced by grinding and polishing according to standard procedures. DVC analysis was performed using LaVision's StrainMaster software.

3. Results and discussion

The studied ductile iron is alloyed with silicon and molybdenum for improved properties at high temperatures. The standard denomination of the material is EN-GJS-SiMo5-1 and the chemical composition is given in table 1. The main microstructural constituents of the studied SiMo material are graphite in a ferritic matrix. Due to the high levels of Si and Mo, the microstructure also contains molybdenum carbides and spheroidized pearlite [5].

Table 1. The chemical composition in weight percent of the studied EN-GJS-SiMo5-1. The iron content is implicit.

C	*Si*	*Mn*	*S*	*P*	*Ni*	*Mo*	*Cu*	*Sn*	*Ti*
3.16	4.33	0.41	0.008	0.014	<0.050	0.91	0.073	<0.01	0.017

These constituents are found together with shrinkage defects in regions corresponding to the last to freeze areas. Figures 2 and 3 show the microstructure as seen in LOM and SEM, respectively. In figure 2 the ferrite grains are clearly seen. Similar structures are observed in tomography images, as shown in figure 4, where the light grey/white areas correspond to the molybdenum carbides seen in figures 2-3.

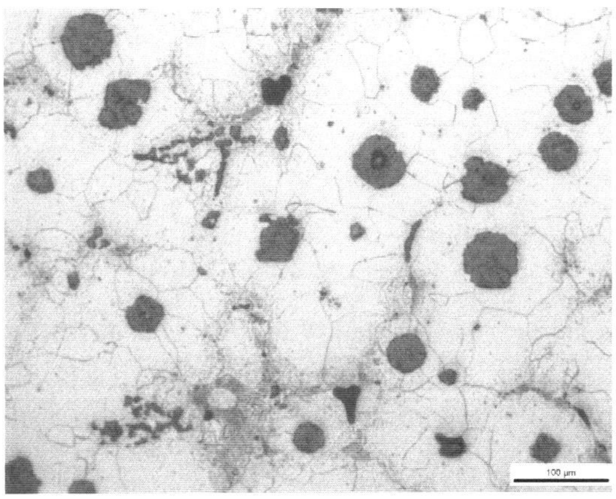

Figure 2. LOM image of the SiMo material.

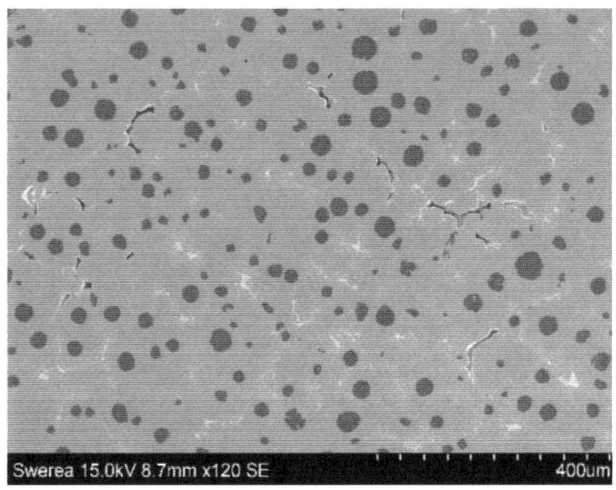

Figure 3. SEM image of the SiMo material.

Through segmentation of the tomography images, using suitable grey level thresholds, it is possible to differentiate between the graphite, carbide and shrinkage porosities as shown in figure 5. In the figure, the graphite phase is shown in light grey, the carbides in light blue and the porosities in green. The volume percentage of each constituent (within the analysed domain) is; 9.4% Graphite, 5.3% Carbide and 0.5%shrinkage porosities. The volume analysed in figure 5 is a cubic sub-volume of extracted from the full image before deformation.

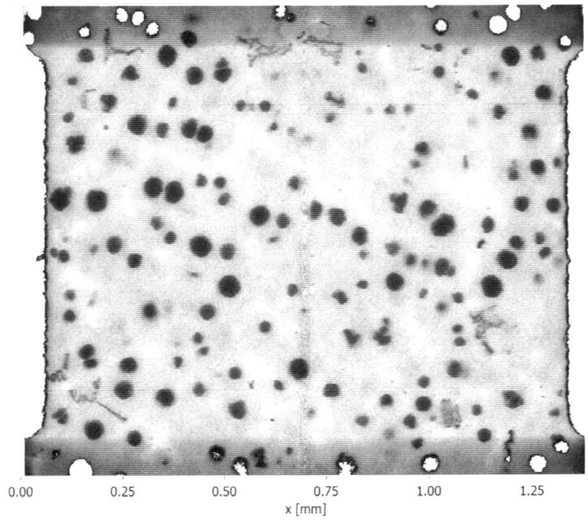

Figure 4. X-ray tomography picture of SiMo sample in unloaded state.

Figure 5. 3D analysis of microstructure.

In the loading experiments, displacement controlled tensile loading was applied, as shown in figure 6. For this specific sample (DCI1) seven load steps were analysed (including the unloaded state) at which tomography and 3DXRD was done, with the displacement held constant, which resulted in some measured relaxation.

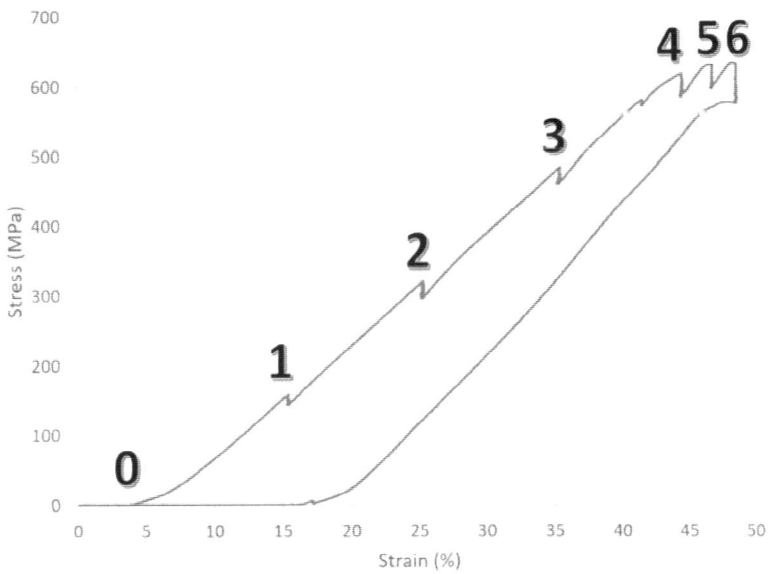

Figure 6. Stress-strain curve for sample DCI1.

By comparing the tomography images acquired at unloaded and loaded states, it is possible to see where in the microstructure that deformation and damage occurred. Comparison of the unloaded state in figure 4 and the deformed state in figure 7 (load step 4), it is observed that cracks occurred in the carbide areas (corresponding to the white areas). It is also observed that debonding occurred at the graphite particles where a gap forms preferentially above and below the graphite particle along the loading axis.

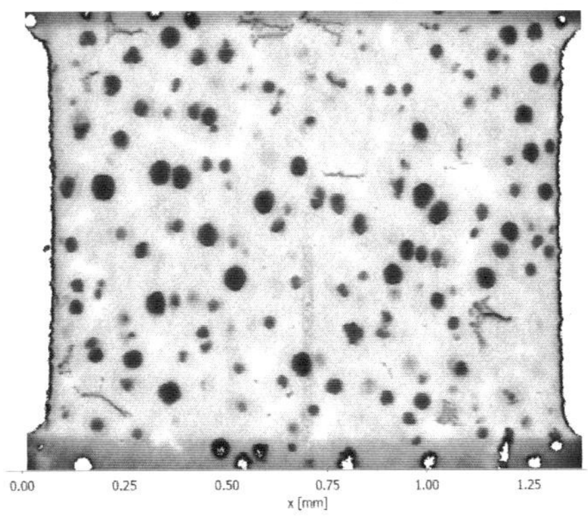

Figure 7. Tomography section at 620 MPa (Step 4).

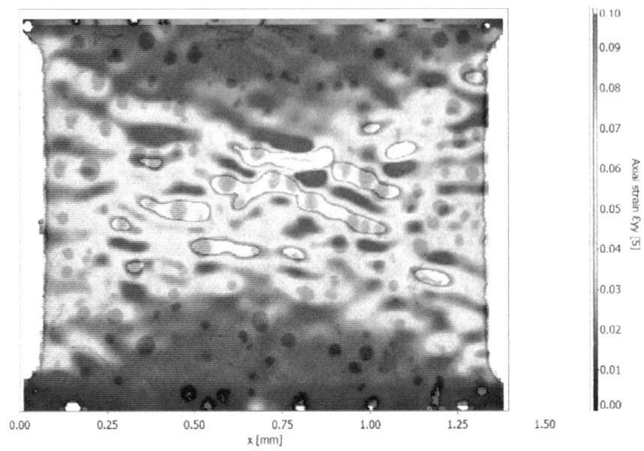

Figure 8. Axial strain DVC data at 620 MPa (Step 4).

DVC analyses of the tomography images allows quantification of the strains that arise within the sample and to further understand the severity of the damages and deformations observed qualitatively in figure 7. Figure 8 shows example DVC results for the SiMo material corresponding to the structure in figure 7 (i.e., the strain field between the unloaded state and step 4). Figure 9 shows a visualisation of the DVC-derived axial strain field. Central planes along the x-, y- and z-axis show the strain distribution. In addition, an ISO-surface, for which the chosen strain is 10%, reveals the structure of the distribution within the material in 3D. In both figure 8 and 9, the strain distribution reveals a shear behaviour that depends on the size and dimension of the sample. It also shows the strains in the vicinity of the discussed cracks and graphite debonding.

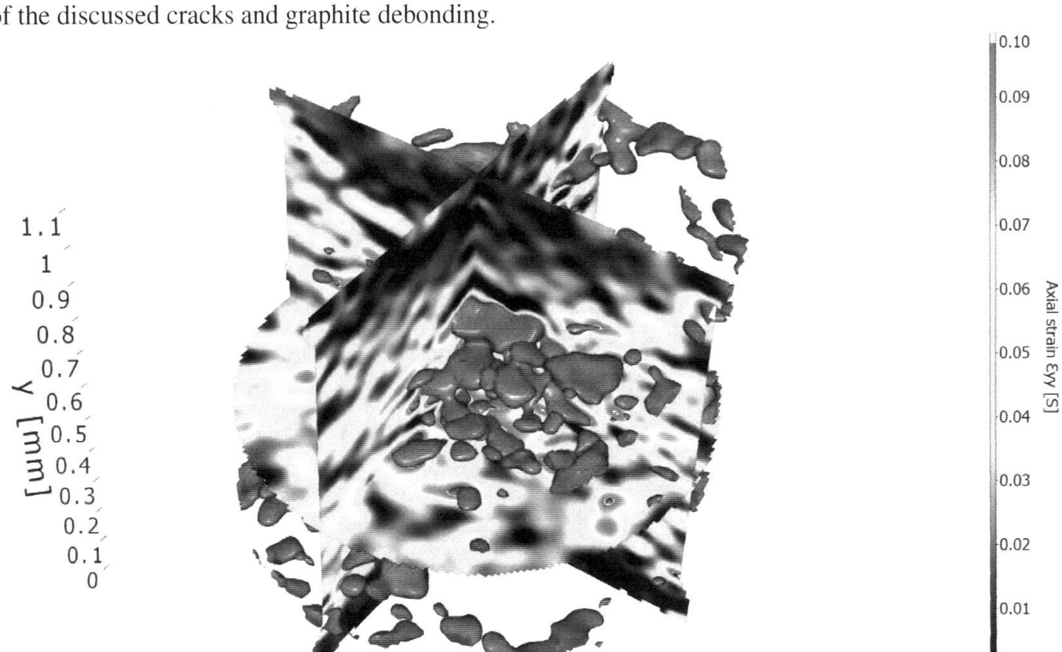

Figure 9. DVC data showing the 3D distribution of axial strain at 620 MPa (Step 4).

Crystallographic unit cell parameters were derived from the 3DXRD data for the individual ferrite grains. The input data for this analysis were 2D diffraction patterns acquired over a 180° rotation of the sample, examples of these are shown in figure 10. Note the smearing of spots at higher loads, indicating an increase in the orientation spread in the grains (inferred as being related to plastic deformation); this leads to reduced number of indexed grains. The number of indexed grains at the centre region drops from 3500 at load step 1 to 250 at load step 4.

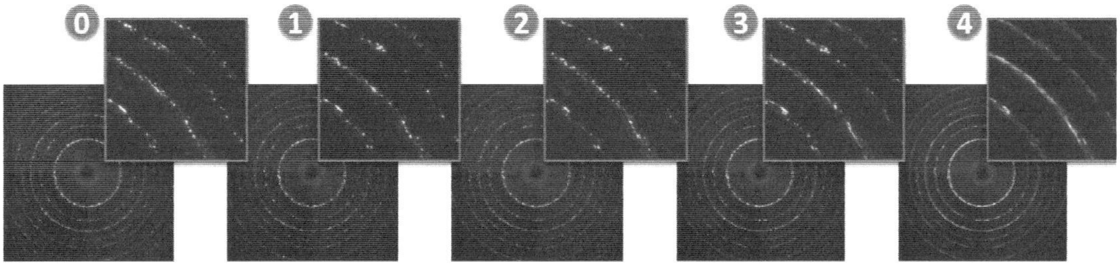

Figure 10. Example diffraction patterns as function of load and position (load steps 0-4). The depicted diffraction patterns are taken from the centre 100-micron slice.

The grain-strains determined from the evolution of the unit cell parameters are shown in figure 11 for the axial (ε11) and horizontal (ε22, ε33) directions. The grain-strains along the loading axis, for the indexed grains, are around 0.2% at load step 4. Since the grains subjected to high strains are not indexed, due to the smearing of the spots, this value reflects only the grains that are elastically deformed.

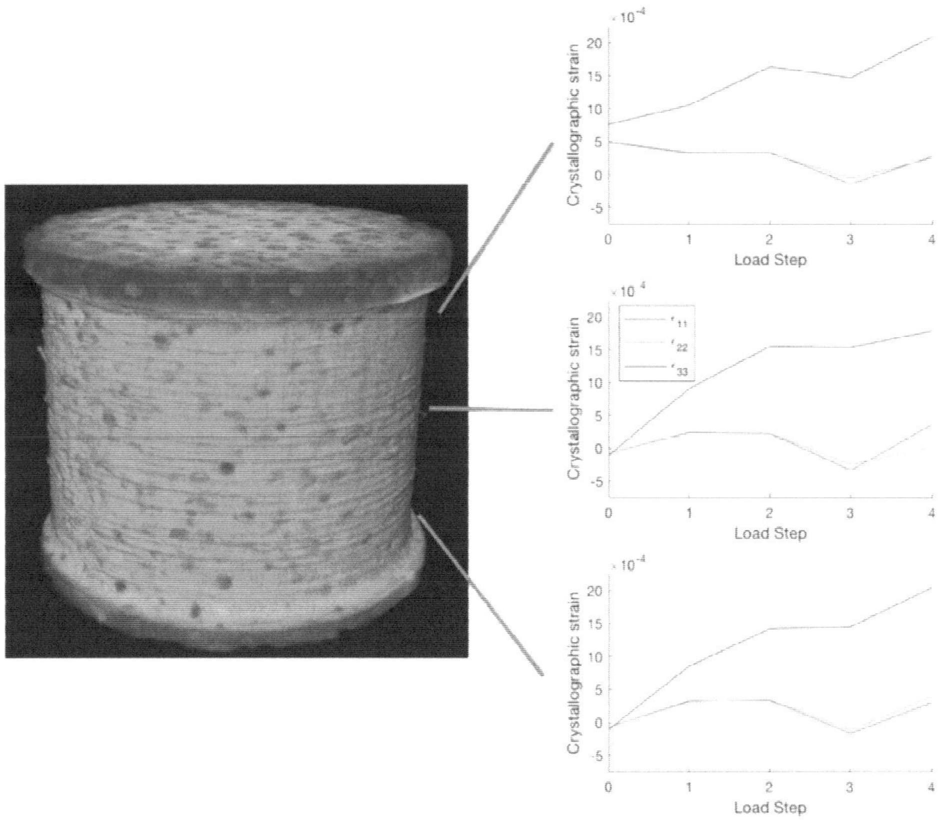

Figure 11. Average grain-strain evolution (load steps 0-4) for the axial (ε_{11}) and horizontal (ε_{22}, ε_{33}) components averaged across 100 μm high slices over the gauge diameter. Data are based on the upper, mid and lower 100-micron slice.

The centre of mass of each ferrite grain was also determined from the 3DXRD data, based on which an estimate of the ferrite grain structure can be made. Results from an initial analysis using an adapted distance transform and watershed algorithm is shown in figure 12. Further work is needed to improve this approach, e.g. to take into account the regions of carbides and porosities, however the grain structure in figure 12 is in good agreement with that in figure 2.

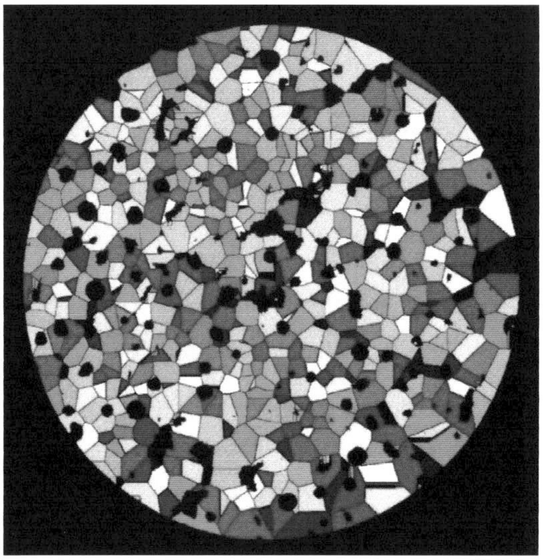

Figure 12. Tesselation of grain centres from 3DXRD overlain on the graphite distribution from the tomography images. The grains are coloured by index

4. Conclusions

An overview of 3D analyses of the microstructure and micromechanics of cast irons using 3D x-ray tomography and 3DXRD has been presented and shows the usefulness of such synchrotron experiments, with in-situ mechanical tests, for the study of microstructural influence on mechanical behaviour in such materials. Combining 3DXRD, x-ray tomography, 3D image analysis and DVC enables analysis of deformation across multiple scales. With full-field tomography, the microstructure can be analysed at high resolution in 3D in combination with 4D evolution. In addition, DVC gives a continuum strain fields based on the tomography images. 3DXRD data provides additional information for modelling (e.g. initial grain positions and parameters) although further work is required to produce grain maps taking into account structural information as well as crystallographic data. For the analysis of texture evolution alternative approaches are needed (high grain deformation).

Acknowledgments

The authors acknowledge European Synchrotron Radiation Facility (ESRF) in Grenoble, France, for hosting the experiment which was facilitated through an application for academic beam time (experiment ref.no. MA3615).

References

[1] Norman V, Skoglund P and Moverare J 2016 Damage evolution in compacted graphite iron during thermomechanical fatigue testing, *Int. J. Cast Metal Res.*, **29** (1-2), 26-33
[2] Norman V, Skoglund P, Leidermark D and Moverare J 2016 The effect of superimposed high-cycle fatigue on thermo-mechanical fatigue in cast iron, *Int. J. Fatigue*, **88**, 121-31
[3] Nadot Y, Mendez J and Ranganathan N 2004 Influence of casting defects on the fatigue limit of nodular cast iron, *Int. J. Fatigue*, **26** (3), 311-19
[4] Hervas I, Thuault A, and Hug E 2015 Damage Analysis of a Ferritic SiMo Ductile Cast Iron Submitted to Tension and Compression Loadings in Temperature *Metals* **5** (4) 2351-69
[5] Ekström M 2013 *Development of a ferritic ductile cast iron for increased life in exhaust applications*, Licentiate thesis, KTH, Stockholm, Sweden

IOP Conf. Series: Materials Science and Engineering **861** (2020) 012040 doi:10.1088/1757-899X/861/1/012040

Modelling macrosegregation modification in dc casting of aluminium alloys in sheet ingots accounting for inlet melt flow, equiaxed grain morphology and transport

A Pakanati[1], M M'Hamdi[1,2], H Combeau[3] and M Založnik[3]

[1] Dept. of Materials Technology, NTNU, N-7491 Trondheim, Norway
[2] SINTEF Materials and Chemistry, N-0314 Oslo, Norway
[3] Université de Lorraine, CNRS, IJL, F-54000 Nancy, France

E-mail: mohammed.mhamdi@sintef.no

Abstract. Macrosegregation is a severe defect present in direct-chill (DC) cast aluminium ingots and billets. In the recent years, experimental studies were conducted to modify and to an extent optimize macrosegregation formation by modifying the inlet melt flow. Due to several limitations, the grain settling behavior and corresponding liquid flow pattern is difficult to analyze using experiments. Simulations on the other hand can provide this insight. However, conducting 2D sheet ingot simulations, as has been previously done, provides an incomplete description of flow pattern. To avoid this and as a first qualitative study, full scale 3D sheet ingot simulation results with two different inlets are presented in this paper. A simplified three-phase multiscale solidification model accounting for solidification shrinkage, natural convection and equiaxed grain growth and transport is used to conduct this study. We show that modification of inlet flow results in modification of grain settling and eventually leading to modification of macrosegregation. The impact of grain morphology is also additionally analyzed.

1. Introduction

The root cause of irregularity in solute content at the scale of the cast product, referred to as macrosegregation, lies in the relative movement between the solid and liquid phases [1]. This relative movement can manifest in many forms in DC casting of aluminium alloys: solidification shrinkage, natural and forced convection, equiaxed grain transport and thermal deformation of the mushy zone. Of these, movement of solute lean equiaxed grains is considered to be a major cause of negative segregation at the center of the ingot [2]. Recent experimental work on Al-4.5wt%Cu sheet ingots by Wagstaff and Allanore [3, 4] focussed on modifying and to an extent eliminating negative segregation by introducing a vertical jet at the center of the ingot. This vertical jet resulted in resuspension of the grains leading to macrosegregation modification. Insight into the complex flow pattern of the solid and liquid phase is not possible using experimental methods and can be gained using numerical simulations. At the same time, sheet ingots have an asymmetric geometry and using 2D simulations, as has been done by several researchers [5–8], provides an incomplete description of the flow phenomena. The flow becomes even more complex with the introduction of the inlet jet. To avoid this problem, 3D simulations of sheet ingots needs to be conducted.

3D simulations are computationally expensive, especially keeping in mind the multiphase, multiscale solidification process. To have realistic simulation time, coarse meshes need to be used. But Al-4.5wt%Cu used by Wagstaff and Allanore [3, 4] has a thin mushy zone and its reasonably accurate numerical resolution might increase the computational cost to prohibitive level. Pakanati *et al.* [8]

recently conducted numerical study and showed that Al-4.5wt%Cu and Al-8.375wt%Zn show similar macrosegregation behaviour in 2D sheet ingots. Al-8.375wt%Zn has a thick mushy zone and its accurate resolution is possible with coarse mesh. Taking all these aspects into account, in the current paper we qualitatively study the impact of inlet flow on macrosegregation formation and modification for Al-8.375wt%Zn in a sheet ingot. We modify the inlet flow by considering a simplified open inlet and an inlet with a vertical jet. Consequently we analyze the interaction of the inlet melt flow with equiaxed grains resulting in their resuspension and macrosegregation modification. Additional influence of the grain morphology on macrosegregation formation is also discussed. This study is conducted using the simplified three-phase, multiscale solidification model proposed by Tveito *et al.* [9].

2. Numerical model

The simplified three-phase, multiscale numerical model of equiaxed solidification, proposed by Tveito *et al.* [9], is based on the volume-averaging approach of Wang and Beckermann [10] and employs the operator splitting algorithm proposed by Založnik and Combeau [11]. For a detailed description of the model the reader is referred to these papers. Only the main features are described here. The Euler-Euler volume averaged model has two parts – macroscopic transport and microscopic growth. The macroscopic transport accounts for solute and heat transfer coupled with liquid flow induced by solidification shrinkage and thermosolutal convection. The densities of liquid and solid are assumed to be constant but different and the Bousinessq approximation is used for the liquid density in the buoyancy term accounting for thermal and solutal effects. For the solid phase, the flow pattern depends on envelope fraction (g_{env}). For envelope fractions smaller than the packing fraction (g_{pack} , a model parameter) the solid (equiaxed) grains are freely floating. The interfacial drag term C_D is modeled in the same manner as in [12] for spherical particles but by considering the envelope fraction instead of the solid fraction. The measure of grain morphology can be obtained by taking the ratio of the solid fraction and the envelope fraction, called as the internal solid fraction: $g_{intern} = g_s/g_{env}$. The grain is globular as the g_{intern} approaches 1 and is dendritic if $g_{intern} \ll 1$. For envelope fractions greater than the packing fraction, grains are assumed to form a rigid porous solid matrix moving with the casting velocity, $\vec{V}_{cast}$. The interfacial drag in the porous medium is modeled by a Darcy term, where the hydrodynamic permeability is calculated by the Kozeny Carman relation, using a the characteristic size of the porous structure, l_{KC}.

The microscopic part is treated locally within each volume element. The model accounts for finite diffusion in both solid and liquid phases and local thermal equilibrium is assumed. Nucleation of grains is assumed to occur on grain-refiner (inoculant) particles. According to the athermal nucleation theory of Greer *et al.* [13], the critical undercooling for free growth of a grain on an inoculant particle of diameter *d* is given by $\Delta T_c = 4\Gamma_{GT}/d$ where Γ_{GT} is the Gibbs-Thompson coefficient. The number of activated particles then depends on the size distribution of the particle population, which can be represented by an exponential distribution density function This representation holds for the largest particles, which are activated at small undercoolings and therefore successful as nuclei. This size distribution is then discretized into *m* classes of inoculants. Each class *i* is represented by a volumetric population density, N_{nuc}^i, and a critical undercooling, ΔT_c^i . When the local undercooling reaches the critical undercooling of class *i*, its local inoculant density, N_{nuc}^i, is instantaneously added to the grain density, N_g, and N_{nuc}^i becomes locally zero.

3. Problem description

The current study is conducted on the DC cast ingot used in [7]. The ingot geometry with dimensions is illustrated in figure 1(a). We can reduce the computational domain to a quarter of the ingot (represented in red) due to summetry. We consider two types of inlet: a simplified open inlet where liquid metal enters the domain from the top surface (figure 1(b)) and an inlet with a vertical jet directed towards the center of the ingot (figure 1(c)). The jet is constructed by reducing the inlet to a 30 mm x 30 mm orifice (marked in black in figure 1(c)). From here on, the simplified open inlet is referred to as Inlet 1 and the inlet with vertical jet is referred to was Inlet 2. The liquid metal is assumed to enter the domain at casting temperature T_{cast}, nominal solute concentration C_o and inoculant density N_{nuc}^i. The inlet velocity is calculated from the global mass balance. The solidified metal is assumed to leave the domain from the bottom with a predefined casting velocity $\vec{V}_{cast}$ of 60 mm/min. The heat is extracted by primary cooling through the mold and by secondary cooling directly to the falling water film flowing over the ingot surface. Primary cooling consists of three zones: meniscus, mold, and air gap. The boundary conditions are specified in [7]. The heat transfer coefficient due to secondary cooling is modelled using Weckmann and Niessen [14]. A binary Al-8.375%Zn alloy is used, similar to Založnik *et al.* [7] emulating the AA7449 alloy. The thermal and physical properties (including packing fraction 0.3) are also obtained from [7]. The inoculation data is taken from [8].

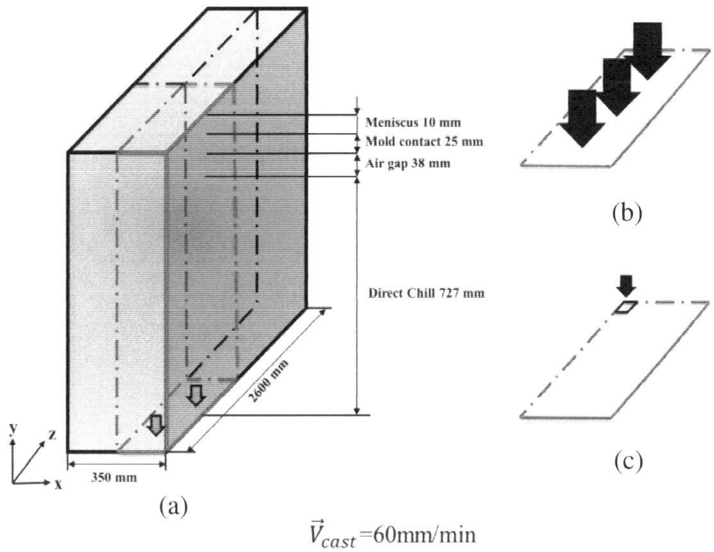

Figure 1. (a) The full isometric view of the sheet ingot. Symmetry axes is represented by dashed lines. The quarter of ingot marked in red is the simulation domain. (b) Simplified open inlet (Inlet 1) and (c) inlet with vertical jet at the center of the ingot (Inlet 2).

The transport equations are solved with a Finite Volume Method and the SIMPLE-algorithm for staggered grid is used for pressure-velocity coupling. The convective terms are discretized with a first-order upwind scheme and for time discretization a fully implicit first-order scheme is used. For all simulations a structured grid of 524,288 cells ($N_x \times N_y \times N_z$ =32x128x128) is employed. A constant time step of 5 ms is used and the calculations are run until steadty state is reached.

4. Results and discussion

All studied cases are summarized in table 1. A total of four cases (1-4) based on the inlet mechanism and grain growth model are considered. Globular morphology is imposed ($g_{env} = g_s$) for Case 1 and Case 3. Grain morphology is simulated for Case 2 and 4. All the cases have the same transport mechanisms: shrinkage induced flow, natural convection and equiaxed grain motion. Case 3 and Case 4 additionally have forced convection due to the inlet jet.

The macrosegregation contour along the horizontal x-z plane for all the cases are shown in figure 2. The main features of the contours are discussed starting with Case 1. Negative segregation along the ingot width in z-direction is due to settling of solute lean grains and shrinkage induced flow. However, the formation of concentrated negative segregation region (marked in red) can be explained by taking a look at figure 3. An isosurface of packing fraction is overplotted with solid grains trajectory along the inclined mushy zone in the x-y plane (black arrows) and z-y plane (white arrow). The region marked in red in figure 2 for Case 1 has grains arriving from both the x and z direction. For the other parts of the ingot, the grain settling is predominantly along the x-direction. This additional contribution results in more grains settling in that region, leading to more pronounced negative segregation. This can be further corroborated by examining figure 4, where the grain densities along the x-z plane are plotted for Case 1 and 3. For Case 1, the region marked in red shows high grain density, which corresponds to the negative segregation zone marked in red for Case 1 in figure 2.

Table 1. Simulation Cases.

Description	Case
Inlet 1 – Globular Grain Growth Model	1
Inlet 1 – Dendritic Grain Growth Model	2
Inlet 2 – Globular Grain Growth Model	3
Inlet 2 – Dendritic Grain Growth Model	4

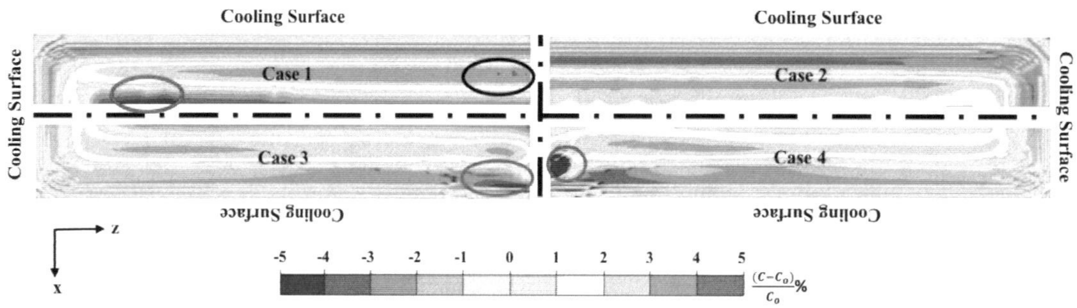

Figure 2. Relative macrosegregation contour plots along the horizontal x-z plane for all the cases. Only quarter ingot plots for each case are shown and the cooling sides are marked. The central axes are also marked.

The region marked in black for Case 1 in figure 2 shows more pronounced positive segregation. In fact, if we observe along the ingot width, the intensity of positive segregation gradually increases and peaks at the center of the ingot. This asymmetric positive segregation is due to the combination of the thermal-solutal convection and grain motion and can be explained using the illustration in figure 5. A 2D sheet ingot with packing front and relative liquid velocity flow loop in the x-y plane is shown in figure 5(a). Due to the grain settling towards the center of the ingot, solute rich liquid is expulsed up and enriches the slurry region. The solute rich liquid is carried into the liquid pool and towards the mid-section of the ingot resulting in positive segregation [7]. In 3D, apart from the flow loop in the x-y plane (light grey to black), there is contribution from the z-y plane (red) and are illustrated with the packing front isosurface in figure 5(b) and figure 5(c), respectively. Similar to the 2D model, grain settling along the x-y plane expulses solute rich liquid up. The additional driving force along the z-y plane pushes this solute rich liquid towards the ingot center. Hence, the extent of enrichment varies along the z-direction. At the position of the light grey loop in figure 5(b), lower amount of solute enrichment can be observed. As we move further and reach the position where the black flow loop is marked, more enriched solute is available due to the driving force from z-y plane. This results in the enriched region marked in black for Case 1 in figure 2. The isometric view of the quarter ingot is shown in figure 6. The flow vectors for relative liquid velocity in black and solid settling velocity in red are marked. Due to the uniform open inlet, the inlet flow has little to no influence on the grain settling. This can be observed from the consistent settling of solid grains along the ingot width.

For Case 2, we have dendritic grains and the macrosegregation contour along the horizontal plane is shown in figure 2. As mentioned in the previous section, due to dendritic nature of the grains, slightly positive segregation is observed along the ingot width. Similar to Case 1, we also see solute accumulation towards the ingot center for Case 2. The mechanisms through which macrosegregation formation remains the same for Case 1 and 2 but the intensity of segregation is different to due the presence of dendritic grains. This indicates that for a given inlet, macrosegregation depends on the morphology of the equiaxed grains, an inference previously reported in [9, 15].

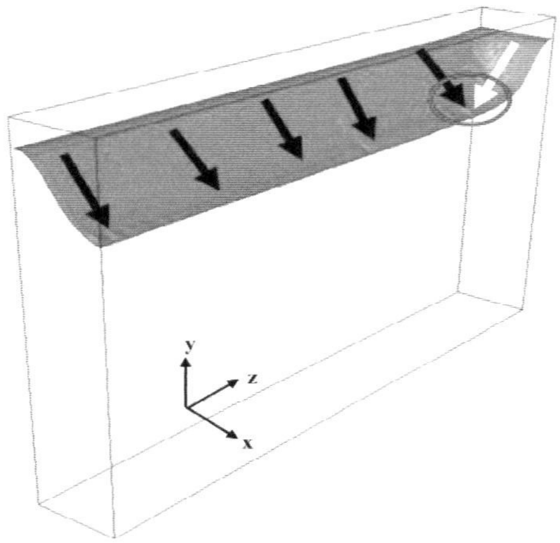

Figure 3. Illustration of Case 1 isosurface of solid fraction at the packing fraction. The black and white arrows indicate the solid grains trajectory in the x-y plane and z-y plane.

With the introduction of inlet jet in Case 3, macrosegregation is modified when compared with Case 1 for the same globular morphology. The inlet jet is known to create a flow loop which could potentially wash away solute along the inclined mushy zone. In figure 2, we can see this clearly in the region marked in green. In addition to this, the asymmetric negative and positive segregation observed in Case 1 is not present and an almost uniform distribution of segregation can be observed. The inlet flow results in redistribution of the grains and this can be seen in figure 4 for Case 3 and when we compare with Case 1, near uniform grain density can be seen along the ingot width. The interaction of the inlet flow with grain settling can be seen in the isometric view of the quarter ingot in figure 7. Similar to Case 1, the relative liquid velocity vectors are marked in black and the solid settling velocity vectors are marked red. The inlet jet ploughs through the mushy zone in the central part of the ingot, creating a small crater. In this region, there is no grain motion (unlike Case 1 in figure 6) as the inlet jet acts like a barrier and resuspends the grains away from the center of the ingot. This is the central theme of the research done by Wagstaff and Allanore [3, 4, 18]. The pushing of the grains from the center of the ingot can be clearly seen by the solid velocity vectors (red) in figure 7. Especially along the z-y plane, the grains are moving away from the center. Along the x-y plane, some degree of resuspension is seen as a fraction of grains have managed to override the inlet jet and settle towards the center. This is a consequence of having a relatively weak inlet jet. Stronger inlet jets can cause stronger barriers resulting in complete resuspension of the grains. However, the important aspect to be noted is the recirculation of the grains due to the inlet jet and the corresponding modification of macrosegregation.

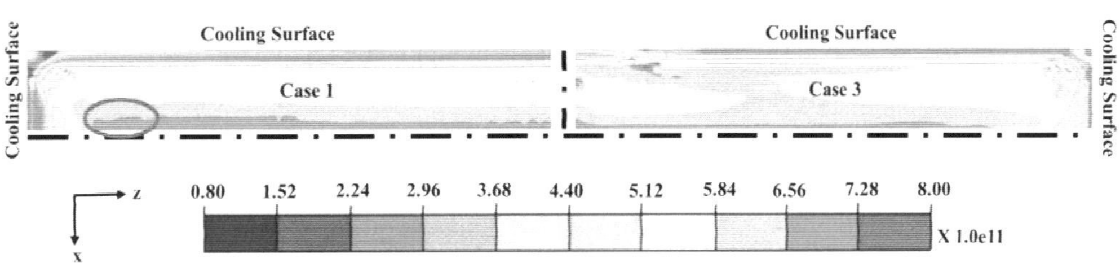

Figure 4. Grain density contours on the horizontal x-z plane for Case 1 and Case 3.

Solving of grain morphology and inlet jet requires proper treatment of the mushy zone due to the strong gradients in the velocity and envelope fraction along the inclined mushy zone. Since this work employs a relatively coarser mesh, it is possible that the island of negative segregation for Case 4, marked in green in figure 2, is caused due to numerical issues. However, solute wash up due to strong forced convection is nothing new in DC cast process [18]. The extent of solute depletion can depend on the morphology of the grain, as discussed in the previous section. Compared to Case 2 with the same dendritic morphology (g_{intern}~0.5), macrosegregation in Case 4 is different due to the presence of the inlet jet. Compared with Case 3 with the same inlet jet, macrosegregation in Case 4 is again different due to the presence of dendiritcg grains in the latter case. This comparison establishes that inlet jet can modify macrosegregation by resuspension and recirculation of the grains. The extent of modification further depends on the morphology of the grain: globular or dendritic.

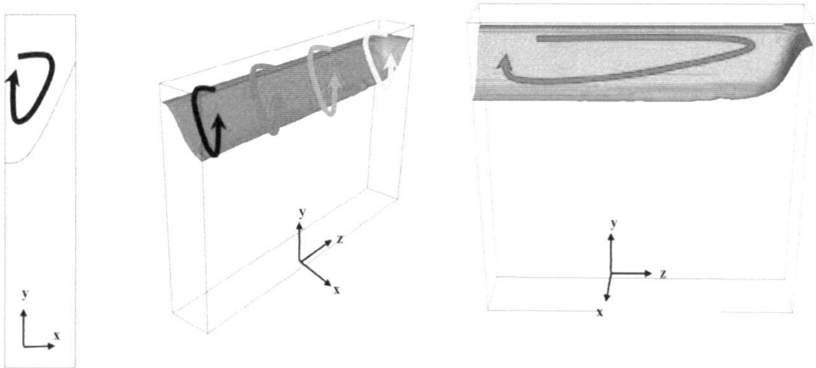

(a) (b) (c)

Figure 5. Illustration of the relative liquid velocity for (a) a 2D sheet ingot model and (b) & (c) a quarter of the 3D sheet ingot model.

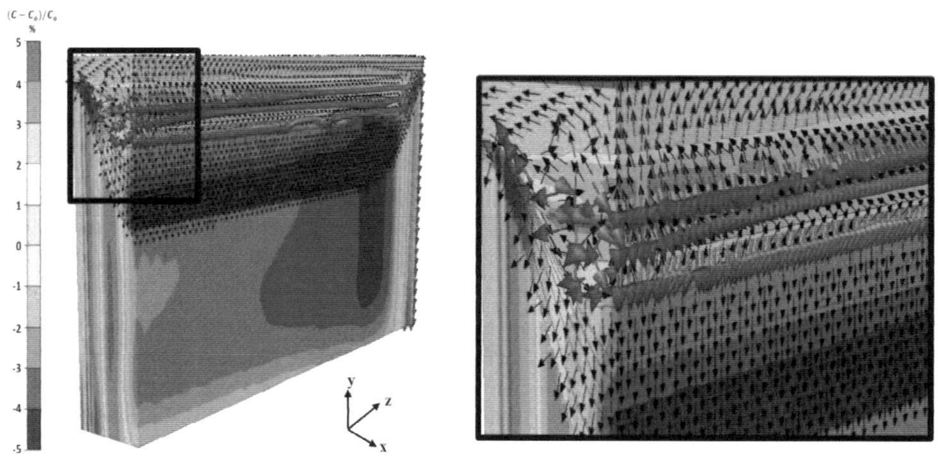

(a) (b)

Figure 6. Isometric view of Case 1 plotted with solid fraction isolines, relative macrosegregation contours, relative liquid velocity vectors in black and solid settling velocities in red. (a) Full quarter ingot and the region marked in black square is enlarged in (b).

As previously mentioned, this study is to qualitatively complement the work done by Wagstaff and Allanore [3,4] where an Al-4.5wt%Cu alloy is used. Owing to numerical issues, this study is conducted using Al-8.375wt%Zn alloy which has a thick mushy zone. However, Pakanati *et al.* [8] indicated that both the alloys behave similarly when it comes to the interplay of various transport mechanisms. A direct comparison of these simulation cases with experiment is not possible as the ingot size and casting conditions are different. Nonetheless, this study provides a first qualitative study in this direction.

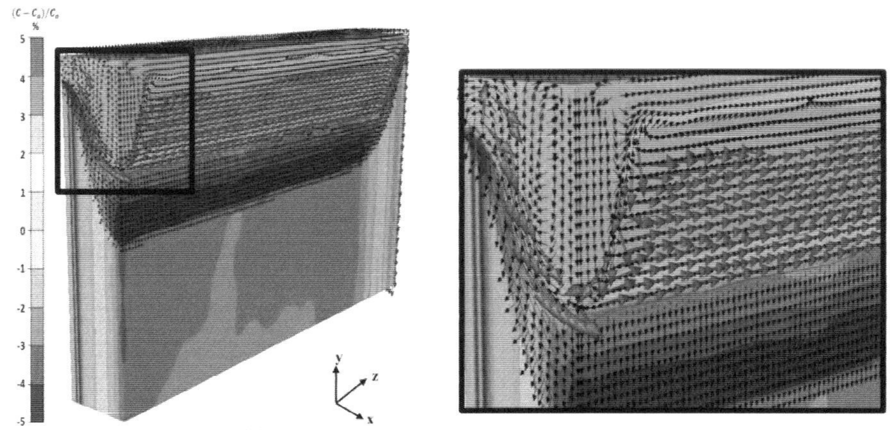

(a) (b)

Figure 7. Isometric view of Case 3 plotted with solid fraction isolines, relative macrosegregation contours, relative liquid velocity vectors in black and solid settling velocities in red. (a) Full quarter ingot and the region marked in black square is enlarged in (b).

5. Conclusions

Modification of macrosegregation in DC casting by modification of the inlet melt flow has been analyzed in the current study. By using a simplified three-phase multi-scale solidification model, we have also analyzed the extent of the modification by accounting for both globular and dendritic grain morphology. The interaction of the inlet melt flow and the solid grains has been discussed by using visual representation of the flow, something that is not possible in experimental studies. The current model is sensitive enough to study macrosegregation modification and maybe optimization for future studies. Since this is a qualitative study, focus should also be brought on complementing the experimental work done on macrosegregation modification. Direct comparison with experiment is feasible only if we are able to keep the error induced due to mesh resolution, especially in the mushy zone, to minimum. Constructing uniform fine grids is an option but for 3D cases, it can result in prohibitive computational time. It can be avoided by using mesh adaptation to locally resolve the mushy zone. This could be the focus of future work.

Acknowledgement

This work is conducted within the framework of PRIMAL project of which Hydro Aluminium ASA, Alcoa Norway ANS, Aleris Rolled Products Germany GmbH, Institute of Energy Technology (IFE), NTNU and SINTEF are the partners. This project is supported by the Research Council of Norway. AP and MM acknowledge the support of NOTUR High Perfomance Computing program.

References

[1] Nadella R, Eskin D G, Du Q and Katgerman L 2008 Macrosegregation in direct-chill casting of aluminium alloys *Prog. Mater. Sci.* **53** 421–80

[2] Yu H and Granger D A 1986 Macrosegregation in aluminum alloy ingot cast by the semicontinuous direct chill method *Aluminium Alloys: Their Phyiscal and Mechanical Properties* (Sheffield, UK) pp 17–29

[3] Wagstaff S R and Allanore A 2016 Modification of Macrosegregation Patterns in Rolling Slab Ingots by Bulk Grain Migration *Williams E. (eds) Light Metals 2016* pp 715–9

[4] Wagstaff S R and Allanore A 2016 Minimization of Macrosegregation in DC Cast Ingots Through Jet Processing *Metall. Mater. Trans. B* **47** 3132–8

[5] Jalanti T, Swierkosz M, Gremaud M and Rappaz M 2000 Modelling of macrosegregation in continuous casting of aluminium *DGM Conf.* 1–7

[6] Bedel M 2014 Étude de la formation des structures de solidification et des macroségrégations en coulée semi-continue d'aluminium (PhD Thesis Université de Lorraine, Nancy,France)

[7] Založnik M, Kumar A, Combeau H, Bedel M, Jarry P and Waz E 2011 Influence of Transport

Mechanisms on Macrosegregation Formation in Direct Chill Cast Industrial Scale Aluminum Alloy Ingots *Adv. Eng. Mater.* **13** 570–80
[8] Pakanati A, M'Hamdi M, Combeau H and Založnik M 2018 Investigation of Macrosegregation Formation in Aluminium DC Casting for Different Alloy Systems *Metall. Mater. Trans. A*
[9] Tveito K O, Pakanati A, M'Hamdi M, Combeau H and Založnik M 2018 A Simplified Three-Phase Model of Equiaxed Solidification for the Prediction of Microstructure and Macrosegregation in Castings *Metall. Mater. Trans. A* **49** 2778–94
[10] Wang C Y and Beckermann C 1996 Equiaxed dentritic solidification with convection: Part I. multiscale/multiphase modeling *Metall. Mater. Trans. A* **27A** 2754–64
[11] Založnik M and Combeau H 2010 An operator splitting scheme for coupling macroscopic transport and grain growth in a two-phase multiscale solidification model: Part I - Model and solution scheme *Comput. Mater. Sci.* **48** 1–10
[12] Založnik M, Kumar A and Combeau H 2010 An operator splitting scheme for coupling macroscopic transport and grain growth in a two-phase multiscale solidification model: Part II – Application of the model *Comput. Mater. Sci.* **48** 1–10
[13] Greer A L, Bunn A M, Tronche A, Evans P V. and Bristow D J 2000 Modelling of inoculation of metallic melts: application to grain refinement of aluminium by Al-Ti-B *Acta Mater.* **48** 2823–35
[14] Weckman D and Niessen P 1982 A Numerical Simulation of the D.C. Continuous Casting Process Including Nucleate Boiling Heat Transfer *Metall. Trans. B* **13** 593–602
[15] Heyvaert L, Bedel M, Založnik M and Combeau H 2017 Modeling of the Coupling of Microstructure and Macrosegregation in a Direct Chill Cast Al-Cu Billet *Metall. Mater. Trans. A* **48** 4713–34
[16] Wagstaff S R and Allanore A 2016 Centerline Depletion in Direct-Chill Cast Aluminum Alloys: The Avalanche Effect and Its Consequence for Turbulent Jet Casting *Metall. Mater. Trans. B* **47** 3139–43
[17] Wagstaff S R and Allanore A 2017 Jet Mixing in Direct-Chill Casting of Aluminum: Crater Effects and its Consequence on Centerline Segregation *Metall. Mater. Trans. B* **48** 2114–22
[18] Zhang L, Eskin D G, Miroux A, Subroto T and Katgerman L 2012 Effect of Inlet Geometry on Macrosegregation During the Direct Chill Casting of 7050 Alloy Billets: Experiments and Computer Modelling *IOP Conf. Ser. Mater. Sci. Eng.* **33** 1–8

Numerical investigation of an in-situ observed flow regimes during solidification of an NH_4Cl – 70 wt% H_2O solution

M Wu, M Stefan-Kharicha, A Kharicha and A Ludwig

Chair for Simulation and Modelling of Metallurgical Processes, Dept. of Metallurgy, University of Leoben, A-8700 Leoben, Austria

E-mail: menghuai.wu@unileoben.ac.at

Abstract. A volume-average mixed columnar-equiaxed solidification model, where the dendritic morphology of crystals is described with the dendrite envelope, was developed by the current authors. As an evaluation effort, this model is used to simulate a solidification experiment of ammonium chloride water (NH_4Cl – 70 wt%H_2O) solution. The solution was cooled in a water cooled copper mould (10×8×1 cm^3), and the solidification process and the flow dynamics during solidification were in-situ observed/recorded with a Particle Imagine Velocimetry technique. The striking features of solidification in this experiment are that the crystal morphology is dominant in columnar structure, and typical 4 flow regimes in the bulk solution during solidification appear: pure thermal convection (TH), unstable (turbulent) flow caused by double diffusive convection (TU), turbulent-stratified flow (TU-ST), meandering flow (MF). These flow regimes correspond to different cooling/solidification stages, and reflect their interactions with the developing mushy zone. The calculated solidification sequence and the flow dynamics (4 flow regimes) agree with the experimental ones. The experimentally observed phenomena were successfully explained by the simulation.

1. Introduction

Volume-average based multiphase approach was introduced by Beckermann and co-workers to model the solidification problem by bridging the transport phenomena at different length scales [1-2]. A concept of "dendrite envelope", as introduced by Rappaz and Thevoz [3-4], was also incorporated into the volume-average approach by Wang and Beckermann to consider the dendritic crystal morphology during solidification [5]. Recently, this work was extended by Wu and Ludwig to consider the mixed columnar-equiaxed solidification in a five-phase solidification model [6-7]. This model was evaluated by comparison with in-situ solidification experiments on organic materials [8] and also on metal castings [9-10]. A primary goal of this conference paper is to further evaluate the model by comparison with solidification experiment of ammonium chloride water (NH4Cl – 70 wt%H2O) solution. A more ambitious goal is to explain the experimentally observed flow phenomena, by correlating each of them with the solidification structure in the advancing mushy zone, hence to achieve deeper understanding of the flow-solidification interaction.

2. Experimental setup and numerical model

2.1. Experimental setup and material data

An in-situ experiment on solidification of the NH_4Cl – 70 wt%H_2O solution was made [11]. The solution was cooled in a mould (10×8×1 cm^3) from an initial temperature 314.15 K. The mould top is open to the air (heat transfer coefficient 12 W/m2/K); the side/bottom walls are water-cooled with imposed temperature ($T_{\rm Wall} = 2.69\times10^{-6}t^2 - 2.488\times10^{-2}t + 314.15$) until reaching the lowest limit of 276.6 K;

the front/back walls are glass (adiabatic). One key solidification feature with this experiment setting is the columnar crystal morphology. A Particle Imagine Velocimetry (PIV) technique was used to record the flow dynamics. The columnar tip front was recorded by a light camera, whose image (darkness) indicated the solid volume fraction in the mushy zone. Material properties are listed in table 1.

Table 1. Material properties and other parameters

	Symbol	Units	Value	Sources
Thermodynamic parameter (local linearized phase diagram)				
Nominal concentration	c_0	wt% H_2O	70.43	[12]
Liquidus temperature	T_{liq}	K	304.15	[12]
Melting point of solvent	T_f	K	638.7	extrapolation
Eutectic composition	c_{eu}	wt%	0.803	[12]
Eutectic temperature	T_{eu}	K	257.267	[12]
Liquidus slope	m	K $(wt\%)^{-1}$	-4.75	[12]
Equilibrium partition coefficient	k	-	0.0	[12]
Gibbs Thomson coefficient	Γ	K.m	4.0×10^{-8}	[13]
Thermo-physical properties				
Reference density	ρ_ℓ , ρ_s	kg·m^{-3}	1073	[14]
Specific heat	c_p^ℓ , c_p^s	J kg^{-1} K^{-1}	3249	[15]
Thermal conductivity of liquid	k_ℓ	W m^{-1} K^{-1}	0.468	[15]
Thermal conductivity of solid	k_s	W m^{-1} K^{-1}	2.7	[15]
Latent heat	Δh_f	J kg^{-1}	3.18×10^5	[15]
Viscosity	μ_ℓ	Kg m^{-1} s^{-1}	1.3×10^{-3}	[15]
Diffusion coefficient (solid)	D_s	m^2 s^{-1}	8.0×10^{-13}	estimation
Diffusion coefficient (liquid)	D_ℓ	m^2 s^{-1}	4.8×10^{-9}	[15]
Boussinesq parameters				
Liquid thermal expansion coefficient	β_T	K^{-1}	3.9×10^{-4}	[14]
Liquid solutal expansion coefficient	β_C	wt%$^{-1}$	2.87×10^{-3}	[14]
Dendrite morphology				
Primary dendritic arm spacing	λ_1	m	6.0×10^{-4}	[11]
Secondary dendritic arm spacing	λ_2	m	1.0×10^{-4}	[11]

2.2. Brief model description

The mixed columnar-equiaxed solidification model as developed by the current author has considered five phases [6-7]. As dominant columnar dendritic structure is observed in this experiment, a simplified version with pure columnar growth structure is applied. Two hydrodynamic phases are considered: the primary liquid melt (ℓ) and columnar phase (c), and their volume fractions sum up to unit ($f_\ell + f_c = 1.0$). Only one set of Navier-Stokes equations are solved for the velocity of the primary liquid melt ($\vec{u}_\ell$); the velocity of the columnar phase ($\vec{u}_c$) is set as zero (stationary). The flow permeability (K) of the mushy zone, derived from the Carman-Kozeny law, refers to Ramirez *et al.* [16]; and it is a function of f_ℓ and λ_1.

As shown in figure 1, a smooth hypothetical contour connecting primary (or primary, secondary and tertiary) dendrite tips, called dendrite envelope, is constructed to separate the inter-dendritic melt and primary (extra-dendritic) melt. With this concept, the system includes three thermodynamic phases: the solid dendrites (f_s^c), the inter-dendritic melt inside the envelope (f_d^c), and the extra-dendritic melt (identical to the primary melt f_ℓ), where $f_s^c + f_d^c = f_c$. The shape of the columnar dendrite envelope is further simplified as a cylinder with cross-section-area equivalent to the columnar tree trunk envelope (figure 1). Two morphological parameters are necessarily defined to describe the geometry of the dendrite envelope: the shape factor Φ_{env}^c and circularity factor Φ_{circ}^c. Assuming that the growth kinetics of the dendrite tips follows the LGK model [17], where the stability constant is $\sigma^* = 1/(4\pi^2)$, from the tip growth velocities ($v_{tip^*}^c$) one can calculate the lateral growth velocity of the cross-section-area equivalent cylinder ($v_{env}^c = \Phi_{env}^c \cdot v_{tip^*}^c$). Φ_{circ}^c is defined to relate the true diffusion area of the dendrite envelope with the area of the equivalent cylinder, $S_{env,D}^c = S_{env,M}^c / \Phi_{circ}^c$. Both shape factors are here assumed and further verified for this experiment per numerical parameter studies: Φ_{env}^c =0.7979, Φ_{circ}^c =0.33. The columnar primary dendrite tip front is tracked according to the growth kinetics of LGK model [6-7, 17].

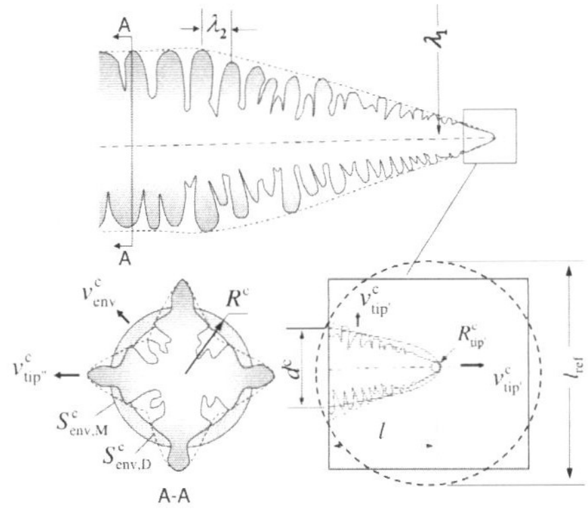

Figure 1. The shape of the columnar tree trunk envelope is simplified as a cylinder with cross-section-area equivalent to the columnar tree trunk envelope (dashed line), which connects the secondary and tertiary dendrite tips. The contour of the columnar envelope near the primary dendrite tip (dashed line connecting the primary and secondary dendrite tips) is simplified as a paraboloid.

3. Simulation – experiment comparison

The simulation was made in full 3D with average mesh size of 0.7 mm. The calculated solidification sequence and flow patterns, in comparison with the experimental results, are shown in figure 2. According to the experimental report [11], 4 typical flow patterns corresponding different cooling/solidification stages are observed. They are successfully reproduced by the numerical simulation, despite of the slight difference in the timing of the starting point for each pattern/stage. The applied thermal boundary conditions for the simulation are not perfectly in light with the experiment. In order to explain the mechanism of each flow pattern, some key solidification quantities, which mostly influence the flow, are shown in figure 3. They are the temperature (T_ℓ), solute (H_2O) concentration in the melt (c_ℓ in mass fraction), density of the melt (ρ_ℓ), volume fraction of the solid dendrite (f_s^c). The density of the melt is calculated as $\rho_\ell = \rho_{ref}\left(1-\beta_T\left(T_\ell - T_{ref}\right)-\beta_c\left(c_\ell - c_{ref}\right)\right)$.

During the early cooling stage before the start of solidification, figure 2(a), it is the thermal convection regime (TH). The flow is laminar and symmetrical: downward along the side wall and upward in the mould centre. Note that the PIV technique is not able to detect the velocity near the wall, hence the largest velocity along the side walls as predicted by the simulation was missing in the experiment.

As solidification starts, NH_4Cl crystals grow from the side/bottom walls. The mushy zone is very porous, and the maximum f_s^c in the mush is about 0.12. The rejection of solute (H_2O) in the mushy region leads to the "density reversion" [18], i.e. the melt density along the side wall becomes lighter than that in the bulk region. To the moment of 900 seconds, as shown in figure 2(b), the flow direction along the side wall is reversed, namely upwards. As shown in figure 3(a1)-(a4), temperature along the side wall is lower, while the solute concentration is higher. Balancing the thermal and solutal effect, the melt density there is smaller than the bulk region, leading to the upward melt flow, which brings less-denser melt from lower region along the wall (through porous mushy region) to the top region. The solute rejection from the mushy zone along the bottom wall tends to rise as well, driving many plumes (or freckle trails [18]). A typical double diffusive convection pattern occurs. To this moment the flow pattern becomes chaotic. This flow regime is called turbulent regime (TU). The previous study demonstrated that the Rayleigh number due to solute-buoyancy could reach the order of magnitude of 10^{10}, larger than the turbulence criterion 10^9, but most regions were observed as laminar with very large and coherent structures.

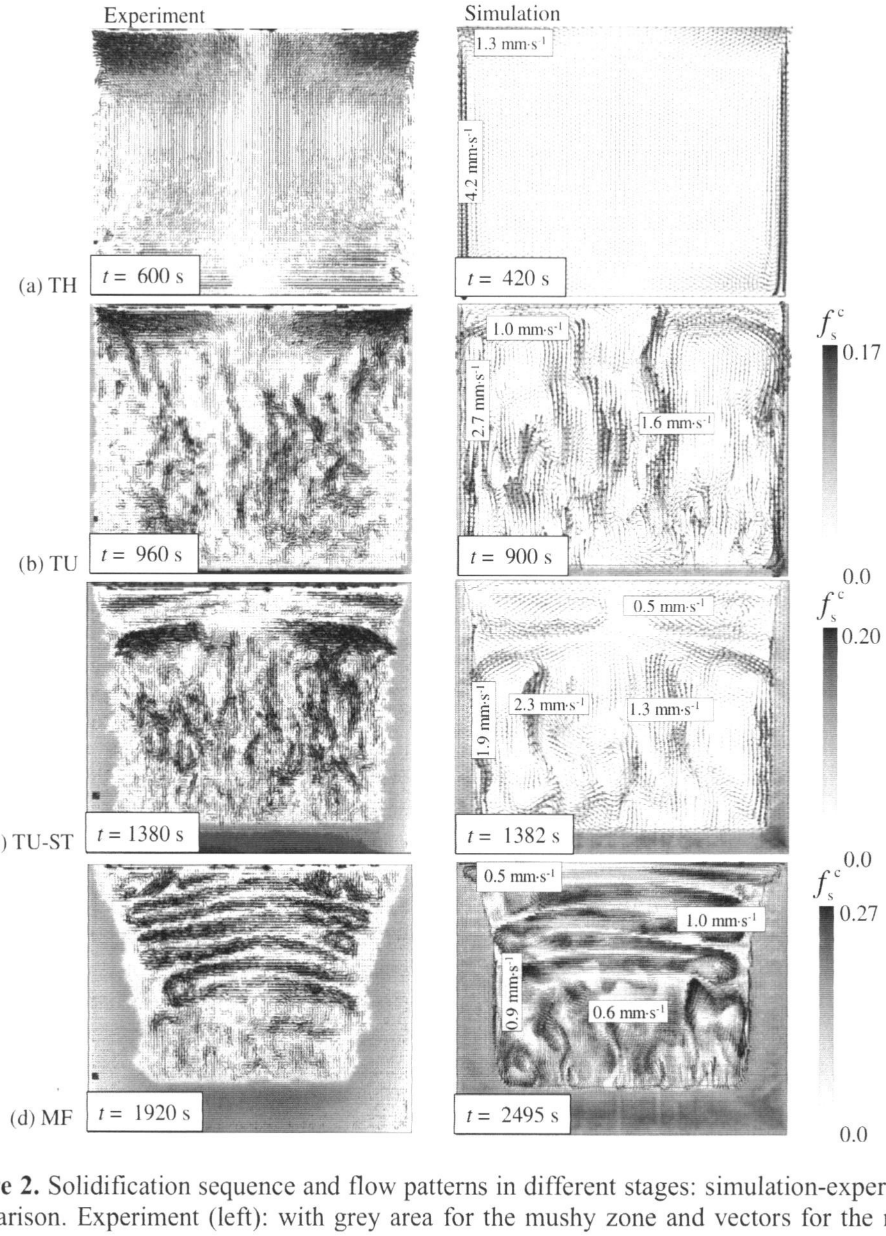

Figure 2. Solidification sequence and flow patterns in different stages: simulation-experiment comparison. Experiment (left): with grey area for the mushy zone and vectors for the measured velocity field with PIV technique. Simulation (right): with grey scale for f_s^c (volume fraction of solid dendrites) and vectors for the liquid velocity $\vec{u}_\ell$. 4 different flow patterns, which correspond to 4 different cooling/solidification stages, are: (a) TH (thermal convection)

(b) TU (turbulence) (c) TU-ST (turbulence – stratified flow) and (d) MF (meandering flow). All simulation results and the flow measurement were taken from the symmetry section.

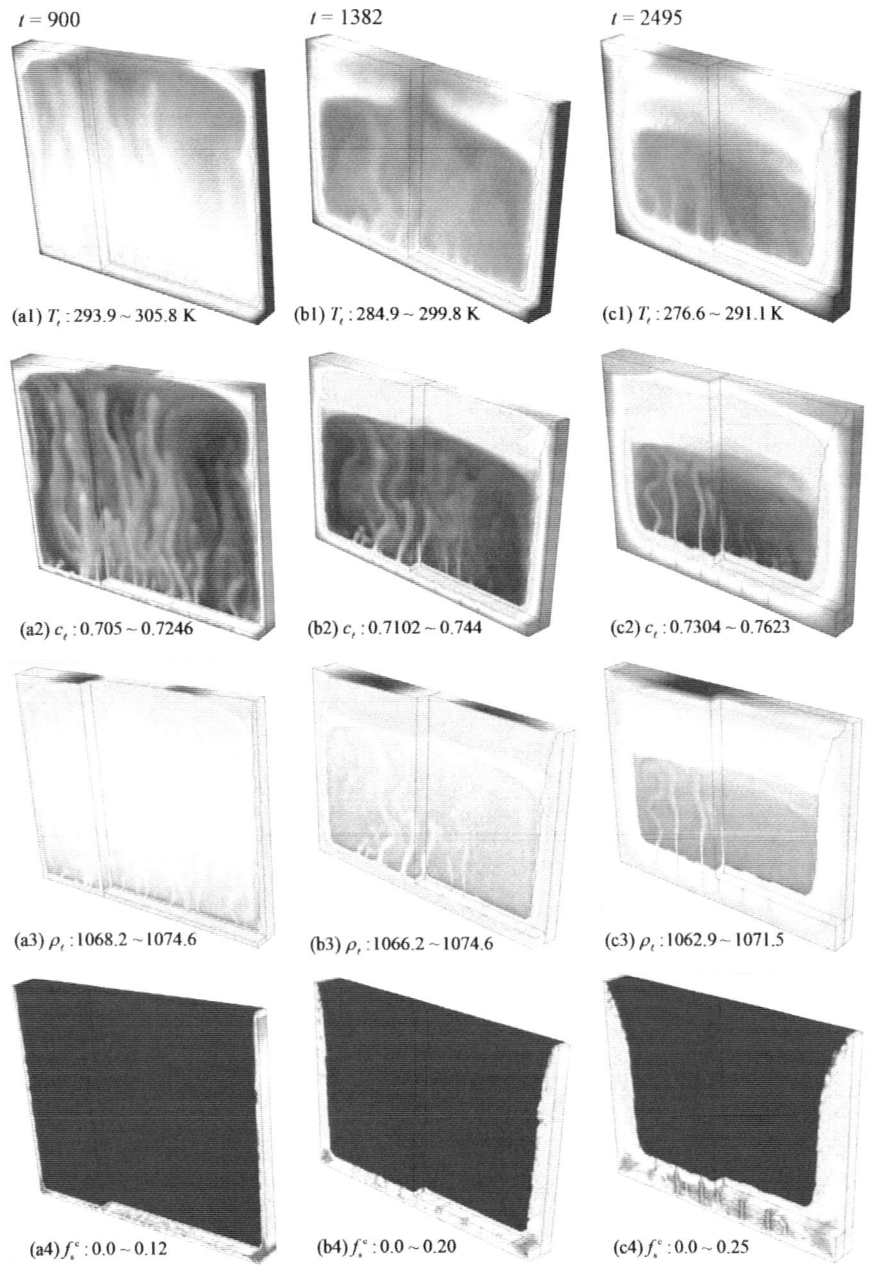

Figure 3. Some key solidification quantities (T_ℓ, c_ℓ, ρ_ℓ, f_s^c), which mostly influence the flow, are shown in a sequence at the time 900, 1382 and 2495 s. The results are shown in 3D, therefore, one can only see the distribution of each quantity in selected sections. All quantities are shown in colour scale with blue for the minimum value and red for the highest value, which are given below each figure.

To the moment ca. 1380 s, figure 2(c), the flow pattern in the mould seems divided into two parts: the lower part is similar to the previous TU region, i.e. chaotic with coherent structures, while the upper part starts a flow stratification with two symmetrical circulation loops. This flow regime is called turbulent – stratified flow region (TU-ST). As shown in figure 3(b1)-(b4), the upward melt flow in the mushy zone along the side wall brings continuously the lighter melt (lower temperature and higher concentration of H_2O) from lower part to the upper part, leading to a separation of the bulk melt region into two parts, namely the upper part with lighter melt and the lower part of denser melt. Along the bottom wall, plumes continue to develop. The distribution of f_s^c is non-uniform, which hints the onset of freckle trails [18]: lower solid fraction inside the channel and high solid fraction in the rim regions. From the experiment, however, no stable channel was observed. This is probably due to the narrow range of f_s^c distribution, which is not detectable by the light camera.

During the late stage of solidification, figure 2(d) and figure 3(c1)-(c4), a so-called meandering flow regime (MF) is developed. The stratification flow pattern in the upper part of the mould is gradually organized horizontally in several layers in a structure of snake. Figure 4 shows more details of the flow structure for the moment of 2495 s. Generally, both simulation and experiment agree with each other very well. The reason for the snake-like MF is as follows. With the continuing of the TU-ST regime, the upper part of lower-density region becomes broader, more circulation loops are built from both sides of the mould. Gradually, the circulation loops from two sides join together, and re-arrange the flow from both sides, leading to the snake-like MF. The experiment seems to show more flow layers than those of the simulation, which is probably due to that the numerical grid resolution is not high enough. Most freckle trails develop vertically upwards; some of them become inclined and later on merged with the neighbouring ones; and some of them disappear.

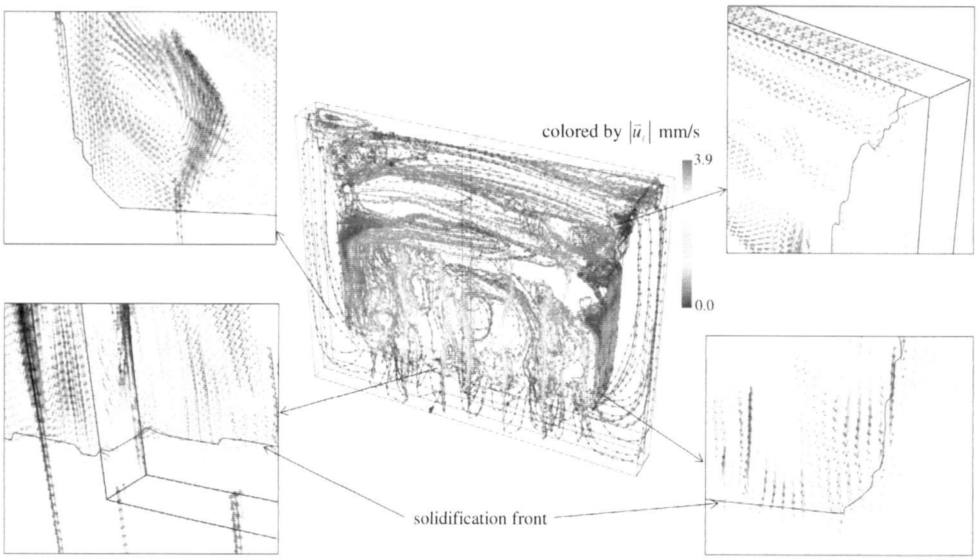

Figure 4. Flow pattern at the moment 2495 s is shown in streamlines ($|\vec{u}_\ell|$ in mm/s) and vectors as zoomed in some regions. The solidification front (position of columnar tips) is also marked with thin line. Flow in the mush is active, but with less intensity. In the bottom part of the mushy zone, the flow in some locations is still relatively strong, where plume develops.

4. Discussions and outlook

Volume-average approach was developed to model the solidification process at the macroscopic scale when the formation of as-cast structure (columnar, equiaxed zones) and macrosegregation is of primary interest [1-2, 19]. The global multiphase flow/transport phenomena are key important at the process scale. To find an experiment, which is able to record the flow at the process scale and in the meantime to record the dendrite grow at the microscopic scale, as a purpose to evaluate above volume-average model is extremely difficult. The in-situ solidification experiment based on X-ray radioscopy, which has been used to detect both flow and crystal growth during solidification [20], is not suitable for evaluate the volume-average model as the experiment is limited to a sample thickness of ca. 100 μm (Hele-Shaw cell). Checking the as-cast structure and final macrosegregation profile with the post-mortem analysis of engineering/laboratory castings [9-10] is the most frequently used method to evaluate the volume-average model, but the global multiphase flow/transport phenomena, the key information for the formation of as-cast structure and macrosegregation, cannot be verified. Therefore, the experiment with in-situ observation of solidification of the NH_4Cl – 70 wt%H_2O solution provides the most efficient way to evaluate volume average model with following merits: (1) it can record of the flow quantitatively; (2) the sample (mould) size of the experiment is at the similar scale of engineering castings; (3) the solidification front (partial mushy zone) can be tracked.

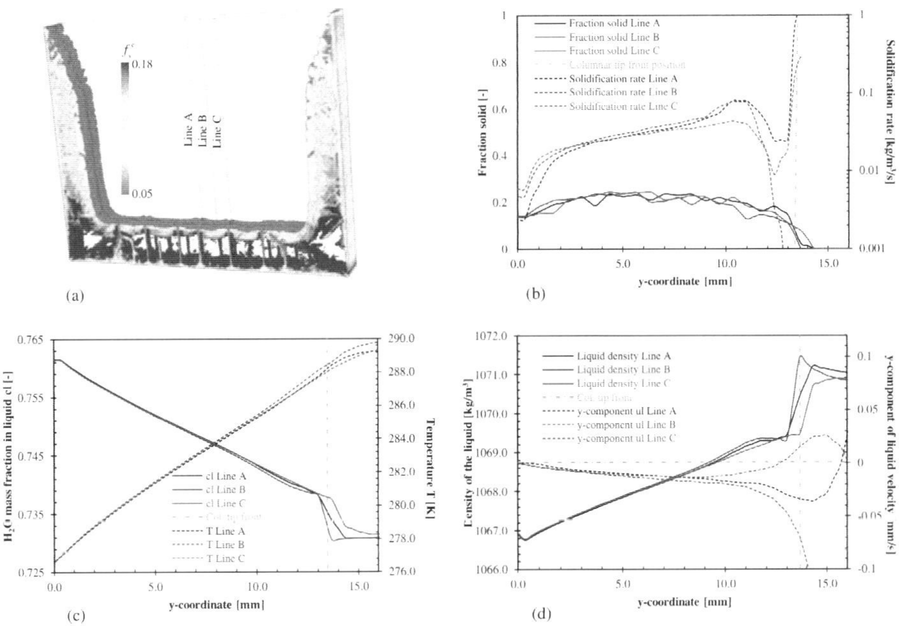

Figure 5. Formation of freckle trails at the moment 2495 s. (a) Iso-surface of f_s^c (= 0.18) indicating the position of solidification front and the formation of freckle trails; (b)-(d) distributions of 6 different quantities along 3 vertical lines (Lines A, B, C, as marked in figure (a)) in y-coordinate direction are shown: fraction solid f_s^c [-], solidification rate M_{ds}^c (kg/m^3/s), H_2O mass fraction in liquid [-], temperature [K], density of the liquid ρ_ℓ [kg/m^3], velocity component in y-coordinate direction $u_{\ell,y}$ [m/s].

All experimental data as provided in [11] were proven to be reproducible by the current volume average model, e.g. figure 2; and the embedded solidification mechanisms can be explained by analysing the modelling results in detail, figures 3-4. Furthermore, the modelling results can also be used to explain some interdendritic flow phenomena during the formation of freckle trail, as experimentally measured by Chen [21] based on a similar experiment with NH_4Cl – 74 wt%H_2O and NH_4Cl – 70 wt%H_2O solutions. Chen used dye ($KMnO_4$) injection and X-ray tomography methods to observe the flow pattern and fraction solid distribution in the mushy zone during solidification. Following findings/statements were made by Chen [21]: the upward plume in the chimneys (channels) induced a downward flow into the mush; with the development of the chimney, solid fraction around the rim of the chimney increased; however, the solid fraction at the bottom of the mushy layer remained low *through an unknown mechanism*; as a consequence, the flow in the chimney was maintained by drawing liquid downward through the mush, then in the horizontal streams along the bottom. Freckle trails in the current experimental setting are simulated and shown in figure 4 and figure 5(a). In order to explain the experiment of Chen's with the current model, further solidification quantities/parameters along 3 vertical lines (A, B and C, in figure 5(a)) are shown in figure 5(b)-(d). The simulation confirms the experimental phenomena that (1) the liquid velocity component in the bottom mushy zone is downward (negative value), except for those which is exactly located in the freckle trails; (2) the solid fraction increases from the solidification front into the middle of the mush, then it decreases towards the bottom wall. The downward flow in the bottom mushy zone can reduce the solidification driven force (constitutional undercooling), leading a reduction of the solidification rate and allowing the flow more easy in the horizontal streams along the bottom mushy zone to feed the upward flow in the freckle trails. "Density reversion" as reported by [18] was also verified near the solidification front.

Further numerical analyses and parameter studies were carried out, e.g. the competing growth between neighboring freckle trails (seen in figure 5(a)); the influence of the crystal morphological parameters (the shape factor Φ^{c}_{env} and circularity factor Φ^{c}_{circ}) on the flow dynamics in the interdendritic region and in the bulk; the effect of the mushy permeability, or the description of permeability by different laws; the formation of crystal fragments (source of equiaxed) as favoured by the flow (plumes/chimneys). These studies would lead to a deeper understanding of the embedded mechanisms of the formation of as-solidified structure and freckle trails. However, they cannot be included in this proceeding article due to the page limit.

Acknowledgement

This work was financially supported by the Austrian Science Fund (FWF, I 4278-N36), FFG Austrian Space Application Program (ASAP, 859777), FFG Bridge I (868070).

References

[1] Beckermann C and Viscanta R 1993 *Appl. Mech. Rev.* **46** 1-27
[2] Beckermann C and Wang C 1995 *Annu. Rev. Heat Trans.* **6** 115-98
[3] Rappaz M and Thévoz P 1987 *Acta Metall.* **35** 1478-97
[4] Rappaz M and Thévoz P 1987 *Acta Metall.* **35** 2929-33
[5] Wang C and Beckermann C 1996 *Metall. Trans. A* **27** 2754-64
[6] Wu M, Fjeld A and Ludwig A 2010 *Comp. Mater. Sci.* **50** 32-42
[7] Wu M, Ludwig A and Fjeld A 2010 *Comp. Mater. Sci.* **50** 43-58
[8] Sturz L, Wu M, Zimmermann G, Ludwig A and Ahmadein M 2015 *IOP Conf. Series: Mater. Sci. Eng.* **84** 012086
[9] Ahmadein M, Wu M, Li J, Schumacher P, Ludwig A 2013 *Metall. Mater. Trans. A* **44** 2895-903
[10] Ahmadein M, Wu M and Ludwig A 2015 *J. Crystal Growth* **417** 65-74
[11] Stefan-Kharicha M, Kharicha A, Wu M and Ludwig A 2014 *Fluid Dyn. Res.* **46** 041424
[12] Stefan-Kharicha M, Kharicha A, Mogeritsch, Wu M and Ludwig A 2018 *J. Chem. Eng. Data* **63** 3170-83
[13] Ramani A and Beckermann C 1997 *Scripta Materialia* **36** 633-8
[14] Perry R, Green D and Maloney J 1999 *Perry's Chem. Engineers' Handbook*, 7th ed. The McGraw-Hill Companies: New York
[15] Beckermann C and Wang C 1996 *Metall. Mater. Trans. A* **27** 2784-95
[16] Ramirez J and Beckermann C 2003 *Metall. Mater. Trans. A* **34** 1525-36
[17] Lipton J, Glicksman M and Kurz W 1984 *Mater. Sci. Eng.* **65** 57-63
[18] Copley S, Giamei A, Johnson S, Hornbecker M 1970 *Metall. Trans.* **1** 2193-203
[19] Wu M, Ludwig A and Kharicha A 2019 *Metals* **9** 229
[20] Shevchenko N, Boden S, Gerbeth G and Eckert S 2013 *Metall. Mater. Trans. A* **44** 3797-808
[21] Chen CF 1995 *J. Fluid Mech.* **293** 81-98

IOP Conf. Series: Materials Science and Engineering **861** (2020) 012042 doi:10.1088/1757-899X/861/1/012042

Impact of crystal sedimentation and viscoplastic semi-solid dynamics on macrosegregation

C M G Rodrigues[1], A Ludwig[1], M Wu[1], A Kharicha[1,2] and A Vakhrushev[1]

[1] Chair of Simulation and Modelling of Metallurgical Processes, Metallurgy Department, Montanuniversitaet Leoben, Franz-Josef Street 18, 8700 Leoben, Austria
[2] Christian-Doppler Laboratory for Metallurgical Application of Magnetohydrodynamics, Montanuniversitaet of Leoben, Franz-Josef Street 18, 8700 Leoben, Austria

E-mail: christian.gomes-rodrigues@unileoben.ac.at

Abstract. A numerical investigation of melt convection with solid-phase transport during equiaxed solidification is performed to examine the structural and compositional evolution of an Al-Si alloy. Nucleation and growth kinetic laws are coupled with the general conservation equations employed in fluid dynamics to predict the development of the solidification structure. Besides crystal sedimentation due to buoyancy-induced forces in the low solid fraction range, the model covers also the viscoplastic semi-solid dynamics at high solid fractions. A test case is devised such that two viscoplastic structures grow from the vertical walls towards the centre of the domain and eventually merge at an arbitrary distance from the bottom of a simple cavity setup. This introduces difficulties for the solidification-driven feeding flow which has been associated with pressure drops across the domain. Dilatation of the mush is also predicted in the lower part of the solidifying viscoplastic mush, which means that additional melt is drawn to this area. Interestingly, besides liquid feeding, also a significant amount of solid feeding has been detected to compensate for solidification shrinkage. Such synergetic behaviour between the phases is analysed in terms of macrosegregation.

1. Introduction

In the last decades, solidification has evolved from a purely industrial-driven activity to a science-based field. Numerous fundamental developments have been achieved in recent years and the understanding of the underlying physics during a phase change has increased greatly from both analytical and experimental studies. However, industry has also coped with today's technological competitiveness by building up greater reliance on computer-aided engineering, and more particularly on numerical simulations. This paradigm shift has driven the development of computational models aimed at replicating definite phenomena.

In the present paper, a two-phase Eulerian-Eulerian volume-average model is presented that simulates the transport, nucleation and growth of equiaxed crystals during solidification. The solidification spectrum is separated into two distinct regimes. At low solid fractions, the behavior of the solid-liquid mixture is described as suspensions [1], and the rheology of the mixture is mainly dictated by the transporting character of the liquid phase. At high solid fractions, on the other hand, the crystals create a coherent structure that is capable of sustaining significantly higher stresses [2]. In this regime (referred as the viscoplastic regime), the solid phase is characterized by having a dynamic behavior at high temperatures. It is also often characterized as a solid skeleton saturated with liquid [3, 4]. Furthermore, it can also deform by shrinkage, convection or due to the application of external loads. As a result, the mush either densifies (by expelling liquid from its core) or dilates (by draining liquid from its surroundings) [3], which alters the composition of the alloy [5, 6]. Based on the experimental work of Nguyen [7], who studied the mechanical behavior of an alloy in the semi-solid state, the transition between the two regimes has been set at a solid fraction (g_s^r) of 57%.

In a previous work [8], it was found that densification or dilatation of the mush in simple cavity setups was not strong enough to induce solute redistribution. In order to gain more insight into the semi-solid dynamics in the viscoplastic regime, the configuration in the present work has been modified so that viscoplastic structures start to interact between each other more intensely, and solidification feeding requirements across the domain accentuates the effect of the semi-solid dynamics and mush deformation during the casting process. It is worth noting that the formation of porosity, which would limit locally the pressure drop, is neglected in this work.

2. Model description

The two-phase Eulerian-Eulerian volume-average model is based on the *twoPhaseEulerFoam* solver from OpenFOAM toolbox [9]. However, the original code has been greatly altered mainly to incorporate the viscoplastic regime, which was not supported by the solver. This includes (among other developments) the addition of the bulk viscosity in the viscous stress term, the adaptation of the Navier-Stokes equation to cope with the different rheological material behaviors depending on the regime, and the calculation of the effective solid viscosity as a quantity that is function of the solid fraction or the equivalent strain rate (depending on the regime). These quantities are defined in the following.

The volume-average conservation equations of mass, momentum, species, and enthalpy are solved for the liquid and the solid phases. They are presented in table 1, where g denotes the volume fraction, ρ the density, $\mathbf{v}$ the velocity vector, p the shared pressure, Σ_i^{eff} the effective deviatoric stress tensor, $\mathbf{g}$ the gravity vector, c the species concentration, h the enthalpy per unit mass, $\mathbf{q}$ the heat flow vector, n the grain number density, and the subscripts i refer to either liquid (ℓ) or solid (s). In each computational cell, the sum of solid and liquid volume fractions (g_s and g_ℓ, respectively) is equal to one.

Table 1. Volume-Averaged Conservation Equations.

Mass conservation:	$\frac{\partial g_i \rho_i}{\partial t} + \nabla \cdot (g_i \rho_i \mathbf{v}_i) = \mp M_{\ell s}$	(1)
Momentum conservation:	$\frac{\partial g_i \rho_i \mathbf{v}_i}{\partial t} + \nabla \cdot (g_i \rho_i \mathbf{v}_i \mathbf{v}_i) = -g_i \nabla p + \nabla \cdot g_i \mathbf{\Sigma}_i^{eff} + g_i \rho_i \mathbf{g} \mp \mathbf{U}_{ls}^D \mp \mathbf{U}_{ls}^M$	(2)
Species conservation:	$\frac{\partial g_i \rho_i c_i}{\partial t} + \nabla \cdot (g_i \rho_i \mathbf{v}_i c_i) = \nabla \cdot (g_i \rho_i D_i \nabla c_i) \mp C_{\ell s}$	(3)
Enthalpy conservation:	$\frac{\partial g_i \rho_i h_i}{\partial t} + \nabla \cdot (g_i \rho_i \mathbf{v}_i h_i) = -\nabla \cdot \mathbf{q}_i \mp H_{\ell s}$	(4)
Grain transport:	$\frac{\partial n}{\partial t} + \nabla \cdot (\mathbf{v}_s n) = N_{\ell s}$	(5)

The coupling between the phases is achieved through the single pressure field, p, and the different interphase exchange terms $M_{\ell s}$, $\mathbf{U}_{\ell s}$, $C_{\ell s}$, $H_{\ell s}$ and $N_{\ell s}$ which are presented in table 2. The sign of the source terms depends on the phase, with "∓" being positive when $i=s$, and negative when $i=\ell$.

The quantities v_r, $S_{\ell s}$, Φ_{imp}, $K_{\ell s}$ and u^* are the crystal growth velocity, specific surface area, impingement factor, momentum exchange coefficient due to drag and momentum exchange velocity coefficient due to phase change. The enthalpy exchange term is used to satisfy the condition of thermal equilibrium between phases (with $h_c = 10^9$ W/m²/K). The diffusion coefficient, D, in equation (3) is equal to 1.63×10^{-9} m²/s for the liquid and 8×10^{-13} m²/s for the solid, whereas the effective thermal diffusivity for enthalpy, α, in equation (10) is equal to 0.058 m²/s for the liquid and 0.166 m²/s for the solid. Further details on the conservation equations, source terms and auxiliary equations considered in the development of the model have been described in detail in a companion paper [5].

Table 2. Exchange terms and other closure laws for the conservation equations.

mass transfer:	$M_{\ell s} = v_r S_{\ell s} \rho_s \Phi_{\text{imp}}$	(6)
momentum transfer:	$\mathbf{U}_{\ell s} = \mathbf{U}_{\ell s}^M + \mathbf{U}_{\ell s}^D = K_{\ell s}(\mathbf{v}_\ell - \mathbf{v}_s) + u^* M_{\ell s}$	(7)
species transfer:	$C_{\ell s} = c_s^* M_{\ell s}$	(8)
enthalpy transfer:	$H_{\ell s} = h_c (T_\ell - T_s)$	(9)
heat flux:	$\mathbf{q}_i = \alpha_i \nabla h_i$	(10)

The important novelty of this model is the implementation of the viscoplastic regime, which is imposed when $g_s > g_s^t$. The change in material behavior is transmitted through the effective deviatoric stress tensor included in equation (2), which is defined differently depending on whether the solid fraction is in the viscoplastic regime.

Below the viscoplastic threshold, the equation is defined by the general Newtonian viscous stress tensor ($\mathbf{\Sigma}_s^{eff} = \boldsymbol{\tau}_s = 2\mu_s \text{dev}(\dot{\boldsymbol{\varepsilon}}_s)$). This expression is analogous to the one adopted in the liquid phase with the strain rate tensor being defined as $\dot{\boldsymbol{\varepsilon}}_i = 1/2\left(\nabla \mathbf{v}_i + (\nabla \mathbf{v}_i)^T\right)$. The solid viscosity is expressed according to the power-law viscosity model of Ishii [10]. Assuming the validity of the mixture rule, it yields:

$$\mu_s = \mu_\ell / g_s \left(\left(1 - g_s / g_s^p\right)^{-2.5 g_s^p} - (1 - g_s) \right). \quad (11)$$

where the packing limit is set as $g_s^p = 0.585$, according to the loose random packing of spheres proposed by Olmedilla *et al.* [11].

In the viscoplastic regime, on the other hand, experimental studies on the behavior of metallic alloys at high temperatures show that the effective deviatoric stress tensor of the solid phase can be described by the following expression [7]:

$$\mathbf{\Sigma}_s^{eff} = \frac{\mu_s^{\text{app}}}{g_s} \left(\frac{2}{A} \text{dev}(\dot{\boldsymbol{\varepsilon}}_s) + \frac{1}{9B} \text{tr}(\dot{\boldsymbol{\varepsilon}}_s) \mathbf{I} \right) \quad (12)$$

where A and B are two rheological parameters proposed by Nguyen *et al.* [7], which affect the deformability of the mush in pure shear and hydrostatic states, respectively. They can be written as follows:

$$A = \frac{3}{g_s^{6.47}} \text{ and } B = \frac{1}{g_s^{6.94}} - 1 \quad (13)$$

and the apparent solid viscosity takes the form of the Norton-Hoff model:

$$\mu_s^{app} = 3K_v \left(\sqrt{3} \dot{\varepsilon}_s^{eq} \right)^{m-1} \quad (14)$$

In the above expression, the apparent solid viscosity can be seen to depend on the equivalent strain rate, which is given by the following expression:

$$\dot{\varepsilon}_s^{\text{eq}} = \sqrt{\frac{2}{A}(\dot{\boldsymbol{\varepsilon}}_s : \dot{\boldsymbol{\varepsilon}}_s) - \left(\frac{2}{3A} - \frac{1}{9B}\right) \text{tr}(\dot{\boldsymbol{\varepsilon}}_s)^2} \quad (15)$$

which is again function of the rheological parameters defined in equation (13). The quantities K_v and m presented in equation (14) are the viscoplastic consistency and the strain-rate sensitivity, respectively. The latter parameter is usually assumed as a fixed value which depends on the material properties. In the present work, it is set as 0.213. On the other hand, the viscoplastic consistency can be calculated by combining the strain-rate tensor proposed by Nguyen *et al.* [7] and the Norton-Hoff stress-strain-rate relation. It takes the value of 6.31×10^6 Pa s.

3. Results and discussion

In a previous work [8], a simple cavity test case was analyzed in terms of melt convection, crystals dynamics, and macrosegregation. A comprehensive picture of the flow dynamics was described and details concerning the flow inside the viscoplastic regime and the eutectic reaction were also discussed. However, although the semi-solid dynamics at high solid fractions could be captured by the model, no clear evidences of solute redistribution driven by densification or dilatation of the mush were detected. In the present manuscript, the setup has been altered so that further interactions between the solidifying structures take place. In addition, the present results are compared against the outcome of a simulation that contemplates a rigid solid body formulation instead of the viscoplastic regime.

A vertical column test case is simulated, as represented in figure 1. The setup is characterized by an intensified cooling region that is enforced by assuming a very high heat transfer coefficient (HTC) defined in a specific section of both vertical walls. This simple setup provides a good opportunity for experimental validation in the future.

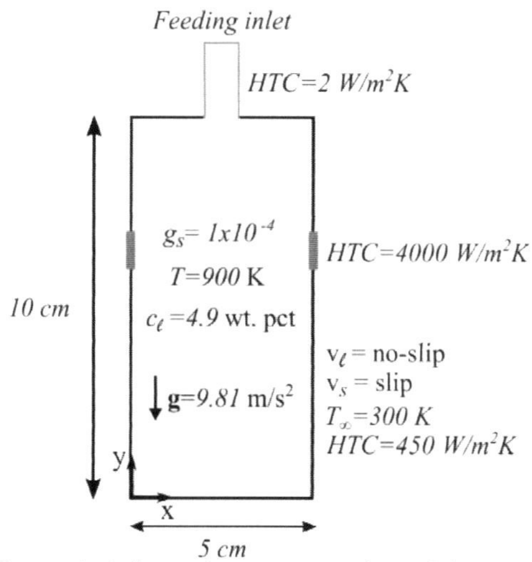

Figure 1. Schematic representation of the test case.

The presented simulation results are for an inoculated Al-5wt.%Si alloy. An inlet tube (with fixed pressure, $p=10^5$ Pa) is designed so that shrinkage can be compensated by mass/liquid feeding. No-slip and slip boundary conditions are applied to the liquid and solid phases, respectively. Along the walls, a convection heat transfer boundary condition is imposed with a heat transfer coefficient (HTC) of 450 W/m^2/K along most of the walls, except at two locations. The first is along the feeding tube where HTC = 2 W/m^2/K so that almost no heat transfer is allowed along that portion of the wall. The second is in the already mentioned intensified cooling region where HTC = 4000 W/m^2/K (see figure 1). A very large heat transfer is adopted in this portion of the domain so that the two viscoplastic structures grow faster than elsewhere and eventually interact between each other during the solidification.

The simulation starts with a solid fraction of 1×10^{-4} (for computational reasons, a finite but very small value is considered), initial temperature of 900 K (liquidus line), and initial species concentrations of $c_l = 4.9$wt.% and $c_s = 0.68$wt.%. The present model only considers globular equiaxed crystals. The solid and liquid densities are 2535 kg/m^3 and 2430 kg/m^3, respectively, and the constant liquid viscosity is taken as 1.09×10^{-3} kg/m/s. A two-dimensional orthogonal mesh with 50×100 cells is used (discounting the inlet tube). The simulation is performed with a variable time-step which remains in the order of 10^{-4} s a few moments after the start of the simulation.

Figure 2 illustrates the solidification and sedimentation of the Al-Si alloy, from a starting point where it is totally liquid (dark blue color). The feeding tube is not shown for simplicity and the black line signals the interface at which the viscoplastic model is triggered. Dark red regions correspond to eutectic structures. Initially, due to the slip boundary condition imposed to the solid phase, the equiaxed crystals immediately sink as they nucleate and grow. Because of the coupling between the two phases (through the momentum exchange term), the melt is also affected by the sinking crystals. This gives rise to a symmetric flow pattern with downward motion near the vertical walls and an upward motion at the center of the domain.

While the solid fraction is below the viscoplastic limit, the equiaxed crystals sink along the vertical boundaries and then, coming from both lateral walls, they collide at the bottom center of the domain. Once the viscoplastic regime is entered, the viscosity of the material increases substantially and the mush becomes stiffer. This can be seen near the walls in figure 2(a), where a coherent structure develops near the sides of the domain. Nevertheless, a more dynamic motion of the solid phase at lower solid fractions is kept between this two viscoplastic structures.

As the simulation proceeds, equiaxed crystals continue to sediment at the bottom and along the vertical walls (figure 2(b)). Once the stiffer formations reach the intensified cooling region (first near the walls), they provide a physical structure where the solid phase can accumulate. As a result, the crystals stop sinking immediately down and start to solidify also along the intensified solidification direction, i.e., towards the center of the domain (figure 2(c)). At the bottom, the coherent structure develops a strong curved interface as a result of the preferential accumulation of equiaxed crystals near the lateral walls and the established flow map [8].

With further solidification, the two separate structures that grow in the intensified cooling region start to approach each other near the center of the domain and, eventually, end up by forming a continuous coherent structure spanning the entire width of the domain (figure 2(d)). This generates an obstacle for the flow to move longitudinally. Notice also that a small region with solid fraction below the viscoplastic limit is still visible in the center of the configuration. In the last snapshot (figure 2(e)), the eutectic formations (dark red color) have merged at the center in the intensified cooling region, which at this point totally precludes the flow to move towards the bottom.

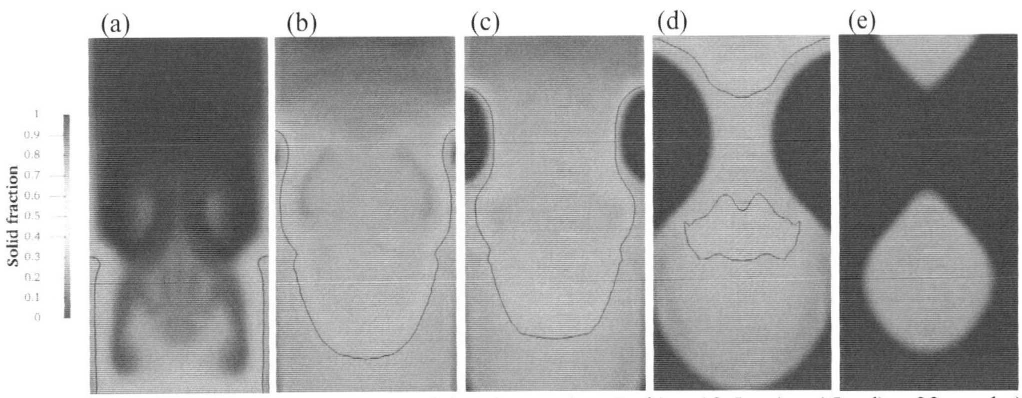

Figure 2. Qualitative results for solid fraction at a) t=5s, b) t=12.5s, c) t=15s, d) t=22s and e) t=28s.

While the two solidifying structures are separated, sedimentation and solidification feeding can progress from top to bottom with no substantial resistance, although macrosegregation deviations occur due to the induced cross-sectional change. However, once they approach each other, material has to flow through the mush to feed solidification shrinkage at the bottom of the domain. While liquid feeding is generally expected, the viscoplastic formulation introduces a dynamic character to the solid phase that come into play here.

Figure 3 illustrates the vertical component of the solid velocity at t = 22 s. According to figure 2(d), it corresponds to a stage where a continuous viscoplastic structure extends between both vertical walls, and separates the solidifying material located at the bottom from the inlet located at the top. Although eutectic structures are identified near the sides, they still are separate enough to allow for material to flow along the centre. Furthermore, in figure 3, the results of the present model are compared against a simulation that considers the common simplification of assuming a rigid solid body (given by an arbitrary high constant solid viscosity) instead of the viscoplastic regime. The snapshot is divided in half longitudinally: the left half corresponds to the outcome of the current model, whereas the right half shows the outcome of the rigid solid body formulation.

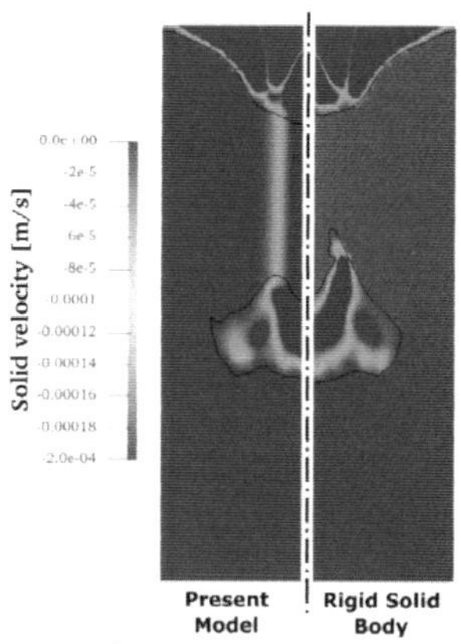

Figure 3. Vertical component of the solid velocity at t = 22 s for present model (left half) and rigid solid body formulation (right half).

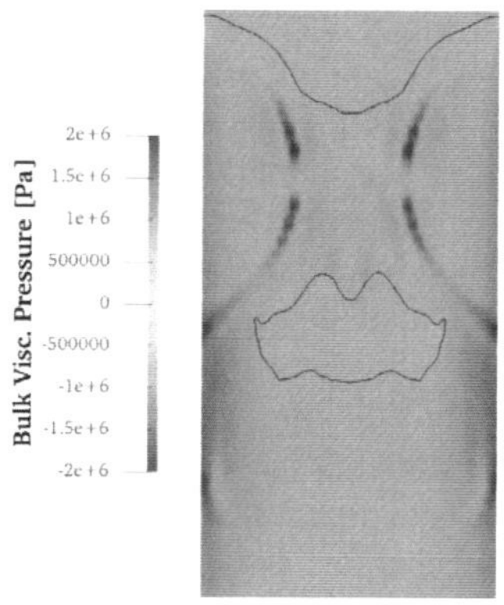

Figure 4. Calculated pressure induced by the bulk viscosity term in the viscoplastic regime at t = 22 s. The red color corresponds to dilatation of the mush, whereas the blue color corresponds to densification.

Because the solid viscosity in the viscoplastic regime is dynamic (i.e. it depends on the strain rate and solid fraction, as shown in equations (13) to (15), the downward solid velocity through the mush also varies across the horizontal cross section of the intensified cooling region. Interestingly, the maximum absolute solid velocity (found in the centre) is of the same order of magnitude as the liquid velocity in that region. This downward flow only occurs in the viscoplastic region, whereas in the eutectic structure the solid velocity is essentially zero. On the other hand, in the rigid solid body formulation, the velocity is roughly non-existent (dark red color in the snapshot) across the whole width of the domain. These results suggest that in the latter approach, the solidification taking place at the bottom of the domain is fed purely with melt (liquid feeding), whereas in the present model the solid phase also contributes to feed solidification (mass feeding).

Such mechanism is the results of a dynamic viscosity assumed in the viscoplastic regime but is also driven by the extra solid pressure that is derived from the second term on the right-hand side of equation (12). This latter quantity is shown in figure 4 at t = 22 s (referred in the results as the bulk viscosity pressure). Dilatation of the mush is observed in the lower part of the solidifying viscoplastic mush, which means that additional melt is drawn to this area. As a consequence, further material has to compensate for the solidification feeding, which comes not only as liquid but partly also in the form of solid phase (as indicated by the previous figures).

For reference, the largest absolute values found in figure 4 are roughly one order of magnitude larger than the shared pressure across the domain (in absolute terms) during most of the simulation. This finding highlights the significance of such parameter in the results, even though it only appears at very specific locations where densification or dilation occurs. Nevertheless, it is important to emphasize that the shared pressure field is also highly influential on the flow dynamics.

Although not directly shown here, the pressure field has been found to be an important driving force for the solidification-driven feeding flow, with larger pressure gradients along the vertical axis being observed as the feeding difficulties increase during that process. However, right after the two eutectic structures form a continuous solid structure in figure 2(e), the lower limit of the pressure field rapidly drops to around 10 kPa. In reality, this scenario would most probably lead to feeding related defects, such as porosity, and hot-tearing formations, as their appearance has been associated with large pressure drops at the end of the casting processes [12, 13]. In the simulation, however, since such mechanisms are not accounted for in the present model and the feeding flow through the solidified material becomes practically impossible, the simulation ends up by crashing.

The present results have shown the presence of local dilatation of the mush in certain regions. This leads to an additional inflow of melt that supplements the feeding flow necessities required to compensate solidification shrinkage. However, besides the melt, also solid feeding has been observed during relevant stages of the simulation. These findings have both a direct effect on the relative motion between the phases, which eventually affects the composition deviation of the alloy. Figure 5(a) illustrates the macrosegregation formation at t = 22s across the domain for both the present model and the rigid solid body formulation. Qualitatively, similar distributions can be seen. This means that the rigid solid body formulation might be employed in test cases where a simple estimation of the composition deviations is needed.

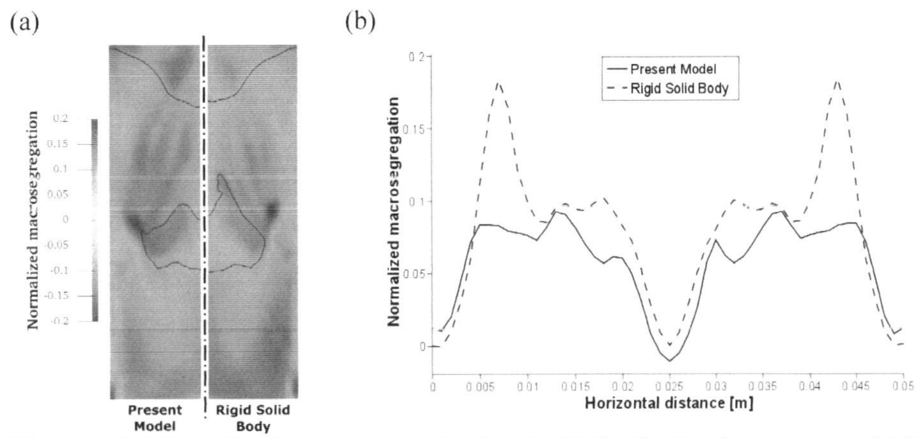

Figure 5. (a) Normalized macrosegregation (c_{mix}/c_0-1) distribution for present model (left half) and rigid solid body model (right half) at t = 22 s. (b) Quantitative results for normalized macrosegregation along a horizontal line located in the middle of the domain.

However, in quantitative terms, some discrepancies can be identified. Figure 5(b) presents the composition deviation along a horizontal line located in the middle of the domain. It exhibits a significant positive deviation spike in the rigid solid body formulation at x=0.007m and x=0.043m, whereas the present model presents a more uniform distribution in this region (due to the reasons discussed above). Both models have a slight negative deviation along the centerline mostly because the inflow of melt to feed solidification in this area mainly comes from the inlet, which possesses the initial melt composition. Therefore, although the general macrosegregation distribution looks similar, the extra complexity of the viscoplastic regime ends up by stabilizing the spikes observed in the macrosegregation results.

IOP Conf. Series: Materials Science and Engineering **861** (2020) 012042 doi:10.1088/1757-899X/861/1/012042

4. Conclusion

In the present paper, a comprehensive model has been described that can be employed to simulate the entire solidification process. A simple test case has been devised so that solidifying structures strongly interact with one another and impose obstacles to the solidification-driven feeding flow across the domain. The simple setup has been considered so that it can be replicated experimentally for proper validation of the computational results.

The results show that the viscoplastic regime employed in the present formulation provides for a more dynamic version of the solid phase, even for high solid fractions. Besides the expected liquid feeding, mass feeding has also been observed in a scenario where a strong inflow was required. The results contrasted with the common general rigid solid body formulation, which completely neglects such mechanism. Also, areas of densification and dilatation have been observed, which influenced the flow behavior during the simulation. Such mechanisms provide for more active relative motion between the phases, which end up by affecting the final macrosegregation formations.

Acknowledgments

This work was financially supported mainly by the FWF Austrian Science Fund (P28785-N34), and by the Austrian Federal Ministry of Economy, Family and Youth and the National Foundation for Research, Technology and Development within the framework of the Christian Doppler Laboratory for Metallurgical applications of Magneto-hydrodynamics.

References

[1] Guazzelli E and Pouliquen O 2018 *Journal of Fluid Mechanics* **852** 1-62
[2] Drezet J M, Mireux B, Szaraz Z and Pirling T 2014 *Miner.* Met. Mater. Soc. **66** 1425–30
[3] Martin C L, Favier D and Suéry M 1999 *Int. J. Plast.* **15** 981–1008
[4] Geindreau C and Auriault J L 1999 *Mech. Mater.* **31** 535–51
[5] Rodrigues C M G, Ludwig A, Wu M, Kharicha A and Vakhrushev A 2019 *Metall. Mater. Trans. B* **50** 1334–50
[6] Fachinotti V D, Le Corre S, Triolet N, Bobadilla M and Bellet M 2006 *Int. J. Numer. Methods Eng.* **67** 1341–84
[7] Nguyen T G, Favier D and Suery M 1994 *Int. J. Plast.* **10** 663–93
[8] Rodrigues C M G, Ludwig A, Wu M, Kharicha A and Vakhrushev A *Computational Materials Science* (submitted)
[9] Greenshields C J, OpenFOAM – The OpenFOAM Foundation – User Guide, 2018. http://foam.sourceforge.net/docs/Guides-a4/Op enFOAMUserGuide-A4.pdf
[10] Ishii M 1977 *Report No. 77-47* (Argonne National Laboratory, ANL, Illinois)
[11] Olmedilla A, Zaloznik M and Combeau H 2017 *Powders and Grains 2017* **140** 1-4
[12] Eskin D G and Katgerman L 2007 *Metall. Mater. Trans. A* **38** 1511–9
[13] M'hamdi M, Mo A and Martin C L 2002 *Metall. Mater. Trans. A* **33** 2081–93

IOP Conf. Series: Materials Science and Engineering **861** (2020) 012043 doi:10.1088/1757-899X/861/1/012043

Modelling the evolution of MnS inclusions and macrosegregation during solidification using a three-phase model

D Cai[1,2], J Li[1], H Dong[1] and J Li[2]

[1] School of Engineering, University of Leicester, Leicester, LE1 7RH, UK
[2] School of Materials Science and Engineering, Shanghai Jiao Tong University, Shanghai, 200240, China

E-mail: li.jun@le.ac.uk, h.dong@le.ac.uk

Abstract. Manganese sulphide (MnS) is one of the major non-metallic inclusion in steel which greatly impacts the properties of steels. Previous models have been developed to simulate the dynamics of MnS during solidification without considering the effects of macrosegregation formation. A comprehensive model incorporating the formation kinetics of MnS with a ternary macrosegregation model is presented to investigate the evolution of MnS inclusions and its effects on C and Mn macrosegregation. Classical nucleation theory and a diffusion-controlled growth model are applied to describe the evolution of MnS, which is fully coupled with a two-phase ternary alloy solidification model. The model considers the growth of columnar dendrite trunks, thermosolutal convection of the melt, solute transport by convection and inclusion floatation. It has been applied to a benchmark case, and reasonable results are obtained with a good agreement with the reported experimental ones. The effects of MnS behaviours on the segregation of different solute elements are discussed.

1. Introduction

Macrosegregation and non-metallic inclusions in casting ingot have been the focus of metallurgists for nearly half a century, since they cannot be alleviated during the down-stream hot processing and may severely limit the application of ingots. Non-metallic inclusion like MnS, which is a ubiquitous inclusion formed in almost every steel grade and is detrimental to the properties of steel by acting as a potential starting point for crack formation or corrosion [1]. Both numerical [2, 3] and experimental [4] researches have been carried out to study the formation mechanism of macrosegregation and the precipitation, cohesion and remove of inclusions during solidification.

Solute heterogeneity occurs as soon as solidification starts when partition coefficient brings the difference between both sides of liquid/solid interface [5]. Consequently, when solidification starts, the solute rejected from the solid phase will accumulate in the solidification front, and the relative motion of solid and melt will turn the microscope solute segregation into macroscale inhomogeneity, i.e. macrosegregation. Several inner mechanisms are believed to contribute to this pivotal relative motion between phases, namely the thermal-solutal convection [6], crystal sedimentation [7], shrinkage-induced flow [8, 9] and motions of gas bubbles or inclusions [10] that develop during solidification. Modern computational ability makes it possible to directly observe and analyse the above mechanisms individually to guide experimental study. The multi-phase macrosegregation models have been greatly improved during the last decades since the first mushy zone model was developed by Fujii *et al.* [11]. Wang *et al.* [12] established a model that considered two principal solidification characteristics namely solid-phase transport and melting in the 1990s. More solidification phenomena such as equiaxed motion [13], gas-induced flow [14] and dendrite morphology [15] were coupled into the sophisticated model.

For macrosegregation modelling, nevertheless, only a few works [16] payed attention on the inclusions, though the oxide-based inclusions have been the focus of Cao *et al.* [17], who found that more inclusions exist in or at the adjacent onset site of long-fastigiate A-segregation, one of the most typical macrosegregation in large-scale ingot. Their simulation results also showed that the inclusions which nucleate before solidification would disturb the solidification front and alter the local melt flow then trigger the macrosegregation. The non-metallic inclusion generated by deoxidation reactions during the solidification, such as MnS, has less been considered in the macrosegregation model, although thermodynamic method [18] was widely employed to calculate the composition and evolution of such inclusions. In addition, classic precipitation and growth theories [19] have also been studied and verified to calculate the nucleation condition. It is quite important to establish a model that consider the interaction between the evolution of MnS inclusion and the formation of macrosegregation.

In this study, a comprehensive model incorporating the nucleation and growth of MnS inclusion with a ternary alloy (Fe–0.45 wt%C–1.06 wt%Mn) macrosegregation model is presented to investigate the evolution of MnS inclusions and its effects on C and Mn macrosegregation in a benchmark case.

2. Model description and configuration

The calculated domain is a 2D 50×50 mm2 grid, half of which is shown in figure 1. The other half of figure 1 indicates the melt (l) solidifies in the form of columnar (c) solidification and the MnS (p) inclusions (orange spheres) participate between the dendrites considered in this model. The sketch also gives the initial and boundary conditions in such solidification process. The whole domain is fill with liquid at the beginning.

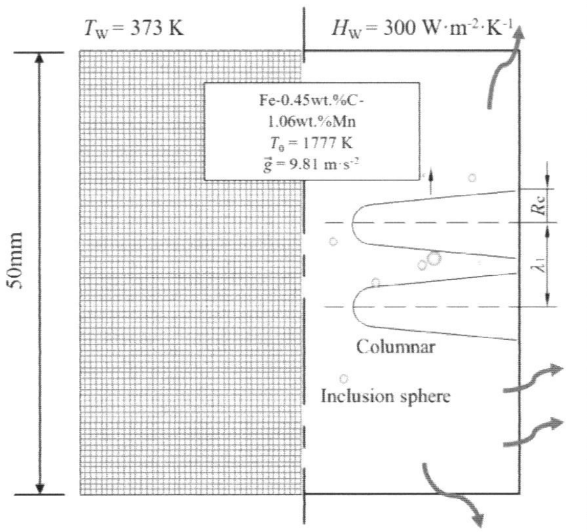

Figure1. A schematic of columnar solidification model with the formation of MnS inclusion in a 2D (50×50 mm2) domain. Thermal boundary conditions are also shown in the left half.

2.1. *Eulerian volume average macrosegregation model*

To elucidate the effect of inclusions on the macrosegregation of Mn and C elements, a Eulerian volume averaged ternary alloy model is established, which is basically extended from the ternary approach established by Wu [20]. More details about the model can be found in the previous literatures [21, 22]. A brief outline of the model is given below:

- Columnar dendrites are approximated by growing cylinders starting from the mould wall towards the casting centre by the heat transfer and each cell in the left half of figure 1 contain uniform volume fraction of columnar. The solidification proceeds by means of the mass transfer from the original liquid to this columnar.
- Growth of the columnar trunk is governed by diffusion with the neglection of back diffusion.
- Enthalpy equations for all phases are solved. Due to the fact that thermal diffusion is faster than solute diffusion, we assume that only one temperature (T) represents each volume element. Therefore, a large inter-phase volume heat transfer coefficient between the phases is applied to balance the temperatures among the phases.

- The permeability in the columnar mush zone is modelled according to Blake-Kozeny [23].
- The densities of the solid and liquid are considered constant and equal. The Boussinesq approach is employed to model thermo-solutal convection. Feeding flow due to density change in the mushy zone can is not considered.

2.2. Nucleation and evolution of MnS inclusion
MnS is normally generated from the enrichment of Mn and S in the residual liquid during solidification. Thus, the abovementioned macrosegregation model alone is beyond qualified to simulate the MnS formation process since the enriched solute in the melt can be withdrawn every time-step during the calculation, in other words, the inclusion-combined model should be fully coupled. To simplify the calculations, the physical assumptions of the model are as follows:

- Only MnS formed in the residual liquid is considered.
- MnS particles are independent with each other and the collision is neglected.
- The morphology of the particles is spherical.
- Diffusion-controlled growth is used and the equilibrium content of Mn on the surface of MnS can be taken as zero since the S concentration in the surface is high [24].

It is believed that MnS is thermodynamically stable when the real concentration product, K, is larger than the solubility product, K_{eq}, with the real concentration and solubility products be calculated by $K=c_s \cdot c_{Mn}$ and $\log K_{eq}=-8750/T+4.63$, respectively. Classical nucleation theory is applied to the model once the critical thermodynamic condition is achieved. For simplicity, homogeneous nucleation is considered in the current case. The growth of the spherical particle is assumed to be controlled by diffusion of the solutes in liquid steel [24, 25].

3. Simulation results and discussions

3.1. The evolution of C and Mn macrosegregation
Such columnar solidification model has been verified in previous published work, and the macrosegregation formation mechanisms have been studied in detail [21]. The focus of the current work is on the interaction between the macrosegregation and the formation of MnS. In addition, the calculated results of MnS volume fraction in the steel matrix is compared with published data which is discussed in the following parts. In this calculated case, the melt takes around 100 s of physical time for the square domain to fully solidify. The segregation of Mn and C is evaluated by segregation index (SI=(cmix-c0)/c0). cmix is the local solute concentration after solidification, and c0 is the initial solute concentration. The solidification sequence and the evolution of C and Mn SI are shown in figure 2, which are the distribution of such properties in each volume-averaged cell in the mesh shown in figure 1. The melt velocities are also indicated in the C macrosegregation contours (upper part). The black isoline of f_l=0.9 indicates the solidification front. Due to the combination of the solute and thermal driving force, a counterclockwise melt flow forms in the bulk melt. With this melt flow, solute depleted zones, i.e. negative macrosegregation zone, has been formed in the upper left region of the ingot in the early stage of solidification (i.e. at 10 s). The solute originally present in the negative segregation region is brought to the centre of the domain by the counterclockwise flow, thereby forming a solute enriched zone here. Both C and Mn components have similar shape of the solute depleted and enriched zones. As for C component, the negative segregation zone remains almost unchanged as the solidification progressed, but the solute enrichment degree in the solute enriched zone gradually increases, and eventually a severe positive segregation zone is formed. For the Mn component, the area of the negative segregation zone gradually increases with the progress of solidification, which results from the consumption of Mn due to the formation of MnS in the later solidification process.

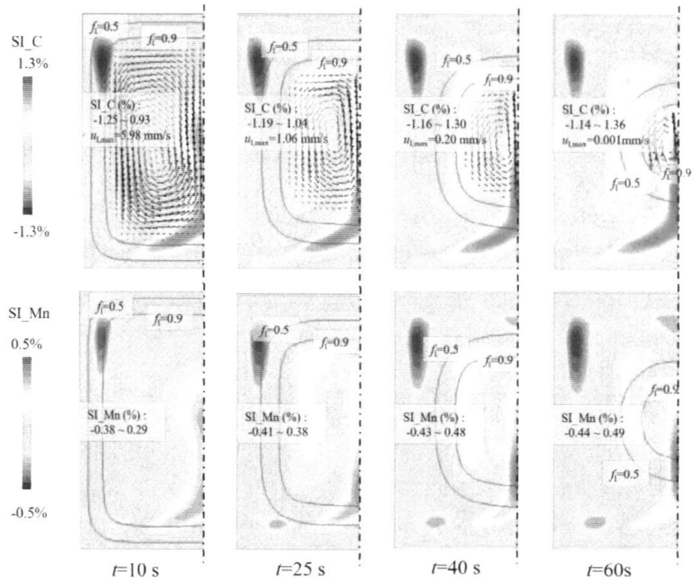

Figure 2. Evolution of C and Mn macrosegregation at different time. The vectors indicate the melt velocity, the black lines indicate the liquid fraction isoline.

3.2. *The effect of MnS formation on macrosegregation*

The macrosegregation patterns of C and Mn without and with the combination of MnS formation are shown in figure 3(a) and (b). By considering the inclusion formation, the range of segregation index of C is slightly enlarged from [-0.93%~+1.35%] to [-1.12%~+1.45%]. The area of negative macrosegregation zone is reduced while the positive macrosegregation zone is obviously enlarged.

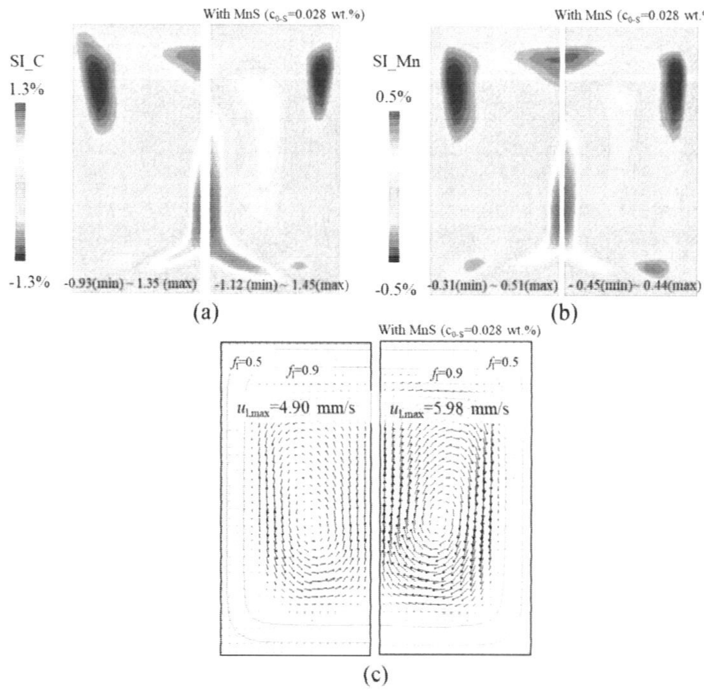

Figure 3. Effect of MnS formation on (a) C macrosegregation, (b) Mn macrosegregation, and (c) melt flow at t=10s.

The reason for these changes is that the melt flow faster, as shown in figure 3(c), due to the formation of MnS inclusion. Under such flow change, the same changes occur in the Mn component, as shown in figure 3(b) with reduced negative segregation area and increased positive segregation area. However, the consumption of Mn during the nucleation and growth of MnS is the dominant influence on Mn positive macrosegregation. As shown in figure 3(b), the positive macrosegregation index is significantly reduced from 0.51% to 0.44%.

3.3. The effect of S composition on MnS formation

It is believed that the S composition also plays an important role in the evolution of MnS. The distribution and volume fraction of MnS with two different S composition are displayed in figure 4. It should be noted that the image in figure 4 and the embedded image in figure 5 are both based on the 50×50 mm2 calculated domain shown in figure 1. The contour in figure 4 also shows the volume fraction of MnS in each cell, so the results give the preferential location for such inclusion to precipitate under such solidification condition. The distribution of inclusion shows a central symmetry characteristic, because the cooling conditions at all boundaries are uniform as 300 W·m-2·K-1. Both contours reveal that the MnS inclusions mainly located at a quarter of the height of the benchmark grid. With the varies of S composition from 0.0028 wt% to 0.3 wt%, the maximum volume fraction of MnS is at a range of 0.0013%-0.0022%. The CT experiment of inclusions in large scale ingot from reference [17] is used to validate the predicted volume fraction. The results from experiment show one or two order of magnitude higher than the predicted results from the perspective of volume fraction, because the experiment inspected all kinds of inclusion including the more common alumina. The consistency between simulation and experiment indicates that it is reasonable to estimate the effect of the formation of MnS on the macrosegregation as discussed above.

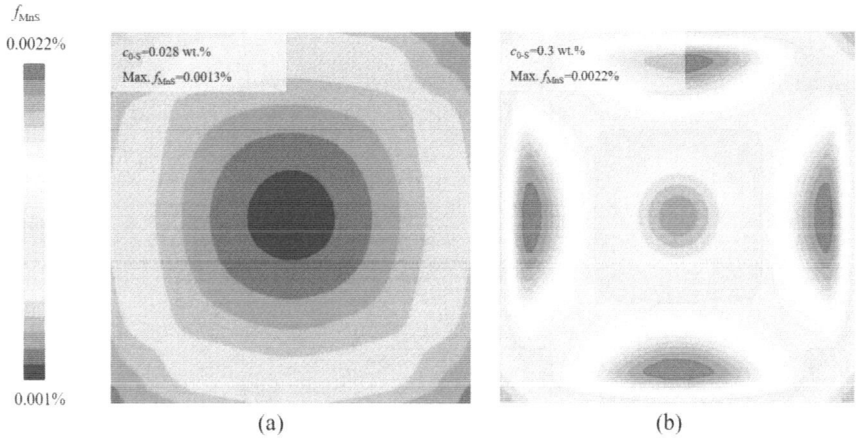

Figure 4. Predicted distribution and final volume fraction of MnS with initial S composition of (a) 0.028 wt% and (b) 0.3 wt%.

Another two cases with different S composition are conducted to investigate its dependency with the thermodynamic condition of MnS formation. The solidification process is barely affected by the S composition, thus at t = 70 s, for all the cases, around 86% of the melt has solidified, and the temperature and f_l line along the A-A line(cross central line on left half part of the benchmark) are uniform, as depicted in figure 5 with red and black lines, respectively. The other four colourful lines represent the mass transfer rate (Mlp) from liquid to MnS, which denote the one-to-one correspondence between the MnS formation start point and critical condition in each case. By increasing the S composition, both of the critical liquid fraction and the temperature to precipitate inclusion MnS are magnified.

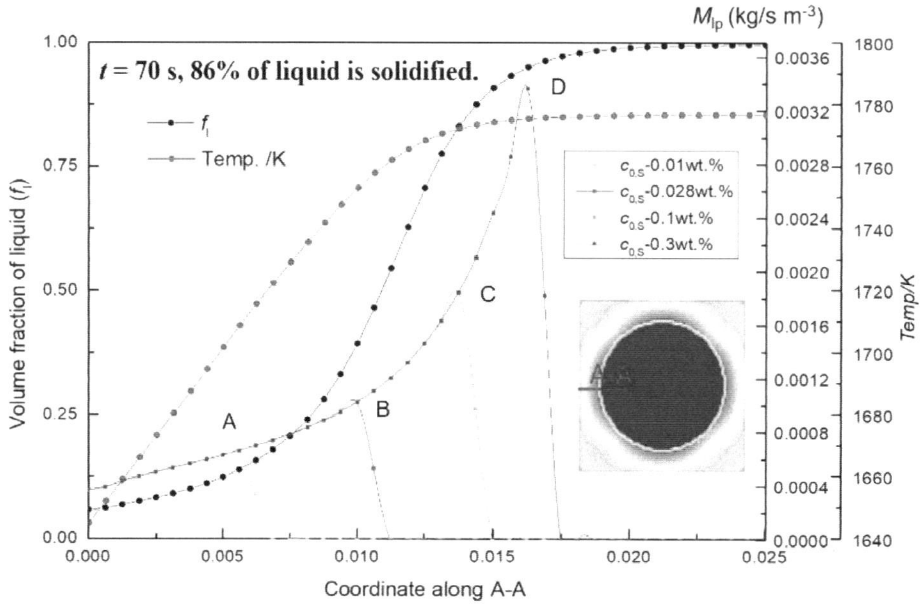

Figure 5. Effect of S composition on the formation condition (liquid volume fraction and temperature) of MnS during solidification. The embedded image is the distribution contour of mass transfer rate form liquid to MnS phase within the square domain when t = 70 s.

4. Conclusion

A comprehensive model combining nucleation and growth of MnS inclusion with a ternary alloy (Fe-0.45wt%C-1.06wt%Mn) columnar solidification model has been established and applied to a benchmark case study. Reasonable results were obtained with a qualitative agreement with the reported experimental results which demonstrates that the current inclusion-combined solidification model has the ability to predict the formation of MnS inclusion and macrosegregation at a certain given alloy composition. It was revealed that the formation of MnS will increase melt flow which finally alter the distribution of both C and Mn macrosegregation. However, the inclusion-correlated solute Mn is mainly affected by the consumption itself.

Acknowledgments

This work is sponsored by Innovate UK (Project Number. 104015), the Joint Funds of the National Natural Science Foundation of China (No. U1660203) and Jiangsu International Science and Technology Cooperation Project (No. BZ2017050) as well as China Scholarship Council (CSC).

References

[1] Ray P, Jarman A and Thomas J 1994 *J. Mater. Sci.* **29** 47-53
[2] Ludwig A, Wu M and Kharicha A 2015 *Metall. Mater. Trans. A* **46** 4854-67
[3] Choudhary S K and Ghosh A 2009 *ISIJ Int.* **49** 1819-27
[4] Mehrabian R, Keane M and Flemings M 1970 *Metall. Mater. Trans. B* **1** 3238-41
[5] Kurz W and Fisher D J 1989 *Fundamentals of Solidification* (Aedermannsdorf: Trans Tech Publications)
[6] Worster M G 2003 *Ann. Rev. Fluid Mech.* **29** 91-122
[7] Wu M and Ludwig A 2006 *Metall. Mater. Trans. A* **37** 1613-31
[8] Wu M, Ludwig A and Kharicha A 2016 Appl. Math. Model. 41
[9] Ge H, Ren F, Li J and Li J 2017 *Metall. Mater. Trans. A* **48** 1139-1150
[10] Li D, Chen X Q, Fu P and Ma X 2014 *Nat. Commun.* **5** 5572
[11] Fujii T, Poirier D R and Flemings M C 1979 *Metall. Mater. Trans. B* **10** 331-9
[12] Wang C Y and Beckermann C 1996 *Metall. Mater. Trans A* **27** 2754-64
[13] Li J, Wu M, Ludwig A and Kharicha A 2012 *IOP Conf. Ser.: Mater. Sci. Eng.* **33** 012091
[14] Ren F, Ge H, Cai D, Li J, Hu Q and Xia M 2018 *Metall. Mater. Trans. A* **49** 6243-54
[15] Combeau H, Založnik M and Hans S 2009 *Metall. Mater. Trans. B* **40** 289-304

[16] Cai D, Ren F, Ge H, Kim H S, Li J and Li J 2019 *Metall. Mater. Trans. A* **50** 1323-32
[17] Cao Y F, Chen Y and Li D Z 2016 *Acta Mater.* **107** 325-36
[18] Nurmi S, Louhenkilpi S and Holappa L 2010 *Steel Res. Int.* **80** 436-40
[19] Perez M, Dumont M and Acevedo-Reyes D 2008 *Acta Mater.* **56** 2119-32
[20] Wu M, Li J, Ludwig A and Kharicha A 2014 *Comp. Mater. Sci.* **92** 267-85
[21] Li J, Wu M, Hao J and Ludwig A 2012 *Comp. Mater. Sci.* **55** 407–18
[22] Li J, Wu M, Hao J and Ludwig A 2012 *IOP Conf. Ser.: Mater. Sci. Eng.* **27** 012055
[23] Bird R B, Stewart W E and Lightfoot E N 1960 *Transport phenomena* (New York: John Wiley and Sons)
[24] Valdez M E, Wang Y and Sridhar S 2004 *Steel Res. Int.* **75** 247-56
[25] You D, Michelic S K Wieser G and Bernhard C 2017 *J. Mater. Sci.* **52** 1797-1812

IOP Conf. Series: Materials Science and Engineering **861** (2020) 012044 doi:10.1088/1757-899X/861/1/012044

Sensitivity of macroscopic transport calculations to uncertain microscale relationships during metal alloy solidification

J Coleman and M J M Krane

School of Materials Engineering, Purdue University, West Lafayette IN, 45707, USA

E-mail: colemaj@purdue.edu

Abstract. The formation of grains during metallic solidification results in a multiphase system consisting of many moving interfaces. Models have been formulated in terms of variables describing the microscopic details of interface motion during solidification, but the computational cost prohibits their extension to the length scale of industrial castings. For this reason, macroscopic transport equations are mathematically formulated using averaging methods to replace the instantaneous description of interfaces. The most common approach is the volume-averaged, Eulerian approach relying on both the proper derivation of macroscopic field equations as well as uncertain interfacial conditions used to account for the average behavior of the interphase transfer occurring on the sub-grid length scale. In this work, common models used to describe the microsegregation occurring during solidification are evaluated in terms of the uncertainty of parametric inputs invoked in each model and their effect on macroscale predictions. We show that the qualitative macroscale predictions are not significantly changed by using more sophisticated microsegregation models, however significant uncertainty in input parameters used to describe the unresolved sub-grid interface is accumulated. Therefore, the use of simplified models is preferred until the uncertainty in model inputs can be reduced.

1. Introduction

Macrosegregation during alloy solidification is initiated by solute partitioning at the solid-liquid interface (microsegregation) and propagated by the relative motion between solid and liquid phases. Two different numerical modeling approaches have been used for calculating macrosegregation in alloy castings. The multiphase approach attempts to resolve the microscopic features of solidification in significant detail by solving partial differential equations for each phase transport term at the expense of computational efficiency and the introduction of highly uncertain model inputs [1, 2]. The continuum mixture approach [3] instead solves transport equations in terms of mixture quantities using explicit relationships to describe the individual phase transport when needed. While the microsegregation relations used in mixture models have historically been limited to simplified analytical expressions neglecting liquid undercooling (e.g. Gulliver-Scheil and Lever rule), this is not a general requirement.

In order to justify the use of more complicated microsegregation models it must be first demonstrated that these models more accurately capture the physical phenomena occurring during equiaxed solidification. Perhaps the weakest area of the volume-averaged microsegregation models lies in the interfacial relationships used to describe the evolution of equiaxed grain morphologies. Unfortunately for intermediate grain radii, of critical importance in industrial castings, there are currently no analytical solutions to fully describe the growth kinetics. Since this class of microsegregation models are based on incomplete closure relations with uncertain input parameters, simplification is often desired. Beckermann and co-workers [4] recently showed that the qualitative macrosegregation predictions during columnar solidification do not significantly change when a volume-averaged microsegregation model is replaced with the simplified Gulliver-Scheil equation. Instead, only the prediction of mesosegregates changed which is now known to be limited by the length scale of the numerical grid rather than the interfacial features which produce such variations in the physical system [5]. Nevertheless, a simplified truncated-Scheil model which includes liquid undercooling at the primary dendrite tips was able to reproduce the full volume-average model results in about half the computational time. Motivated by this study of columnar solidification, the primary goal of this work is to investigate the degree to which macrosegregation predictions change using a volume-average microsegregation model compared to a Gulliver-Scheil model during equiaxed solidification.

2. Microsegregation modeling

A modified form of the volume-averaged microsegregation model of dendritic solidification formulated by Tveito *et al.* [6] is used here. The average solid and liquid volume fractions (g_s, g_l) and compositions (C_s, C_l) are calculated using the conservation equations:

$$\frac{\partial}{\partial t}(\rho_s g_s) + \nabla \cdot (\rho_s g_s U_s) = \dot{m}_s \tag{1}$$

$$\frac{\partial}{\partial t}(\rho_s g_s C_s) + \nabla \cdot (\rho_s g_s \boldsymbol{U}_s C_s) = \dot{m}_s k_p C_l^* \tag{2}$$

$$\frac{\partial}{\partial t}(\rho_l g_l C_l) + \nabla \cdot (\rho_l g_l \boldsymbol{U}_l C_l) = -\dot{m}_s k_p C_l^* \tag{3}$$

where ρ_k is the phase density, $\boldsymbol{U}_k$ the phase velocity vector, $\dot{m}_s$ the mass phase change rate due to solidification, k_p the equilibrium partition coefficient, C_l^* the composition of the interdendritic liquid, and $g_s + g_l = 1$. Due to the highly coupled nature of these nonlinear equation, splitting methods, where the contributions of phase transport and growth are considered separately, are commonly adopted [7, 8].

2.1. *Transport stage*

In the transport stage equations (1)-(3) are solved without the influence of growth. The left-hand side of these equations are solved explicitly using the following expressions:

$$g_s^{tr} = g_s^o - \Delta t \nabla \cdot (g_s^o \boldsymbol{U}_s^o) \tag{1a}$$

$$C_s^{tr} = \frac{1}{g_s^t + \varphi_s}\left((g_s^o + \varphi_s)C_s^o - \Delta t \nabla \cdot (g_s^{tr} \boldsymbol{U}_s^o C_s^o)\right) \tag{2a}$$

$$C_l^{tr} = \frac{1}{g_l^{tr} + \varphi_l}\left((g_l^o + \varphi_l)C_l^o - \Delta t \nabla \cdot (g_l^{tr} \boldsymbol{U}_l^o C_l^o)\right) \tag{3a}$$

where φ_k is a stabilization coefficient used to bound the solution in the dilute limits of phase k, i.e. $\varphi_k = \max(10^{-6} - g_k, 0)$, the superscript o denotes the values at the previous time step Δt, and the superscript tr denotes the resulting transport values calculated in this stage. The MULES algorithm (multi-dimensional limiter for explicit solution) currently implemented in OpenFOAM [9] is used for solving equation (1a). Using the updated values of g_s^{tr} and g_l^{tr}, the solute advection equations are solved (equations (2a) and (3a)), and the resulting transport fields are stored for reference in the subsequent growth stage. Solving the transport equations explicitly introduces a strict Courant number limit, and hence time step limit, however no time step sub-cycling is required in the transport stage leading to an overall decrease in computation time.

2.2. Growth stage

In the growth stage equations (1)-(3) are solved using the updated values from the transport stage. A simplified and efficient growth model is obtained by solving the energy equation for temperature and using the semi-implicit source method of Swaminathan and Voller [10] to couple to the solid fraction evolution to the temperature field. This is the same approach used for the analytical expressions, and therefore, should limit the numerical differences attributed to the solution procedure when comparing the predictions of the microsegregation models. Details of this method can be found in the original text, the most important detail being that phase fraction is updated using a truncated Taylor Series expansion of the solid fraction. Therefore, expressions for the change in solid fraction with respect to temperature (dg_s/dT) and the temperature evaluated using the previous solid fraction value T^* are required. Using this method, the following explicit equations are used to update the phase fraction and composition fields due to the influence of growth:

$$g_s^{n+1} = g_s^n + \frac{dg_s}{dT}(T^{n+1} - T^*) \tag{1b}$$

$$C_s^{n+1} = \frac{1}{g_s^{n+1} + \varphi_s}\left((g_s^{tr} + \varphi_s)C_s^{tr} + \frac{\Delta t}{\rho_s}\dot{m}_s k_p C_l^*\right) \tag{2b}$$

$$C_l^{n+1} = \frac{1}{g_l^{n+1} + \varphi_l}\left((g_l^{tr} + \varphi_l)C_l^{tr} - \frac{\Delta t}{\rho_l}\dot{m}_s k_p C_l^*\right) \tag{3b}$$

where the superscript n denotes the time step sub-iteration level. The change in solid fraction with respect to temperature (dg_s/dT) is obtained applying Leibniz's rule to equation 1 assuming a constant phase density, i.e.:

$$\frac{dg_s}{dT} = \frac{\partial}{\partial T}\left(\frac{\partial \rho_s g_s}{dt}\right) \cong \frac{\Delta t}{\rho_s}\left(\frac{d\dot{m}_s}{dT}\right) \tag{4}$$

The mass phase change rate $\dot{m}_s$ is given by a solute mass balance at the solid-liquid interface neglecting diffusion into the solid phase:

$$\dot{m}_{s,l} = \left(\frac{1 - g_s}{1 - g_{env}}\right)\frac{\frac{S_{env}\rho_l D_l}{\delta_{env}}(C_l^* - C_l)}{(1 - k_p)C_l^*} = f\frac{a_l(C_l^* - C_l)}{(1 - k_p)C_l^*} \tag{5}$$

where g_{env} is the volume fraction of the grain envelope, S_{env} is the interfacial area of the grain envelope, δ_{env} is the liquid diffusion length, $a_l = S_{env}\rho_l D_l/\delta_{env}$ is the diffusion rate coefficient characterizing the length and time-scales of solute diffusion to the extradendritic liquid, and $f = (1 - g_s)/(1 - g_{env})$ is a modification factor to formulate the solute balance in terms of the average liquid composition in place of the extradendritic liquid composition. Equation (5) can be specified as a differentiable function with respect to temperature following the condition of interfacial equilibrium:

$$T = T_m + m_l C_l^* \tag{6}$$

where T_m is the melting temperature of the pure solvent and m_l is the linear liquidus slope take from the binary phase diagram. The terms required for the semi-implicit source-based method are found following the algebraic rearrangement of equations 5 and 6:

$$\frac{dg_s}{dT} = \frac{\Delta t}{\rho_s}\left(f\frac{m_l C_l a_l}{(1 - k_p)(T - T_m)^2}\right) \tag{7}$$

and

$$T^* = T_m + m_l C_l \left(f \frac{a_l}{(k_p - 1)\dot{m}_{s,l} + a_l} \right) \tag{8}$$

It is noted that $k_p C_l^*$ in equations (2b) and (3b) is replaced with C_s^{tr} for remelting (i.e. $\dot{m}_{s,l} < 0$) and C_l^{tr} during a eutectic reaction. In order to close the numerical model auxiliary relationships are required for determining g_{env}, S_{env}, and δ_{env}. These relationships are taken from Tveito *et al.* [6] and are calculated at the end of a given time step once the solid fraction and phase compositions have converged to a relative tolerance of 10^{-6}. Details of the full numerical model can be found in upcoming papers from the same authors; however, it is noted that the microsegregation model was verified against the recalescence benchmark of Rappaz and Thevoz for an Al-5wt%Si alloy [11].

3. Results and discussion

Before investigating the sensitivity of macrosegregation predictions to different microsegregation model inputs, the controlling transport phenomenon occurring during solidification are illustrated using the Gulliver-Scheil microsegregation model. The test case is based on SMACS benchmark [12] for a Sn-10wt%Pb alloy modified for equiaxed solidification using the following expression for the relative phase velocity in the slurry:

$$(\boldsymbol{U}_s - \boldsymbol{U}_l) = \frac{4}{3} \frac{(1 - g_{env}) g_{si}^c (\rho_s - \rho_l) \bar{d}_g^{\,2}}{C_D Re \mu_l} \mathbf{g} \tag{9}$$

where C_D is the drag coefficient taken from drag model of Wang and Beckermann [13], Re the grain Reynolds number, $\mu_l$the dynamic viscosity of the liquid, and $\bar{d}_g$ the variable grain diameter. A constant envelope packing fraction is assumed ($g_{env}^c = 0.3$), the grains are assumed to be fully globular ($g_{si}^c = 1$) assuming a value of $\bar{d}_g(g_{env} = 0.3) = 75\mu m$ at packing, and a constant relative phase density is used ($\rho_s - \rho_l = 280\ kg/m^2$).

Contour plots of the composition field, packing front, and streamlines are shown in figure 1. Since no superheat is provided and the microsegregation model assumes zero liquid undercooling, the slurry region quickly occupies the entire domain upon cooling from the right wall. Buoyancy forces near the chill drive a clockwise flow cell and negative segregation develops near the bottom, left corner as the depleted grains separate from the mixture flow due to settling. At the onset of packing, the grains become immobilized and the penetration of the buoyancy-driven flow transports enriched solute from the rigid mush to the slurry further depleting the rigid mush and enriching the slurry. By 200 s, the entire domain is occupied by the rigid mush and the clockwise cell of buoyancy-driven flow transports enriched interdendritic liquid from the top right to the bottom right corner. The final composition field due to this combined transport is shown in figure 1(c).

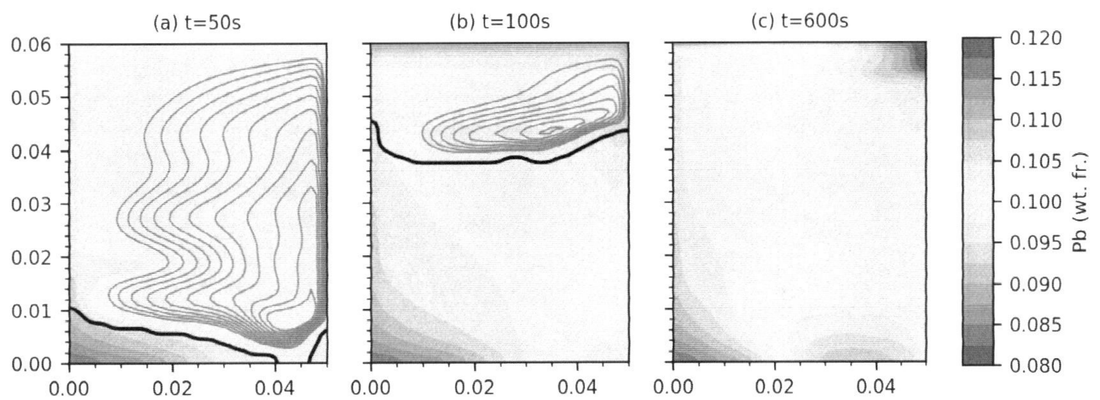

Figure 1. Lead composition distribution with clockwise rotating flow at (a)=50s, (b)=100s, and (c)=600s.

IOP Conf. Series: Materials Science and Engineering **861** (2020) 012044 doi:10.1088/1757-899X/861/1/012044

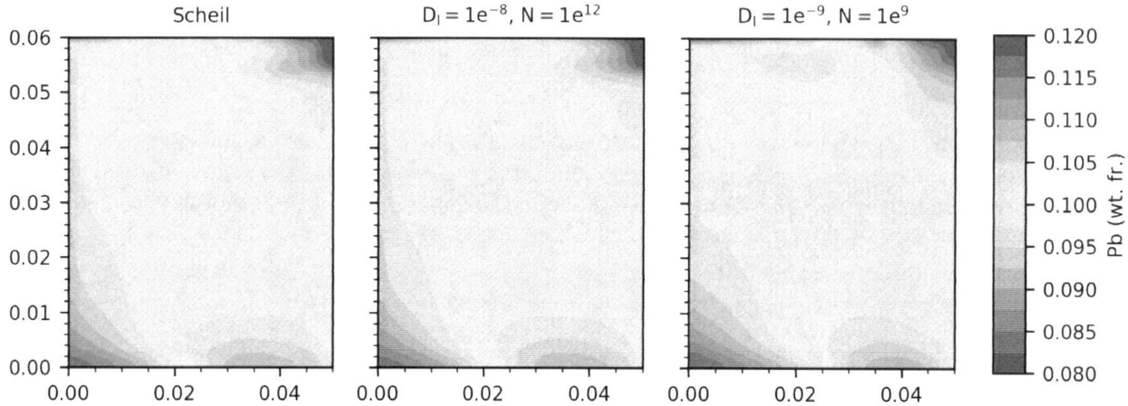

Figure 2. Final macrosegregation fields for the Scheil model and the kinetics models at extremes of the parameter space for D_l and N.

3.1. Influence of microsegregation model

The total possible effect of liquid undercooling on the macrosegregation predictions during equiaxed solidification is demonstrated by varying the two primary model inputs to the microsegregation model: the liquid mass diffusivity D_l and the grain number density n_g. These inputs are varied over the range expected for equiaxed solidification, $D_l = (10^{-9}, 5.5 \times 10^{-9}, 10^{-8})\ m^2/s$ and $n_g = (10^9, 10^{`0}, 10^{11}, 10^{12})\ m^{-3}$, for a total of 12 simulations. The liquid undercooling increases with decreasing D_l and n_g, and therefore, the highest deviation from the assumption made in the Gulliver-Scheil equation should occur for $D_l = 10^{-9}\ m^2/s$ and $n_g = 10^9 m^{-3}$. Contour plots of the final Pb composition fields are shown for the two extremes of microsegregation model inputs and show excellent agreement with the predictions using the Gulliver-Scheil equation (figure 2). This result indicates the liquid undercooling does not significantly affect the final macrosegregation pattern and can be explained by the fact that the bulk liquid phase becomes solutally well-mixed at the solid fractions upon which equiaxed grains typically pack (0.1-0.5) and the more complicated microsegregation model becomes functionally equivalent to the Gulliver-Scheil solution. Therefore, the only change in macrosegregation is due to the relative motion of the solid, interdendritic liquid, and extradendritic liquid in the slurry.

The change in the macrosegregation prediction using the volume-averaged microsegregation model from the previous Gulliver-Scheil results is quantified using the mean absolute percentage error

$$MAPE = \frac{100}{n} \sum_{i=1}^{n} \left| \frac{F_i - A_i}{A_i} \right| \tag{10}$$

where F_i is the local composition at each control volume i and A_i is the local composition predicted using the Gulliver-Scheil model. Shown in figure 3, the difference is less than 1.5% over the entire physical parameter space indicating that solutal undercooling alone does not play a significant role in the macrosegregation formation for the current case of equiaxed solidification. For reference, it was previously shown by Schneider and Beckermann [14] that assumption of linear solidus and liquidus slopes taken from binary phase diagrams and used in most solidification models is of a similar difference.

3.2. Influence of grain hydrodynamics

To investigate the effect of grain morphology on the macrosegregation, the internal solid fraction upon packing (g_{si}^c) is varied over a physically reasonable range for dendritic grains 0.75-1.0. An octahedral grain envelope is assumed, characterized by a packing fraction of $g_{env}^c = 0.4$, therefore the grains become immobilized between a solid fraction of 0.3-0.4. It is assumed that both the inter- and extradendritic liquid have the same velocity. A response surface of the global macrosegregation index (GMI) is shown in figure 4 for the grain number density and internal solid fraction. It is clear that the grain number density has a negligible effect on final macrosegregation prediction compared to internal solid fraction of the grain. This is further demonstrated by quantifying the GMI sensitivities for various model inputs using the elementary effects (EE) method in figure 4(b).

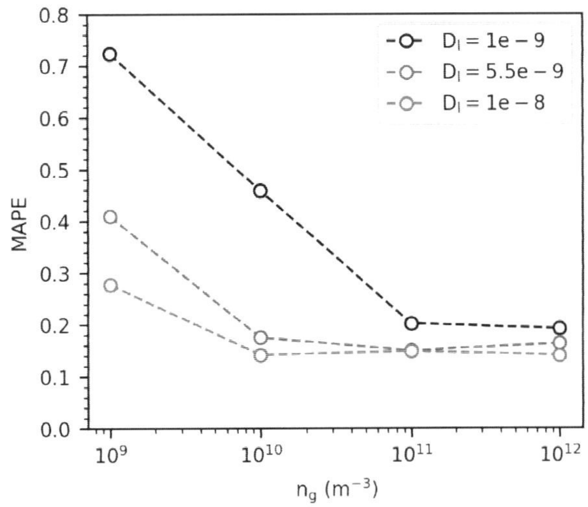

Figure 3. The mean absolute percent error (MAPE) measuring deviation of the macrosegregation predictions using the VA microsegregation models from the Gulliver-Scheil predictions for different values of liquid mass diffusivity and grain number density.

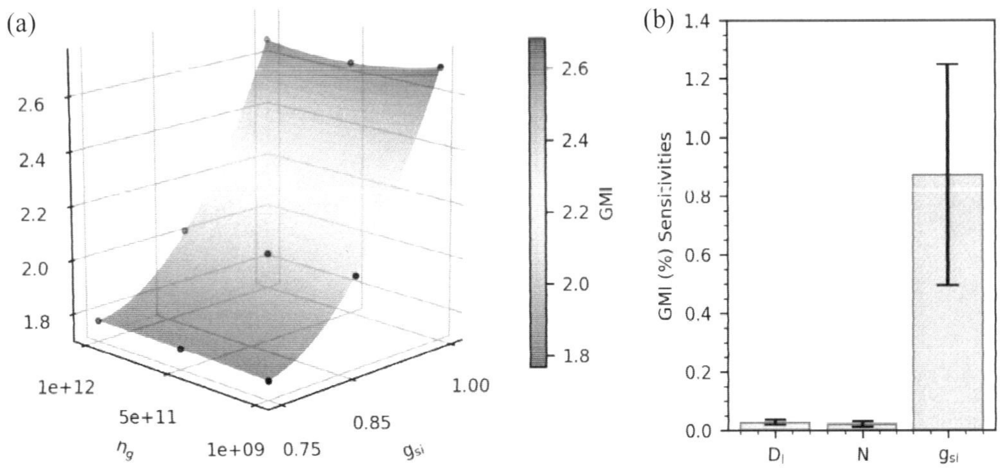

Figure 4. (a) The response surface of the mean absolute percent error (MAPE) measuring deviation of the macrosegregation predictions from those predicted using the Gulliver-Scheil equation for different values of liquid mass diffusivity and grain number density. (b) GMI sensitivities for the inputs.

4. Conclusions

First, the degree to which the macrosegregation prediction varies due to the microsegregation changes of the complicated volume-averaged models can vary from those of the Gulliver-Scheil equation are investigated. The mean absolute percent error (MAPE) over the entire physical parameter space is shown. From these results, it appears that previous studies directing the need for complicated equiaxed dendrite models were primarily motivated by the observed change in hydrodynamic behavior compared to globular grain morphologies and not the need to capture the amount of liquid undercooling. Further effort in modeling solidification should focus more on the closure relationships used for the grain structure predictions. If the uncertainty in model inputs and/or sensitivity of the grain structure predictions to said inputs cannot be reduced using these kinetics-based models, then more pragmatic relationships should be investigated.

5. Acknowledgements
The authors thank Robert Wagstaff and Novelis Inc. for the financial contribution that funded this work.

References

[1] Ni J and Beckermann C 1991 *Metall. Trans. B* **22** 349–61
[2] Wang C Y and Beckermann C 1996 *Metall. Mater. Trans. A* **27A** 2754–83
[3] Bennon W D and Incropera F P 1987 *Int. J. Heat Mass Transf.* **30** 2161–70
[4] Torabi Rad M and Beckermann C 2019 *Materialia* **7** 100364
[5] Plotkowski A and Krane M J M 2016 *Int. J. Heat Mass Transf.* **100** 11-23
[6] Tveito K O, Pakanati A, M'Hamdi M, Combeau H and Založnik M 2018 *Metall. Mater. Trans. A* **49** 2778–94
[7] Guo J and Beckermann C 2003 *Numer. Heat Tr. A-Appl.* **44** 559–76
[8] Založnik M and Combeau H 2010 *Comput. Mater. Sci.* **48** 1–10
[9] OpenFOAM®, OpenCFD 2016
[10] Voller V R and Swaminathan C 1991 *Numer. Heat Transf.* **19** 175–90
[11] Rappaz M and Thevoz P 1987 *Acta Metall.* **35** 1487–97
[12] Bellet M *et al.* 2009 Int. J. Therm. Sci. **48** 2013–16
[13] Wang C Y, Ahuja S, Beckermann C and de Groh H C 1995 *Metall. Mater. Trans. B.* **26B** 111–9
[14] Schneider M C and Beckermann C 1995 *Metall. Mater. Trans. A* **26** 2373–88

IOP Conf. Series: Materials Science and Engineering **861** (2020) 012045 doi:10.1088/1757-899X/861/1/012045

Microsegregation model including Soret effect: an application for Sn-Bi alloy

J E Santos Filho[1] and R Heringer[2]

[1] Universidade Federal da Paraíba, CCAE/DCX, 58297-000 - Rio Tinto-PB, Brazil
[2] Universidade Federal de Santa Catarina, EGR, 88040-970 - Florianópolis-SC, Brazil

E-mail: elias@dcx.ufpb.br, romulo.heringer@ufsc.br

Abstract. We present an application of a microsegregation model that takes into account the Soret effect. This model, summarized here, lies in the context of the evaluation of thermodiffusion influence in alloys solidification. Initially, we deal with the thermodiffusion effect on the microsegregation scale. Subsequently, the Soret effect on macrosegregation shall be analyzed. The present model was conceived with similar considerations compared to the Gulliver-Scheil approach. A dimensionless parameter emerged from the model, which is potentially useful for assessing the Soret effect on the solidification of binary alloys. In the present work, this parameter was applied to an Sn-Bi alloy. The microsegregation model including the Soret effect can be widely used in the study of solute redistribution.

1. Introduction

The Soret effect is the diffusion of chemical species in a mixture driven by thermal gradients. This kind of transport has been systematically neglected on solidification models [1]. This is done based on the typically low value of thermodiffusion coefficient in solid solutions compared to chemical driven diffusion coefficient. A parameter commonly used to show that is the Soret coefficient, S_T, defined by $S_T = D^T/D$, where D^T is the thermodiffusion coefficient, D is the diffusion coefficient. However, this parameter does not includes other important phenomena inherent to solidification process, such as thermal conduction or energy and solute supplying due to phase change.

In fact, recent results such as those of references [2] and [3] suggest that thermodiffusion can significantly influence relevant parameters of solidified materials in the cases studied. Accordingly, since the temperature gradient in the liquid is often high, the thermodiffusion may have a considerable or influence on the composition of the liquid and the solidified final product.

The present work is part of an effort to analysis the influence of thermodiffusion in solidification processes [4]. Initially, the objective is to obtain parameters to evaluate the effect of thermodiffusion on the microsegregation scale. The present work fits this context. Subsequently, the Soret effect on macrosegregation will be analyzed.

In this paper, we present an application of the microsegregation model, presented elsewhere, which includes the Soret effect. The present model was conceived from similar considerations made by Gulliver and after by Scheil in their well known microsegregation model [1, 5], hereafter called G-S model in this text. But here the thermodiffusion across the system boundary was taken into account. The development of the mentioned model allowed the definition of a potentially useful dimensionless parameter for evaluating Soret effect on the solidification of binary alloys. In the present work, this parameter was applied to a Sn-35wt%Bi binary alloy. Lack of space, here we will summarize the development of the model used in this text that is focused on a specific result. A more comprehensive explanation will be done elsewhere [4]. The microsegregation model including Soret effect can be widely used in the study of solute redistribution, with the advantage of being less restrictive than G-S model concerning the physical phenomena inherent to the solidification process.

In the section 2 we summarize the model developed for describe microsegregation on binary alloys including thermodiffusion. This model exposes a potential useful dimensionless parameter, which is presented in the section 3. Some results for several values of the mentioned parameter show the range of influence of thermodiffusion on microsegregation. Finally, a case study is presented for the Sn-35wt%Bi alloy.

2. Summary of microsegregation model including thermodiffusion

The present authors have proposed a solute balance equation to describe microsegregation in a binary alloy where the component can flow through the system boundary, driven by thermal gradient (thermodiffusion) [4]. This flow was taken into account via source term affecting the composition of the whole system over time.

The model was based on the following assumptions: a) an one-dimensional system between secondary dendrite arms, similar to that established for the G-S model [1, 5]; b) the solute diffusion in the solid is null ($D_s = 0$); c) the solute diffusion in the liquid is fast enough to have a uniform profile in this phase; d) local equilibrium at the solid/liquid interface and $C_s(t) = k_0 C_l(t)$; e) a binary alloy with equal and constant densities for liquid and solid phases, so that the mass and volume solid fractions are coincident; f) no convection in the liquid; g) solute may flow across the system boundary due to the effect of thermodiffusion (Soret effect); h) the secondary arm spacing remains constant over time.

2.1. One-dimensional solute balance

The solute balance in a control volume defined in accordance to the constraints *a*), *e*), *f*) and *g*) listed above may be expressed as follows:

$$\int_0^{x^*(t)} C_s(x,t)\,dx + \int_0^{x^*(t)} C_l(x,t)\,dx = X(t)\bar{C}(t) \tag{1}$$

In the present text, the subscript *l* refers to the liquid phase, while *s* indicates the solid phase. X is the whole length of the system (half the secondary arm spacing); C is the total solute composition for the entire control volume, $0 \le x \le X(t)$; the integration limit x^* is the location of the solid-liquid interface measured from the center of dendrite arm (the amount of solid in our one-dimensional analysis).

Applying the simplifications c) and d) to equation (1), dividing the result by X, making $(X - x^*) = x_l^*$ and recognizing that $g_l = x_l^*/X$, we have

$$\frac{1}{X}\int_0^{x^*} C_s\,dx + g_l\,C_l^* = \bar{C} \tag{2}$$

where $C_s^* = C_l^*(x,t)$. We have omitted (x,t) function notation to simplify the writing. Differentiating the previous equation with respect to time, using Leibniz's rule for the first term,

$$\left(\int_0^{(1-g_l)} C_s\,dg_l\right)\frac{1}{X}\frac{dX}{dt} - \int_0^{(1-g_l)} \frac{dC_s}{dt}\,dg_l + g_l\frac{dC_l^*}{dt} + (1-k_0)C_l^*\,\frac{dg_l}{dt} = \dot{\bar{C}} \tag{3}$$

where $\dot{\bar{C}} = d\bar{C}/dt$. In the last result we have also used the partition coefficient defined as $k_0 = C_l^*/C_s^*$, along with $g_s = 1 - g_l$, resulting in $C_s^*\,dg_s/dt = -k_0 C_l^*\,dg_l/dt$. In addition, X does not depend directly on *x*, allowing changing variables: $x/X = dg_s = -\,dg_l$.

Equation (3) express the solute balance in a one-dimensional control volume, where the total amount of solute can change. In the present work we deal with a particular case of (3) by applying simplifications

b) (no diffusion in the solid phase) and h) (constant secondary arm spacing). That is, $dX/dt = dC_s/dt = 0$, and equation (3) reduces to

$$g_l \frac{dC_l^*}{dt} + (1-k_0)C_l^* \frac{dg_l}{dt} = \dot{\bar{C}} \tag{4}$$

2.2. Thermodiffusion across the boundary as a source term

The total mass of solute crossing the boundary S surrounding V, per unit of time, is

$$F_p = -\int_S \vec{j}\,.\,\hat{n}\,dS \tag{5}$$

where the flux of solute $\vec{j}$ due to thermodiffusion is given by [6],

$$\vec{j} = -\frac{L_{1q}}{T^2}\vec{\nabla}T = -\rho C_A C_B D^T \vec{\nabla}T \tag{6}$$

C_A and C_B are the solvent and solute mass fraction, respectively. D^T is the thermodiffusion coefficient ($[D^T] = m^2\,K^{-1}\,s^{-1}$). L_{1q} is the Onsager's phenomenological cross coefficient and T is the temperature.

To consider solute flux crossing the boundary as a source term acting at each point in the volume,

$$\dot{\bar{C}} = \frac{1}{\rho V}F_p = -\frac{1}{\rho V}\int_{S_l}\vec{j}\,.\hat{n}\,dS = \frac{1}{\rho V}\int_{S_l}\frac{L_{1q}}{T^2}\vec{\nabla}T\,.\,\hat{n}\,dS \tag{7}$$

Note that V is the total volume (solid plus liquid), since $\bar{C}$ is the average concentration over the entire volume. However, we consider that all thermodiffusion occurs only through the liquid boundary, S_l. Here, it should be noted that the control volume V is small enough so that the integrand can exit the above integral. However, for convenience, $\vec{\nabla}T$ will be left in the operator. In fact, the is not computed within V, but occurs between V and the neighborhood via S_l.

For what follows it will be useful to remember that, for a single phase region of volume V, subjected only to heat conduction, we have that,

$$\frac{1}{\alpha}\int_V \frac{\partial T}{\partial t}dV = \int_S \vec{\nabla}T\,.\,\hat{n}\,dS \tag{8}$$

where the divergence theorem was applied, with S being the surface containing the arbitrary volume V, and the thermal diffusivity α was considered a constant.

Extracting L_{1q}/T^2 from the integral in equation (7) and using equation (8) with $V = V_l$ and $S = S_l$,

$$\dot{\bar{C}} = \frac{1}{\rho V}\frac{L_{1q}}{T^2}\int_{V_l}\frac{1}{\alpha}\frac{\partial T}{\partial t}dV = \frac{L_{1q}}{\rho\alpha}\frac{g_l}{T^2}\frac{dT}{dt} \tag{9}$$

For the last result, both time derivative of temperature, $\partial T/\partial t$ $(= dT/dt)$, and thermal diffusivity α were considered uniform in V. In addition, it was made $V_l/V = g_l$.

2.3. Solute balance including thermodiffusion

Replacing from equation (9) in the equation (4),

$$g_l \frac{dC_l^*}{dt} + (1-k_0)C_l^* \frac{dg_l}{dt} = \frac{L_{1q}}{\rho\alpha}\frac{g_l}{T^2}\frac{dT}{dt} \tag{10}$$

Assuming that the *liquidus* curve can be approximated locally by a linear coefficient line m_l, the *liquidus* slope, we obtain $T - T_f = m_l C_l^*$ and $dC_l^*/dt = (1/m_l)dT/dt$. Here, T_f is a reference temperature for which the solute composition is null (the melting temperature of pure solvent). Using this linearization of the *liquidus* line and some rearrangement, we rewrite (10) as follows.

$$\frac{1}{g_l}\frac{dg_l}{dt} = \frac{1}{k_0 - 1}\left[\frac{1}{\mathrm{T} - \mathrm{T}_f} - \frac{\mathrm{m}_l L_{1q}}{\rho\alpha}\frac{1}{\mathrm{T}^2(\mathrm{T} - \mathrm{T}_f)}\right]\frac{d\mathrm{T}}{dt} \tag{11}$$

The expression (11) is a separable first order ordinary differential equation subject to the following boundary conditions:

$$g_l = 1 \Leftrightarrow \mathrm{T} = \mathrm{T}_0 \tag{12}$$

where T_0 is the *liquidus* temperature corresponding to the nominal (initial) concentration of the alloy, C_0. That is, $\mathrm{T}_0 = \mathrm{m}_l\mathrm{C}_0 + \mathrm{T}_f$.

The resolution of the above boundary value problem (11) and (12) provided the main result for the microsegregation model in a binary alloy including Soret effect [4]:

$$g_l = \left(\frac{\mathrm{T} - \mathrm{T}_f}{\mathrm{T}_0 - \mathrm{T}_f}\right)^{\Theta}\left(\frac{\mathrm{T}_0}{\mathrm{T}}\right)^{\Lambda}\left[exp\left(\frac{\mathrm{T}_0}{\mathrm{T}} - 1\right)\right]^{\Phi} \tag{13}$$

with

$$\Theta = -\frac{1}{1 - k_0} + \frac{\mathrm{C}_{A_0}\mathrm{C}_{B_0}\mathrm{m}_l\mathrm{D}^{\mathrm{T}}}{(1 - k_0)\alpha}; \quad \Lambda = \frac{\mathrm{C}_{A_0}\mathrm{C}_{B_0}\mathrm{m}_l\mathrm{D}^{\mathrm{T}_f}}{(1 - k_0)\alpha}; \quad \Phi = \frac{\mathrm{C}_{A_0}\mathrm{C}_{B_0}\mathrm{m}_l\mathrm{D}^{\mathrm{T}_r}}{(1 - k_0)\alpha} \tag{14}$$

where $\mathrm{T}_\mathrm{r} = \sqrt{\mathrm{T}_f\mathrm{T}_\mathrm{r}}$ is the geometric mean between T_f and T_r. The phenomenological coefficient L_{1q} is not directly available. To get around this fact, the exponents (14) were written in terms of thermodiffusion coefficient, D^T, evaluated for each temperature T_f and T_r:

$$\mathrm{D}^{\mathrm{T}_f} = \frac{L_{1q}}{\rho \mathrm{C}_{A_0}\mathrm{C}_{B_0}\mathrm{T}_f^2} \quad \text{and} \quad \mathrm{D}^{\mathrm{T}_r} = \frac{L_{1q}}{\rho \mathrm{C}_{A_0}\mathrm{C}_{B_0}\mathrm{T}_r^2} \tag{15}$$

where C_{A_0} and C_{B_0} are the initial mass fraction of the alloy components.

Note that if $\mathrm{D}^\mathrm{T} = 0$ (no thermodiffusion), equation (13) reduces to the Gulliver-Scheil microsegregation model given by

$$g_l = \left(\frac{\mathrm{T} - \mathrm{T}_f}{\mathrm{T}_0 - \mathrm{T}_f}\right)^{-\frac{1}{1-k_0}} \tag{16}$$

3. A dimensionless parameter for Soret effect on solidification

Observing the exponents of equation (13) presented at equations (14), we recognize the following dimensionless parameter [4]:

$$\mathcal{H} = \frac{\mathrm{C}_{A_0}\mathrm{C}_{B_0}\mathrm{m}_l\mathrm{D}^{\mathrm{T}}}{(1 - k_0)\alpha} \tag{17}$$

This parameter relates thermodiffusion, heat transfer and the amount of solute segregated due to solidification.

Applied to (13), $\mathcal{H}$ shall be evaluated for T_f and T_r using equations (15), so that

$$g_s = 1 - \left(\frac{\mathrm{T} - \mathrm{T}_f}{\mathrm{T}_0 - \mathrm{T}_f}\right)^{-\frac{1}{(1-k_0)} + \mathcal{H}_f}\left(\frac{\mathrm{T}_0}{\mathrm{T}}\right)^{\mathcal{H}_\mathrm{r}}\left[exp\left(\frac{\mathrm{T}_0}{\mathrm{T}} - 1\right)\right]^{\mathcal{H}_\mathrm{r}} \tag{18}$$

where, for convenience, we have used the solid fraction g_s with $g_s = 1 - g_l$. Note that if $\mathcal{H}$ vanishes, we have the G-S equation from the above expression.

IOP Conf. Series: Materials Science and Engineering **861** (2020) 012045 doi:10.1088/1757-899X/861/1/012045

4. Results and discussion

4.1. Preliminary results

We have generated some results from equation (18) using values $\mathcal{H} = \{10^{-4}, 10^{-3}, 10^{-2}, 10^{-1}\}$ and $\mathcal{H} = \{-10^{-4}, -10^{-3}, -10^{-2}, -10^{-1}\}$, considering data of table 1. These results are shown in the figures 1 and 2, plotted along with both G-S and lever-rule (LR) equations. The values in these figures are arbitrary, but table 1 are valid for Sn-35*wt%*Bi. For the case $\mathcal{H} > 0$ (figure 1), the curves from equation (18) lie below that of G-S model. For $\mathcal{H} < 0$ (figure 2), the curves from equation (18) are above the G-S curve. In our case, the figures 1 and 2 show that for $|\mathcal{H}| < 10^{-2}$ the contribution of Soret effect for (g_S, T) profile is negligible. Thermodiffusion gets relevance around $|\mathcal{H}| \sim 10^{-1}$, when the curve for the present model is quite different from that of G-S model.

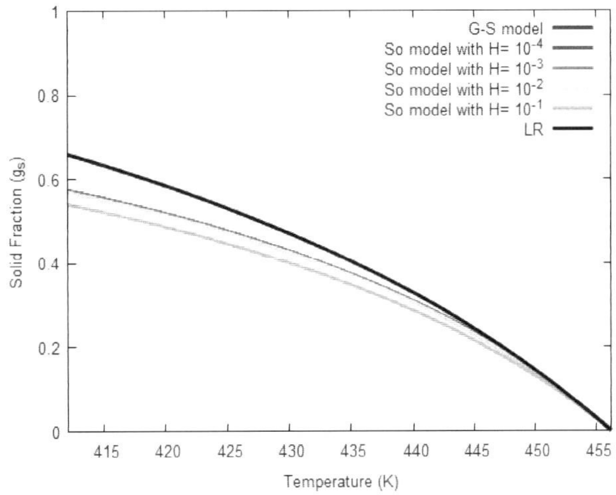

Figure 1: Solidification path for some values of $\mathcal{H} > 0$.

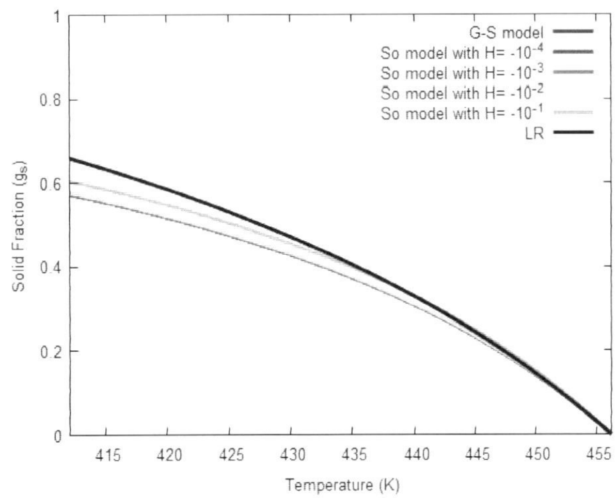

Figure 2: Solidification path for some values of $\mathcal{H} < 0$.

4.2. An application of microsegregation model with Soret effect

We present here an application of the microsegregation model including the Soret effect. We have based our calculation on the data from Jafar-Salehi [3] (see table 1).

Considering the Sn-Bi phase diagram from reference [7], the *liquidus* and *solidus* lines have been locally fitted by linear approximations. For values of (C,T) around $\mathrm{C}_0 = 35wt\%\mathrm{Bi}$, this approximation would be as follows

$$\mathrm{T} = -1.6314\mathrm{C}_l + 513.34 \tag{19}$$

$$\mathrm{T} = -4.4280\mathrm{C}_s + 505.14 \tag{20}$$

$$\mathrm{m}_l = -1.6314 \tag{21}$$

$$\mathrm{m}_s = -4.4280 \tag{22}$$

$$k_0 = \frac{\mathrm{m}_l}{\mathrm{m}_s} = 0.368 \tag{23}$$

Table 1: Physical properties of Sn-Bi alloy [3].

Property	Symbol	Value
Nominal solute mass fraction (wt % Bi)	C_0	35
Thermal diffusivity ($\mathrm{m}^2\,\mathrm{s}^{-1}$)	α	2.622×10^{-5}
Eutectic mass fraction (wt % Bi)	C_{eut}	57
Diffusion coefficient ($\mathrm{m}^2\,\mathrm{s}^{-1}$)	D	1.89×10^{-9}
Thermal diffusion coefficient ($\mathrm{m}^2\,\mathrm{s}^{-1}\,\mathrm{K}^{-1}$)	D^{T}	4.697×10^{-10}

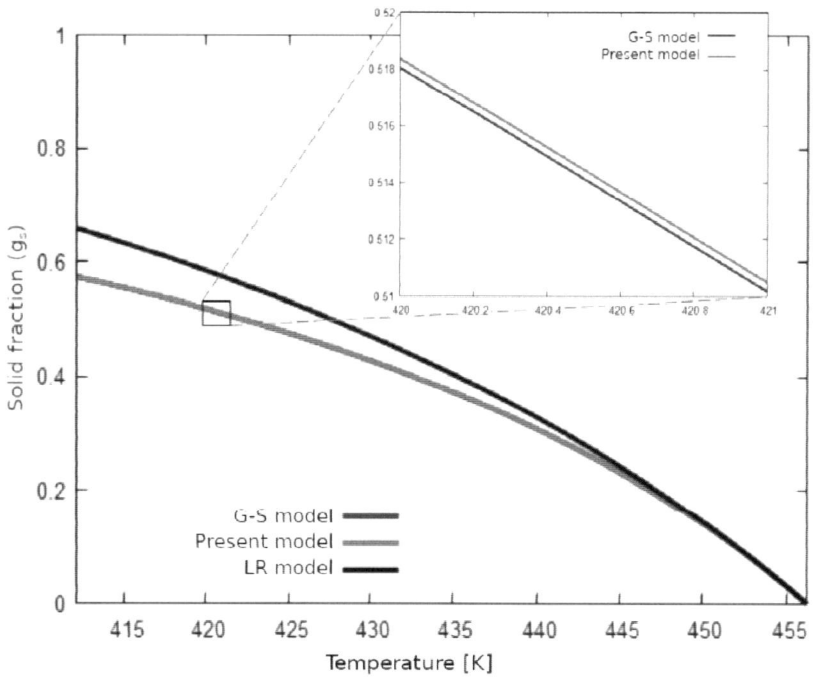

Figure 3: Solidification path of Sn-35%wtBi. Comparison between G-S model, the present model and the lever rule (LR). An enlarged detail of the present model and the G-S model curves is shown.

To employ the microsegregation model, we must calculate the dimensionless number $\mathcal{H}$. According to the physical properties listed in the table 1, the equation (17) results in

$$\mathcal{H} = -1.053\times10^{-3} \tag{24}$$

This result allows us to calculate $g_S(T)$ from equation (18). However, based on the analysis made in the previous section, we can expect that thermodiffusion can be neglected in this case. To verify this, we calculate the solidification path $g_S(T)$ to compare it with G-S results.

Figure 3 shows the results of equation (18) for the present case. The curve of G-S and that of the present model are indistinguishable on the scale of the whole path. This fact is confirmed in figure 4, which shows the difference between the models. The maximum difference found was 3.24×10-4, confirming the previous conclusion made only by evaluating the parameter.

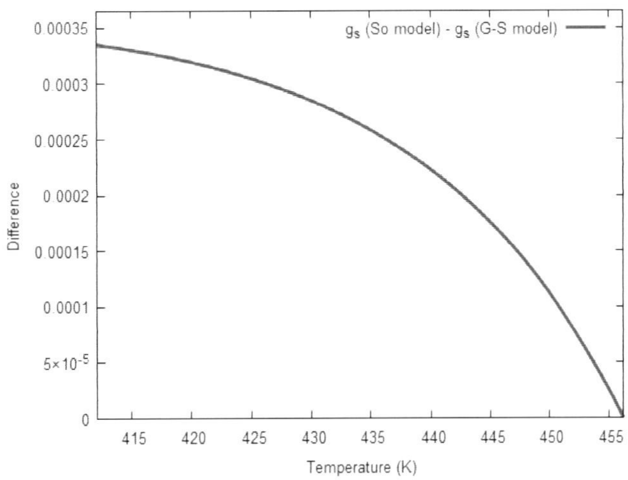

Figure 4: Difference between the curves 'present model' and 'G-S model' from the figure 3.

Despite the seemingly negligible influence of thermodiffusion on microsegregation (for the present alloy), it does not mean that it has no effect on other phenomena present in solidification. On this subject Platten and Legros [8, p. 576] stated: "The mass fraction gradient established under the effect of thermal diffusion is very small (...), but our theme is that such a small concentration gradient disproportionately influences hydrodynamics relative to its contribution to the buoyancy of the fluid. Therefore thermal diffusion cannot be ignored". Convection is not taken into account in the present model. However, microsegregation models can be coupled in multi-scale solidification modeling, where convection can be a major player.

The solidification paths presented in figures 1 and 2 were calculated for a specific alloy. For now, we cannot conclude that parameter $\mathcal{H}$ is better than the Soret coefficient to evaluate the thermodiffusion influence on solidification. However, this parameter groups all relevant transport properties presented at solidification process. This prevents us that it can be more suitable for evaluation of the thermodiffusion on solidification than Soret coefficient is.

5. Conclusions

In this text we have summarized an analytical model for microsegregation during the solidification of a binary alloy, including thermodiffusion effect. A new parameter emerges naturally from the model, which correlates thermodiffusion, heat transfer, and the amount of solute segregated due to solidification. Equation (17) suggests that, in addition to thermodiffusion coefficient itself, the solute content, the *liquidus* slope and the partition coefficient are also parameters indicating the thermodiffusion relevance in solidification phenomena.

Depending on the value of the parameter $\mathcal{H}$, we can attempt a non-negligible difference between the solidification paths calculated using the present model and those from G-S equation. The present model was applied to calculate the solidification path of an Sn-35$wt\%$Bi alloy. The results were compared with those from G-S equation. The calculation shown no significant difference between the two models, indicating that thermodiffusion is not a relevant phenomenon for solid fraction path in the case studied. This conclusion could also be made inspecting the parameter $\mathcal{H} \sim 10^{-3}$ for the case considered.

More studies are necessary to verify the usefulness of $\mathcal{H}$ as a similarity parameter. In addition, a simplified version of the result (18) may be possible without major losses.

References

[1] Dantzig J A and Rappaz M 2009 *Solidification* (EPFL press)
[2] Satheesh A and Mohan C 2017 *Front. Heat Mass Transf.* **8** 36
[3] Jafar-Salehi E, Eslamian M and Saghir M 2016 *Eng. Sci. Techn. Int. J.* **19** 511-517
[4] Heringer R and Santos Filho J E *To be published*
[5] Kurz W and Fisher D J 1998 *Fundamentals of Solidification* (Trans Tech Publications Ltd)
[6] De Groot S and Mazur P 2013 *Non-Equilibrium Thermodynamics* (Dover Publications)
[7] Abdelaziz S M, Zahran H and El-Rehim A A 2017 *Int. J. Adv. Eng. Technol.* **10** 73
[8] Platten J and Legros J 2012 *Convection in Liquids* (Springer Verlag)

IOP Conf. Series: Materials Science and Engineering **861** (2020) 012046 doi:10.1088/1757-899X/861/1/012046

Cellular automaton modelling of M_7C_3 carbide growth during solidification of Fe-C-Cr alloy

H W Zhang[1,2], S Zhang[1,2], Y C Wang[1,2], Y Z Hao[1,2], M Miao[1,2], K Nakajima[3], H Lei[1,2], Q Wang[1,2] and J C He[1,2]

[1] Northeastern University, Key Laboratory of Electromagnetic Processing of Materials, Shenyang, 110004, P.R. China
[2] Northeastern University, School of Metallurgy, Shenyang, 110004, P.R. China
[3] KTH Royal Institute of Technology, Department of Materials Science and Engineering, SE-100 44 Stockholm, Sweden

E-mail: hongweizhang@epm.neu.edu.cn

Abstract. A microscopic cellular automaton combined with macroscopic heat and solute transport was developed to simulate the mutual growth and evolution of austenite and M_7C_3 carbide grains during Fe-C-Cr ternary alloy solidification process. The diffusion of solute C and Cr are contributed together to the constitutional undercooling, together with curvature undercooling, for obtaining the grain growth rate of austenite and M_7C_3 carbide. Results show that, the absorption of solute C and Cr by M_7C_3 grain and rejection by austenite grain promote the two grains' cooperative growth. Once they approach to each other, the austenite grain quickly overgrows towards the M_7C_3 grain till finally envelops it. The simulated solidification morphology of the Fe-4wt%C-17wt%Cr alloy, predicted averaged grain size of M_7C_3 carbides and C and Cr concentration in austenite grains agree with the experimental measurements and solidification path prediction. The predicted average liquid concentration curve fits with the LR, GS and PE prediction at the initial M_7C_3 precipitation regime and after austenite grains fully enveloping towards the M_7C_3 grains, returns to overlap the LR, GS and PE prediction curves.

1. Introduction

Fe-C-Cr based alloys have been widely applied in metallurgical, mineral and mechanical fields due to their great wear resistance [1], in which the carbides, especially M_7C_3 carbide's amount, size, morphology and distribution in the microstructure contribute much to the final properties. As reported, the coarse M_7C_3 carbide precipitated in Fe-C-Cr melt spoils the continuity of the matrix, speeds up the formation and extension of the crack and consequently shortens the operation span of the alloys [2].

The theoretical and experimental researches have shown that M_7C_3 carbide grows in a faceted hexagonal close-packed (HCP) structure, which behaviours as an irregular hexagonal shape in 2D section [3]. The austenite grows in a dendritic structure when growing isolatively. With solidification proceeding, the precipitated austenite and M_7C_3 carbide grains grow up mutually, while the growing details are not very clear. Besides the theoretical and experimental studies, the numerical simulation gives great help in revealing the solidification mechanics. And among the field, the microscopic cellular automaton has gained great favour [4] in quantitatively evaluating the interaction between concentration evolution and grain growth, combining with macroscopic melt flow, heat and solute transport, and describing the detailed dendritic structure with an actual calculation size and economic calculation cost.

Thus the present paper studied the mutual growth of M_7C_3 carbide and austenite grains during solidification of a Fe-4wt%C-17wt%Cr ternary alloy with microscopic cellular automaton model. The study is expected to clarify their mutual growth mechanics and further control the distribution of M_7C_3 carbide grains in austenite matrix for obtaining better mechanical properties of the alloy.

2. Cellular automaton model

2.1. Assumptions

A) The heat diffusion of the molten alloy and the release of the latent heat during the solidification are ignored. *B)* The melt flow is omitted. *C)* The interface is always kept as thermodynamic equilibrium with ignoring the thermodynamic undercooling and kinetic undercooling. *D)* The physical properties of the alloys are kept constant.

2.2. Temperature field

The temperature of the melt is uniformly cooled down with a constant cooling rate $\dot{T}$,

$$\frac{\partial T}{\partial t} = \dot{T} \tag{1}$$

Then it is updated through the explicit scheme.

2.3. Concentration field

The diffusion of solute i (C or Cr) is governed by the following equation with zero flux condition at the region boundary,

$$\frac{\partial \bar{C}_i}{\partial t} = D\left(\frac{\partial^2 P_i}{\partial x^2}\right) + D\left(\frac{\partial^2 P_i}{\partial y^2}\right) \tag{2}$$

where $\bar{C_i}$ is the average solute concentration of solute i, expressed as $\bar{C_i} = f_\gamma C_{\gamma,i} + f_M C_{M,i} + f_L C_{L,i}$. f_γ, f_M and f_L are the mass fraction of austenite, M_7C_3 carbide and liquid; $C_{\gamma,i}$, $C_{M,i}$ and $C_{L,i}$ are the concentration of solute i in austenite, M_7C_3 carbide and liquid, respectively.

P_i is defined as an equivalent liquid concentration to unify the discontinuous solute concentration in phase and at interface, which is expressed as $C_{L,i} f_L + \sum_n \Delta f_M^n C_{L,i}^n + f_\gamma C_{\gamma,i}/k_{p,\gamma/L,i}$ at interface, $C_{L,i}$ in liquid, $\sum_n \Delta f_M^n C_{L,i}^n$ in carbide M and $C_{\gamma,i}/k_{p,\gamma/L,i}$ in austenite cell, respectively. Here Δf_M^n is the increase of M_7C_3 phase fraction at time step n, $C_{L,i}^n$ is the liquid concentration before M_7C_3 carbide's new formation at step n. $k_{p,\gamma/L,i}$ is partition coefficient of solute i at austenite/liquid interface, valued as 0.407 for C and 0.576 for Cr through solidification path prediction with partial equilibrium approximation [5, 6] and TCFE6 steel database [7].

The diffusion coefficient D is expressed as $D = 0.5[f_\gamma D_{\gamma,i} + f_M D_{M,i} + f_L D_{L,i}]|_j + 0.5[f_\gamma D_{\gamma,i} + f_M D_{M,i} + f_L D_{L,i}]|_{j+1}$. $D_{\gamma,i}$, $D_{M,i}$ and $D_{L,i}$ are the diffusion coefficient of solute i in austenite, M_7C_3 and liquid, valued as $D_{\gamma,C}$=2.57×10^{-10} m^2/s, $D_{\gamma,Cr}$=3.67×10^{-14} m^2/s, $D_{M,C}$=$D_{M,Cr}$=0, $D_{L,C}$=9.6×10^{-10} m^2/s, $D_{L,Cr}$=8.23×10^{-10} m^2/s through retrieving from MOB2 steel mobility database [8].

2.4. Solidification path and nucleation

Upon the nominal concentration of Fe-4wt%C-17wt%Cr ternary alloy, the precipitation temperatures of austenite and M_7C_3 carbide are T_γ=1266 °C and T_M=1304 °C through solidification path prediction with partial equilibrium approximation.

The nucleation sites of M_7C_3 and austenite grains are initialized according to the experimental morphology [3]. At each phase's nucleation site, the nucleation undercooling is randomly chosen among those nucleation undercoolings at the prescribed nucleation sites for the phase, which is set based on the idea of reference [9]. That is, the number of the nucleation sites, N_V^{nucl}, among the total volume V is determined by ($n_{\max} V dn$) at a certain nucleation undercooling, ΔT_v^{nucl}, sequently chosen within the given range, here is 0 ~ -20K. dn is the increase of grain density by the increase of undercooling d(ΔT) as indicated in Gaussian distribution, dn/d(ΔT), which is dependent of three parameters: the maximum nucleation density, $n_{\max}$, the maximum nucleation undercoolig, $\Delta T_{\max}$, and its deviation, ΔT_σ. The data for M_7C_3 and austenite are $n_{\max, M_7C_3} = 2 \times 10^8/m^3$, $\Delta T_{\max, M_7C_3} = 0.4K$, $\Delta T_{\sigma, M_7C_3} = 0.1K$ and $n_{\max,\gamma} = 2 \times 10^8/m^3, \Delta T_{\max,\gamma} = 0.2K$, $\Delta T_{\sigma,\gamma} = 0.1K$, respectively. These nucleation sites are randomly positioned in region V with the same ΔT_v^{nucl}. Through this way each phase is provided with a set of prescribed nucleation undercoolings at its nucleation sites.

Once the local undercooling of the chosen nucleation site is larger than the prescribed nucleation undercooling of a certain phase, it nucleates. The state of the cell then changes to interface, and a random crystal orientation is given to the nucleation site. As for a nucleated austenite site, a random orientation ranged within [-π/4, π/4] is given since it is growing in a four-fold symmetrical dendrite shape. Similarly, the random crystal orientation for a M_7C_3 carbide nucleus is ranged at [-π/6+2, π/6+2] due to its six-fold symmetrical faceted hexagonal shape in 2D cross section, here +2 is just to distinguish its orientation with that of the austenite.

2.5. *Interface equilibrium concentration*

The total undercooling at the grain interface is the contribution of constitutional and curvature undercooling when omitting the thermodynamic and kinetic effects,

$$T = T_L - \Delta T_R - \Delta T_C \tag{3}$$

where T_L is liquidus temperature. The curvature undercooling ΔT_R is expressed as

$$\Delta T_R = \Gamma K f(\varphi, \theta) \tag{4}$$

where the Gibbs-Thomson coefficient Γ is valued as 3.7×10^{-7}m·K for austenite/liquid interface and 1.9×10^{-7} m·K for M_7C_3 carbide/liquid interface [10]; interface curvature K is expressed through counting - cell technique [11] of the solid fractions around; $f(\varphi, \theta)$ is an expression describing the anisotropy of the interface energy according to Gibbs-Thomson equation [12], $f(\varphi, \theta) = 1 + (1 - k^2)\varepsilon\cos[k(\varphi - \theta)]$. The crystal symmetry parameter k is valued as 4 for FCC and 6 for HCP structure [13]; the anisotropy coefficient ε is chosen as 0.04 for austenite and 0.025 for M_7C_3 carbide; θ is crystallographic orientation; φ is the angle between the normal of the interface and x-axis, $\varphi = \cos^{-1}[(\partial f_S/\partial x)/\sqrt{(\partial f_S/\partial x)^2 + (\partial f_S/\partial y)^2}]$.

The constitutional undercooling, ΔT_C, in a Fe-C-Cr alloy is the contribution of solute C and Cr,

$$\Delta T_C = m_{L,C}(C_{0,C} - C^*_{L,C}) + m_{L,Cr}(C_{0,Cr} - C^*_{L,Cr}) \tag{5}$$

where $m_{L,i}$ is the liquidus slope of solute i, its value is retrieved from solidification path prediction with partial equilibrium approximation for Fe-4wt%C-17wt%Cr alloy. They are $m_{L,\gamma,C}$= -95.49 K/wt% and $m_{L,\gamma,Cr}$ =6.14 K/wt% at austenite-liquid regime and $m_{L,M,C}$=90.37 K/wt% and $m_{L,M,Cr}$=15.29 K/wt% at M_7C_3 carbide-liquid regime. $C^*_{L,i}$ is interface equilibrium concentration, expressed as $C^*_{L,\gamma,i}$ for austenite and $C^*_{L,M,i}$ for M_7C_3 carbide. $C_{0,i}$ is initial concentration.

The intrinsic relation of the two solutes' diffusion for one unique interface growth rate is obtained from the definition of solute supersaturation [14]. In pure diffusion process, it is expressed as an Ivantsov function of Peclet number Pe_i,

$$\Omega = (C^*_{L,i} - C_{0,i})/(C^*_{L,i} - C^*_{S,i}) = Iv(Pe_i) = Pe_i \exp(Pe_i)E_1(Pe_i) \tag{6}$$

and $C^*_{L,i}$ can be written for austenite ($k_{p,\gamma/L,i}$=const) and carbide interface ($C^*_{M,i}$=const) as

$$C^*_{L,i} = \frac{C_{0,i}}{1-(1-k_{p,\gamma/L,i})\Omega} \quad \text{and} \quad C^*_{L,i} = \frac{C_{0,i} - \Omega C^*_{M,i}}{1-\Omega} \tag{7}$$

where $E_1(Pe_i)$ is the exponential integral function, $E_1(Pe_i) = \int_{Pe_i}^{\infty} \exp(-\tau)/\tau d\tau$. Peclet number Pe_i is defined as

$$Pe_i = V_n R/2D_{L,i} \tag{8}$$

and the solute C and solute Cr 's Peclet number has the relation of

$$Pe_C = Pe_{Cr} D_{L,Cr} / D_{L,C} \tag{9}$$

Thus the constitutional undercooling ΔT_{C} (equation (5)) can be expressed as an unique function of Pe_{Cr}, i.e. $\Delta T_{\mathrm{C}} = f(Pe_{Cr})$. Substituting ΔT_{C} (equation (5)) and ΔT_{R} (equation (4)) in to equation (3), the unknown Pe_{Cr} can be obtained. Then the interface equilibrium concentration $C_{L,i}^*$ is got from equation (7).

2.6. *Interface growth rate and solid fraction*

The interface growth rate, $\vec{\mathrm{V}}_{\mathrm{n}}$, is calculated through the solute conservation at solid/liquid interface,

$$\vec{V}_n(C_{L,i}^* - C_{S,i}^*) = \left[-D_{L,i}\left(\frac{\partial C_{L,i}}{\partial x} + \frac{\partial C_{L,i}}{\partial y}\right) + D_{S,i}\left(\frac{\partial C_{S,i}}{\partial x} + \frac{\partial C_{S,i}}{\partial y}\right)\right] \cdot \vec{n} \tag{10}$$

Then the solid fraction is updated,

$$f_S^{n+1} = f_S^n + \Delta f_S \tag{11}$$

where

$$\Delta f_S = \frac{V_x \Delta t \Delta s + V_y \Delta t \Delta s - V_x V_y \Delta t^2}{\Delta s^2} = \frac{\Delta t}{\Delta s}\left(V_x + V_y - V_x V_y \frac{\Delta t}{\Delta x}\right) \tag{12}$$

Upon the austenite growth, a disturbance function, $1 + A(1 - 2\mathrm{Rand}(\,))$, is multiplied with Δf_S to describe the influence of interfacial disturbance on dendrite growth, where A is chosen as 0.1; Rand( ) is a random number within [0,1]. As for M_7C_3 carbide, a shape factor G is adopted to eliminate the mesh anisotropy, which is related to the state of the nearest, second, third and fourth order adjacent neighbour group, S_m^I, S_m^{II}, S_m^{III} and S_m^{IV},

$$G = \min\left[1, \frac{1}{4}\left(\sum_{m=1}^{4} S_m^{I} + \frac{1}{\sqrt{2}}\sum_{m=1}^{4} S_m^{II} + \frac{1}{2}\sum_{m=1}^{4} S_m^{III} + \frac{1}{\sqrt{5}}\sum_{m=1}^{4} S_m^{IV}\right)\right], \quad S_m = \begin{cases} 0 & (f_{s,m} < 1) \\ 1 & (f_{s,m} = 1) \end{cases} \tag{13}$$

2.7. *Solid concentration*

An austenite cell's concentration is updated through the accumulation of the newly solidified solutes to the existing ones,

$$C_{S,i} = (C_{S,i}^n f_S^n + k_{p,i} C_{L,i}^n \Delta f_S)/(f_S^n + \Delta f_S) \tag{14}$$

The M_7C_3 cell's concentration is set fixed.

2.8. *Solute redistribution*

With the grain growth, the solutes are apt to be absorbed from or rejected to the liquid interface due to their solubility in solids. The lacked or accumulated concentration at the liquid interface needs to be evened up, that is, solute redistribution.

2.8.1. Neighbourhood relations. According to the crystal morphologies, a Moore relation is adopted for austenite grain, and a hexagonal-like neighbourhood relation [15] is adopted for M_7C_3 grain.

2.8.2. Solute redistribution for M_7C_3 carbide. With M_7C_3 phase fraction increasing and the high solubility of C and Cr in M_7C_3 phase, i.e. the local partition coefficients $k_{p,M/L,i} > 1$, the solutes keep be absorbed from the liquid interface and spread over the neighbour cells. Assuming $C_{M,i}^n$ and $C_{L,M,i}^n$, f_M^n and f_L^n are the concentration of solute i and phase fraction of M_7C_3 and liquid for the present cell at moment n. Once there is an increase Δf_M of M_7C_3 at n+1 moment, the composition of the residual liquid updates accordingly:

(1) If $\Delta f_M < f_L^n$, the interface cell absorbs the solute from its residual liquid, then its liquid concentration decreases to $C_{L,M,i}^{n+1} = (C_{L,M,i}^n f_L^n - \Delta f_M C_{M,i})/(f_L^n - \Delta f_M)$.

If $C_{L,M,i}^{n+1}$<0, the lacked concentration, $\Delta C_i = C_{L,M,i}^{n+1}(f_L^n\text{-}\Delta f_M)$, will be compensated evenly by the available neighbour cells which have remaining liquid, then $C_{L,M,i}^{n+1}$=0; if $C_{L,M,i}^{n+1}$>$C_{L,M,i}^*$, the excess solute $\Delta C_i = (C_{L,M,i}^{n+1}\text{-}C_{L,M,i}^*)(f_L^n\text{-}\Delta f_M)$ will be re-distributed evenly to the available neighbour cells containing liquid, then set $C_{L,M,i}^{n+1} = C_{L,M,i}^*$.

(2) If $\Delta f_M \geq f_L^n$, setting $\Delta f_M = f_L^n$. The cell changes its state to M_7C_3 solid. The solid concentration is set $C_{M,i}$=const. The absorbed solute $\Delta C_i = (C_{M,i}\text{-}C_{L,M,i}^n)\Delta f_M$ is provided evenly by the available neighbour cells. Then set $C_{L,M,i}^{n+1} = 0$.

2.8.3. Solute redistribution for austenite. The solute redistribution around austenite grain is opposite to that around M_7C_3 carbide since $k_{p,\gamma/L,i} < 1$. With the growth of austenite grains, the C and Cr solutes are rejected from the austenite and accumulated in the residual liquid at the interface, which need to be redistributed around. The solute redistribution for austenite follows the similar routine as those for M_7C_3.

2.9. Determination of time step

$$\Delta t \leq 0.2\min(\Delta x^2/D_{L,C}\,,\ \Delta x^2/D_{L,Cr}\,,\ \Delta x/V_{\max}\,,\ \Delta t_0) \tag{15}$$

where $D_{L,i}$ is the diffusion coefficient of solute i in liquid, Δt_0 is the initial time step.

2.10. Solution

A) Setting the initial concentration $C_{0,i}$ and the initial temperature for the total region. *B)* Determining the time step (equation (15)). *C)* Cooling down the melt uniformly with a constant cooling rate (equation (1)). *D)* Once the temperature of a certain cell approaches the nucleation undercooling of M_7C_3 or austenite, the cell is nucleated and labelled as an interface one. *E)* Obtaining the unique Peclet number of solute Cr through coupling the expressions of the local undercooling (equation (3)), curvature undercooling (equation (4)), constitutional undercooling (equation (5)) and interface equilibrium concentration (equation (6)) with each solute Peclet number Pe_i (equation (8)). *F)* Obtaining the interface equilibrium concentration of solute Cr, i.e. $C_{L,Cr}^*$, through equation (7). *G)* Updating the grain growth rate and the solid fraction through equation (10) and equation (11). *H)* Repeating steps *E)* - *G)* for iteration until the interface curvature is converged. *I)* Updating the solid concentration through equation (14). *J)* Redistributing solute concentration through section 2.8. *K)* Solving the concentration field (equation (2)). *L)* Going back to step *B)* until the calculation region is completely solidified.

3. Solidification microstructure evolution of Fe-4wt%C-17wt%Cr alloy

According to the experimental cooling condition and obtained microstructure morphology of Fe-4wt%C-17wt%Cr alloy [3], a 700 μm ×520 μm 2D region with CA mesh size 2 μm × 2 μm is adopted. The initial undercooling of the melt is set as -4 K from the liquidus temperature 1304 °C and the constant cooling rate $\dot{T}$ is chosen as -1.3 K/s corresponding to the graphite mould cooling condition. The austenite and M_7C_3 carbide nucleation sites were set according to the microstructure morphology [3]. The nucleation undercooling was set randomly.

The mutual growth of M_7C_3 grain and adjacent austenite grains during solidification can be seen from austenite phase fraction evolution in figure 1. The envelope of a single austenite grain B to a M_7C_3 grain A can be found at the bottom centre of the region. The M_7C_3 grain A keeps absorbing solutes while the adjacent austenite grains B keep rejecting them as they grow up. Once the two kinds of grains grow up to approach each other, the mutual complement of the solutes at the interface accelerate their growth till finally the austenite grains B totally envelope the M_7C_3 grain A.

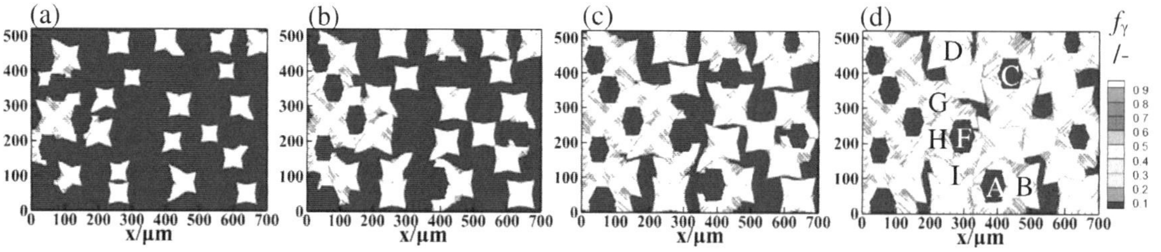

Figure 1. Enveloping of austenite grains to M_7C_3 grain A and F during solidification: (a) t=27.00 s (b) t=27.26 s (c) t=27.53 s and (d) t=27.88 s.

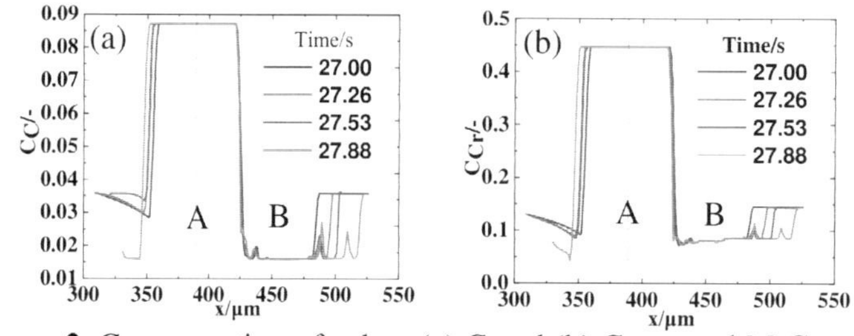

Figure 2. Concentration of solute (a) C and (b) Cr around M_7C_3 grain A and austenite grain B during enveloping.

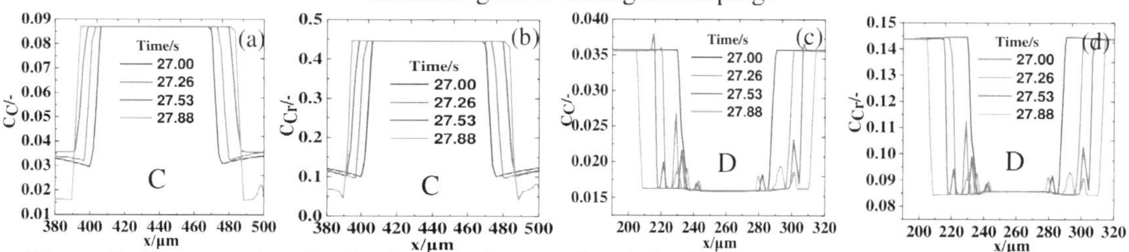

Figure 3. Concentration distribution of solute (a) C and (b) Cr around a separated M_7C_3 grain C and (c) C and (d) Cr around austenite grain D as labelled in figure 1.

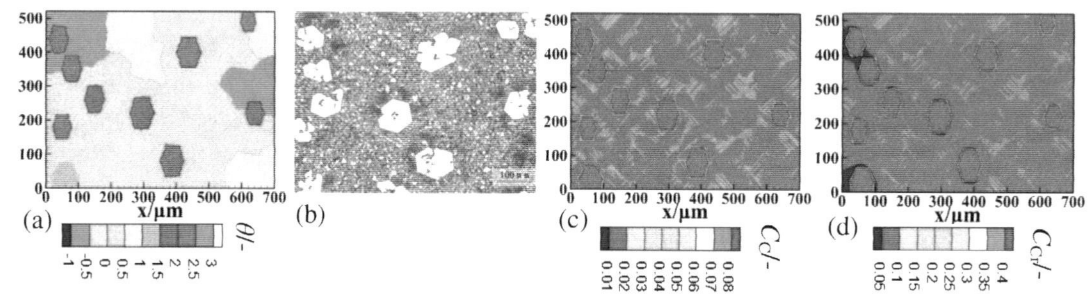

Figure 4. Final solidification microstructure (a) simulated and (b) experimental [3] and simulated solute concentration of (c) C and (d) Cr for Fe-4wt%C-17wt%Cr alloy.

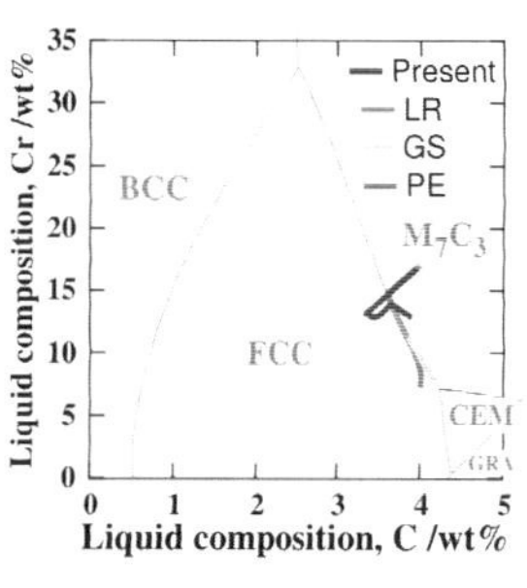

Figure 5. Predicted average liquid composition of Cr with respect to that of C for Fe-4wt%C-17wt%Cr alloy by the present model, comparing with the solidification path prediction by LR, GS and PE approximations. The thin grey curves delimit the domains of the liquidus surfaces of different phases in the system.

Figure 2 shows the solute concentration evolution along the horizontal centerline of austenite grain B and M_7C_3 grain A during this envelope. Due to the partition of solutes, the lacked solute concentration is over double of the liquid concentration at M_7C_3/liquid interface, while the accumulated concentration at austenite grain interface is merely one half of its interface liquid concentration. As seen, at 27.00 s, the liquid concentration at the right-side interface of grain A, i.e. in between grain A and B, which is 1.6wt%C and 7.4wt%Cr, is lower than the left-side one, which is 2.9wt%C and 9.2wt%Cr, since at that moment the right-side part of grain A grows a little faster. Then, grain B successively envelop grain A and complement the solutes around it. Thus at 27.88 s, the interface concentration at the right-side of A is almost unchanged since there is no space there to grow up due to the touching between grain A and B, while is merely 1.6wt%C and 4.2wt%Cr at left-side of A due to its free growth. Thus among 27.00-27.88 s, the right-side part of grain A almost keeps 31 μm from the grain's centreline while its left-side part grows up to 39 μm, with the lacked liquid concentration there. At 27.88 s, the adjacent austenite dendrite B branches up to complete the circling around the grain A, with its dendritic structure changing to a deformed pliers-shaped cluster structure. Besides the single grain's enveloping, several austenite grain's enveloping can also be found in figure 1, as illustrated, the austenite grains G-I contribute together to circle the adjacent M_7C_3 grain F.

On the contrary, a separate M_7C_3 carbide grain C and austenite grain D 's growth are shown in figure 3 by tracking the solute concentration distribution along the horizontal centerline of them. As illustrated in figure 3(a)-(b), the solute C and Cr concentrations in M_7C_3 carbide, which are 8.71wt%C and 44.7wt%Cr, are much higher than those in adjacent liquid. With M_7C_3 carbide absorbing the solutes during the growth, the concentration at the interface is the lowest during the solute absorption process. Contrastingly, figure 3(c)-(d) shows the solute concentration distribution around a separate austenite grain D, as labelled in figure 1. In this case, the solute concentration in solid is lower than that in liquid, which is consistent with the partition coefficient $k_{p,\gamma/L,i} < 1$ for austenite/liquid interface. The austenite rejects the solutes during its growth. Due to the inner dendrite structure of austenite, the uneven concentration distribution is tracked along the horizontal centerline of grain D.

Figure 4 compares the predicted final solidification morphology with the experimental one [3]. Based on the orientation group set for M_7C_3 carbide and austenite (in section 2.4), the orange to pink-coloured grains represent the M_7C_3 carbide ones, and the blue to green-coloured grains represent the austenite ones. As seen, the hexagon-shaped M_7C_3 carbide grains distribute scattered in the austenite matrix, the predicted morphology (figure 4(a)) shows similar to the experimental one (figure 4(b)). Figure 4(c) and 4(d) show the solute concentration profile, the interdendritic structure can be seen there. The predicted average size of M_7C_3 carbides, measured by Image-pro Plus 6.0, is 71.20 μm, which is fit with the experimental one, 72 μm. And the predicted average solute concentrations in austenite, i.e. the summation of austenite solute mass in all cells divided by the total austenite mass fraction in the region, are 1.61wt%C and 7.87wt%Cr, which are also fit with the predicted concentrations in austenite by solidification path prediction, i.e. 1.64wt%C and 6.63wt%Cr from Gulliver Scheil (GS), 1.65 wt%C and 6.47wt%Cr from lever rule (LR), and 2.04wt%C and 6.87wt%Cr from partial equilibrium (PE) approximation.

Finally, the predicted average liquid concentration evolution during solidification of the Fe-4wt% C-17wt%Cr alloy was checked in Fe-C-Cr alloy liquidus surface phase diagram (figure 5). The averaged liquid concentration is the summation of liquid solute mass in all cells divided by the total liquid mass fraction in the region. As seen, the solidification starts from the liquid+M_7C_3 regime, then goes directly to FCC regime and finally finishes near the valley between liquid+M_7C_3 and liquid+FCC regime. The predicted liquid concentration curve by the present model, i.e. the blue-coloured one, shows three turning points, corresponding respectively to the start of the austenite grain precipitation, the touching of the austenite grains with almost all the M_7C_3 ones, and the accomplishment of the austenite grains enveloping the M_7C_3 ones. The solidification path prediction by LR, GS and PE prediction are also plotted in the phase diagram. The present prediction fits with them at the initial M_7C_3 precipitation regime and after the austenite grains' complete enveloping towards the M_7C_3 grains, returns to overlap the LR, GS and PE prediction curves.

4. Conclusions

A microscopic cellular automaton combined with macroscopic heat and solute transport was developed to simulate the mutual growth and evolution of austenite and M_7C_3 carbide grains during Fe-4wt%C-17wt%Cr ternary alloy solidification process. Results show that,

1) The diffusion of solute C and Cr are contributed together to the constitutional undercooling, together with curvature undercooling for obtaining the grain growth rate of austenite and M_7C_3.

2) Absorption of solute C and Cr to M_7C_3 carbide grain makes its interface the lowest liquid concentration in the region; contrastingly, rejection of solutes from austenite grain makes its interface the higher one. Due to the partition of solutes, the lacked solute concentration is over double of the liquid concentration at M_7C_3/liquid interface, while the accumulated concentration at austenite grain interface is merely one half of its interface liquid concentration.

3) The absorption of solute C and Cr by M_7C_3 grain and rejection of the solutes by austenite grain promote these two grains' cooperative growth. Once they approach to each other, the austenite grain quickly overgrows towards the M_7C_3grain till finally envelops it.

4) The simulated solidification morphology of the Fe-4wt%C-17wt%Cr alloy, the predicted averaged grain size of M_7C_3 carbides and the C and Cr concentration in austenite grains all agree with the experimental measurements and solidification path prediction. The predicted average liquid concentration curve fits with the LR, GS and PE prediction at the initial M_7C_3 precipitation regime and after the austenite grains' complete enveloping towards the M_7C_3 grains, returns to overlap the LR, GS and PE prediction curves.

Acknowledgements

The work is supported by National Natural Science Foundation of China [51574074] and [51425401], National Natural Science Foundation of China and Shanghai Baosteel [U1460108] and [U1560207], and Natural Science Foundation of Liaoning Province, China [L20150183]. The work is also supported by JERNKONTORET, Prytziska fonden 2, Sweden.

References

[1] Mendez P F, Barnes N, Bell K, Borle S D, Gajapathi S S, Guest S D, Izadi H, Gol A K and Wood G 2014 *J. Manuf. Process.* **16** 4-25
[2] Qi X W and Jia Z N 2011 *Surf. Coat. Technol.* **205** 5510-14
[3] Liu Q, Zhang H W,Wang Q, Zhou X K, Jönsson P G and Nakajima K 2012 *ISIJ Int.* **52** 2210-19
[4] Zhu M F, Zhang L, Zhao H L and Stefanescu D M 2015 *Acta Mater.* **84** 413-25
[5] Chen Q and Sundman B 2002 *Mater. Trans.* **43** 551-59
[6] Zhang H W, Nakajima K, Gandin Ch-A and He J C 2013 *ISIJ Int.* **53** 493-501
[7] Shi P 2008 *TCS steels/Fe-alloys database V6.0*, Thermo-Calc Software AB, Stockholm, SE
[8] Helander T 1998 *MOB2-Mobility database,* Thermo-Calc software AB, Stockholm, SE
[9] Gandin Ch-A, Desbiolles J L, Rappaz M and Theroz P 1999 *Metall. Mater. Trans.*A **30** 3153-65
[10] Burbelko A A, Fraś E, Gurgul D, Kapturkiewicz W and Sikora J A 2011 *Surf. Coat. Technol.* **205** 5510-14
[11] Nastac L 1999 *Acta Mater.* **47** 4253-62
[12] Beltran-Sanchez L, Stefanescu D M 2004 *Metall. Mater. Trans.* A **35** 2471-86
[13] Zhao D W, Li Q S and Hao W X 2011 *Foundry Technol.* **32** 318-20
[14] Shi Y F, Xu Q Y and Liu B C 2012 *Acta Phys. Sin.* **61** 1008101-1-7
[15] Fu Z N, Xu Q Y and Xiong S M 2007 *Chin. J. Nonferrous. Met.* **17** 1567-73

IOP Conf. Series: Materials Science and Engineering **861** (2020) 012047 doi:10.1088/1757-899X/861/1/012047

Effect of Al-V-B grain refiner on refining aluminium alloys: estimation from *ab initio* calculations

S Ma[1], N Zong[1], H Dong[2] and T Jing[1]

[1] School of Materials Science and Engineering, Tsinghua University, Beijing 100084, China
[2] School of Engineering, University of Leicester, Leicester, LE1 7RH, UK

E-mail: jingtao@mail.tsinghua.edu.cn

Abstract. Ab initio calculations based on density functional theory were conducted to estimate the effect of Al-V-B grain refiner on refining aluminium alloys. Al-Ti-B, an industrial grain refiner, was investigated simultaneously to act as a reference. Six interfaces formed by (111) Al with (112) Al3V(Ti), V(Ti)-terminated (0001) V(Ti)B2 and B-terminated (0001) V(Ti)B2 were studied in detail. Interfacial lattice mismatch, work of adhesion and stain-free interfacial energy were calculated for each of these interfaces, respectively. Analysis of interfacial energy demonstrated that Al3V overlayers are less likely to form on the surface of (0001) VB2 surface compared with the Ti-related system. By combination of interfacial lattice mismatch and work of adhesion, it was concluded that VB2 is much more potent in refining aluminium alloys than TiB2, while Al3V is much weaker in nucleant potency compared with Al3Ti.

1. Introduction

Al foundry alloys have a broad application in the automotive and aerospace industries due to their superior corrosion and wear resistance, particularly high strengths after heat-treatment [1, 2]. It is known that the achievement of fine grain structure is beneficial to the soundness and mechanical properties of Al alloy castings [3]. In recent years, the focus of attention is on grain refining efficiency of Al-Ti-B master alloys since the master alloys are the most frequently used in the foundry [4-6]. However, in the cast Al-Si alloys the silicon-poisoning phenomenon significantly limits the degree of grain refinement achieved by adding Al-Ti-B master alloys to Al-Si casting alloys [7]. Therefore, researches were performed to overcome the Si poisoning effect and reduce the grain size of solidified Al-Si alloys [8, 9].

Ti element in Ti-based master alloy has a higher affinity with Si than Al, which leads to the formation of Ti silicide, and the aluminium melt gradually is depleted of Ti, reducing TiB_2 and Al_3Ti substrates' potency at pouring temperature of Al-Si alloy [10]. The Nb-based master alloys have seldom occurred the phenomenon of Si poisoning [11], but there is a serious precipitation phenomenon due to the sedimentation problem. Zhang *et al.* [12] reported that a Al-V-B master alloys can overcome the Si poison effect on grain refinement and refine the grain structure of the Al-7wt%Si alloy effectively through the experimental research. The mechanism of this phenomenon was not investigated clearly, and it is widely accepted that the *ab initio* calculations with density functional theory [13, 14] can provide fundamental information at atomic or even electronic levels to investigate the heterogeneous nucleation mechanism. However, published papers about this issue are rarely retrieved.

The atomic geometry, interfacial adhesion and interfacial bonding nature play an important role in affecting nucleation potency of a heterogeneous substrate. Therefore, heterogeneous nucleation interfaces of Al/V(Ti)B_2 and Al/Al_3V(Ti) were extensively studied by using *ab initio* calculations, this work with the aim of estimating the potential and mechanism of Al-V-B as a grain refiner for aluminium alloys. Interfacial structure, adhesion strength and interfacial energy are obtained, which will be discussed in detail in the following sections.

2. Establishment of interfacial models of Al with Al₃V(Ti) and V(Ti)B₂

Atomistic geometries of the interfaces of Al with Al_3V, VB_2, Al_3Ti and TiB_2 were first determined. From previous experiments [15], orientation relationship (OR) of $(0001)_{V(Ti)B2}$ // $(111)_{Al}$ was confirmed. And on the surface of (0001) TiB_2, (112) Al_3Ti atomic layers were likely to appear which showed a relatively small mismatch with (111) Al [15]. Considering the similarity between Al_3Ti and Al_3V, (112) Al_3V is expected to form adhering to the (0001) VB_2 surface, which then may promote the nucleation of Al with (111) plane parallel to the substrate surface. Therefore, the OR of $(112)_{Al3V(Ti)}$ // $(111)_{Al}$ is necessary to be examined. In-plane crystal orientation matching for the above two types of interfaces is $<11\bar{2}0>_{V(Ti)B2}$ // $<1\bar{1}0>_{Al}$ and $<20\bar{1}>_{Al3V(Ti)}$ (or $<1\bar{1}0>_{Al3V(Ti)}$) // $<1\bar{1}0>_{Al}$, respectively [15]. Following the above ORs, interfacial models between Al and the four nucleant substrates were established (figure 1). For all the substrates and solid Al, 7-layer slabs were used to build the supercell, which have been tested to be sufficiently large to ensure the interaction between two surfaces to be negligible. A vacuum region with length larger than 15 Å was inserted into each supercell above the solid Al to separate it from the substrate. For the (0001) V(Ti)B_2 surface, two different terminations (V(Ti)-termination and B-termination) exist, which were both studied in this work, as shown in figure 1(b), (c), (e) and (f). We note that for each interfacial model, several stacking modes (ST) with high symmetry were considered as depicted by the solid circles that are connected by dashed lines in figure 2. This is helpful to find the interface structure with the lowest energy.

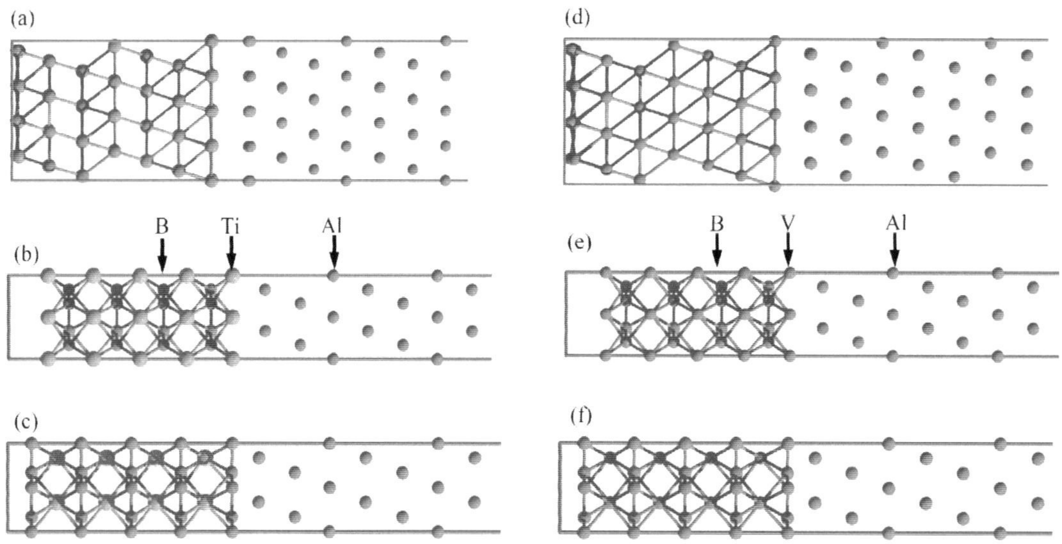

Figure 1. Atomistic models of the six studied interfaces in this work. (a) to (f) are (112) Al_3Ti/(111) Al, Ti-terminated (0001) TiB_2/(111) Al, B-terminated (0001) TiB_2/(111) Al, (112) Al_3V/(111) Al, V-terminated (0001) VB_2/(111) Al, and B-terminated (0001) VB_2/(111) Al interfaces, respectively.

The above established initial interfacial models (totally 12 models) were relaxed within the framework of density functional theory. Vienna *Ab initio* Simulation Package (VASP) was adopted to conduct these calculations [16, 17]. Projected-augmented waves (PAW) method [18, 19] was employed to deal with the relationship between valence electrons and the ionic cores. The Perdew-Burke-Ernzerhof functional with generalized gradient approximation (GGA-PBE) [20] was selected to approximate the exchange-correlation energy. We chose a plane wave cutoff energy of 400.0 eV for all the calculations. Gaussian smearing method with a smearing width of 0.01 eV was used in the self-consistent calculations to solve the ground state electronic density. The ionic relaxation continued until all forces were smaller than 0.01 eV·Å^{-1}, which was used to determine ground state energies and atomic geometries for each structure. $3 \times 3 \times 1$ and $7 \times 7 \times 1$ **Γ**-centered **k**-point meshes were used for $Al_3V(Ti)$ and $V(Ti)B_2$ related interface calculations, respectively.

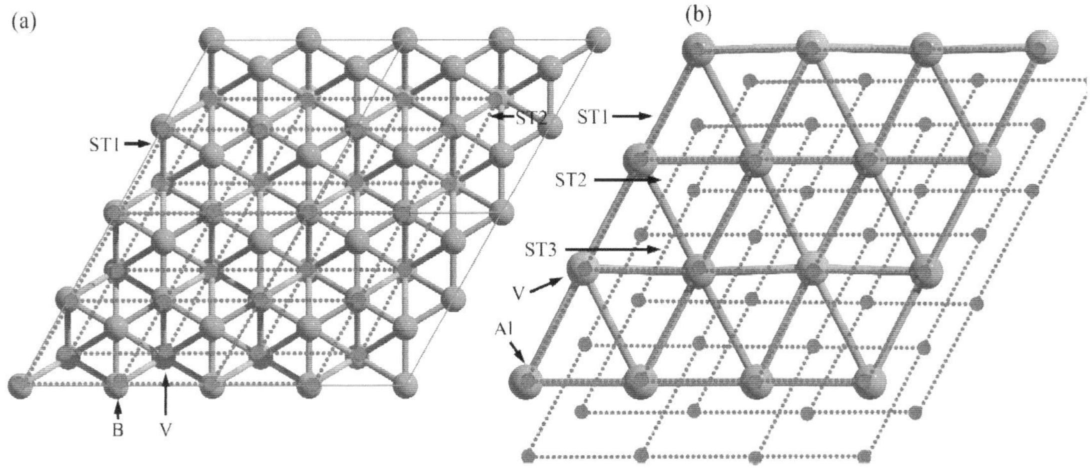

Figure 2. Top view of (a) (0001) VB_2 surface and (b) (112) Al_3V surface with the possible stacking modes for (111) Al depicted by dashed lines and solid circles.

To estimate the nucleant potencies of $Al_3V(Ti)$ and $V(Ti)B_2$ particles, lattice mismatch f at the interfaces and interfacial energies were calculated based on the aforementioned interfacial models, which will be discussed in detail in the following sections.

3. Calculation of interfacial mismatch

Although (112) $Al_3V(Ti)$, (0001) $V(Ti)B_2$ and (111) Al all show six-fold symmetry, their lattice parameters (or interatomic distances) are different which renders interfacial strain when the substrates are combined with the solid Al. During heterogeneous nucleation process of solid Al, interfacial strain hinders the formation of nuclei. Therefore, a smaller interfacial strain is expected for the substrate with high nucleant potency. Here, lattice mismatch f was used to quantify the interfacial strain.

For the $V(Ti)B_2$/Al interfaces, f was calculated using the following equation:

$$f = \frac{d<11\bar{2}0>_{V(Ti)B2} - d<1\bar{1}0>_{Al}}{d<1\bar{1}0>_{Al}} \times 100\%, \tag{1}$$

where $d<11\bar{2}0>_{V(Ti)B2}$ and $d<1\bar{1}0>_{Al}$ are interatomic distances along the $<11\bar{2}0>_{V(Ti)B2}$ and $<1\bar{1}0>_{Al}$ directions, respectively. For the $Al_3V(Ti)$/Al interface, f has the following formula:

$$f = \left(\frac{1}{3}\frac{d<1\bar{1}0>_{Al3V(Ti)} - d<1\bar{1}0>_{Al}}{d<1\bar{1}0>_{Al}} + \frac{2}{3}\frac{d<20\bar{1}>_{Al3V(Ti)} - d<1\bar{1}0>_{Al}}{d<1\bar{1}0>_{Al}}\right) \times 100\%. \tag{2}$$

In the above equation, $d<1\bar{1}0>_{Al3V(Ti)}$ and $d<20\bar{1}>_{Al3V(Ti)}$ denote the interatomic distances in the $<1\bar{1}0>$ and $<20\bar{1}>$ directions of the $Al_3V(Ti)$ substrate. The prefactors 1/3 and 2/3 in the above equation are determined due to the fact that for the OR between (112) $Al_3V(Ti)$ and (111) Al, 1/3 of the in-plane crystal orientation matches are $<1\bar{1}0>_{Al3V(Ti)}$ // $<1\bar{1}0>_{Al}$, while the rest are $<20\bar{1}>_{Al3V(Ti)}$ // $<1\bar{1}0>_{Al}$.

The calculated interatomic distances based on *ab initio* simulations and the corresponding interfacial lattice mismatches are listed in table 1.

Table 1. Interfacial mismatch at the studied interfaces.

Parameters	**Al**	**VB_2**	**TiB_2**	**Al_3V**		**Al_3Ti**	
Direction	$<1\bar{1}0>$	$<11\bar{2}0>$	$<11\bar{2}0>$	$<1\bar{1}0>$	$<20\bar{1}>$	$<1\bar{1}0>$	$<20\bar{1}>$
***d* (Å)**	2.86	3.00	3.03	2.66	2.80	2.72	2.89
***f* (%)**	—	4.9	5.9	-3.7		-0.93	

In table 1, lattice mismatches of Al with the four substrates are compared. A positive value of *f* means the solid Al is stretched, while a negative value suggests the solid Al is compressed. Clearly illustrated by table 1 is that Al_3Ti has the smallest lattice mismatch with Al while TiB_2 exhibits the largest *f*. This finding has already been verified for a number of times by previous experimental and theoretical analyses, and it explains why the appearance of an Al_3Ti layer on the surface of TiB_2 substrate can significantly enhance the nucleant potency of TiB_2. More interesting are the *f* values of Al with V-based compounds. It is found that the absolute values of *f* for VB_2 and Al_3V approach, which implies that effect of Al_3V on enhancing the nucleant potency of VB_2 may be much weaker than the Ti-based system. Additionally, a slight decrease of *f* is found when replacing TiB_2 by VB_2. The above discussion based on interfacial lattice mismatch predicts that for the Al-V-B master alloys, VB_2 alone may play an indispensable role in refining the aluminum alloys even without the help of Al_3V overlayers.

4. Quantification of interfacial affinity

The affinity between the substrate and Al also plays a vital role in affecting nucleant potency. Interfacial affinity describes the tendency to form an interface and embodies the stability of it. Work of adhesion W_{ad} and strain-free interfacial energy γ (or interfacial chemical energy) were together used to quantify the interfacial affinity. W_{ad} is expressed as follows:

$$W_{ad} = \frac{1}{A}(E_{inter} - E_{sub} - E_{Al}). \tag{3}$$

In equation (3), A is the total area of the interface; E_{inter}, E_{sub} and E_{Al} are the ground state total energies of the interface, substrate ($Al_3V(Ti)$ or $V(Ti)B_2$) and Al slab, respectively. W_{ad} has another expression which connects γ_{sf} with the surface energies of the substrate σ_{sub} and Al σ_{Al}:

$$W_{ad} = \gamma_{sf} - \sigma_{sub} - \sigma_{Al} \tag{4}$$

Using equation (4), γ_{sf} can be calculated through $\gamma_{sf} = W_{ad} + \sigma_{sub} + \sigma_{Al}$.

Surface energies of (111) Al and (112) $Al_3V(Ti)$ are easy to acquire since these surfaces are stoichiometric, while those of the non-stoichiometric (0001) $V(Ti)B_2$ surfaces are influenced by the chemical potential of the individual element. We have taken the (0001) VB_2 surface as an example to demonstrate how the non-stoichiometric surface energy was calculated, which was readily applied for the calculation of (0001) TiB_2 surface energy. Surface energy of (0001) VB_2 surface is derived as

$$\sigma_{\mathrm{VB2}} = \frac{1}{A}(E_{\mathrm{slab}} - N_{\mathrm{V}}\mu_{\mathrm{V}}^{\mathrm{VB2,s}} - N_{\mathrm{B}}\mu_{\mathrm{B}}^{\mathrm{VB2,s}}), \tag{5}$$

where A is the total surface area of (0001) VB_2 surface; E_{slab} denotes the ground state total energy of the (0001) VB_2 slab; N_{V} and N_{B} are the number of V and B atoms in the slab, respectively; $\mu_{\mathrm{V}}^{\mathrm{VB2,s}}$ and $\mu_{\mathrm{B}}^{\mathrm{VB2,s}}$ indicate the chemical potentials of V and B in the surface slab, respectively. Considering the equilibration between surface and bulk VB_2, it is assumed that $\mu_{\mathrm{V}}^{\mathrm{VB2,s}}=\mu_{\mathrm{V}}^{\mathrm{VB2,b}}$ and $\mu_{\mathrm{B}}^{\mathrm{VB2,s}}=\mu_{\mathrm{B}}^{\mathrm{VB2,b}}$, where $\mu_{\mathrm{V}}^{\mathrm{VB2,b}}$ and $\mu_{\mathrm{B}}^{\mathrm{VB2,b}}$ are the chemical potentials of V and B in bulk VB_2 [21]. For bulk VB_2, an equation can be established:

$$\mu_{\mathrm{VB2}}^{\mathrm{b}} = \mu_{\mathrm{V}}^{\mathrm{VB2,b}} + 2\mu_{\mathrm{B}}^{\mathrm{VB2,b}} = \mu_{\mathrm{V}}^{\mathrm{VB2,s}} + 2\mu_{\mathrm{B}}^{\mathrm{VB2,s}}, \tag{6}$$

in which $\mu_{\mathrm{VB2}}^{\mathrm{b}}$ is the chemical potential of a unit of VB_2, which is readily calculated from the *ab initio* simulations. Combining equation (5) and (6), σ_{VB2} can be written as:

$$\sigma_{\mathrm{VB2}} = \frac{1}{A}[E_{\mathrm{slab}} - \frac{1}{2}N_{\mathrm{B}}\mu_{\mathrm{VB2}}^{\mathrm{b}} + (\frac{1}{2}N_{\mathrm{B}} - N_{\mathrm{V}})\mu_{\mathrm{V}}^{\mathrm{VB2,s}}] \tag{7}$$

Now, $\mu_{\mathrm{Ti}}^{\mathrm{TiB2,s}}$ is still needed to be determined for the calculation of σ_{VB2}. According to Han *et al.* [21], $\mu_{\mathrm{Ti}}^{\mathrm{TiB2,s}}$ is indeed in a range determined by the following formula:

$$\Delta H_f \le \mu_{\mathrm{V}}^{\mathrm{VB2,s}} - \mu_{\mathrm{V}}^{\mathrm{V,b}} \le 0 . \tag{8}$$

In equation (8), $\mu_{\mathrm{V}}^{\mathrm{V,b}}$ is the chemical potential of a V atom in bulk V and ΔH_f is the formation enthalpy of VB2. Both $\mu_{\mathrm{V}}^{\mathrm{V,b}}$ and ΔH_f were obtained from *ab initio* calculations.

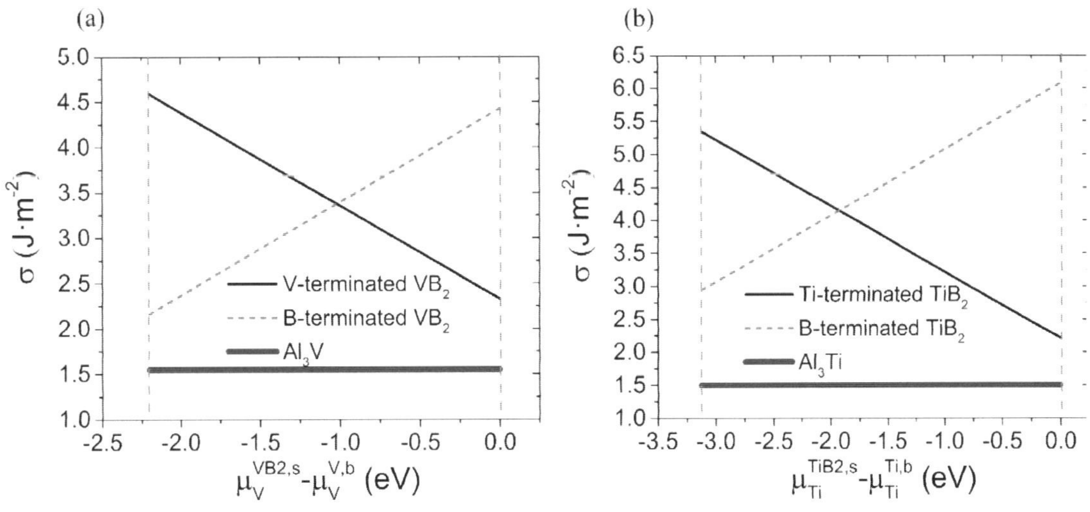

Figure 3. Surface energies of (a) (0001) $V(Ti)B_2$ and (b) (112) $Al_3V(Ti)$ surfaces.

Figure 3 depicts the calculated surface energies of the (0001) $V(Ti)B_2$ and (112) $Al_3V(Ti)$ surfaces. A prominent feature shown in figure 3 is that the surface energies of the (112) $Al_3V(Ti)$ surfaces are much lower than those of the (0001) $V(Ti)B_2$ surfaces. This finding well explains why $Al_3V(Ti)$ overlayers are likely to form on the (0001) $V(Ti)B_2$ surface, because such structures significantly reduce the surface energies of the borides. Another finding from figure 3 is that surface energies of the (0001) $V(Ti)B_2$ are closely related to the chemical potential of the V(Ti) atoms in the surface slabs. More importantly, surface energies of V(Ti)-terminated and B-terminated surfaces are in the same range, and thus both of these two types of surfaces are likely to exist.

Prior to the calculation of interfacial energy, W_{ad} for the six interfaces were computed firstly, as shown in figure 4, in which terminations of the (0001) V(Ti)B_2 surfaces are indicated by the element symbols in parentheses. To compute W_{ad}, ground state total energies E_{inter} of the six interfaces need to be determined. As the first step, the equilibrated interface structures were obtained using the approach introduced in section 2. At the (112) Al_3V(Ti), we found that (111) Al stacking in the ST2 manner (figure 2) was energetically more favorable. At the (0001) V(Ti)B_2 surface, ST2 (atop V(Ti) atoms) and ST1 (atop B atoms) are the stacking modes with the lowest energy for B and V(Ti)-terminated substrates, respectively. E_{inter} were extracted from these interfacial structures and then put into equation (3) to calculate W_{ad}. Contrary to σ (figure 3), W_{ad} for the (112) Al_3V(Ti) substrate is obviously larger than that for the (0001) V(Ti)B_2 substrate nevertheless of Ti(V) or B termination. Basically, W_{ad} quantifies the energy releasing from combining two slabs (in this work, substrate and Al slabs) to form an interface. A smaller W_{ad} suggests that it is energetically more favorable to form the corresponding interface. Therefore, (0001) Al slab is more readily to form an interface with V(Ti)B_2 substrates, especially when it is terminated by V(Ti), compared with the Al_3V(Ti) substrate.

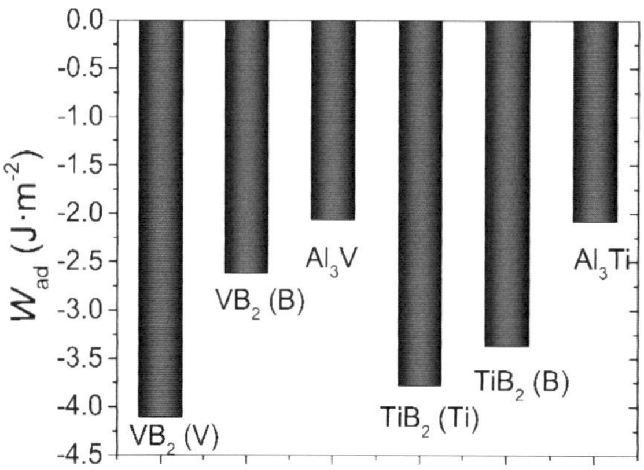

Figure 4. Work of adhesion for the interfaces of (111) Al with (0001) V(Ti)B_2 and (112) Al_3V(Ti). Contents in the parentheses indicate the surface termination of the V(Ti)B_2 substrates.

Equation (4) was used together with the already obtained results of σ (figure 3) and W_{ad} (figure 4) to calculate the interfacial energy γ. Surface energy of (111) Al is also required for this calculation, which was computed to be 0.904 J • m^{-2} based on the *ab initio* method. The calculated γ for the six interfaces (figure 1) are exhibited in figure 5. γ embodies the stability of an interface. An interface presenting a smaller γ is more likely to exist compared with its counterpart with a larger γ. This is important when analyzing which of the terminations (V(Ti) or B) is more likely to appear when Al is present. From figure 5, it is observed that with $\mu_V^{VB2,s}$ ($\mu_{Ti}^{TiB2,s}$) increasing, γ of the V(Ti)-terminated (0001) V(Ti)B_2/Al interface decreases gradually which even reaches zero when $\mu_V^{VB2,s}$ ($\mu_{Ti}^{TiB2,s}$) is near $\mu_V^{V,b}$ ($\mu_{Ti}^{Ti,b}$). Meanwhile, γ of the interface with the B-terminated substrate increases. Obviously, $\mu_V^{VB2,s}$ ($\mu_{Ti}^{TiB2,s}$) is influenced by the concentration of V(Ti) at the interface, which is difficult to measure. However, some experiment results may provide some hints on the more exact $\mu_V^{VB2,s}$ ($\mu_{Ti}^{TiB2,s}$) range. Normally, an excess of solute Ti atoms is guaranteed when designing the Al-Ti-B master alloys to ensure an excellent grain refining effect. As observed in previous experimental investigations [22], these free Ti solute atoms partially segregate to the surface of the TiB_2 resulting an increase of local Ti concentration, which then renders the increase of $\mu_{Ti}^{TiB2,s}$ to approach $\mu_{Ti}^{Ti,b}$. When $\mu_{Ti}^{TiB2,s}$ is larger than A_{Ti} in figure 5(b), the Ti-terminated TiB_2 surface becomes more stable. And if $\mu_{Ti}^{TiB2,s}$ is beyond B_{Ti} (figure 5(b)), formation of the Al_3Ti layer is not possible. In the experiments of Fan *et al.* [22], a Ti-terminated TiB_2 was observed which was covered by an Al_3Ti monolayer at the TiB_2/Al interface. This finding suggests that $\mu_{Ti}^{TiB2,s}$ is actually between A_{Ti} and B_{Ti}. For the V-related interfaces, interval between A_V and B_V is obviously smaller than that between A_{Ti} and B_{Ti}. This makes $\mu_V^{VB2,s}$ easier to pass B_{Ti} as the concentration of V increases at interface when formation of Al_3V overlayer is energetically unfavorable. In fact, some experimental results can evidence such a conjecture. Zhang *et al.* [12] found in their pioneering investigation of Al-V-B grain refiners that VB_2 alone seems to be the potent nucleants instead of the Al_3Ti particles.

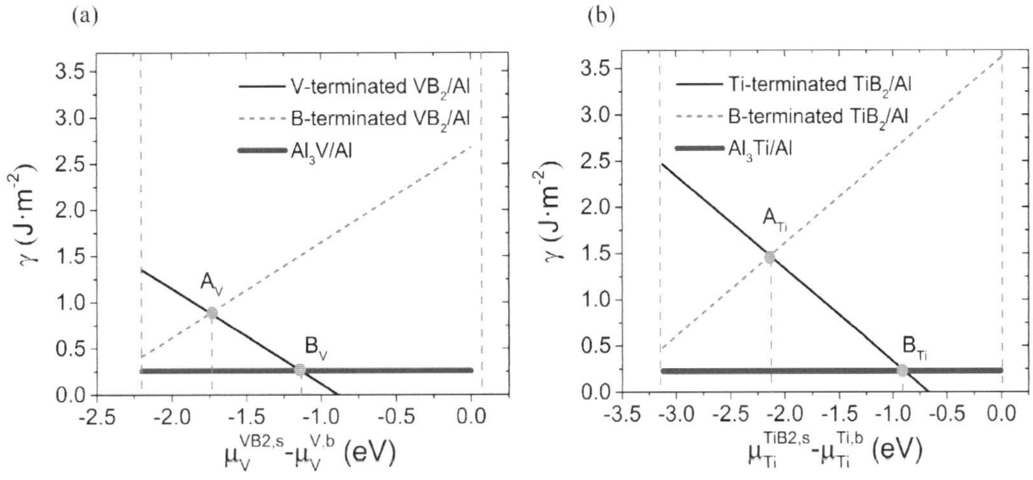

Figure 5. Interfacial energies of the interfaces of (111) Al with (a) (0001) V(Ti)B2 and (b) (112) Al3V(Ti).

5. Discussion

According to classical nucleation theory, nucleant potency for heterogeneous nucleation can be quantified by contact angle θ, which is defined as follows:

$$\cos\theta = \frac{\gamma_{nl} - \gamma_{ns}}{\gamma_{sl}} . \tag{9}$$

In this equation, γ_{sl}, γ_{nl} and γ_{ns} are the interfacial energies of solid/liquid, nucleant/liquid, and nucleant/solid interfaces, respectively. In the case of estimating potency of nucleant substrates, γ_{sl} is the same for all the substrates. Therefore, the larger $(\gamma_{nl} - \gamma_{ns})$ is, the higher the potency will be. A substantial difference between nucleant/liquid and nucleant/solid interfaces is the existence of interfacial strain at the nucleant/solid interface, which is effectively quantified by *f*. Another difference is embodied in the chemical bonding feature [23, 24] that can be qualitatively reflected by W_{ad}. Therefore, by combination of *f* and W_{ad}, the nucleant potency of the substrates in Al-V(Ti)-B master alloys can be estimated. Al_3V show much worse nuleant potency than Al_3Ti, because W_{ad} of for these two substrates are nearly the same while interfacial strain of Al_3V/Al is much larger than the interface containing Al_3Ti. On the contrary, VB_2 is more potent than TiB_2 with the former presenting smaller *f* and W_{ad} than the latter.

6. Conclusion

Ab initio calculations were conducted to estimate the effect of Al-V-B grain refiner on refining aluminium alloys with a comparison to the Al-Ti-B grain refiner. We found VB_2 in Al-V-B can play a much more important role in refining Al than TiB_2 in Al-Ti-B, and nucleant potency of Al_3V is much weaker than Al_3Ti. These findings theoretically explain the previous experimental observation that VB_2 alone instead of together with Al_3V refines an aluminium alloy.

Acknowledgments
This work was supported by the National Natural Science Foundation of China (Grants 51674153 and 51320105003) and the National Key Research and Development Programme of China (2017YFB1103700). This research used the supercomputing facilities of National Laboratory for Information Science and Technology at Tsinghua University.

References

[1] Easton M A and Stjohn D H 1999 *Metall. Mater. Trans. A.* **30** 1613
[2] Mohanty P S and Gruzleski J E 1995 *Acta Metall. Mater.* **43** 2001
[3] Murty B S, Kori S A and Chakrabort M 2002 *Int. Mater. Rev.* **47** 3
[4] Greer A L, Bunn A M, Tronche A, Evans P V and Bristow D J 2000 *Acta Mater.* **48** 2823
[5] Quested T and Greer A L 2005 *Acta Mater.* **53** 2683
[6] Xu Y, Casari D, Du Q, Mathiesen R H, Arnberg L and Li Y 2017 *Acta Mater.* **140** 224
[7] Sigworth G K 1996 *Scr. Mater.* **34** 919
[8] Chen Z, Kang H, Fan G, Li J, Lu Y, Jie J, Zhang Y, Li T, Jian X and Wang T 2016 *Acta Mater.* **120** 168
[9] Li Y, Hu B, Gu Q, Liu B and Li Q 2019 *Scr. Mater.* **160** 75
[10] Spittle J.A and Sadli S 1995 *Mater. Sci. Technol.* **11** 533
[11] Bolzoni L and Hari Babu N 2019 *J Mater Res Technol.* **8** 5631
[12] Zhang Y H, Ye C Y, Shen Y P, Chang W, StJohn D H, Wang G and Zhai Q J 2019 *J. Alloys Compd.* **812** 152022
[13] Han Y, Dai Y, Shu D, Wang J and Sun B 2006 *J. Phys.: Condens. Matter.* **18** 4197-205
[14] Finnis M W 1996 *J. Phys.: Condens. Matter.* **8** 5811
[15] Schumacher, P, Greer, A L, Worth, J, Evans, P V, Kearns, M A, Fisher, P, Green, A H 1998 *Mater. Sci. Tech.* **14** 394-404
[16] Kresse G and Hafner J 1994 *Phys. Rev. B* **49** 14251
[17] Kresse G and Furthmuller J 1996 *Phys. Rev. B* **54** 11169-86
[18] Blöchl P E 1994 *Phys. Rev. B* **50** 17953-79
[19] Kresse G and Joubert D 1999 *Phys. Rev. B* **59** 1758-75
[20] Perdew J P and Burke K 1996 *Phys. Rev. Lett.* **77** 3865-68
[21] Han Y, Dai Y, Shu D, Wang J and Sun B 2006 *Appl. Phys. Lett.* **89** 144107
[22] Fan Z, Wang Y, Zhang Y, Qin T, Zhou X R, Thompson G E, Pennycook T, Hashimoto T 2015 *Acta Mater.* **84** 292-304
[23] Ma S, Yan R, Jing T and Dong H 2018 *Metals* **8** 521
[24] Ma S, Brown A, Yan R, Davidchack R, Howes P, Nicklin C, Zhai Q, Jing T and Dong H 2019 *Commun. Chem.* **2** 1

IOP Conf. Series: Materials Science and Engineering **861** (2020) 012048 doi:10.1088/1757-899X/861/1/012048

Influence of crystal fragmentation on the formation of microstructure and macrosegregation during directional solidification under forced convection condition

H Zhang, M Wu, Z Zhang, A Ludwig and A Kharicha

Chair for Simulation and Modeling of Metallurgy Processes, Department of Metallurgy, University of Leoben, A-8700 Leoben, Austria

E-mail: menghuai.wu@unileoben.ac.at

Abstract. Directional solidification experiment under forced convection condition was conducted. The AlSi7 alloy was solidified in an alumina cylindrical crucible (⌀8 mm) in a Bridgeman furnace; and forced convection was induced by applying rotating magnetic field (RMF). The RMF induced flow in the sample during solidification leads to the formation of equiaxed crystals by the mechanism of crystal fragmentation (assumption). The current study is to use a mixed columnar-equiaxed solidification model to simulate this experiment by considering the crystal fragmentation as sole origin of equiaxed crystals. An inward flow (Ekman effect) forms in the front of the (columnar) mushy zone under the RMF. Solute-driven remelting, as enhanced by the interdendritic flow, leads to fragmentation near the columnar tip front. Some fragments are transported by the forced convection to the sample centre and remelted there, while many of them are captured by the columnar structure near the sample centre. The modelling result on the mixed columnar-equiaxed structure agrees with the post-mortem analysis of as-solidified sample. As conclusion following impacts of the crystal fragmentation on solidification are suggested: (1) it widens the central segregation channel and promotes the formation of side-arms; (2) it leads to the formation of relatively high volume fraction of equiaxed crystals near the sample centre.

1. Introduction

To investigate the effect of the forced convection during alloy solidification on the evolution of microstructure and macrosegregation, a series of unidirectional solidification experiments on Al-alloys were conducted in a Bridgeman furnace as equipped with the rotating magnetic field (RMF) [1-4]. Under weak RMF or natural convection condition, the sample solidified normally as columnar structure. With the increase of RMF, a transition of microstructure from columnar to equiaxed (CET) occurred in the center part of a sample, whereas in the out-of-center region columnar dendrites still existed. Under strong RMF, a CET occurred in the whole cross section. Formation of fragments under forced convection was believed to be the main source of equiaxed crystals [5-7]. The RMF induced flow has also strong influence on the macrosegregation.

This paper will use a three-phase mixed columnar-equiaxed solidification model [8-10] to simulate the unidirectional solidification of a binary Al-7.0 wt%Si alloy under the RMF. The model was recently extended by incorporating the fragmentation of columnar dendrites as the source of equiaxed crystals [11, 12]. One goal of this study is to verify the numerical model by comparison with the experiments; a more ambitious goal is to study the influence of crystal fragmentation on the formation of microstructure and macrosegregation.

2. Model descriptions

A three-phase mixed columnar-equiaxed solidification model has been described elsewhere [8-10]. The three phases refer to the liquid melt, equiaxed crystals and columnar dendrite trunks, and their amounts are quantified by their volume fractions, f_ℓ, f_e and f_c, respectively. Both columnar dendrites and equiaxed crystals are treated as two separate solid phases. The primary and secondary arm spaces are taken from the experiment (λ_1 = 300 μm and λ_2 = 50 μm). The columnar phase is stationary, i.e. $\vec{u}_c \equiv 0$; while the motion of equiaxed phase, $\vec{u}_e$, is calculated by solving the corresponding momentum conservation equation.

In this study crystal fragmentation is considered as sole origin of equiaxed crystals; heterogeneous nucleation is ignored. The method to treat crystal fragmentation in the mixed columnar-equiaxed solidification model was described previously [11, 12]. The fragmentation-induced mass transfer rate from the columnar phase to the equiaxed phase (M_{ce}) can be calculated by:

$$M_{ce} = -\gamma \cdot \vec{u}_\ell \cdot \nabla c_\ell \cdot \rho_e \tag{1}$$

where $\vec{u}_\ell$ is velocity of liquid; ∇c_ℓ is the liquid concentration gradient; and ρ_e is the density of equiaxed phase. A fragmentation coefficient γ is assigned to bridge the unknown contributions to M_{ce}. Those unknown contributions to the fragmentation could be the curvature effect of the dendrites, latent heat induced thermal fluctuation, solute diffusion in the interdendritic liquid, etc. The value of γ should be estimated and modified according to some available experimental results. The initial diameter of the fragment ($d^0_{e,frag}$) is assumed to be proportional to λ_2 and local volume fraction of columnar phase (f_c).

$$d^0_{e,frag} = \lambda_2 f_c \tag{2}$$

As the ideal sphere is assumed for the morphology of equiaxed crystals [12], the production rate of fragments can be calculated with the following equation:

$$N_{frag} = \frac{M_{ce}}{\rho_e \cdot \frac{\pi}{6}\left(d^0_{e,frag}\right)^3} \tag{3}$$

In this paper, the mixture concentration, as calculated by equation (4), was used to analyse the macrosegregation.

$$c_{mix} = \frac{c_\ell \cdot \rho_\ell \cdot f_\ell + c_e \cdot \rho_e \cdot f_e + c_c \cdot \rho_c \cdot f_c}{\rho_\ell \cdot f_\ell + \rho_e \cdot f_e + \rho_c \cdot f_c} \times 100\% \tag{4}$$

where, c_ℓ, c_e, c_c and c_{mix} are concentrations of liquid, equiaxed, columnar, and mixture phase respectively; ρ_ℓ, ρ_e and ρ_c are corresponding densities of liquid, equiaxed and columnar phase.

3. Simulation settings

The numerical model is in accordance with the experiments [1], as shown in figure 1. The simulation is carried out in 2D axisymmetry. The cylindrical sample of AlSi7 alloy with a diameter of 8 mm solidifies in an Al_2O_3 crucible unidirectionally. A Dirichlet thermal boundary condition is imposed on the top (T_{Top}) and bottom (T_{Bottom}) of the sample. The outer wall of the sample is considered to be adiabatic. The fragmentation coefficient in equation (1) γ is equal to 0.2 for current study. It was determined through numerical parameter study by fitting the modelling results with the available experimental results. An RMF inductor is installed outside the sample. As shown by the dotted line in figure 1, the RMF is activated at $t = 210$ s. An analytical approximation of the azimuthal component of the electromagnetic force ($\vec{F}_\theta$) is valid (equation (5)) [13]:

$$\vec{F}_\theta = \frac{1}{2}\sigma\omega B^2 r\left(1-\frac{u_\theta}{\omega R}\right)\vec{e} \tag{5}$$

where σ is the electrical conductivity of the melt, $\omega = 2\pi f$ is angular frequency (f = 50 Hz), B (= 10 mT) is the magnetic induction, r and R (= $d/2$) is a radial coordinate and radius of the sample, u_θ is the azimuthal velocity of the melt at a radial coordinate r, and $\vec{e}$ is the tangential unit vector. The material properties and other parameters used in this study are referred to Zhang *et al.* [14].

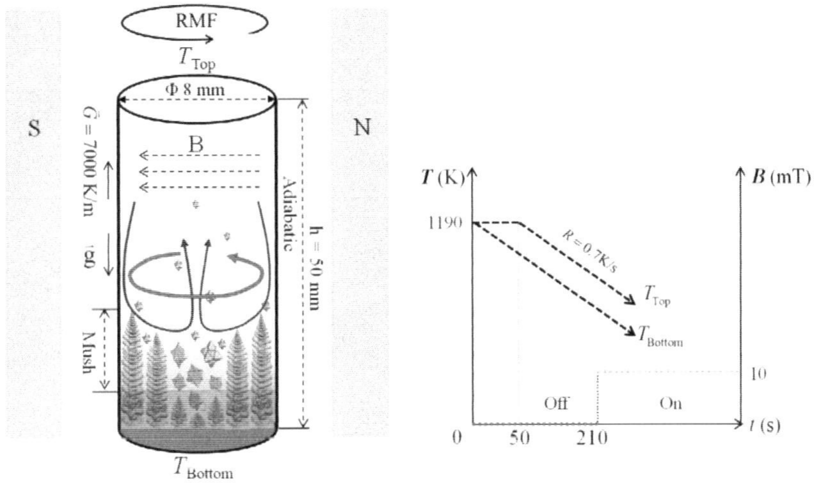

Figure 1. Geometry configuration of the simulation domain and applied boundary conditions.

4. Simulation results and discussions

4.1. Modelling results of solidification process

Before the start of the RMF, the thermo-solutal convection is very weak and the alloy solidifies in columnar dendritic structure. Once the RMF is activated at 210 s, some fragments form and develop into equiaxed crystals. Simulation results under the RMF at t = 350 s are shown in figure 2. To assist in the result explanation, the solidification and melt flow pattern are schematically drawn in figure 2(a). Due to the axisymmetry, only the modeling results on the left half of the sample are shown. The pink dotted line represents the axis.

During the solidification process, the solute element (Si) is rejected into the interdendritic melt region, resulting in an increase of solute concentration in the deep mushy region (figure 2(b)). Under the electromagnetic stirring of the RMF, the buoyancy effect is negligible. At the solidification front, the so-called Ekman effect [15] leads to a meridional circulation with the velocity magnitude of 0.014 m/s (vectors in figure 2(c)). The flow pattern in the mushy zone is similar to that in the bulk liquid, but the flow intensity is 2~4 orders of magnitude weaker [14]. This kind of flow transports the solute-enriched interdendritic liquid (figure 2(b)) from the periphery of the sample to the centre, causing the formation of a central segregation channel which is filled with solute-enriched liquid (figure 2(h)).

Figure 2(d) shows the regions where fragments mostly form. Maximum formation rate of fragments reaches a value of 10^9 (1/m^3/s) at the solidification front. In the side-arms of central segregation channel, some fragments also form, but the formation rate is relatively low ($10^5 \sim 10^7$ ($1/m^3/s$)). The meridional motion of equiaxed crystals (vectors) and the number density of equiaxed crystals (isolines) are shown in figure 2(e). In the bulk liquid region, equiaxed crystals move almost with the same velocity as the liquid.

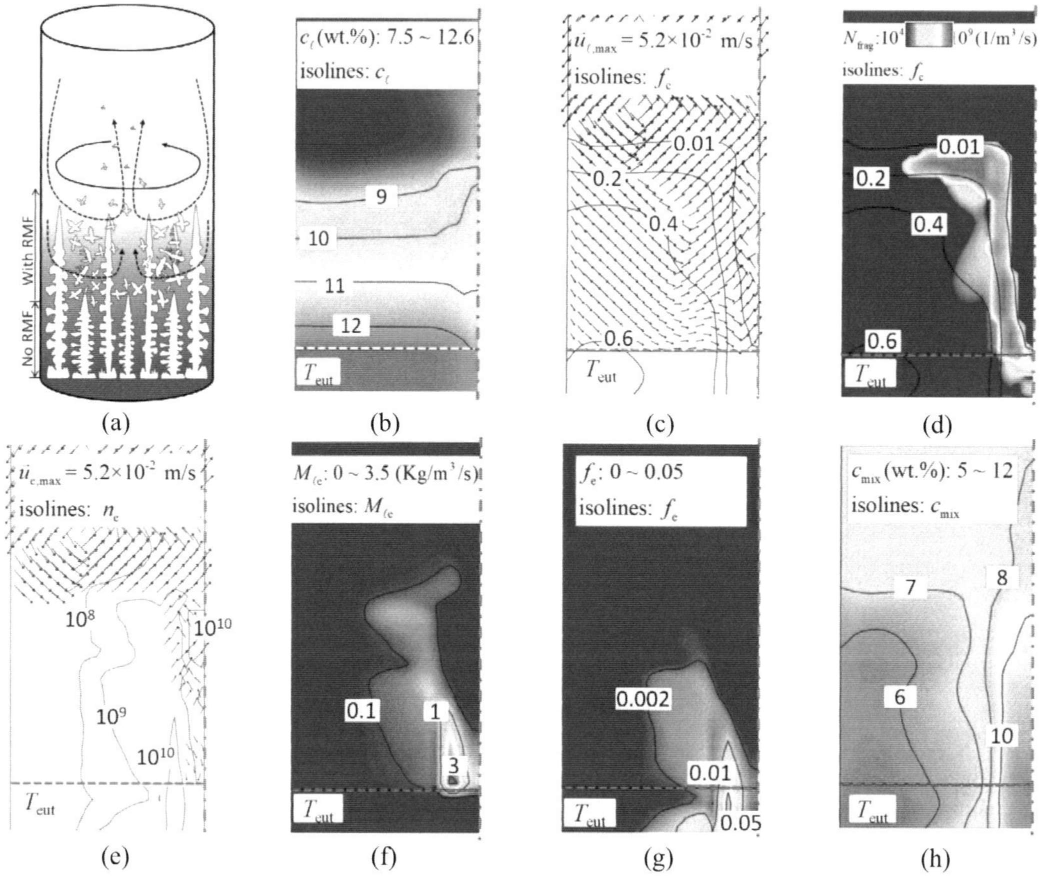

Figure 2. Solidification process of the sample under the RMF. (a) Schematic of the solidification process and melt flow pattern. (b) Contour of liquid concentration (c_ℓ). (c) Meridional flow of liquid ($\vec{u}_\ell$) overlaid with isolines of volume fraction of columnar phase (f_c). (d) Formation rate of fragments (N_e) overlaid with f_c. (e) Meridional motion of equiaxed crystals ($\vec{u}_e$) overlaid with isolines of number density of equiaxed crystals (n_e). (f) Contour of mass transfer rate from the liquid phase to the equiaxed phase ($M_{\ell e}$). (g) Contour of volume fraction of equiaxed phase (f_e). (h) Contour of mixture concentration (c_{mix}). To better display the velocity of liquid and equiaxed phase both in bulk liquid region and in mushy zone, length of vectors in (c) and (e) has been scaled by a log function. Contours in (b), (d), (f)-(h) are also shown in colour scale with red for maximum and blue for minimum values.

Near the solidification front, the velocity of equiaxed crystals is rapidly reduced. When the columnar phase reaches a critical volume fraction (assumption 0.2), the equiaxed crystals are captured by columnar dendrites. There is no motion of equiaxed crystals seen in lower left corner of figure 2(e). From the isolines, it can be seen that most of the fragments (10^9~10^{10} 1/m^3) are captured by the columnar dendrites near the sample centre. The equiaxed crystals, as transported to the sample centre, lead to an accumulation of equiaxed crystals (~10^{10} 1/m^3) initially; and they are then further transported into bulk liquid region and remelted there. The same fate of remelting occurs to those fragments which form in the side-arm regions of central segregation channel.

If the captured fragments survive, they can continue to grow and interact with the columnar dendrites during the subsequent solidification. The mass transfer (solidification) rate from the liquid phase to the equiaxed phase ($M_{\ell e}$) is shown in figure 2(f). The maximum $M_{\ell e}$ is observed near the solidification front with a value of 3.5 kg/m^3/s. From figure 2(e), large number of equiaxed crystals are observed near the sample centre, and $M_{\ell e}$ is also relatively large there.

The final distribution of equiaxed phase fraction (f_e) is shown in figure 2(g). Based on the current simulation results, equiaxed crystals with a volume fraction of approximate 0.05 form near the sample centre, whereas there are almost no equiaxed crystals at the periphery of the sample. Figure 2(h) exhibits the distribution of mixture concentration (c_{mix}). A strong positive segregation channel forms along the axis of the sample.

4.2. *Model validation*

The effect of the RMF on as-solidified structure and macrosegregation can been seen in figure 3. The distribution of the eutectic phase fraction (f_{eut}) is shown in figure 3(a). Before the RMF is activated, the eutectic phase uniformly distributes between columnar dendrites. After the activation of the RMF, an accumulation of the eutectic phase in the sample centre forms a shape of "fishtail". It is then followed by a tube-shape. Figure 3(b) demonstrates the mixture concentration (c_{mix}), which takes the same distribution as the eutectic phase. At the solidification front, the RMF induced flow sweeps the solute-enriched melt to the sample centre, leading to the formation of this kind of segregation pattern. Figure 3(c) displays the distribution of equiaxed phase fraction (f_e). With the effect of the RMF, some equiaxed crystals form and then are captured by the columnar dendrites. The maximum value of f_e is equal to 0.05. Figure 3(d) shows the as-solidified microstructure in the experiment. The current numerical study cannot quantitatively reproduce the experimental results. However, both numerical and experimental results agree with each other qualitatively.

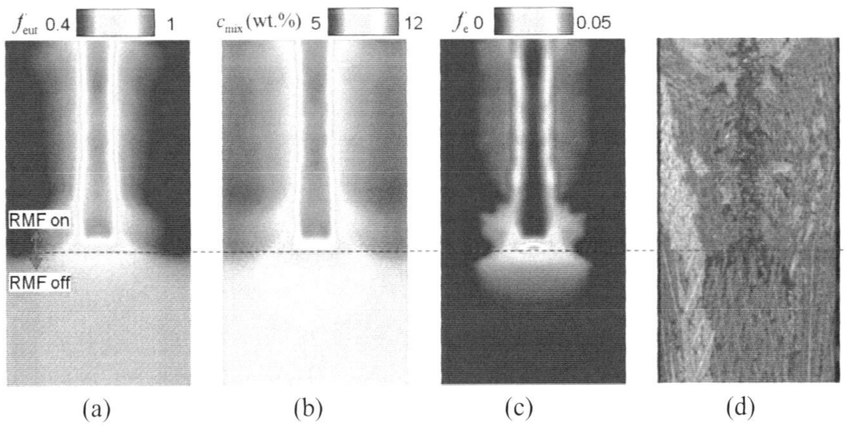

Figure 3. Microstructure and macrosegregation in the as-solidified sample near the section of RMF switch-on (a) eutectic phase fraction (f_{eut}), (b) mixture concentration (c_{mix}), (c) equiaxed phase fraction (f_e). (d) Experimentally measured microstructure.

4.3. *Influence of fragmentation on microstructure and macrosegregation*

In order to study the influence of crystal fragmentation on the microstructure and macrosegregation, an additional case of simulation is performed by assuming no fragmentation. Two cases are compared: one with the fragmentation and one without fragmentation. As shown in figure 4(a), without consideration of fragmentation there is no formation of equiaxed crystals, and the alloy solidifies as columnar dendrites with rest eutectics embedded in the inter columnar dendritic region and in the sample centre. When fragmentation is considered, some equiaxed phase forms near the sample centre; at the periphery of the sample columnar dendrites are dominant. The macrosegregation distribution patterns of both cases are quite similar, as shown in figure 4(b). A strong positive segregation channel form in the centre of the sample, and mixture concentration (c_{mix}) reaches 12.0, which is close to the eutectic composition (12.6); at the periphery, c_{mix} for both cases is equal to 5.5. The main difference in macrosegregation between these two cases can be seen in the side-arm regions of the central segregation channel. Without consideration of fragmentation, a thinner segregation channel forms in comparison to the case with fragmentation. It seems that the crystal fragmentation promotes the formation of side-arms (oscillations in figure 4(b)). As shown in figure 2(d), some fragments form in the side-arm regions. If these fragments are not captured by columnar dendrites, they can be transported by the liquid flow to the bulk liquid region. This fragmentation phenomenon makes more space for the flow and promote the formation of the side-arms. c_{mix} along the radial direction, as drawn by the dash line, is plotted in figure 4(c). Consideration of fragmentation during solidification increases c_{mix} in the central segregation channel.

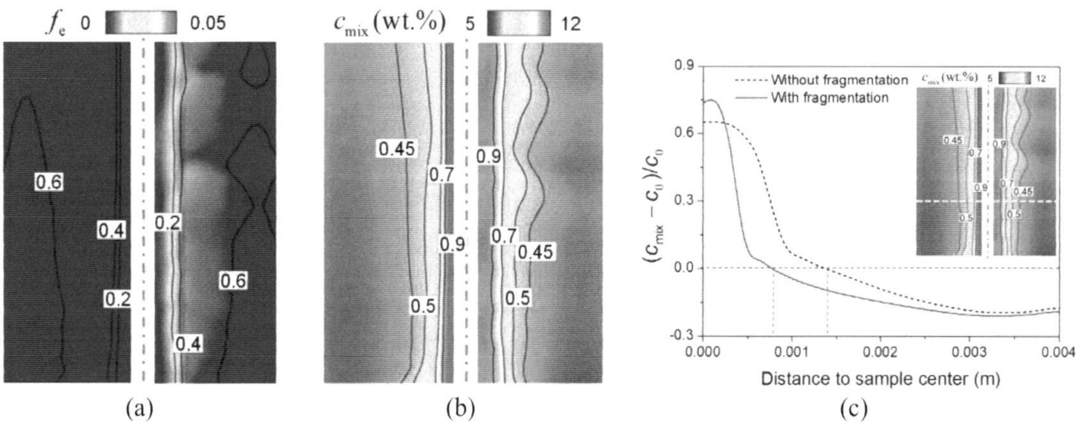

Figure 4. Numerical parameter study of the influence of fragmentation event on the as-solidified structure and macrosegregation. Left-half figures of (a) and (b) shows the modelling result for the case without considering the fragmentation; right-half figures of them shows the modelling result for the case with the fragmentation. (a) Contour of f_e overlaid with isolines of f_e. (b) Contour of c_{mix} overlaid with isolines of the volume fraction of rest eutectic phase (f_{eut}). (c) c_{mix} along the radial direction of the sample.

4.4. *Discussion on the formation mechanism of fragments*

Figure 5 shows the simulation results on the fragmentation rate and fragmentation-induced columnar-to-equiaxed mass transfer rate at $t = 350$ s. According to the current model, equation (3), the production rate of fragments (N_{frag}) should be proportional to fragmentation-induced mass transfer rate from the columnar phase to the equiaxed phase (M_{ce}). As shown in figure 5(a) and (b), in the regions where M_{ce} is positive, fragments form. Most fragments form at the solidification front, and a small amount of fragments form in the side-arms of central segregation channel. Based on equation (1), if the direction of $\vec{u}_\ell$ is opposite to

∇c_ℓ, the interdendritic flow leads to local increase of c_ℓ, which promotes the formation of fragments. This can be seen from the zoom-in views of zone A and zone B in figure 5(c). In regions where the angle between two vectors ($\vec{u}_\ell$ and ∇c_ℓ) is larger than 90°, M_{ce} is positive and some fragments form.

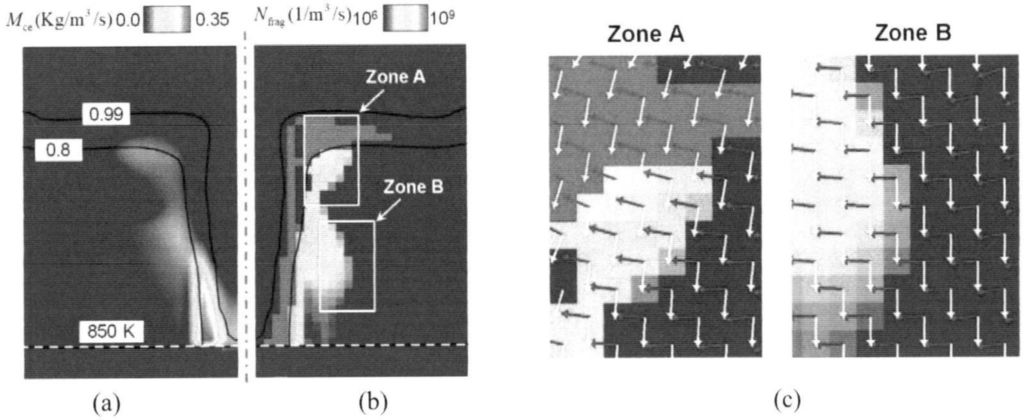

Figure 5. Analysis of the formation mechanism of fragments by solute-driven remelting due to interdendritic flow (t = 350 s). (a) Fragmentation-induced mass transfer rate from the columnar phase to equiaxed phase (M_{ce}), overlaid by isopleths of liquid volume fraction f_ℓ (solid lines) and eutectic temperature (dash line). (b) Contour of production rate of fragments (N_{frag}). (c) Zoom-in views of zone A and zone B in (b) with vectors indicating the directions of liquid velocity $\vec{u}_\ell$ (dark red) and liquid concentration gradient ∇c_ℓ (yellow).

It is well known that both heterogeneous nucleation and crystal fragmentation are two main origins of equiaxed crystals during alloy solidification. In current study, we only investigated the influence of crystal fragmentation on the formation of microstructure and macrosegregation during unidirectional solidification under forced convection condition. Although fragmentation is considered to be the most important one for the formation of equiaxed crystals under forced convection condition [16], the role of another mechanism by heterogeneous nucleation in the current experiment is not clear. It needs further investigation.

5. Conclusions

A mixed columnar-equiaxed solidification model was used to study the influence of crystal fragmentation on the formation of microstructure and macrosegregation in a unidirectional solidification experiment under the forced convection condition (RMF: rotating magnetic field). Although a simple formulation [11, 12] which is based on the solute-driven remelting as enhanced by the interdendritic flow was used to estimate the fragmentation rate the modelling results showed a very promising agreement with the experiment. New knowledge was obtained. An inward flow forms above the mush under the RMF, leading to the formation of central segregation channel. Solute-driven remelting leads to fragmentation near the columnar tip front and in the central segregation channel. Some fragments are transported by the fluid flow to the sample centre, while some fragments are captured by columnar dendrites near the sample centre. Two important impacts of the fragmentation on solidification are suggested: (1) it widens the central segregation channel and promotes the formation of side-arms; (2) it leads to the formation of relatively high volume fraction of equiaxed crystals near the sample centre.

Acknowledgements
The authors acknowledge the financial support from Austrian Research Promotion Agency (FFG) - Austrian Space Application Program (ASAP) through the project FLOWSICONS (No. 859777), as well as the support from European Space Agency (ESA) through the project MICAST. The authors also acknowledge MICAST-Hungary Team (leader Prof. András Roósz), University of Miskolc, Hungary, for providing the experimental data.

References

[1] Rónaföldi A, Kovács J and Roósz A 2009 *Trans. Indian Inst. Met.* **62** 475-7
[2] Zimmermann G, Weiss A and Mbaya Z 2005 *Mater. Sci. Eng.* A **413** 236-42
[3] Steinbach S and Ratke L 2005 *Mater. Sci. Eng. A* **413-414** 200-4
[4] Roósz A, Kovács J, Rónaföldi A and Kovács Á 2013 *Mater. Sci. Forum* **752** 57-65
[5] Zimmermann G, Pickmann C, Hamacher M, Schaberger-Zimmermann E, Neumann-Heyme H, Eckert K and Eckert S 2016 *Acta Mater.* **126** 236-50
[6] Hellawell A, Liu S and Lu S 1997 *JOM* **49** 18-20
[7] Boden S, Eckert S, Willers B and Gerbeth G 2008 *Metall. Mater. Trans. A* **39** 613-23
[8] Wu M and Ludwig A 2006 *Metall. Mater. Trans. A* **37A** 1613-31
[9] Wu M, Ludwig A and Kharicha A 2019 *Metals* **9** 229
[10] Wu M and Ludwig A 2007 *Metall. Mater. Trans. A* **38** 1465-75
[11] Zheng Y, Wu M, Kharicha A and Ludwig A 2017 *Model. Simul. Mater. Sci. Eng.* **26** 015004
[12] Zheng Y, Wu M, Karimi-Sibaki E, Kharicha A and Ludwig A 2018 *Int. J. Heat Mass Transfer* **122** 939-53
[13] Roplekar J and Dantzig J 2001 *Int. J. Cast Met. Res.* **14** 79-95
[14] Zhang H, Wu M, Kharicha A and Ludwig A 2020 *Mater. Today Commun.* **22** 100842
[15] Willers B, Eckert S, Nikrityuk P A, Räbiger D, Dong J, Eckert K and Gerbeth G 2008 *Metall. Mater. Trans. B* **39** 304-16
[16] Griffiths G and McCartney D 1996 *Mater. Sci. Eng. A* **216** 47-60

IOP Conf. Series: Materials Science and Engineering **861** (2020) 012049 doi:10.1088/1757-899X/861/1/012049

Three-dimensional needle network model for dendritic growth with fluid flow

T Isensee [1,2] **and D Tourret** [1]
[1] IMDEA Materials Institute, Getafe, 28906 Madrid, Spain
[2] Polytechnic University of Madrid, 28040 Madrid, Spain

E-mail: damien.tourret@imdea.org

Abstract. We present a first implementation of the Dendritic Needle Network (DNN) model for dendritic crystal growth in three dimensions including convective transport in the melt. The numerical solving of the Navier-Stokes equations is performed with finite differences and is validated by comparison with a classical benchmark in fluid mechanics for unsteady flow. We compute the growth behavior of a single equiaxed crystal under a forced convective flow. As expected, the resulting dendrite morphology differs strongly from the case of the purely diffusive regime and from similar two-dimensional simulations. The resulting computationally efficient simulations open the way to studying mechanisms of microstructure selection in presence of fluid flow, using realistic alloy and process parameters.

1. Introduction

Dendritic microstructures are common in solidification-processed metals and alloys [1,2]. Their morphology has a crucial influence on the thermo-mechanical properties of these materials [3]. Dendritic patterns are shaped by the interaction of individual dendritic branches and hence arise from mechanisms on different length scales, i.e. phenomena on the scale of the microscopic solid-liquid interface, and macroscopic heat and solute transport. For decades, the aim to bridge scales has motivated the development of a wide range of multiscale simulation approaches, e.g. using dedicated formulations for enthalpy terms [4], volume-averaged conservation equations [5], cellular automata coupled with finite elements [6] or finite differences [7], or dynamics of grain envelopes [8], to name a few. Macroscopic heat and solute transport can be strongly influenced by advective currents. Since buoyancy can be caused by gravity [9-11], it is nearly impossible to carry out solidification experiments in homogeneous conditions. Experiments with reduced gravity have provided important insight into the effect of buoyancy onto microstructure selection [12-14], however only at great expense, which is why the modeling of dendritic growth under fluid flow is of tremendous interest.

In contrast to growth in the purely diffusive regime, in the presence of fluid flow, growing crystals are asymmetric. The problem has been investigated analytically, e.g. using a diffusive boundary layer ("stagnant film") around the dendrite [15-17], and computationally, e.g. with models that explicitly track the solid-liquid surface [18-20] or phase-field (PF) simulations [21-26]. However, these models still require a reasonably accurate resolution of the dendrite tip morphology, which limits their use in the case of concentrated alloys that usually solidify with thin needle-like branches with a tip radius much lower than the scale of macroscopic transport of heat and solute in the melt.

To address this multiscale challenge, we developed a dendritic needle network (DNN) model, in which the dendrite branches are represented as thin paraboloids. The model quantitatively predicts the dynamics of individual branches in hierarchical networks during solidification at a scale much larger than the diffusion length. It is particularly adapted to solidification at low supersaturation, i.e. low Péclet number, when phase-field becomes prohibitively costly. The key computational advantage of the method comes from the fact that the numerical grid spacing can be chosen of the same order as the dendrite tip radius, which is typically one order of magnitude coarser than the requirement for quantitative phase-field simulations. This results in simulations faster by several orders of magnitude, with minimal loss in accuracy [27]. The DNN model, developed in two dimensions (2D) [28] and three dimensions (3D) [29], was already verified against classical theories and phase-field simulations of dendritic growth, and validated against directional solidification experiments in a predominantly diffusive transport regime [29-31]. A first incorporation of fluid flow in 2D was also achieved [32]. However, since flow patterns strongly differ in 2D and 3D, quantitative simulations require a 3D implementation. Here, we present first results of the DNN model including fluid flow in 3D.

2. Model

2.1. Sharp-interface problem

Considering a binary alloy at a temperature $T = T_0$ below its liquidus temperature T_L, we introduce a dimensionless form of the solute concentration field c, i.e. the supersaturation

$$U = \frac{c_0 - c}{(1-k)\, c_0}, \tag{1}$$

with c_0 and k the liquid equilibrium concentration at T_0 and the interface solute partition coefficient, respectively. Neglecting diffusion in the solid phase, the solute concentration in the liquid close to the solid-liquid interface (e.g. within a diffusive boundary layer) follows

$$\partial_t U = D\nabla^2 U\,, \tag{2}$$

with D the solute diffusivity. Solute mass conservation at the interface yields a Stefan condition

$$v_n = D\partial_n U|_i \tag{3}$$

that relates the interface normal velocity v_n to the normal solute concentration gradient $\partial_n U|_i$. Neglecting kinetic undercooling, the Gibbs-Thomson relation for interfacial equilibrium is

$$U_i = d_0 f_\gamma(\theta)\,\kappa\,, \tag{4}$$

where κ is the interface curvature and $d_0 = \Gamma_{sl} / [|m|(1-k)\, c_0]$ is the solutal capillary length, with the liquidus slope $m < 0$ and the Gibbs-Thomson coefficient Γ_{sl}. The anisotropy function $f_\gamma(\theta)$ represents the dependence of the interface stiffness $\gamma(\theta) + \partial_\theta^2\gamma(\theta)$ upon the interface orientation θ, where $\gamma(\theta)$ is the excess free energy of the solid-liquid interface [33,34]. Equations (2)-(4) describe the sharp interface problem for a growing solid-liquid interface, combined with an imposed supersaturation

$$\Omega = \frac{c_0 - c_\infty}{(1-k)\, c_0} \tag{5}$$

far away from the interface, with c_∞ the nominal solute concentration.

2.2. Dendritic Needle Network model

The DNN model is designed to model solidification at low supersaturation, i.e. at a Péclet number $\mathrm{Pe} = RV/(2D) \ll 1$, with R and V the dendrite tip radius and velocity, respectively. With a tip radius much smaller than the diffusion length $l_D = D/V$, conservation equations can be derived at different length scales, such that: (i) mass conservation on an intermediate scale between the tip radius R and the diffusion length provides the product RV (or RV^2 in 2D), and (ii) microscopic solvability theory at the scale of R prescribes the constant value of R^2V. The combination of these conditions, described below, enables to obtain the instantaneous growth conditions $R(t)$ and $V(t)$ for each dendritic tip.

2.3. Microscopic solvability condition at the scale of the dendrite tip

The well-established solvability theory [35-38] states that the sharp-interface problem described by equations (2)-(4) has a steady solution only if

$$R^2V = \frac{2Dd_0}{\sigma}, \tag{6}$$

where the selection parameter σ is uniquely determined by the strength of the interface anisotropy. The theory was confirmed using PF simulations [39-41], which even showed that the product R^2V relaxes to a constant very early during growth, while $R(t)$ and $V(t)$ still undergo significant variations [41]. Analytical [42] and phase-field [22,24,25] studies have also shown that σ remains constant up to fluid velocities about one order of magnitude higher than the tip velocity. Thus, the DNN model considers equation (6) to be valid at all times.

2.4. Conservation of solute on an intermediate scale

On a scale larger than the tip radius, dendritic branches appear sharp and curvature effects can be neglected, so that the equilibrium concentration $U = U_i \approx 0$ is considered along the interface. Assuming a paraboloid with its tip at x_t and a cross section of radius $r_i(x) = [2R(x_t - x)]^{1/2}$ growing in a shape-preserving manner with a velocity V, the mass conservation (equation (3)) integrated over a contour Γ_0 that spans until a length a behind the tip (see figure 1) yields

$$D\int_{\Gamma_0} (\partial_{n^*}U)\, \mathrm{d}\Gamma_0 = \int_0^{2\pi}\int_0^{r_a} V\, r\, \mathrm{d}r\, \mathrm{d}\theta = 2\pi aRV, \tag{7}$$

where r_a is the paraboloid cross section at $x = x_t - a$. By introducing the flux intensity factor (FIF)

$$\mathcal{F} = \frac{1}{2\pi a}\int_{\Gamma_0} (\partial_{n^*}U)\, \mathrm{d}\Gamma_0, \tag{8}$$

the product RV can be expressed as

$$RV = D\mathcal{F}. \tag{9}$$

To integrate over Γ_0, we use the divergence theorem and the assumption of a Laplacian field in a moving frame with velocity V, i.e. $D\nabla^2U = -V\,\partial_xU$, within the FIF integration region, yielding

$$\int_{\Gamma_0} (\partial_{n^*}U)\, \mathrm{d}\Gamma_0 = \int_{\Gamma_i} (\partial_nU)\, \mathrm{d}\Gamma_i + \frac{V}{D}\int_{\Sigma_i} (\partial_xU)\, \mathrm{d}\Sigma_i. \tag{10}$$

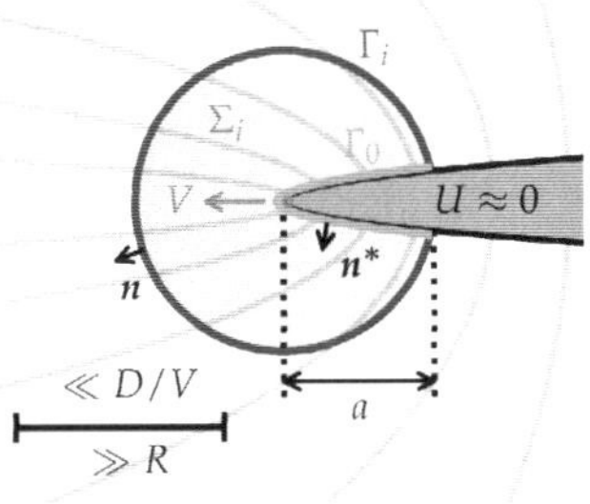

Figure 1. Conservation of solute on the needle tip moving with velocity V on an intermediate scale larger than the tip radius R but smaller than the diffusion length $l_D = D/V$.

2.5. Navier-Stokes equation in the liquid phase

We consider the liquid phase to be an incompressible and Newtonian fluid, in which the mass conservation results in the incompressibility condition $\nabla \cdot \boldsymbol{u} = 0$. The conservative Navier-Stokes equations as a statement of momentum conservation read

$$\rho\, [\partial_t \boldsymbol{u} + \nabla \cdot (\boldsymbol{uu})] = \boldsymbol{F} - \nabla p + \eta\nabla^2\boldsymbol{u}, \tag{11}$$

where $\boldsymbol{u}$ is the fluid velocity field of (x,y,z)-components (u,v,w), p is the pressure field, η the dynamic viscosity, ρ the fluid density, and $\boldsymbol{F}$ represents external body forces, e.g. gravity-induced buoyancy. The transport of solute within the liquid is described by the advection-diffusion equation

$$\partial_t U + \nabla \cdot (\boldsymbol{u} U) = \nabla \cdot (D \nabla U) \,. \tag{12}$$

2.6. Implementation

Equations are solved using a method previously presented in 2D [32]. We use a projection method to solve the momentum equation, with an iterative successive over relaxation method for the incompressibility condition $\nabla \cdot \boldsymbol{u} = 0$. Space is discretized using finite differences on a homogeneous grid of cubic elements with staggered velocity components. Time stepping uses an explicit Euler method. The flux intensity factor $\mathcal{F}$ is integrated over a sphere centered around the tip location, similarly as done with a circle in 2D [32]. The code is implemented in C-based cuda programming language, which allows for parallelization on Nvidia graphic cards.

3. Code validation

3.1. Unsteady flow past an obstacle

First, we validate the implementation of the Navier-Stokes equation for an unsteady flow. To do so, we simulate a von Kármán vortex street instability past a cylindrical obstacle with different Reynolds numbers, which was extensively studied theoretically, numerically, and experimentally [43-48]. An inflow velocity of $u_i = 1$ is imposed at $x = 0$ in a cuboidal domain with dimensions $L_x \times L_y \times L_z$ with $25.5 \leq L_x \leq 34.7$, $10.1 \leq L_y \leq 13.4$, and $0.21 \leq L_z \leq 0.56$, using a discrete grid element size $\Delta x = 0.04$ (for Re ≤ 150) or 0.035 (for Re > 150). A cylindrical obstacle of axis z and diameter $d = 1.75$ is set at $x \approx 9$. The boundaries parallel to the flow are set with free slip conditions, while the outflow at the upper x boundary is free. The resulting oscillatory flow can be characterized by the Strouhal number St $= f\,d/u_i$, that uniquely depends on the Reynolds number, with f being the outflow frequency. For $49 <$ Re < 180 the Strouhal number follows a universal law [43], while for higher Reynolds numbers the flow becomes three-dimensional and follows a different law [44]. Figure 2(a)-(b) show a snapshot slice through the three-dimensional domain, illustrating iso-values of the vorticity magnitude for a Reynold number of 50 (a) and 175 (b). In figure 2(c) the predictions for St(Re) compare well with the universal laws assessed experimentally, with a deviation by less than 3%.

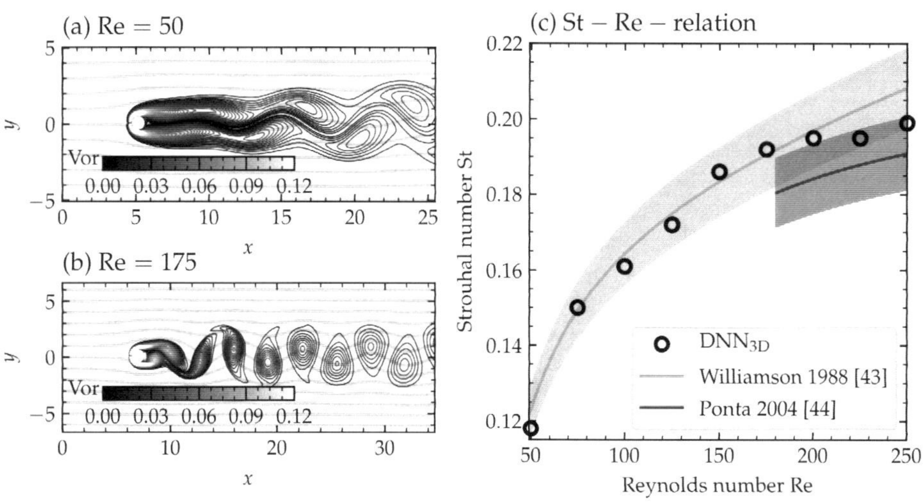

Figure 2. (a) and (b) Contour plots of the absolute value of the vorticity in a flow with Re = 50 and Re = 175, respectively, past a cylindrical obstacle with the maximum value shown in white and the minimum value in black. The gray streamlines indicate that the domain is large enough for the boundary conditions to not come into play. The cross section was taken in the center of the three-dimensional domain. (c) Strouhal versus Reynolds number in comparison with universal laws from the literature, i.e. [43] for $49 <$ Re < 180 and [44] for Re ≥ 180. The shaded area indicates a 5% uncertainty range.

3.2. Steady state growth of an isolated dendrite

Next, we test the growth of a free dendrite compared to the analytical Ivantsov solution [49], which relates the Peclét number to the solute supersaturation as

$$\Omega = \mathrm{Iv}_{2\mathrm{D}}(\mathrm{Pe}) = (\pi \mathrm{Pe})^{1/2} \exp(\mathrm{Pe})\, \mathrm{erfc}(\mathrm{Pe}^{1/2}) \quad \text{with} \quad \mathrm{erfc}(\zeta) = 1 - \frac{2}{\sqrt{\pi}} \int_{\zeta}^{\infty} \exp(-s^2)\, \mathrm{d}s\,, \tag{13}$$

$$\Omega = \mathrm{Iv}_{3\mathrm{D}}(\mathrm{Pe}) = \mathrm{Pe} \exp(\mathrm{Pe})\, \mathrm{E}_1(\mathrm{Pe}) \quad \text{with} \quad \mathrm{E}_1(\zeta) = \int_{\zeta}^{\infty} \frac{\exp(-s)}{s}\, \mathrm{d}s\,, \tag{14}$$

in 2D and 3D, respectively.

We simulate the growth of a single dendrite at a given imposed supersaturation until a steady state is reached, and compare the resulting Péclet number with the theoretical solution. We use a grid size of $230 \times 230 \times 230$ with a grid spacing Δx that yields $\Delta x = l_D/15$ over a time of at least $50D/V_s^2$. A unique needle is set to grow in the $x+$ direction and no-flux conditions are applied on all boundaries. The needle is located either at the center (for $\Omega \leq 0.02$) or at the edge (for $\Omega > 0.2$) of the domain in y and z. The domain is shifted progressively to keep the location of the tip at 1/3 of the domain length in x. The FIF integration radius is $r_{\mathrm{FIF}} = 5\Delta x$.

Figure 3 illustrates the comparison with the Ivantsov solution, showing that, while Pe is slightly overestimated at low Ω, the model essentially reproduces the expected steady growth velocity across several orders of magnitude in Péclet number.

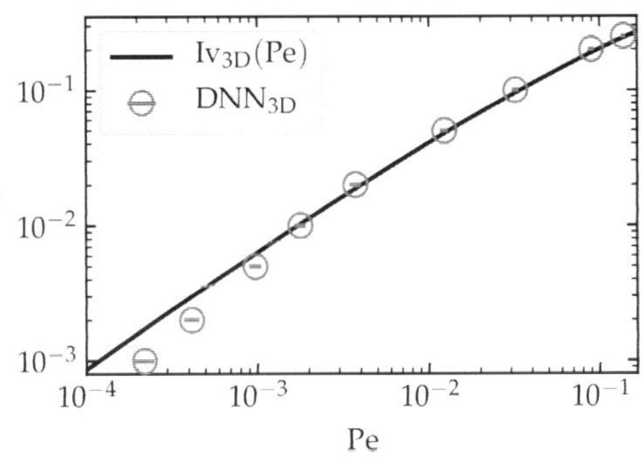

Figure 3. Imposed supersaturation Ω versus Péclet number Pe of a single isolated dendrite predicted by 3D DNN simulations in comparison with the analytical Ivantsov law $\mathrm{Iv}_{3\mathrm{D}}(\mathrm{Pe})$ [49]. The grid spacing Δx was chosen so that the diffusion length $l_D = 15\Delta x$. Each value is an average over a time range that covers several oscillations. The error bars correspond to the standard deviation.

4. Equiaxed crystal growth in a forced flow

4.1. Simulations

Finally, having independently verified the implementation of the Navier-Stokes equations and dendritic growth in the diffusive regime, we simulate the growth of an equiaxed grain under forced flow. We perform simulations in both 2D and 3D for conditions close to those studied in 2D in [32]. We consider a model alloy of Schmidt number $\mathrm{Sc} = \nu/D = 20$, with $\nu = \eta/\rho$ the kinematic viscosity, at a solute supersaturation $\Omega = 0.5$. We use a tip selection parameter $\sigma = 0.06$ in 2D and $\sigma = 0.04$ in 3D, which both correspond to a solid-liquid interface energy anisotropy of amplitude ≈ 0.01 for a one-sided model [37].

For the 3D simulation, we use a domain with a size $L_x \times L_y \times L_z = 51.0\, R_s \times 25.4\, R_s \times 25.4\, R_s$ with a grid spacing of $\Delta x = 0.1 R_s$. An inflow with velocity $(u_i,0,0)$ is imposed at the left boundary ($x = 0$), while the right boundary is set to allow free outflow, and the remaining boundaries hold free-slip conditions for the velocity (i.e. $v = \partial v/\partial y = 0$ for y-boundaries and $w = \partial w/\partial z = 0$ for z-boundaries). Mirror symmetry conditions are applied for the solute field U at all boundaries. Exploiting the symmetry of the domain and the expected laminar flow, we initiate a single equiaxed grain along the ($y = z = 0$) edge of the domain, centered at ($25.5\, R_s$,0,0). The grain consists of four branches (with directions $x+$, $x-$, $y+$, and $z+$) that have an initial radius $0.4R_s$ and length $2R_s$. For the 2D simulation, we use similar boundary conditions and grid spacing but with a size of $L_x \times L_y$ = 204.6 $R_s \times$ 102.2 R_s. The grain is generated at (102.2 R_s, 0).

For similar physical parameters, the expected steady state velocities V_s in 2D and 3D differ, such that the scaled inflow velocities u_i/V_s also differ. Combining microscopic solvability (6) with the definition of the Péclet number, Pe = $RV/(2D)$, one can write $V_s d_0/D = 2\sigma\mathrm{Pe}^2$. For $\Omega = 0.5$, since the Péclet number is 0.1873 in 2D and 0.6101 in 3D, the steady growth velocity V_s in 3D is about 7 times higher than in 2D. The inflow velocity is set to $u_i = 0.0421\ D/d_0$, which corresponds to $u_i/V_s = 10.0$ in 2D and $u_i/V_s = 1.41$ in 3D.

The total simulation time is chosen as 20 R_s/V_s, which was found to be sufficient to achieve steady-state growth. For both 2D and 3D simulations, the numerical parameters are $K_{\Delta t} = 0.6$, $\omega_{up} = 0.9$, $\omega_{SOR} = 1.7$ and $r_{SOR} = 10^{-3}$ (see Ref. [32] for details). The FIF integration domain has a radius $r_{FIF} = 5\Delta x$. Side branches are generated every time a needle grows by 10 R_s.

4.2. Results

The 3D simulation was performed in under 20 h with a single Nvidia RTX 2080Ti GPU. The results are illustrated in figure 4. Figure 4a shows the grain shape in the 3D simulation after a time $20R_s/V_s$, with iso-surfaces of the solute concentration and streamlines of the flow. Figure 4b shows the evolution of the upstream (US), downstream (DS), and transverse (T) tips in both 2D and 3D simulations. Figure 4(c)-(d) show the interface, solute field, and streamlines in 2D (c) and along the (z = 0) plane in 3D (d).

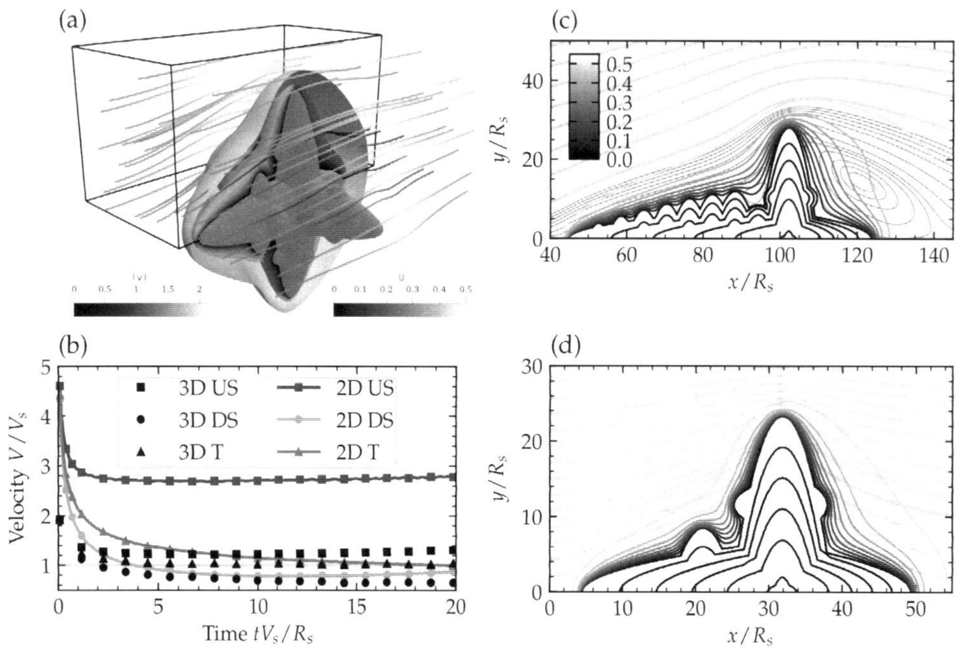

Figure 4. (a) Dendritic grain (gray) under forced flow predicted by the DNN model at a time $20R_s/V_s$. The iso-surfaces indicate the dimensionless solute field U from 0 to 0.5 with steps of 0.05 and the streamlines show the laminar flow around the crystal. (b) Velocities of the upstream (US), downstream (DS), and transverse (T) tips versus time. The values correspond to average values over a short time range in order to smoothen the curves. (c) and (d) show the cross section along the fluid flow direction. The colored contour lines indicate the field U; the black lines show the location of the solid-liquid interface at tV_s/R_s = 0, 4, 8, 12, 16, and 20. The gray streamlines represent iso-values of the stream function. Plots do not show the complete simulation domain. Plot (c) shows a higher density of thinner streamlines close to the dendrite in order to highlight the convective vortices next to the downstream tip.

Qualitatively, the effect of fluid flow on the tip velocities is similar, namely the upstream tip velocity is increased, the downstream tip velocity is decreased, and the steady transverse tip velocity is barely affected. However, the amount of change differs substantially between 2D and 3D simulations. The upstream velocity increase is much higher in 2D (about 3 times higher than the transverse tip velocity) than in 3D (increased by about 40%). This is not only due to the different scaling of V_s in 2D and 3D, but also to the resulting flow pattern. As already pointed out in previous studies [23-25,50], in 3D the flow can easily pass by the transverse arm, whereas in 2D the arm acts as a wall that the flow has to entirely go around. As a result, the flow velocities are overestimated in 2D, and flow patterns are also importantly affected. An important consequence on the flow pattern is the formation of convective vortices around the downstream arm, which only appear in 2D. An outcome of these 2D vortices is that the downstream tip accelerates as the vortices feed it solute, and hence the tip cannot reach a steady state. Meanwhile, in 3D, the downstream tip seems to reach a well-defined steady growth velocity.

5. Summary and outlook

We presented the first three-dimensional implementation of the Dendritic Needle Network (DNN) model for binary alloy isothermal solidification with fluid flow. The code implementation was validated for unsteady oscillatory flow past an obstacle, and verified for steady state growth in the diffusive regime. Then, we performed simulations of equiaxed growth of a single grain in a forced flow and compared results of 2D and 3D simulations. The results show that the growth dynamics significantly deviates from diffusive solidification. Furthermore, the acceleration of the upstream tip and deceleration of the downstream tip differ significantly in the 2D and 3D cases, in agreement with previous studies [23,24]. These results further highlight the importance of 3D simulations in order to produce results that can be compared to experiments or solidification processes on a quantitative basis.

In the future, we expected the DNN model to provide computationally efficient and spatially extended simulations of solidification in a low Péclet number regime that remains challenging to established simulation methods such as phase-field. This should allow exploring the mechanisms of microstructure selection at the scale of tens of thousands of dendritic branches [29,31]. Ongoing and upcoming developments from this work include: quantitative comparison to phase-field simulations [23,24,27], study of the effect of the relative orientations of crystal and flow [50,51], and the extension to directional solidification conditions [28,29] in order to study, for instance, the selection of dendritic spacings in the presence of micro- or hyper-gravity conditions [52,53] or the selection of grain boundary orientations during columnar grain growth competition [54,55].

Acknowledgements

This work was supported by the European Union's Horizon 2020 research and innovation programme through DT's Marie Skłodowska-Curie Individual Fellowship (Grant Agreement 842795).

References

[1] Langer J 1980 *Rev. Mod. Phys.* **52** 1
[2] Trivedi R and Kurz W 1994 *Int. Mater. Rev.* **39** 49
[3] Dantzig JA and Rappaz M 2009 *Solidification* (Lausanne, Switzerland: EPFL Press)
[4] Voller VR and Prakash C 1987 *Int. J. Heat Mass Tran.* **30** 1709
[5] Wang C and Beckermann C 1996 *Metall. Mater. Trans.* A **27** 2754
[6] Gandin CA, Desbiolles JL, Rappaz M and Thevoz P 1999 *Metall. Mater. Trans.* A **30** 3153
[7] Wang W, Lee PD and Mclean M 2003 *Acta Mater.* **51** 2971
[8] Steinbach I, Beckermann C, Kauerauf B, Li Q and Guo J 1999 *Acta Mater.* **47** 971
[9] Mehrabian R, Keane M and Flemings M 1970 *Metall. Trans.* **1** 1209
[10] Nguyen-Thi H, Billia B and Jamgotchian H 1989 *J. Fluid Mech.* **204** 581
[11] Dupouy M, Camel D and Favier J 1989 *Acta Metall.* **37** 1143
[12] Glicksmann M, Koss M and Winsa E 1994 *Phys. Rev. Lett.* **73** 573

[13] Nguyen-Thi H, Dabo Y, Drevet B, Dupouy M, Camel D, Billia B, Hunt J and Chilton A 2005 *J. Cryst. Growth* **281** 654
[14] Nguyen-Thi H, Reinhart G and Billia B 2017 *CR Mécanique* **345** 66
[15] Cantor B and Vogel A 1977 *J. Cryst. Growth* **41** 10
[16] Li Q and Beckermann C 2002 *J. Cryst. Growth* **236** 482
[17] Sekerka R, Coriell S and McFadden G 1995 *J. Cryst. Growth* **154** 370
[18] Udaykumar H, Marella S and Krishnan S 2003 *Int. J. Heat Mass Tran.* **46** 2615
[19] Al-Rawahi N and Tryggvason G 2002 *J. Comput. Phys.* **180** 471
[20] Zhao P, Heinrich J and Poirier D 2005 *Int. J. Numer. Meth. Fl.* **49** 233
[21] Beckermann C, Diepers HJ, Steinbach I, Karma A and Tong X 1999 *J. Comput. Phys.* **154** 468
[22] Tong X, Beckermann C, Karma A and Li Q 2001 *Phys. Rev.* E **63** 61601
[23] Jeong JH, Goldenfeld N and Dantzig J 2001 *Phys. Rev.* E **64** 416021
[24] Jeong JH, Dantzig J and Goldenfeld N 2003 *Metall. Mater. Trans.* A **34** 459
[25] Lu Y, Beckermann C and Ramirez J 2005 *J. Cryst. Growth* **280** 320
[26] Rojas R, Takaki T and Ohno M 2015 *J. Comput. Phys.* **298** 29
[27] Tourret D, Sturz L, Viardin A and Založnik M 2020 *IOP Conf. Ser.: Mater. Sci. Eng.* ***current issue***
[28] Tourret D and Karma A 2013 *Acta Mater.* **61** 6474
[29] Tourret D and Karma A 2016 *Acta Mater.* **120** 240
[30] Tourret D, Karma A, Clarke AJ, Gibbs PJ and Imhoff SD 2015 *IOP Conf. Ser.: Mater. Sci. Eng.* **84** 012082
[31] Tourret D, Clarke AJ, Imhoff SD, Gibbs PJ, Gibbs JW and Karma A 2015 *JOM* **67** 1776
[32] Tourret D, Francois M and Clarke A 2019 *Comput. Mater. Sci.* **162** 206
[33] Haxhimali T, Karma A, Gonzales F and Rappaz M 2006 *Nature Mater.* **5** 660
[34] Dantzig JA, Di Napoli P, Friedli J and Rappaz M 2013 *Metall. Mater. Trans.* A **44** 5532
[35] Langer J 1987 *Lectures in the theory of pattern formation* (Les Houches, session XLVI) Chance and Matter, Eds Souletie J, Vannimenus J and Stora R (New York: North-Holland)
[36] Langer, J 1989 *Science* **243** 1150
[37] Barbieri A and Langer J 1989 *Phys. Rev.* A **39** 5314
[38] Ben Amar M and Brener E 1993 *Phys. Rev. Lett.* **71** 589
[39] Karma A and Rappel WJ 1998 *Phys. Rev.* E **57** 4323
[40] Provatas N, Goldenfeld N and Dantzig J 1998 *Phys. Rev. Lett.* **80** 3309
[41] Plapp M and Karma A 2000 *Phys. Rev. Lett.* **84** 1740
[42] Bouissou P and Pelce P 1989 *Phys. Rev.* A 40 6673
[43] Williamson CHK 1988 *Phys. Fluids* **31** 2742
[44] Ponta F and Aref H 2004 *Phys. Rev. Lett.* **93** 084501
[45] Von Kármán T 2004 *Aerodynamics: selected topics in the light of their historical development* (Dover)
[46] Williamson CHK 1988 *Phys. Fluids* **31** 3165
[47] Williamson CHK 1989 *J. Fluid Mech.***206** 579
[48] Williamson CHK 1996 *Annu. Rev. Fluid Mech.* **28** 477
[49] Ivantsov G 1947 *Dokl. Akad. Nauk. SSSR* **58** 567
[50] Badillo A, Ceynar D and Beckermann C 2007 *J. Cryst. Growth* **309** 197
[51] Sakane S, Takaki T, Ohno M, Shibuta Y, Shimokawabe T and Aoki T 2018 *J. Cryst. Growth* **483** 147
[52] Steinbach I 2009 *Acta Mater.* **57** 2640
[53] Viardin A, Zollinger J, Sturz L, Apel M, Eiken J, Berger R and Hecht U 2020 *Comput. Mater. Sci.* **172** 109358
[54] Tourret D and Karma A 2015 *Acta Mater.* **82** 64
[55] Tourret D, Song Y, Clarke AJ and Karma A 2017 *Acta Mater.* **122** 220

IOP Conf. Series: Materials Science and Engineering **861** (2020) 012050 doi:10.1088/1757-899X/861/1/012050

Comparison of solidification kinetics of compacted and lamellar cast irons

J Lacaze[1], U de la Torre[2], J Sertucha[2] and A Regordosa[2]

[1] CIRIMAT, Université de Toulouse, France
[2] Fundación AZTERLAN, Durango, Spain

E-mail: jacques.lacaze@ensiacet.fr

Abstract. Thermal analysis is largely used in cast-iron foundry shops as a means to check melt preparation before casting. It has been suggested that the shape of the cooling curves could be related to graphite form, either spheroidal, lamellar or else compacted. The present work is part of a larger study intended to improve melt control for compacted graphite castings by using controlled additions of sulfur and magnesium. Experiments showing lamellar graphite and their counterpart with compacted graphite have been selected for analysis and simulation of the thermal analysis records. Simulation makes use of the fact that compacted graphite differs from lamellar graphite by the limitation of graphite branching in the former. Solidification of both types of irons may thus be simulated using the same basic growth law for irregular eutectics, though changing the parameters describing the branching capability of the faceted graphite phase. The increase of the undercooling of the eutectic plateau during solidification of compacted cast iron when compared to that of lamellar cast iron could be verified and reproduced by simulation. The limitations of the present approach which assumes the temperature in thermal cups is homogeneous at any time are also pointed out.

1. Introduction

Compacted graphite irons (CGI) have been used since the 1960s [1] but are now breaking markets previously occupied by lamellar graphite irons. The most usual way of preparing CGI consists to under-treat a melt with usual spheroidizing elements, namely magnesium and rare earths [1]. However, the range of residual spheroidizing elements for getting the proper microstructure is quite limited and highly sensitive to casting conditions. Amongst the process parameters that greatly affect graphite shape are the cooling rate and inoculation. It may be worth noting that Subramanian *et al.* [2] suggested it should be much easier to control graphite shape by addition of cerium rather than of magnesium for spheroidizing but this does not appear to be applied. Other means have been looked for such as balancing spheroidizing elements with addition of sulfur. In this line, Bazdar *et al.* [3] investigated the transition from SG to CG and then LG by adding sulfur to a melt containing quite a high level of magnesium at 0.057 wt%. The graphite shape was characterized by the aspect ratio, i.e. graphite length/graphite width, and it was found that compacted graphite prevails in a large range of sulfur content.

An extensive study was carried out to investigate the coupled effect of magnesium and sulfur levels, inoculation and cooling rate. Out of the whole series of experiments, samples with lamellar graphite were sometimes obtained. Some of the experiments were thus selected as an experimental basis for checking the growth process of compacted graphite as compared to lamellar graphite. Indeed, it has been previously suggested [4, 5] that compacted graphite cells grow as lamellar graphite cells but with a limited capability of graphite for branching, leading to the coarser structure seen in metallographic sections of CGI when compared to LGI. An approach based on the model developed by Jones and Kurz [6] to describe LGI solidification was thus considered. The present work compares predictions for lamellar and compacted graphite irons to experimental cooling curves recorded during standard thermal analysis.

2. Experimental details

All melts were prepared in a medium frequency induction furnace (250 Hz, 100 kW) 100 kg in capacity from metallic charges composed of low-alloyed steel scrap and low alloyed ferritic cast iron returns. These charges also contained high purity graphite (C > 99.0 wt%), a FeSi alloy (wt% Si = 74.0, C = 0.15, Al = 1.06 and balance Fe) and extra pure sulfur powder. Once melting process was completed, the composition of the batches was checked and carbon, silicon and sulfur contents were then adjusted according to the required targets. After this adjustment stage, about 50 kg of melt was transferred from the furnace to a 60 kg capacity ladle at a temperature between 1500 °C and 1510 °C. The nodularization treatment was carried out following the sandwich method by adding a commercial FeSiMg alloy (wt% Mg = 6.3, RE = 1.01, Si = 45.5, Ca = 0.97, Al = 0.97 and balance Fe). The nodularizer was added in the reaction chamber at an amount of 0.35-0.40 % of the total weight of the treated melt and it was then properly covered with steel scrap. Once the nodularizing reaction finished, a sample (medal) was taken out for determining the chemical composition of the batch.

The melts were then cast in standard thermal analysis (TA) cups which were empty or contained inoculant and/or some other addition as sulfur powder. For the present report, samples from four melts were selected which were all without inoculant. Also, they were without sulfur addition and labelled not-inoc, or with either 0.005 wt% (12.5 mg) or 0.010 wt% (25 mg) sulfur and labelled +0.005% S or +0.010% S, respectively. These sulfur additions were made to form lamellar graphite particles by decreasing the free magnesium content of the liquid poured in the cups. The cooling curves obtained from these samples were recorded using the Thermolan® software.

Carbon and sulfur contents were determined on medals by using combustion analysis (LECO CS300) while spark spectrometry was used for the other elements (SPECTRO LAB). The chemical analyses of the four selected alloys are listed in table 1 in which the sulfur additions performed on the TA cups haven't been considered. Table 1 also contains the carbon equivalent CE and the temperature of the stable eutectic T_{EUT}, in Celsius, which were calculated using the following formulas [7]:

$$CE = w_C + 0.28 \cdot w_{Si} + 0.007 \cdot w_{Mn} + 0.303 \cdot w_P \tag{1}$$

$$T_{EUT} = 1154.02 + 4.246 \cdot w_{Si} - 5.00 \cdot w_{Mn} \tag{2}$$

where w_i is the content of the alloy in element i (wt%). It is seen that the CE values are either slightly lower or slightly higher than 4.34 wt%, meaning the alloys are either hypo- or hyper-eutectic.

Table 1. Composition of the selected alloys (wt%), carbon equivalent, CE, and temperature of the stable eutectic, T_{EUT}.

Trial	C	Si	Mn	P	S	Ti	Mg	CE	T_{EUT} (°C)
4.1	3.64	2.39	0.15	0.012	0.018	<0.010	0.011	4.31	1163.4
8.1	3.73	2.43	0.16	0.021	0.014	<0.010	0.021	4.42	1163.5
9.1	3.68	2.40	0.15	0.013	0.015	<0.010	0.016	4.36	1163.5
9.2	3.73	2.38	0.15	0.012	0.016	<0.010	0.012	4.40	1163.4

Metallographic inspections were performed on samples taken out from the geometrical centre of the TA cups, i.e. close to the thermocouple junction, and their surface was properly polished. Micrographs at 100x magnification of three different fields were obtained from each sample and the shape of the graphite particles was then evaluated by determining the relative area fraction of class III graphite which is related to compacted shape, f_{III_A}.

3. Experimental results

Table 2 summarizes the observations made on the samples obtained from the 4 alloys, and figures 1, 2 and 3 present the cooling curves and typical microstructures. Lamellar graphite was obtained by addition in the quick cup of 0.005 wt% sulfur to alloy 4.1 and 0.01 wt% to alloy 8.1, see figure 1. In both alloys, part of the lamellar graphite was of the undercooled type D. Note in the micrograph of figure 1 the presence of ex-austenite dendrites. Alloy 9.1 with 0.01 wt% of sulfur addition in the quick cup and alloy 9.2 with 0.005 wt% sulfur showed a lamellar graphite type D, see figure 2. Here also, ex-austenite dendrites are observed. All other five alloys showed compacted graphite at a level varying between 40% and 68%, the remaining being spheroidal graphite. Three of the five cooling curves are shown in figure 3 together with a typical microstructure.

As expected, the cooling curves for samples showing undercooled graphite present a eutectic plateau at lower temperature than those with lamellar graphite. On the contrary, comparing the graphite distribution in figures 1 and 3 confirms that compacted graphite is much coarser than lamellar graphite when solidified in the same casting conditions, even though it grows at a higher undercooling. A more detailed analysis of the curves is provided later when comparing calculated to experimental records.

Table 2. Microstructure observed for all studied samples.

Trial	not-inoc	+0.005% S	+0.010% S
4.1	65% CG	40% lamellar graphite type A 60% lamellar graphite type D	
8.1	59% CG	68% CG	70% Lamellar graphite type A 30% lamellar graphite type D
9.1	56% CG	51% CG	Mostly lamellar graphite type D
9.2	40% CG	Mostly lamellar graphite type D	

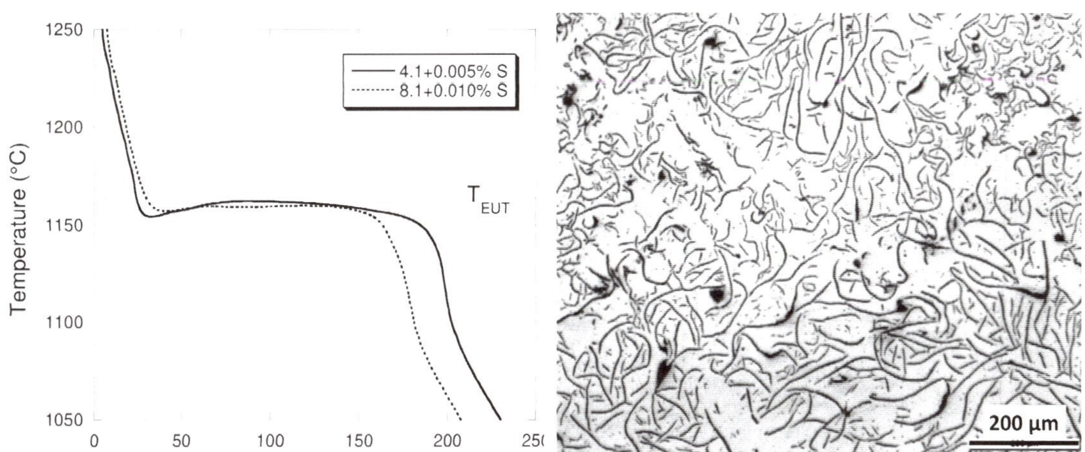

Figure 1. Cooling curves for samples showing lamellar graphite and typical microstructure.

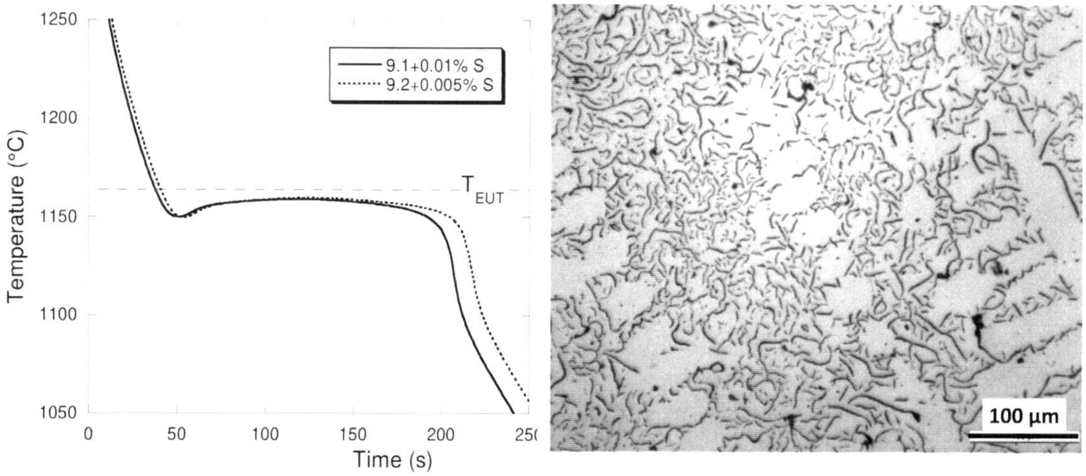

Figure 2. Cooling curves for samples showing undercooled graphite and typical microstructure.

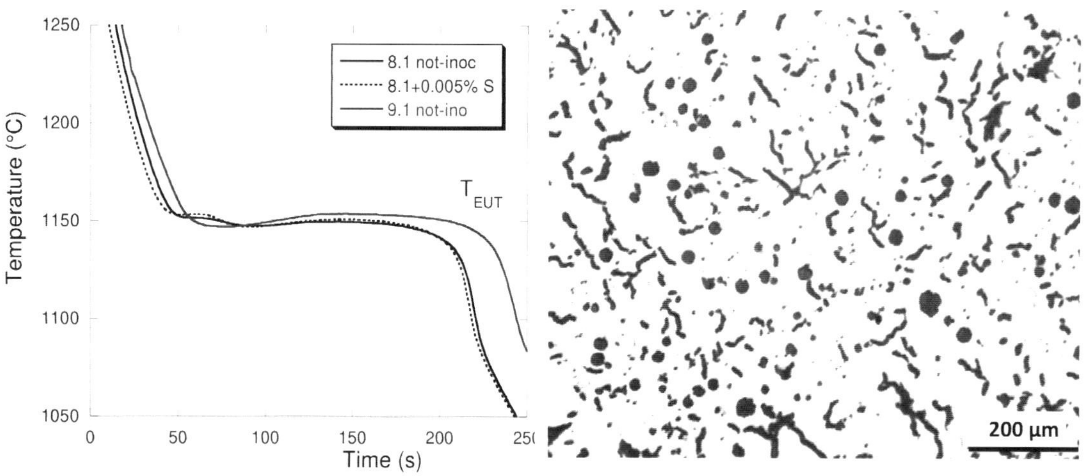

Figure 3. Cooling curves for samples showing compacted graphite and typical microstructure.

4. Simulation approach

Solidification of cast irons is quantitatively described by writing the appropriate mass balances following the work by Lesoult *et al.* [8] on spheroidal graphite cast irons. The whole details of the present model will be presented elsewhere. For describing the eutectic reaction in the present study, spheroidal graphite eutectic cells are replaced by lamellar or compacted graphite cells of radius R whose growth law is written according to Jones and Kurz [6] as:

$$\frac{dR}{dt} = \frac{(\Delta T)^2}{\left(\frac{1}{\varphi} + \varphi\right)^2 \cdot a \cdot b} \tag{3}$$

in which t is time, ΔT is the undercooling expressed with respect to the equilibrium eutectic temperature, T_{EUT}, a and b are constants evaluated by Jones and Kurz as a=2.3 µm·K and b=0.080 K·s·µm^{-2}, and φ characterizes the capability of the faceted graphite phase to branch.

For an ideal regular eutectic between two non-faceted phases, φ should be 1, while it has been proposed it is 2.5 for Fe-C graphite/austenite eutectic [9]. Jones and Kurz could reproduce their experimental results of directional solidification of lamellar graphite iron with φ set to 3.9 while Zou Jie [10] found a value of 6.5 for equiaxed solidification of an Fe-C-Si alloy. Zou Jie suggested that part of the difference with directional solidification is due to the expanding nature of the eutectic cells in equiaxed solidification, and one may even think of a value of φ changing during growth in this case. In the case of CG cells, the protuberances that form on the primary graphite precipitates then develop without much branching. Thus, the distance between them increases as the size of the CG cells increases. This suggested writing $\varphi=\varphi^0+R/R_{GE}$, where R_{GE} is the initial size of the cell, and to limit the increase of φ to some maximum value lower than 10 at which branching of graphite lamellae is anyway expected because of the high undercooling developed between the lamellas.

After primary deposition of austenite in the case of a hypo-eutectic alloy, a volume density of N_V eutectic cells were nucleated when T_{EUT} was reached. During eutectic solidification, off-eutectic austenite may continue to grow as during primary deposition or may dissolve in order for the solidification path to stick to the extrapolation of the austenite liquidus. In the case of hyper-eutectic alloys, the solidification path during primary deposition was calculated assuming spheroidal graphite precipitates following a nucleation law given as $N_V=A_1\cdot(\Delta T_{LG})$ mm^{-3}, where ΔT_{LG} is the undercooling with respect to the graphite liquidus [8]. When the extrapolation of the austenite liquidus is reached, the number of graphite particles is set constant at the value which has been reached. An equivalent diameter of the graphite particles is calculated which is then used as initial size for the compacted graphite cells, R_{GE}. Solidification proceeds further as in the case of hypo-eutectic alloys, with possible growth and dissolution of off-eutectic austenite together with growth of eutectic cells.

The solidification process of the thermal cups was described assuming their temperature is homogeneous at any time during the cooling process. For each time step of calculation, the calculated change of the solid fraction, V^S, was thus introduced in the following heat balance:

$$\rho\cdot V/A\cdot\left(C_p\cdot\frac{dT}{dt}-\Delta H\cdot\frac{dV^S}{dt}\right)=-\sqrt{\Delta}\cdot\left(T-T^0\right)\cdot t^{-0.5} \qquad (4)$$

where ρ and C_p are, respectively, the density and the heat capacity of the metal at temperature T, ΔH is the latent heat of fusion of the metal, Δ is a quantity characteristic of the mould and T^0 is the ambient temperature. V is the volume of metal having an outer surface A, and V/A is the so-called casting modulus. All data used in the present calculations are listed in table 3. During solidification, the specific heat C_p and the density ρ were calculated as a weighted average of the solid and liquid values.

Table 3. Values of the parameters that were used for calculations.

V/A [m]	$\sqrt{\Delta}$ [J·m^{-2}·K^{-1}·s$^{-0.5}$]	C_p^{liquid} [J·K^{-1}·kg^{-1}]	C_p^{solid} [J·K^{-1}·kg^{-1}]	ρ^l [kg·m^{-3}]	ρ^γ [kg·m^{-3}]	ρ^g [kg·m^{-3}]
0.009	728 before solidification 1215 after solidification	920	750	6800	7000	2200

5. Simulation results

It has been recently observed that soon after the nodularizing treatment, not inoculated alloys behave much alike inoculated ones [11]. This suggests that this treatment leads to the precipitation of magnesium oxides or sulfides that serve as graphite nuclei. Accordingly, the number of eutectic cells was set at the same value for all alloys at the start of the eutectic reaction. To determine this value, preliminary calculations were performed on sample 8.1+0.01%S with φ set constant at the value of 2.5 which was expected to apply for lamellar graphite. Figure 4 compares the experimental and calculated curves for N_V finally set to 2585 mm^{-3}. It is seen that the predicted and experimental eutectic undercoolings are the same during most of the plateau, and this N_V value was then used for all calculations.

On the graph in figure 4 are also plotted the evolution of the fraction of off-eutectic austenite, of eutectic cells and of the whole solid. It is seen that off-eutectic austenite forms at the beginning of the eutectic reaction, but then dissolves to reappear when the temperature decreases at the end of the eutectic plateau. On the whole, this simulation appears thus quite satisfactory but it is noticeable that the predicted maximum undercooling at the beginning of the eutectic reaction is much larger than the experimental one, this is further discussed below.

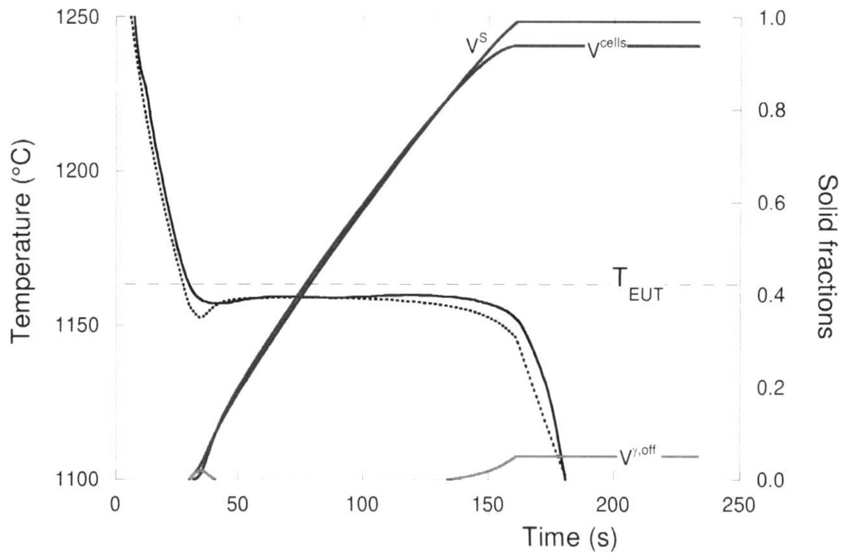

Figure 4. Comparison of the experimental (solid line) and calculated (dotted line) cooling curves for alloy 8.1+0.01%S. The predicted evolution of the fraction of off-eutectic austenite, V^{off}, lamellar graphite cells, V^{cells}, and total solid, V^S, is also shown.

Figure 5 compares the calculated and experimental curves for samples 8.1 with lamellar and compacted graphite. For this latter, φ was varied from 2.5 to a maximum value of 7.5 which was found appropriate for retrieving the undercooling during the eutectic plateau. It is thus seen that even with a relatively high number of eutectic cells, their growth law may significantly affect the predicted eutectic plateau undercooling. However, the main difference between calculated and experimental curves for compacted graphite eutectic (sample 8.1+0.005% S) appears at the beginning of the bulk solidification where it is seen experimentally a short plateau while a deep undercooling is predicted. Nevertheless, it is noticed that the predicted recalescence is in quite good agreement with the measured one.

Finally, figure 6 compares the calculated cooling curve to the experimental one for sample 4.1 not-inoc which is very slightly hypo-eutectic. The origin of the discrepancy at the beginning of the eutectic reaction is then made evident. While austenite is predicted to precipitate as soon as the austenite liquidus temperature is reached, the corresponding thermal arrest is shifted by about 15°C on the experimental record. This undercooling has sometimes been simulated by assuming it is due to a nucleation barrier for austenite [12]. However, thermal cups are quite small castings and it is expected austenite nucleates at the wall of the cup and grows inwards in a columnar way. Accordingly, an alternative possibility is that the shift in the thermal arrest corresponds to the delay needed by the dendrites formed at the surface of the sample to reach the thermocouple located in the middle of the thermal cup. This would agree with the fact that this arrest is not associated with recalescence. When the thermal arrest associated to austenite growth reaches the centre of the cup, the cooling rate decreases giving more time to the eutectic cells to grow, hence the lower maximum undercooling experienced by the material upon the start of the eutectic reaction.

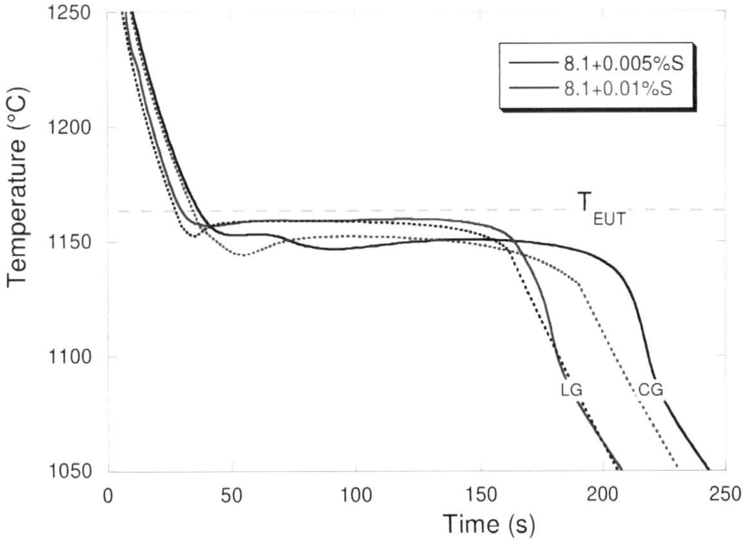

Figure 5. Comparison of calculated and experimental curves for alloy 8.1 with lamellar (LG) and compacted (CG) graphite. Experimental and calculated curves are in solid and dotted lines respectively.

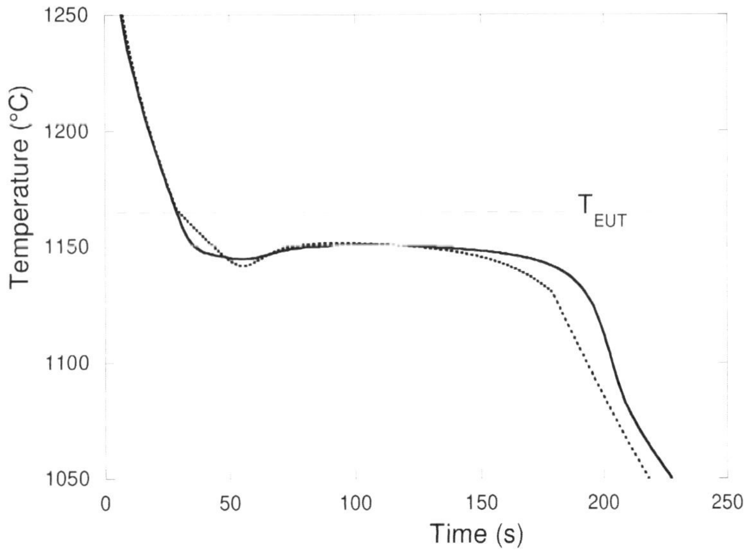

Figure 6. Comparison of calculated and experimental curves for alloy 4.1-not-inoc with compacted graphite. Experimental and calculated curves are in solid and dotted lines respectively.

6. Conclusion

Focusing on comparing lamellar and compacted graphite, it has been shown that it is possible to reproduce the change in eutectic plateau undercooling by simply considering the associated difference in branching capability of graphite during eutectic cell growth. It was also noticed that the predicted recalescence for compacted graphite is well reproduced.

Improvement of the present work could be achieved by accounting properly of the temperature distribution in the thermal cups during solidification instead of considering their temperature is homogeneous. This would certainly allow a better description of the cooling curves when austenite precipitation starts being sensed by the thermocouple. Also, no distinction was made in the modelling approach between flake and undercooled graphite while describing the transition from one to the other would certainly be of interest.

References

[1] Nechtelberger E, Puhr H, Nesselrode J B and Nakayasu A 1982 Cast iron with vermicular/compacted graphite – State of the art. Development, production, properties, applications, *Proc. Int. Foundry Congress, CIATF (Chicago)* p 1-37
[2] Subramanian S V, Kay D A R and Purdy G R 1982 Compacted graphite morphology control *AFS Trans.* **90** 589-603
[3] Bazdar M, Abbasi H R, Yaghtin A H and Rassizadehghani J 2009 Effect of sulfur on graphite aspect ratio and tensile properties in compacted graphite irons *J. Mater. Proc. Tech.* **209** 1701-5
[4] Lacaze J and Sertucha J 2018 Some paradoxical observations about spheroidal graphite degeneracy *China Foundry* **15** 457-63
[5] Lacaze J, Connétable D and Castro de Roman M J 2019 Effects of impurities on graphite shape during solidification of spheroidal graphite cast ions *Materialia* **8** 100471
[6] Jones H and Kurz W 1981 Relation of interphase spacing and growth temperature to growth velocity in Fe-C and $Fe\text{-}Fe_3C$ eutectic alloys *Z. Metallkunde* **72** 792-7
[7] Castro M, Herrera M, Cisneros M M, Lesoult G and Lacaze J 1999 Simulation of thermal analysis applied to the description of the solidification of hypereutectic SG cast irons, *Int. J. Cast Metals Research* **11** 369-74
[8] Lesoult G, Castro de Roman M J and Lacaze J 1998 Solidification of spheroidal graphite cast irons – I. Physical modelling *Acta Mater.* **46** 983-95
[9] Trivedi R and Kurz W 1988 Microstructure selection in eutectic binary systems, *in Solidification Processing of Eutectic alloys*, Ed. D.M. Stefanescu, G.J. Abbaschian and R.J. Bayuzick, The Metallurgical Society, pp. 3-34
[10] Zou Jie 1989 Simulation de la solidification eutectique, PhD thesis #774, EPFL
[11] Regordosa A, de la Torre U, Loizaga A, Sertucha J and Lacaze J Microstructure changes during solidification of cast irons - Effect of chemical composition and inoculation on competitive spheroidal and compacted graphite growth *International Journal of Metalcasting*, DOI : 10.1007/s40962-019-00389-y
[12] Suli Chang, Dongkai Shangguan and Stefanescu D M 1992 Modeling of the liquid/solid and the eutectoid phase transformations in spheroidal graphite cast iron *Metall. Trans. A* **23** 1333-46

IOP Conf. Series: Materials Science and Engineering **861** (2020) 012051 doi:10.1088/1757-899X/861/1/012051

In situ observation of austenite coarsening induced by massive-like transformation during solidification in Fe–C alloys

H Yasuda[1], T Suga[1], K Ichida[1], T Narumi[1] and K Morishita[1, 2]

[1] Department of Materials Science and Engineering, Kyoto University, Sakyo, Kyoto 606-8501, Japan
[2] Department of Materials Science and Engineering, Kyushu University, Nishiku, Fukuoka, 819-0395 Japan

E-mail: yasuda.hideyuki.6s@kyoto-u.ac.jp

Abstract. A massive-like transformation, in which δ-ferrite massively transforms to γ-austenite in the solid state, was examined by time-resolved X-ray imaging and X-ray diffractometry on a four-dimensional–computed tomography setup. In the unidirectional solidification of a hyperperitectic steel (0.3 mass% C) at 50 μm/s, δ dendrites, fine γ grains that are produced through the massive-like transformation, and coarse γ grains grow together in the steady state. The massive-like transformation occurred commonly in Fe-based alloys with a peritectic reaction in equilibrium. Time resolved X ray diffraction measurements for a peritectic steel (0.18 mass% C) showed that the massive-like transformation induced strains in the γ grains. The fine γ grains coarsened and/or vanished and new γ grains formed. The induced strains were released during coarsening. The results obtained by the observations contribute to building of a physical model for γ grain coarsening.

1. Introduction

In high-temperature steel processing, it is essential to avoid coarse γ grains (referred to as γ) because the coarse grains cannot be refined in subsequent hot working [1–3]. However, abnormal γ grain growth is produced frequently in as-cast microstructures [4]. Details of γ grain coarsening was examined by an interrupted and isothermal solidification method [1]. According to the experimental results [1], γ grain suppression needed the coexistence of a second phase, such as δ-ferrite (referred to as δ) or liquid phase and only occurred in a limited temperature range below the peritectic temperature. Subsequent cooling led to coarse γ grains of as large as 5 mm in diameter. Annealing for 600 s yielded γ grains of ~1 mm at 1473K and 200 μm at 1673 K. The γ grain size was smaller than that observed in abnormal coarsening. Thus, abnormal coarsening after solidification is not explained simply by the grain growth in the specimen that was heated from the α-ferrite region (referred to as α). Ambiguities exist in understanding the coarsening of γ grains during or after solidification and causes difficulties to understand and to build models for predicting the microstructure formation after the transformation from ferrite (δ) to austenite (γ). It is desired to obtained experimental data which show how γ grains coarsen for building physical model.

Previous work on the coarsening of γ grains was performed by quenching unidirectionally solidifying specimens [5]. The as-cast coarse columnar γ grain structure (referred to as CCG) in hyper-peritectic steel was investigated by a rapid unidirectional solidification method [5]. In the quench structure, fine columnar γ grains (referred to as FCG) always existed ahead of the CCG region. The FCG disappeared owing to CCG growth. This study indicates that discontinuous changes in γ grain size differed from conventional coarsening between grains.

It is difficult to analyse the microstructure evolution in Fe-based systems with phase transformations from δ to γ and from γ to α. Thus, solidification should be observed in the Fe-based alloys in situ. Two techniques exist for in situ observation, namely, X-ray imaging using synchrotron radiation X-rays and high-temperature laser-scanning confocal microscopy [6, 7]. X-ray imaging allows for internal specimen observation and for a combination of other techniques, such as X-ray diffraction (XRD) and florescence X-ray analysis (EDS). However, access to synchrotron radiation facilities is limited.

Solidification of Sn-, Al-, and Mg-based alloys [8–12] was observed in situ by X-ray transmission imaging (two-dimensional, 2D imaging). Later, the technique allowed for an observation of solidification in Cu-, Ni-, and Fe-based alloys above 1300 K [13–18]. Time-resolved computed tomography (three-dimension plus time, referred to as 4D-CT) has been developed and used to observe dendrite evolution in Al–Cu alloys (pink X-ray beam, rotation of 0.1 rps) [19, 20] and Al–Mg–Si–Y_2O_3 alloys [21]. 4D-CT was extended to observe a semi-solid deformation and transgranular liquation cracking in Al–Cu alloys [22, 23]. Time-interlaced model-based iterative reconstruction (TIMBIR) was developed to improve temporal resolution [24]. A filtering technique using a phase field model was proposed to improve image quality even at a relatively high temporal resolution [25]. In addition to the three-dimensional observation, 4D-CT and XRD were combined to observe the microstructure and crystallographic orientation simultaneously [25, 26].

Two-dimensional and 3D observations using X-ray imaging proved that massive-like transformation occurred in Fe–C alloys [16–18, 27–29] and Fe–Cr–Ni alloys [30]. In this transformation, δ transforms to γ in the solid state and multiple γ grains are produced. The massive-like transformation occur in preference to the peritectic reaction, in which diffusion-controlled growth of γ occurs and γ grains are crystallographically continuous along δ. Laser-scanning confocal microscopy has also shown that the δ phase massively transforms into the γ phase [7]. In the unidirectional solidification of a hyper-peritectic carbon alloy [17, 18], the massive-like transformation was selected even at a growth rate of 5 μm/s and a coarse γ grain existed at 200 μm behind the δ dendrite tips. Therefore, the γ grain growth could be influenced by the massive-like transformation. Recently, in situ XRD analysis before and after the massive-like transformation was achieved by using a 4D-CT setup [25]. Fine γ grains were produced through the massive-like transformation and the crystallographic orientation was distributed widely.

In this study, the steady state growth of the massive-like transformation behind the δ dendritic growth in a hyper-peritectic 0.3 mass% C alloy was confirmed and grain growth of the γ grains after the massive-like transformation was examined by XRD on a 4D-CT setup.

2. Experiments

Two-dimensional imaging of the unidirectional solidification was performed for Fe – 0.3 mass% C – 0.6 mass% Mn – 0.3 mass% Si alloys (referred to as 0.3C steel) at an imaging beamline BL20B2 at SPring-8. The preliminary results were presented previously [17, 18]. The setup for the 2D observation, which was developed by the authors' group, is shown in figure 1(a). A monochromatized 28 keV X-ray was passed from the left to the right of figure 1(a). A beam monitor with a pixel size of 2.5 μm × 2.5 μm for the transmission images and a panel-type detector with a pixel size of 50 μm × 50 μm for the XRD images was used. A specimen was pulled down at 50 μm/s under a temperature gradient of approximately 10 K/mm. Positions of the liquid-δ, the δ–γ (fine), and the γ (fine) –γ (coarse) interfaces were measured. The diffraction spots were analysed in terms of γ grain size.

XRD data for 0.18 mass% C – 0.6 mass% Mn – 0.3 mass% Si alloys (1.18C steel) on a 4D-CT setup [25, 26], as shown in figure 1(b), were obtained at beamline BL20XU of SPring-8. X-rays with an energy of 37.7 keV were used. The specimen was inserted into a sintered Al_2O_3 pipe with inner and outer diameters of 0.8 mm and 2 mm, respectively, and placed in a graphite furnace. XRD spots were observed with a flat-panel sensor with a format of 1008 × 682 pixels and a pixel size of 100 μm × 100 μm at a

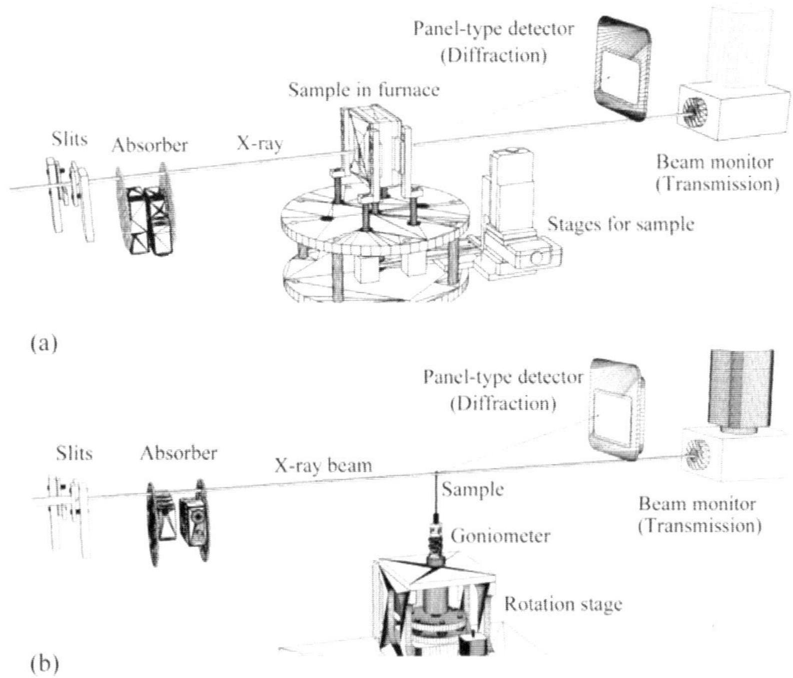

Figure 1. (a) Setup for the transmission imaging (2D observation) with X-ray diffraction and (b) setup for time-resolved tomography with X-ray diffraction. transformation.

frame rate of 30 fps. The sample was rotated at 0.25 rps and XRD spots over a 360°-rotation were obtained every 4 s. Distributions of diffracting planes were analysed.

3. Results and discussion

3.1. Unidirectional solidification

Figure 2 shows the unidirectional solidification in 0.3C steel at a pulling rate of 50 μm/s [17, 18]. Initially (0 s), liquid and δ were observed and γ exited behind the liquid-δ interface (out of the observation area) as shown in figure 2(a). The tip position of δ on the yellow axis was defined to be zero as a vertical position. As shown in figure 2(b), the solidification was completed at the δ–γ interface, which indicates that γ solidification occurred after the δ–γ transformation. Dark spots, where the Bragg condition was satisfied, were observed within the γ region at 80 s. Divergence of the X-ray beam at the beamline BL20B2 was 1.5 mrad in the vertical direction and 0.06 mrad in the horizontal direction [31]. Thus, many spots in the γ region indicated that fine γ grains with different crystallographic orientations were

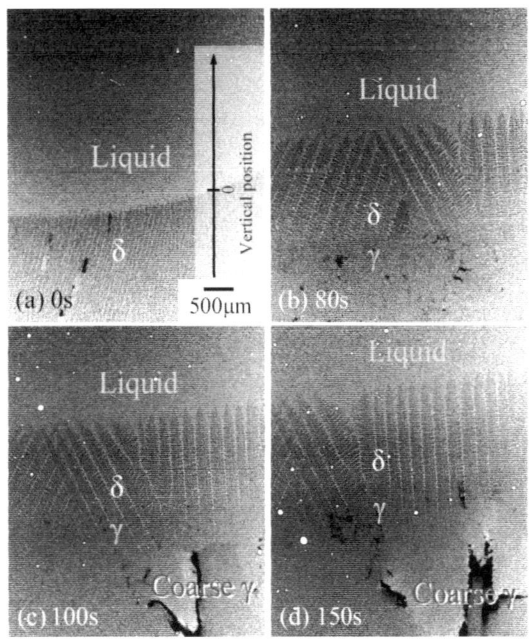

Figure 2. Unidirectional solidification of Fe-0.30C-0.6Mn-0.3Si alloy at a pulling rate of 50μm/s. (a) 0s, (b) 80s, (c) 100s and (d) 150s.

produced by the massive-like transformation. In addition to the fine γ grains behind the liquid-δ interface, a coarse γ grain, which had a front at 500 μm from the tip, appeared at 100 s, as shown in figures 2(c) and (d). The configuration of fine γ grains and a coarse γ grain is essentially the same as the microstructure from quenching during solidification [5]. The grain growth rate of the coarse γ grain is larger than the reported rate [1]. Thus, the growth rate of coarse γ grains during cooling from the melt does not agree with that observed in the specimen that was heated from the α region to the γ region.

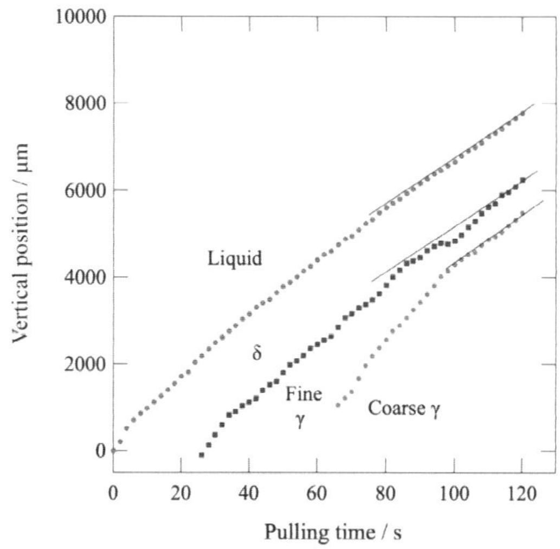

Figure 3. Positions of liquid-δ interface, δ-γ(fine) interface and γ(fine)-γ(coarse) interface during the unidirectional solidification at 50 μm/s.

Recently, it was reported that a massive-like transformation from δ to γ occurred at a cooling rate above 1 K/s [32]. In the massive-like transformation during the unidirectional solidification at 5 μm/s [17, 18], the estimated cooling rate was 0.005 K/s. The observation showed that the massive-like transformation occurred even when the temperature was kept constant below the peritectic temperature [16]. The discrepancy suggests that the massive-like transformation may occur as a transient phenomenon. Thus, it is of interest to confirm the massive-like transformation during unidirectional solidification, either in the transient state or in the steady state. Figure 3 shows the position of the δ tip, the δ–γ (fine) interface, which is the front of the massive-like transformation, and the γ (fine)–γ (coarse) interface, which is the front of the crystallographically continuous γ growth, on the vertical axis in figure 2. The growth of δ into liquid and the growth of fine γ reached a steady state at 100 s. The growth of coarse γ also reached a steady state. The positions showed that the front of coarse γ did not overtake the front of fine γ. The configuration of FCG and CCG in the quenched specimen [5] is consistent with the observation result in figure 2. We conclude that the massive-like transformation is selected even at a growth rate of 5 μm/s.

Figure 4 shows a snapshot in the unidirectional solidification and an XRD image. The dark regions in the transmission images range from less than 100 μm to several 100 μm. Dark regions correspond to

(a)
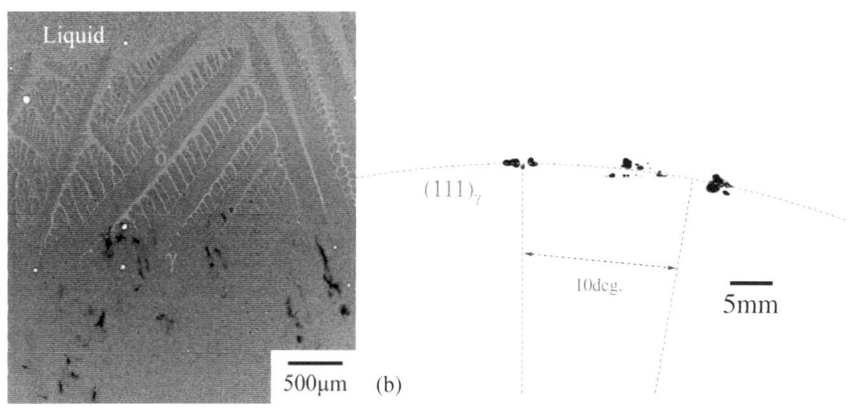

Figure 4. (a) Snapshot of the unidirectional solidification and (b) close-up view of (111) spots at the time.

the grain shape where the Bragg condition is satisfied, because the diffraction angle, 2θ, was approximately 10°. The XRD spots of the (111) plane in figure 4(b) showed that the fine γ grains were produced through the massive-like transformation. Some spots tended to segregate in a certain direction, which indicates that γ grains exist with similar crystallographic orientations and/or a large strain was induced in a single γ grain.

3.2. Coarsening of γ grains

Figure 5 shows examples of XRD images over a 3°-rotation, obtained from the panel-type detector. The rotation angles of specimen are the same. The diffraction spots varied after the massive-like transformation. As indicated by a red arrow, the XRD spots varied gradually after the massive-like transformation and some spots vanished. The change in spots indicates that a coarsening of γ grains commenced after the massive-like transformation.

Normal vectors of (111), (200), and (220) are plotted in stereo projections [33] in figure 6. The stereo projections draw the intensities of XRD spots over a 360°-rotation. Orange solid lines indicate the measurable limits at the low- and high-angle sides. Time was designated to be zero when the massive-like transformation was detected.

The spots of δ before the massive-like transformation correspond to those obtained from a single crystal, although the spots spread within 5°. As shown in figure 6(b), the crystallographic orientations of the γ grains spread widely at 0–4 s, which indicates that multiple γ grains were produced even in a single δ grain. The production of fine γ grains in single δ grains was observed in 0.45 mass% C steel and a single

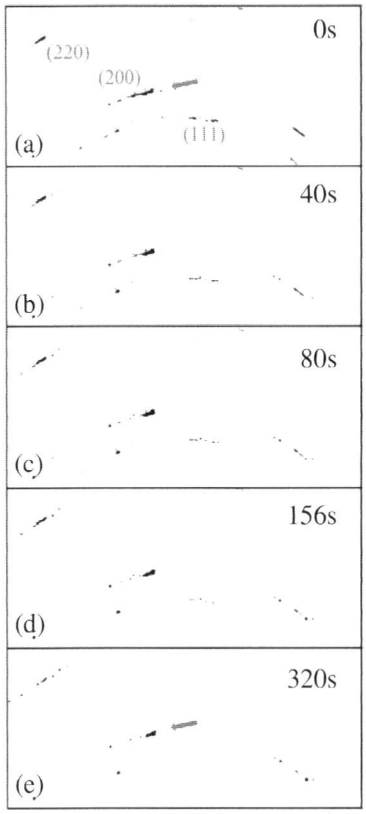

Figure 5. Diffraction spots after the massive-like transformation. Sample rotation angles are the same each other. (a) 0s, (b) 40s, (c) 80s, (d) 156s and (e) 320s.

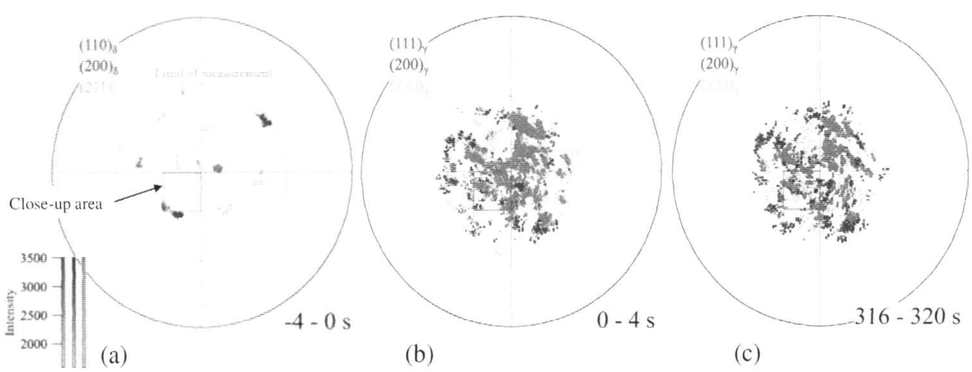

Figure 6. Crystallographic orientations. (a) δ phase before the massive-like transformation, (b) γ phase at 0-4 s after the transformation and (c) γ phase at 316-320 s.

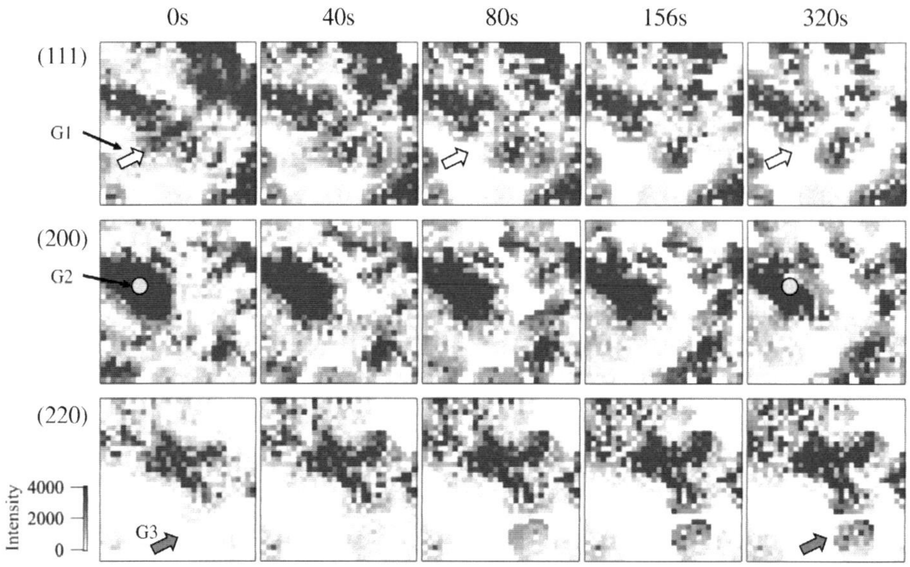

Figure 7. Close-up views of diffraction spots. The close-up area indicated in figure 6 is depicted. Yellow arrows: spots disappearing, green circles: spot s sharpened and red arrows: spots newly appearing during coarsening of γ grains.

dendrite was divided into pieces owing to melting at the γ grain boundaries [16]. Thus, the observation of multiple γ grains in this study is consistent with the observation for different conditions [16]. Multiple γ grain formation in a single γ grain is a required condition for defining the massive-like transformation. Figure 6(c) shows the distribution at 316–320 s. Compared with figure 6(b), the scatter of each plane tended to decrease with an increase in time after the massive-like transformation. Some spots vanished during coarsening. It would be interesting to examine how the coarsening proceeded after the massive-like transformation.

Figure 7 shows close-up views of the stereo-projections, with an area indicated by a rectangle in figure 6(a). The distributions of (111), (200), and (222) planes are drawn separately. The "G1" intensity, which is indicated by yellow arrows in the (111) plane decreased immediately after the massive-like transformation and vanished at 320 s. Therefore, γ grains with the direction of "G1" vanished or changed their crystallographic orientation gradually during coarsening. The "G2" intensities, which are indicated by green circles, spread widely immediately after the massive-like transformation and converged gradually to a narrower region during coarsening. The gradual change in crystallographic orientation is attributed to the release of strains that are induced in a γ grain; the strains that caused a distribution of

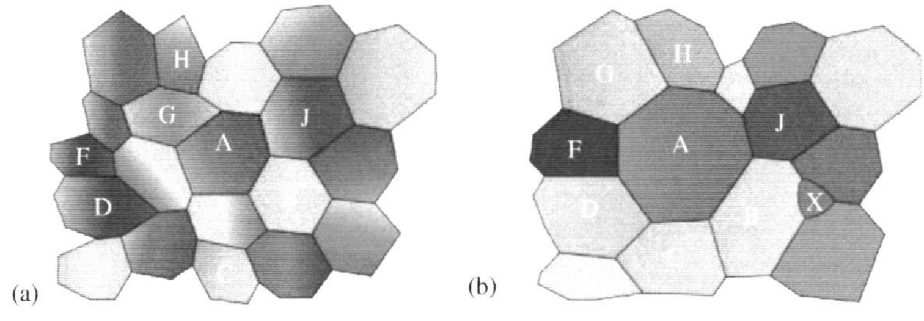

Figure 8. Schematic illustration of coarsening of γ grains after the massive-like transformation. (a) γ grains just after the massive-like transformation and (b) γ grains during coarsening. Colors in the grains correspond to crystallographic orientation.

several degrees in the crystallographic orientation were introduced in the massive-like transformation. The strains in this study are consistent with large strains that were observed by time-resolved Laue diffraction using white X-rays [28]. Although it is difficult to estimate the distribution by XRD of white X-rays, this study using a monochromatized X-ray to confirm the induced strains. The volume decreased 0.5 % by the transformation from δ to γ [18]. The volume change can induce strains within γ grains even above 1673K. "G3" is indicated by red arrows formed, which suggests that new grains were produced during coarsening. The production of new γ grains may correspond to strains within γ grain strain produced by the massive-like transformation.

Figure 8 shows schematic illustrations of γ grains immediately after the massive-like transformation and during coarsening to summarize the observation results. The fine γ grains in which strains are induced are produced through the massive-like transformation. The crystallographic orientation changes within a γ grains can reflect lattice defects in the γ grains. After the massive-like transformation, γ grains coarsened and/or vanished. The strains that were induced in the γ grains were released and this process can influence the kinetics of γ grain coarsening. The influence of the massive-like transformation modifies the coarsening kinetics in subsequent cooling after solidification and is different, compared with the kinetics [1–4].

4. Summary

The massive-like transformation in Fe–C alloys was examined by time-resolved X-ray imaging (2D observation) and XRD on a 4D-CT setup.

The observation results for the unidirectional solidification were analysed extensively. In the unidirectional solidification in 0.3C steel at 50 μm/s, δ dendrites, fine γ grains that were produced through the massive-like transformation, and coarse γ grains can grow together in the steady state. The XRD images indicated that fine γ grains formed during unidirectional solidification. Based on detailed analysis, we concluded that the massive-like transformation was selected even at low growth rates.

Changes in the crystallographic orientations of γ grains during coarsening were analysed. Strains in the γ grains were induced by the massive-like transformation. The fine γ grains coarsened and/or vanished after the massive-like transformation and new γ grains formed. The strains were released during coarsening. The massive-like transformation can modify the kinetics of γ grain coarsening.

Acknowledgments

In situ observations (2D and 3D observations) using synchrotron radiation X-rays were performed as general projects (2016B1458, 2017B1463, 2018A1380, 2019A1400) at the BL20B2 and the BL20XU of SPring-8 (JASRI), Japan. The study was initiated by "Heterogeneous Structure Control: Towards Innovative Development of Metallic Structural Materials" in the Industry–Academia Collaborative R&D Program (JST). The study was supported by a Grant-in-Aid for Scientific Research (S) (No. 17H06155). We thank Laura Kuhar, PhD, from Edanz Group (www.edanzediting.com/ac) for editing a draft of this manuscript.

References

[1] Pottore N S, Garcia C I and DeArdo A J 1991 *Metall. Trans. A* **22** 1871
[2] Cuddy L J and Raley J C 1983 *Metall. Trans. A* **14** 1989
[3] Yoshida N, Kobayashi Y and Nagai K 2004 *Tetsu-to-Hagane* **90** 198
[4] Manohar P A, Dunne D P, Chandra T and Killmore C R 1996 *ISIJ Int.* **36** 194
[5] Tsuchiya S, Ohno M, Matsuura K and Isobe K 2011 *Acta Mater.* **59** 3334
[6] Shibata H, Arai Y, Suzuki M and Emi T 2000 *Metall. Mater. Trans. B* **31** 981
[7] Griesser S, Reid M, Bernhard C and Dippenaar R 2014 *Acta Mater.* **67** 335
[8] Mathiesen R H, Arnberg L, Mo F, Weitkamp T and Snigirev A 1999 *Phys. Rev. Lett.* **83** 5062
[9] Yasuda H, Ohnaka I, Kawasaki K, Sugiyama A, Ohmichi T, Iwane J and Umetani K 2004 *J. Cryst. Growth* **262** 645
[10] Li B, Brody H and Kazimirov A 2004 *Phys. Rev. E* **70** 062602

[11] Schenk T *et al.* 2005 *J. Cryst. Growth* **275** 201
[12] Wang T, Xu J, Xiao T, Xie H, Li J, Li T and Cao Z 2010 *Phys. Rev. E* **81** 42601
[13] Yasuda H *et al.* 2009 *Int. J. Cast Met. Res.* **22** 15
[14] Yasuda H, Nagira T, Yoshiya M, Nakatsuka N, Sugiyama A, Uesugi K and Umetani K 2011 *ISIJ Int.* **51** 402
[15] Kareh K M, O'Sullivan C, Nagira T, Yasuda H and Gourlay C M 2017 *Acta Mater.* **125** 187
[16] Yasuda H, Morishita K, Nakatsuka N, Nishimura T, Yoshiya M, Sugiyama A, Uesugi K and Takeuchi A 2019 *Nat. Commun.* **10** 3183
[17] Nishimura T, Morishita K, Yoshiya M, Nagira T and H Y 2020 *ISIJ Int.* in press (doi:10.2355/ isijinternational.ISIJINT-2019-6)
[18] Nishimura T, Morishita K, Yoshiya M, Nagira T and Yasuda H 2019 *Tetsu-To-Hagane* **105** 168
[19] Ludwig O, Dimichiel M, Salvo L, Suéry M and Falus P 2005 *Metall. Mater. Trans. A* **36** 1515
[20] Aagesen L K, Fife J L, Lauridsen E M and Voorhees P W 2011 *Scr. Mater.* **64** 394
[21] Daudin R, Terzi S, Lhuissier P, Salvo L and Boller E 2015 *Mater. Des.* **87** 313
[22] Cai B, Karagadde S, Yuan L, Marrow T J, Connolley T and Lee P D 2014 *Acta Mater.* **76** 371
[23] Karagadde S *et al.* 2015 *Nat. Commun.* **6** 8300
[24] Gibbs J W, Mohan K A, Gulsoy E B, Shahani A J, Xiao X, Bouman C A, De Graef M and Voorhees P W 2015 *Sci. Rep.* **5** 11824
[25] Yasuda H, Hashimoto T, Sei N, Morishita K and Yoshiya M 2019 *IOP Conf. Ser. Mater. Sci. Eng.* **529** 012013
[26] Nakano K, Narumi T, Morishita K and Yasuda H 2020 *Mater. Trans.* **61** in press
[27] Yasuda H *et al.* 2011 *IOP Conf. Ser. Mater. Sci. Eng.* **27** 012084
[28] Yasuda H *et al.* 2012 *IOP Conf. Ser. Mater. Sci. Eng.* **33** 012036
[29] Nishimura T, Morishita K, Nagira T, Yoshiya M and Yasuda H 2015 *IOP Conf. Ser. Mater. Sci. Eng.* **84** 012062
[30] Nishimura T, Matsubayashi R, Morishita K, Yoshiya M, Nagira T and Yasuda H 2019 *ISIJ Int.* **59** 459
[31] Goto S *et al.* 2001 *Nucl. Instruments Methods Phys. Res. Sect. A Accel. Spectrometers, Detect. Assoc. Equip.* **467–468** 682
[32] Liu T, Long M, Chen D, Huang Y, Yang J, Duan H, Gui L and Xu P 2020 *Metall. Mater. Trans. B* **51** 338
[33] Yoshiya M, Nakajima K, Watanabe M, Ueshima N, Nagira T and Yasuda H 2015 *Mater. Trans.* **56** 1461

3D cellular automaton modelling of silicon crystallization including grains in twin relationship

A Pineau[1], G Guillemot[1], G Reinhart[2], G Regula[2] N Mangelinck-Noël[2] and Ch-A Gandin[1]

[1] MINES ParisTech, PSL Research University, CEMEF UMR CNRS 7635, CS10207, 06904 Sophia Antipolis, France
[2] Aix Marseille Univ, Université de Toulon, CNRS, IM2NP, Marseille, France

E-mail : Charles-Andre.GANDIN@mines-paristech.fr

Abstract. Production of silicon for solar cells in photovoltaic systems is mainly based on directional casting processes. Twin nucleation is favoured during silicon growth due to the low-level twin energy of formation. As a consequence, in all solidification process, a large amount of grain boundaries (GB) are in twin relationship. A 3D cellular automaton (CA) model has been recently developed for the growth of multi-crystalline silicon including facet formation and nucleation of new grains in twin relationship. Activation of facets is based on an undercooling parameter assigned to each grain and for each of the <111> crystal directions. The model also considers nucleation and growth of grains on <111> facets corresponding to Σ3 twin relationships between twin grains. This model is first applied to comparison with experimental observations. It is found that impingement of growing grains that nucleated in Σ3 twin relationships meet during growth and form Σ3, Σ9 and Σ27 GB, in good agreement with experimental observations. Finally, the model is applied at a larger scale to generate grain structures representative of industrial practice. While quantitative experimental data is missing for comparison at such scale, the model is promising and its implementation in heat and mass transfer models should be considered for assistance to production of silicon for solar cells dedicated to photovoltaic systems.

1. Introduction

Industrial photovoltaic panels are mainly based on the use of silicon material. Crystalline silicon has the largest market share due to: the availability of the raw material, non-toxicity, efficiency that could reach 15% and a high level of technology readiness [1]. The photoelectric properties of the solar cells are strongly linked to the production chain, including purity of the raw material, crystallization, wafering, cell processing, and module integration. During crystallization, the grain structure and associated defects are generated, defining density, crystallographic orientations and twin-relationships. Similarly, impurities are developed and dislocations are formed in material. The crystallization features need to be controlled in order to optimize final electrical properties because they can act as recombination centres of charge carriers, thus lowering the efficiency of the photoelectric panels [2-5].

During the manufacturing of silicon ingots in crucibles, twin grain boundaries are commonly observed in any solidification process and strongly influence the final grain structure. Voigt *et al.* [6, 7] analysed the orientation of each grain on a 5×5 cm^2 wafer. They observed that most of the grains are in twin relationship and more than 50% of the grain boundaries correspond to Σ3, Σ9, Σ27 and Σ81 twins boundaries. The most frequent twins in silicon are of Σ3 type and can be described as an odd number of 60° rotation of the crystal lattice around a <111> direction. New grains in Σ3 relationship nucleate on {111} plane [3], their formation being enhanced by the low energy needed for twinning process [8]. According to Jackson's criterion [9], facets may correspond to the single {111} crystallographic planes in silicon material. Being coherent boundaries, Σ3 twins exhibit no dangling bonds and do not directly lower the final efficiency of solar cells [10]. However, the nucleation of a grain in Σ3 twin relationship can be at the origin of the formation of higher order incoherent, symmetric or asymmetric twin boundaries such as Σ9 or Σ27 [11] and hence responsible for part of the dislocation emission during growth [3, 12] and for impurity segregation thus degrading the photoelectric properties.

CA aims at modelling complex phenomena through the use of local simple laws. This approach was firstly proposed by Gandin and Rappaz [15] to model solidification processes. The development of grain envelopes is simulated rather than the full dendritic microstructure, with the advantage to be applicable to large simulation domains. The present work aims at modelling silicon growth upon directional solidification using a 3D CA method. The proposed model is firstly introduced. It is based on phenomenological and geometrical considerations, so as to retrieve {111} facet growth and nucleation of grains in twin relationship as commonly observed experimentally. Finally, the model is applied at a macroscopic scale to highlight the influence of facet growth and new twinned grains nucleation on the final ingot grain structure.

2. Modelling

The description of the CA algorithm is presented in [17] with details and associated schematics. The strategies implemented in the CA model to allow modelling of polycrystalline silicon growth are here briefly summarized. The model used in the present work is divided into three steps: (i) {111} facet activation and growth, (ii) grain in twin relationship nucleation and (iii) twin growth. In the following subsections, index μ refers to a liquid cell.

2.1. Facet activation and growth

The standard growth algorithm developed for cubic materials is used [17]. However, dedicated developments are introduced to model faceted growth.

2.1.1. Facet activation. Eight facets are defined for each grain, g, and tracked with virtual planes using Cartesian equations $D_g^{\{111\}} : a_g^{\{111\}}x + b_g^{\{111\}}y + c_g^{\{111\}}z + d_g^{\{111\}} = 0$. An activation undercooling, $\Delta T_g^{\{111\}}$ is associated to each plane. Its value is randomly chosen using a given normal distribution law of mean value $\Delta T_{g,m}^{\{111\}}$ and standard deviation $\Delta T_{g,s}^{\{111\}}$. If a liquid cell is captured by a growing cell belonging to grain g, its undercooling ΔT_μ is computed. This undercooling is compared to the activation undercooling, $\Delta T_g^{\{111\}}$, associated to each {111} plane leading to faceted growth for grain g. Facet activation occurs if the capture undercooling is lower than the activation undercooling ($\Delta T_\mu < \Delta T_g^{\{111\}}$). This stochastic approach is necessary to reproduce the size distribution between twin grain boundaries. The mean value and standard deviation parameters are adjusted based on comparison with experimental observations.

2.1.2. Facet growth. Facets of a grain g being described by its Cartesian equations $D_g^{\{111\}}: a_g^{\{111\}}x + b_g^{\{111\}}y + c_g^{\{111\}}z + d_g^{\{111\}} = 0$, their growth is simply modeled by the time integration of the crystal growth velocity of the {111} facets $v_c^{\{111\}}$ defined along the normal $\boldsymbol{n}^{\{111\}}$. In practice it is sufficient to update $d_g^{\{111\}}$ with time using

$$d_g^{\{111\}} = v_c^{\{111\}}(t - t_0) + d_{g,t_0}^{\{111\}} \tag{1}$$

where $v_c^{\{111\}}$ is taken constant following measurements available elsewhere [16]. The CA standard algorithm is modified [17] and referred as conditioned capture algorithm. This is achieved by conditioning the capture of a cell μ by a neighboring growing cell belonging to grain g considering its location with respect to the facets described by their Cartesian equations. The coordinates (x_μ, y_μ, z_μ) of the center, C_μ, of cell μ are used in the equation associated to activated facets of grain g to compute the quantity $a_g^{\{111\}}x_\mu + b_g^{\{111\}}y_\mu + c_g^{\{111\}}z_\mu + d_{g,t_0}^{\{111\}}$. The capture is considered only if a negative value is obtained and inhibited otherwise. This condition corresponds to the facet overpassing the cell center.

2.2. Nucleation of a new grain in twin relationship
Nucleation undercooling of twinned grains on {111} facets is defined at sample edges, $\Delta T_{N,T}^E$, and at grain boundary grooves, $\Delta T_{N,T}^B$ (i.e. in the bulk material) in accordance with experimental observations. Indeed twin formation upon directional solidification is only of Σ3 type and $\Delta T_{N,T}^B$ has a lower value compared with $\Delta T_{N,T}^E$ [3, 16]. Upon capture of a liquid cell μ by its neighboring growing cell ν, location of cell μ is tested to know if it is located on a sample edge or as part of the bulk. For an edge cell (respectively a bulk cell), if the test $\Delta T_\mu > \Delta T_{N,T}^E$ (respectively $\Delta T_\mu > \Delta T_{N,T}^B$), is verified, a nucleation event takes place. The Euler angles of the new grain formed in cell μ correspond to the crystallographic orientations of the capturing grain to which a Σ3 twin rotation (60°) around the normal to the capturing facet is applied.

2.3. Twins growth
Twin boundaries between grains are locked during growth [3, 16]. To maintain the twin boundary while pursuing growth of the solid-liquid interfaces, the conditioned capture algorithm is used. The twin grain boundary remains static in space and develops as a result of growth of the grains on each side of the corresponding {111} facet. Note that this is equivalent to keep a constant value for $d_{g,t_{N,T}}^{\{111\}}$ in equation (1), defined at the time $t_{N,T}$ of the grain nucleation in twin relationship. Let us consider two grains, g_1 and g_2, in twin relationship, their common twin plane is denoted $D^{\{111\}}$ and also defined by its cartesian equation. The position of the twin plane $D^{\{111\}}$ remains unchanged. For a given liquid cell μ located in the neighbourhood of $g_1$and g_2, the capture liquid the cell μ is allowed for only g_1 or g_2, depending on the spatial position of its centre with respect to $D^{\{111\}}$. If its centre is located beneath $D^{\{111\}}$, only the capture by grain g_1 will be authorized. Capture by g_2 will occur otherwise. The reader will find schematized details of the algorithm in [17].

3. Results and discussion

3.1. Comparisons with a benchmark experiment
The 3D CA model described above is applied to simulate an experiment corresponding to observations obtained by X-ray imaging [18]. This benchmark experiment is fully detailed in [17, 18]. Numerical and process parameters as well as material properties are provided in [17]. Figure 1 shows the comparison between the simulation and experimental results. Figure 1(a) is the inverse pole figure (IPF) map along the growth direction and displayed on the longitudinal section (left) simulated and (right) measured. Figure 1(b) shows the coincidence site lattice (CSL) map (left) simulated and (right) measured. The two sets of Σ3 twin relationships with a common $(1\bar{1}1)$ plane on the right-hand side (A/B type grain boundaries) and $(11\bar{1})$ plane on the left-hand side (A/C type grain boundaries) are made of grains with various band widths. This is directly controlled by the value of the standard deviation to activate the {111} facets, $\Delta T_{g,s}^{\{111\}}$ (see [17] for more details).

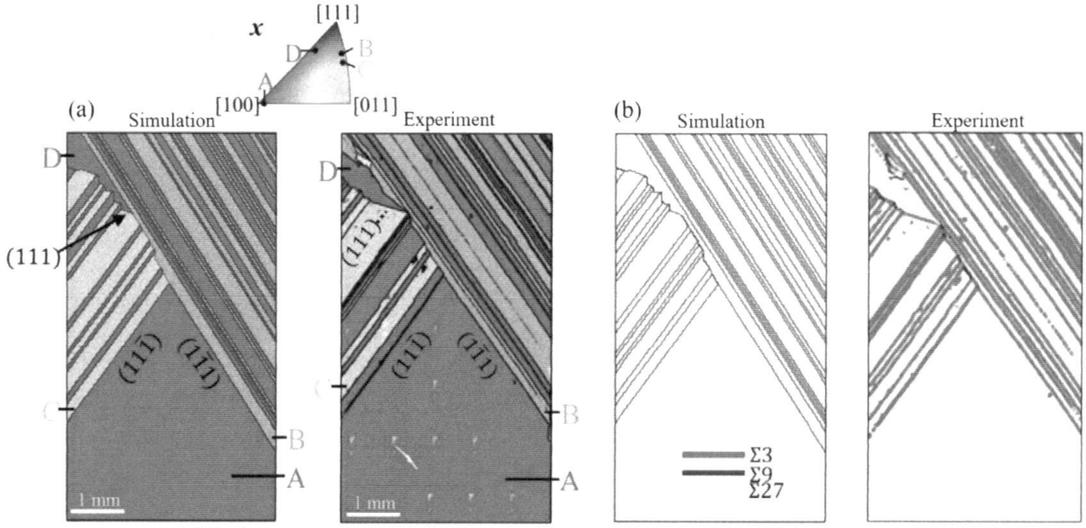

Figure 1. Qualitative comparison between the simulation results and the benchmark experiment: (a) IFP maps along the growth direction on longitudinal sections and (b) CSL maps revealing the different types of twinned GBs.

Unlike grains of B-type and C-type, the grain labelled D nucleates in the bulk at a faceted/faceted groove on a {111} plane of a C-type grain. The Σ9 twin boundaries are observed whenever B-type and C-type or D-type and A-type grains meet, whereas a Σ27 twin boundary originates from D-type and B-type. Twin grain boundaries apart from the one due to Σ3 twin nucleation are the consequences of grain encounters during growth of A-, B-, C- and D-type grains. This result is consistent with the experimental observations. Prior to grain D nucleation, a total of 5 grooves between B-type and C-type grains were formed but did not lead to nucleation of the D-type grain on the {111} plane of a C-type grain.

Quantitative comparisons between simulation and experiment are summarized in figure 2. The area fraction, the number of grains of types A–D and the mean twin band width of grains of type A–C were determined. For a given grain, the latter is defined by the width between two twinning events of a given grain type in a direction perpendicular to the twinning plane. The area of the initial seed is not considered when computing the grain area fraction since we focus on the grains that formed as a result of twinning nucleation events: this is why for band width computations, only edge twin nucleation is considered. One can notice the overall good agreement between simulations and experimental observations. Area fraction of A-type grains obtained in the simulation (41.3%) is very close the area fraction measured in the experiment (39.8%). The number of grains in the computation is also higher (24) than the one in the experiment (22), the mean band width being consequently a little higher, 0.11 mm in the simulation and 0.10 mm in the experiment. For B-type grains, the area fraction is 4.7% lower in the simulation (33.4%) than in the experiment (38.1%). The number of grains is almost identical and the mean band width in the simulation is lower (0.1 mm) compared to the experiment (0.12 mm). Similar observations can be done for C-type grains.

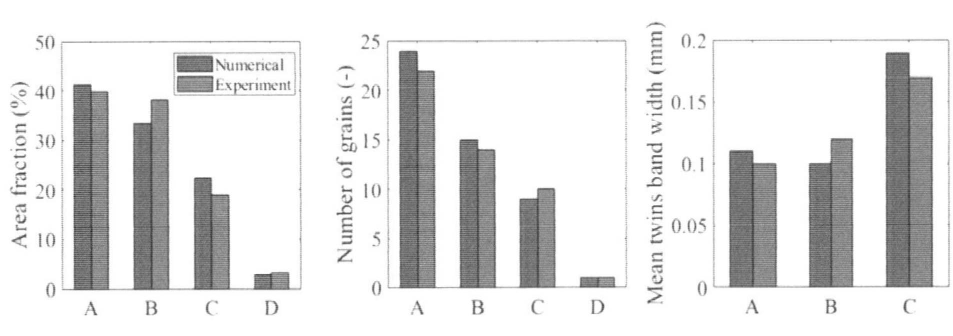

(a) (b) (c)

Figure 2. Statistical analyses showing (a) the area fraction, (b) the number of grains and (c) the mean band width between twin boundaries for the simulation results and the benchmark experiment shown in figure 1.

3.2. Application to a cube geometry

The model is now applied at a larger scale, *i.e.* to a cube of 5×5×5 cm^3. Three simulation cases are proposed: directional growth in an imposed and vertical temperature gradient with (i) no facet formation nor grains in twin relationship nucleation, (ii) facet formation without grains in twin relationship nucleation and (iii) facet formation and grains in twin relationship nucleation. For the sake of simplicity, these simulations are denoted I, II and III, respectively. Simulation I is similar to dendritic grain growth situation since no facet or twin nucleation are modelled in this case. The purpose of such simulations is to demonstrate the influence of the facet formation solely and facet formation coupled with twins nucleation on the final grain structure when compared to a dendritic one. In all cases, heterogeneous nucleation takes place on the bottom surface of the cube, the crystallographic orientations of the initial equiaxed grains being randomly chosen.

3.2.1. Simulation I. We firstly disable facet formation. Since twin nucleation can occur only on facets, these specific events are also disabled. The corresponding results are shown in figure 3(a). The final grain structure in displayed on the left part of figure 3(a) with a cold colour related to the misorientation, α_x, with respect to the x direction (growth direction). It corresponds to the lowest angle formed between all <100> crystallographic directions of the grains and the x axis. The centred part of figure 3(a) shows the IPF map in the x direction and the right one gives the CSL map. The warm colours in figure 3(a) correspond to highly misoriented grains as shown at the very beginning of the simulation at nucleation stage. The growth of highly misoriented grain is blocked in all <100> crystallographic directions. Cold colours are the only one to remain at the end of the solidification which indicates that well orientated grains, with respect to the temperature gradient, can survive. This is simply due to the grain growth competition which commonly occurs between columnar grains during directional solidification.

3.2.2. Simulation II. Facet formation is now enabled and nucleation of grains in twin relationship is disabled. The corresponding results are shown in figure 3(b). In comparison with figure 3(a), less grain remains at the end of the solidification. Regarding the misorientation, the colour code indicates that more misoriented grains survived (warm colours). This comes from the fact that some of the grains develop {111} facets and become constrained during their growth. Neighbouring grains, that grow without developing {111} facets, can solidify into the free liquid present in the facet and potentially block growth of side grains that form {111} facets.

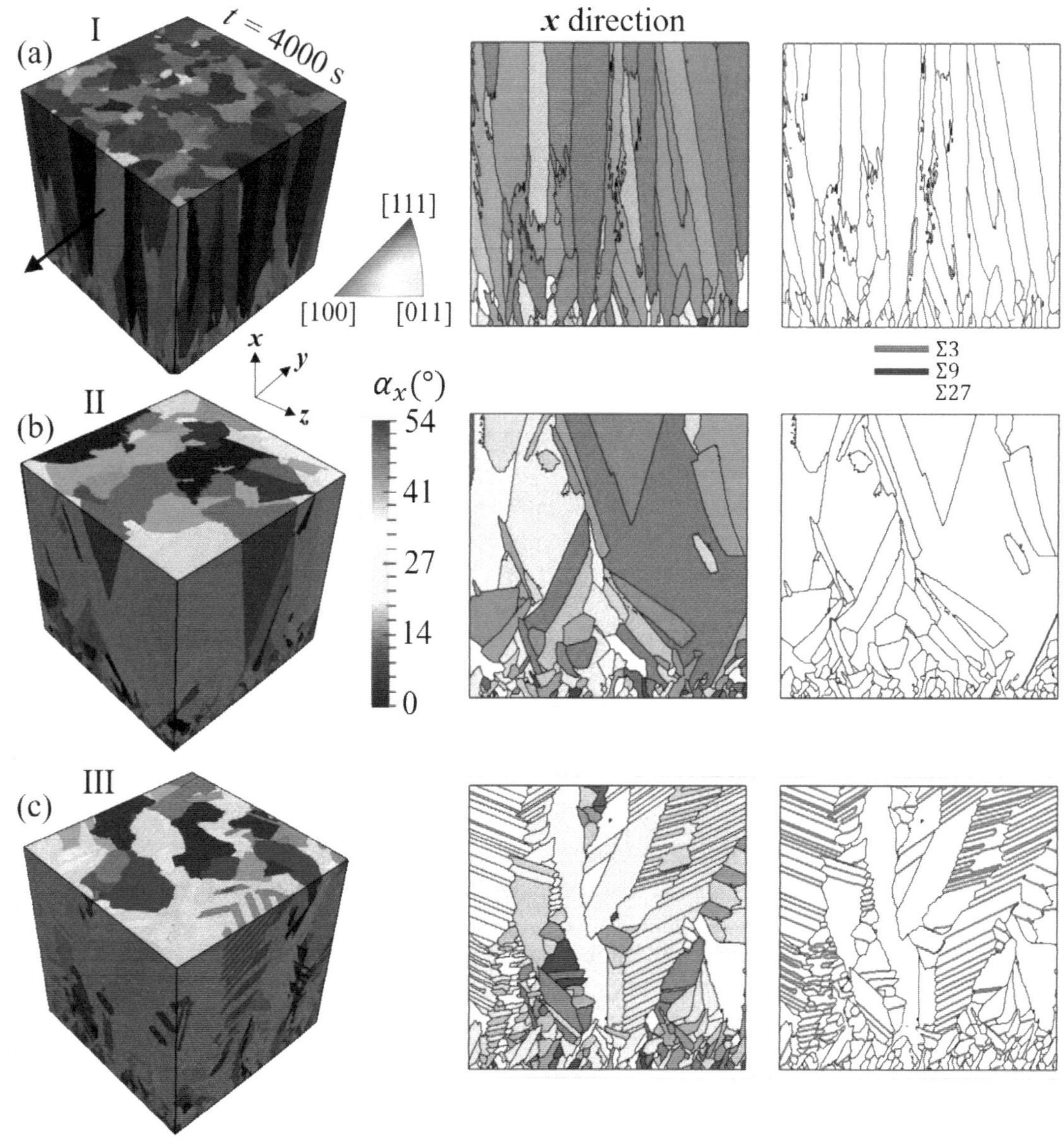

Figure 3. Results of simulations corresponding to (a) simulation I with a dendritic growth regime, (b) simulation II with growth restricted to facet formation, (c) simulation III with facet formation and grain nucleation in twin relationship. From left to right: final grain structure observed at time 4000 s showing the misorientation, α_x, the centered figure is the IFP map along the growth direction, x, and the CSL map is on the right.

3.2.3. Simulation III. Both facet formation and grains in twin relationship nucleation are enabled. The corresponding results are shown in figure 3(c). Grains in twin relationship can be clearly observed in final structure. Their presence being also shown in IPF and CSL maps. As it is commonly observed in industrial casting ingots [6, 7], most of the grains are in Σ3 twin relationship. Similar explanations can be developed (figure 4) regarding the evolutions of the grain area density, n_g (figure 4(a)), and mean misorientation, $\bar{\alpha}_x$ (figure 4(b)) with respect to the solidification time, t.

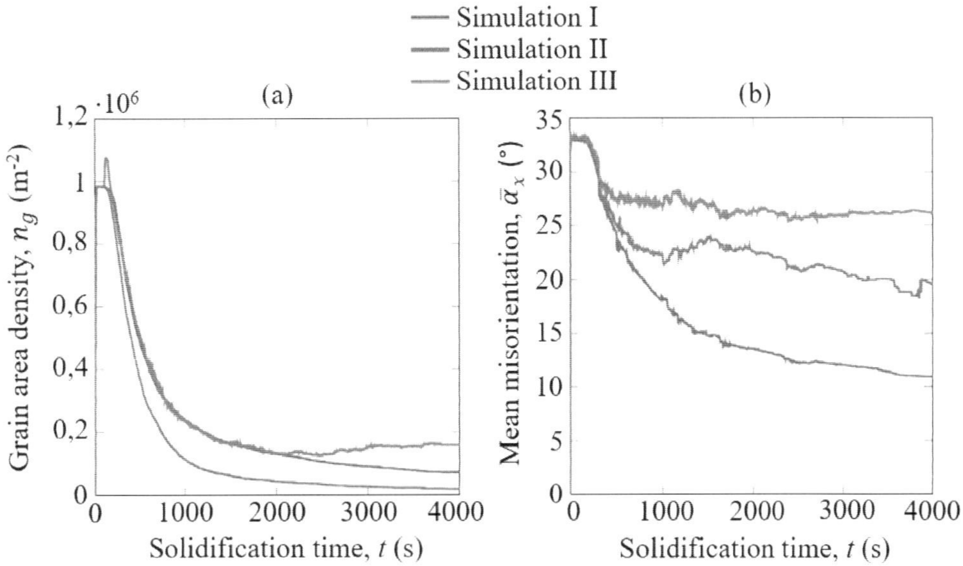

Figure 4. Evolution of (a) the grain area density, n_g, and (b) the mean misorientation, $\bar{\alpha}_x$, as a function of the solidification time, t for the three simulations cases.

For simulations I and II, the grain density n_g decreases continuously. However, the value at time $t =$ 4000 s is lower in simulation II than in simulation I demonstrating that grains development is stopped due to the formation of $\{111\}$ facets on some of the grains. The value of $\bar{\alpha}_x$ at $t = 4000$ s is higher in simulation II when compared to simulation I. This is in agreement with figure 3 where more misoriented grains survive as shown in case II. Simulation III includes nucleation of grains in twin relationship. An increase in grain density at $t \approx 2600$ s is shown. It indicates that rate of twin grains nucleation overcomes grain elimination induced by grain growth competition.

4. Conclusion

A 3D CA model dedicated to the modelling of silicon solidification in multi-crystalline structure is developed. IPF and CSL maps are simulated and compared with their counterpart produced on a fully solidified sample from a benchmark experiment. Qualitative and quantitative comparisons show good agreements and similar tendencies in microstructure evolution. Application to a larger scale demonstrates that facet formation and grains in twin relationship drastically modify the final structure compared to dendritic development. Such numerical tool is suitable to follow microstructure evolution in casting process dedicated to silicon material. Future development should consequently permit grain structure optimization by direct simulation at the process scale. Further developments of the model could include a thermomechanical analysis during solidification of the grain structure, thus assessing the additional stress field that may play a role on the formation of the twin grain boundaries and of dislocations.

Acknowledgements

The present work was funded by the project CrySaLID (ANR-14-CE05-0046-01) of the French National Research Agency (ANR).

References

[1] Lacerda JS and van der Bergh JC 2016 Diversity in solar photovoltaic energy: implications for innovation and policy *Renew. Sust. Energ. Rev.* **54** 331

[2] Trempa M, Reimann C, Friedrich J, Müller G, Krause A, Sylla L and Richter T 2014 Defect formation induced by seed-joints during directional solidification of quasi-mono-crystalline silicon ingots *J. Cryst. Growth* **405** 131

[3] Tsoutsouva M G, Riberi-Béridot T, Regula G, Reinhart G, Baruchel J, Guittonneau F, Barrallier L and Mangelinck-Noël N 2016 In situ investigation of the structural defect generation and evolution during the directional solidification of ⟨110⟩ seeded growth Si *Acta Mater.* **115** 210

[4] Kakimoto K 2009 Crystallization of Silicon by a Directional Solidification Method, *Crystal Growth of Si for Solar Cells*, eds Nakajima K and Usami N (Springer Berlin Heidelberg) p 55

[5] Ouaddah H, Périchaud I, Barakel D, Palais O, Di Sabatino M, Reinhart G, Regula G and Mangelinck-Noël N 2019 Role of impurities in silicon solidification and electrical properties studied by complementary in situ and ex situ methods *Physica Status Solidi* A 1900298

[6] Voigt A, Wolf E and Strunk H P 1997 *Proc 14th Europ. PV Solar Energy Conf.* (Barcelona) p 774

[7] Voigt A, Wolf E and Strunk H P 1998 Grain orientation and grain boundaries in cast multicrystalline silicon *Mat. Sci. Eng. B* **54** 202

[8] Bristowe P D 1999 *Properties of Crystalline Si*, ed R Hull EMIS 299308

[9] Jackson K A 2004 Constitutional supercooling surface roughening *J. Cryst. Growth* **264** 519

[10] Duffar T 2010 Comprehensive review on grain and twin structures in bulk photovoltaic silicon, *Recent Res. Dev. Cryst. Growth* **5** 61

[11] Ribéri-Béridot T, Mangelinck-Noël N, Tandjaoui A, Reinhart G, Billia B, Lafford T, Baruchel J and Barralier L 2015 On the impact of twinning on the formation of the grain structure of multi-crystalline silicon for photovoltaic applications during directional solidification *J. Cryst. Growth* **418** 38

[12] Ryningen B, Stokkan G, Kivambe M, Ervik T and Lohne O 2011 Growth of dislocation clusters during directional solidification of multicrystalline silicon ingot, *Acta Mater.* **59** 7703

[13] Tandjaoui A, Mangelinck-Noël N, Reinhart G, Furter J-J, Billia B, Lafford T, Baruchel J and Guichard X 2012 Real time observation of the directional solidification of multicrystalline silicon: X-ray imaging characterization *Energy Procedia* **27** 82

[14] Tandjaoui A, Mangelinck-Noël N, Reinhart G, Billia B and Guichard X 2013 Twinning occurrence and grain competition in multi-crystalline silicon during solidification, *C. R. Physique* **14** 141

[15] Gandin Ch-A and Rappaz M 1994 A coupled finite element-cellular automaton model for the prediction of dendritic grain structures in solidification processses, *Acta Metall. Mater.* **42** 2233

[16] Stamelou V, Tsoutsouva MG, Riberi-Béridot T, Reinhart G, Regula G, Baruchel J and Mangelinck-Noël N 2017 {111} facet growth laws and grain competition during silicon crystallization *J. Cryst. Growth* **479** 1

[17] Pineau A, Guillemot G, Reinhart G, Regula G, Mangelinck-Noël N and Gandin Ch-A 2020 Cellular automaton modelling of silicon crystallization with grains in twin relationship, *Submitted for publication*

[18] Riberi-Béridot T, Tsoutsouva MG, Regula G, Reinhart G, Guittoneau G, Barrallier L and Mangelinck-Noël N 2019 Strain building and correlation with grain nucleation during silicon growth, *Acta Mater.* **177** 141

IOP Conf. Series: Materials Science and Engineering **861** (2020) 012053 doi:10.1088/1757-899X/861/1/012053

Molecular dynamics simulation of heterogeneous nucleation from concave cavity at surface of grain refiner

T Fujinaga and Y Shibuta

Department of Materials Engineering, The University of Tokyo, 7-3-1, Hongo, Bunkyo-ku, Tokyo 113-8656, Japan

E-mail: shibuta@material.t.u-tokyo.ac.jp

Abstract. Grain refiners are often used in practical production processes of metallic materials to promote and control nucleation. Although morphology of the grain refiner strongly affects nucleation behaviour, it is not well understood how unevenness of surface of the grain refiner affects nucleation from atomistic viewpoint. In this study, effect of concave cavity at surface of the grain refiner on heterogeneous nucleation is investigated by molecular dynamics (MD) simulation. HCP-Ti cube with a concave cavity is employed in the MD simulation as a grain refiner of the solidification of undercooled Al melt as a model study. HCP-Al solid is firstly formed at the tip or on the side wall of cavity and FCC-Al growth is followed on the exposed (0001) surface of HCP-Al for most of the cavities, whereas FCC-Al solid is directly formed on HCP(0001) terraces at the side wall of the cavity with large-obtuse apex angle.

1. Introduction

Since most metallic materials are formed through solidification from undercooled melt, it is indispensable to control nucleation at the initial stage of solidification [1]. However, it is not straightforward to directly observe nucleation process in experiments and therefore molecular dynamics (MD) simulations have contributed to the understanding of atomistic details of nucleation [2-4]. At present, there are many MD studies of nucleation [2] including nucleation rate [5], transient structure at onset of nucleation [6], non-classical behavior [7] and so on. Recently, homogeneous nucleation from undercooled melt in a micrometer-scale system with 10 billion atoms are successfully achieved owing to parallel computation on a GPU-rich supercomputer [8, 9]. From practical viewpoint, heterogeneous nucleation happens in general during the production process of metallic materials. Moreover, grain refiners are often used to control and enhance heterogeneous nucleation from undercooled melt in the production process [10]. For example, TiB_2 and Al_3Ti particles [11-13] are often employed as nucleants for solid grains of Al alloys. However, most of current MD studies on nucleation are limited to focus on homogeneous nucleation for simplicity

We have performed MD simulations of heterogeneous nucleation at the surface of grain refiners [14, 15]. Especially, we have directly derived threshold undercooling temperature, which divides free growth and stagnation from MD simulations [14]. This agrees with theoretical model of athermal nucleation (i.e., a free growth model) [10]. On the other hand, surface morphology is not taken into account in these previous works. Recently, Ueshima and coworkers [16] reported that notched substrate enhances nucleation and crystal growth from experimental observation. Therefore, nucleation and crystal growth at the surface with unevenness such as notch and cavity should be different from that from flat surface. To this end, the effect of concave cavity at surface of the grain refiner on heterogeneous nucleation is investigated by MD simulation in this study.

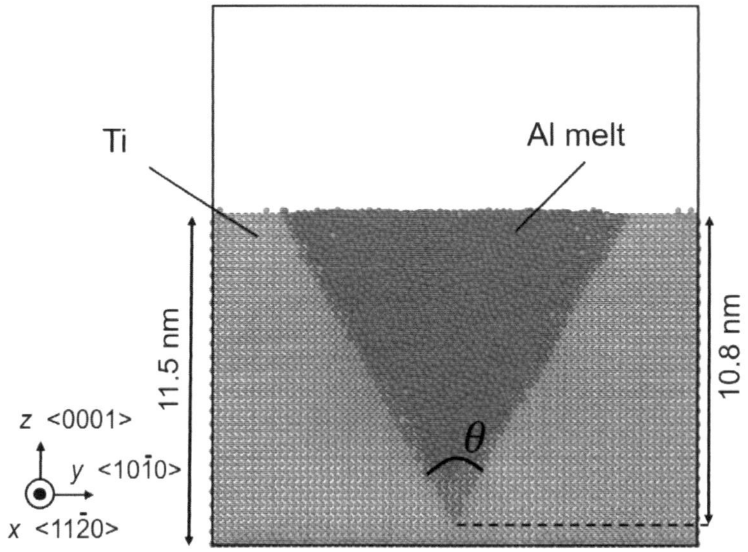

Figure 1. Initial configuration of calculation system.

2. Methodology

In this study, MD simulations are performed using LAMMPS (Large-scale Atomic/Molecular Massive Parallel Simulator) [17]. An EAM (embedded atom method) potentials fitted by Zope and Mishin [18] is employed as an interatomic potential of Al-Ti interaction. Parameter files for the EAM potential from the database of IPR (Interatomic potential depository) at NIST (national institute of standard and technology) [19] are utilized for simulations by the LAMMPS. The velocity-Verlet method is used to integrate the classical equation of motion with a time step of 1.0 fs. The Nose-Hoover thermostat and barostat [20, 21] are employed to control temperature and pressure. OVITO (Open visualization tool) [22] is used for visualization of atomic configuration and post-analyses of calculation results using adaptive common neighbor Analysis (a-CNA) [23].

Figure 1 shows the initial configuration of calculation system, which is prepared by cutting out a cone shape (i.e., the cavity) from an HCP-Ti crystal and filling Al melt inside the cavity. Various systems with a cavity of apex angle 30°, 60°, 90°, 105°, 120° and 150° are prepared. HCP (hexagonal close packed)-Ti crystals consist of 60 × 35 × 40 unit cells (17.6 × 17.8 × 19.0 nm^3) for 30°and 60°, 180 × 105 × 40 unit cells (53.0 × 53.6 × 19.0 nm^3) for 90°, 105° and 120°, and 360 × 210 × 40 unit cells (106.0 × 107.2 × 19.0 nm^3) for 150°, respectively. The height of the cone is fixed to be 10.8 nm for all cases. The vacuum layer is set at the upper side of the cell. The periodic boundary condition is employed for all boundaries of the calculation system. Initial configurations are then annealed isothermally in the main calculation for 300 ps under zero pressure with the constant-NPT (number of atoms, pressure and temperature) ensemble at 750 K. Note that the melting point of pure Al of this EAM potential is estimated to be 870 K [14] by the convergence temperature technique [24, 25]. Therefore, annealing temperature of 750 K corresponds to undercooling temperature of 120 K.

3. Results and discussion

Figure 2 shows representative snapshots of atomic configuration from 300 ps calculation of heterogeneous nucleation inside the cavity with various apex angles and figure 3 shows time change of the ratio of HCP-Al and FCC (face-centered cubic)-Al atoms to all Al atoms in the cavity. It is confirmed from snapshots that HCP-Al solid is formed at the tip of concave cavity and it gradually piles up in the cavity following the orientation of HCP-Ti surface for all cases except for the case of 150°. On the other

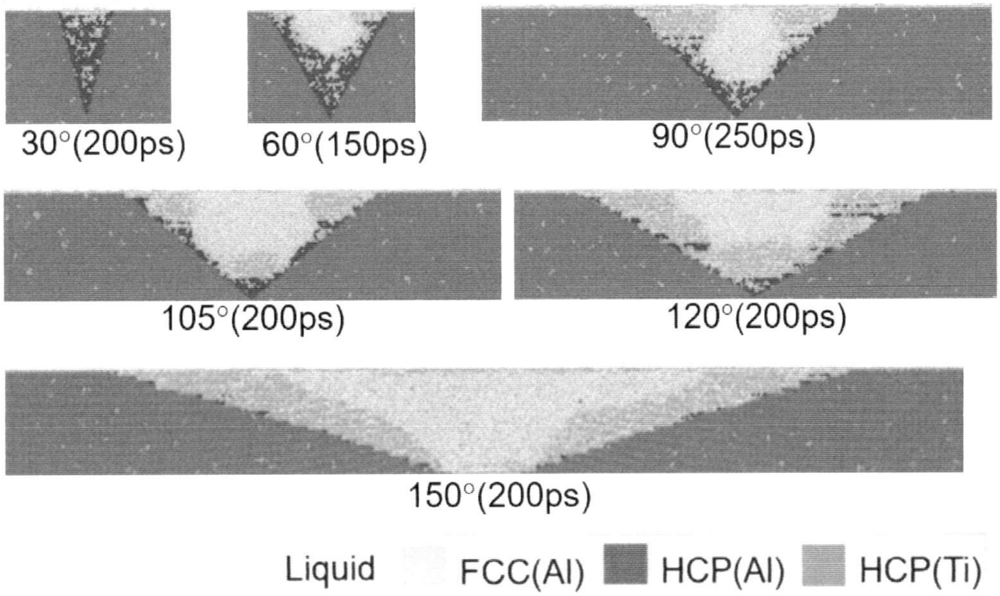

Figure 2. Snapshots of atomic configuration from 300 ps calculation of heterogeneous nucleation in the cavity with various apex angles. White, green and purple atoms represent Al atoms with liquid, FCC and HCP structures. Brown atoms represent Ti atoms with HCP structure.

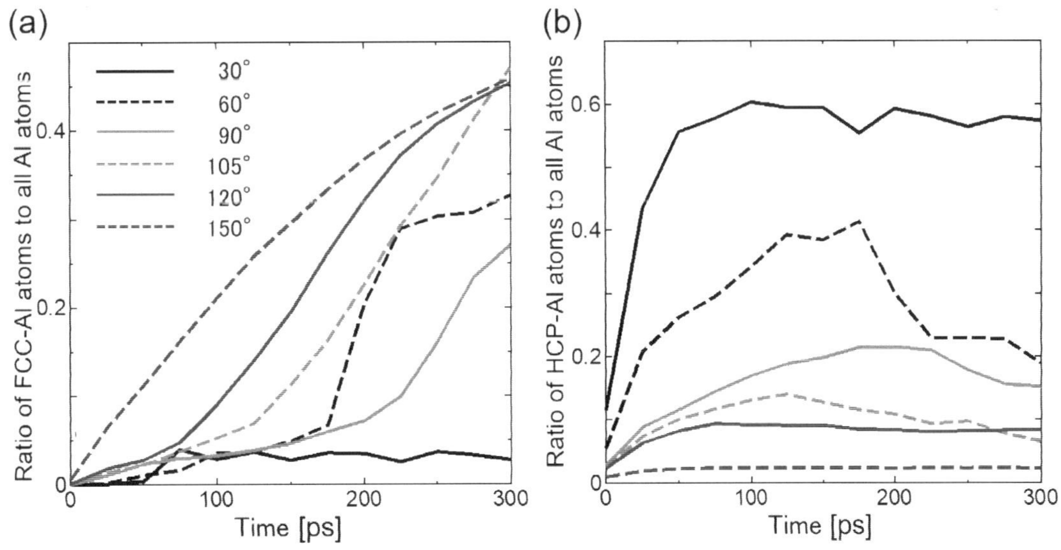

Figure 3. Time change of ratio of (a) FCC-Al and (b) HCP-Al atoms to all Al atoms in the cavity.

hand, FCC-Al solid is directly formed on the side wall of the cavity for the case of 150°. Therefore, the ratio of FCC-Al atoms to all Al atoms in figure 3(a) for 150° increases from the beginning of simulation whereas there is some waiting time until the increase of FCC atoms for all other apex angles. This difference is discussed layer from the atomistic viewpoint. The speed of pile up of HCP-Al solid becomes lower with increasing the apex angle, which is confirmed both from snapshots in figure 2 and

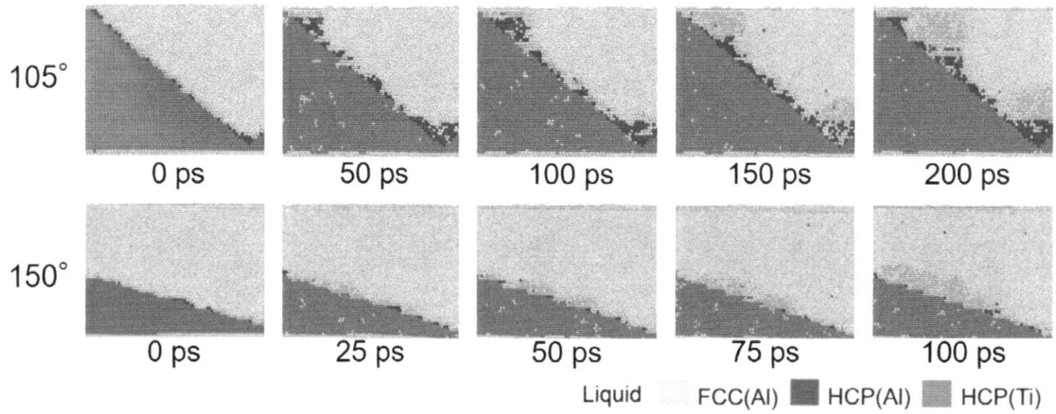

Figure 4. Snapshots of the side wall of cavities with apex angle of 105° and 150° during the simulation of heterogeneous nucleation. White, green and purple atoms represent Al atoms with liquid, FCC and HCP structures. Brown atoms represent Ti atoms with HCP structure.

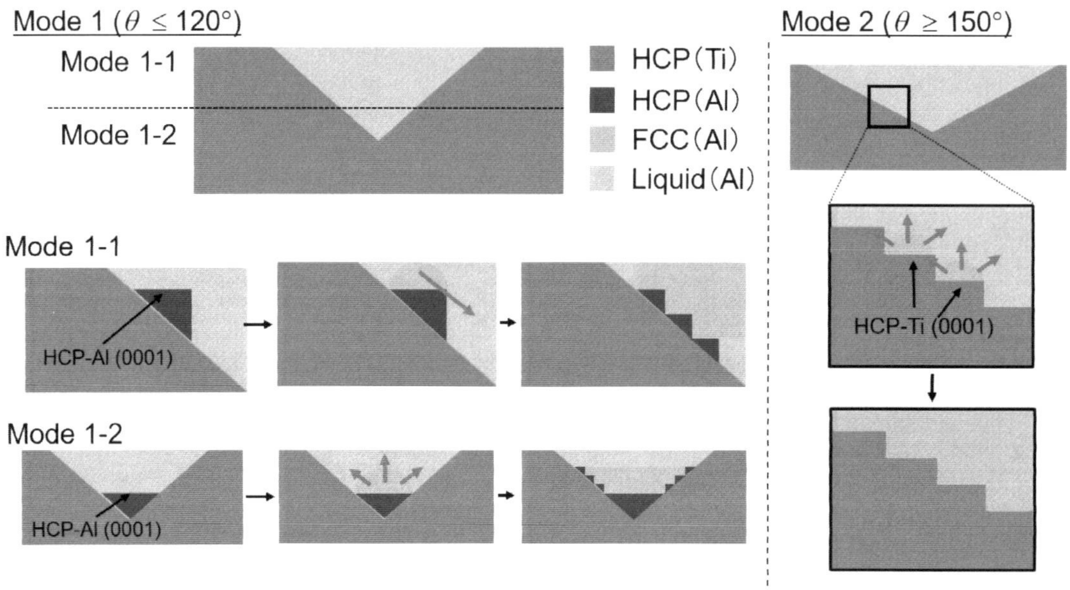

Figure 5. Schematic image of crystal growth behaviour on the side wall of concave cavity.

figure 3(b). Especially, almost all spaces in the cavity is filled with the HCP-Al before 100 ps for the case of 30° and therefore, the ratio of HCP-Al to all Al atoms becomes almost constant after 100 ps. It is considered that the constraint by the side wall of the cavity consisting of HCP-Ti becomes stronger with decreasing the apex angle (i.e., narrower cavity). In addition, several layers of HCP-Al solid are formed on the side wall of the cavity except for the case of 150°. This trend agrees with our previous simulation [14], in which HCP-Al layers were formed on the various Ti surfaces contacting with undercooled Al melt except for HCP(0001) surface. Then, FCC-Al solid is formed from exposed surfaces of HCP-Al solid perpendicular to z-direction, which basically have (0001) orientation in line with the surrounding Ti solid. It is reasonable since HCP(0001) and FCC(111) surfaces are equivalent as a close-packed face. Actually, FCC-Al layers grew continuously at the HCP(0001) surface in our previous simulation [14]. Once FCC-Al solid is generated, growth of the FCC phase proceeds since FCC is the most stable phase for solid Al.

Next, the origin of direct formation on the side wall of cavity of the apex angle of 150° is discussed. Figure 4 shows snapshots of the side wall of cavities with apex angle of 105° and 150° during the simulation of heterogeneous nucleation. In the case of the apex angle of 105°, HCP-Al solid is generated at the tip of the cavity and the side wall of the cavity near the gas-liquid interface. When the (0001) plane of HCP-Al equivalent to FCC(111) plane is exposed, Al-FCC solid is formed on the HCP(0001) plane. A slight amount of HCP-Al is also formed at the side wall of the cavity since epitaxy between HCP-Ti and FCC-Al surfaces is not good except for HCP(0001)-FCC(111) interface. On the other hand, side wall of the cavity with apex angle of 150° has a stepwise structure with HCP(0001) terraces due to the large-obtuse apex angle. Therefore, FCC-Al solid is directly formed on exposed HCP(0001) terraces of the side wall and the FCC-Al solid keeps growing without the appearance of HCP-Al solid since FCC is the most stable phase of solid Al. Figure 5 shows the schematic image of crystal growth behavior on the side wall of concave cavity. As observed in figure 4, there are two modes of growth behavior: FCC-Al growth on the (0001) surface of HCP-Al solid (named Mode 1) and FCC-Al growth on the HCP(0001) terrace of side wall of the cavity consisting of solid Ti (named Mode 2). In our simulation, Mode 1 appears in cases with apex axis of 120° and less and Mode 2 appears in the case with apex axis of 150°. Moreover, Mode 1 is divided into two patterns depending on the place where nucleation of FCC-Al happens. That is, nucleation on the surface of HCP-Al solid at the tip of the cavity (named Mode 1-1) and that on the side wall of the cavity (named Mode 1-2). Mode 1-1 is dominant in the case of apex axis of 90° and less. Both Modes 1-1 and 1-2 are found for 105° and 120°. Then, Mode 2 becomes dominant for the case of the apex axis of 150°.

4. Conclusions

In this study, the effect of the concave cavity at the surface of grain refiner on heterogeneous nucleation of undercooled melt is investigated by MD simulation. Especially, dependence of apex axis of the cavity on the growth behavior of Al melt is systematically examined. It is confirmed that there are two modes in growth behavior of FCC-Al growth from the undercooled melt of Al inside the concave cavity. In the case of apex angle of 120° and less, HCP-Al solid is firstly formed at the tip and on the side wall of cavity and FCC-Al growth is followed on the exposed (0001) surface of previously existing HCP-Al. On the other hand, FCC-Al is formed on the HCP(0001) terrace of side wall of the cavity for the case of cavity of 150°. It is confirmed from this study that the existence of HCP(0001) surface (whichever HCP-Ti or HCP-Al) is a key factor to initiate FCC-Al formation on the surface of HCP-based grain refiner. In the next step, other factors affecting nucleation such as the orientation anisotropy and the temperature dependence will be examined.

Acknowledgements
This work was supported by Grant-in-Aid for Scientific Research (B) [No. 19H02415] from Japan Society for the Promotion of Science (JSPS), Japan.

References

[1] Dantzig J A and Rappaz M, Solidification 2009 (Lausanne: EPFL Press) pp. 249-85
[2] Sosso G C, Chen J, Cox S J, Fitzner M, Pedevilla P, Zen A and Michaelides A 2016 *Chem. Rev.* **116** 7078
[3] Shibuta Y, Ohno M and T. Takaki 2015 *JOM* **67** 1793
[4] Shibuta Y, Ohno M and T. Takaki 2018 *Adv. Theo. Sim.* **1** 1800065
[5] Auer S and Frenkel D 2001 *Nature* **409** 1020
[6] Swope W C and Andersen H C 1990 *Phys. Rev. B* **41** 7042
[7] Cherne F J, Baskes M I, Schwarz R B, Srinivasan S G and Klein W 2004 *Model. Simul. Mater.*

Sci. Eng. **12** 1063
[8] Shibuta Y, Sakane S, Miyoshi E, Okita S, Takaki T and Ohno M 2017 *Nat. Commun.* **8** 10
[9] Shibuta Y, Sakane S, Miyoshi E, Takaki T and Ohno M 2019 *Model. Simul. Mater. Sci. Eng.* **27** 054002
[10] Greer A L 2016 *J. Chem. Phys.* **145** 211704
[11] Greer A L, Bunn A M, Tronche A, Evans P V and Bristow D J 2000 *Acta Mater.* **48** 2823
[12] Watanabe Y, Eryu H and Matsuura K 2001 *Acta Mater.* **49** 775
[13] Fan Z, Wang Y, Zhang Y, Qin T, Zhou X R, Thompson G E, Pennycook T and Hashimoto T 2015 *Acta Mater.* **84** 292
[14] Fujinaga T and Shibuta Y 2019 *Comp. Mater. Sci.* **164** 74
[15] Fujinaga T and Shibuta Y 2019 *IOP Conf. Ser.: Mater. Sci. Eng.* **529** 012047
[16] Ueshima Y, Takebayashi S, Yamamoto K, Miyazaki M and Mizukami K 2018 *J. Jpn. Inst. Met. Mater.* **82** 223
[17] Plimpton S J 1995 *J. Comput. Phys.* **117** 1, http://lammps.sandia.gov
[18] Zope R R and Mishin Y 2003 *Phys. Rev. B* **68** 024102
[19] Becker C A, Tavazza F, Trautt Z T and Buarque de Macedo R A 2013 *Curr. Opin. Solid State Mater. Sci.* **17** 277, http://www.ctcms.nist.gov/potentials
[20] Nose S 1984 *J. Chem. Phys.* **81** 511
[21] Hoover W G 1985 *Phys. Rev. A* **31** 1695
[22] Stukowski A. 2010 *Model. Simul. Mater. Sci. Eng.* **18** 015012
[23] Stukowski A. 2012 *Model. Simul. Mater. Sci. Eng.* **20** 045021
[24] Shibuta Y, Takamoto S and Suzuki T 2008 *ISIJ Int.* **48** 1582
[25] Shibuta Y 2019 *Mater. Trans.* **60** 180

IOP Conf. Series: Materials Science and Engineering **861** (2020) 012054 doi:10.1088/1757-899X/861/1/012054

The integration of structural mechanics into microstructure solidification modelling

P Soar[1], A Kao[1], G Djambazov[1], N Shevchenko[2], S Eckert[2] and K Pericleous[1]

[1] Computational Science and Engineering Group (CSEG), University of Greenwich, Old Royal Naval College, Park Row, London SE10 9LS, UK
[2] Helmholtz-Zentrum Dresden-Rossendorf, Institute of Fluid Dynamics, Bautzner Landstrasse 400, 01328 Dresden, Germany

E-mail: p.m.soar@gre.ac.uk

Abstract. *In situ* structural mechanics are an often neglected area when modelling alloy microstructure during solidification, despite the existence of practical examples and studies which seem to indicate that the interaction between thermal or mechanical stresses and microstructure can have a significant impact on its evolution and hence the final properties at a macroscopic level. A bespoke structural mechanics solver using the finite volume method has been developed to solve the linear elasticity equations, with design choices being made to facilitate the coupling of this solver to run *in situ* with an existing solidification model. The accuracy of the structural mechanics solver is verified against an analytic solution and initial results from a fully coupled system are presented which demonstrate in a fundamental example that the interaction between structural mechanics and a solidifying dendrite can lead to a significant change in growth behaviour.

1. Introduction

When casting metal alloys, the material properties at the macroscopic level are dependent on the underlying morphology of the microstructure as defined by the arrangement of the dendrites that form as the liquid metal solidifies. Consequently, the ability to understand and manipulate the development of dendrites by controlling the environment in which the solidification occurs has the potential to allow for the creation of alloys that possess superior material properties. There are numerous practical experiments and numerical models which demonstrate the utility of these principles, the understanding of which has been a subject of research for decades as highlighted in the review of the field by Asta *et al.* [1]. More recent work by Kao *et al.* [2] explores the use of magnetic fields to induce fluid flow to further understand defect formation and alter dendritic development.

Numerical models simulating the process of solidification at the microstructure level can currently consider many of the parameters influencing the growth behaviour of the dendrites. However, one area of the modelling seemingly neglected so far is the impact of including *in situ* structural mechanics as a factor influencing dendritic growth. This *in situ* interaction requires that the deformation and accumulation of stress under external forces is modelled at the same time as the solidification itself, with the processes interacting throughout. Asta *et al.* [1] identified the interaction between structural mechanics and the microstructure solidification causing casting defects as being one of the open questions requiring further research, with numerous experimental examples further underlining the importance of this currently neglected area. Reinhart *et al.* [3] demonstrate in an experiment at ESRF that stress and gravity cause defects to occur and the dendrite arms to be impacted mechanically. More recently an experiment undertaken by He *et al.* [4] investigated the impact of pulling, directional solidification and temperatures on the microstructural development and mechanical properties of a solidified aluminium alloy, finding it to impact both the microstructure and defect formation. A further example of structural mechanics behaviour influencing the microstructure has been identified by experiments run on a solidifying Gallium 25wt% Indium alloy by N. Shevchenko and S. Eckert from Helmholz-Zentrum Dresden-Rossendorf, where the experimental radiograph in figure 1(a) shows that the dendrites begin by growing in a seemingly fixed straight orientation, but figure 1(b) shows that later in the experiment the highlighted dendrite has deformed under its own weight, leaving it with a clear deviation in growth orientation when compared to the start. Any attempts to model this scenario using current methods which neglect mechanical effects would ignore this behaviour, leading to the dendrite continuing to grow along the initial orientation with no deformation, consequently causing a significantly different microstructure.

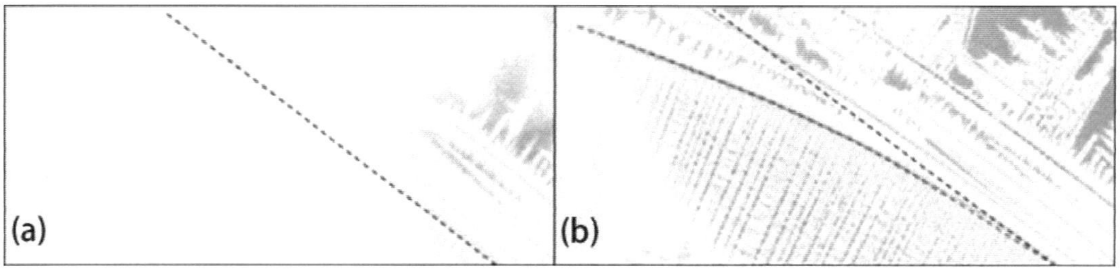

Figure 1. Dendrite growth showing structural mechanics impact.

The impact of mechanical interaction during the solidification process has been identified as an important modelling consideration in many studies, however many such cases take a post processing approach where the stresses will have no effect on the development of the solidification of the cast structure [5, 6]. Furthermore, the modelling that does consider these interactions largely restrict themselves to macroscopic models where the only mechanical behaviour considered is temperature driven [7, 8]. A solidification model created by Fackeldey, Ludwig and Sahm [9] considers these *in situ* effects, however this is achieved by modelling the process macroscopically and accounting for the microstructure in a more general sense using parameters to describe the primary and secondary arm spacing and eutectic fraction within a region of the macrostructure, as opposed to directly modelling the dendrites comprising the microstructure as this research intends.

Numerical models developed within this research group by Kao *et al.* [10] have been used to investigate how external parameters such as fluid flow and electromagnetics can be used to control the microstructural evolution. This work can be expanded further with the introduction of structural mechanics as an additional parameter in the microstructural solidification process, with the potential to lead to more accurate simulations of the dendritic growth behaviour.

The aim of this research will be to design a structural mechanics model that can be coupled with the ThermoElectric Solidification Algorithm (TESA), a bespoke code developed in house by CSEG which runs large scale simulations of microstructural growth in parallel [11]. This code has been designed to simulate the processes of solidification, fluid flow and electromagnetics within a single model capable of solving complex multi-physics phenomena such as casting [2], freckle defect formation [12] and undercooled growth [13]. However, to avoid some of the complications of coupling straight to this complex code, a more straightforward solidification code using the enthalpy method [14] was utilised for the initial coupling to demonstrate the fundamental changes in behaviour from the *in situ* interaction.

2. Numerical modelling

2.1. Solidification model

The solidification model that has been chosen for the initial tests of coupling with *in situ* structural mechanics is the enthalpy based method described by Voller [14]. This approach to solidification introduces the order parameter f to formulate an enthalpy-based method for a diffuse interface, where f represents the liquid fraction where $f = 1$ is fully liquid and $f = 0$ is fully solid.

The governing equations for this approach relate f to the volumetric enthalpy H, by defining H as the sum of the latent heats:

$$H = c_p T + fL \tag{1}$$

where c_p is the specific heat, T is the temperature and L is the latent heat of the volume. With the conservation of enthalpy being given by

$$\frac{\partial H}{\partial t} = \nabla.(K \nabla T) \tag{2}$$

which assumes that the thermal conductivity K is constant. The interface is undercooled to the temperature T^i:

$$T^i = T_m - \frac{\Gamma(\theta)}{L} T_m \kappa \tag{3}$$

where κ is the curvature, T_m is the melting temperature and the surface stiffness is defined via an often utilised fourfold symmetry model:

$$\Gamma(\theta) = d_0(1 - 15\epsilon_4 \cos(4\theta)) \tag{4}$$

The numerical approach Voller took was to convert this equation set into a dimensionless system which was discretised onto a Cartesian grid. In this paper the same approach was taken, so for the sake of brevity the detail of this approach has not been repeated.

2.2. Governing equations & assumptions for structural mechanics

Due to the fundamental nature of the research being undertaken, it was decided the linear elasticity equations would make for an appropriate choice of material model. This describes the behaviour of materials where the following assumptions are true: the deformations are 'small', there is a linear relationship between the stress and the strain and that the results of the problem show no plastic behaviour. Furthermore, as the model is currently only working in two dimensions, plane strain has been assumed. These simplifying assumptions will not hold for all solidification problems, but due to the paucity of existing published work exploring this phenomenon, it makes intuitive sense to begin by ensuring the most fundamental material model provides meaningful results. For example, effects such as thermal contraction and solidification driven density changes would introduce shrinkage flows, but the inclusion of these hydrodynamical effects is currently beyond the scope of the work. The material properties also have temperature dependence, but aside from interfacial undercooling the dendrite can be considered predominantly isothermal at the melting temperature with little spatial variation in material properties.

The equations for linear elasticity in two dimensions are described as follows:

$$\rho \frac{\partial^2 u_x}{\partial t^2} = \frac{\partial \sigma_x}{\partial x} + \frac{\partial \tau_{yx}}{\partial y} + F_x \tag{5}$$

$$\rho \frac{\partial^2 u_y}{\partial t^2} = \frac{\partial \sigma_y}{\partial y} + \frac{\partial \tau_{xy}}{\partial x} + F_y \tag{6}$$

where ρ is the density of the material, u_x and u_y the displacements in the x and y directions respectively, σ_x and σ_y the stresses in the x and y direction, τ_{xy} is the shear stress and F_x or F_y represent the x and y components of a body force acting upon the material.

The target application of his work is ultimately directional solidification in casting, where growth velocities are typically on the order of 10μm/s, with computational grid sizes of 10μm and time steps of 1ms giving a characteristic solidification length of 0.01μm per time step. For structural mechanics to have any noticeable effect the displacements per time step should be on a similar order of magnitude or notably longer, allowing the structural mechanics to be treated as a quasi-stationary process for a broad range of problems. This assumption simplifies the implementation, however as each solution is steady state the force component is changed from the full body force components to ΔF_x and ΔF_y, representing the change in the force exerted between the current time and the last timestep the structural mechanics was resolved.

Using Hooke's law equations (5) and (6) can be transformed into a displacement formulation, leading to the following pair of partial differential equations which are then solved simultaneously:

$$(2\mu + \lambda)\frac{\partial^2 u_x}{\partial x^2} + (\lambda + \mu)\frac{\partial^2 u_y}{\partial x \partial y} + \mu\frac{\partial^2 u_x}{\partial y^2} + \Delta F_x = 0 \quad (7)$$

$$(2\mu + \lambda)\frac{\partial^2 u_y}{\partial y^2} + (\lambda + \mu)\frac{\partial^2 u_x}{\partial x \partial y} + \mu\frac{\partial^2 u_y}{\partial x^2} + \Delta F_y = 0 \quad (8)$$

where μ and λ are the two Lamé constants defined as follows:

$$\lambda = \frac{vE}{(1 + v)(1 - 2v)}; \mu = \frac{E}{2(1 + v)} \quad (9)$$

defining E and v are respectively as the Young's Modulus and Poisson's ratio of the material.

2.3. *Structural solver development*

When solving structural mechanics problems, it is generally taken for granted that the Finite Element Method (FEM) will be the most appropriate choice of model. However, the process of meshing the structure intrinsic to the FEM approach can present a problem in the framework of coupled solidification with structural mechanics since every change in the structure due to solidification, would require computational power to constantly be diverted to re-meshing the structure to remain accurate. Furthermore, FEM generally models the geometry using unstructured grids, presenting a complication in coupling to TESA as the interpolation required to pass information between solvers to the Cartesian grid would increase complexity.

Instead a Finite Volume Method (FVM) approach has been taken as it synergises with solidification processes that are typically represented as a change in volume on a fixed grid, this change in volume providing the mechanical model with a straightforward means of calculating the change in forces exerted as solidification progresses. FVM is often associated with Computational Fluid Dynamics due to the volume conserving approach. However this is mere convention, as the attainable accuracy of results is comparable when modelling structural mechanics problems, as shown by Oñate, Cervera and Zienkiewicz [15] where for a simple structure under constant loads both the FEM and FVM produce the same results, with other groups supporting this position by finding modelling using the FVM to be in good agreement with analytic solutions [16, 17].

The FVM can be found applied to structural mechanics in many existing multi-physics problems such as fluid-structure interaction, both in a general sense [18] and how it can be related to solidification fronts [19] and welding [20]. In this related area, while models using different numerical schemes can certainly be coupled together, it is generally more straightforward and efficient for a similar numerical method to be used for both the fluid and solid equations.

For the purposes the future coupling to TESA, the FVM employed uses a Cartesian grid, which simplifies the discretisation of the governing equations down to relatively straightforward numerical schemes, which synergise well with the staggered grid Finite Difference approach being used to model solidification. As both approaches store material properties in the volume centred nodes with displacements and velocities on the corresponding cell faces due to the staggered mesh, meaning that less work is required to pass any quantities required by both solvers between the corresponding grids without requiring any complicated interpolations.

In order to validate the accuracy of the structural mechanics solver, a problem was set up that could be modelled by the Euler Bernoulli Beam theory, which provides an analytic solution for the deflection of a beam with a point load applied to the end of a cantilever beam. The results of this validation can be observed in figure 2 and show the FVM is returning results in close agreement with the analytic solution, clearly within the 3% interval beam theory is held to be accurate for.

However, as some assumptions of the Euler-Bernoulli beam theory limit the scenarios that can be modelled, this validation was further verified by modelling several 2D scenarios violating these assumptions. Here the FVM results were found to be in close agreement with solutions obtained using the commercial code COMSOL. This is reassuring to see as COMSOL, aside from being used widely both in academia and industry, also uses the FEM to obtain its answers, so the close agreement that can be found here seems to further support that the use of FVM to solve structural mechanics will not have any negative implications on the accuracy of the model.

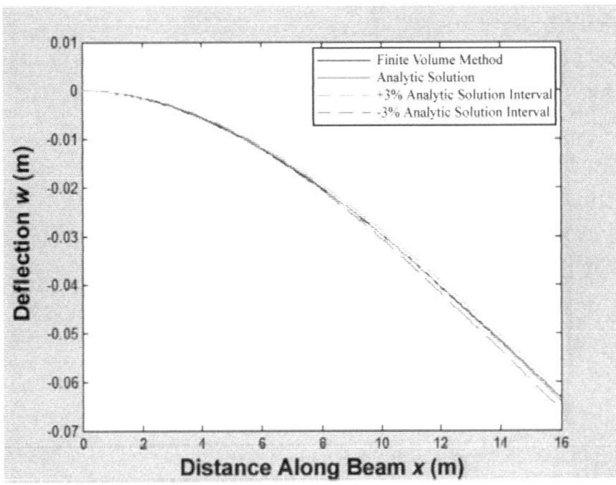

Figure 2. Comparison of deflections calculated using beam theory analytic solution and finite volume structural mechanics solver along 16m×4m×4m cantilever beam with a -1N Point load applied at the free tip using grid spacing $\Delta x = 0.1$m and material properties $E = 1000$Pa, $\nu = 0.33$ and $\rho = 8960$kgm^{-3}.

2.4. Advection of solid material

After the displacements are obtained, an advection-based process is used to account for changes in the solid fraction profile after the change in external forces have been accounted for. This process can be represented by the following equation:

$$\Delta f + \nabla \cdot \boldsymbol{u} f = 0 \qquad (10)$$

where Δf is the change in solid fraction and $\boldsymbol{u}$ is the vector of displacement values obtained from the structural mechanics code. To avoid the results showing a bias caused by a dependence on the order of the x and y sweep, an operator split scheme was written to discretise this equation. The scheme implemented utilises a Volume of Fluid (VOF) method with a Simple Line Interface Calculation (SLIC) approach as described by Noh and Woodward [21] to reconstruct the interface so that unrealistic diffusion of the solid material from the advection process is avoided.

2.5. In situ coupling process

In order to bring the previously outlined elements together in a loosely coupled solver they must be able to pass information between them. This begins by identifying which areas of the domain, represented by a liquid fraction profile f are to be considered solid by marking areas with a sufficiently low f value. Once the solid domain is identified, the structural mechanics code runs to obtain $\boldsymbol{u}$ which contains the displacements in x and y at the faces of each cell for the current timestep. These displacements are then used to drive an advection process causing solidified material to move into the surrounding cells to generate an updated profile for f. This f can then be used as an initial condition for the solidification solver, itself producing an updated f after a period of solidification has occurred. This represents a complete iteration of the coupled solver, which can be repeated for as many steps as the scenario being modelled requires with the final f profile after solidification acting as the initial condition for the next iteration of the structural mechanics solver. As the structural mechanics solver is steady state, the displacement is calculated on the difference between the previous force and the updated force from f. With the body force is dependent on f, if no solidification occurs the solution would remain the same as the previous calculation as it remains in steady state. However, as f is modified by the solidification solver, the change in force gives a new displacement. The magnitude of the change in force is dependent on the time interval between successive calls to the structural mechanical solver. Frequent calls give small displacements, while less frequent calls will give larger displacements. Thus, while the structural mechanical solver is steady state, this inherently introduces a time scale dependent on solidification.

3. Results

To understand the behaviour of the fully coupled model a simple test case of columnar growth with a body force applied to the solid cells was conducted. In this problem set up all of the domain boundaries are considered adiabatic, with the top surface where solidification initiates being a solid wall and the other boundaries considered to be Neumann conditions which allow no flux. A dimensionless approach is taken where the variables $\Delta x = \rho = E = 1$ which are scaled using base SI scaling factors to obtain the values in m, kg and s allowing the system to be governed by the dimensionless force. A consequence of this approach is that for a given set of material properties i.e. ρ and E, the dimensionless force scales with the meter SI unit. The dimensionless temperature is scaled by $T^* = (T - T_m)/(L/c)$, where L is the latent heat of fusion, c the specific heat and T_m the melting temperature. The initial conditions of the domain area assume dimensionless undercooling of ΔT = -0.643. In this context, high undercooled growth gives very small dendrites with high solidification velocities, while conversely in low undercooled growth the dendrites are much larger and grow more slowly. However, dendrites exhibit self-similarity across a wide range of undercooling and so while the real force required to influence solidification will vary significantly, this dimensionless force should be similar in magnitude.

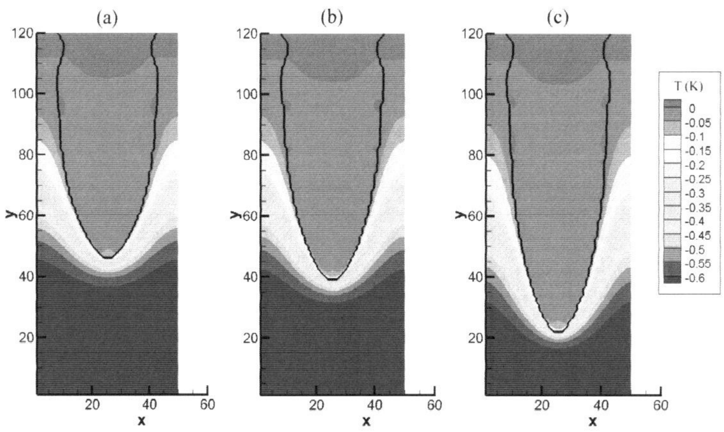

Figure 3. Dendrite behaviour under forces with thermal field T. (a) Positive force opposing growth direction, (b) no force, (c) negative force in growth direction.

In figure 3 there are three cases displaying the solid fraction and thermal field behaviour under different body force conditions. The cases considered are zero force, a body force acting in the direction of solidification and a body force acting against the direction of solidification. These cases were chosen as large deformations tangential to the growth direction will modify the preferential growth of the dendrite, which is behaviour currently unaccounted for in the solidification model.

Using the no force case in figure 3(b) as a reference, it is clear that when the force is applied in the direction of solidification as in figure 3(c) the dendrite arm is extended, while conversely when the body force is applied against the direction of solidification as in figure 3(a) the structure becomes shorter and wider. The components of the normal and shear strains which accumulate during the *in situ* solidification occurring under these positive and negative force conditions have been provided in figure 4 and figure 5 respectively. For growth under the positive force, it can be observed in figure 4 that for all types of strain the largest component seems to be running through the centre of the dendrite, caused as the tip of the dendrite is the region of the highest growth of new solid material, meaning that this is the region where the largest deformations occur with each run of the structural mechanics solver. In the negative force case in figure 5, while there is a corresponding high strain region running through the centre for the same reasons, the largest components of the strain can be found near the wall which is accumulating strain due to being stretched throughout the process by a dendrite increasing in solid mass at a greater rate.

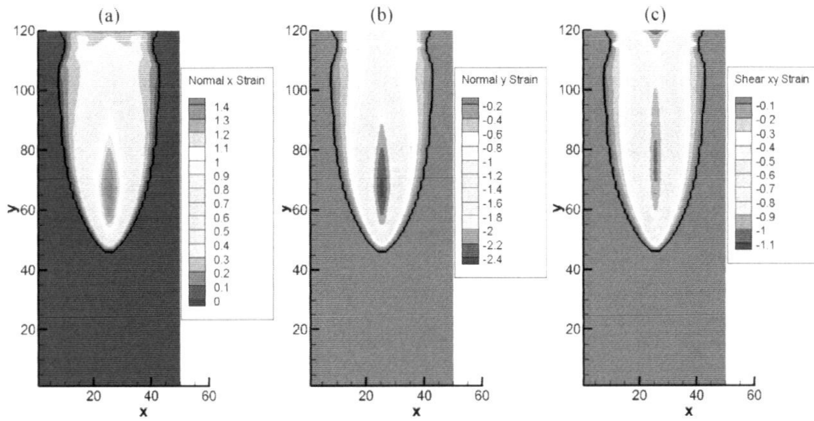

Figure 4. Accumulated strains in dendrite grown under positive body force. (a) Normal x strain, (b) normal y strain and (c) shear xy strain.

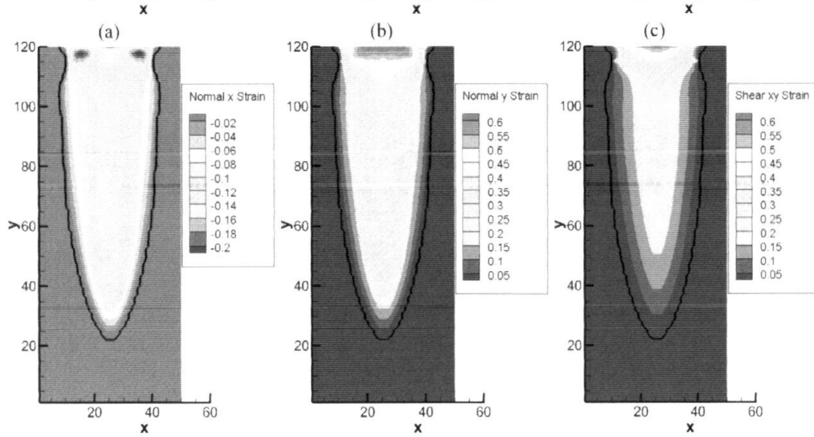

Figure 5. Accumulated strains in dendrite grown under negative body force. (a) Normal x strain, (b) normal y strain and (c) shear xy strain.

Although the cases represent the unrealistic situation of high undercooled growth and large body forces, these results are consistent with expectations and provide further qualitative understanding of the potential mechanisms for more realistic systems. When the force is in the direction of solidification structural deformations cause the dendrite to extend into the bulk. The deformation per unit time, which represents an added velocity allows the dendrite tip to extend further into the melt condensing isotherms which provides additional heat transfer. Conversely, the resisting force in figure 3(a), has the opposite effect where the slowing of the tip allows the thermal boundary to extend decreasing heat transfer. In more realistic casting situations of alloys, structural mechanics will have much less effect on the thermal problem compared to the solute problem, where mass transfer has very similar mechanisms, except there is a large disparity in both time and length scales, with a typical Lewis number on the order of 10000. However, with the focus on understanding the coupling of structural mechanics to solidification this simplified problem removes the necessity to also consider advection of solute through the displacement field, but requires very large forces to have a significant morphological change.

4. Conclusion

In summary, a coupled system modelling microstructure solidification while considering *in situ* structural mechanics was described which utilised a FVM approach due to constraints endemic to the nature of solidification modelling. This was done by creating a purpose-built structural mechanics solver to allow for coupling with the existing numerical code TESA, which was then validated against an analytic solution and coupled to an enthalpy-based solidification model. This coupled system was used to explore the fundamental case of a dendrite growing under the influence of a body force, both following and opposing the growth orientation of the dendrite. This demonstrates the behaviour that would be intuitively expected and is supported by observations of the analogous case of crystal growth under flow, where complementary forces induce comparatively elongated and slender dendrites, whilst opposing forces induce dendrites which appear compressed. Two key aspects still requiring inclusion are to account for the spatial variation as the dendrite deforms which alter the growth orientation and for implementing for alloy systems a similar advective process capable of redistributing f within a partitioned solute. These promising results highlight that the *in situ* effect of structural mechanics on the solidification process can be captured using the described approach and act as proof of concept, justifying the further development, where the structural mechanics code will be ultimately be fully coupled to the main TESA code enabling studies to be run which allow the impacts of these interactions to be better understood.

Acknowledgments

Peter Soar gratefully acknowledges the financial assistance offered by the University of Greenwich VC scholarship for his PhD studies.

References

[1] Asta M *et al.* 2009 *Acta Mater.* **57** 941-71
[2] Kao A, Cai B, Lee PD and Pericleous K 2017 *J. Cryst. Growth* **457** 70-4
[3] Reinhart G *et al.* 2008 *Met. and Mater. Trans. A* **39** 865-74
[4] He T, Min L, Ming Huo Y, Liu H J and Jie Yi X 2018 *Mechanics* **24** 149-54
[5] Gandin C A, Brechet Y, Rappaz M, Canova G, Ashby M and Shercliff H 2002 *Acta Mater.* **50** 901-27
[6] Palumbo G, Piccininni A, Piglionico V, Guglielmi P, Sorgente D and Tricarico L 2015 *J. Mat. Pro. Tech.* **217** 253-61
[7] Thorborg J, Klinkhammer J and Heitzer M 2012 *IOP Conf. Ser.: Mater. Sci. Eng.* **33** 1-8
[8] Srinivasan M, Karuppasamy P, Ramasamy P and Barua A K 2016 *Elec. Mater. letters* **12** 431-8
[9] Fackeldey M, Ludwig A and Sahm P R 1996 *Comp. Mater. Sci.* **7** 194-9
[10] Kao A, Shevchenko N, Roshchupinka O, Eckert S and Pericleous K 2015 *IOP Conf. Ser.: Mater. Sci. Eng.* **84** 012018
[11] Kao A, Shevchenko N, Alexandrakis M, Krastins I, Eckert S and Pericleous K 2019 *Phil. Trans. Royal Soc. A* **377** 20180206
[12] Kao A and Pericleous K 2012 *J. Alg. Comp. Tech.* **6** 173-201
[13] Kao A, Krastins I, Alexandrakis M, Shevchenko N, Eckert S and Pericleous K 2019 *JOM* **71** 48-58
[14] Voller V R 2008 *Int. J. Heat Mass Trans.* **51** 823-34
[15] Oñate E, Cervera M and Zienkiewicz O C 1994 *Int. J. Num. Meth. Engin.* **37** 181-201
[16] Oliveira P J and Rente C J 1999 *Proc. NAFEMS World Congr. Effect. Eng. Anal.* **1** 297-309
[17] Cavalcante M A and Pindera M J 2012 *J. App. Mech.* **79** 051006
[18] Xia G and Lin C L 2008 *Comp. Struct.* **86** 684-701
[19] Udaykumar H S, Mittal R and Rampunggoon P 2002 *Com. Num. Meth. Eng.* **18** 89-97
[20] Taylor G A, Hughes M, Strusevich N and Pericleous K 2002 *App. Math. Mod.* **26** 311-22
[21] Noh W F and Woodward P 1976 *Proc. Fifth Int. Conf. Num. Meth. Fluid Dyn.* 330-40

IOP Conf. Series: Materials Science and Engineering **861** (2020) 012055 doi:10.1088/1757-899X/861/1/012055

Calphad-based phase-field study of the interplay between spheroidal graphite growth and chemical segregation in ductile cast iron

J Eiken

Access e.V., Intzestraße 5, 52072 Aachen, Germany

E-mail: j.eiken@access.rwth-aachen.de

Abstract. Spheroidal graphite in ductile cast irons forms in a divorced eutectic solidification, controlled by interstitial diffusion of carbon through the enveloping austenitic shell. As technical cast irons are multicomponent alloys, carbon diffusion can strongly be affected by content and chemical gradient of other solute elements. During growth of a eutectic cell, all solute elements are partitioned at the austenite/liquid interface. Substitutional diffusion within the austenite is almost negligible, but expansion-induced spatial displacement of the iron-rich matrix and solute atoms during graphite growth significantly complicates multidimensional numerical simulation. A recently developed volumetric multi-phase-field approach, implemented in the Micress software, now accounts for the volume expansion and related matter and solute transport. In the present case study, this approach is applied to investigate the evolving chemical gradients and their potential effect on carbon diffusion during growth of a three-dimensional eutectic cast iron cell. Phase and concentration dependent thermodynamic data, partial molar volumes and diffusion coefficients are evaluated from linked Calphad-databases. The interface properties of the polycrystalline spheroidal graphite nodule are modelled based on a dedicated effective multi-facetted anisotropy function. A special target is to investigate whether increase in Si content may result in a significant reduction of the carbon flux, eventually favoring graphite degeneration and chunky growth.

1. Introduction

Microstructure evolution in ductile cast irons is challenging because of the complex interplay between interstitial and substitutional diffusion and the non-negligible role of volumetric expansion. The spheroidal graphite growth is mainly controlled by interstitial diffusion of carbon through the austenite shell. While being interstitially dissolved in austenite, the carbon atoms merely contribute to the material's volume, but significantly increase in partial volume when transforming to graphite. As a consequence, the solidifying material locally expands in volume during graphite growth. This transition-related expansion can be exploited to counteract thermal shrinkage and minimize shrinkage porosity [1], but is also of importance for the kinetic control of the divorced eutectic transition. Despite their slow solid-state diffusion, the substitutional elements Si, Mn and Mg hardly exert a growth restricting effect on the austenite to graphite transition, because they are passively transported together with the displaced fcc-(Fe) matrix during local expansion [2]. Figure 1 illustrates the complex interplay between graphite growth, solute transport and expansion.

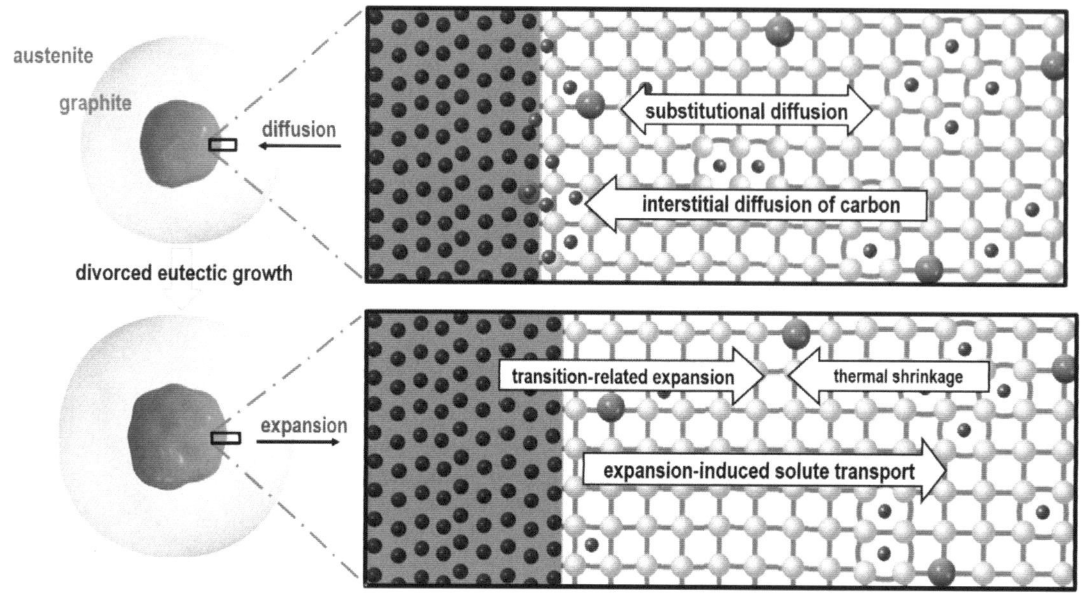

Figure 1. Scheme of the divorced eutectic growth of graphite. Carbon atoms diffuse from the liquid/austenite to the austenite/graphite interface driven by chemical gradients. The transition of the interstitial carbon atoms to graphite results in local volumetric expansion. Substitutional diffusion is of negligible importance and does not slow down the carbon-controlled transition, since the substitutional elements are passively transported together with the displaced fcc-(Fe) matrix.

Numerical 3D simulation of microstructure evolution has become feasible thanks to an extended volumetric multicomponent multiphase-field (Vol-MMPF) model [3, 4] which accounts for the changing partial molar volume of the individual elements and the related expansion-induced solute transport based on Calphad data. In a previous study [4], solute distribution maps were predicted for a hypoeutectic Fe-C-Si-Mg-Mn cast iron and validated against experimental as-cast data. The present study is based on the same approach, but aims at a better understanding of the interplay between spheroidal graphite growth and evolving chemical gradients during the divorced eutectic solidification. It has been suggested [5, 6] that chemical gradients of substitutional elements might essentially affect carbon transport by cross diffusion. Counteracting carbon cross diffusion fluxes were hypothesized to be a major reason for graphite degeneration. The here presented phase-field simulations shall serve as a basis to discuss the role of the initial Si concentration and related cross diffusion on the growth of a single three-dimensional eutectic cell from a slightly hypereutectic Fe-C-Si-Mn-Mg melt,

2. Microstructure simulations

Simulations were performed using the Vol-MMPF model [3] implemented in the Micress® software [7, 8] under assumption of two hypereutectic compositions with different Si content: (a) Fe - 3.8 C - 2.0 Si - 0.2 Mn - 0.045 Mg and (b) Fe - 3.8 C - 2.5 Si-0.2 Mn - 0.045 Mg (all in wt%). All phase-specific thermodynamic, volumetric and diffusion data was evaluated from the databases TCFe8 [9] and MobFe3 [10] via the TQ interface of the ThermoCalc software [11] as illustrated in figure 2. The kinetic equations were corrected in the thin-interface limit [12], however without anti-trapping currents at the graphite/austenite interface, inconsistent with the volumetric approach. A dedicated finite-differences formulation [13] ensured high accuracy for a diffuse interface width of 1.6 μm = 4 Δx. The side length of the cubic simulation domain with periodic boundary conditions was set to 40 μm (figure 3).

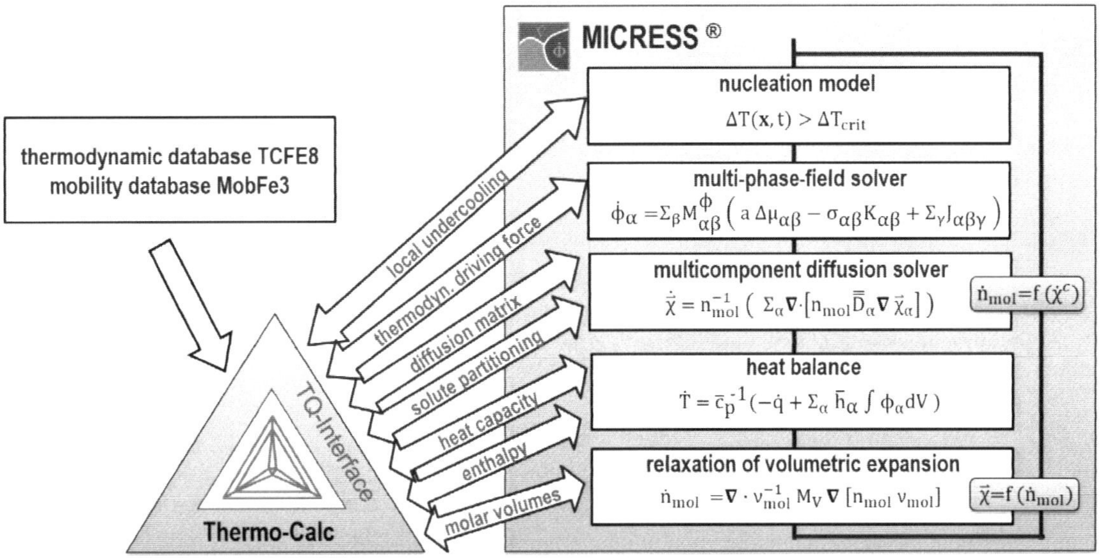

Figure 2. Illustration of the volumetric multicomponent multi-phase-field model (Vol-MMPF) [3, 4] and its implementation in the Micress® software. Phase-specific thermodynamic, volumetric and diffusion data is evaluated from Calphad databases.

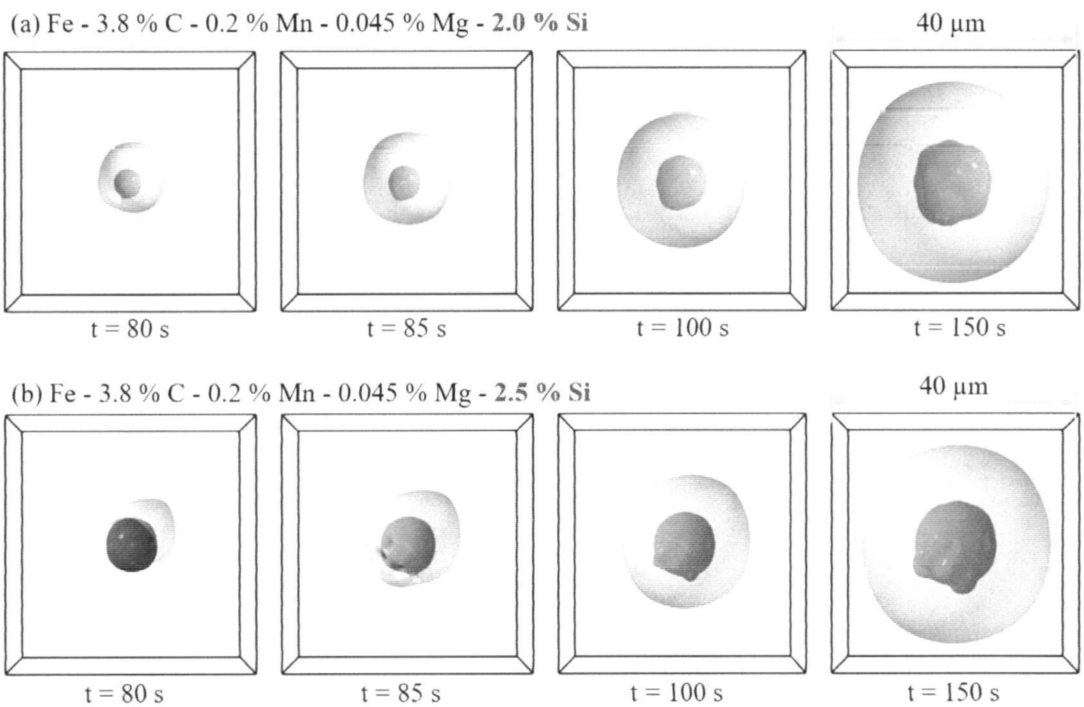

Figure 3. Phase-field simulations of the divorced growth of a single eutectic cell in comparison for two hypereutectic cast iron alloys with different initial Si contents (a) 2.0% and (b) 2.5%. Both simulations were started at a temperature of 1266.85 °C and cooled by a constant heat extraction rate $\dot{q} = 7 \cdot 10^6$ $Js^{-1}m^{-3}$ (White: liquid, yellow: austenite, violet: graphite).

IOP Conf. Series: Materials Science and Engineering **861** (2020) 012055 doi:10.1088/1757-899X/861/1/012055

Both simulations started at a temperature of 1266.85 °C. The temperature was computed homogeneously throughout the simulation domain from the balance between growth related latent heat release and constant heat extraction ($\dot{q}$ = 7·10^6 Js^{-1}m^{-3}), resulting in a solidification time of approximately 300s. Figure 4 shows the predicted temperature and phase fractions in comparison for both alloys. Due to the effect of Si on the heat capacity, small temperature differences can be observed even before nucleation. Graphite nucleation was modelled with a critical undercooling ΔT = 2 °C in the center of the domain. In alloy (a), this undercooling was reached after 70s at a temperature of T = 1170 °C, in (b) already after 35.25 s at a significantly higher temperature of T = 1218.25 °C. Austenite nucleation was modelled with a critical undercooling ΔT = 3 °C in the melt with some distance to the graphite interface. This undercooling was reached almost at the same time for both alloys: (a) after 77.4 s at a temperature of T = 1160.15 °C and (b) after 78.7 s at a temperature of T = 1163.15 °C.

Interfacial energies were modelled based on estimated data from literature [14, 15] under assumption of a weak cubic anisotropy for austenite and an effective multi-facetted anisotropy for the multi-crystalline graphite [4]. Table 1 summarizes the data specified for the individual interfaces. The assumed high interfacial mobility of the liquid/austenite ensures complete diffusion-control, whereas the lower mobility of the graphite interface was defined at the limit between diffusion- and interfacial-control. This limit is defined by the constraint that a further increase of the mobility value hardly affects the simulation results, while a reduction results in kinetic undercooling of graphite. It may be noteworthy that a reduced austenite/graphite mobility was found to trigger coupled eutectic growth (liquid→austenite + graphite) and multidimensional branching, proceeding at a higher temperature than the divorced eutectic growth (liquid → austenite, austenite → graphite), obtained in the here investigated diffusion-controlled limit.

Table 1. Interfacial properties

interface	anisotropic interfacial energy		anisotropic interfacial mobility	
liquid/austenite	0.17 Jm^{-2} (mean)	0.03 (anisotropy)	10^{-7}m^{4}J^{-1}s^1(mean)	0.05 (anisotropy)
liquid/graphite	1.46 Jm^{-2} (facet)	1.80 Jm^{-2} (edge)	3·10^{-14}m^{4}J^{-1}s^1(facet)	5·10^{-14}m^{4}J^{-1}s^1(edge)
austenite/graphite	1.21 Jm^{-2} (facet)	1.50 Jm^{-2} (edge)	3·10^{-15}m^{4}J^{-1}s^1(facet)	5·10^{-15}m^{4}J^{-1}s^1(edge)

Figure 3 shows the evolving eutectic cell at t = 80 s, 85 s, 100 s and 150 s for both alloys. The most striking difference is that the size of the graphite nodule before encapsulation by austenite is much higher for alloy (b) than for alloy (a). At t = 100 s, the eutectic cells of both alloys exhibit a spheroidal geometry of comparable size. Although the whole solidification lasted about 300 s, only the first 150 s were evaluated in order to avoid artifacts from the cubic shape of the box and impingement of the eutectic cell. The molar fraction of solid at t = 150 s was approx. f_s = 0.3 for both alloys with the ratio between austenite and graphite being slightly higher for alloy (a) f_A/f_G = 11.03 then for alloy (b) f_A/f_G = 9.16. The corresponding intrinsic volume expansion, resulting from the balance between transition-related expansion and thermal shrinkage without consideration of melt feeding or pore formation, was (a) V/V_o = 0.983 and (b) V/V_o = 0.982. The ratio between austenite and graphite volumes was (a) V_A/V_G = 14.29 and (b) V_A/V_G = 11.83, corresponding to mean radii ratios of (a) r_A/r_G = 2.48 and (b) r_A/r_G = 2.34.

The major difference in the divorced eutectic growth of the two alloys is that alloy (b) solidifies at a higher temperature and also shows a slightly different recalescence behavior (figure 4). Despite this difference, the graphite morphology of both alloys is very similar, except for some surface perturbations inherited from the moment of encapsulation. This difference can be explained by the fact that the austenite exhibited a smooth, stable front when covering the graphite surface in alloy (a), but formed some kind of surface dendrites in alloy (b) and as a consequence, the shell did not close in a single, but in several spots. This detail is here hardly visible, but was found to be more pronounced for further increased Si contents. It is still unclear whether the observed instability of the evolving three-phase-line can be explained by the larger nodule size only, or whether it is additionally affected by the increased Si segregation.

3. Chemical segregation

During the growth of the eutectic cell, the solute elements are partitioned at the phase interfaces and redistributed by diffusion and expansion-induced transport. Graphite was modelled with 100% carbon content and zero solubility for other elements. Figure 5 shows the evolution of the average contents of C, Si, Mn and Mg in the melt evaluated as function of time. Primary growth of graphite was found to reduce the carbon content, but hardly affect the Si, Mn and Mg content. After nucleation of austenite, the melt becomes enriched in Mn and Mg and depleted in Si. The C content of the melt rapidly increases during encapsulation of graphite, but temporarily decreases, before going up again. This change is less accentuated for alloy (b), where the carbon content continuously increases during encapsulation and subsequent divorced growth. Furthermore, the increased Si content of alloy (b) slightly reduces the enrichment in Mn, but increases the enrichment in Mg. The Si content itself seems to be simply constantly shifted by 0.5%.

The line profiles in figure 6, evaluated in horizontal direction through the center of the simulation domain, illustrate the spatial solute distribution at three moments in time (t = 80 s, 100 s and 150 s). In the liquid regions, all solute concentrations appear homogenous. The fact that the three-phase-line becomes morphologically instable in alloy (b) provides some evidence that the growth of austenite from liquid might be limited by diffusional transport of Si, Mn or Mg in the melt, but this is not visible on the scale of the depicted solute profiles. In contrast, the chemical gradients in austenite, representing the driving forces for diffusion from liquid to graphite, are clearly visible. They are caused by the difference in composition between the austenite/graphite interface on the one side and the austenite/liquid interface on the other side. The corresponding solute partitioning was evaluated from the thermodynamic database [9] based on quasi-equilibrium calculations [8], comprehensively accounting for thermal, solutal, curvature and kinetic undercooling. It is important to note, that the quasi-equilibrium concentrations of the individual elements are not independent of each other. Figure 6 shows that the higher initial Si content of alloy (b) affects the profiles of all elements, with Si and Mn contents in austenite being increased and C and Mg being decreased.

Apart from their effect on the thermodynamic quasi-equilibrium conditions, the substitutional elements affect the carbon flux by cross-diffusion. In agreement with previous findings [4], the profiles of substitutional elements in austenite exhibit Scheil-type progression with negative gradients of Si and positive gradients of Mn and Mg in normal direction to the austenite/graphite interface. These gradients represent additional driving forces for the carbon flux, which was explicitly considered in the phase-field simulations based on chemical mobility data from the multicomponent database [9]. Especially the cross-diffusional contribution of the Si gradients to the carbon flux was found to be non-negligible, with an off-diagonal coefficient approximately half the size of the diagonal carbon diffusion coefficient and chemical gradients of the same order of magnitude. It is important to note that the impact of the negative Si gradient on the carbon diffusion flux counteracts the impact of the positive carbon gradient.

It has been supposed that an increased content of negatively segregating elements, such as Ni or Si, increases the negative chemical gradients and the related counteracting cross diffusion [5, 6], which might then result in an essential reduction of the carbon diffusion flux. However, this was not observed in the present simulations. Figure 6 reveals that the carbon and silicon profiles in austenite are significantly affected by the increased initial Si-content, but the corresponding change in chemical gradients is hardly visible. Moreover, a detailed evaluation revealed that any increase in the chemical gradient of Si is related to a simultaneous increase in C gradient, such that the resulting changes in carbon diffusion flux directly compensate each other. Hence, the diffusion fluxes and the transition kinetics remain almost unaffected by the increased initial Si content, while the thermodynamic conditions, i.e. the temperature and interface composition during the divorced eutectic transformation are significantly changed. Whether this may favour repeated nucleation of graphite in the melt and/or trigger transition to chunky growth at a critical Si content is still unclear and shall be the subject of future studies. In this context, it may also be of interest to compare alloys with different Si and C contents, but equal carbon equivalent.

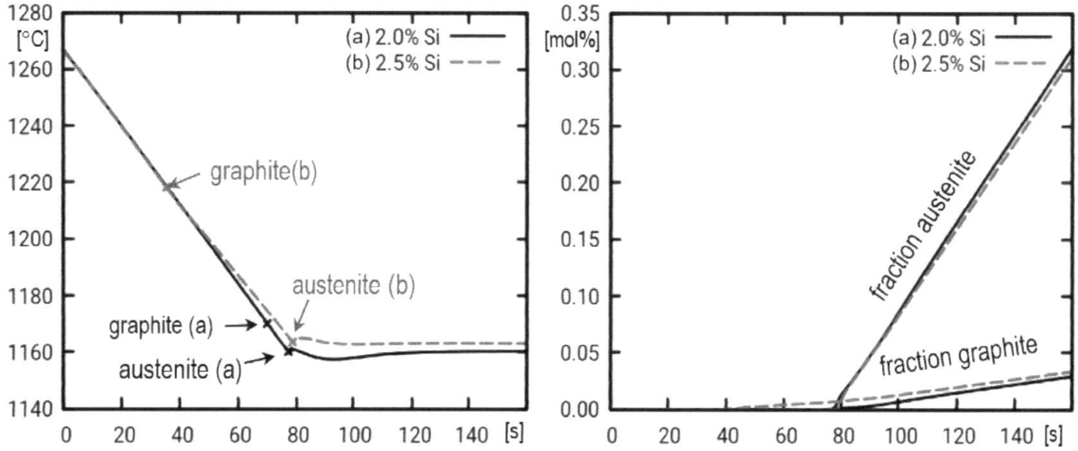

Figure 4. Temperature evolution and phase fractions evaluated as function of time from two simulations with different initial Si contents.

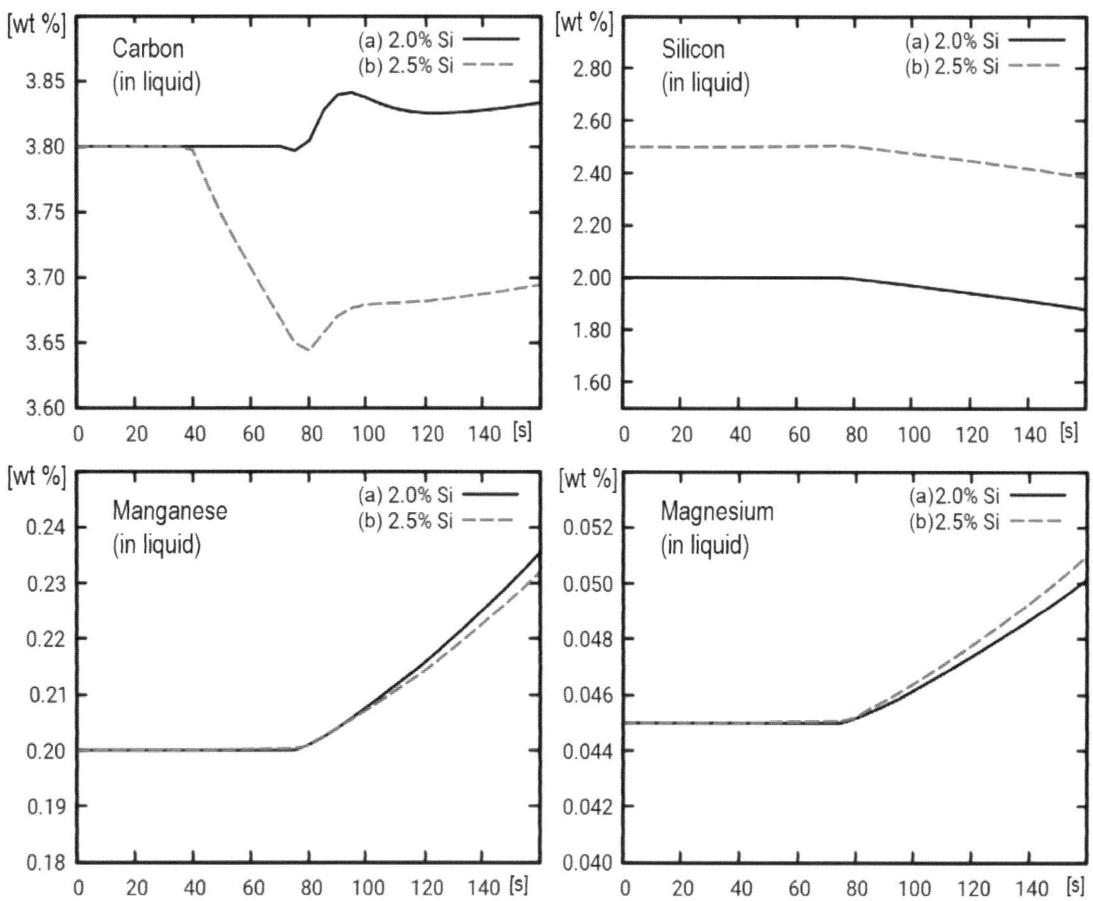

Figure 5. Average contents of carbon, silicon, manganese and magnesium in the melt evaluated as function of time from two simulations with different initial Si contents.

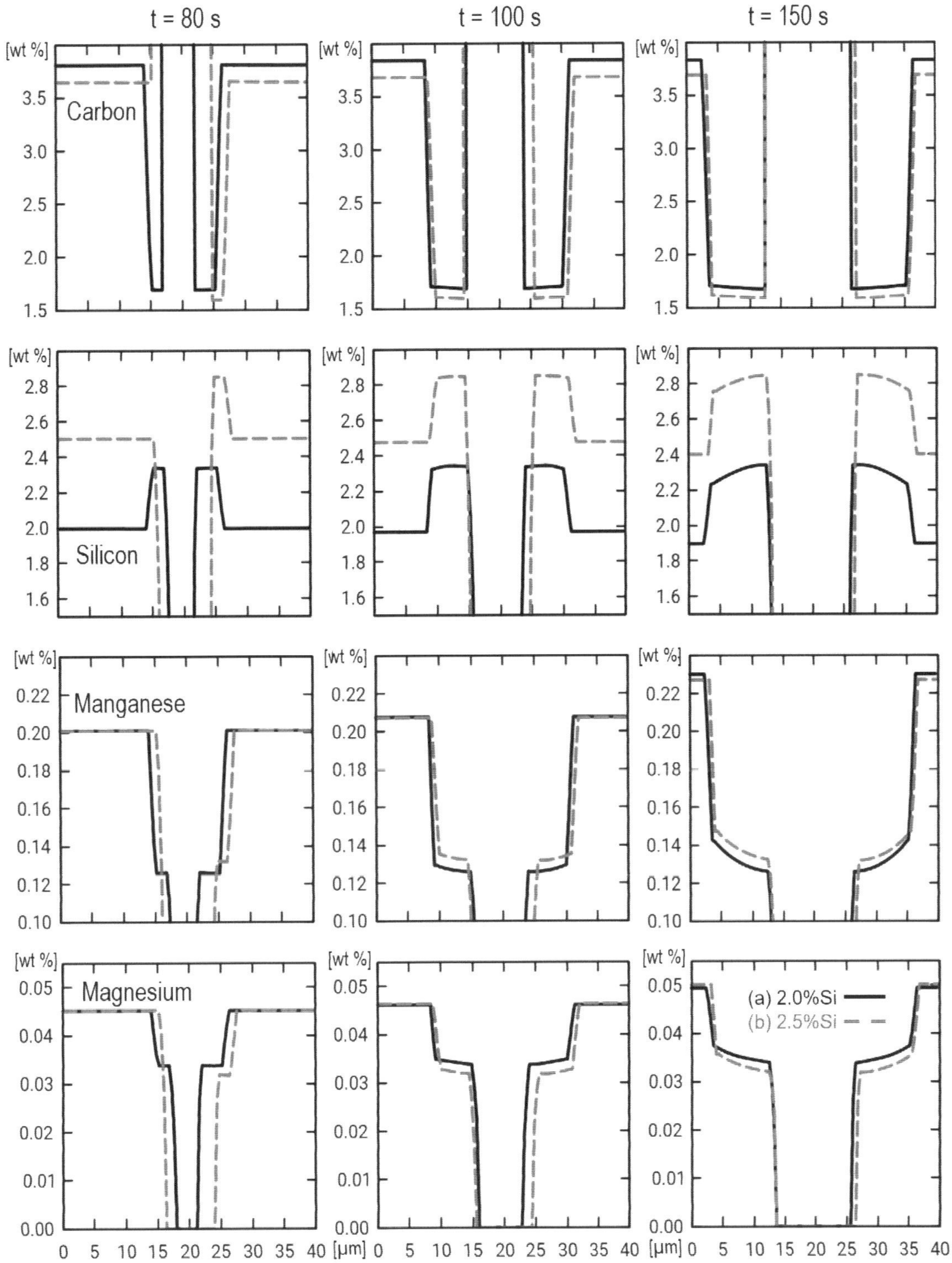

Figure 6. Horizontal line profiles of carbon, silicon, manganese and magnesium evaluated at three different times t= 80 s, 100 s and 150 s from two simulations with different initial Si contents.

4. Conclusions

The interplay between spheroidal graphite growth and chemical segregation during the solidification of a single eutectic cast iron cell was simulated using a volumetric multi-phase-field approach. Thermodynamic and diffusion data, as well as element and phase specific partial molar volumes of the multicomponent Fe-C-Si-Mn-Mg alloy were evaluated from Calphad databases. As expected Mn and Mg segregated positively and Si negatively into the melt, while carbon segregation changes from negative to positive after encapsulation of the graphite by austenite. Carbon diffusion through the austenite shell was found to be mainly driven by the positive chemical gradient of C and the negative gradient of Si. The counteracting contribution of the silicon gradient to the carbon flux was found to be non-negligible, approximately of half the size of the contribution of the carbon gradient.

However, against previous expectations, no reduction of the carbon flux was observed when increasing the initial Si content from 2.0 to 2.5 wt%. The predicted carbon and silicon profiles in austenite were considerably shifted, but the observed change in chemical gradients was rather small. Moreover, any change in Si gradient was found to be directly compensated by a simultaneous change in C gradient. While the increased initial Si content hardly affected the diffusion flux, it essentially changed the nucleation and thermodynamic conditions. Graphite nucleated much earlier, which favored the primary growth of graphite from the melt and affected not only the graphite size, but also the morphology of the three-phase-line during encapsulation by austenite. Furthermore, the temperature and the interface compositions during the divorced eutectic were significantly changed. These effects are expected to be more pronounced for further increased Si contents. An interesting future task will be to investigate whether the modification of the nucleation and thermodynamic conditions or rather a simultaneous modification of the interfacial properties can explain the often reported impact of the initial solute concentrations on graphite degeneration.

References

[1] Lekakh S and Hrebec B 2016 *Int. J. Metalcast.* **4** 389-400
[2] Lesoult G, Castro M and Lacaze J 1998 *Acta Mat.* **46(3)** 983
[3] Eiken J and Böttger B 2018 *T Indian I Metals* **71(11)** 2725-9
[4] Eiken J, Subasic E and Lacaze J 2020 *Materialia* **9** 100538
[5] Pustal B and Bührig-Polaczek A 2017 *Acta Mat.* **124** 137-42
[6] Domeij B 2019 Compacted graphite iron: On solidification phenomena related to shrinkage defects. PhD thesis, *Jönköping University, School of Engineering*
[7] MICRESS®, the microstructure evolution simulation software *www.micress.de*
[8] Eiken J, Böttger B and Steinbach I 2006 *Phys. Rev. E* **73(6)** 066122
[9] TCFe8, TCS Steels/Fe-Alloys Database version 8.0
[10] MobFe3, TCS Steels/Fe-Alloys Mobility Database, version 3.0
[11] Thermo-Calc software, *www.thermocalc.com*
[12] Carré A, Böttger B, Apel M 2013 *J. Cryst. Growth* **380** 5
[13] Eiken J 2012 *IOP Conf. Ser. Mater Sci. Eng.* **33** 012105
[14] McSwain RH, Bates CE and Scott WD 1974 *AFS Trans.* **82** 85
[15] De La Torre U, Loizaga A, Lacaze J and Sertucha J 2014 *Mater. Sci. Tech.* **30(12)** 1425-31

Effects of Bi addition on Si features, tensile properties and wear resistance of hypereutectic Al-15Si alloy

C L Pereira[1], L F Gomes[2] and J E Spinelli[1]

[1] Department of Materials Engineering, Federal University of São Carlos, São Carlos, SP 13565-905, Brazil
[2] Federal University of São Carlos, Graduate Program in Materials Science and Engineering, 13565-905 - São Carlos, SP, Brazil

E-mail: spinelli@ufscar.br

Abstract. Al-Si alloys are very important in the automotive industry. This is thanks to their high strength to weight ratio, suitable fluidity, good corrosion resistance and high productivity. The shape of the eutectic Silicon (Si) has been widely studied since it is considered an index of quality regarding many application properties. Understanding the effects of alloy additions on the morphology of Si has drawn researcher's attention. One of these elements of interest is Bismuth (Bi), which may be part of the composition of Al-Si alloys either as an addictive or a contamination. Hence, the microstructure aspects related to the presence of Bi are important requiring further consideration. In the present study, Al-15wt% Si hypereutectic alloy is modified with 1.0 wt% Bi. The goal is to compare the effect of this addition on the morphology of the phases forming the microstructure and on tensile/wear properties. In order to pursue that comparison, previously available data for the binary Al-15wt% Si alloy will be used. Various samples characterized by distinct cooling rates were generated by transient directional solidification of the ternary Al-Si-Bi alloy. The experimental variations of the eutectic spacing will be compared to each other based on their trends. It was found that the addition of minor Bi content (~1%) permitted tensile strength and ductility of 195 MPa and 14%. Furthermore, wear resistance was improved thanks to Bi. The reasons for that will be outlined.

1. Introduction

Al-Si alloys are widely used to produce shaped castings for a variety of applications as a result of their high specific strength, wear resistance, corrosion resistance and low thermal expansion coefficient. The usage of these alloys as an alternative to gray cast iron in automotive parts such as cylinder heads, brake discs and pistons aims to reduce automobile weight and improve fuel efficiency. The distribution of the primary and eutectic Si phases, as well as their size and morphology have a directly influence on the mechanical properties of Al-Si hypereutectic alloys. Additionally, adding silicon into the aluminum acts to improvement of the fluidity and hot tear resistance [1–3].

As potential substitutes for some applications where steels are still used, mainly in view of weight reduction and higher corrosion resistance, aluminum-based alloys have been more and more used during the years. In relation to the expanding applications of these alloys in the aeronautical and automotive industries, energy efficiency is directly associated to the weight factor, in which Al is about three times less dense than steel [4, 5].

Al-Bi alloys, may have immiscible phases and microstructures defined by a minority soft phase (Bi), following the morphology of fibers and/or droplets embedded into the Al-rich matrix. Practical applications include the fabrication of porous materials, electrical contact and self-lubricated bearings [6]. Tribological applications can be achieved in automotive components using aluminum alloys dispersed with bismuth. Such dispersions of low melting temperature elements decrease hardness and flow easily under sliding conditions [7].

During the wear process, while the Al-rich matrix provided some mechanical strength, the Bi-rich droplets were shown to act as self-lubricating agents, for Al-Bi alloys. The wear volume increases with an increase in the microstructural spacing (λ_{Bi}), as recently reported, for a same sliding distance for a monotectic Al-Bi alloy. It means that in order to induce a higher wear resistance, a better dispersion of Bi droplets is required [8].

Nowadays, Al industries have been progressing towards using recycled, or secondary, Al to reduce the initial energy output of the Al productive chain. The use of secondary Al causes some drawbacks mostly associated with elemental contamination. Due to that, understanding the effects of these additions has drawn researcher's attention. Bismuth (Bi) may be part of the composition of Al-Si alloys either as an addictive or a contamination. Hence, the microstructure aspects related to the presence of Bi are an important requiring further consideration.

A bearing material with lower coefficient of friction as well as higher wear resistance is required for new car developments [9]. Systematic studies regarding ternary metallic alloys in comparison to the binary monotectic Al-base alloys remain scarce. To increase the ability of these materials to resist loads without fracturing, the addition of elements to Al alloys can contribute positively, as for instance Si. In order to realize the potential of a hypereutectic Al-Si alloys with Bi doping as bearing material, a deeper microstructure examination is required.

The present study aims to contribute to a better comprehension of the effects of cohesion, size and distribution of the hard and soft phases on the mechanical properties and wear behavior of the ternary Al-15wt%Si-1wt%Bi alloy. Upward directional unsteady-state solidification (DS) experiment was performed. This study focuses on the influence of solidification thermal parameters (growth rate and cooling rate) on the microstructural evolution. Quantitative measurements of the Bi fraction were performed. Samples from Al-Si-Bi alloy were subjected to advanced SEM-EDS technique to determine the main evolution of Bi and Si compositional profiles along the axial length of the DS casting. SEM images of the worn surfaces were further characterized. Experimental interrelations between the wear volume and the Si spacing will be outlined.

2. Experimental Procedure

To permit a transient heat flow condition to be achieved, a directional solidification system was used as detailed in previous articles [10, 11]. Heat is directionally extracted through a water-cooled low carbon steel bottom, promoting vertical upward directional solidification. A stainless steel split mold was used having an internal diameter of 60 mm, a height of 157 mm and a wall thickness of 5 mm. To minimize radial heat losses, the vertical inner mold surface was covered with a layer of insulating alumina. A thin (3 mm) 1020 carbon steel was designed to close the bottom part of the mold.

The Al-Si-Bi alloy was prepared by melting weighed quantities of aluminum, silicon and bismuth of high purity elements in a silicon carbide crucible inserted into an induction furnace. The alloy was poured in the mold and the electric heaters were disabled so that directional solidification could be initiated, and at the same time, the controlled water flow was opened. During solidification, continuous temperature measurements in the casting were performed by fine type K thermocouples, with 1.6 mm diameter and 304 stainless steel sheath, placed along the casting length. The determination of solidification rates of cooling down was based in the passage of two isotherms along the length of the DS casting, which are liquidus and eutectic isotherms.

The cylindrical ingot was sectioned along its axial axis, sanded with a 1200 mesh sandpaper and etched with an acid solution (Poulton's reagent: 5 mL H_2O, 30 mL HNO_3, 60 mL HCl and 5 mL HF) to reveal the macrostructure. A solution of HCl in water was used to reveal the silicon particles for metallography. The Si and Bi spacings (λ_{Si}, λ_{Bi}) were measured using optical microscopy's images by examining longitudinal sections of the casting. Furthermore, representative micrographs were obtained with an optical image processing system [12]. The intercept method was employed to quantify the microstructural spacings with an average of 40 counting intercepts per position. This method is typically employed in the examination of longitudinal sections in the direction of heat flow, measuring the length between the center of a particular particle to another center aligned with the first [13].

In addition, microstructural characterization was completed with a scanning electron microscopy (SEM), equipped with an energy dispersive spectrometer (EDS). An image processing software (ImageJ) was used to measure the fraction of Bi particles. Transverse sections along the axial length of the DS Al-Si and Al-Bi-Si samples were used in the micro-adhesive (ball crater) wear tests. A schematic representation of the used wear tester is shown in [8]. To produce a wear crater, during the tests, a hard spherical bearing steel ball (AISI 52100, diameter of 25.4 mm and hardness of 818 HV) was rotated against the sample. The rotating ball is pressed into the sample from the top by test loads placed directly above the sample. The applied normal contact load was 0.9 N and the used ball speed was 0.33 ms^{-1} (250 RPM). Equation 1 was used to calculate the wear volume V, where d is the crater diameter and R is the ball radius [8].

$$V = \frac{\pi d^4}{64R} \tag{1}$$

The tests were carried out under dry sliding wear conditions to prevent any interfacial element from causing an influence. Features of the worn surfaces were assessed by using SEM. Chemical composition of the worn surfaces was examined by EDS. Transverse specimens were machined from different positions along the length of the DS casting. As such, tensile test samples according to specifications of ASTM Standard E 8M were prepared. A strain rate of about 3×10^{-3} s^{-1} was adopted.

3. Results and discussion

Characteristic microstructures of the examined alloy are shown in figure 1. The lighter areas in each micrograph are the α-Al phase, while the dark grey regions are the Si crystals forming eutectic structures. From the left to the right the images correspond to positions (P) with increasing distance from the chill interface, that is P = 2.5 mm, P = 10mm and P = 44 mm. The morphology of Si shifted from fibers to lamellae at approximately P = 3.0 mm. From the thermocouple data in our directional solidification system the liquidus cooling rates related to these positions were determined as being 21.6 K/s, 11.6 K/s and 1.1 K/s, respectively. At positions further from the chill interface (i.e., for lower cooling rates) it is worth noting that the size of the Si particles increases. This is thanks to the recurrent decrease of the liquidus growth velocity as a function of the length of the casting, i.e. 2.3 mm/s, 1.7 mm/s and 0.5 mm/s, respectively.

The SEM images in figure 2 were taken at a higher magnification and with the SEM in back-scatter electron (BSE) mode to show detailed aspects regarding either Bi or Si morphologies. The bright spots spread over the images in figure 2 are Bi particles. Bi addition remains in its elemental form within the alloys. Near globular Bi droplets prevail along the examined microstructures. The shape of the Si particles can be better evaluated through the deep etched SEM images in figure 2. From this visualization, the microstructures of the Al-15 wt%Si-1.0 wt%Bi alloy displayed a (αAl + Si) eutectic structure in association with Bi. Although Si morphology of the eutectic structure for the tested alloy was found to be predominantly lamellar across the casting length, higher cooling rates induced structural modifications in the eutectic Si phase, which transitioned to fibers for cooling rates higher than 22 K/s. This key value is roughly two-fold higher than that observed in the Al-15 wt%Si alloy, whose presence of fibers + globules was observed for rates exceeding 9 K/s during the formation of eutectic Si [14].

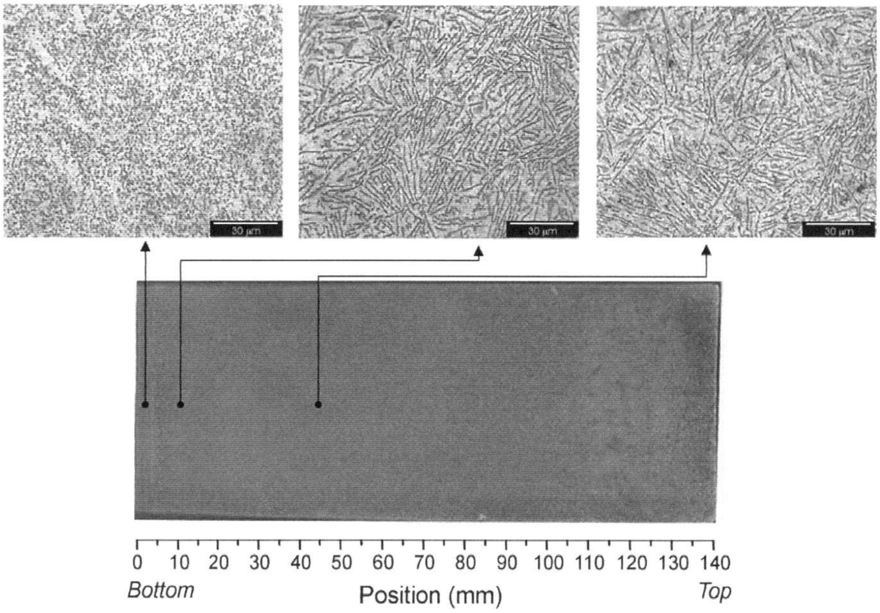

Figure 1. Directionally solidified macrograph of the Al-15 wt%Si-1.0 wt%Bi alloy with light micrographs related to specific positions along its length.

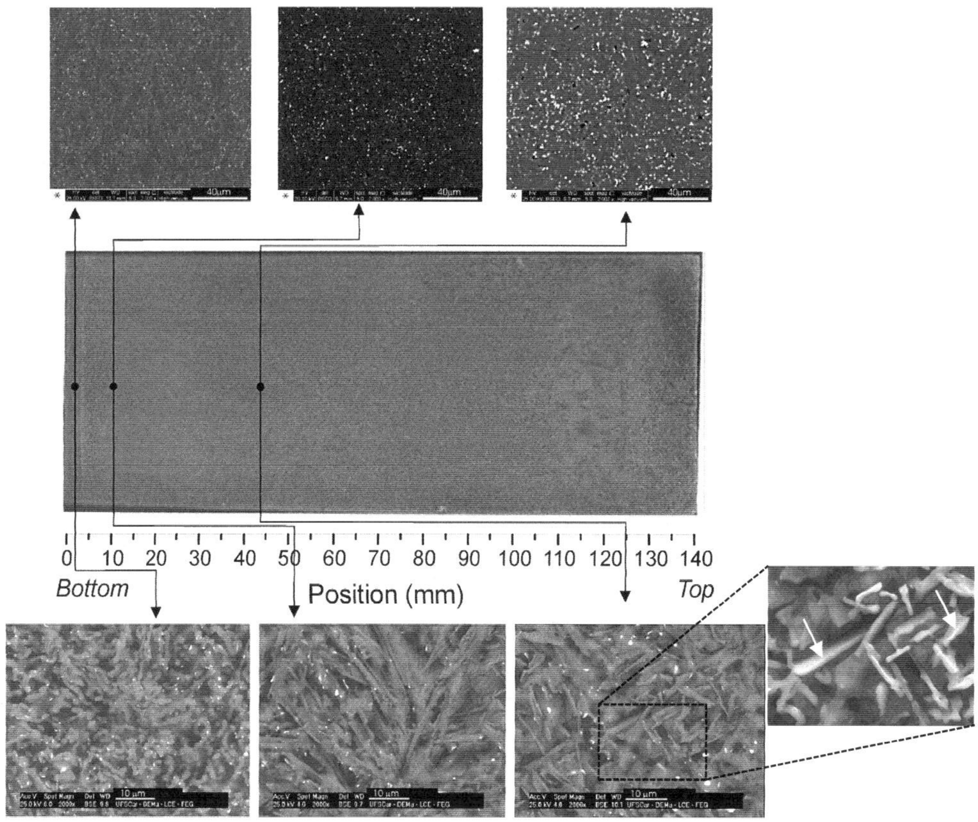

Figure 2. SEM images showing details of the microstructure found in the Al-15 wt%Si-1.0 wt%Bi alloy. The upper row shows the Bi droplets while the lower set of images highlights the Si particles.

IOP Conf. Series: Materials Science and Engineering **861** (2020) 012056 doi:10.1088/1757-899X/861/1/012056

Figure 3 shows the variations found for both λ_{Bi} and λ_{Si} along the casting length. For the variations of λ_{Bi} and λ_{Si} in the ternary alloy, single relationships following -1/2 function exponent for the eutectic velocity (V_E) are appropriate to represent the experimental data generated, as can be seen in figure 3 (a) and (b). The experimental expression reported by Reyes *et al.* [14] for the binary Al-15wt%Si alloy is $\lambda_{Si}=1.3(v_E)^{-1/2}$, which is very close to that found for the Bi-containing alloy in figure 3(a). It appears that doping with Bi is not able to change significantly the Si spacing. Both microstructure spacings, λ_{Bi} and λ_{Si}, vary with growth rate. The best distributions of either Si or Bi can be seen for higher growth velocities. This is explained due to the lower spacing values under such conditions.

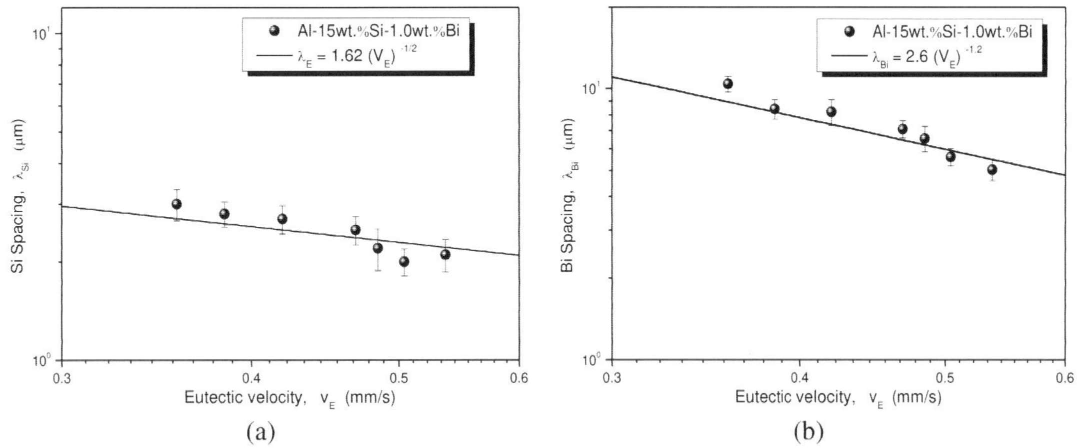

Figure 3. (a) λ_{Si} and (b) λ_{Bi} as a function of the eutectic growth velocity v_E.

Both Bi area fraction and Bi content increase with the transient upward movement of the solidification front while the Si content remains practically unaltered, as can be seen in figure 4. It seems that Bi is pushed to the Al/Si front during solidification enriching the interdendritic region with Bi. The accumulation of Bi promoted the development of perfectly rounded edges as shown by the arrows in the magnified inset image of figure 2. The complex skeleton arrangement is mainly formed by lamellar Si as also described by Farahany *et al.* [15].

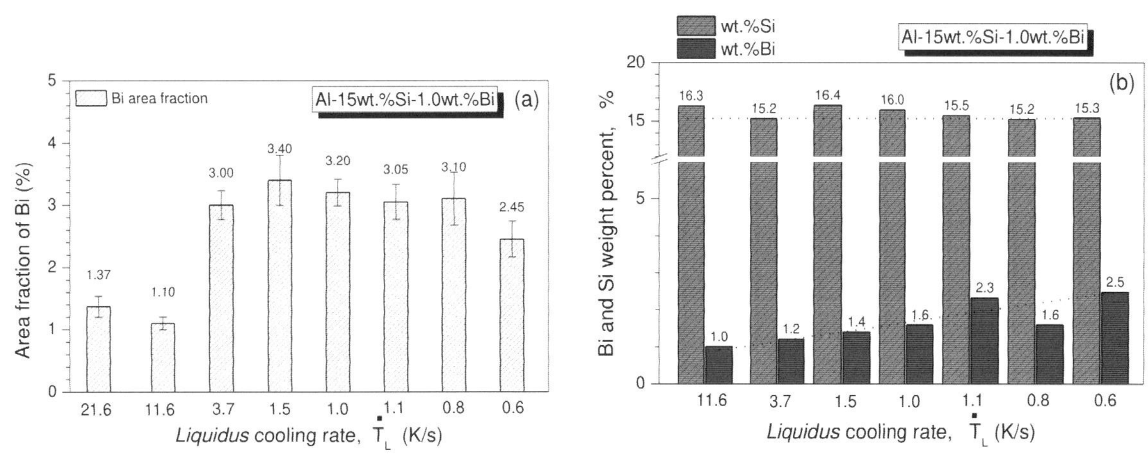

Figure 4. (a) Area fraction of Bi and (b) Bi / Si EDS macrosegregation profiles for the Al-15 wt.%Si-1.0 wt%Bi alloy.

As can be seen in figure 5, the ultimate tensile strength (σ_u) and the strain to failure (δ) vary along the length of the DS Al-Si-Bi alloy casting (square and circle points, respectively) with unaltered values for $\lambda_{Si}^{-1/2} < 0.6$. As previously demonstrated, the quantified aspects of the microstructure (λ_{Si}, λ_{Bi}) are related to the various cooling rates and growth rates determined for the solidified samples (see figure 3). Only lamellar Si samples were considered for tensile tests. This is because the region containing fibers was very confining to the very beginning of the casting (first 2.5 mm), which avoided its examination by using the current tensile test configurations.

Considering that the major microstructural characteristic constituting the samples is the spacing of the eutectic structure, λ_{Si}, a method was required to describe the scale variation analysis of the spacing affecting the strength and ductility. Therefore, the application of "Hall-Petch" relationships to compare the present results with the previous results for binary the Al-15 wt%Si was adopted [14].

In general, the tensile properties of the alloy containing Al-15 wt% Si -1.0 wt%Bi are higher. The behavior of the strength here is in agreement with that reported by Biswas [16]. This means that the addition of Bi in hypereutectic Al-Si alloys promotes increase in the alloy strength. The positive effect of Bi is very clear if one compares the Hall-Petch plots for ternary and binary compositions in figure 5. The prevalence of lamellar instead of flaky structure is the main reason for increasing strength.

The highest σu and δ of 195 MPa and 14 % for the ternary Al-15 wt%Si -1.0 wt%Bi alloy casting are associated with fine lamellar microstructural spacings. Better distribution of either Si or Bi resulted in augmentation of the tensile properties. The average Si spacing, λ_{Si}, related to those properties is 2.2 µm for the ternary alloy. In order to achieve the same properties for the same alloy without Bi addition a λ_{Si} of 1.3 µm is required. Furthermore, Bi inhibit the propagation of cracks through the grain boundaries of α-Al phase, reducing the premature fracture of the Si particles. This could improve the ductility.

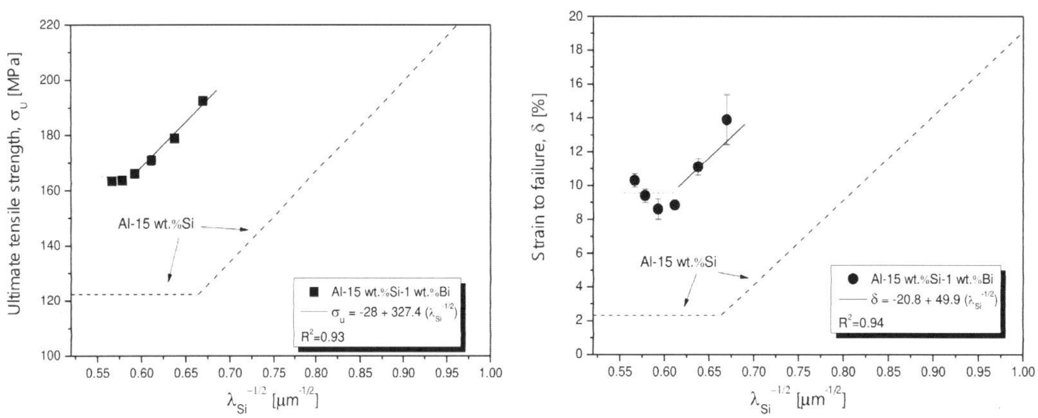

Figure 5. Evolutions of ultimate tensile strength and strain to failure along the length of the DS Al-15 wt%Si -1.0 wt%Bi alloy as a function of the inverse of the square root of the Si spacing, $\lambda_{Si}^{-1/2}$. R^2 is the coefficient of determination related to the adjusted curves.

Figure 6 shows a representative SEM image of the surface morphology after wear test. Two points were chosen to be examined by using SEM-EDS instrument. The Al/Si eutectic constituent (Spectrum #1) is subjected to distortion while the spread Bi-rich layers (brighter areas - Spectrum #2) behave as a solid lubricant on the friction surface. As the test develops, soft particles of Bi tend to form a solid lubricant layer on the sliding metal contact. However, the wear contact may change from adhesive to abrasive as soon as Fe and Al oxides form, as can be inferred examining the EDS data of the spectra. Fe comes from the used steel ball for the tests.

The experimental results of wear volume, V, as a function of λ_{Si} for both tested alloys can be seen in figure 7 for 30 minutes of wear tests. This time corresponds to a sliding distance of 598.2 m during the wear test of each sample. The points are experimental results of each alloy while the lines represent the best fittings to the experimental points.

The variations of wear volume can be considered significant in both cases with higher volumes being related to the alloy without Bi doping. For such range of microstructures, it can be seen that the wear volume, V, decreases with the decrease in λ_{Si}. A more homogeneous distribution of Si particles appears to be beneficial to the wear. It seems that the higher strength provided by the addition of Bi interferes in the wear behavior. The same was demonstrated by Kang *et al.* [17]. As a consequence, higher wear resistance is attained for Bi-modified samples. Furthermore, a more homogeneous distribution of Bi droplets in the microstructure is beneficial to lower the wear volume.

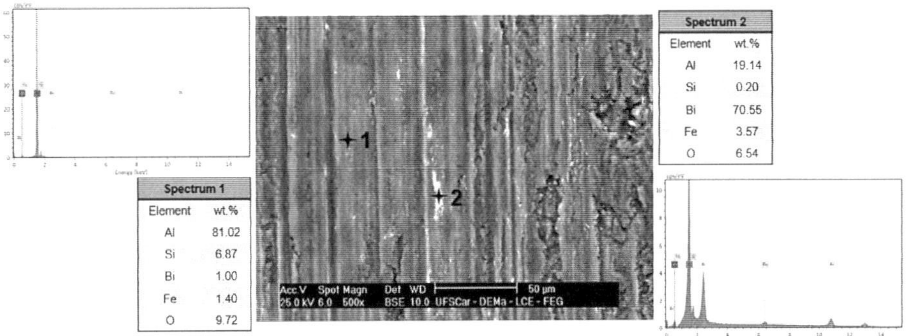

Figure 6. SEM image of the worn surface and EDS compositions of Bi-rich (Spectrum 2) and Al-rich (Spectrum 1) areas. It shows the action of the solid lubricant (brighter areas) which has been spread by the rotating ball.

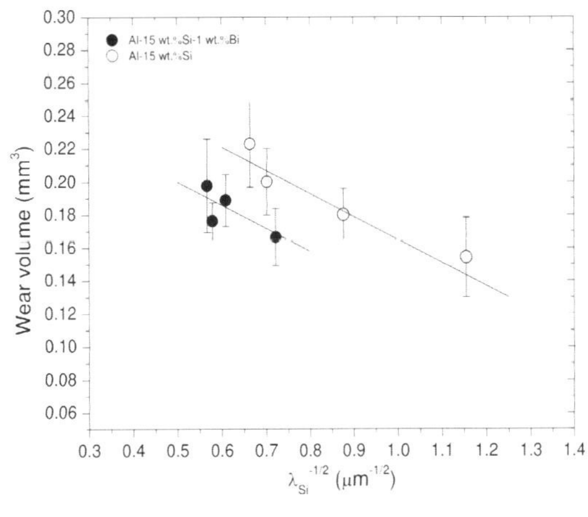

Figure 7. Experimental variation of wear volume with the Si spacing measured in the Al-15 wt%Si and Al-15 wt%Si-1 wt%Bi alloys samples for a test time of 30 minutes (i.e. 598.2 m).

4. Conclusions

After a complete characterization of the microstructure associated with the directional solidification of the ternary Al-15 wt%Si-1 wt%Bi alloy for a range of cooling rates, the following major conclusions were attained:

- The Si morphology of the eutectic mixture of the Bi containing alloy was found to be fine lamellar with particles having rounded edges. Furthermore, Bi segregates to the Al/Si front resulting in enrichment of the farther positions from the cooled surface of the casting.

- The addition of Bi in hypereutectic Al-Si alloy was shown to promote important increase in the alloy tensile properties as well as in the wear resistance.

Acknowledgements
The authors are grateful to FAPESP- São Paulo Research Foundation, Brazil (grant 2017/12741-6); Capes- Coordenação de Aperfeiçoamento de Pessoal de Nível Superior, Brazil (Funding Code 001); and CNPq- National Council for Scientific and Technological Development, Brazil.

References

[1] Cruz K S, Meza E S, Fernandes F A P, Quaresma J M V, Casteletti L C and Garcia A 2010 *Metall. Mater. Trans. A Phys. Metall. Mater. Sci.* **41** 972–84
[2] Jeon J H, Shin J H and Bae D H 2019 *Mater. Sci. Eng. A* **748** 367–70
[3] Lv G, Bao Y, Zhang Y, He Y, Ma W and Lei Y 2018 *Mater. Sci. Semicond. Process.* **81** 139–48
[4] Kamberovic, Z, Romhanji, E, Filipovic, M, Korac M 2009 *J. Metall.* **15** 189–200
[5] Miller W, Zhuang L, Bottema J, Wittebrood A, De Smet P, Haszler A and Vieregge A 2000 *Mater. Sci. Eng. A* **280** 37–49
[6] Freitas E S, Silva A P, Spinelli J E, Casteletti L C and Garcia A 2014 *Tribol. Lett.* **55** 111–20
[7] Silva A P, Spinelli J E and Garcia A 2009 Al–Bi alloys *J. Alloys Compd.* **480** 485–93
[8] Reyes R V., Pinotti V E, Afonso C R M, Casteletti L C, Garcia A and Spinelli J E 2019 *J. Mater. Eng. Perform.* **28** 1201–12
[9] Ratke L, Diefenbach S, Drees S, Alkemper J, Prinz B, Romero A and Ahlborn H 1995 *Adv. Sp. Res.* **16** 185–90
[10] Costa T A, Freitas E S, Dias M, Brito C, Cheung N and Garcia A 2015 *J. Alloys Compd.* **653** 243–54
[11] Canté M V., Spinelli J E, Cheung N and Garcia A 2010 *Met. Mater. Int.* **16** 39–49
[12] Brito C, Costa T A, Vida T A, Bertelli F, Cheung N, Spinelli J E and Garcia A 2015 *Metall. Mater. Trans. A* **46** 3342–55
[13] Gündüz M and Çadırlı E 2002 *Mater. Sci. Eng. A* **327** 167–85
[14] Reyes R V., Kakitani R, Costa T A, Spinelli J E, Cheung N and Garcia A 2016 *Philos. Mag. Lett.* **96** 228–37
[15] Farahany S, Ourdjini A, Bakar T A A and Idris M H 2014 *Metall. Mater. Trans. A* **45** 1085–8
[16] Biswas P, Prasadu K D and Mondal M K 2019 *Mater. Res. Express* **6** 1165b9
[17] Kang N, Coddet P, Liao H, Baur T and Coddet C 2016 *Appl. Surf. Sci.* **378** 142–9

IOP Conf. Series: Materials Science and Engineering **861** (2020) 012057 doi:10.1088/1757-899X/861/1/012057

Applications of pattern recognition for dendritic microstructures

J Strickland, B Nenchev and H B Dong

School of Engineering, University of Leicester, Leicester, LE1 7RH, UK

E-mail: jcjs2@leicester.ac.uk

Abstract. The Primary Dendrite Arm Spacing (PDAS) is the most important length scale in directionally solidified single crystal alloys. It determines the propensity for defect formation, solution heat treatment times and mechanical properties of the material. In this work a CMSX4 single crystal sample was imaged under a Scanning Electron Microscope (SEM). An automatic dendritic mapping (DenMap) algorithm using Normalised Cross-Correlation (NCC) is combined with Shape-Limited Primary Spacing (SLPS) to determine the local nearest neighbour dendrites and the corresponding dendritic packing. The algorithm located the dendritic centres, calculated the local PDAS, packing pattern, and relationship between PDAS and packing pattern for 256 dendrites in 1 minute 10 seconds. This is the first fully automatic method to produce a clear Gaussian distribution of local PDAS and packing pattern; thus, enabling rapid data gathering potential for single-crystal microstructures.

1. Introduction

Dendritic structures are one of the most complex forms of crystallisation in nature and technology. The primary dendrite arm spacing (PDAS) is known to control the segregation properties [1,2], which controls the chances for defect formation [3] and the mechanical properties of the material [4].

It is of vital importance to be able to characterise this length scale in a way that is repeatable, accurate and rapid. Current methods of determining mean primary spacing utilise the counting method [5-7]:

$$\bar{\lambda} = B\sqrt{\frac{A}{N_d}} \tag{1}$$

where A is area, N_d is number of dendrites within that area and B is a fitting coefficient for the array. However, this method provides no way of determining the local PDAS or the PDAS distribution. Other authors have attempted to characterise the local PDAS [8-12]. Warnken and Reed [12] developed an algorithm based on calculating the closest three neighbouring dendrites then iteratively adding neighbours. Their upper limit was determined by first calculating the local PDAS then adding the standard deviation of PDAS multiplied by a statistically determined parameter, α. No physical basis for α was provided, only that values between 1.4 and 1.6 yielded reasonable results when analysing directionally solidified alloys. However, local spacing outliers inflate the standard deviation, causing the algorithm to continue adding neighbours beyond the true nearest neighbours (TNN) with no upper limit in place. Since dendrites are only influenced by the local diffusion fields of their true neighbours, an accurate description of the effect of the local solidification conditions on the microstructure is fruitless without first identifying accurately the TNN. Tschopp *et al.* [8] modified the Warnken-Reed technique by applying Voronoi tessellation which provided a method to limit the maximum nearest neighbours, however, this provided no means to determine true nearest neighbours.. Tschopp *et al.* [13] attempted to automate the dendritic core selection process in optical images using a four-fold symmetry algorithm. However, this method required removing of background and manual thresholding, returned only an 87.7% detection accuracy, and required extensive computational time.

In this work, we utilise a Normalised Cross-Correlation technique [14] to detect dendritic cores within a single crystal CMSX4 sample, then apply Voronoi tessellation combined with a novel Shape-Limited Primary Spacing (SLPS) algorithm to calculate true local nearest neighbours. Thereby creating a map of packing pattern and local PDAS distribution over the sample surface.

2. Application of pattern recognition

A 2nd generation CMSX4 single crystal bar sample, with a diameter of 9.4mm, was provided by Rolls-Royce Plc. The bar was sectioned using normal metallographic techniques and imaged on a FEI 650 Quanta SEM at the University of Leicester.

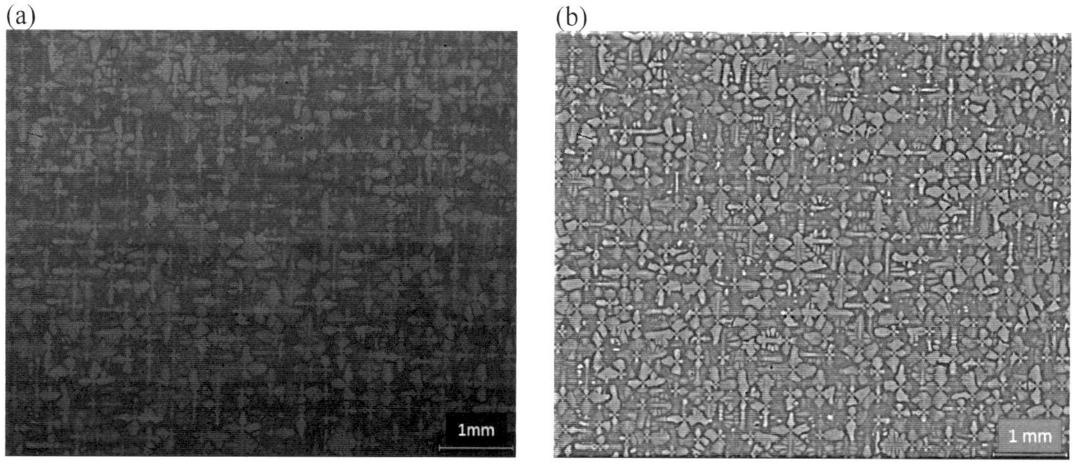

Figure 1. (a) Backscattered electron image of CMSX4 ni-base superalloy, (b) CMSX4 imaged with DenMap, dendritic centres shown as red rots. 256 dendrites are determined.

Dendritic core detection classically requires time intensive laborious manual determination of the dendrite cores. In this work, automatic, rapid and accurate core detection is achieved using a Normalised Cross Correlation algorithm (NCC), DenMap. First the image is pre-processed first with a Fast Fourier Transform Band Pass Filter to suppress or enhance unwanted features followed by a transformation back into real space using an inverse Fourier transform. This step is followed by histogram equalisation to remove blurring and shift the contrast to sharpen all the features, especially the dendritic core. Once image pre-processing is completed, scale and rotation are calculated to generate a template image. The template image is inputted into NCC and scanned across the sample surface. NCC provides a measure of similarity between the two data sets, with outputs being between 1 and -1, 1 being an exact match with the template. Following this, automatic thresholding is applied, with the maximum value of the NCC algorithm corresponding to the exact location of the dendritic core, as shown in figure 1(b). The total processing time for dendritic core determination from figure 1(a) is exactly one minute, and achieved 100% detection. Now a method to determine the local PDAS is required.

First the x and y pixel positions are sorted in a clockwise direction, the first data point taken from the top left of the image and the last data point the centre. Then the same method as Tschopp *et al.* [8] is applied, whereby, the image is split by Delaunay triangulation. Each dendrite has a corresponding region where all points of the plane are closer to that dendrite than to any other. To overcome the problems associated with determining the true nearest neighbours, a Shape-Limited Primary Spacing (SLPS) algorithm is developed, derived from observed single-crystal packing patterns in nature. The method suggested herein applies the Modified-Warnken-Reed method as suggested by Tschopp *et al.* [8], however, the coefficient alpha is replaced with a SLPS term, K_{SLPS} (equation (2)). The purpose of K_{SLPS} is to permit the algorithm to automatically remove any incorrectly captured FNN from Voronoi tessellation; thus, providing an accurate description of the influence of the solidification conditions on the local dendritic neighbourhood. K_{SLPS} is geometrically derived from the square counting method, equation (1), combined with the formula for a regular polygon as

$$K_{SLPS} = \frac{2\lambda}{a}\sqrt{\tan\left(\frac{\pi}{N}\right)} \tag{2}$$

where a is side length, λ is PDAS and N is coordination number. The N is determined from the number of local dendrites bound to a central dendrite. In order to set systematic rules and comparatively relate packing of dendrites of different N, each central dendrite is bound by a regular polygon, with a fixed λ from the centre to a neighbouring dendrite. With a unique K_{SLPS} determined for each N it is now possible to determine true nearest neighbours and thus the local PDAS distribution. K_{SLPS} is applied to the Modified-Warnken-Reed method [10] (equation 3)

$$R_{SLPS} = \bar{\lambda}_{Local} + K_{SLPS}\lambda_{\sigma_{Array}} \tag{3}$$

where, $\bar{\lambda}_{Local}$ is the average local PDAS, $\lambda_{\sigma Array}$ is the PDAS standard deviation for the whole array and R_{SLPS} is the radius of the shape limiting circle about the central dendrite. First the nearest neighbours are determined by Voronoi tessellation and then R_{SLPS} is calculated. If a dendrite is outside of R_{SLPS} then the algorithm removes the outlier and recomputes based on the new determined nearest neighbours and updated K_{SLPS}. The algorithm follows the same process for each point in the dataset. A full description of the method will be published in the future.

3. Application of pattern recognition

The total processing time for dendritic core determination combined with SLPS for the CMSX4 micrograph with 256 dendrites is exactly one minute and 10 seconds on an Intel® Core™ i5-6500 Desktop PC with a 3.2 GHz CPU and 8GB RAM. The results from SLPS are shown in figure 2.

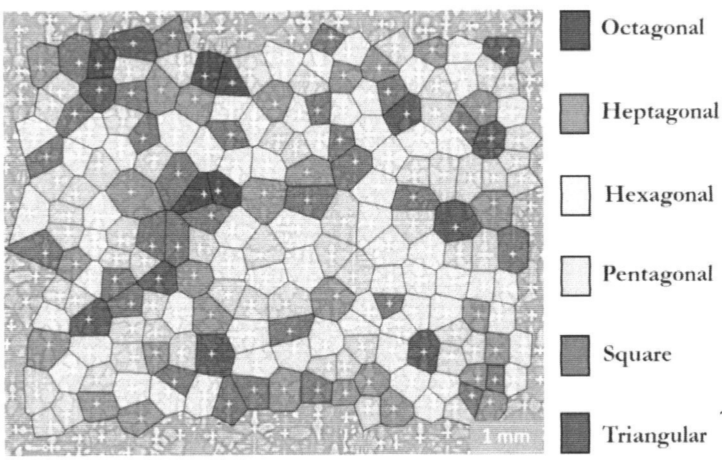

Figure 2. SLPS algorithm applied to CMSX4 single-crystal microstructure. 241 Voronoi shapes are calculated using SLPS.

The results from the application of shape-limited primary spacing (SLPS) to the CMSX4 superalloy single-crystal microstructure (figure 1) are shown in figure 2. A variety of packing formations have been determined within the microstructure. The results demonstrate combined sections of hexagonal packing (N_6) on the right hand side of figure 2 and a less homogenous distribution of triangular (N_3), square (N_4) and pentagonal (N_5) packing on the left.

The local PDAS variation results are shown in figure 3(a). The algorithm captures a Gaussian distribution of local PDAS and determines the $\bar{\lambda}$ as 419 μm. The classical counting method (equation (1)) calculated the $\bar{\lambda}$ as 423 μm using a B coefficient (equation (1)) equal to 1.075. This small discrepancy of 4 μm is likely due to taking $B = 1.075$ for an ideally packed hexagonal array [6]. The smooth Gaussian distribution and the close match with the classical method implies SLPS can successfully remove incorrectly captured first nearest neighbour dendrites included in the Voronoi tessellation. Therefore, the algorithm can successfully capture the influence of solidification conditions on the local dendritic neighbourhood. As a result, the model can provide new information in regard to the types and quantity packing patterns within figure 2. It should be clarified that a sample section provides only a snapshot of solidification at a moment in time and packing patterns may change when viewed in 3D due to a change in solidification conditions [11].

Figure 3(b) quantifies the frequency and types of packing pattern determined from figure 2. For the array studied here, the largest frequency of detected packing patterns are pentagonal (N_5) and hexagonal (N_6). The lowest frequency packed dendrites are the triangular (N_3) and octagonal (N_8) arrangements.

a) b)

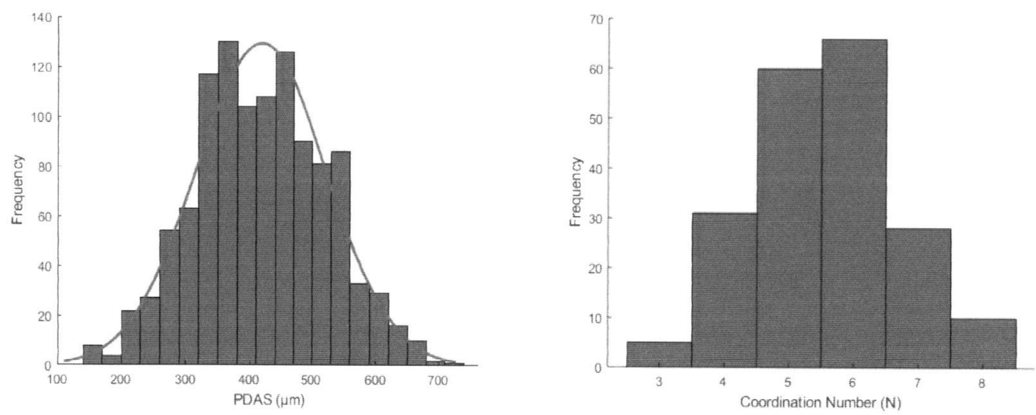

Figure 3. (a) PDAS vs frequency for CMSX4 superalloy, $\bar{\lambda} = 419$ μm, (b) histogram of total packing patterns within the CMSX4 dendritic array.

The relationship between PDAS and determined packing pattern is shown in figure 4. The N_7 and N_8 packing patterns identify shapes where the local PDAS is large in comparison to the spacing between true nearest neighbour dendrites (figure 4). Large PDAS indicate the interdendritic regions, as Ni, Ti, Ta, and Al preferentially partition to these regions, high coordination number shapes indicate areas rich in these elements. These high coordination numbers are typically surrounded by N_3, N_4 and N_5 packing patterns with small local PDAS (figure 2). As Re, W, Co and Mo partition to the dendritic core, areas with increased core density indicate regions with a high concentration in these elements. As N_3 and N_8 possess the smallest and largest PDAS on average, they are related to the areas with the most severe microsegregation levels. As the N_6 patterns are between the two segregation extremes they form under a composition closer to nominal. Interestingly, the $\bar{\lambda}$ for the N_6 patterns is 430 μm which is the closest out all determined patterns to the $\bar{\lambda}$ for the bulk array. One could draw the conclusion that as the N_6 pattern frequency is the highest (figure 3(b)), forms under a composition closer to nominal and the local and bulk $\bar{\lambda}$ are similar, that it could be the preferential packing formation within a single-crystal array. This conclusion is supported by what is observed within experiment [15].

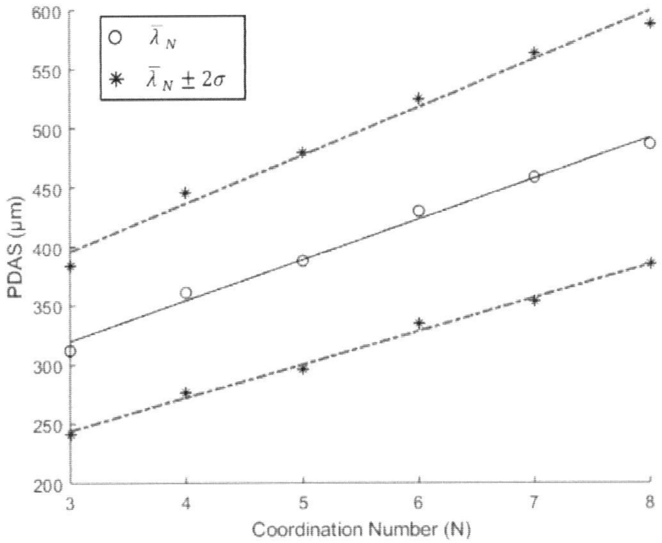

Figure 4. Line Graph of mean PDAS, $\bar{\lambda}$, and standard deviation, 2σ, is plotted for each coordination number N. The N_3, N_4 and N_5 shapes demonstrate smaller average PDAS and are related to the preferential segregation of Re, W, Co and Mo. The N_7 and N_8 shapes with larger average PDAS are related to the preferential segregation of Ni, Ti, Ta, and Al.

4. Conclusions

DenMap is developed to automatically detect dendritic cores using Normalised Cross-Correlation. Auto detection is combined with a dendritic pattern detection algorithm, Shape-Limited Primary Spacing (SLPS). SLPS calculated the local PDAS distribution in a CMSX4 alloy with 256 dendrites in 1 minute and 10 seconds. The model provides a Gaussian distribution of local PDAS variation and removes any large outliers. The technique can quantify the relationship between packing patterns and microsegregation and the results suggest single-crystal arrays have a natural tendency to form hexagonally packed structures.

DenMap-SLPS enables the rapid, accurate and quantitive data gathering. Consequently, this new model enables the standardisation of single-crystal characterisation and confidence in comparison between different experimental results. It follows, that the new approach will elucidate a deeper understanding between single-crystal formation and the processing parameters.

Acknowledgements

We gratefully acknowledge the financial support from the Centre for Doctoral Training in Innovative Metal Processing (IMPaCT) funded by the UK Engineering and Physical Sciences Research Council (EPRSC) and Rolls-Royce Plc for providing financial support and supplying the single-crystal CMSX4 samples.

References

[1] Matache G, Stefanescu D, Puscasu C and Alexandrescu E 2016 *Int. J. Cast Metal Res.* **29** 303-16
[2] Giamei A F and Kear B 1970 *Metall. Trans.* **1** 2185-92
[3] Ridolfi M R, Tassa O and De Rosa G 2017 *Mater. Sci. Forum* **879** 1582-7
[4] Santos G, Goulart P R, Couto A A and Garcia A 2017 *Properties and Characterization of Modern Materials* vol 33 ed Öchsner A and Altenbach H (Singapore: Springer) pp 215-29

[5] McCartney D and Hunt J 1981 *Acta Metall.***29** 1851-63
[6] Flemings M C 1974 *Metall. Trans.* **5** 2121-34
[7] Jacobi H and Schwerdtfeger K 1976 *Metall. Trans. A* **7** 811-20
[8] Tschopp M A, Miller J D, Oppedal A L and Solanki K N 2014 *Metall. Mater. Trans. A* **45** 426-37
[9] Tschopp M A, Miller J D, Oppedal A L and Solanki K N 2015 *Metall. Mater. Trans. A* **46** 4610-28
[10] Tiryakioğlu M 2019 *Mater. Sci. Tech.* **35** 509-11
[11] Takaki T, Sakane S, Ohno M, Shibuta Y, Shimokawabe T and Aoki T 2016 *Acta Mater.* **118** 230-43
[12] Warnken N and Reed RC 2011 *Metall. Mater. Trans. A* **42** 1675-83
[13] Tschopp M, Groeber M, Fahringer R, Simmons J, Rosenberger A and Woodward C 2010 *Scr. Mater.* **62** 357-60
[14] Luo J and Konofagou E E 2010 *IEEE T. Ultrason. Ferr.* **57** 1347-57
[15] Tewari S N, Weng Y-H, Ding G and Trivedi R 2002 *Metall. Mater Trans A* **33** 1229-43

IOP Conf. Series: Materials Science and Engineering **861** (2020) 012058 doi:10.1088/1757-899X/861/1/012058

Modelling of intermetallic layers formation during solid-liquid joining of dissimilar metallic materials

V Glück Nardi[1], T Greß[2], B Tonn[1,3] and W Volk[2]

[1] Institute of Metallurgy, Clausthal University of Technology, Robert-Koch-Str. 42, 38678, Clausthal-Zellerfeld, Germany
[2] Chair of Metal Forming and Casting, Technical University of Munich, Walther-Meißner-Str. 4, 85748, Garching, Germany
[3] Simulation Science Center Clausthal / Göttingen, Clausthal University of Technology, Arnold-Sommerfeld-Straße 6, 38678 Clausthal-Zellerfeld, Germany

E-mail: vanessa.glueck.nardi@tu-clausthal.de

Abstract. Joining of dissimilar metallic materials gives rise to the production of components with improved properties. At the joint interface, the formation of brittle intermetallic layers might deteriorate the component application. Therefore, its formation and growth must be well understood and carefully controlled. Numerical simulations assist in the determination of processes parameters to obtain the desired bond characteristics. In solid-liquid joining processes, the melt movement plays an important role und must be accounted in the simulations. In this paper, a mathematical model at the continuum scale is used for the description of the interface formation during solid-liquid joining. The description of convection and advection of the molten alloy and the dissolution of the solid substrate are of interest. The substrate dissolution into the melt alters the local composition of the molten alloy, leading to precipitation of intermetallic compounds on the substrate surface during the solidification process. To account for these phenomena, convection heat transfer with phase transformation and diffusion and reaction at the solid-liquid interface are considered in the mathematical model that is based on conservation equations of mass, momentum, energy and species. The different scale problem is solved by applying a dynamic mesh refinement at the interface .to accurately solve the intermetallic layer growth. The model was implemented in OpenFoam® and simulation results for the intermetallic layer thickness as a function of solid substrate temperature (*Ts*) for compound casting between initially liquid Al and a solid Cu substrate are presented. In general, for *Ts* initial value ranging from 400 to 700 °C while keeping the initial Al melt temperate at 750 °C, the intermetallic layers thickness increases as *Ts* increases.

1. Introduction

The use of materials with improved properties is essential to attend the technological and industrial advances. However, the material may not have the desired properties to perform a specific function on its own. Therefore, the possibility of joining different materials to obtain a component with better mechanical, physical and chemical properties can overcome this problem. Depending on the required application of the component, the combination of different materials can be achieved by different techniques, such as coating, welding or casting, among others. The interest of this paper is on the joining of dissimilar materials by compound casting. In general, it consists of liquid metal being cast onto a solid metallic substrate. The high temperatures promote the dissolution of the solid substrate into the melt that starts to solidify. Under the right process parameters, a controlled dissolution of the solid material into the melt occurs and a sound metallurgical bond is achieved. However, if the materials are not completely soluble in each other, intermetallic compound layers (ICL) may form in between, weakening the bond and, consequently, the performance of the component. Over the years, the compound casting has become sophisticated and new challenges have been faced, such as its application in continuous casting [1, 2], which makes this technique economically competitive.

Much effort is put into determining the correlation between process parameters and formation and growth of ICL so that an optimum metallurgical bond for a given component can always be achieved. In most cases, this optimization is done via several trial-and-error experimental investigations and researches focus on the characterization of the interface, while its formation mechanism is still controversial. To overcome the necessity of expensive experimental work and to help to understand the interface formation, numerical simulations are performed. Nevertheless, numerical techniques have also been improved over the years to better describe the physical phenomena and to help to define the optimal process parameters for general metallurgical processes. One of the applications of numerical modelling is to help understand the behaviour that is not yet fully understood through an analysis of experiments, such as the formation mechanism of ICL during solid-liquid joining, as occurs in compound casting. Among the physical phenomena associated with compound casting, convection heat transfer with phase transformation and diffusion-reaction at the interface between materials are highlighted, which makes a comprehensive modelling of this complex phenomena challenging.

The modelling of compound casting resembles the modelling of solidification problems with the addition of interfacial diffusion and reaction effects. The necessity of realistic models applicable to high scale processes is therefore of interest. Depending on the degree of detail that is desired, the problem can be studied in different scales. In order to understand the application of different scales for the study of solidification, the reader is referred to the work of Stefanescu [3]. In this article, we focus on the formation and growth of ICL and how it affects and is affected by the solidification process during the compound casting.

In the literature, dissimilar joint problems are approached in different scales, such as molecular simulations [4] or mesoscale simulations, where usually phase-field techniques [5–7] or lattice-Boltzmann simulations [8] are applied. These methods, however, lack in providing the correlation between process parameter and the ICL in the compound casting framework due to the usual very small numerical domain. Thus, its applicability to industrial scale is impractical. In a trial to overcome these problems, we present a continuum approach for the compound casting with the prediction of thickness and composition of the ICL. For instance, the conventional conservation equations are applied and a local dynamic mesh refinement at the interface is used to properly solve the diffusion and reaction problem.

The modelling of formation and growth of intermetallic phases at the solid-liquid interface during solidification based on a continuum-scale requires a deep understanding of the relevant physical mechanisms in order to be described into model equations. Different methods must be combined to obtain a comprehensive model for ICL formation and growth. These methods involve thermodynamic modelling, reaction kinetics, heat transfer and fluid flow. This multifaceted approach is presented in this paper within the mathematical description. The current model is, however, a simplified approach to assist in the determination of input parameters, such as pouring temperature and preheating of the substrate, for the compound casting process. Figure 1 schematically shows the problem address in this paper. The solid-liquid diffusion and reaction are coupled with the enthalpy-porosity method to simulate the compound casting process. For this, the complex phenomena observed in the process are translated into a simplified mathematical model that is based on conservation equations of mass, momentum, energy, and species. The model description is an adaption from [9].

2. Mathematical model

The compound casting is physically understood as a solidification process with diffusion and reaction taking place at the interface between both base materials. For the modelling of the solidification process, the enthalpy-porosity method was used to track the solid-liquid interface [10]. To track the formation and growth of the ICL, a source term is added to the species transport equation. The equations are presented under the assumption of laminar Newtonian fluid flow in Cartesian coordinates. To comprehension of the model, the solidification modelling and the formation and growth of ICL modelling are discussed separately, as follows.

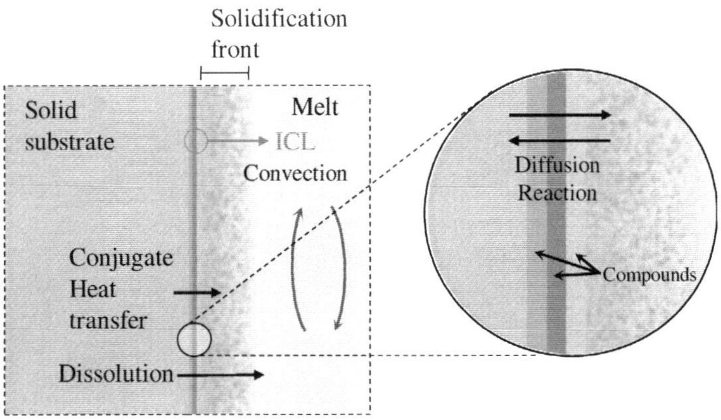

Figure 1. Schematic compound casting problem. Substrate dissolution at the moment melt wets the solid surface and diffusion and reaction are considered under the influence of heat transfer and fluid flow.

2.1. Solidification modelling

In the enthalpy-porosity formulation, it is considered that the solidified structures adjacent to the solid-liquid interface form a mushy zone that is represented as a porous medium, where the porosity of the medium is given by the liquid fraction (f_l) and it decreases as solidification progresses. The governing equations are the conservation of mass, momentum and energy, respectively given by [11]:

$$\frac{\partial(\rho)}{\partial t}+\nabla\cdot(\rho\boldsymbol{u})=0 \tag{1}$$

$$\frac{\partial(\rho\boldsymbol{u})}{\partial t}+\nabla\cdot(\rho\boldsymbol{uu})=-\nabla p+\nabla\cdot(\mu\nabla\boldsymbol{u})+\boldsymbol{b}+\boldsymbol{S_M} \tag{2}$$

$$\frac{\partial(\rho h)}{\partial t}+\nabla\cdot(h\rho\mathbf{u})=\nabla\cdot(k\nabla T)+\boldsymbol{S_H} \tag{3}$$

where ρ is the density, $\boldsymbol{u}$ is the velocity field, μ is the viscosity, h is the enthalpy, T is the temperature and $\boldsymbol{S_M}$ and $\boldsymbol{S_H}$ are source terms that include the effects of solidification in the model. The natural convection effect is accounted for in the body force term, $\boldsymbol{b}$, in which the Oberbeck-Boussinesq approximation is applied [12], given by:

$$\boldsymbol{b}=\rho\boldsymbol{g}\left[1\text{-}\beta\left(T\text{-}T_{ref}\right)\right] \tag{4}$$

where $\boldsymbol{g}$ is the gravity vector, β is the thermal expansion coefficient and T_{ref} is the reference temperature.

$\boldsymbol{S_M}$ is added in the momentum equation, equation (2), to account for the fluid velocity drop due to the decrease of liquid fraction during the evolution of the mushy zone and it is given by:

$$\boldsymbol{S_M}=\frac{\mu}{K}\boldsymbol{u} \tag{5}$$

where K is the permeability of the porous media, formulated here from the Karman-Cozeny equations, as follows:

$$K=B\frac{(1\text{-}f_s)^3}{f_s^{\,2}+\varepsilon} \tag{6}$$

where B is a constant that describes the morphology of the porous medium, ε is a small number to avoid division by zero and f_s is the solid fraction, related with f_l by equation (7). It can be seen from equations (5) and (6) that $\boldsymbol{S_M}$ is modelled so that tends to zero in the liquid region ($f_s = 0$), while becoming dominant in the mushy zone ($0 < f_s < 1$), in such a way that the momentum equation approaches D'Arcy's law, which is the equation that governs fluid flow in porous media. Because of its function, $\boldsymbol{S_M}$ is also called D'Arcy source term [13].

$$f_s = 1 - f_l \tag{7}$$

In the energy equation, conveniently written in the enthalpy form [14], the added source term $\boldsymbol{S_H}$, represents the release of latent heat during solidification, given by:

$$\boldsymbol{S_H} = \frac{\partial(L\rho f_s)}{\partial t} + \nabla\cdot(L\rho \boldsymbol{u} f_s) \tag{8}$$

where L is the latent heat.

To determine the evolution of the mushy zone the solid fraction must be calculated. In the current model, the solid fraction is determined with the help of the software Pandat, using Scheil-Gulliver calculations, as described in [15]. The results depend on the local temperature and solute distribution in the mushy zone. The composition distribution near the interface is highly affected by the dissolution of the substrate material into the melt and must be considered. Therefore, the redistribution of the chemical solution near the interface is accounted in the model of ICL formation and growth, described below, while details for the enthalpy-porosity method can be found in [10] and [14].

2.2. Intermetallic compound layer formation and growth modelling

The chemical composition is a compulsory quantity to predict not just the solid fraction required in the solidification model, but it is also an input parameter to determine if a reaction might take place for the formation of a new compound. The local mass fraction of each species, Y_i, evolved in the system is predicted within the conservation of species equation [11], given by:

$$\frac{\partial(\rho Y_i)}{\partial t} + \nabla\cdot(\rho \mathbf{u} Y_i) = -\nabla\cdot(\boldsymbol{J}_i) + R_i \tag{9}$$

where $\boldsymbol{J}_i$ is the diffusion flux of species i, and R_i is the reacting source term. In a binary diffusion system, $\boldsymbol{J}_i$ can be related to the species concentration using Fick's first law [16]. However, for a multi-component mixture, the Maxwell-Stefan equations are recommended [11]. Regardless, the diffusion flux depends on the diffusion coefficient, D. This is an important parameter that describes the diffusivity of one species into another and it is usually determined via experimental analyses. It is important to note that the summation of equation (9) over all the i elements recovers equation (1), which in the case is the equation of continuity for the mixture [11].

The formation and growth of a compound are considered in equation (9) in the reaction source term R_i, which for a non-reversible reaction is given by [17]:

$$R_i = M_{w,i} \sum_{r=1}^{N_R} \left(v''_{i,r} - v'_{i,r}\right) k_{f,r} \prod_{j=1}^{N_R} \left[C_{j,r}\right]^{\eta'_{j,r}} \tag{10}$$

where $M_{w,i}$ is the molecular weight of species i, N_R is the number of reactions that the species participate in. $v''_{i,r}$, $v'_{i,r}$, are the stoichiometric coefficients for the reactant and the product of the i specie, M_i, in the reaction r, as generally given by equation (12) [17]. $k_{f,r}$ is the forward reaction rate constant, computed using the Arrhenius expression, given by equation (11).

$$k_{f,r} = A_r T^{\beta_r} \exp\left(-\frac{E_r}{RT}\right) \tag{11}$$

$$\sum_{i=1}^{N_R} v'_{i,r} M_i \underset{k_{f,r}}{\rightarrow} \sum_{i=1}^{N_R} v''_{i,r} M_i \tag{12}$$

where A_r is the pre-exponential factor, β_r is the temperature power exponent, E_r is the activation energy, and R is the universal gas constant.

In metallurgy, the phases to precipitate are closely related to the phase diagram of the representative components of the liquid melt and solid substrate, as demonstrate in [18]. As an example, figure 2 shows the Al - Cu binary phase diagram with a diffusion scheme and the specific reactions written in the form of equation (12). In the present model, the precipitation and evolution of each phase is tracked by equation (9).

2.3. Coupling between models

Although the ICL modelling and the solidification modelling were presented separately, they are naturally coupled together. The composition information obtained with the ICL model is used as input to determine the solid fraction in the solidification model, while the velocity and temperature fields calculate in the solidification model are used as input in the species transport equation. The interface between materials is coupled considering that at the interface between the solid and the liquid/semi-solid, the temperature must be the same, as well as the heat flow between them, which are usual boundary condition applied in conjugate heat transfer problems [14, 19] . At the solid state, the diffusion reaction problem is simplified to a one-dimentional problem.

3. Numerical implementation

For the numerical simulation, the conservation equations were discretized using the finite volume method. The gradient compilation was done via least squares method. For the approximation of advective terms, the QUICK scheme was applied, while the discretization of diffusion fluxes was done via Gauss method with linear interpolation scheme. A first order implicit scheme was applied for time discretization.

The implementation was done using OpenFOAM® environment, as described in [9]. Due the very small scale needed to solve the ICL model, mesh refinement is applied to achieve greater accuracy at the interface to evaluate the different compound layers. In this manner, the size of the most refined volume at the interface reaches the order of 0.1 μm.

4. Model application to Cu-Al compound casting

The model presented in this paper was applied to determine the ICL between Cu and Al. The problem consists of pouring Al melt over a solid Cu substrate. Dissolution from Cu substrate surface into the Al melt and the precipitation reaction of Cu-Al compounds are simulated. In the continuum modelling, the composition profile cannot be calculated, unless all the diffusion coefficients are known as a function of temperature and concentration. For the present work, we consider the diffusion of Al and Cu between each other and within the compounds. The diffusion of Al melt into the solid Cu substrate is neglected.

Another important simplification is regarding the order of phase's appearance. Following [18], the phases to appear are considered as in the phase diagram, going from the solid substrate in the direction of higher temperatures. Based on these assumptions, the reactions are listed in figure 2(c). They are based on the phase diagram of figure 2(a). Since the δ_1 phase is usually not observed in experimental results, [20, 21] it was omitted in the calculation for simplification. Also, adjacent the Al_2Cu layer, a eutectic layer composed of Al_2Cu and Al solid solution is usually observed [20]. The model, however, does not consider the eutectic reaction.

For the simulation, diffusion coefficients and activation energies were taken from the literature. The values not available were determined experimentally as described in [9]. The simulation boundary conditions and geometry resembles the experimental work described in [21]. The influence of the initial temperature of the solid substrate is investigated. The liquid Al initial temperature is kept at 750 °C while the solid substrate temperature varies from 400 to 700 °C.

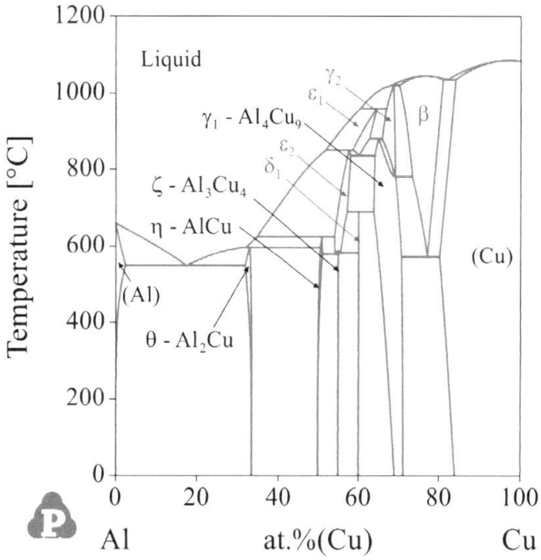

(a)

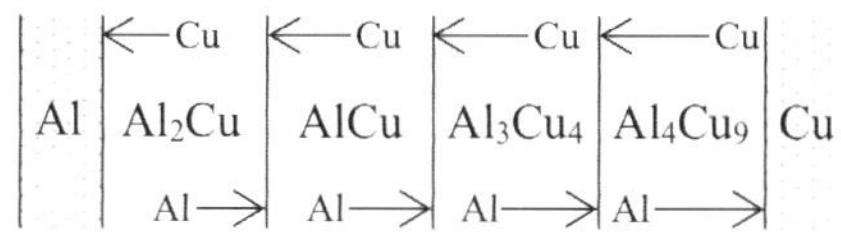

(b)

$$9Cu+4Al \rightarrow Al_4Cu_9$$
$$11Al+4Al_4Cu_9 \rightarrow 9Al_3Cu_4$$
$$1Al+1Al_3Cu_4 \rightarrow 4AlCu$$
$$Al+AlCu \rightarrow Al_2Cu$$

(c)

Figure 2. (a) Al-Cu binary phase diagram and (b) a diffusion scheme and (c) the specific reactions.

4.1. Results

The phases on the interface are expressed in terms of their volume fraction, φ_i. Figure 3 shows the result of φ_i's profile on the interface obtained for Ts = 600 °C. The curves in figure 3 overlap, so the diffusion and reaction model used does not reproduce a clear interface between the precipitated phases. To obtain a quantitative result, the length of a phase was determined considering only the length where this phase has its largest volume fraction, as exemplified in figure 3 for the AlCu phase.

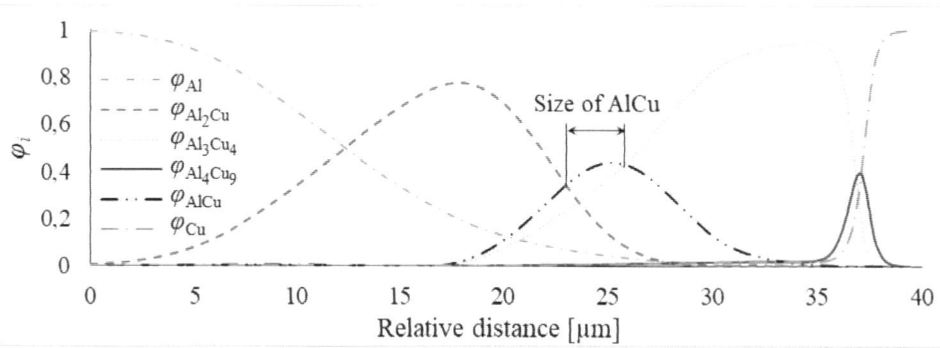

Figure 3. Volume fraction of the phases at the interface and representation of the calculated phase size.

For the reaction order considered in figure 2(c), first Al_4Cu_9 is formed at the interface and it is rapid consumed for the formation of Al_3Cu_4. It is considered then, that as soon as Al_3Cu_4 forms, it reacts with Al and forms AlCu phase. With the formation of AlCu, the Al_2Cu phase starts to form. This reaction order is a significant simplification to the model. However, it proved to be enough to determine the size of each intermetallic layer.

The model was validated using experimental data from the literature, as shown in table 1. The numerical results are within the value range of the experimental results. This shows that the model applied, and the interpretation of the results presented here, are quantitively comparable with experimental data. Simulations for initial substrate temperature, *Ts*, higher them the pouring temperature, *Tp*, were unstable and no convergence was achieved. The model is considered successful for *Ts* < *Tp*.

Table 1. Numerical results for intermetallic layers size obtained for comparison with experimental data from the literature.

Phase	Phase length [μm] *Tp* = 800 °C / *Ts* = 506°C	
	Numerical results (present work)	Experimental [20]
Al_4Cu_9	0.2	<0.5
Al_3Cu_4	8.1	7.2–8.1
AlCu	1.7	1.8–2.4
Al_2Cu	6.2	5.7–11.3

For different substrate preheating temperatures, the calculated size of the intermetallic layers is shown in figure 4. In general, the intermetallic layers thickness increases with increasing Ts. The Al_4Cu_9 layer is the least present at the interface. Its size barely exceeds 0.2 μm for *Ts* = 700 °C. Al_2Cu and Al_3Cu_4 layers are of comparable sizes for *Ts* = 400, 500 and 600 °C. For *Ts* = 700 °C Al_2Cu is significantly larger than Al_3Cu_4. It shows how the Al_2Cu layer is the most affected by substrate temperature, it increases much faster as the temperature rises compared to the layers of other compounds. These finds corroborate experimental results from the literature [20, 21].

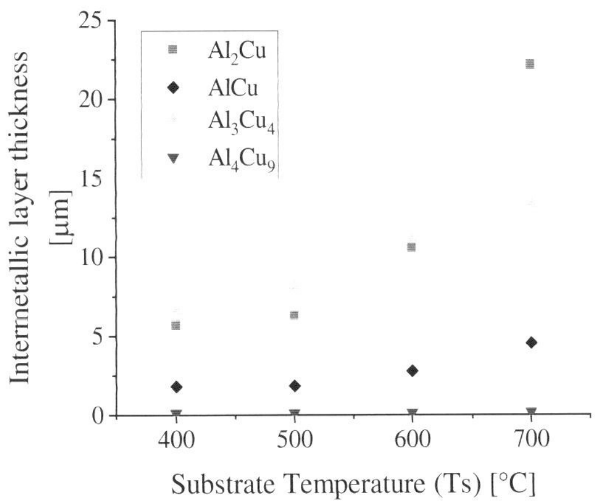

Figure 4. Intermetallic layers thickness as a function of the Cu substrate temperature. Initial Al liquid temperature was kept at 750 °C.

5. Conclusion

This paper presented a numerical model for compound casting simulation. The model considers the formation of intermetallic layers during the contact between a solid substrate and the melt with subsequent solidification. It was implemented in the OpenFoam environment, adjusting existing solvers.

The continuum model ability to predict the intermetallic layers thickness in solid-liquid joining problems, depends on the quality of the mobility data and the thermodynamic properties available, which makes necessary an adjustment of the model for each system of interest. Results applied for the Al-Cu compound casting problem demonstrates the use of the model.

Improvements in modelling are still needed, especially regarding nucleation of a compound at the interface, so that the order in which the compounds appear in the intermetallic layers is not an input, but rather a result. This would assist in understanding the reason for the lack of a compound in the ICL, which stills a controversial subject in the literature. The model application into higher-order systems, such as ternary systems, are also necessary to obtain more realistic results.

Acknowledgements
The present work is part of the project "Continuous compound casting of rotationally symmetrical Cu-Al semi-finished products" with the conveyor No. TO-271/24-1, founded by the German Research Foundation (DFG). The authors gratefully acknowledge the financial support from DFG.

References

[1] Wu L, Wang T, Fu Y, Li J, Cao Z and Li T 2012 *IOP Conf. Ser.: Mater. Sci. Eng.* **33** 12020
[2] Greß T, Glück Nardi V, Mittler T, Schmid S, Buchberger P, Tonn B and Volk W 2019 *Int. J. Metalcast.* 1–16
[3] Stefanescu D M 2015 *Science and engineering of casting solidification* (Cham: Springer)
[4] Yi P, Falk M L and Weihs T P 2018 *J. Appl. Phys.* **124** 165303
[5] Attari V and Arroyave R 2016 *J. Phase Equilib. Diff.* **37** 469–80
[6] Hong K K and Huh J Y 2006 *J. Electron. Mater.* **35** 56–64
[7] Park M S, Gibbons S L and Arróyave R 2012 *Acta Mater.* **60** 6278–87
[8] Min T, Gao Y, Chen L, Kang Q and Tao W-Q 2016 *Int. J. Heat Mass Tran.* **92** 370–80
[9] Starykov O, Wölck J, Pintore M, Tonn B and Volk W 2017 In *Proc. 6th Decennial Int. Conf. on Solidification Processing (Beaumont Estate, Old Windsor, UK)*
[10] Voller V R and Prakash C 1987 *Int. J. Heat Mass Tran.* **30** 1709–19
[11] Bird R B, Stewart W E and Lightfoot E N 2007 *Transport phenomena* 2nd edn (New York, Chichester: J. Wiley)
[12] Glück Nardi V and Franco A T 2015 In *Proc. 23rd ABCM Int. Cong. of Mechanical Engineering (Rio de Janeiro, Brazil)* (Rio de Janeiro: ABCM)
[13] Miehe A Numerical Investigation of Horizontal Twin-Roll Casting of the Magnesium Alloy AZ31 Technische Universität Bergakademie Freiberg
[14] Viswanath R and Jaluria Y 1995 *Numer. Heat Tr. A-Appl.* **27** 519–36
[15] Pelton A D 2018 *Thermodynamics, phase diagrams and thermodynamic modeling of solutions* 1st edn (Waltham MA: Elsevier)
[16] Theodore L. Bergman, Adrienne S. Lavine, Frank P. Incropera, David P. DeWitt 2011 *Fundamentals of heat and mass transfer* 7th edn (Hoboken NJ: John Wiley & Sons)
[17] André Bakker *Modeling chemical reactios with CFD: Reacting Flows*
[18] Dybkov V I 1987 *J. Mater. Sci.* **22** 4233–9
[19] Lugarini de Souza A, Franco A and Glück Nardi V 2017 *Proc. 24th ABCM Int. Cong. of Mechanical Engineering (Curitiba, Brazil)* (Rio de Janeiro: ABCM)
[20] Pintore M, Starykov O, Mittler T, Volk W and Tonn B 2018 *Int. J. Metalcast.* **12** 79–88
[21] Pintore M, Wölck J, Mittler T, Greß T, Tonn B and Volk W 2020 *Int. J. Metalcast.* **14** 155–66

Cellular automaton modelling to predict multi-phase solidification microstructures for Fe-C peritectic alloys

J Ogawa[1] and Y Natsume[2]

[1] Graduate School of Engineering Science, Akita University, 1-1 Tegatagakuen-machi, Akita 010-8502, Japan
[2] Department of Material Science, Faculty of Engineering Science, Akita University, 1-1 Tegatagakuen-machi, Akita 010-8502, Japan

E-mail: natsume@gipc.akita-u.ac.jp

Abstract. In this study, we developed a cellular automaton (CA) model to simulate the multi-phase microstructural evolution of the solidification of Fe-C binary peritectic alloys. Three phases were considered in the model: δ, γ, and liquid (L). To simulate microstructures formed by the peritectic transformation, moving interfaces at δ-γ, δ-L, and γ-L transformations were introduced for diffusion-controlled growth. The coarsening growth of γ grains after solidification of peritectic alloys was modelled using the CA method. Numerical simulations of multi-phase microstructure evolution for hyper-peritectic and hypo-peritectic alloys were performed in two and three dimensions. In these simulations, the nuclei of γ phases were formed at the δ/L interface below the peritectic temperature. After nucleation of the γ phase in hyper-peritectic alloys, the δ phase in the calculation domain disappeared before γ solidification was completed. Conversely, in hypo-peritectic alloys, the L phase disappeared before the δ-γ transformation was complete. Moreover, the coarsening of γ grains occurred for both alloys after the end of the peritectic transformation. These outcomes agree with the fact that the phase transformation of alloys involves the solidification of peritectic alloys. Thus, we confirmed that the proposed model was a valid approach to simulating multi-phase microstructures formed by the solidification of peritectic alloys.

1. Introduction

Solidification microstructure has a great influence on the quality of metal products. In Fe-C binary alloys, the δ phase solidifies at the liquidus temperature, and the δ-γ transformation, or γ solidification, occurs at the peritectic temperature. It is well known that the phase transformation from the δ phase to the γ phase causes volume shrinkage that is closely related to various defects such as the formation of uneven shell and surface cracks [1, 2]. Thus, it is important to understand the microstructure formation that occurs during the solidification of Fe-C binary alloys. It is well known that the peritectic reaction and transformation occur at the peritectic temperature. In the solidification of peritectic alloys, the nuclei of γ phase are formed on the δ/liquid (L) interface, and the γ phase grows along the δ phase boundary (peritectic reaction). When δ phases are enclosed by the γ phase due to the peritectic reaction, the γ phase grows into the sides of the δ and L phases (peritectic transformation). After the peritectic transformation completes, the coarsening of the γ grains begins. The longer the time required for the peritectic transformation, the larger the size of the γ grains just after the peritectic transformation [3].

Although it has been commonly stated that the coarsening of γ grains occurs just after the disappearance of the δ and L phases [4], new behaviour related to δ/γ transformation and γ-grain coarsening has recently been reported [5-8]. Tsuchiya *et al.* reported that the coarsening of γ grains does not occur immediately after peritectic transformation under the cooling conditions of continuous casting [5, 6], and Yasuda *et al.* reported that the massive-like transformation occurs as a new transformation at approximately the peritectic temperature [7, 8]. However, the relationship between the end of peritectic transformation and the start of γ-grain coarsening is not yet clear. Thus, it is necessary to investigate the relationship between peritectic transformation and the coarsening of γ grains in detail.

Numerical simulations have been used as an effective tool for understanding the mechanism of microstructure formation during solidification. Cellular Automaton (CA) models are one type of numerical simulation model describing the microstructure formation of alloys. The majority of CA models for microstructure formations have been proposed or developed for dendritic growth [9-15], which is a single-phase model that simulates the growth of a single solid phase in a liquid phase. We previously developed a single-phase CA model to simulate dendritic growth [11, 13]. However, a multi-phase model to simulate the solidification microstructures of eutectic or peritectic alloys has not been sufficiently developed. Although several multi-phase CA models have been proposed for the solidification of eutectic alloys [14, 15], a model combined with the solidification of peritectic alloys and the coarsening of γ grains is still required.

In this study, to investigate the relationship between the peritectic transformation and coarsening of γ grains, we modified the single-phase CA model that we previously proposed, and developed a multi-phase CA model to simulate microstructure formation with peritectic reaction and transformation, which can simulate the growth of the δ, γ, and L phases of an Fe-C binary alloy. Moreover, we modeled the coarsening of γ grains after the peritectic transformation and combined this with the multi-phase CA model of peritectic alloys. The development of three-dimensional (3D) model is important to evaluate quantitatively the multi-phase transition behaviour. However, the computational cost of 3D simulations is very high. Thus, two-dimensional (2D) simulations are also very important to evaluate their behaviour. We performed 2D and 3D numerical simulations of multi-phase microstructural evolution for hyper-peritectic and hypo-peritectic alloys using the proposed model, and the validity of the model was evaluated in terms of modeling the solidification of peritectic alloys and the coarsening of γ grains after the peritectic transformation.

2. Model description

2.1. Phase fractions

In the present model, the calculation domain is divided into a regular Cartesian grid referred to as a CA cell, and the five variables of solute concentration, temperature, grain orientation, state of cell, and phase fraction, are provided for each CA cell. The fractions of the L, δ, and γ phases are constrained according to the following condition.

$$\sum_{i=L,\delta,\gamma} f_i = 1 \quad \left(0 \le f_i \le 1\right) \tag{1}$$

where f_i is the fraction of the i phase (i = L, δ, and γ). The state of the CA cell is defined as follows: L phase ($f_L = 1, f_\delta = f_\gamma = 0$), δ phase ($f_\delta = 1, f_L = f_\gamma = 0$), γ phase ($f_\gamma = 1, f_L = f_\delta = 0$), δ/L interface ($0 \le f_L, f_\delta \le 1, f_\gamma = 0$), γ/L interface ($0 \le f_L, f_\gamma \le 1, f_\delta = 0$), and γ/δ interface ($0 \le f_\delta, f_\gamma \le 1, f_L = 0$).

2.2. Solute concentration and temperature fields

The solute concentration field is calculated using the expression for solute diffusion and redistribution proposed by equation (2).

$$\frac{\partial C_i}{\partial t} = \nabla \cdot \left(D_i \nabla C_i\right) + \sum_{j=L,\delta} C_j \left(1 - k_{i/j}\right) \frac{\partial f_i^{i/j}}{\partial t} \quad (i = \delta, \gamma, L) \tag{2}$$

where C_i is the solute concentration of the i phase, D_i is the solute diffusion coefficient in the i phase, and $k_{i/j}$ is the partition coefficient of the i/j interface. The first and second terms on the right-hand side of equation (2) represent the solute diffusion and redistribution, respectively. Local equilibrium at each interface is assumed as $C^*_i = k_{i/j} C^*_j$, where C^*_i is the equilibrium concentration of the i phase at the i/j interface. As a condition of solute diffusion between interfaces, only solute diffusion between the δ/L and γ/L interfaces is considered, ignoring solute diffusion between other types of interface. This is to avoid an extreme complexity of the solute partition near the triple junction. The time step (Δt) is $\Delta x^2/4.5D_L$ in 2D and $\Delta x^2/8.5D_L$ in 3D. Temperature is provided for each cell, and the temperature field is calculated by $T^{t+\Delta t} = T^t - R_c\Delta t$, where R_c is the cooling rate. In this study, the heat flow equation is not solved.

2.3. Change in phase fractions

In the solidification of peritectic alloys, the δ/L, γ/L, and γ/δ interfaces must be considered to simulate the solidification of the δ and γ phases and the γ-δ transformation. The change in phase fractions of the i/j interface cell at every time step ($\Delta f_i^{i/j}$) is calculated using equation (3).

$$\Delta f_i^{i/j} = \frac{C_j^* - C_j}{C_j^*\left(1-k_{i/j}\right)} \tag{3}$$

C^*_j is given by

$$C_j^* = C_0 + \frac{1}{m_{i/j}}\left(T_{i/j}^* - T + \Gamma_{i/j}\cdot wmc_{i/j}\right) \tag{4}$$

where C_0 is the initial concentration, $m_{i/j}$ is the slope of the j phase at the i/j interface, $T^*_{i/j}$ is the equilibrium temperature of the j phase at the i/j interface for the initial composition, T is the local temperature, $\Gamma_{i/j}$ is the Gibbs-Thomson coefficient at the i/j interface, and $wmc_{i/j}$ is the weighted mean curvature for 3D simulation at the i/j interface [10].

$$\begin{aligned} wmc_{i/j} = {} & \left(3\varepsilon_{i/j}-1\right)\cdot\left(\partial_x n_x + \partial_y n_y + \partial_z n_z\right) - 48\varepsilon_{i/j}\left(n_x^2\partial_x n_x + n_y^2\partial_y n_y + n_z^2\partial_z n_z\right) \\ & + 12\varepsilon_{i/j}Q\left(\partial_x n_x + \partial_y n_y + \partial_z n_z\right) + 12\varepsilon_{i/j}\left(n_x\partial_x Q + n_y\partial_y Q + n_z\partial_z Q\right) \end{aligned} \tag{5}$$

where $Q = n_x^4 + n_y^4 + n_z^4$, $n_x = \partial_x f_i / |\nabla f_i|$, and $\varepsilon_{i/j}$ is the degree of anisotropy of the surface energy of the i/j interface. In the present model, the state of the CA cell is varied according to the local rule of Moor's neighborhood, and updated using the values of the phase fractions of the interface cells. After the change in phase fractions is calculated, the values of each phase fraction and the state of the cell are updated. For example, in the δ/L interface cell, if the fraction of the δ phase exceeds one ($f_\delta \geq 1$), the state of the cell is updated from the δ/L interface to the δ phase. As a result, the cells of the L phase around the updated δ phase are changed to an δ/L interface. The same CA algorithm is used for updating the cell states of γ/L and γ/δ interfaces. The δ/L interface cell with the all phases in Moor's neighbourhood is captured as the γ/L interface.

2.4. Grain coarsening model

The grain coarsening model used in this study is based on a model proposed by Li *et al.* [16]. The growth rate of a grain boundary (v) is generally described by the expression $v = M\sigma\kappa$, where M is the grain boundary mobility, σ is the grain boundary energy, and κ is the curvature of the grain boundary. In this model, κ is determined using equation (6).

$$\kappa = \frac{A}{\Delta x}\frac{Kink - N_i}{N} \tag{6}$$

where A is the shape factor, Δx is the cell size, $Kink$ is the number of cells required to create a flat grain boundary, N_i is the number of cells belonging to the same grain within Moor's neighborhood, and N is the number of all cells within Moor's neighbourhood. A commonly used value for A is 1.28 in 2D simulations [17]. In 3D simulations, the value of A is unknown (as discussed in Section 3.4). When the cell adjoins a cell belonging to a different grain, if $\kappa > 0$, the cell properties (grain orientation) may change to those of the cell belonging to the different grain. The probability (p_g) of this event is

$$p_g = \sum_{(n,m)\in I} \Delta t \frac{v}{a_{(n,m)}} \tag{7}$$

where $a_{(n,m)}$ is the distance between two neighbouring cells, and I is the set of cells belonging to the different grain within Moor's neighbourhood. When the random number (0–1) that is generated at each cell is less than p_g, the cell changes to the cell match the neighbouring cell that belongings to the other grain. Since $p_g \le 1$, Δt must be carefully chosen. The Δt used in the calculation of solute diffusion is sufficient small for both 2D and 3D simulations, as mentioned in Section 2.2. In the grain growth theory of Hillert [18], the grain growth rate (dr/dt) in 2D is calculated with equation (8), while that in 3D is calculated using equation (9), where r is the grain radius, and $\langle r \rangle$ is the average grain radius.

$$\frac{2r}{M\sigma}\frac{dr}{dt} = \left(\frac{r}{\langle r \rangle} - 1\right) \tag{8}$$

$$\frac{r}{M\sigma}\frac{dr}{dt} = \left(\frac{\langle r^2 \rangle}{\langle r \rangle^2}\frac{r}{\langle r \rangle} - 1\right) \tag{9}$$

2.5. Material properties and calculation parameters

Table 1. Physical and thermodynamic properties for Fe–C binary alloy

Name of parameter	Symbol	Values	Unit	Ref.
Melting point of pure δ-Fe	$T_{m,\delta/L}$	1811	K	[19]
Melting point of pure γ-Fe	$T_{m,\gamma/L}$	1801	K	[19]
δ/γ transformation temperature	$T_{m,\gamma/\delta}$	1667	K	[19]
Peritectic temperature	T_p	1768.4	K	[19]
Peritectic composition	C_p	0.778	mol%$^{-1}$	[19]
Liquidus slope of δ phase	$m_{\delta/L}$	-18.28	K mol%$^{-1}$	[19]
Liquidus slope of γ phase	$m_{\gamma/L}$	-13.99	K mol%$^{-1}$	[19]
Partition coefficient of δ/L interface	$k_{\delta/L}$	0.179	-	[19]
Partition coefficient of γ/L interface	$k_{\gamma/L}$	0.334	-	[19]
Partition coefficient of γ/δ interface	$k_{\delta/\gamma}$ (= $k_{\gamma/L}$ / $k_{\delta/L}$)	1.866	-	[19]
Frequency factor of diffusion in L phase	$D_{0,L}$	5.2×10^{-7}	m^2 s^{-1}	[19]
Activation energy of diffusion in L phase	Q_L	5.0×10^{4}	J mol^{-1}	[19]
Frequency factor of diffusion in δ phase	$D_{0,\delta}$	1.27×10^{-6}	m^2 s^{-1}	[19]
Activation energy of diffusion in δ phase	Q_δ	8.3×10^{4}	J mol^{-1}	[19]
Frequency factor of diffusion in γ phase	$D_{0,\gamma}$	7.61×10^{-6}	m^2 s^{-1}	[19]
Activation energy of diffusion in γ phase	Q_γ	13.7×10^{4}	J mol^{-1}	[19]
Gibbs-Thomson coefficient of L/δ interface	$\Gamma_{\gamma/L}$	1.9×10^{-7}	K m	[20]
Gibbs-Thomson coefficient of L/γ interface	$\Gamma_{\gamma/L}$	3.7×10^{-7}	K m	[21]
Gibbs-Thomson coefficient of δ/γ interface	$\Gamma_{\gamma/\delta}$	5.3×10^{-7}	K m	Estimated
Anisotropy of the surface energy	$\varepsilon_{\delta/L}$, $\varepsilon_{\gamma/L}$, $\varepsilon_{\gamma/\delta}$	0.03	-	[10]
Grain boundary energy of γ grain	σ	0.79	J m^{-2}	[6]

Table 1 shows the physical and thermodynamic properties used in the simulations performed in this study. The diffusion coefficient is provided in the form $D_i = D_{0,i}\exp(-Q_i/(RT))$. Although the majority of material property values are obtained from references [6, 19, 20, 21], the Gibbs-Thomson coefficient of the γ/δ interface ($\Gamma_{\gamma/\delta}$) and the anisotropy parameters cannot be obtained from the data in the literature [10]. The $\Gamma_{\gamma/\delta}$ is estimated using the γ/δ interface energy (0.56 J m^{-2} [22]) and the melting entropy of pure Fe (1.07×10^{6} JK^{-1}m^{-3} [20]), while the value for the anisotropy parameter that is often used in simulations of dendritic growth are adopted.

3. Results and discussion

3.1. Model validation for γ-δ and γ-L transformations

To evaluate the validity of the proposed model in terms of the δ-γ and γ-L transformations, 1D simulations were performed under isothermal conditions. The calculation domain was set to 100 μm using a cell of 0.5 μm, and was divided into two equal parts as an initial setting. The δ and L phases were set in the left and right halves, respectively, and the γ phase was set in one cell between the δ and L phases. The initial concentrations of the δ, γ, and L phases were $C^{*}_{\delta/\gamma}$, $C^{*}_{\gamma/\delta}$, and $C^{*}_{L/\gamma}$, respectively. The relationship between the change in positions of the γ/δ and γ/L interfaces ($\Delta S_{\gamma/\delta}$ and $\Delta S_{\gamma/L}$) and the transformation time (t) was evaluated by comparison with equation (10).

$$\Delta S_{i/j} = B_{i/j} t^{0.5} \tag{10}$$

where $\Delta S_{i/j}$ is the change in position of the i/j interface. $B_{i/j}$ is the material parameter of the i/j interface, and is temperature dependent. The values of $B_{i/j}$ were determined by comparing the simulation results and equation (10), as shown in figure 1(a). Figure 1(b) shows the $B_{i/j}$ obtained from the simulation results at different temperatures. For comparison, the experimental results reported by Matsuura *et al.* [23, 24] are plotted on the figure. The simulation results were in good agreement with the experimental results. From these results, we confirmed that the present model could appropriately simulate the solidification of peritectic alloys (γ-δ and γ-L transformations).

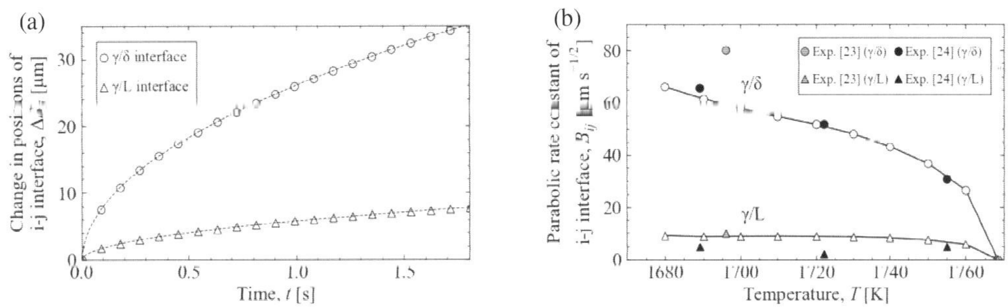

Figure 1. (a) Change in position of γ/δ and γ/L interfaces at 1760 K, and (b) the relationship between temperature and the $B_{i/j}$ of the γ/δ and γ/L interfaces.

3.2. 2D simulations of peritectic reaction, peritectic transformation, and grain coarsening

We performed 2D simulations of microstructure formation in the hypo-peritectic and the hyper-peritectic alloys. The calculation domain was set to 250 μm × 250 μm using a cell size of 0.5 μm. The Neumann boundary condition was used for all variables. As an initial condition, the calculation domain was cooled at R_c = 10 K/s from an initial temperature of an undercooling of 5 K, and the forty small solids of the δ phase were randomly set in the calculation domain. The nuclei of the γ phase were provided at the δ/L interface, below the peritectic temperature. Grain boundary mobility (M) was determined using the formula $M = 2.53\times10^{-3}\exp(-177\times10^{3}/(RT))$ m^{4}J^{-1}s^{-1} [4]. Figures 2 and 3 show snapshots of simulated microstructures and the change in fractions of each phase in the Fe-0.7mol%C hypo-peritectic and the Fe-1.0mol%C hyper-peritectic alloys, respectively. In the hypo-peritectic alloy, the δ dendrites grew in the domain at 1793.0 K (figure 2(a)). Below the peritectic temperature, the γ phases were nucleated at the δ/L interface and the grains of δ phase were immediately enclosed by the γ phases (figures 2(b), (c)). Although the amount of δ phase decreased due to δ/γ transformation, the L phase disappeared before the δ/γ transformation finished (figure 2(e)). Moreover, the coarsening of γ grains occurred after peritectic transformation had ended (figure 2(f)). Conversely, in the hyper-peritectic alloy, although the microstructure formation behaviour before the peritectic reaction was similar to that of the hypo-peritectic alloy (figures 3(a)-(d)), the variations in each phase fraction differed. Although the L phase decreased due to the γ solidification, the δ phase disappeared before γ solidification finished (figure 3(e)). The coarsening of γ grains occurred after γ solidification ended (figure 3(f)). These simulation results provided reasonable description of the phase transformations of peritectic alloys. Thus, it is found that the proposed 2D model effectively simulated the multi-phase microstructures formed by the peritectic reaction, peritectic transformation, and grain coarsening.

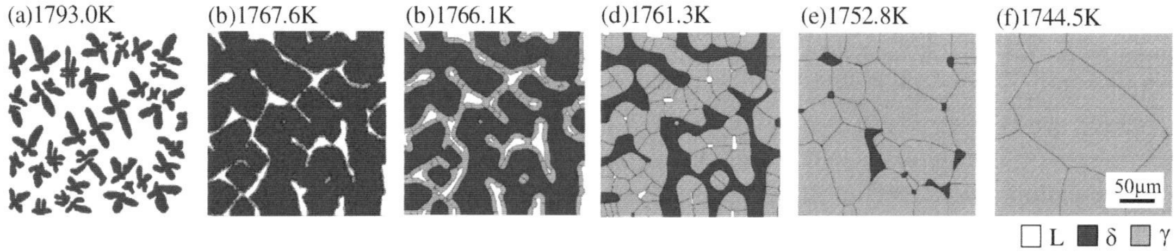

Figure 2. (a)-(f) Microstructure formation of the Fe-0.7mol%C hypo-peritectic alloy at different temperatures, simulated using the proposed 2D model.

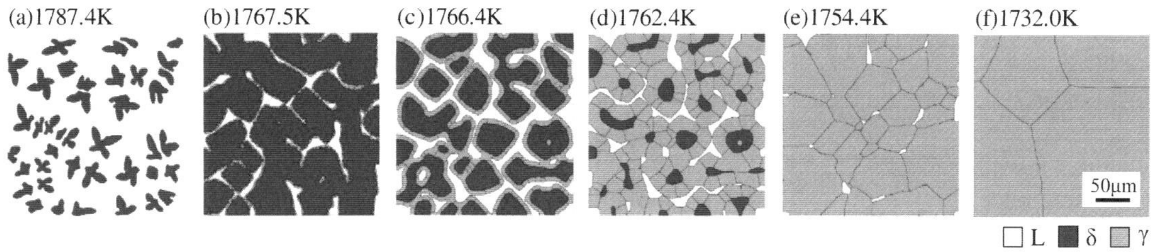

Figure 3. (a)-(f) Microstructure formation of the Fe-1.0mol%C hyper-peritectic alloy at different temperatures, simulated using the proposed 2D model.

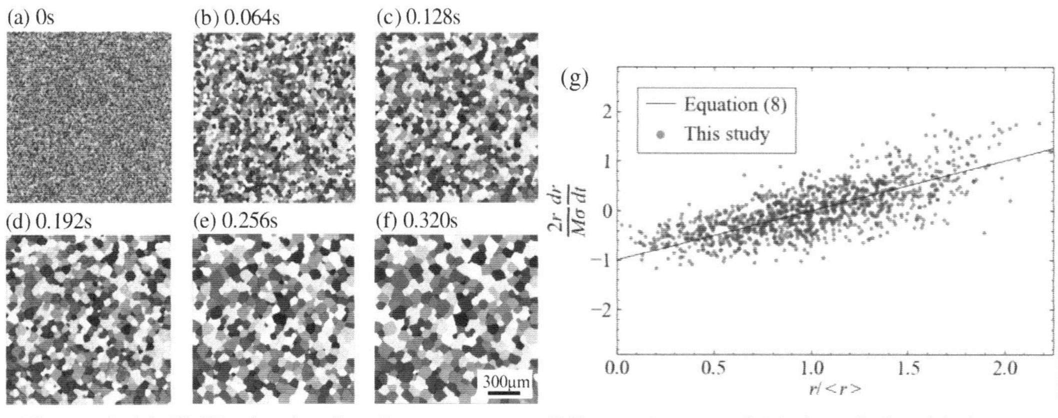

Figure 4. (a)-(f) 2D simulated grain structures at different times, and (g) the relationship between normalized grain radius and grain growth rate obtained from the 2D simulations.

Next, to evaluate the validity of the γ-grain coarsening model in 2D, we performed 2D simulation of polycrystalline structure evolution under isothermal conditions of 1768.4 K for an Fe-2.1mol%C peritectic alloy. The calculation domain of the Neumann boundary condition was set to 1500 μm × 1500 μm using a cell size of 0.5 μm. In the initial γ-grain distribution, 40000 grains with various orientations were randomly set in the calculation domain. Solute diffusion was not considered in this simulation. Figure 4 shows the relationship between the normalized grain radius and the normalized grain growth rate for each grain. The solid line on the figure shows the results that were calculated using equation (8). Although dispersions in the data were observed, there was good agreement between the tendencies of the simulated and analytical (obtained from equation (8)) results. Thus, we confirmed that the proposed model produced a reasonable description of the coarsening of γ grains in 2D.

3.3. 3D simulations of peritectic reaction, peritectic transformation, and grain coarsening

We performed 3D simulations of microstructure formation for unidirectional solidification in the Fe-2.1mol%C hyper-peritectic alloy. The calculation domain was set to 625 μm × 125 μm × 500 μm using a cell size of 1.0 μm, and was cooled at R_c = 5 K/s under a temperature gradient of 1.0 K/mm. The Neumann boundary condition was used for all variables. Calculation of γ-grain coarsening after peritectic transformation was not performed in 3D because of a huge computational time requirement in 3D simulation for many grains as performed in the 2D simulation. At an undercooling of 3 K, five small solids of δ phase were initially added to the bottom of the calculation domain. When the temperature on the bottom of the calculation domain was below the peritectic temperature, the γ phase was set up around the δ dendrites on the bottom of the domain. Figure 5 shows snapshots of the simulated microstructure. After the δ columnar dendrites grew from the bottom, the γ phase grew and enclosed the δ phase from the bottom (figures 5(a)-(c)). The δ phase then disappeared due to the end of the δ/γ transformation (figure 5(d)). From this result, we confirmed that the proposed 3D model was able to simulate the peritectic reaction and transformation.

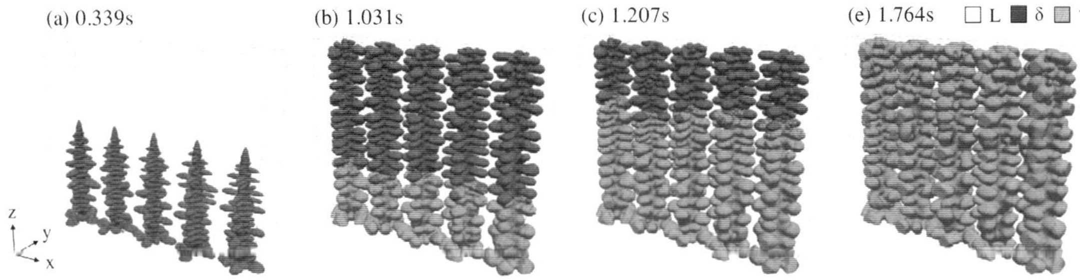

Figure 5. Microstructures formation of the Fe-2.1mol%C hyper-peritectic alloy simulated using the proposed 3D model.

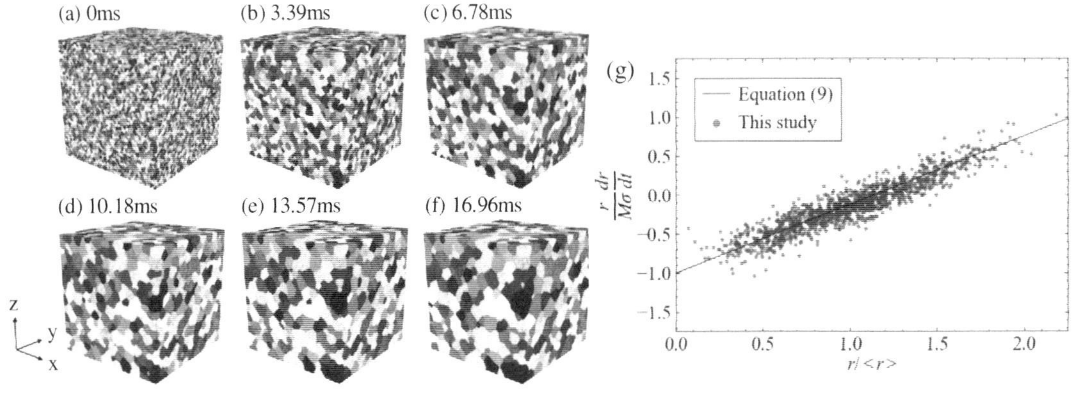

Figure 6. (a)-(f) The 3D simulated grain structures at different times, and (g) the relationship between normalized grain radius and grain growth rate that was obtained from the 3D simulations.

Finally, to evaluate the validity of the γ-grain coarsening model in 3D, we performed a 3D simulation of polycrystalline structure evolution under isothermal conditions of 1768.4 K for the Fe-2.1mol%C peritectic alloy. The calculation domain of the Neumann boundary condition was set to 225 μm × 225 μm × 225 μm using a cell size of 0.5 μm. In the initial γ-grain distribution, 0.12 million grains with various orientations were randomly set in the calculation domain. As before, solute diffusion was not considered in this simulation. Figure 6 shows the relationship between the normalized grain radius and the normalized grain growth rate for each grain. The solid line in the figure shows the result that was calculated using equation (9). Since the value of A in equation (6) for the 3D case was unknown, this value was obtained by comparing the simulated results with the analytical line calculated using equation (10). From this comparison, a value of A = 0.394 was estimated. The simulated results were in good agreement with the result of phase-field simulations performed by Kim *et al.* [25]. Thus, we confirmed that the present model could reasonably simulate the coarsening of γ grains in 3D.

4. Conclusion

In this study, we developed a multi-phase CA model to simulate the microstructural evolution of the peritectic solidification of an Fe-C binary peritectic alloy, combining a multi-phase model for solidification with a model for γ-grain coarsening after the peritectic reaction and transformation. First, 2D and 3D simulations of microstructure formations appearing with the solidification of peritectic alloys were performed using the proposed model. Reasonable 2D and 3D simulation results were obtained that describe the solidification of peritectic alloys. Next, 2D and 3D simulations of isothermal grain growth were performed, and simulation results were compared with the analytical results obtained using Hillert's equation. Both 2D and 3D simulation results were in good agreement with those obtained with Hillert's equation. From those results, we confirmed that the proposed model can accurately simulate both the microstructures involved in the solidification of peritectic alloys and the grain coarsening that occurs after peritectic transformation. In the near future, we plan to modify and develop our model to simulate microstructure formations such as the massive-like transformation, which has been recently reported by Yasuda's group as a new transformation model for steel [7, 8].

References

[1] Saeki T, Ohguchi S, Yamamoto T, Misumi H and Tsuneoka S 1982 *Tetsu-to-Hagané* **68** 1773
[2] Matsumiya T, Saeki T, Tanaka J and Ariyoshi T 1982 *Tetsu-to-Hagané* **68** 1782
[3] Maehara Y, Yasumoto K, Sugitani Y and Gunji K 1985 *Tetsu-to-Hagané* **71** 1534
[4] Bernhard C, Reiter J and Presslinger H 2008 *Metall. Mater. Trans.* **39**B 885
[5] Tsuchiya S, Ohno M, Matsuura K and Isobe K 2011 *Acta Mater.* **59** 3334
[6] Ohno M, Tsuchiya S and Matsuura K 2011 *Acta Mater.* **59** 5700
[7] Yasuda H *et al.* 2012 *Mater. Sci. Eng.* **33** 012036
[8] Nishimura T, Morishita K, Yoshiya M, Nagira T and Yasuda H 2019 *Tetsu-to-Hagané* **105** 290
[9] Wang W, Lee P D and McLean M 2003 *Acta Mater.* **51** 2971
[10] Pan S and Zhu M F 2010 *Acta Mater.* **58** 340
[11] Natsume Y and Ohsasa K 2015 *Mater. Sci. Eng.* **84** 012050
[12] Natsume Y 2016 *ISIJ Int.* **56** 1376
[13] Kao A, Krastins I, Alexandrakis M, Shevchenko N 2019 *JOM* **71** 48
[14] Wei L, Cao Y, Lin X, Wang M and Huang W 2019 *Comp. Mater. Sci.* **156** 157
[15] Dobravec T, Mavrič B and Šarler B 2017 *J. Comp. Phys.* **349** 351
[16] Li Z, Wang J and Huang H 2019 *J. Alloys and Comp.* **791** 411
[17] Lan Y J, Li D Z and Li Y Y 2006 *Metall. Mater. Trans. B* **37** 119
[18] Hillert M 1965 *Acta Metall.* **13** 227
[19] Ohno M and Matsuura 2010 *Acta Mater.* **58** 6134
[20] Kurz W and Fisher D J 1998 *Fundamentals of Solidification* Appendix 14 p.294
[21] Zhu M, Zhang L, Zhao H and Stefanescu D M 2015 *Acta. Mater.* **84** 413
[22] Yoshiya M, Watanabe M, Nakajima K, Ueshima N, Hashimoto K, Nagira T and Yasuda H 2015 *Mater. Trans.* 2015 **56** 1467.
[23] Matsuura K, Itoh Y and Narita T 1993 *ISIJ Int.* **33** 583
[24] Matsuura K, Maruyama H, Itoh Y, Kudoh M and Ishii K 1995 *ISIJ Int.* **35** 183
[25] Kim W T and Park Y B 2006 *Phys. Rev E* **74** 061605

IOP Conf. Series: Materials Science and Engineering **861** (2020) 012060 doi:10.1088/1757-899X/861/1/012060

Phase field modelling of dendritic solidification by using an adaptive meshless solution procedure

T Dobravec[1], B Mavrič[1,2] and B Šarler[1,2]

[1] Institute of Metals and Technology, Lepi pot 11, 1000 Ljubljana, Slovenia
[2] Faculty of Mechanical Engineering, University of Ljubljana, Aškerčeva cesta 6, 1000 Ljubljana, Slovenia

E-mail: bozidar.sarler@fs.uni-lj.si

Abstract. A novel numerical procedure is developed for modelling two-dimensional dendritic solidification in dilute binary alloys. The evolution of the phases and the solute concentration is described by the partial differential equations, obtained from the phase field model. The meshless radial basis function-generated finite difference (RBF-FD) method is used for the spatial discretisation of the partial differential equations. The forward Euler scheme is used for the time-stepping. In order to reduce the computational cost, an adaptive procedure is developed, based on the quad-tree strategy, ensuring the highest density of the computational nodes at the solid-liquid interface. In the procedure, the computational domain is divided into overlapping sub-domains which can be dynamically refined or coarsened. The regular or scattered node distribution with constant node density is used for discretisation of each sub-domain. The h-adaptive procedure is ensured by the constant product between the area of a sub-domain and the computational node density. The accuracy and speedup in comparison to the solution on a uniform node distribution are assessed by solving the benchmark problem for dendritic solidification in dilute binary alloys. The main originality of the model represents the first use of RBF-FD method for the spatial discretisation of the PF equations in combination with adaptive solution procedure. The RBF-FD method can be used on unstructured node distributions, which is especially advantageous in the solution of PF model for dendritic growth, since the solution is very sensitive to the regularity of the node distribution. The developed spatial-temporal-adaptive numerical model represents an accurate and computationally efficient tool for the prediction of the dendrite morphology and micro-segregation during the solidification in binary alloys.

1. Introduction

Dendritic morphology is one of the most commonly observed microstructures in the industrial casting of metallic alloys. The microstructure evolution significantly determines the material properties of the final product [1,2]. The theoretical and computational studies of dendritic growth can therefore be used to better understand and optimise the casting procedure.

The phase field (PF) approach [3, 4] represents a one-domain formulation for the simulation of dendritic solidification. Different phases are formulated by different values of PF, a continuous variable, which is constant in the bulk of the phases and smoothly varies over the interfaces between them. In the first step of the derivation of a PF model [5], a free-energy functional is constructed, describing the thermodynamics of a phase change and the capillary effects. Next, the kinetic equations for PF and other thermodynamic variables are derived by minimising the free-energy functional. Finally, appropriate free PF model parameters [6] have to be set in order to correctly capture the underlying physics of the phase change. Additionally, in some cases, the kinetic equations have to be modified [7] in order to ensure quantitatively correct simulation results. In this paper, a PF model of dendritic solidification in dilute binary alloys [8] is considered. The model consists of two partial differential equations (PDEs), one for the PF and one for the dimensionless supersaturation.

In the numerical solution of the PF model, the spacing between the neighbouring computational nodes has to scale with the PF interface thickness. Although the interface thickness can be increased by using the thin-interface limit of the PF model [6], the ratio between the size of the computational domain and the interface thickness usually remains to be in the range of a few orders of magnitude, making the PF simulations computationally extremely costly. In the last few decades, a spectra of different numerical techniques [9, 10] have been applied for the solution of the PF models. One of the most popular techniques for PF model is the adaptive mesh refinement [11, 12], where the node density is dynamically increased near the solid-liquid interface and dynamically decreased in the bulk of the solid and liquid phase. This reduces the computational efforts while sustaining the accuracy of the numerical solution.

The PF model is in the present paper solved by a novel adaptive meshless solution procedure, based on the numerical technique for modelling of solidification by using a novel next-generation adaptive mesh refinement [12]. The computational domain is divided into the computational sub-domains with different computational node densities by using a quad-tree data structure [13]. Computational sub-domains are dynamically refined/de-refined in order to ensure the highest node density in the areas with the largest gradients of PF and dimensionless supersaturation. The meshless RBF-FD method [14] is used for spatial discretisation of the PF equations. The solution of PF equations is propagated in each sub-domain independently, while the boundary conditions are set by the weighted least squares method (WLS) [15] interpolation from the overlapping neighbouring sub-domains. The forward Euler scheme is used for temporal discretisation of PDEs on each of the sub-domains, where a suitable stable time step depends on the node density in a sub-domain. The synchronisation between the neighbouring sub-domains with different node densities only occurs when they are integrated to the same simulation time.

In the paper, firstly, the PF model for solidification of dilute binary alloys is presented. Next, the adaptive solution procedure and the RBF-FD method are described. The benchmark problem for solidification in binary alloys is used for the assessment of the accuracy and the efficiency of newly developed numerical algorithm.

2. Governing equations

We study solidification of a dilute binary alloy [8] by solving a coupled system of two partial differential equations, one for the PF ϕ and one for the dimensionless supersaturation U. The values $\phi=1$ and $\phi=-1$ represent the solid and the liquid phase, respectively. The dimensionless supersaturation is defined as $U=[\exp(u)-1]/(1-k_p)$, where u measures the dimensionless departure of the chemical potential from the equilibrium value and k_p stands for the partition coefficient, respectively. The departure is defined as

$$u=\ln\left(\frac{2c/c_l^0}{1+k_p-(1-k_p)\phi}\right), \tag{1}$$

where c and c_l^0 stand for the concentration and the equilibrium concentration in the liquid phase, respectively.

It is convenient for numerical analysis to write PF model in dimensionless form by introducing the dimensionless spatial coordinates, the dimensionless time, and the dimensionless diffusivity of the solute in the liquid phase

$$(\bar{x},\bar{y})=(x/W_0, y/W_0), \quad \bar{t}=t/\tau_0, \quad \bar{D}=D\tau_0/W_0^2, \tag{2}$$

where D is the diffusivity of the solute in the liquid phase and W_0 and τ_0 stand for the characteristic interface thickness and the characteristic time of attachment of the atoms, respectively. The dimensionless governing equations in case of a negligible interface kinetics [8] are given as

$$\left(1+(1-k_p)\bar{R}_c\bar{t}\right)a^2(\mathbf{n})\partial_{\bar{t}}\phi = \\ =\bar{\nabla}\cdot\left(a^2(\mathbf{n})\bar{\nabla}\phi\right)+\sum_{\xi=x,y}\partial_{\bar{\xi}}\left(|\bar{\nabla}\phi|^2\, a(\mathbf{n})\frac{\partial a(\mathbf{n})}{\partial(\partial_{\bar{\xi}}\phi)}\right)-\phi-\phi^3-\lambda(1-\phi^2)^2\left(U-\bar{R}_c\bar{t}\right), \tag{3}$$

$$\left(\frac{1+k_p}{2}-\frac{1-k_p}{2}\phi\right)\partial_{\bar{t}}U = \\ =\bar{\nabla}\cdot\left(\bar{D}\frac{1-\phi}{2}\bar{\nabla}U+\frac{1}{2\sqrt{2}}(1+(1-k_p)U)\partial_{\bar{t}}\phi\mathbf{n}\right)+(1+(1-k_p)U)\frac{1}{2}\partial_{\bar{t}}\phi, \tag{4}$$

where λ is the coupling constant and $\mathbf{n}=(n_x,n_y)$ stands for the normalized gradient of PF. The dimensionless cooling rate is defined as

$$\bar{R}_c=R_c\tau_0/\Delta T_0, \quad \Delta T_0=|m_l|(1-k_p)c_l^0, \tag{5}$$

where R_c and m_l stand for the cooling rate and the liquidus slope, respectively. In the case of the solid phase with a cubic crystal structure, the following anisotropy function $a(\mathbf{n})$ is in use

$$a(\mathbf{n})=1-3\varepsilon_4+4\varepsilon_4(n_x^4+n_y^4), \tag{6}$$

where ε_4 is the strength of anisotropy.

The thin-interface limit of the PF model [6-8], yields the following relations between the physical and the PF parameters

$$W_0=\lambda d_0/a_1, \quad \tau_0=a_2\lambda W_0^2/D, \tag{7}$$

where a_1=0.8839 and a_2=0.6267. The constant $d_0=\Gamma/\Delta T_0$ is the chemical capillary length, where Γ is the Gibbs-Thomson coefficient. The thin-interface limit yields only one free parameter in the PF model. It has to be set consistently with the validity condition $|\mathbf{v}|<<1$, where $\mathbf{v}$ is the dimensionless interface velocity. As stated in [8], the convergence of the model actually starts to break at $|\mathbf{v}|\approx 0.2$, yielding more freedom in the selection of the free parameter.

3. Solution procedure

3.1. Adaptive discretisation of computational domain

In the solution procedure, the computational domain Ω is decomposed into N_p overlapping computational sub-domains Ω_i^*, i=1,2,3,…N_p. The boundaries of Ω and Ω_i^* are denoted as $\partial\Omega$ and $\partial\Omega_i^*$, respectively. $\partial\Omega_i^*$ can lie partially in $\partial\Omega$ and partially in the neighbouring Ω_j^*, $j\neq i$. In the latter case, the boundary condition on $\partial\Omega_i^*$ is set according to the solution in Ω_j^*, $j\neq i$.

In our case, we consider a rectangular Ω which is recursively divided into smaller rectangular sub-domains Ω_i^0 by using the balanced quad-tree data structure [13] as seen in figure 1(a). A quad-tree is balanced if the level difference between any two neighbouring leafs is less than or equal to one. Ω_i^* is constructed by extending Ω_i^0 in order to obtain the overlapping sub-domains. For each Ω_i^*, uniform, regular or scattered node distribution is generated as seen in figure 2(b). The details of algorithm for construction of scattered node distribution are given in [16]. The boundary condition in computational nodes from Ω_i^*, lying on $\partial\Omega_i^*\cap\partial\Omega$ is defined by the physical problem, while the Dirichlet boundary condition is assumed in computational nodes on $\partial\Omega_i^*\cap\Omega_j^*$, $j\neq i$. The values of Dirichlet boundary condition in such nodes are interpolated from Ω_j^*.

In order to accurately capture the details of the solution during simulation, i.e. if the refinement condition is fulfilled, the domain Ω_i^* is divided into four new sub-domains with twice reduced spacing h_i^*. Additional Ω_i^* have to be refined too in order to maintain the balanced quad-tree. Similarly, four Ω_i^* can be merged into one Ω_i^* with twice increased spacing h_i^*, if de-refinement condition is fulfilled. De-refinement can only occur, if the quad-tree is still balanced after the potential merger of four Ω_i^*.

The computational node density in Ω_i^* is determined as $\rho_i^*=1/h_i^{*2}$, while the spacing is given as $h_i^*=2^{N-M}h_{min}$, where h_{min} is the minimum spacing, and N and M stand for the level of Ω_i^0 and the maximum level of the quad-tree, respectively. The length and height of Ω_i^0 are given as $L_i^0=n_L h_i^*$ and $H_i^0=n_H h_i^*$, respectively, where n_L and n_H are the number of regularly distributed nodes per length and height, respectively. The length and height of Ω_i^* depends on the position of Ω_i^0 in Ω; in the case from figure 1(a), they are given as $L_i^*=(n_L+n_{buff})h_i^*$ and $H_i^*=(n_H+2n_{buff})h_i^*$, where the parameter n_{buff} controls the area of overlapping.

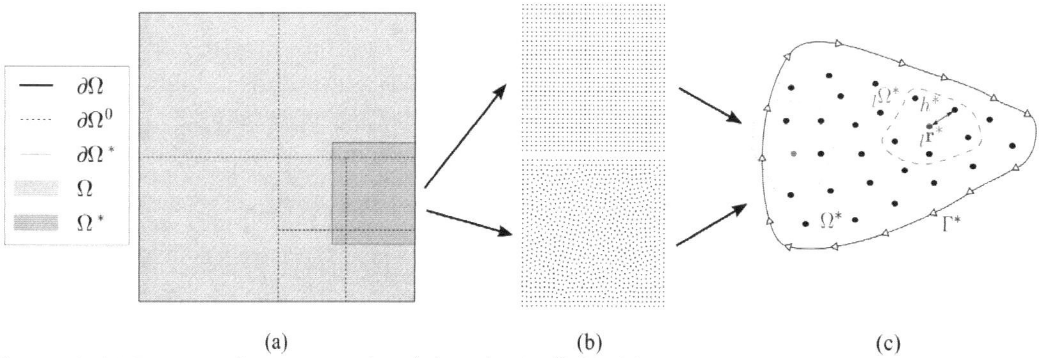

Figure 1. (a) Rectangular computational domain Ω, divided by the quad-tree algorithm into sub-domains Ω^0, which are extended in order to obtain the overlapping sub-domains Ω^*. A quad-tree subdomain is extended by constructing boundary of computational domain $\partial\Omega^*$ which lies inside the neighbouring sub-domains. (b) Regular or scattered node distribution is used for discretisation of a Ω^*. (c) Illustration of a computational domain Ω^* and boundary Γ^*. The solid circles and empty triangles represent the interior and the boundary computational nodes, respectively. The dashed line represents the boundary of a local sub-domain, containing only interior nodes. The boundary of a local sub-domain containing the interior and the boundary nodes is represented by the dotted line.

3.2. Discretisation of PDEs on overlapping sub-domains

The spatial differential operators in each Ω_i^* are evaluated by the use of RBF-FD method. In the section, notation Ω^* is used instead of Ω_i^*, since the procedure is the same for every Ω_i^* in Ω. The method is based on the interpolation of the field value over a local sub-domain ${}_l\Omega^*$, centred at a node ${}_l\mathbf{r}^*$ from Ω^*, by the use of N_{loc} radial basis functions (RBFs) and N_{aug} augmentation monomials. ${}_l\Omega^*$ consists of ${}_l\mathbf{r}^*$ and its N_{loc}-1 nearest neighbours as seen in figure 1(c). The details of the method are given in our very recent work [14]. The RBF-FD has a finite-difference-like structure, i.e., a general spatial differential operator Θ at ${}_l\mathbf{r}^*$ is approximated as

$$\Theta\eta({}_l\mathbf{r}^*) \approx \sum_{k}^{N_{loc}} {}_l w_k^* \, {}_l\eta_k^*, \tag{8}$$

where η is an arbitrary scalar field, ${}_l w_k^*$ is a finite-difference-like coefficient, and ${}_l\eta_k^*$ field value at k-th node in ${}_l\Omega^*$. Similar approximation can be written for discretisation of Θ acting on a vector or tensor field. WLS method [15] is used for interpolation of field values to the boundary nodes in the neighbouring sub-domains. WLS has also a finite-difference-like structure, however, the weighted least squares minimisation is used instead of collocation in the calculation of the finite-difference-like coefficients.

The forward Euler scheme is used for the temporal discretisation of the partial differential equations. The interpolation of boundary values between neighbouring sub-domains only occurs, when they are at the same simulation time, since the explicit time step in Ω_i^* is calculated as

$$\Delta t_i^* = \frac{h_i^{*2}}{12\max(D,1/\min(a(\mathbf{n})))}. \tag{9}$$

4. Numerical results

4.1. Benchmark definition and model parameters
The solution procedure is verified by solving the benchmark problem for solidification in dilute binary alloys proposed by Karma [8]. In the benchmark, growth of a single dendrite into a supersaturated melt is considered while the cooling rate is set to zero. The growth velocity and the concentration profile in the solid phase are used to verify the numerical model.

A square Ω with the side length $L/d_0=740$ is used in the simulation, while a small circular nucleus with radius $R_0/d_0=22$ is put in the south-west corner of Ω. Initial supersaturation in the whole Ω is set to $U=-0.55$. Zero flux Neumann boundary conditions are used for ϕ and U on Γ. The partition coefficient and the strength of anisotropy are set to $k_p=0.15$ and $\varepsilon_4=0.02$, respectively. The simulation is run till $tD/d_0^2=26000$. PF free parameter is set to $\lambda=a_1/0.277$. As stated in [7], this value fulfil the validity condition of the PF model for the benchmark problem.

The parameters of the adaptive discretisation of Ω are set to: $h_{min}/W_0=0.4$, $n_L=n_H=16$, $M=5$, and $n_{buff}=1$. A Ω_i^* is refined, if $|\phi({}_l\mathbf{r}_i)|<0.95$ at each ${}_l\mathbf{r}_i$ from Ω_i^*. A Ω_i^* can be de-refined if $|\phi({}_l\mathbf{r}_i)|>0.99$ at each ${}_l\mathbf{r}_i$ from Ω_i^*. A regular node distribution is used for discretisation of Ω_i^* in the present work. Irregular distribution can be easily employed too, because of the meshless concept. Second-order accurate RBF-FD method [14] is used by setting augmentation with monomials up to the second order while $N_{loc}=13$ fifth-order poly-harmonic splines are used as RBFs. The interpolation of boundary values in the neighbouring sub-domains is performed by WLS [15] with $N_{loc}=6$.

4.2. Results
The rescaled concentration profile c/c_l^0 at time $tD/d_0^2=26000$ is shown in figure 2(a). Non-overlapping computational sub-domains Ω_i^0, determined by the quad-tree algorithm are also shown. The actual computation is done on extended overlapping sub-domains Ω_i^*. The adaptive algorithm dynamically ensures that the solid-liquid interface lies in sub-domains with $h_i^*=h_{min}/W_0=0.4$ while larger values of h_i^* are used in the bulk of the solid and the liquid phase. The concentration profile in the middle of the trunk is shown on the right in figure 2(b). A plot of the dendrite's tip velocity as a function of time is shown in figure 3(c). Both, c_s/c_l^0 and Vd_0/D, are very close to the benchmark values [7].

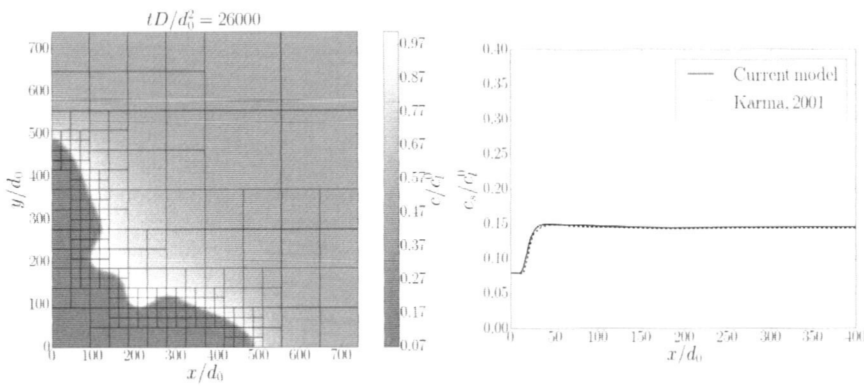

(a) (b)

Figure 2. (a) Concentration profile and quad-tree subdomains at the end of the simulation. Each extended sub-domain in the interior of Ω contains $(n_L+2n_{buff})\times(n_H+2n_{buff})=324$ computational nodes. (b) Concentration profile in the middle of the dendrite's trunk.

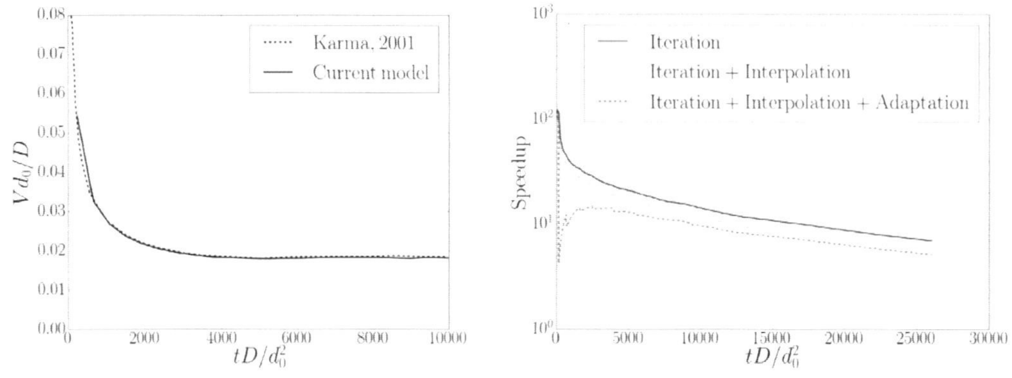

(a) (b)

Figure 3. (a) Dendrite's tip velocity as a function of time. (b) Speedup of adaptive solution procedure as a function of time. The speedup is defined as a ratio between ERTs on adaptive and uniform node distribution.

The same simulation is run also on uniform node distribution with parameters h_{min}/W_0=0.4, n_L=n_H=512, and M=0 in order to assess the speedup of adaptive solution procedure. In the adaptive procedure, we reduce the number of the computational nodes and the number of time iterations, however, we introduce some additional computational work due to the interpolation of boundary values and the adaptation of the sub-domains. The speedup of the adaptive numerical model in comparison with the solution on the fixed node distribution is shown in figure 3(b). The speedup is divided into three parts: we compare the iteration elapsed real time (ERT) on uniform node distribution to ERT needed only for iteration, sum of ERTs for iteration and interpolation of boundary values, and sum of ERTs for iteration, interpolation, and adaptation. As expected, the speedup is a decreasing function of the solid-liquid interface area. Each extra term in the sum of ERTs reduces speedup. The highest speedup is observed at the beginning of the simulation. The highest growth velocity is observed initially, hence, a lot of time is spent for adaptation of computational sub-domains. The adaptation time is reduced as the velocity decreases.

5. Conclusions

An adaptive meshless solution procedure for the PF simulation of dendritic growth is developed. In the procedure, a rectangular computational domain is dynamically refined or de-refined into sub-domains with different node density by the use of quad-tree data structure. The procedure ensures that the solid-liquid interface is in the sub-domains with the highest node density. RBF-FD method and WLS are used for discretisation of PDEs and interpolation of boundary values to the neighbouring sub-domains, respectively. Simple forward Euler scheme is applied for time stepping.

The newly developed model is verified by solving the PF benchmark problem for solidification in dilute binary alloys. We have compared the segregation profile in the solid phase and the dendrite's growth velocity. Both quantities are in good agreement with results from the benchmark problem [7]. We have also assessed the speedup of the adaptive algorithm by comparing ERTs on adaptive and uniform node distributions. The speedup drops with time as more and more sub-domains are created as the size of the solid-liquid interface increases. Naturally, the communication between neighbouring sub-domains and refinement/de-refinement procedure reduce the speedup of the method.

In the current version of the method, the finite-difference-like coefficients in the RBF-FD and WLS method are calculated whenever a new sub-domain is created. The finite-difference-like coefficients will only be calculated at the beginning of the simulation and then copied and appropriately rescaled as new sub-domain is created in the future foreseen version of the numerical implementation. This improvement will decrease the adaptation ERT, however, speedup will remain reduced due to communication. In the future, the PF model for solidification in dilute multi-component alloys [17] will be numerically solved by the developed numerical method and integrated into the simulation system for direct-chill casting of aluminium alloys [18] and continuous casting of steel [19].

Acknowledgements
This work was supported by the Slovenian Research Agency (J2-1718 Advanced Meshless Modelling and Simulation of Multiphase Systems; L2-9246 Multiphysics and Multiscale Modelling for Competitive Continuous Casting; P2-0162 Transient Two-phase Flows).

References

[1] Dantzig J A and Rappaz M 2009 *Solidification* (Lausanne, Switzerland: EFPL Press)
[2] Glicksman M E 2011 *Principles of Solidification* (New York, USA: Springer New York)
[3] Chen L-Q 2002 Phase-field models for microstructure evolution *Annu. Rev. Mater. Res.* **32** 113–40
[4] Boettinger W J, Warren J A, Beckermann C and Karma A 2002 Phase-field simulation of solidification *Annu. Rev. Mater. Res.* **32** 163–94
[5] Provatas N and Elder K 2010 *Phase-Field Methods in Materials Science and Engineering* (Weinheim, Germany: Wiley-VCH)
[6] Karma A and Rappel W-J 1998 Quantitative phase-field modeling of dendritic growth in two and three dimensions *Phys. Rev. E* **57** 4323
[7] Karma A 2001 Phase-Field Formulation for Quantitative Modeling of Alloy Solidification *Phys. Rev. Lett.* **87** 115701
[8] Echebarria B, Folch R, Karma A and Plapp M 2004 Quantitative phase-field model of alloy solidification *Phys. Rev. E* **70** 061604
[9] Karma A and Tourret D 2016 Atomistic to continuum modeling of solidification microstructures *Curr. Opin. Solid State Mater. Sci.* **20** 25–36
[10] Dong X, Xing H, Weng K and Zhao H 2017 Current development in quantitative phase-field modeling of solidification *J. Iron. Steel. Res. Int.* **24** 865–78
[11] Provatas N, Goldenfeld N and Dantzig J 1998 Efficient computation of dendritic microstructures using adaptive mesh refinement *Phys. Rev. Lett.* **80** 3308 11
[12] Greenwood M, Shampur K N, Ofori-Opoku N, Pinomaa T, Wang L, Gurevich S and Provatas N 2018 Quantitative 3D phase field modelling of solidification using next-generation adaptive mesh refinement *Comput. Mater. Sci* **142** 153–71
[13] Dobravec T, Mavrič B and Šarler B 2017 A cellular automaton – finite volume method for the simulation of dendritic and eutectic growth in binary alloys using an adaptive mesh refinement *J. Comput. Phys* **349** 351–75
[14] Dobravec T, Mavrič B and Šarler B 2020 Reduction of discretisation-induced anisotropy in the phase-field modelling of dendritic growth by meshless approach *Comput. Mater. Sci* **172** 109166
[15] Hatić V, Mavrič B and Šarler B 2018 Simulation of a macrosegregation benchmark with a meshless diffuse approximate method *Int. J. Numer. Method. H.* **28** 361–80
[16] Hatić V, Mavrič B, Košnik N and Šarler B 2018 Simulation of direct chill casting under the influence of a low-frequency electromagnetic field *Appl. Math. Model.* **54** 170–88
[17] Ohno M 2012 Quantitative phase-field modeling of nonisothermal solidification in dilute multicomponent alloys with arbitrary diffusivities *Phys. Rev. E* **86** 051603
[18] Šarler B, Dobravec T, Glavan G, Hatić V, Mavrič B, Vertnik R, Cvahte P, Gregor F, Jelen M and Petrovič M 2019 Multi-Physics and Multi-Scale Meshless Simulation System for Direct-Chill Casting of Aluminium Alloys *J. Mech. Eng.* **65** 658–70
[19] Vertnik R, Mramor K and Šarler B 2019 Solution of three-dimensional temperature and turbulent velocity field in continuously cast steel billets with electromagnetic stirring by a meshless method *Eng. Anal. Bound. Elem.* **104** 347–63

IOP Conf. Series: Materials Science and Engineering **861** (2020) 012061 doi:10.1088/1757-899X/861/1/012061

A cellular automaton approach for the prediction of grain size in grain refined alloys

A Jacot[1,2]

[1] Calcom ESI SA, Route Cantonale 105, 1025 St-Sulpice, Switzerland
[2] Brunel University London, BCAST, Uxbridge UB8 3PH, UK

E-mail: alain.jacot@esi-group.com

Abstract. Grain refinement by inoculation is widely practised in metal casting, in particular for aluminium alloys. During the last decades several modelling techniques have emerged to quantitatively predict solidification microstructures. The prediction of grain size remains however a challenge due to the complex competition between nucleation and growth, and physical phenomena taking place at very different length scales. Models often address the thermal recalescence and sometimes the suppression of nucleation by the solute boundary layers. The possible clustering tendency of the nucleant particles is however usually not considered, although it can play an important influence on the final grain size. This effect is particularly difficult to quantify because it requires a small-scale approach to describe the clusters, while a large computational domain is needed for grain size statistics. A cellular automaton (CA) model tracking the grain envelopes was developed to address this challenge. The spatial distribution of the nucleant particles and the solute diffusion field in the intergranular liquid were represented directly on the CA grid. Predicted grain sizes were first compared successfully with published experimental and modelling data. An investigation of the role of clustering of nucleant particles was then carried out, which demonstrated the potential of the model to address this topic.

1. Introduction

Grain refinement by inoculation is widely practiced in metal casting, in particular for aluminium alloys. The goal is to obtain small equiaxed grains which bring several benefits such as reduced defects, easier processing and improved mechanical properties. The added particles favour the nucleation of many equiaxed grains at low undercooling, avoiding also the formation of undesired columnar structures. During the last decades there has been great interest to quantitatively predict solidification microstructures. Several modelling techniques have emerged such as multiphase-models [1-3] and cellular automata [4] which can provide fields of various microstructural quantities on the scale of an entire cast component. On a smaller scale, the phase-field method can now address the growth of columnar or equiaxed grains with a very high level of details about the solid morphology and the distribution of the solute elements in the microstructure including the effect of fluid flow [5,6].

In spite of this progress, the prediction of grain size remains a modelling challenge. The reason is that the average grain size in a microstructure is the result of a complex competition between nucleation and growth and is governed by phenomena taking place at very different length scales. A major progress in the understanding of grain refinement and quantification of the grain size has been the free growth theory of Greer [7]. One of the outcomes of the theory is that the particle size distribution of the inoculant can be translated into a distribution of nucleation undercoolings. Combining this with a proper growth model, the grain size can be predicted [8]. Models predicting the grain size normally also include a calculation of a thermal recalescence, which will stop nucleation, leaving the less potent particles unused. Recently, Du *et al.* developed a mean-field model that also applies to continuous cooling conditions, where no recalescence occurs and nucleation can only be stopped by site capture or solutal effects [9]. The problem was addressed by incorporating the effect of the solute boundary layers of the growing grains, which can stifle nucleation on less potent particles in their vicinity. This phenomenon, known also as the solute suppressed nucleation zone (SSNZ), was well described already earlier by Easton and StJohn [10].

Another factor that can influence the grain size is the spatial distribution of the nucleant particles. If they tend to agglomerate, a large proportion of them will be inefficient due to the SSNZ effect or simple capture. This situation is particularly difficult to quantify in a model. It requires on one hand a small-scale approach to describe the geometrical arrangement of the clusters and to address the SSNZ formed by the overlapping diffusion fields inside the clusters and in their vicinity. On the other hand, the computational domain needs to be sufficiently large to allow for sufficient statistics on grain sizes. In addition, to correctly calculate the recalescence, the approach should address heat transfer, either by a direct resolution of the heat conservation equation or through a micro-macro coupling scheme. The model presented here addresses these challenges by incorporating all the important mechanisms that determine the final grain size: the size and spatial distribution of nucleant particles, the solute diffusion field around growing grains, thermal recalesence and site capture. Although it is not fully coupled yet, the model has been designed to be compatible with a resolution of heat flow on the process scale.

2. Model description

Predicting the grain size using a direct representation of the individual grains requires a relatively large computation domain to have enough grains to carry out meaningful statistics. For grain refined alloys, computation domains must typically be in the order of 1 mm to correctly predict the grain size in a given region of the casting. While state-of-the art implementations of the phase-field method [5] would probably allow to address such volumes, the computation cost of the technique remains very high, and prevents any implementation in numerical schemes where such local microstructure calculations would be coupled with heat transfer on the process scale. Another choice would be mean-field approaches, which have been applied with success [9]. A direct description of the grains is however desired when non-random spatial distributions of nucleant particles are analysed, including the complex multiple interactions through their diffusion fields. For these reasons, an envelope model was considered. In this approach, only the grain envelope is directly represented whilst the internal solid morphology is simplified.

2.1. Nucleation and growth

The model is based on the CAFE model of ProCAST, originally developed at EPFL [4]. The model was recently renovated to support parallel computations (and renamed CAFE2G), before being extended for the current study. The model uses a grid of cells to track the nucleation and growth of grains nucleating either at the mould surface or in the bulk. Nucleation is assumed to be heterogeneous and athermal. The number of nucleated grains is obtained by integration of a distribution of nucleation undercooling, using either a Gaussian distribution function, as in the original model of [4], or a Log-normal function:

$$D(\Delta T)=\frac{dn}{d(\Delta T)}=\frac{n_0}{\sigma\Delta T\sqrt{2\pi}}\exp\left[-\frac{1}{2}\left(\frac{\mathrm{Ln}\left(\frac{\Delta T}{\Delta T_0}\right)}{\sigma}\right)^2\right] \tag{1}$$

where ΔT is the undercooling, ΔT_0 is the median of the distribution, σ is a dimensionless parameter and n_0 is the total number of sites per unit volume.

During the resolution the local undercooling is calculated based on the local concentration c_l :

$$\Delta T = T_m + m_l c_l - T \tag{2}$$

where T_m is the melting temperature of the pure metal and m_l is the liquidus slope.

Grains can be either dendritic or globular. For dendritic grains, only the grain envelopes are tracked using a dendrite tip kinetic model based on the classical LGK theory [11]:

$$v(\Delta T)=\frac{D_l m_l c_l^*(k\text{-}1)}{\pi^2\Gamma}\left[\mathrm{Iv}^{-1}(\Omega)\right]^2 \tag{3}$$

where c_l^* is the liquid concentration at the envelope obtained from the phase diagram at the current temperature (local equilibrium), k is the partition coefficient, Γ is the Gibbs-Thomson coefficient, and D_l is the diffusion coefficient in the liquid. The supersaturation is defined as:

$$\Omega=\frac{c_l^* - c_f}{c_l^*(k - 1)} \tag{4}$$

where c_f is a far-field concentration obtained by averaging c_l over the liquid cells situated in a region centred on the growing cell. The size of that region is a parameter of the model. For isothermal calculations and small domains, c_f, is simply obtained by averaging c_l over all the liquid cells.
Iv^{-1} is the inverse Ivantsov function, which is approximated by the following expression [12]:

$$\mathrm{Iv}^{-1}(\Omega) = 0.4567\left(\frac{\Omega}{1\text{-}\Omega}\right)^{1.195} \tag{5}$$

The envelopes are tracked on the CA grid following the same procedure as in the original model of [4]. The growth rate of globular grains is obtained with a Zener type model, $v = D\Omega/R$, where R is the radius of the globular grain, as proposed by Wu *et al.* [13].

2.2. Solute distribution

In the original CAFE model, the far-field concentrations are assumed to stay at the nominal concentration, and the grain envelopes can only be stopped by mechanical contact with other grains or the border. The model was extended to include the influence of the solute diffusion field in the intergranular liquid associated with solid growth. The solute boundary layers can affect the undercooling at the remaining nucleation sites and the growth rate of neighbouring grains (so-called soft impingement).

The distribution of solute in the intergranular liquid is obtained by solving numerically the solute conservation equation:

$$\frac{\partial c_l}{\partial t} = \nabla\cdot D_l\,\nabla c_l \tag{6}$$

The resolution is performed on the CA grid using a finite volume method and a forward Euler time-discretization scheme. The resolution is restricted to the liquid cells of the CA grid, imposing c_l^* as a boundary conditions at the grain envelopes. Numerically this is achieved by prescribing c_l^* on the cells containing a section of a grain envelope at each resolution step before solving the diffusion problem.

2.3. Thermal coupling

To correctly predict the nucleation arrest, the model must be coupled with a resolution of heat transfer and predict a possible recalescence. The model can be coupled with a finite element resolution of the heat transfer on the process scales as described in the original article [4]. However, owing to the relatively small domain sizes considered in the present study (1-10 mm^3), a simplified version of thermal coupling was used, which considers isothermal solidification conditions (no thermal gradients). Two options were implemented regarding the calculation of the temperature. The first one is to consider the temperature as an input, which is a suitable assumption for Bridgman-type conditions. The second option is to obtain the temperature from a global heat balance using the solid fraction calculated with the CA method and a microsegregation model. More specifically, the cooling rate is obtained from:

$$\frac{\mathrm{d}T}{\mathrm{d}t}=\frac{1}{\rho Cp}\left(L\frac{\mathrm{d}g_s}{\mathrm{d}t}\text{-}\dot{H}\right) \tag{7}$$

where $\dot{H}$ is a prescribed heat extraction rate per unit volume, ρC_p is the volumetric specific heat and L is the volumetric latent heat.

The solid fraction, g_s, is the volume fraction of solid in the calculation domain. It is obtained from the volume fraction occupied by the grains, g_g, and a model for the internal solid fraction. Whilst the latter is simply the unity for globular grains, a microsegregation model is required for dendritic grains.

Lever rule and Scheil approximations were implemented. A set of non-linear equations is finally obtained, and iterations should in principle be carried out to determine the temperature to be used for the CA resolution. As this procedure would be too computationally expensive a linearization technique is used to avoid iterating on the CA resolution.

2.4. *Initialization and algorithms*

During initialization, the nucleant particles are randomly distributed in space and receive random nucleation undercoolings with a probability determined by the Gaussian or log-normal distribution (equation (1)). All cells are initialized at the liquid state and the temperature is usually set at the liquidus temperature of the alloy. The liquid concentration is initialized at the nominal concentration of the alloy, $c_l = c_0$, in all cells. The time-step is taken as half of the maximum value given by the Fourier stability criterion for explicit time discretization schemes.

Nucleation arrest usually occurs at a very low solid volume fraction (typically between 0.01% and 0.1%). Hence, the calculation of the grain and solid fractions can be quite inaccurate if only based on the number of captured CA cells, because at this stage grain radii are still very small in comparison with the cell size. For a better accuracy, the contribution of the small grains to the average grain fraction is obtained analytically from the size of their growing sphere/octahedron, until they can be discretized with sufficient accuracy on the CA grid. The grain envelope is then no longer described with a single octahedron or sphere centered on the nucleation sites, but by a series of off-centered shapes. A cell grain fraction is calculated in the cells that contain a section of the grain envelope. The calculation is based on the position of the grain envelope with respect to the cell centre. This method offers a better accuracy than simply considering a binary state (captured/not captured) as it is often done in CA methods.

3. Results and discussion

3.1. *Model validation*

A potential difficulty with CA methods is to make them quantitative and guarantee a correct accuracy of the resolution. To address this difficulty, the model was first validated against the well-established models of Rappaz-Thévoz [14] and Wang-Beckermann [1]. As in [14] and [1], the model was applied to the solidification of an Al-5wt%Si alloy, considering a single dendritic grain in a spherical domain and a cooling rate at liquidus of 45 K/s. The model was applied to three different final grain radii: 0.1, 1 and 10 mm, using the physical properties listed in [1]. As can be seen in figure 1, a very good agreement is obtained between the present model and the results of [1] and [14]. The results are also compared with the original CAFE model where solute diffusion is not addressed and grain growth is only stopped by the domain border or hard impingement. The better agreement obtained with the present model than the original one already shows the benefits of including solute diffusion into CA solidification models.

3.2. *Comparison with reference X-ray radiography measurements and mean-field models*

A second series of comparisons was made with the excellent investigation done at Sintef, where data have been collected by in-situ microfocus X-radiography during near-isothermal melt solidification experiments [9]. Constant cooling rates and suppressed melt convection conditions were almost achieved, which provides ideal conditions for comparisons. Data were obtained for an Al-10%Cu alloy solidified at different cooling rates without recalescence.

The present CA model was applied to the same conditions and same nucleation parameters as the mean-field model presented in [9] for 0.1 wt% Al-5Ti-1B: $d_0 = 0.77$ µm, $\sigma = 0.5$, $n_0 = 5\times10^{11}$ m^{-3}. The calculations were carried out in a cubic domain of 1.5 mm in size, assuming a globular grain morphology.

Figure 2 shows the concentration field in the liquid at the time of nucleation arrest for different cooling rates. The concentration at the interface is higher when the cooling rate is high, which is associated with a higher undercooling at nucleation arrest.

In figure 3, the grain size and maximum nucleation undercooling are compared with the experimental and modelling data from [9]. The average grain size calculated with the CA model shows very good agreement with the experimental and mean-field results in figure 3(a). Interestingly, the CA model is slightly closer to the experimental data than the mean-field approach. This improvement is more visible in figure 3(b) where a lower maximum nucleation undercooling in substantially better experimental agreement is obtained with the CA method than mean-field. The difference can be explained in terms of overlapping diffusion fields. The CA model can account for the cumulative effect of several neighbouring grains which can stifle less potent nucleant particles in their vicinity. This effect cannot be considered in mean-field approaches where grains and nucleants are not localized.

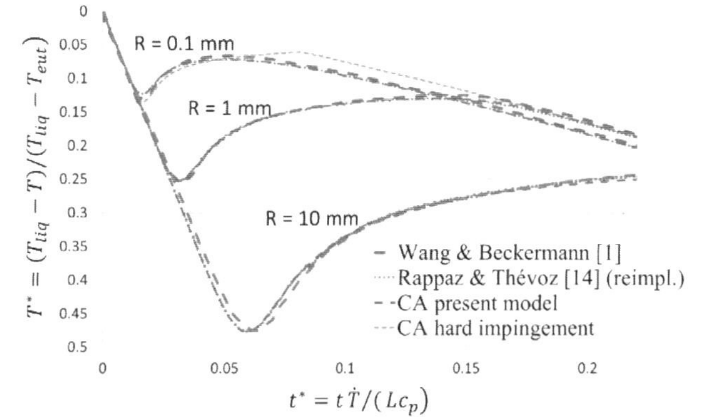

Figure 1. Comparison of the CA model with reference results for a single dendritic grain.

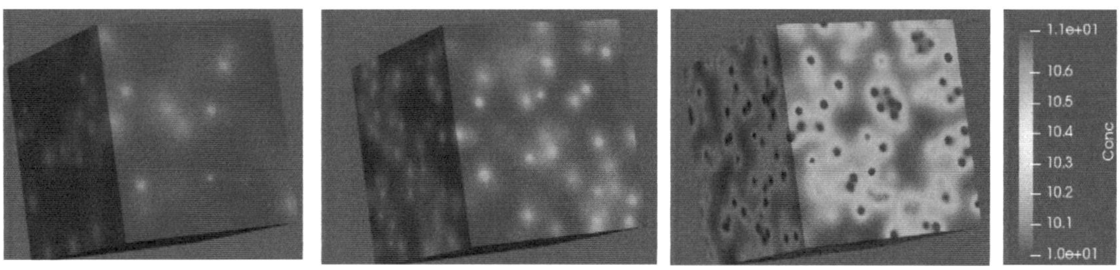

(a) (b) (c)

Figure 2. Concentration in the intergranular liquid at nucleation arrest during globular solidification in Al-10wt%Cu at (a) 0.05, (b) 0.2 and (c) 1 K/s. Solid has the same colour as liquid at the interface.

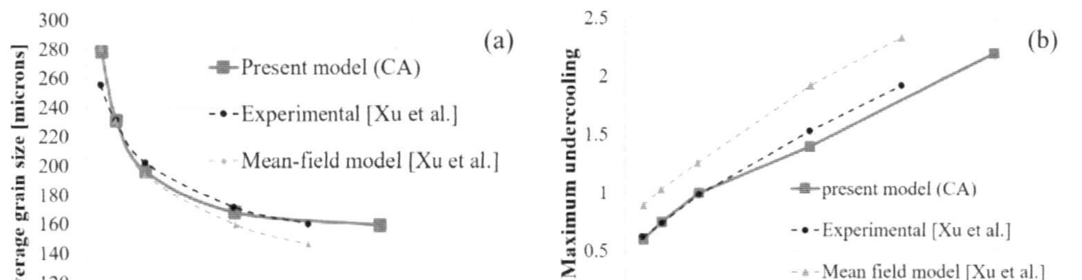

Figure 3. Predicted (a) grain sizes and (b) maximum nucleation undercoolings compared with [9].

3.3. Role of the solute suppressed nucleation zone

The influence of the solute diffusion field on nucleation arrest was analysed. Figure 4 shows the proportion of nucleated particles for different size classes or nucleation potencies. The grey bars simply correspond to the log normal size distribution of nucleation undercoolings introduced in the simulation. As expected, a large proportion of the particles in the most potent classes initiate grains whilst none of the less potent particles are activated due to nucleation arrest occurring at a lower undercooling. The red bars indicate the proportion of particles that had reached their nominal nucleation temperature before they were physically captured by the growing solid but were prevented from initiating a grain due to the local liquid concentration. This data shows the role of the solute suppressed nucleation zone, which is quite substantial under these conditions.

Similar calculations were carried out using a constant heat extraction rate rather than constant cooling, solving equation (7) to calculate the cooling curve. As can be seen in figure 5, a recalescence is obtained, which is deeper as the heat extraction rate (or cooling rate at liquidus indicated in the diagram) increases.

Figure 6 shows that the proportion of sites affected by the solute boundary layers is much lower than in the case of constant cooling. This is obviously due to the recalescence which is now the dominant mechanism for the nucleation arrest. The solute suppressed nucleation mechanism seems to be the most prominent at intermediate cooling rates. This can be understood by considering that slow cooling leads to distant grains and recalescence at low undercooling, whilst fast cooling leads to thin boundary layers.

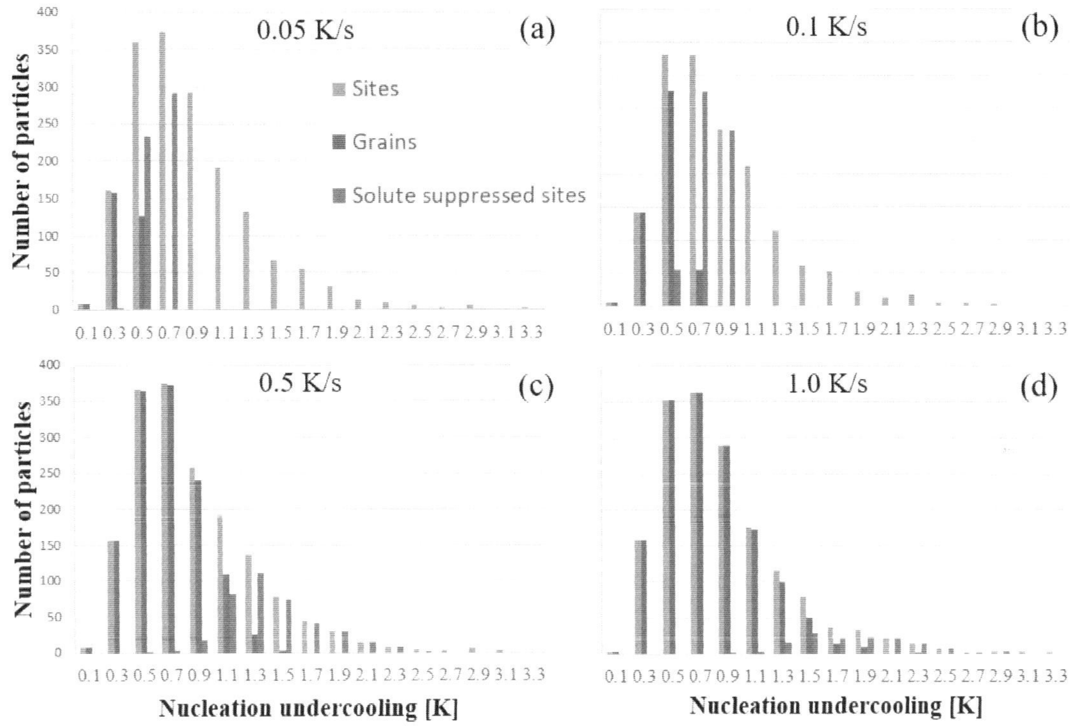

Figure 4. Effectiveness of nucleant particles during solidification in Al-10wt%Cu for various imposed cooling rates.

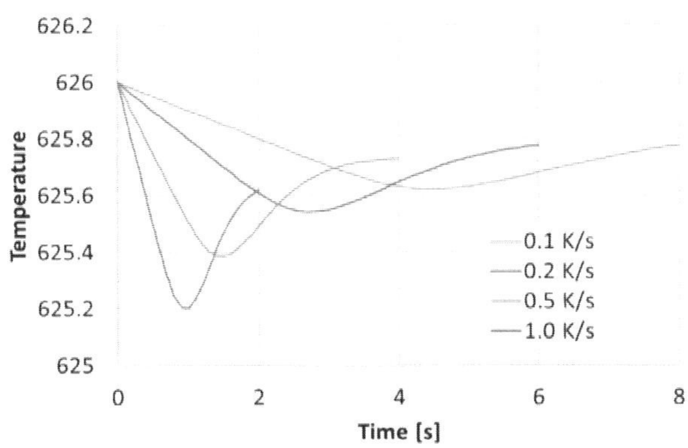

Figure 5. Calculated cooling curves for four different heat extraction rates.

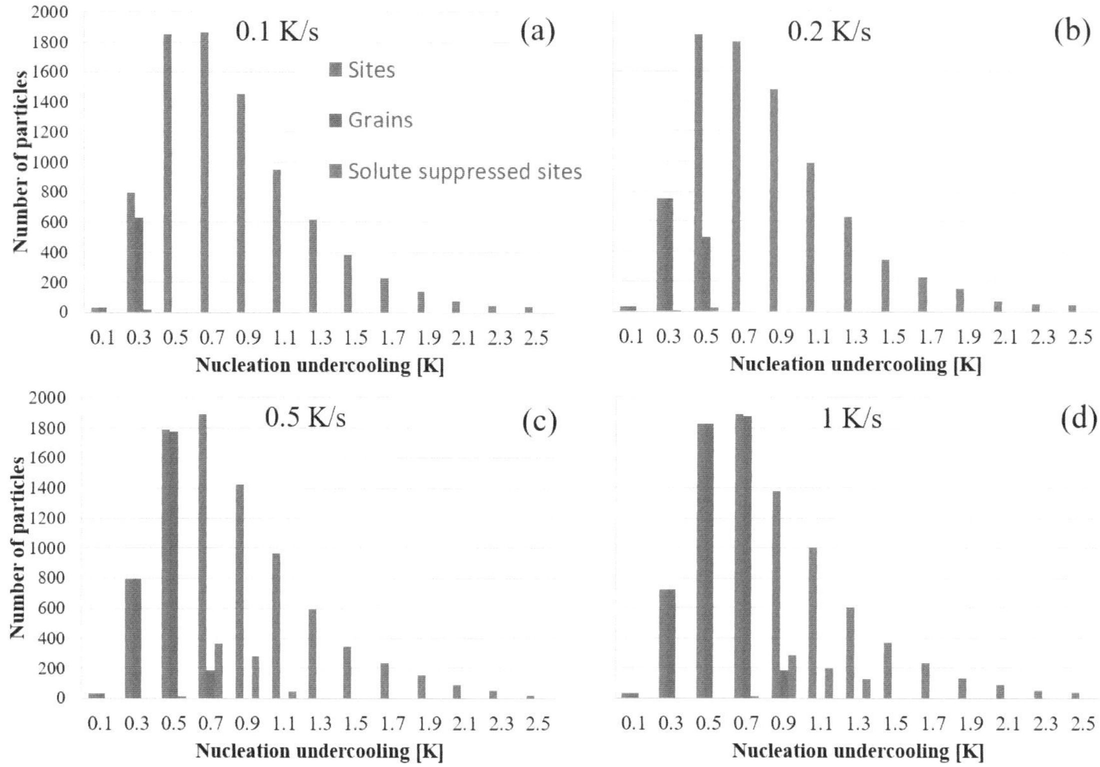

Figure 6. Effectiveness of nucleant particles during solidification in Al-10wt%Cu for various heat extraction rates.

3.4. Introducing clustering

The clustering tendency of the nucleant particles can be an important factor affecting the effectiveness of grain refinement. This phenomenon was investigated with the CA model by introducing nucleant particles according to a non-uniform probability in space. This was achieved by considering clusters of nucleant particles characterized by two parameters: an average number of particles per cluster, $\overline{N}_{cl}$, and a spreading parameter, K. The latter determines the average cluster radius, R_{cl}, obtained from the following expression: $4\pi n_0 R_{\mathrm{cl}}^3/(3\,\overline{N}_{\mathrm{cl}}) = K$. The nucleant particles were then attributed randomly to the clusters with a random position within the average cluster radius.

Calculations were carried out with K=0.1, 0.03, 0.01, 0.003, which corresponds to an increasing clustering tendency. An example of spatial distribution of nucleation sites is shown in figure 7. Figure 8 shows the cooling curves for the different conditions. As anticipated, recalescence occurs at higher undercooling as the clustering tendency increases. This is explained by an increasingly important effect of the solute suppressed nucleation zone. This was confirmed by the quantification of the proportion of nucleant particles in the SSNZ, which is shown in figure 9.

4. Conclusion

A cellular automaton model based on the new CAFE2G module of ProCAST has been developed to address the challenge of predicting the grain size in grain refined alloys without using adjustable nucleation parameters. The model incorporates all the important mechanisms affecting the final grain size: potency of the nucleant particles, competition between nucleation and growth, recalescence, the solute suppressed nucleation zone and the spatial distribution of the nucleant particles. The model has been validated against well-established mean-field models and shows very good agreement with experimental data. A first investigation of the role of clustering of the nucleant particles was carried out to demonstrate the potential of the model to address this topic in a quantitative manner.

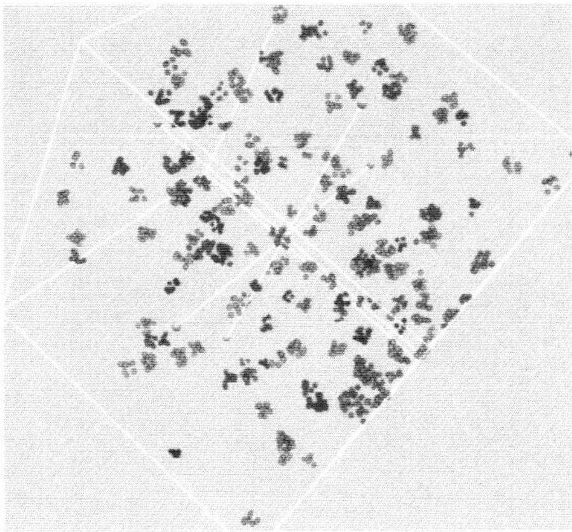

Figure 7. Spatial distribution of nucleant particles considering a clustering tendency (K=0.01).

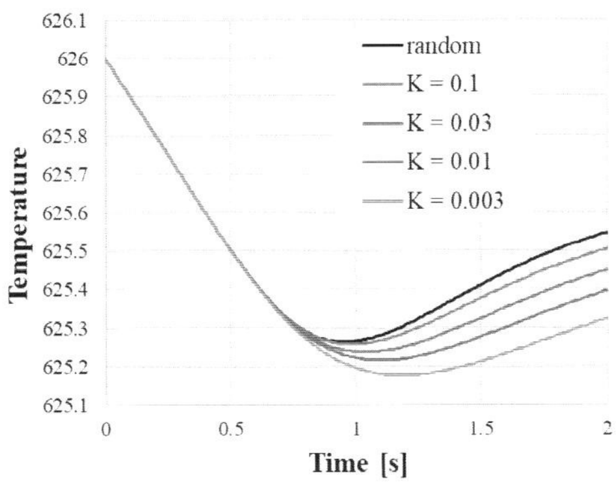

Figure 8. Cooling curves for different clustering tendencies.

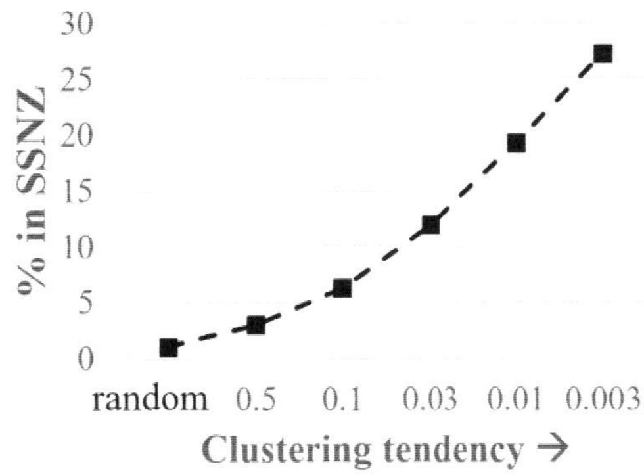

Figure 9. Proportion of nucleant particles in the SSNZ for different clustering tendencies.

Acknowledgments
The author wishes to thank Prof. Z. Fan from BCAST for the inspiring discussions on grain refinement and the useful suggestions.

References

[1] Wang C Y and Beckermann C 1993 *Mat. Sci. Eng. A* **171** 199
[2] Appolaire B, Combeau H and Lesoult G 2008 *Mat. Sci. Eng. A* **487** 33
[3] Wu M and Ludwig A 2009 *Acta Mater.* **57** 5621
[4] Gandin C A, Desbiolles J L, Rappaz M and Thévoz P 1999 *Metal. Mat. Trans. A* **30** 3153
[5] Boettinger W J, Warren JA, Beckermann C and Karma A 2002 *Ann. Rev. Mat. Res.* **32** 163
[6] Takaki T, Sakane S, Ohno M, Shibuta Y, Aoki T and Gandin CA 2018 *Materialia* **1** 104
[7] Greer A, Bunn A, Tronche, A, Evans P and Bristow D 2000 *Acta Mater.* **48** 2823
[8] Quested T and Greer A 2004 *Acta Mater.* **52** 3859
[9] Xu Y, Casari D, Du Q, Mathiesen R, Arnberg L and Li Y 2017 *Acta Mater.* **140** 224
[10] Easton M and StJohn D 1999 *Met Mat. Trans. A* **30** 1613
[11] Lipton J, Glicksman M E and Kurz W 1984 *Mater. Sci. Eng.* **65** 57
[12] Wang C Y and Beckermann C 1996 *Met. Mat. Trans. A* **27** 2754
[13] Wu M and Ludwig A 2009 *Acta Mater.* **57** 5621
[14] Rappaz M and Thévoz P 1987 *Acta Metall.* **35** 1487

IOP Conf. Series: Materials Science and Engineering **861** (2020) 012062 doi:10.1088/1757-899X/861/1/012062

Toward the development of a thermal-stress model of an industrial counter pressure casting process

C Wei[1,2], J Ou[1], P Fan[1], F Mehr[1], D Maijer[1], S Cockcroft[1], Y Zhang[2], Z Chen[2], L A[2] and Z Zhu[2]

[1] The Department of Materials Engineering, The University of British Columbia, 6350 Stores Road, Vancouver, BC, Canada V6T 1Z4
[2] CITIC DICASTAL Co., Ltd, No 185, Long Hai Road, Economic and technological development zone, Qinhuangdao City, Hebei Province, China

E-mail: weichunying@gmail.com

Abstract. It is argued that the Counter Pressure Casting (CPC) process is superior over low-pressure die casting (LPDC) in terms of reducing defects and improving cast product performance. To date, there has been relatively little research conducted on the CPC process to provide reliable data to confirm this argument. In this work, a plant trial has been done on an industrial CPC process using a custom-designed 'H-shaped' die to acquire an extensive amount of quantitative process data. The data acquired includes temperatures obtained from within the die, the casting, the surrounding environment, and within specific die cooling channels. The data has been processed and analysed to support the development of a comprehensive thermal-stress model of the casting process in order to better understand and quantify the essential macro transport processes. This paper presents a methodology of a coupled thermal-stress model development on the CPC process, and the preliminary results obtained from the models. The results to-date clearly show the need for a fully coupled thermal-stress analysis for the particular casting geometry and process conditions examined. Some of the challenges associated with the current modelling approach are also identified and potential solutions presented.

1. Introduction

In the automotive industry, aluminium has been used to replace iron and steel to reduce vehicle weight in order to improve fuel economy and reduce emissions. As a result, there are on-going activities in the research and commercial sectors to develop casting technologies to improve the mechanical properties of cast aluminium-alloy components. One such technology, the Counter Pressure Casting (CPC) process, was developed based on the conventional Low-Pressure Die Casting (LPDC) process. The CPC process is argued to produce improved cast-properties and has been employed in the casting industry to produce aluminium parts -e.g. steering knuckles and control arms. Generally, the CPC process (schematically shown in figure 1) is similar to the LPDC process. The main difference is that the die, in the CPC process, is placed in a pressurized chamber (~2 atm), whereas the die in LPDC is operated under the atmospheric pressure. Due to this difference, it is believed that the CPC process can reduce free surface turbulence during filling and enhance casting/die interfacial heat transfer during solidification. If true, both of these factors can potentially lead to fewer defects and a refined microstructure, thereby improving the mechanical performance of the component [1, 2].

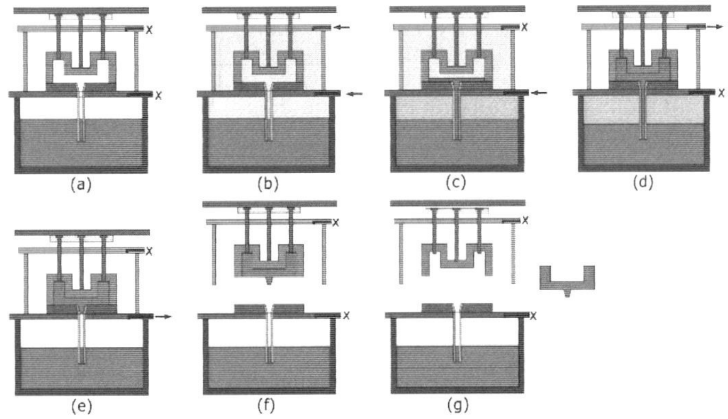

Figure 1. Schematic illustration of the CPC process: (a) the die and chamber are closed and a new casting cycle begins; (b) the furnace and chamber are pressurized with equal pressure; (c) the pressure is increased in the furnace, the liquid metal begins to fill the die cavity under the counter pressure; (d) the pressure in the chamber is released and the casting is solidified under an elevated pressure; (e) the pressure in the furnace is released; (f) the die and chamber are opened; (g) the casting is ejected and the cycle ends.

Numerical modelling has been widely applied in the casting industry as a means of designing and optimizing casting processes in order to reduce defects and improve yields. Typically, the approaches focus on fluid flow during cavity filling and heat transport. Additional areas of simulation have included microstructural evolution during solidification, macro-segregation and stress/strain evolution. As enhanced interfacial heat transfer may be one of the benefits achieved in the CPC process, a detailed, numerical-based, investigation of stress/strain development in the casting and die would appear justified in order to better characterize the benefits of the technology.

Within the casting group at UBC, there has been a sustained effort over several decades to contribute to the knowledge-base to improve casting process simulation, with a particular focus on light metal alloys. Past examples included Roy *et al.* [3] and their characterization of the constitutive behaviour of aluminium alloy A356 in as-cast condition, Thompson *et al.* [4] and their characterization of the evolution of the fraction solid in A356 under different cooling rates and Moayedinia's [5] characterization of the heat transfer within the water-based cooling channels in the LPDC process. Synthesis of this knowledge in comprehensive process models includes the work Zhang *et al.* [6] to develop a thermal model for low-pressure die casting of automotive wheels and the work by Wei [7] to develop a coupled thermal-stress model for low-pressure die casting of automotive wheels. This work has convinced the UBC group of the importance of considering the thermal/stress deformation of the die and casting in understanding interfacial heat transfer in die casting processes. Work outside of the group in this area includes studies by Fackeldey *et al.* [8] and Griffiths *et al.* [9] to name a few.

In the present work, a plant trial has been conducted using an 'H-shaped' casting to acquire quantitative data from a commercial CPC process. The data, in turn, has been used to support the development and to validate a computational model of the process. The model has been developed using the commercial FE package ABAQUS™ owing to its ability to conduct a coupled thermal/stress analysis and its capability with respect to analyses involving contact between ridged surfaces. In this paper, the foundations of the approach are highlighted.

2. Experimental measurements

2.1. Casting and die design

Figure 2 shows the geometry and dimensions of the 'H-shaped' casting that was designed for the research. The intent of the design was to create interfaces between the casting and the die that would form both gaps and pressure, thereby allowing the effect of counter-pressure on interface behaviour and heat transfer to be determined. Figure 3 shows the main die components. The system consists of a top die, a bottom die, a bottom plate, ejector pins and an ejector plate. The die is cooled by water through ten 'finger-cooling' elements and two channels internal to the die structure.

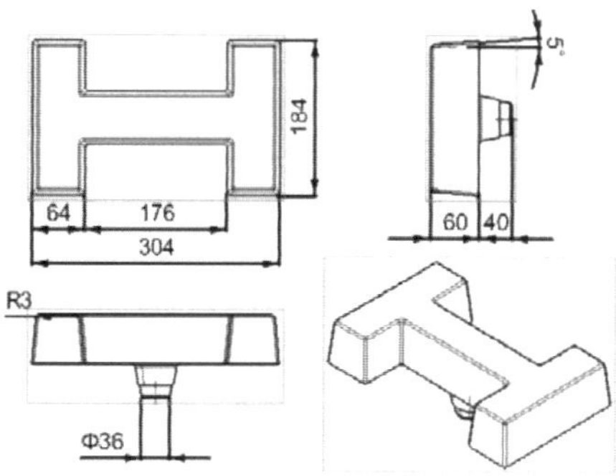

Figure 2. Drawing (in mm) of the "H-shaped" casting

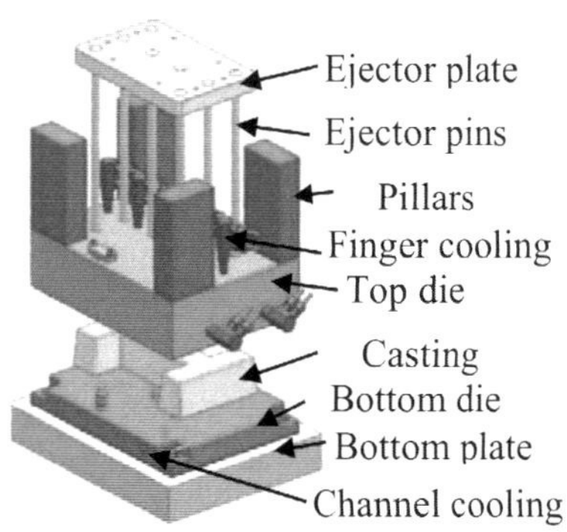

Figure 3. CPC die components

Figure 4. Commercial CPC machine (CPC 1600)

2.2. Plant trial

A plant trial was conducted using a commercial CPC machine (see figure 4) at CITIC Dicastal Co., Ltd. facility in China. In the trial, several sets of process conditions were examined with respect to the Chamber pressure, including typical CPC process conditions and conditions in which the counter-pressure in the die chamber was varied. The data collected from the trial included:

1) Temperature data: Temperature data was obtained from 60+ K-type thermocouples placed within the die, the casting, the cooling channels, on the die surfaces and in the environment surrounding the die.

IOP Conf. Series: Materials Science and Engineering **861** (2020) 012062 doi:10.1088/1757-899X/861/1/012062

2) Process data: Various process data related to the control of the casting machine was obtained including the various chamber pressures, the PLC signals controlling die motion – e.g. die open, die locking, die open and casting ejection – and PLC signals controlling the on- and off-timings for the various water-cooling elements.

3) Cast defects: X-ray and CT-based scans were performed on the castings produced under each process conditions to examine for porosity and to also identify the location of the thermocouple placed within the casting.

3. Model methods

A series of models were developed using the commercial finite element package, ABAQUS, according to the methodology described below:

1. Analysis 1 - To begin, a thermal model of the casting process was developed and run until a cyclic steady-state was achieved. The model considered only conductive heat transport and ignored die filling. The model was run sequentially, which involved input of the temperature distribution from the die at the end of a given cycle as the initial condition for the die temperature at the beginning of the subsequent cycle. The process was started by assuming an initial uniform temperature in the die and continued until the variation in die temperature between subsequent cycles was approximately invariant to within a specified tolerance. See the section below for starting initial conditions.
2. Analysis 2 - As the geometry of the die, and correspondingly the geometry of the casting, would be altered under steady-state operating conditions due to thermal expansion (relative to their room temperature dimensions), a single step thermal-stress analysis was undertaken to correct the geometry of the wheel. This was done by importing the temperature field in the die calculated from Analysis 1 and tying the die interface to the casting interface. For this analysis the Young`s modulus of the H13 die material was set to its nominal value, whereas the cast material (A356) was assigned to a low value to decrease the resistance of the die to deformation, thereby allowing the casting to deform to the geometry of the hot die, without significantly impacting on the deformation of the die. The updated, deformed, casting geometry was then re-meshed and subsequently used as the initial geometry of the casting.

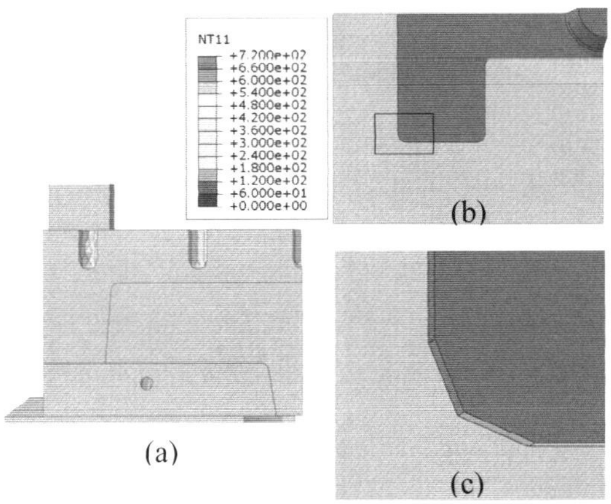

Figure 5. Model of a ¼ symmetric section of the casting showing: (a) The original geometry of the die and casting at 25 °C. (*Note: the legend does not represent the temperature in this image*) (b) The deformed die and the original casting. (Bottom view with the bottom die and bottom plate hidden) (c) Zoom in the square area shown in (b)

3. Analysis 3 - In the final analysis, a fully coupled thermal-stress model was run using the cyclic steady-state temperature field in the die obtained from Analysis 1 as the initial condition for the die, and the updated casting geometry from Analysis 2 as the initial cast geometry. The details of

this analysis have been simplified for the sake of brevity. The model also ignored the die filling stage. This approach allowed the correct contact state between the casting and die -i.e. no gaps or pressure – to be used at the beginning of the steady casting cycle analysis. Moreover, this would allow the coupled thermal-stress model to be able to correctly describe the evolution of the casting/die interface geometry and therefore heat transport across the interface within the casting cycle.

3.1. Geometry and mesh

Figure 6 presents an overview of the model analysis domain and mesh, including the four pillars, the top die, the casting, the bottom die, and the bottom plate. The casting, cooling channels as well as the casting/die interfaces on the die side were seeded in 6 mm increments for meshing, while the rest on the die components were seeded in 10 mm increments. The mesh was changed smoothly between the two regions. The mesh type for the thermal model was the 4-node linear heat transfer tetrahedron (DC3D4), and the mesh type for the coupled thermal-stress model was the 4-node thermally coupled tetrahedron, linear displacement and temperature mesh (C3D4T).

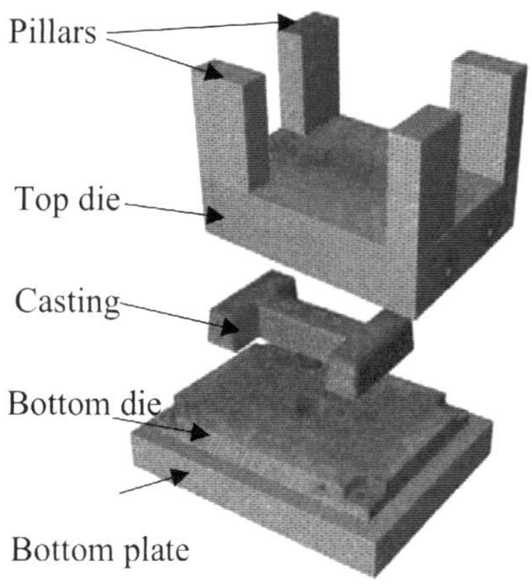

Figure 6. Meshed domain in the models

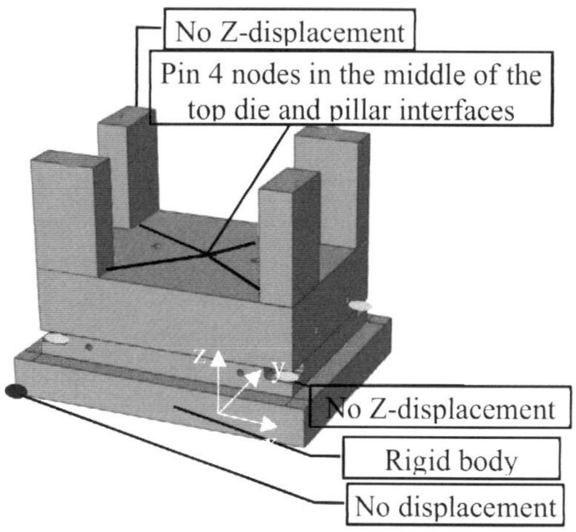

Figure 7. Mechanical boundary conditions in the thermal-stress model

3.2. Part materials and the properties

The casting material was A356 (Al-7Si-0.3Mg), and the die material was H13 tool steel. The thermophysical and constitutive properties of these two materials used in this model were extracted from the literature – see references [4, 6]. These properties include density, specific heat, thermal conductivity, latent heat, Young's modulus, Poisson's ratio, thermal expansion coefficient and yield stress.

3.3. Initial conditions

In the thermal only model (Analysis 1), the initial temperature affects the total number of cycles that are needed to achieve cyclic steady-state. To reduce the execution time needed to reach steady-state, initial temperatures of 200, 350, 350 and 300 °C, were imposed on the pillars, top die, bottom die and bottom plate, respectively. The initial temperature of the casting was set to 720 °C in all the analyses. In Analysis 2, which calculated the deformation of the die and casting at beginning of a cycle the initial temperature of the die was set to 25 °C and ramped to a final temperature of the cyclic steady state temperature distribution at the beginning of a cycle. In the coupled thermal-stress model, the initial temperature of the die components was set to cyclic steady state temperature at the beginning of a cycle.

3.4. Boundary conditions

The necessary mechanical boundary conditions used in the model are described schematically in figure 7 and were specified based on the installation of the die assembly in the trial. There are three different types of thermal boundary conditions used in the model.

3.4.1. Surface film heat transfer. A surface film heat transfer boundary condition was used for the die and casting surfaces that were exposed to the ambient environment either continuously or periodically within a casting cycle. The convective heat flux on surfaces was described using the equation:

$$q = h\left(T_{Surf.} - T_{\infty}\right) \quad (1)$$

where q (W·m^{-2}) is the heat flux, h (W·m^{-2}·K^{-1}) is the surface film coefficient and $T_{Surf.}$ (K) and T_{∞} (K) are the surface and environment temperatures, respectively.

3.4.2. Interfacial heat transfer. An interfacial heat transfer boundary condition was applied in those locations where two surfaces are in contact with each other. The heat flux across the interface was described using the equation:

$$q = h_{int}(T_{Surf.1} - T_{Surf.2}) \quad (2)$$

where h_{int} (W·m^{-2}·K^{-1}) is the interfacial heat transfer coefficient (IHTC), and $T_{Surf.1}$ (K) and $T_{Surf.2}$ (K) are the surface temperature of the surface 1 and surface 2 at the interfaces of the two components, respectively. For the casting/die interface, a temperature-dependent IHTC was applied in Analysis 1 – the thermal model -see figure 8. In the fully coupled thermal-stress model development, gap size and contact pressure-dependent IHTC was employed in the thermal-stress model development – see figure 9.

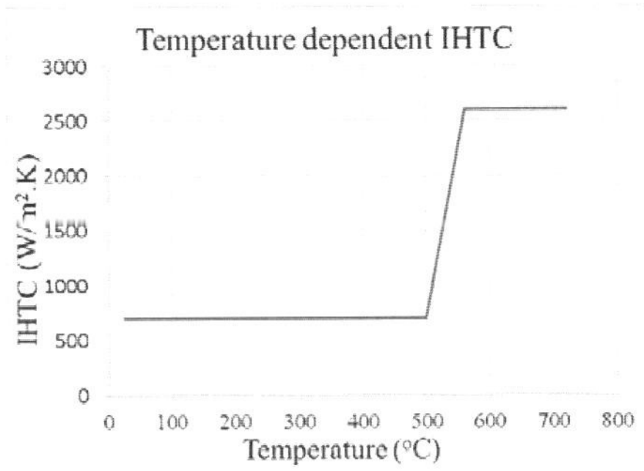

Figure 8. Temperature-dependent IHTC in the thermal model

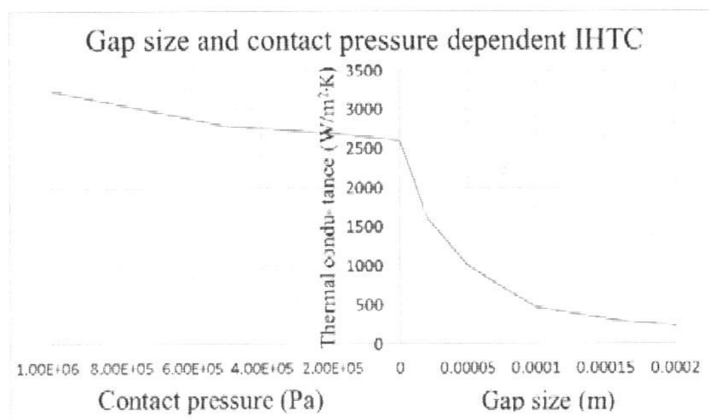

Figure 9. Gap size and contact pressure-dependent IHTC in the coupled thermal-stress model

3.4.3. Water cooling heat transfer. In the CPC process, the die was cooled by water, and the heat flux within the cooling elements can be described by the equation:

$$q = h_{CC}(T_{Die} - T_{Water}) \quad (3)$$

where h_{CC} (W·m^{-2}·K^{-1}) is the heat transfer coefficient associated with the water cooling element and T_{Die} (K), and T_{Water} (K) are the temperature of cooling channel surfaces and water, respectively. Heat transfer within the various cooling elements is complex and may involve boiling-water heat transfer, forced convective-water heat transfer or forced air-convection heat transfer (forced air-cooling is used to purge water from the cooling elements at the end of the cycle). In the model, a temperature-dependent IHTC calculated based on Moayedinia's work [5] was applied when water was on, and a relatively high IHTC was implemented based on the trial-and-error method when water was off.

4. Results and discussion

Figure 10 shows a comparison between the model predictions (both thermal-only and thermal-stress models) and the experimental data at the approximate locations of TC6, TC10, TC13 and TC21 – refer to figure 11 for locations of the four thermocouples. Figure 12 shows the regions on the casting where the two types of interfaces (gap and contact pressure) are distributed. Starting with the measured data, as shown in figure 10, all of the temperature curves obtained from the experiment show the same behaviour – i.e. a brief period of cooling associated with the die closing, a rapid rise in temperature associated with the liquid metal entering the die followed by a relatively slow decline associated with solidification and cooling of the cast component. Note: also the slight increase in cooling rate at ~253 s when the casting was ejected from the top die associated with the inner die faces being exposed to the ambient environment.

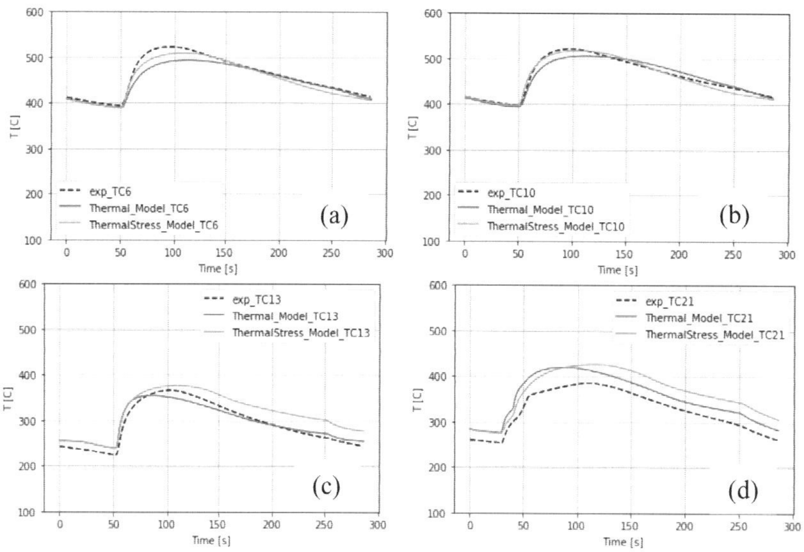

Figure 10. Comparison between model predictions and experiment data at the locations of (a) TC6, (b) TC10, (c) TC13 and (d) TC21.

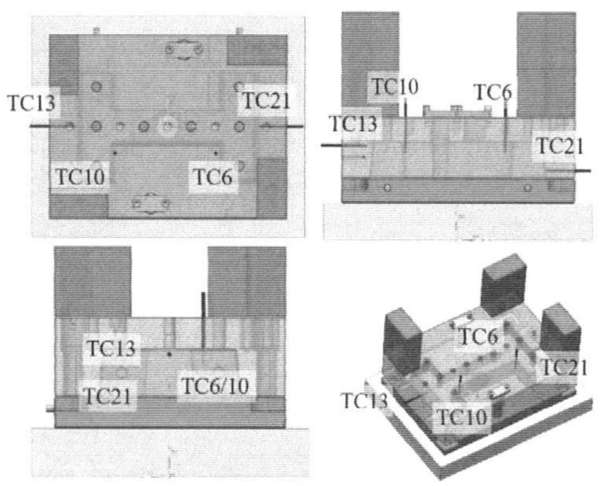

Figure 11. The locations of the TC6, TC10, TC13 and TC21 in the die (4 views)

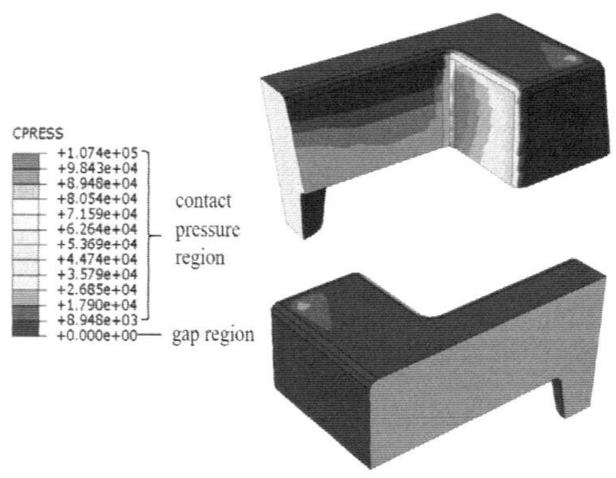

Figure 12. A ¼ symmetric section of the casting showing the regions of gap and contact pressure

Another observation from the measure data was that there was a difference in the time at which the temperature rapidly increases from 30 to 50s. This indicates a difference in the time taken for the metal to reach the location of a given TC within the die. The difference in time is large and is believed, at this stage, to be due to a combination of the back pressure and inadequate venting in the die.

Turning to the comparison between the model predictions and the measurements, it can be seen that both the thermal model and the coupled thermal-stress model predict temperatures close to the experimentally derived data, with the maximum differences around 10-40 °C. On closer inspection, it can be seen that, for the TC's located in the contact pressure region (TC6 and TC10), the model predictions matched the experimental data better than for the TC's located in regions expected to generate a gap (TC13 and TC21) (maximum temperature difference less than 20 °C). Furthermore, for TC6 and TC10, the coupled thermal-stress model fit the empirical data better than the thermal-only model.

For those TCs located in the gap formation region (TC13 and TC21), the prediction results were worse for the coupled thermal stress model indicating that the IHTC needs to be reduced further when a physical gap form. Work is on-going to correlate the IHTC with gap size and contact pressure. A critical assessment of the effects of counter pressure on temperature evolution, defect generation and microstructure will be the subject of forth coming publications.

5. Summary

A new methodology for developing a thermal-stress model was introduced in the paper, which accounts for the steady, expanded, geometry of the die. With this method, the casting/die interfacial behaviour can be better described. Moreover, a large amount of reliable data was acquired in a series of experiments conducted on a commercial CPC machine. Using this data, a thermal model and a coupled thermal-stress model of the CPC process were established and validated in ABAQUS. The preliminary results indicate that the both models can describe the casting process well. However, due to the fact that the thermal-stress model accounts in detail for the effects taking place during the casting process, it is expected that with further refinement of the gap dependent IHTC it will lead to more accurate results.

Acknowledgement

The authors thank CITIC Dicastal Co. Ltd. for the financial support of this work.

References

[1] Viswanathan S 2008 *ASM Handbook Volume 15: Casting* (ASM International)

[2] Vijayaram T R 2012 Counter Pressure Casting Technique for Aluminium Foundries *Metalworld* **12** 24–6

[3] Roy M J, Maijer D M and Dancoine L 2012 Constitutive behavior of as-cast A356 *Mater. Sci. Eng. A* **548** 195–205

[4] Thompson S 2003 *Effect of cooling rate on solidification characteristics of aluminum alloys A356 and AA5182* (University of British Columbia)

[5] Moayedinia S 2014 Quantification of cooling channel heat transfer in low pressure die casting (University of British Columbia)

[6] Zhang B, Maijer D M and Cockcroft S L 2007 Development of a 3-D thermal model of the low-pressure die-cast (LPDC) process of A356 aluminum alloy wheels *Mater. Sci. Eng. A* **464** 295–305

[7] Wei X 2014 Thermal mechanical analysis of interfacial behavior in aluminum alloy wheel casting process (University of British Columbia)

[8] Fackeldey M, Ludwig A and Sahm P R 1996 Coupled modelling of the solidification process predicting temperatures, stresses and microstructures *Comput. Mater. Sci.* **7** 194–9

[9] Griffiths W D and Kawai K 2010 The effect of increased pressure on interfacial heat transfer in the aluminium gravity die casting process *J. Mater. Sci.* **45** 2330–9

IOP Conf. Series: Materials Science and Engineering **861** (2020) 012063 doi:10.1088/1757-899X/861/1/012063

Castings distortion of gas turbine engine parts during solidification and cooling

A Inozemtsev, A Dubrovskaia, K Dongauzer, A Alapanova and A Maleev

JSC «UEC-Aviadvigatel», Perm, Russia

E-mail: dubrovskaya-as@avid.ru

Abstract. The paper studies the matters of numeric simulation of the processes of casting, crystallizing and cooling down to room temperature, and addresses the related problem of the stress-strain state of the casting process in manufacturing the cast parts of a gas turbine engine. The influence of the technological parameters of casting onto the development of the deformation defects was evaluated. The works were performed on incorporating predeformation into the burn-out model for compensating temperature deformations during casting, and on production and control of the above casting with predeformation.

1. Introduction

One of the most important tasks in modern production of aviation and industrial gas turbine engines is improving the quality of the manufactured parts and minimizing spoilage in production, while saving the expensive raw materials, energy and labor resources. At UEC-Aviadvigatel Joint Stock Company this task is being solved through development and implementation of elements of digital manufacturing which are capable of optimizing the parameters of each phase of the technological process for obtaining the required parameters of the final product by using virtual modelling and creating “digital twins”.

Cooling of a casting causes changes of its dimensions following the temperature deformations and shrinkage deformations which show as linear shrinkage and bending of the axes of individual elements or distortions. High heat-shrink deformations result in the distortions of the casting mould, and lead to bending the configuration of the casting itself. Thus, for getting an item of the targeted geometry, the probability of deformation should be considered in advance. Combined method of lost-wax casting for gating system and burn-out casting for parts allows to produce burn-out models with predeformation by 3-D printing for achievement of minimal distortion in casting parts after casting process.

Applying the numerical simulation method allows to predict the casting deformation with high accuracy at all the phases of the technological process, to improve the design of the gating system and register its distortions through no real expensive experiments but using the virtual space of the mathematical model [1-2]. The full contact problem of the stress-strain state determination is resolved in the statement “deformed mould – deformed casting”. Studies are carried out in viscoelastoplastic statement using the Perzyna model allowing to account for linear hardening.

The result of the simulation is the possibility to introduce compensatory predeformation into the burn-out model of the casting produced with 3D printing technology. This allows to significantly increase the accuracy of manufacturing and reduce the number of the subsequent machining operations; this is especially important for the thin-walled items of critical duty having complex configurations.

2. Statement of the problem

A casting of the part named "panel" (figure 1) with nominal geometry was manufactured for the purpose of the mathematical model optimization and calibration.

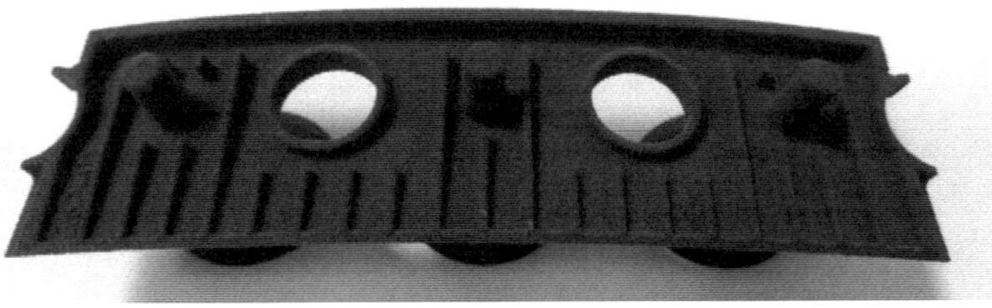

Figure 1. Burn-out model of the casting.

Non-contact optical 3D scanning revealed deviations of the castings geometry from the nominal geometry. Maximum deformation detected at ingates (bosses) exceeded 8%. Maximum deformation within the casting body was about 5.5 % (figure 2).

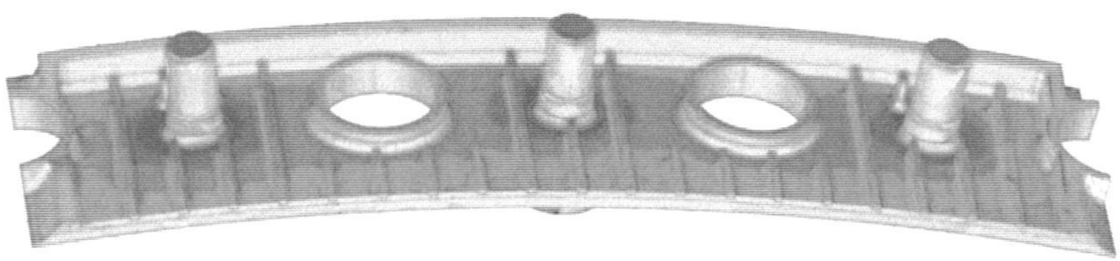

Figure 2. Results of 3D laser scanning of the cast part.

Figure 3 shows the model of the gate system for producing a casting of the "panel" and the finite-element approximation of the model, respectively.

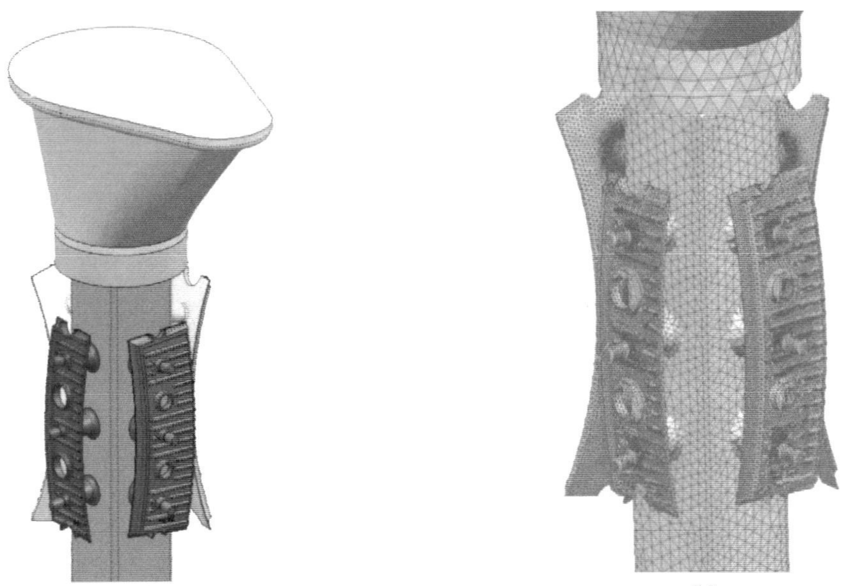

(a) (b)

Figure 3. Gate system unit model: (a) CAD model, (b) finite-element approximation of the model.

3. Mathematical statement of the problem

In the course of the study the full contact problem of the stress-strain state determination is resolved in the statement "deformed mould – deformed casting". Studies are carried out in viscoelastoplastic statement.

Physical relations are taken as follows:

$$\hat{\sigma} = {}^4D \cdot\cdot \hat{\varepsilon}^{el} = {}^4D \cdot\cdot \left(\hat{\varepsilon} - \hat{\varepsilon}^{th} - \hat{\varepsilon}^{vp} - \hat{\varepsilon}^{tr} \right) \tag{1}$$

where ${}^4D$ is quartic elastic constants tensor.

Geometrical relations look as follows:

$$\hat{\varepsilon} = \frac{1}{2}\left[\nabla \bar{u} - \left(\nabla \bar{u}\right)^T\right] \tag{2}$$

The full deformation is taken as the sum of temperature, elastic, viscoplastic deformation, plus additional contribution made by transformations in solid-liquid state [4-5]:

$$\hat{\varepsilon} = \hat{\varepsilon}^{th} + \hat{\varepsilon}^{el} + \hat{\varepsilon}^{vp} + \hat{\varepsilon}^{tr} \tag{3}$$

The temperature deformation is

$$\hat{\varepsilon}^{th} = \alpha\left(T - T_0\right) \tag{4}$$

where $\alpha(t)$ is the temperature expansion coefficient. To compute viscoplastic deformation, the Perzyna model is used that allows to consider linear hardening [3]:

$$\hat{\varepsilon}^{vp} = \frac{1}{n}\left\langle \frac{\hat{\sigma}_{eq}}{\sigma^*} - 1 \right\rangle^p \tag{5}$$

where σ^* is yield stress, σ_{eq} is von Mises equivalent stresses, n is viscous parameter of the material, and p is strain rate sensitivity coefficient.

The phase transformations are considered as follows:

$$\hat{\varepsilon}^{tr} = \frac{\beta_{tr}}{3}\hat{g}_{tr} \tag{6}$$

in which $\hat{g}_{tr}$ is volume fraction of liquid-solid phase, transforming into a new phase, and β_{tr} is the coefficient of volume expansion associated with the phase transformation.

Balance equation is taken as follows

$$\nabla \cdot \hat{\sigma} = 0 \tag{7}$$

At solid and liquid-solid phases, normal and tangential stresses look as follows:

$$\begin{cases} \sigma_n = \left(\langle \sigma \rangle \cdot \bar{n}\right)\bar{n} \\ \sigma_\tau = 0 \end{cases} \tag{8}$$

The conditions of incomplete cohesion with slippage at the contact interfaces of the casting and the ceramic mould are set as follows [2]:

$$\begin{cases} \left[\bar{u}_\tau\right] = 0 \\ \left[\hat{\sigma}_\tau\right] = q\left[\hat{\sigma}_n\right] \end{cases} \tag{9}$$

The mould displacements are limited as the mould is fastened in the casting assembly:

$$\bar{u}_k = 0 \tag{10}$$

4. Results

Numerical simulation of the cast part production was performed using the ProCAST™ simulation software [6]. The Stress module was additionally used for the strain-stress state evaluation.

Thermal fields evolution at pouring is shown in figure 4.

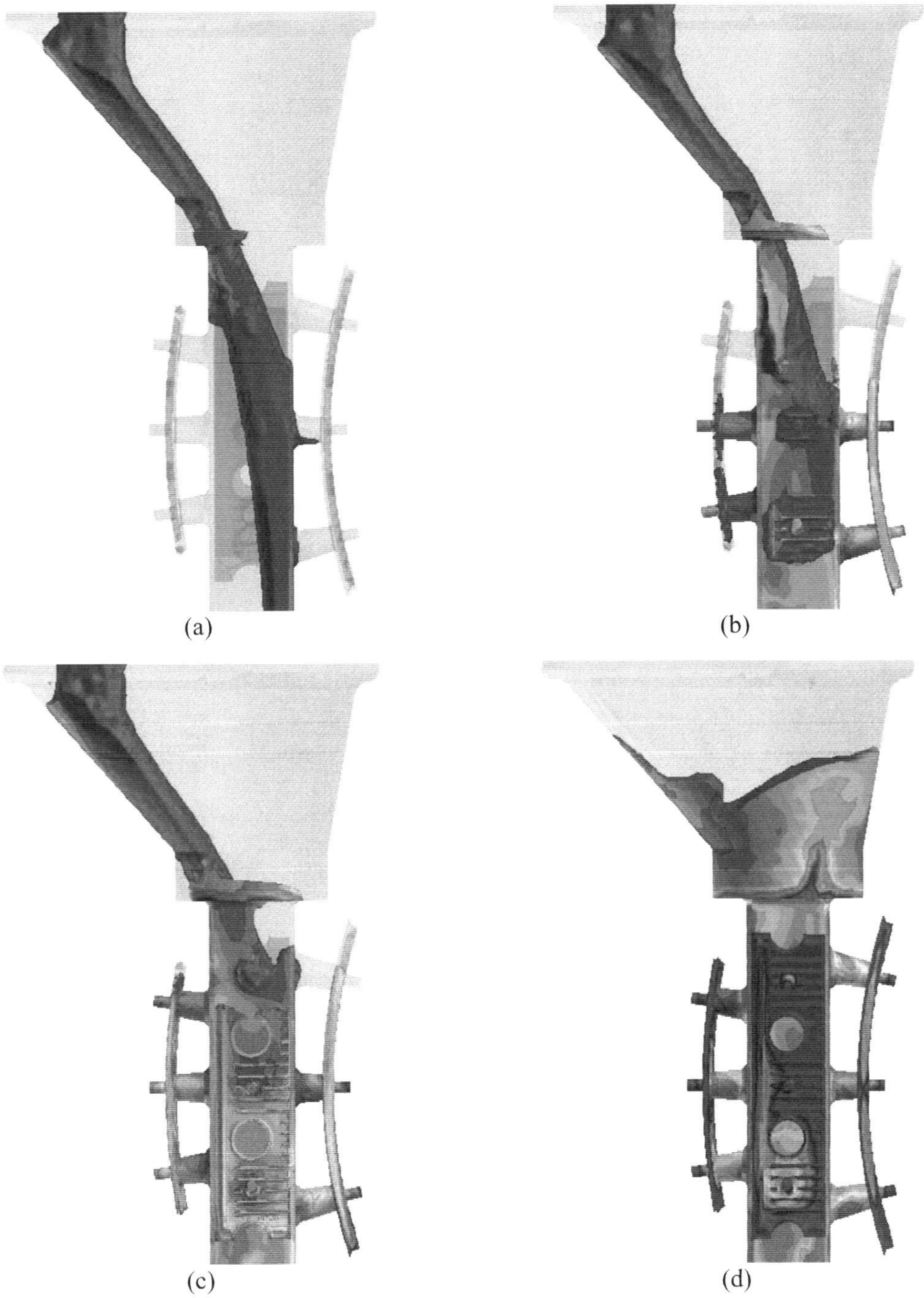

Figure 4. Thermal fields in metal during pouring: (a) initial phase, (b)-(c) middle phase, (d) end phase.

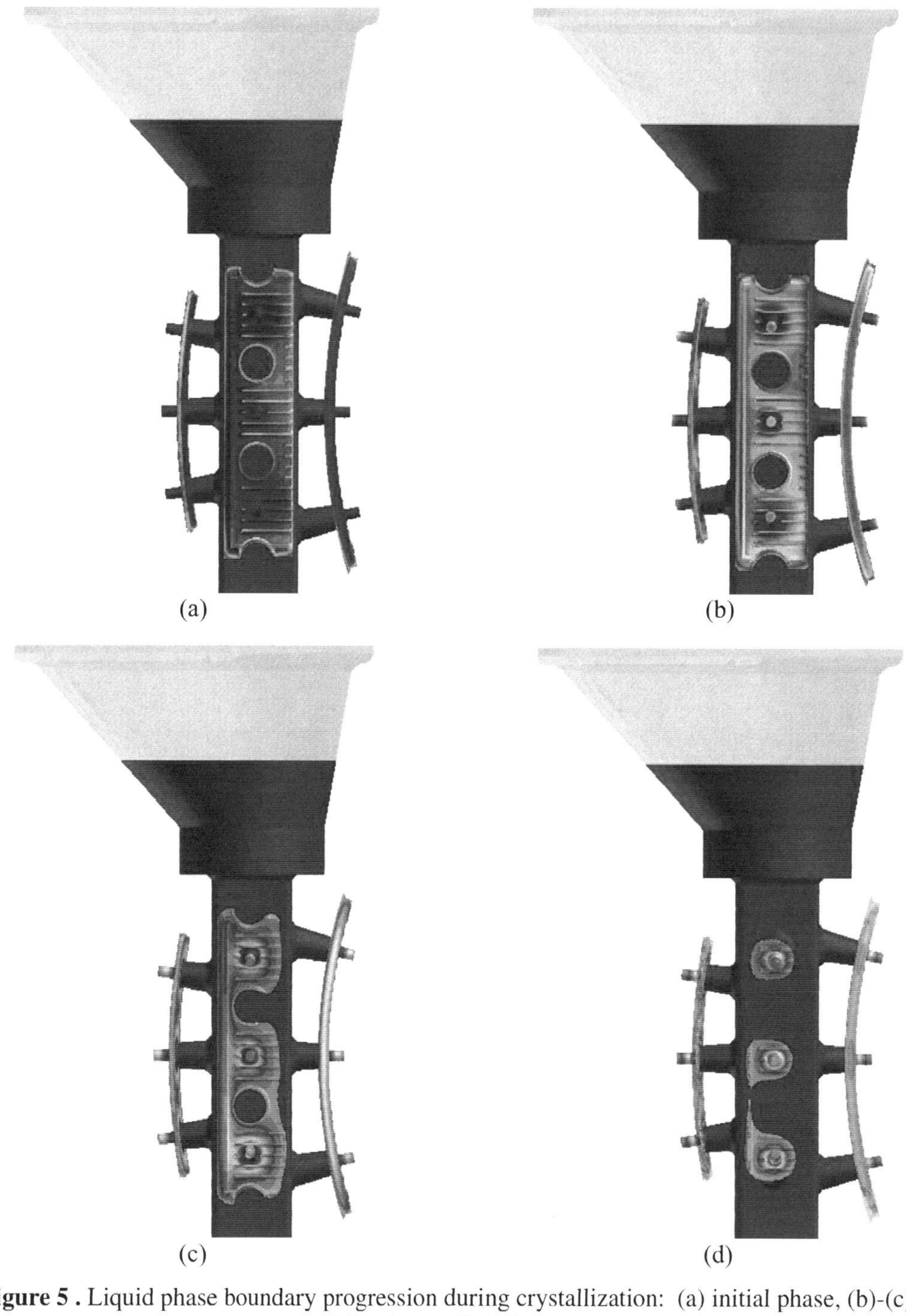

Figure 5 . Liquid phase boundary progression during crystallization: (a) initial phase, (b)-(c) middle phase, (d) end phase.

The evolution of the liquid phase boundary progression during crystallization is shown in figure 5. The degree of distortion and the final shape of the cast panel are shown in figure 6. The main reason for distortion is the difference in the thermal expansion coefficients of the Ni-based alloy and of the ceramic mould, along with the relatively low hardness of the main body of the panels as compared to the massive ingates. In the conditions of high-temperature deformations and mechanical resistance of the ceramic mould material, the localized areas of plastic deformations are formed in the interface zone of bosses and heat-shield panels.

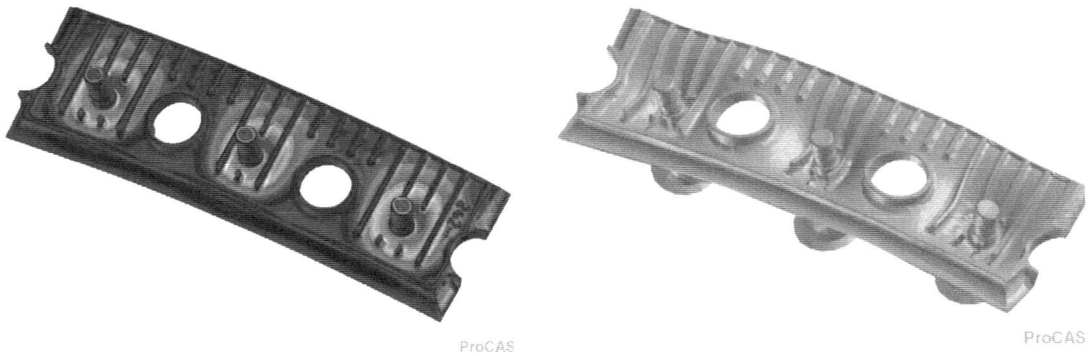

(a) (b)

Figure 6. Simulation results: (a) distortion of the cast panel, (b) the final distortion (magnification x10).

The magnitude of distortion forecasted in numeric simulation process is in good agreement with the measurements results of a real casting shown in figure 1. The scheme of the distortion development is shown in figure 7.

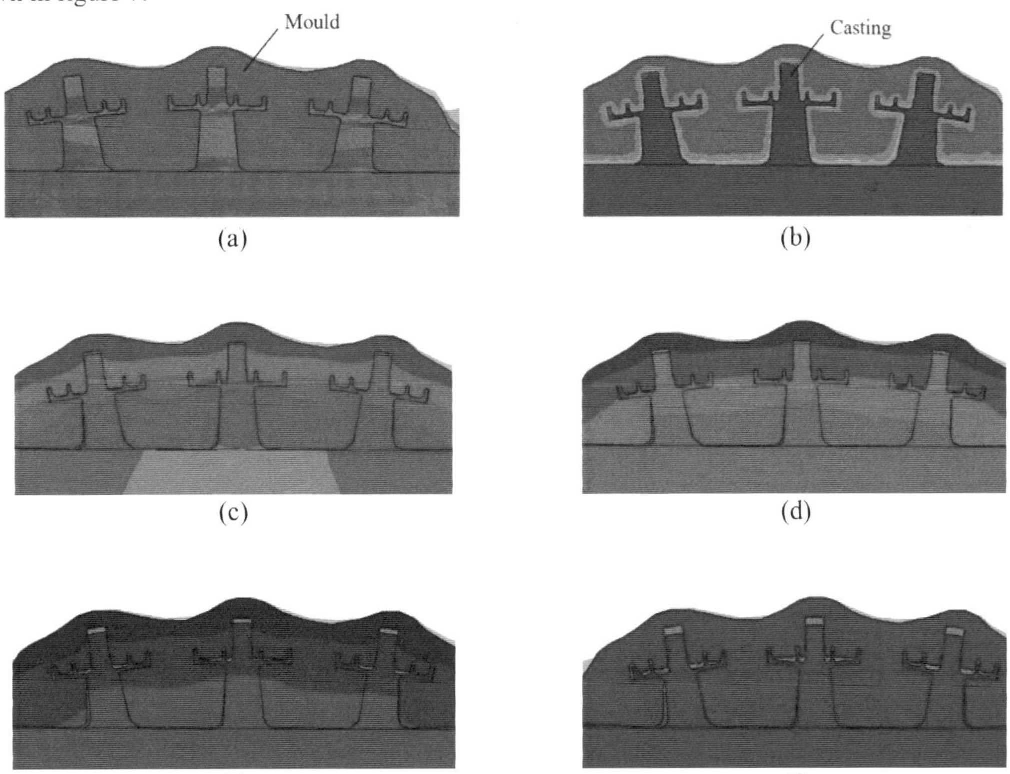

(a) (b)

(c) (d)

(e) (f)

Figure 7. Temperature field °C and deformed state evolution (magnification x5) in the process of cooling down to room temperature, the longitudinal section of part: (a) start of pouring, (b) end of pouring. Time points (c) 0.5 h, (d) 1 h, (e) 2 h, (f) at room temperature.

5. Incorporation of predeformation

Following the results of numerical simulation, predeformation was incorporated into the casting model manually using a CAD system to compensate for the temperature deformations. Figure 8 shows the forecasted magnitude of distortion in the casting with predeformation for compensation, and the shape of the cast panel after cooling, respectively.

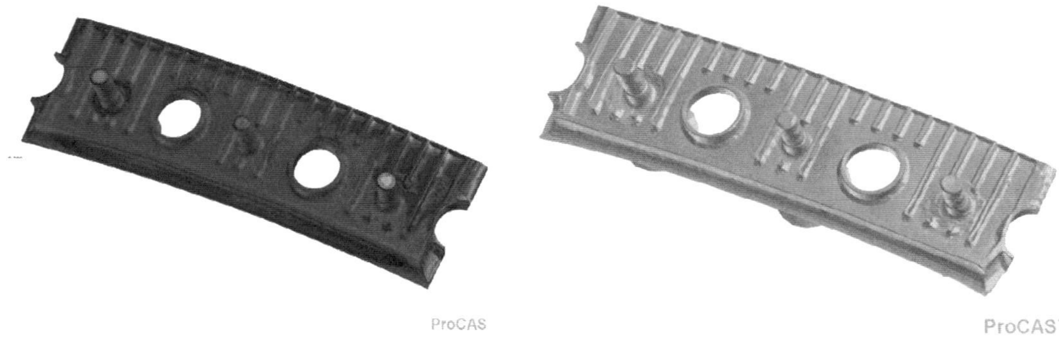

(a) (b)

Figure 8. Simulation results after predeformation was incorporated into the part model: (a) cast panel distortion (b) final distortion (magnification x10).

A casting with incrporated compensating distortions was manufactured in the production shop of UEC-Aviadvigatel JSC. The results of its 3D optical scanning are given in figure 9. The deformation at bosses was 3%, the deformation within the part body was 1%. Thus, the resulting deformations in the casting were more than 2 times lower.

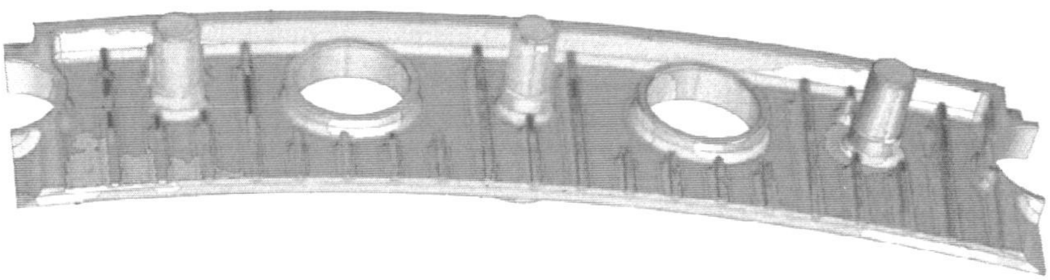

Figure 9. Results of 3D laser scanning of the cast panel with incorporated predeformation.

6. Conclusion

Numeric simulation allows to forecast the formation of residual stress fields and displacements in the process of casting. The results of computation are used for subsequent optimization of the technological process parameters in production, and for incorporating preliminary distortions into the geometric model of the part before its manufacture. In the future, it is planned to develop an automated algorithm for CAD-geometry predeformation.

References

[1] Monastyrsky A V, Monastyrsky V P and Levitan E M 2007 Casting Process Development for Industrial Gas Turbine Large Blades Using Polygon and ProCAST Systems *Cast. Prod.* **9** 29
[2] Inozemtsev A A, Dubrovskaya A S, Dongauser K A and Trufanov N A 2015 Numerical Simulation of Cast Distortion in Gas Turbine Engine Components *IOP Conf. Ser. Mat. Sci. Eng.* **84** 12039
[3] Heeres O M, Suiker A S J and de Borst R 2002 A comparison between the Perzyna viscoplastic model and the Consistency viscoplastic model *Eur. J. Mech. A-Solids* **21** 1-12

IOP Conf. Series: Materials Science and Engineering **861** (2020) 012063 doi:10.1088/1757-899X/861/1/012063

[4] Dantzig J A and Rappaz M 2009 *Solidification* (Lausanne, EPFL Press) 521
[5] Campbell J 2011 Complete Casting Handbook: Metal Casting Processes, Techniques and Design (Oxford, Elsevier) 465
[6] Inozemtsev A A, Dubrovskaya A S, Dongauzer K A and Trufanov N A 2014 Numerical Simulation for Optimizing Temperature Gradients during Single Crystal Casting Process *ISIJ International* **54** 254

IOP Conf. Series: Materials Science and Engineering **861** (2020) 012064 doi:10.1088/1757-899X/861/1/012064

Molecular dynamics study of composition dependence of solid-liquid interfacial energy of Fe-Ni binary alloy

K Ueno and Y Shibuta

Department of Materials Engineering, The University of Tokyo, 7-3-1, Hongo, Bunkyo-ku, Tokyo 113-8656, Japan

E-mail: ueno@mse.mm.t.u-tokyo.ac.jp

Abstract. Interfacial properties strongly affect solidification microstructure of practical alloy materials. Therefore, it is essential to understand interfacial properties including composition dependence for target alloys. In this study, composition dependence of solid-liquid interfacial energy of Fe-Ni alloy is investigated by molecular dynamics simulations in conjunction with a capillary fluctuation method. The solid-liquid interfacial energy of Ni-rich composition taking FCC-solid phase is larger than that of Fe-rich composition taking BCC-solid phase. Moreover, solid-liquid interfacial energy has a high anisotropy at the composition with FCC-solid phase compared to the one with BCC-solid phase.

1. Introduction

It is well known that solid-liquid interfacial properties strongly affect solidification microstructure in practical production process of metallic materials [1, 2]. Therefore, it is essential to estimate solid-liquid interfacial properties precisely. In addition to many experimental approaches, in recent years, numerical approaches such as molecular dynamics (MD) simulations have contributed to the estimation of these properties. For example, there are many reports on solid-liquid interfacial energies of pure metals on the basis of the MD simulation in conjunction with a capillary fluctuation method [3-6] and a classical nucleation theory-based technique [7]. However, there are not so many reports on those of alloys till now since it is not straightforward to reproduce proper solute partition at the solid-liquid interface for alloy systems by the MD simulation with a given interatomic potential. At present, there are several reports focusing on this problem by a semi-grand canonical Monte-Carlo (SGCMC) technique [8-10], an atom swap technique in line with a Metropolis Monte-Carlo (MMC) simulation [11] and a direct monitoring of solute partition at the solid-liquid interface by a long-time MD simulation [11]. These techniques have contributed to the estimation of interfacial properties of various alloys [12-15]. Nevertheless, there are not so many reports on composition dependence of interfacial properties for alloy systems since it takes time and effort. To this end, composition dependence of solid-liquid interfacial energy of Fe-Ni binary alloy is investigated by MD simulations in conjunction with capillary fluctuation method in this study.

Here, Fe-Ni binary alloy is employed as a target alloy material since Fe-Ni alloy has an interesting feature from the scientific viewpoint that the stable solid phase changes from body-centered cubic (BCC) to face-centered cubic (FCC) with increasing Ni composition. Fe-Ni binary alloy is also important from the practical viewpoint since Fe-Ni alloy has unique properties such as a low coefficient of thermal expansion (known as invar alloy) and a large magnetic permeability (known as permalloy).

2. Methodology

The solid-liquid interfacial energy of Fe-Ni alloy is investigated by MD simulation in conjunction with a capillary fluctuation method [3]. Simulation methodology basically follows our previous study [14, 15]. MD simulations are performed by Large-scale Atomic/Molecular Massive Parallel Simulator (LAMMPS) [16]. An embedded atom method (EAM) potential fitted by Bonny *et al.* [17] is employed as the interatomic potential of Fe-Ni alloy. Parameter files of the EAM potential are employed from the database of NIST-IPR project [18] for the MD simulations by LAMMPS. Nose-Hoover thermostat and barostat [19, 20] are employed to control temperature and pressure in MD simulations. Visualization of simulation results and post-analyses are performed by the open visualization tool (OVITO) [21]. The polyhedral template matching (PTM) technique [22] is employed for classifying atoms in crystalline or liquid, which is suitable for the classification of atoms at high temperature. A root-mean-square deviation (RMSD) cutoff parameter of the PTM is set to 0.16.

3. Results and discussion

3.1. Equilibrium temperature for each composition

Prior to derivation of the solid-liquid interfacial energy, equilibrium temperatures at various Ni compositions, at which the solid-liquid interface does not move, are estimated for the potential employed in this study using a convergence temperature technique [23, 24]. Firstly, a solid FCC crystals of Fe-xat%Ni alloys (x = 15, 30, 40, 50, 60, 70, 80, 90 and 100 (i.e., pure Ni)) consisting of 100×25×25 unit cells (250,000 atoms) is annealed at 5000 K for 10 picoseconds (ps) with the number of atoms, volume and temperature (NVT)-constant ensemble to obtain liquid structures. For pure Fe and Fe-rich Fe-Cr alloy (5 at% and 10 at%), solid BCC crystal consisting of 100×25×25 unit cells (125,000 atoms) is employed. A solid crystal of same size and composition is connected with each liquid facing (100) plane at the interface. Then, energy minimization is performed to avoid the unexpected proximity of atoms at the solid-liquid interface. Prepared solid-liquid biphasic systems of various compositions are relaxed with the number of atoms, pressure and temperature (NPT)-constant ensemble for 25 ps at several temperatures near 1700 K. For example, 1700, 1750 and 1800 K are employed for pure Ni. Then, the systems are relaxed with the number of atoms, pressure and enthalpy (NPH)-constant ensemble for subsequent 2975 ps. Pressure is maintained at 0 Pa in all processes. In this technique, the convergence temperature of the system starting from several initial temperatures is regarded as the equilibrium temperature of target composition [24].

Table 1 shows estimated values of the equilibrium temperature of Fe-Ni binary alloy with various compositions for the EAM potential employed in this simulation. The solid FCC structure naturally transforms into bcc one during the relaxation for Fe-15at%Ni alloy. Therefore, we define threshold composition dividing FCC and BCC phases between 15 and 30at%Ni. Note that this value is higher than the value from experimental reports (around 5at%Ni) due to the characteristic of the EAM potential employed in this study [17]. There is almost no solute partition at the solid-liquid interface for all compositions, which basically agrees with the experimental feature.

3.2. Capillary fluctuation method

Next, solid-liquid interfacial energy of Fe-Ni binary alloy is derived by a capillary fluctuation method [3]. A quasi two-dimensional solid-liquid biphasic system is employed to obtain atomic configurations of solid-liquid interface for each composition, which is prepared as follows [14]. Firstly, a liquid structure is prepared by annealing a BCC (259,200 atoms in a cell of 514.8×11.44×514.8 Å^3) for Fe-rich composition and an FCC crystal (345,744 atoms in a cell of 517.44×14.08×517.44 Å^3) for Ni-rich composition at 5000 K at 10 ps with the number of atoms, volume and temperature (NVT)-constant ensemble, respectively. Then, a solid crystal of the same size is connected with the liquid facing (100) plane at the interface, respectively. The short side of the system is set to be [010] direction (notated as (100)[010] orientation hereafter). Figure 1 shows the snapshot of the atomic configuration of pure Ni with the (100) interface. In the same manner, biphasic systems with (110)[001] and (110)[1-10] interfaces are prepared. Details of configuration of the biphasic system are listed in table 2. Then, the energy minimization is performed for the combined structure to avoid unexpected proximity of atoms at the solid-liquid interface. All prepared biphasic systems are relaxed with the number of atoms, pressure and temperature (NPT)-constant ensemble for 100 ps at 0 Pa.

Table 1. Equilibrium temperature and stable solid structure at that temperature for Fe-Ni binary alloy with various compositions estimated with the EAM potential.

Composition	Solid Structure	T_{eq} [K]
Pure Fe	BCC	1768
Fe-5at%Ni	BCC	1751
Fe-10at%Ni	BCC	1740
Fe-15at%Ni	BCC	1722
Fe-30at%Ni	FCC	1726
Fe-40at%Ni	FCC	1724
Fe-50at%Ni	FCC	1743
Fe-60at%Ni	FCC	1749
Fe-70at%Ni	FCC	1740
Fe-80at%Ni	FCC	1719
Fe-90at%Ni	FCC	1681
Pure Ni	FCC	1627

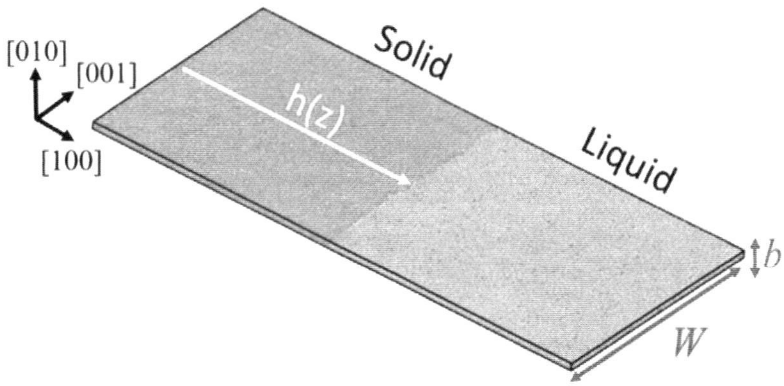

Figure 1. Quasi 2D solid-liquid biphasic system of pure Ni with (100) interface.

Table 2. Size of simulation cells for three interface orientations.

Orientation	Number of Atoms	W [Å]	b [Å]
$(100)[010]_{bcc}$	518,400	514.8	11.44
$(100)[010]_{fcc}$	691,488	517.44	14.08
$(110)[001]_{bcc}$	516,128	513.671	11.44
$(110)[001]_{fcc}$	692,224	517.715	14.08
$(110)[1\text{-}10]_{bcc}$	548,640	514.8	12.134
$(110)[1\text{-}10]_{fcc}$	733,824	517.44	14.9341

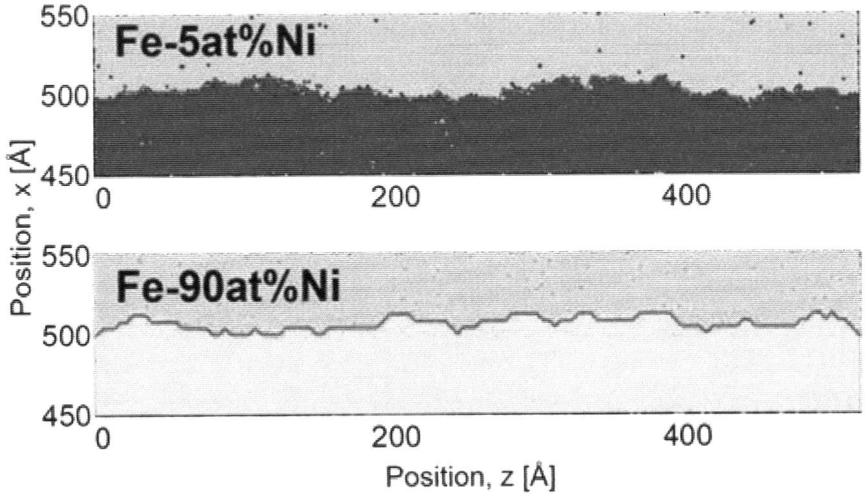

Figure 2. Snapshots of the solid-liquid interface for Fe-5at%Ni and Fe-90at%Ni. Blue, green and white atoms represent atoms with BCC, FCC and liquid configurations, which are defined by the polyhedral template matching (PTM) technique. The solid-liquid interface is drawn by the red curve.

Figure 2 shows the snapshots of the solid-liquid interface for Fe-5at%Ni and Fe-90at%Ni with (100)[010] orientation after 100 ps calculation. Position of the solid-liquid interface is drawn by the red curve, which is then extracted in the same manner of our previous simulations [14, 15]. Coordinates of the solid-liquid interface extracted from snapshots between 60 and 100 ps at 0.1 ps intervals (i.e., 401 snapshots) are used for capillary fluctuation method [3]. In the equilibrium solid-liquid biphasic system, solid-liquid interface is fluctuated in order to minimize the free energy. The interface stiffness can be estimated from the amplitude of the fluctuation of the solid-liquid interface on the basis of the following equation:

$$\langle |A(k)|^2 \rangle = \frac{k_B T_{eq}}{bW\tilde{\gamma}k^2} \tag{1}$$

where $\tilde{\gamma}$ is the solid-liquid interfacial stiffness, $A(k)$ is the Fourier amplitude of the interface height profile, k is the wave number, k_B is Boltzmann's constant and T_{eq} is the equilibrium temperature at which the solid-liquid interface does not move. b and W are the thickness and length of the interface, respectively. The brackets represent averages over the number of sampling times. Therefore, the interfacial stiffness can be estimated from the slope of the relationship between the k^2 and $1/\langle|A(k)^2|\rangle$ in the fluctuation spectrum of a quasi-one-dimensional interface. Here, interfacial stiffness of solid-liquid interfaces with (100)[010], (110)[001] and (110)[1-10] orientations is estimated. Then, solid-liquid interfacial energy is derived using the following relations [14, 25]:

$$\tilde{\gamma} = \gamma_0\left(1 - \frac{18}{5}\varepsilon_1 - \frac{80}{7}\varepsilon_2\right) \quad \text{(for (100)[010] orientation)} \tag{2}$$

$$\tilde{\gamma} = \gamma_0\left(1 + \frac{39}{10}\varepsilon_1 + \frac{155}{14}\varepsilon_2\right) \quad \text{(for (110)[001] orientation)} \tag{3}$$

$$\tilde{\gamma} = \gamma_0\left(1 - \frac{21}{10}\varepsilon_1 + \frac{365}{14}\varepsilon_2\right) \quad \text{(for (110)}[1\bar{1}0]\text{ orientation)} \tag{4}$$

where γ_0 represents the average solid-liquid interfacial energy and ε_1 and ε_2 are the anisotropy parameters. Practically, γ_0, ε_1 and ε_2 are specifically obtained by assigning values of interfacial stiffness for three different orientations into equations (2), (3) and (4). These equations are derived from a low-order expansion consistent with cubic symmetry [25]. Finally, the solid-liquid interfacial energy including its anisotropy can be derived from Fehlner and Vosko's equation [25] expressed as:

$$\gamma = \gamma_0 \left[1 + \varepsilon_1 \left(\sum_{i=1}^{3} n_i^4 - \frac{3}{5}\right) + \varepsilon_2 \left(3\sum_{i=1}^{3} n_i^4 + 66 n_1^2 n_2^2 n_3^2 - \frac{17}{7}\right)\right] \tag{5}$$

where n_i represent the orientation indices of the solid-liquid interface (n_1, n_2, n_3). The solid-liquid interfacial energy of (100), (110) and (111) orientations are derived by assigning γ_0, ε_1 and ε_2 into equation (5). These calculations are performed three times for each configuration using different initial configurations of liquid atoms.

3.3. Solid-liquid interfacial energy of Fe-Ni alloy

Figure 3 shows average solid-liquid interfacial energy as a function of Ni composition for Fe-Ni binary alloy. Error bar represents standard deviation of three calculations. As a whole, solid-liquid interfacial energy at Ni-rich composition taking an FCC solid structure is larger than the one at Fe-rich composition taking a BCC solid structure. This agrees with a general trend of the relationship in the solid-liquid interfacial energy of pure metals between FCC and BCC [4]. In the FCC solid region, solid-liquid interfacial energy basically increases with increasing Ni composition until about 50at%Ni and becomes almost constant with some fluctuation at more than 50at%Ni. On the other hand, solid-liquid interfacial energy basically decreases with increasing Ni composition at the BCC solid region. Figure 4 represents anisotropy parameters ε_1 and ε_2 as a function of Ni composition. It is confirmed that ε_1 in the FCC solid region is larger than that in the BCC solid regions. As a result, the solid-liquid interface for Ni-rich composition with FCC solid structure has a strong anisotropy than that for Fe-rich composition with BCC solid structure. This agrees with previous reports on the comparison of the anisotropy in the solid-liquid interfacial energy between FCC and BCC metals [6].

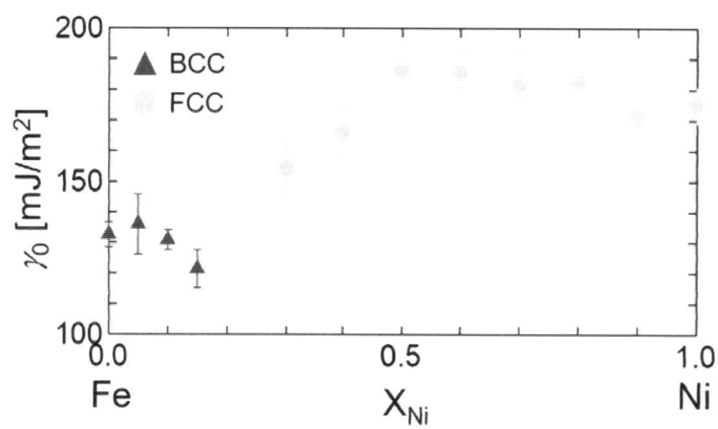

Figure 3. Average solid-liquid interfacial energy, γ_0 of Fe-Ni alloy as a function of Ni composition. Error bar represents standard deviations of three calculations.

IOP Conf. Series: Materials Science and Engineering **861** (2020) 012064 doi:10.1088/1757-899X/861/1/012064

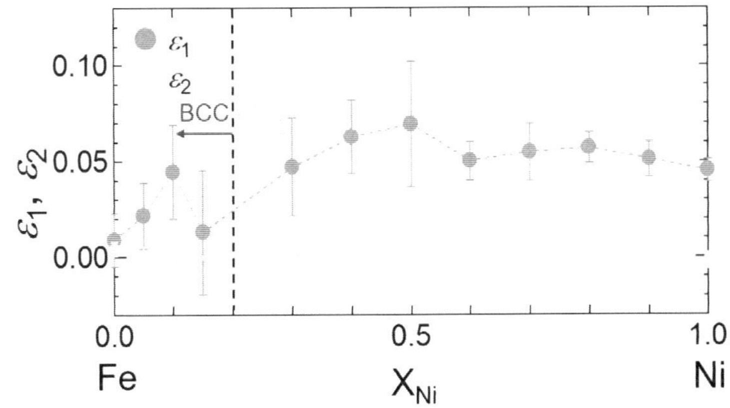

Figure 4. Anisotropy parameters, ε_1, ε_2 for Fe-Ni alloy as a function of Ni composition. Error bar represents standard deviations of three calculations.

Figure 5 shows the solid-liquid interfacial energy of (100), (110) and (111) orientations as a function of Ni composition. It is confirmed that there is a clear tendency, $\gamma_{(100)} > \gamma_{(110)} > \gamma_{(111)}$ at all compositions regardless of its solid structure. This tendency also agrees with previous reports both for FCC and BCC crystals [4, 12, 14]. The difference in the solid-liquid interface among three different orientations becomes larger at Ni-rich composition with FCC solid structure compared to that at Fe-rich composition with BCC solid structure since anisotropy parameters for the former composition is larger than that for the letter composition as shown in figure 4.

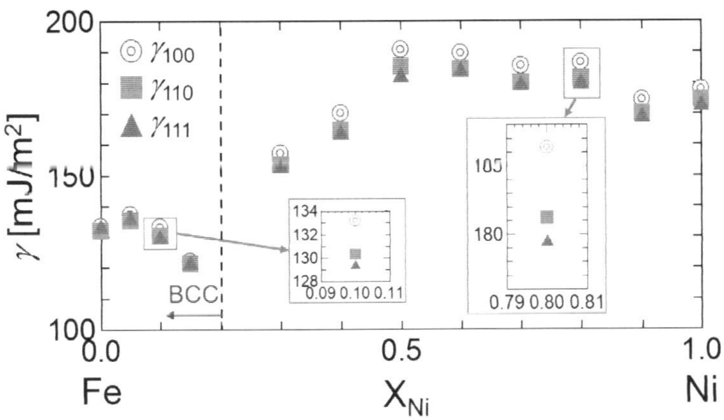

Figure 5. Solid-liquid interfacial energies, γ including the anisotropy effect for Fe-Ni alloy as a function of Ni composition. Insert figures represent enlarged views of representative plots.

4. Conclusions

In this study, solid-liquid interfacial energy of Fe-Ni binary alloy is investigated by MD simulations in conjunction with the capillary fluctuation method. It is revealed that the solid-liquid interfacial energy of Ni-rich composition taking FCC-solid phase is larger than that of Fe-rich composition taking BCC-solid phase. Moreover, the solid-liquid interfacial energy has a higher anisotropy in the FCC solid region than in the BCC solid region. On the other hand, there is a same tendency, $\gamma_{(100)} > \gamma_{(110)} > \gamma_{(111)}$ at all compositions regardless of its solid structure. It is significant in this study to reveal composition dependence of solid-liquid interfacial energy of Fe-Ni alloy, which has a different stable phase of solid depending on the composition. Utilizing this methodology, solid-liquid interfacial energy for many other alloys with various stable solid phase will be examined in the next step.

IOP Conf. Series: Materials Science and Engineering **861** (2020) 012064 doi:10.1088/1757-899X/861/1/012064

References

[1] Dantzig J A and Rappaz M, *Solidification* 2009 (Lausanne: EPFL Press) pp. 287-344
[2] Shibuta Y, Ohno M and Takaki T 2015 *JOM* **67** 1793
[3] Hoyt J J, Asta M and Karma A 2001 *Phys. Rev. Lett.* **86** 5530
[4] Sun D Y, Asta M, Hoyt J J, Mendelev and Srolovitz 2004 *Phys. Rev. B* **69** 020102
[5] Brown N T, Martinez E and Qu J 2017 *Acta Mater.* **129** 83
[6] Kavousi S, Novak B R, Zaeem M A and Moldovan D 2019 *Comp. Mater. Sci.* **163** 218
[7] Watanabe Y, Shibuta Y and Suzuki T 2010 *ISIJ Int.* **50** 1158
[8] Becker C A, Asta M, Hoyt J J and Foiles S M 2006 J. *Chem. Phys.* **124** 164708
[9] Becker C A, Olmsted D L, Asta M, Hoyt J J and Foiles S M 2009 *Phys. Rev. B* **79** 054109
[10] Ueno K and Shibuta Y 2019 *IOP Conf. Ser.: Mater. Sci. Eng.* **529** 012037
[11] Ueno K and Shibuta Y 2018 *Materialia.* **4** 553
[12] Qi C, Xu B, Kong L T and Li J F 2017 *J. Alloys Compounds.* **708** 1073
[13] Hoyt J J, Raman S, Ma N and Asta M 2018 *Comp. Mater. Sci.* **154** 303
[14] Ueno K and Shibuta Y 2019 *Comp. Mater. Sci.* **167** 1
[15] Ueno K and Shibuta Y 2019 *ISIJ Int.* submitted
[16] Plimpton S J 1995 *J. Comput. Phys.* **117** 1, http://lammps.sandia.gov
[17] Bonny G, Pasianot R C and Malerba L 2009 *Model. Simul. Mater. Sci. Eng.* **17** 025010
[18] Becker C A, Tavazza F, Trautt Z T and Buarque de Macedo R A 2013 *Curr. Opin. Solid State Mater. Sci.* **17** 277
[19] Nose S 1984 *J. Chem. Phys.* **81** 511
[20] Hoover W G 1985 *Phys. Rev. A* **31** 1695
[21] Stukowski A 2010 *Model. Simul. Mater. Sci. Eng.* **18** 015012
[22] Larsen P M, Schmidt S and Schiøtz J 2016 *Model. Simul. Mater. Sci. Eng.* **24** 055007
[23] Shibuta Y, Takamoto S and Suzuki T 2008 *ISIJ Int.* **48** 1582
[24] Shibuta Y 2019 *Mater. Trans.* **60** 180
[25] Fehlner W R and Vosko S H 1976 *Can. J. Phys.* **54** 2159

Observation of semisolid deformation by using 4D-CT and 3DXRD

T Narumi, T Nakata and H Yasuda

Department of Materials Science and Engineering, Kyoto University, Yoshida-honmachi, Sakyo, Kyoto, 606-8501 Japan

E-mail: narumi.taka.6n@kyoto-u.ac.jp

Abstract. Time-resolved X-ray tomography (4D-CT) and three-dimensional X-ray diffraction (3DXRD) were performed to investigate semisolid deformation in Al–10mass%Cu alloy. The configurations of solid grains were observed in-situ by 4D-CT and the crystallographic orientations of grains were also measured by 3DXRD. Translation and rotation of the solid grains in the semisolid are quantitatively analysed by the reconstructed 3D images and the diffraction spots. Based on the in-situ observation, we discuss about the motion of the solid grains and the formation of the shear band during the semisolid deformation.

1. Introduction

In various casting processes of metallic alloys, elimination of casting defects such as macrosegregation, porosity and hot tear is required for improving quality of castings and promoting efficiency of casting processes. It is of interest to study semisolid deformation in metallic alloys because the semisolid deformation occurs in various casting processes and is closely related to the formation of casting defects. The semisolid deformation can be driven by solidification shrinkage, external forces and thermosolutal convection during solidification. For example, the semisolid processing and the centrifugal casting generate the stress which influences the semisolid microstructure [1–3]. In common with granular materials, the rheological behaviour such as dilatancy was found in semisolid alloys [1, 2], which provides apparent volume expansion in response to shear deformation. Various studies about the rheology of semisolid alloys such as the mechanical and viscus behaviours, and the deformation of solid grains, have been reported. However, the change in the microstructure during the semisolid deformation has not been fully understood, which is essential to understand the formation mechanism of casting defects.

In-situ observation of solidification phenomena in various alloys has been performed by using X-ray imaging techniques at synchrotron radiation facilities. Thanks to the in-situ observations, quantitative analysis of dendrite growth / fragmentation / coarsening has been achieved [4–9]. In-situ observations proved that the massive-like δ–γ transformation rather than the peritectic solidification dominantly occurred in the conventional steels [10–15]. The growth behaviour of the spheroidal graphite in the ductile cast iron [16–19] were elucidated by the in-situ observation. These imaging techniques have also been applied to observe the semisolid deformation on a solid-grain scale. In the previous study, transmission (2D) observations of the semisolid deformation in thin Al–Cu alloys [20] and carbon steels [21] revealed that the local contraction and the dilation occurred in the semisolid alloys, and the shear deformation was localized during the deformation. The solid grains were mechanically isolated even at the high solid fraction ($f_s > 0.95$), and the rotation and the translation of the solid grains were dominant during the semisolid deformation while the deformation of solid grains was modest. Therefore, to know the motion of the solid grains is essential to elucidate the semisolid deformation. However, by the 2D observation, the grain motion was restricted in the thin specimen (thickness was 200 μm), the liquid film existed between the grains and the sapphire specimen holders and the crystallographic orientation of the grains was difficult to measure. Hence, the 3D observation was necessary to obtain the grain motion during the semisolid deformation in the bulk specimen.

Recently, the 3D observation technique using time-resolved X-ray tomography (4D-CT) has been performed aiming to observe the time evolution and the spatial distribution of the solidifying microstructure in metallic alloys. The microstructural evolution such as the changes in the solid/liquid interfacial area and the curvature along with the solid fraction in specimens was extensively analysed [22–26]. 4D-CT has also worked for the semisolid deformation as well as the solidification phenomena. Spatial and time resolutions of 4D-CT allow to understand the translation of solid grains and the change in the local liquid fraction during the semisolid deformation [27, 28]. The liquation cracking of solid grains during the semisolid deformation was discovered by using 4D-CT, which leads to the structural refinement [29]. Using these techniques, the grain motion has been discussed mainly based on the grain translation during the semisolid deformation. For further development and understanding of the semisolid deformation, measurement of the grain rotation in 3D, which provides the semisolid deformation in addition to the translation, is desired. To analyse the grain rotation, the change in crystallographic orientation of each solid grain during the semisolid deformation at high temperature is required.

3DXRD is a non-destructive method to measure the crystallographic orientation of each grain during deforming and phase-transforming specimens [30–32]. We aimed to analyse the 3D grain motion by measuring the rotation using 3DXRD in addition to the translation using 4D-CT. The observation results are expected to contribute further understanding the 3D grain motion during the semisolid deformation. This paper demonstrates the 3D grain motion in Al–10mass%Cu alloys with diameter of 1 mm during semisolid deformation by using the 4D-CT and 3DXRD technique developed by our group. We also discuss about the motion of the solid grains and formation of the shear band during the semisolid deformation.

2. Experimental method

The measurement of 4D-CT and 3DXRD was performed at beamline BL20XU of SPring-8 (synchrotron radiation facility in Hyogo, Japan). Figure 1 shows the schematic illustrations of 4D-CT and 3DXRD, respectively. To perform 4D-CT, the monochromatized X-ray (37.7 keV, approximately 5 mm in width and 3 mm in height) was applied to a specimen rotated at 4 s/rotation. The transmission images were obtained by a beam monitor, ORCA-Flash (1024 × 1024 pixels, pixel size 6.5 μm × 6.5 μm, Hamamatsu Photonics) at 100 fps. Three-dimensional images were reconstructed by using 200 transmission images over 180 degrees rotation. The voxel size of reconstructed images was 6.5 μm × 6.5 μm × 6.5 μm. To perform 3DXRD with vertical resolution, the slit X-ray whose height was 100 μm was applied repeatedly from the top to the bottom of the rotating specimen divided into 20 layers, which took 15 s in each diffraction measurement. The diffraction spots of the solid grains in the semisolid specimen were detected by a X-ray flat panel sensor (1008 × 682 pixels, pixel size 100 μm × 100 μm, Hamamatsu Photonics) at 30 fps. Diffractions from all solid grains, whose number was 10–20, in the field of view were detected using diffraction images over 360 degrees rotation. In the present research, both diffractions of Al {111} and Al {200} were employed to determine the crystallographic orientations of the solid grains. By the distortion of the diffraction spots from the circles centred at the direct X-ray beam according to the position of the solid grains, the horizontal resolution of 3DXRD was provided.

Al–10mass%Cu alloy was used as the specimen. The specimen which had the rod-like shape (2 mm in diameter and 4 mm in length) with the reduced section in the middle (1.4 mm in diameter) to localize deformation in the observation field, was inserted into a sintered Al_2O_3 pipe (ID 2.0 mm, OD 3.0 mm). The eutectic structure of α-Al and θ-Al_2Cu in the specimen was melted and maintained at above the eutectic temperature for 15 min in a vacuum atmosphere (approximately 1 Pa) to make the semisolid specimen isothermal before applying deformation. The eutectic temperature was determined from the disappearance of the diffraction spots of θ-Al_2Cu {211}. The compression deformation of the specimen was applied intermittently three times by pushing down an Al_2O_3 rod placed on the top of the specimen (each deformation period is described as #1, #2 and #3 hereinafter). The deformation lengths of #1, #2 and #3 were 129 μm, 228 μm and 456 μm, respectively, and the deformation rate in each compression, which was calculated from the specimen length and the velocity of the Al_2O_3 rod, was about 10^{-4} /s. The condition of the specimen during the experiment was continuously observed by the transmission images. At each deformation period, 4D-CT and 3DXRD were performed. Figure 2 shows transmission images of the semisolid specimen of whole observation area for 4D-CT, and the diffraction spots obtained from the represented area (slit area) for 3DXRD as an example. Using the data obtained by the in-situ observation, quantitative analysis of the semisolid deformation including the crystallographic orientation of solid grains was performed.

3. Results and discussion

Figure 3 shows the 3D reconstructed images of the solid grains in the observation area of the specimen from the initial to #3. The colour variation of the solid grains represents the individual grains. The solid grains were distributed uniformly in the semisolid specimen before the deformation. This indicates that no macrosegregation such as the band segregation existed in the specimen. Once deformation was applied to the specimen, the impingement between the solid grains and the rearrangement of the solid grains occurred, resulting in the formation of the shear band indicated as the white arrows in figure 3.

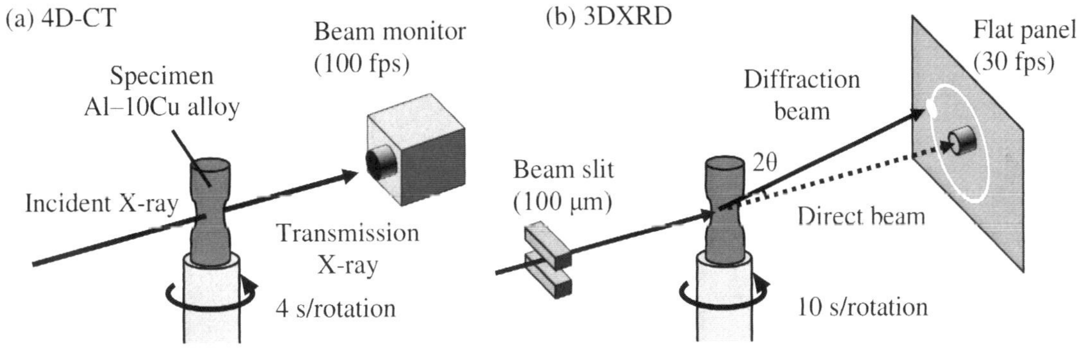

Figure 1. Schematics illustration of (a) 4D-CT and (b) 3DXRD.

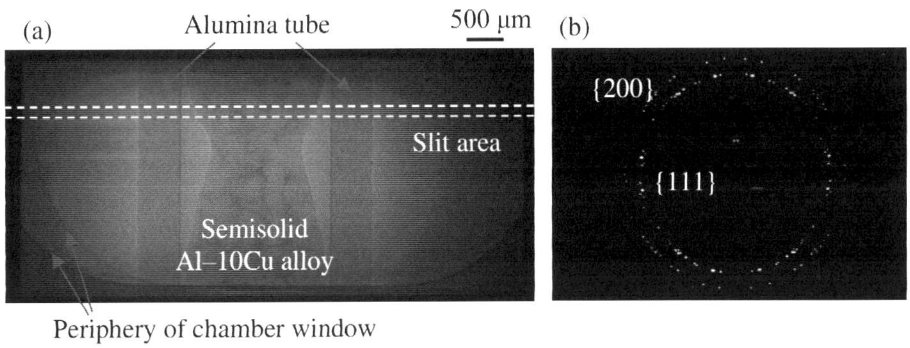

Figure 2. (a) Transmission image of the specimen (the slit beam irradiated the area between the dashed lines), and (b) diffractions obtained from the slit area in (a).

For considering the details of the microstructural evolution during the deformation, the vertical and the horizontal cross-sections which contain the shear band were extracted as shown in figure 4. The bright and dark contrast represent the solid and liquid phase, respectively. Several porosities were observed near the constricted part of the specimen as shown black areas in figure 4. The intergranular liquid films were accumulated with the rearrangement of the solid grains by applying the deformation. The shear band was confirmed in the upper half of the observation area. The shear band tilted from the vertical axis of the specimen (z-axis) whose direction was parallel to the compression direction. This indicates that the solid grains complexly displaced by the impingement between the solid grains instead of the simple translation toward the z-axis. To investigate the correlation of the translation of the solid grains with the formation of the shear band, we define a new coordinate system in which y'-axis is in common with the original y-axis, and x'-axis and z'-axis are perpendicular and parallel to the shear band, respectively. Figure 5 shows the relative translations of the solid grains from the initial to #3 near the shear band using the new coordinate system. Here, the translation was analysed from the movement of centre of mass in the solid grains by using the reconstructed images of 4D-CT. Since those solid grains moved opposite directions, they were classified as group A and group B in accordance with their translational directions. As shown in figure 5, group A (the outer grains) moved larger than group B (the inner grains). The cause of the translation difference between group A and group B was unclear from the grain motion in the observation field. In this observation, there is the possibility that the grains outside of the observation field interacted with group A and group B.

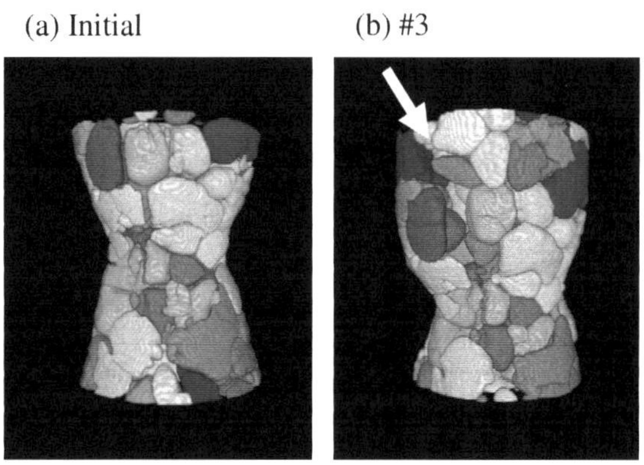

Figure 3. 3D reconstructed images of semisolid Al–10mass%Cu alloy (white arrows indicate the shear band); (a) initial and (b) #3.

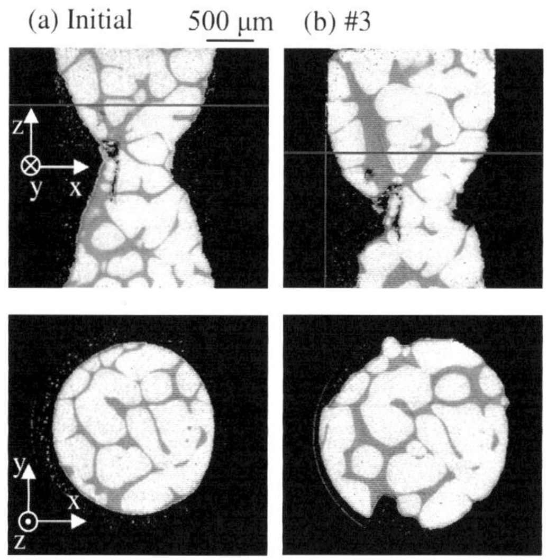

Figure 4. 2D extracted images from 3D images of semisolid Al–10mass%Cu alloy; blue lines in the vertical cross-sections indicate the position of the horizontal cross-sections; (a) initial and (b) #3.

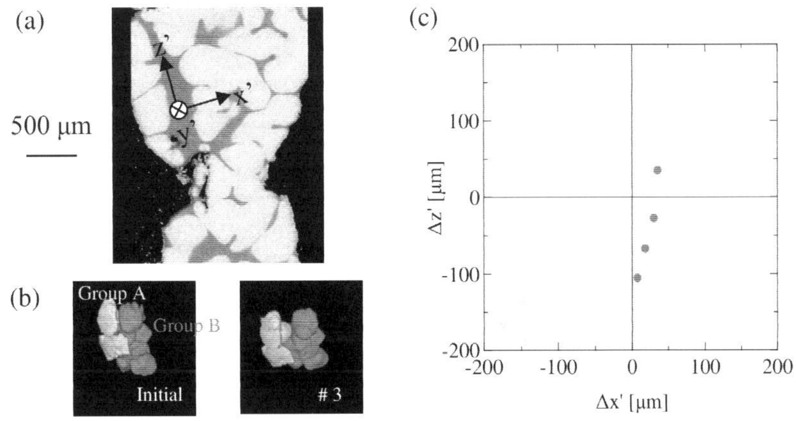

Figure 5. (a) definition of the new coordinate along with the shear band, (b) the solid grains near the shear band at initial and at #3; the number of the solid grains in group A (blue) and group B (red) were three and four, respectively, and (c) the translation length of each grain in group A (blue) and group B (red) from the initial to #3 in the new coordinate of x'-z' plane.

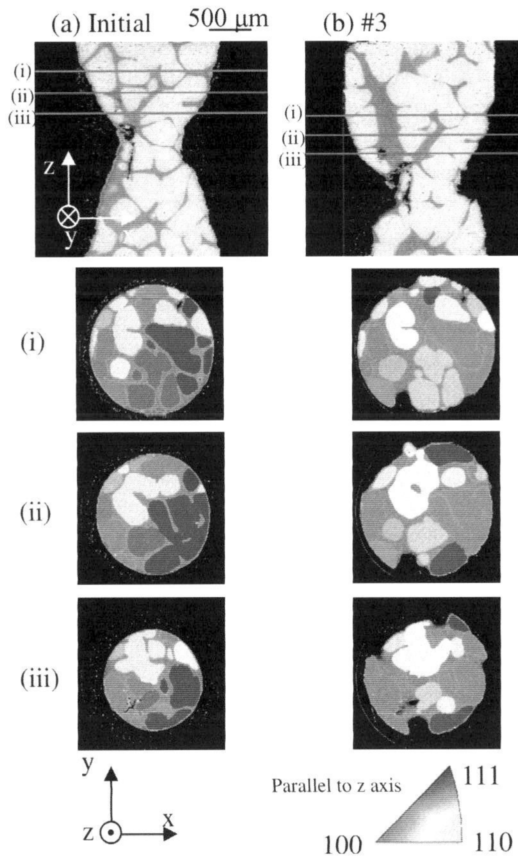

Figure 6. Vertical cross-sections and inverse pole figures of the horizontal cross-sections indicated as (i), (ii) and (iii); (a) initial and (b) #3.

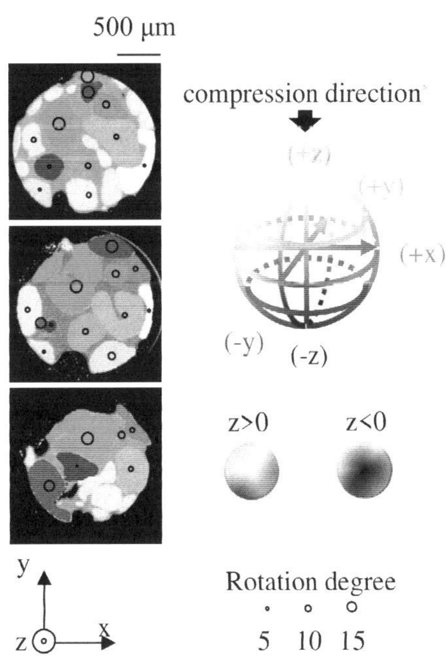

Figure 7. Change in the rotational direction and the degree of rotation from initial to #3; the colour variations and the circles indicates the rotational direction and the degree of rotation, respectively.

Figure 6 shows the inverse pole figure of the solid grains in the horizontal cross section marked in figure 4. The crystallographic orientation analysis of the solid grains in the semisolid alloy was achieved by 3DXRD at each deformation period, though a few grains were unavailable due to their smaller size and weaker diffraction intensity. Using those inverse pole figures, the grain rotation was assessed as shown in Figure 7. Here, the colour variation represents the both of the rotational axis direction and rotative direction obtained by 3DXRD. The rotational axis direction was defined as the tilt angle form z-axis and represented by x-y coordinate. The rotative direction was defined as positive (+z) when a solid grain rotated clockwise as viewed from the top of the specimen. The circles placed in the grains represent the degree of rotation. From figure 7, the solid grains rotated randomly, and no apparent tendency was confirmed. To investigate the correlation among the grain motion and the grain size near the shear band, the translation length, the degree of rotation and the volume of the solid grains from initial to #3 were evaluated based on 4D-CT and 3DXRD, and shown in figure 8. The region near the shear band (figure 8 (a)), two grains were larger than 0.20 mm^3 and the other grains were 0.02–0.06 mm^3. The distribution of the translation length represents that the solid grains moved ununiformly and the grain rearrangement occurred. No apparent tendencies were found between the translation, the rotation and the volume of the solid grains in this study (from figure 8(a)). To elucidate the relationship between the formation of the shear band and the rearrangement of the solid grain, observing the motion of large number of grains from the semisolid specimen and treating those data statistically are necessary.

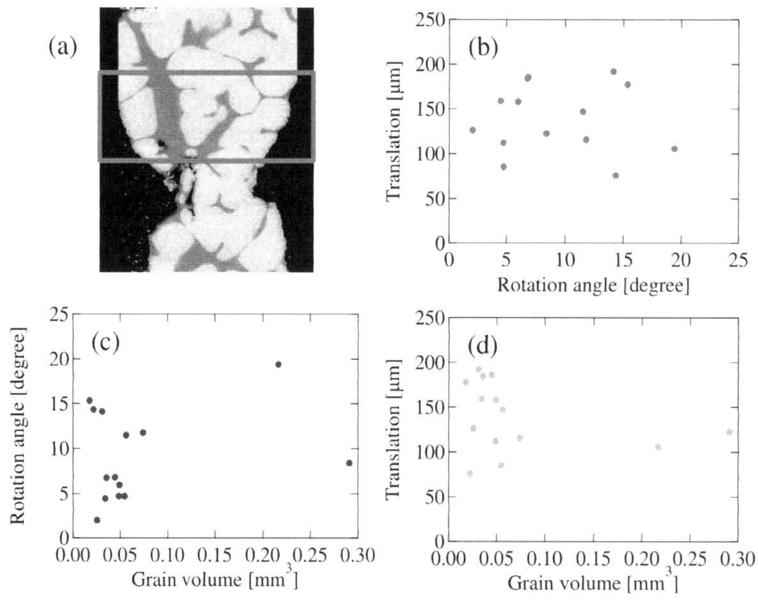

Figure 8. (a) The analysed region in the semisolid specimen, and the correlation between (b) translation and rotation angle, (c) rotation angle and grain volume, and (d) translation and grain volume.

4. Conclusion

In this study, 4D-CT and 3DXRD were performed for in-situ observation of semisolid deformation in a bulk specimen of Al–10mass%Cu alloy. The translation and rotation of the solid grains in the observation area were quantitatively analysed by the 3D reconstructed images and diffraction spots in synchronization with the sample rotation, respectively. As the compression deformation was applied to the semisolid specimen, the formation of the shear band was obtained. The translation of the solid grains was measured by the reconstructed images of 4D-CT. The solid grains in both side of the shear band moved apart each other due to the interaction between the solid grains. The rotation of the solid grains was also measured by 3DXRD. No apparent tendency of the grain rotation was obtained. The interaction between grains essentially caused the translation motions of solid grains and the rotation motion occurred as a result of the translation. The random rotation suggested that the forces causing the translation rather than the rotation should be included in a semisolid-deformation model.

We concluded that the motions of the solid grains during the semisolid deformation are analysed quantitatively based on the in-situ observations using 4D-CT and 3DXRD. For the statistical elucidation of the interaction between the solid grains during the semisolid deformation, it is of interest to study the semisolid deformation in a bulk specimen containing more than 1000 grains.

Acknowledgement

Part of this study was supported by a Grant-in-Aid for Scientific Research (S) (Grant Number 17H06155) from MEXT, Japan. The synchrotron radiation experiments were performed as general projects at a beamline of BL20XU at SPring-8 (JASRI) 2019A1400.

References

[1] Gourlay C M and Dahle A K 2007 *Nature* **445** 70–3
[2] Gourlay C M, Meylan B and Dahle A K 2008 *Acta Mater.* **56** 3403–13
[3] Kang I and Ohnaka I 1997 *J. Japan Foundry Eng. Soc.* **69** 119–26
[4] Mathiesen R H, Arnberg L, Mo F, Weitkamp T and Snigirev A 1999 *Phys. Rev. Lett.* **83** 5062–5
[5] Yasuda H, Ohnaka I, Kawasaki K, Sugiyama A, Ohmichi T, Iwane J and Umetani K 2004 *J. Cryst. Growth* **262** 645 52
[6] Mangelinck-Noël N, Nguyen-Thi H, Reinhart G, Schenk T, Cristiglio V, Dupouy M-D, Gastaldi J, Billia B, Härtwig J and Baruchel J 2005 *J. Phys. D. Appl. Phys.* **38** A28–32
[7] Li B, Brody H D, Black D R, Burdette H E and Rau C 2006 *J. Phys. D. Appl. Phys.* **39** 4450–6
[8] Yasuda H *et al.* 2008 *Int. J. Cast Met. Res.* **21** 125–8
[9] Yasuda H et al. 2009 Int. J. Cast Met. Res. **22** 15–21
[10] Yasuda H *et al.* 2011 *IOP Conf. Ser. Mater. Sci. Eng.* **27** 0–9
[11] Yasuda H *et al.* K 2012 IOP Conf. Ser. Mater. Sci. Eng. **33**
[12] Nishimura T, Morishita K, Nagira T, Yoshiya M and Yasuda H 2015 *IOP Conf. Ser. Mater. Sci. Eng.* **84** 012062
[13] Nishimura T, Morishita K, Yoshiya M, Nagira T and Yasuda H 2019 *Tetsu-to-Hagane* **105** 290–8
[14] Nishimura T, Matsubayashi R, Morishita K, Yoshiya M, Nagira T and Yasuda H 2019 *ISIJ Int.* **59** 459–65
[15] Yasuda H, Morishita K, Nakatsuka N, Nishimura T, Yoshiya M, Sugiyama A, Uesugi K and Takeuchi A 2019 *Nat. Commun.* **10** 3183
[16] Yamane K *et al.* 2014 J. Japan Foundry Eng. Soc. **86** 461–70
[17] Chatchari K, Akira S, Kouhei M and Hideyuki Y 2018 *J. Japan Foundry Eng. Soc.* **90** 602–12
[18] Yasuda H, Sugiyama A, Kiattisaksri C, Morishita K, Nagira T, Yoshiya M, Uesugi K and Takeuchi A 2018 *Mater. Sci. Forum* **925** 104–9
[19] Bjerre M K, Azeem M, Lee P D, Hattel J H and Tiedje N S 2018 **925** 118–24
[20] Gourlay C M, Dahle A K, Nagira T, Nakatsuka N, Nogita K, Uesugi K and Yasuda H 2011 *Acta Mater.* **59** 4933–43

[21] Nagira T *et al.* 2011 *Scr. Mater.* **64** 1129–32
[22] Ludwig O, Dimichiel M, Salvo L, Suéry M and Falus P 2005 *Metall. Mater. Trans. A* **36** 1515–23
[23] Aagesen L K, Fife J L, Lauridsen E M and Voorhees P W 2011 *Scr. Mater.* **64** 394–7
[24] Gibbs J W, Mohan K A, Gulsoy E B, Shahani A J, Xiao X, Bouman C A, De Graef M and Voorhees P W 2015 *Sci. Rep.* **5** 11824
[25] Puncreobutr C, Phillion A B, Fife J L and Lee P D 2014 *Acta Mater.* **64** 316–25
[26] Yasuda H, Kawarasaki T, Tomiyori Y, Kato Y and Morishita K 2019 *IOP Conf. Ser. Mater. Sci. Eng.* **529** 012023
[27] Terzi S *et al.* 2009 *Scr. Mater.* **61** 449–52
[28] Sistaninia M, Terzi S, Phillion A B, Drezet J-M and Rappaz M 2013 *Acta Mater.* **61** 3831–41
[29] Karagadde S *et al.* 2015 *Nat. Commun.* **6** 8300
[30] Jensen D J, Lauridsen E M, Margulies L, Poulsen H F, Schmidt S, Sørensen H O and Vaughan G B M 2006 *Mater. Today* **9** 18–25
[31] Offerman S E, van Dijk N H, Sietsma J, Lauridsen E M, Margulies L, Grigull S, Poulsen H F and van der Zwaag S 2006 *Nucl. Instruments Methods Phys. Res. Sect. B Beam Interact. with Mater. Atoms* **246** 194–200
[32] Toda H, Ohkawa Y, Kamiko T, Naganuma T, Uesugi K, Takeuchi A, Suzuki Y and Kobayashi M 2013 *Acta Mater.* **61** 5535–48

IOP Conf. Series: Materials Science and Engineering **861** (2020) 012066 doi:10.1088/1757-899X/861/1/012066

Modelling of shear bands during solidification

A Ludwig[1], C M G Rodrigues[1], M Wu[1] and A Kharicha[1,2]

[1] Chair of Simulation and Modelling Metallurgical Processes, Metallurgy Department, Montanuniversitaet Leoben, Franz-Josef Street 18, 8700 Leoben, Austria
[2] Christian-Doppler Laboratory for Metallurgical Application of Magnetohydrodynamics, Montanuniversitaet of Leoben, Franz-Josef Street 18, 8700 Leoben, Austria

E-mail: christian.gomes-rodrigues@unileoben.ac.at

Abstract. Major developments have been accomplished in the understanding of the physics involved in granular media and suspensions when subjected to shearing stresses. However, although partially solidified alloys have been reported to exhibit similar characteristics as granular material under certain conditions, such findings have never been fully addressed by the solidification community. Here, the modelling approaches adopted in granular mechanics for the description shear-induced particle migration have been included into a solidification model. It is shown that shear-induced particle migration has a considerable effect on the solid fraction distribution and macrosegregation formation in casting processes where a continuous and significant shear rate exists in a conduit with relatively small geometrical aspect ratio.

1. Introduction

Shear bands, the localization of deformation in certain regions of intense shearing, is a phenomenon that is commonly observed in compacted granular materials. There has been extensive research on the comprehension of the principles involved in such phenomenon through extensive experimental and theoretical studies. A good example of such mechanism is the so-called migration phenomenon. For relatively large particles, measurements taken in different geometrical configurations (e.g., in concentric Couette cells, parallel channels, etc.) have shown that the particles migrate irreversibly from high shear regions towards low shear regions [1-3]. It takes place in suspensions under steady uniform plane shear flows and has been successfully modelled both as a single fluid representing the whole suspension [4, 5] and as a two-phase model where the fluid and the particles are considered as two distinct interpenetrating continua [6, 7]. Although in both case good agreement with experimental data is obtained, the latter approach provides a more detailed description of the multiphase flow dynamics than the mixture models. This is an essential feature when complex phenomena are under scrutiny.

In solidification processes, besides the convection of the equiaxed crystals, a phase transition occurs to the melt which gradually transforms into solid. The rheology of the solid-liquid mixture changes considerably during this process. At low solid fractions, the behaviour of the solid-liquid mixture is described as suspensions and the flow is mainly dictated by the transporting character of the liquid phase. Within this regime, it is commonly accepted that the viscosity of the mixture varies exponentially with solid fraction [4, 8]. At a critical solid fraction, the crystals start to create bridges between one another, and the material develops into a coherent structure which has cohesive properties. This is where strength development of the material slowly begins [9]. With further solidification, the crystals become fully connected and the mush exhibits a more solid-like behaviour, which is often described as a viscoplastic continuous solid skeleton saturated with interstitial liquid [10]. In the latter regime (referred as viscoplastic regime), particular constitutive equations are employed for the deformation of the mush that arises due to shrinkage, convection or imposed external loads. Due to its interest for casting production, both experimental and computational studies have been undertaken to analyse the deformation behaviour of semi-solids at high temperatures [11-13].

Interestingly, experimental research has shown that, under certain conditions, the rheology of equiaxed solidifying alloys exhibits similar characteristics as compacted granular materials at solid fractions below the viscoplastic limit [14]. Such results are particularly important because they explain the formation of casting defects, such as macrosegregation and porosity, in a solid fraction range that can have a major impact on the final solidified microstructure. To our knowledge, such findings have never been applied to models designed to replicate solidification processes.

In the present paper, the modelling approaches adopted in granular mechanics for the description of the rheological behavior of suspensions at moderate solid fractions under steady uniform plane shear flows have been integrated into a thermo-mechanical and macrosegregation solidification model to explore the principles behind shear-induced particle migration during solidification.

2. Model description

The solidification model has been described in detail in previous works [12, 15]. It includes also constitutive equations that replicate the deformation of the mush due to shrinkage, convection or imposed external loads. Under such conditions, the mush can either densify (by expelling liquid from its core) or dilate (by draining liquid from its surroundings), giving rise to local relative motion between the phases, and eventually leading to macrosegregation deviations in the cast products [12, 16, 17]. In the present paper, special attention is drawn to the formulation adopted to account for the shear-induced migration (SIM).

The origin of particle migrations is related to the pressure gradients experienced by the particles. In pressure-driven flows through a channel with homogeneous suspensions, shear rate is usually maximal near the walls, and decreases towards the centreline. This correlates with the higher granular pressure gradient near the walls, where particle-particle collisions are more frequent. As a result, a net body force arises on the solid phase that leads to an inward flux of particles which is compensated by an outwards flux of liquid. This persists until the granular pressure becomes homogenous across the section.

The migration mechanism is taken into account in the present model by mean of a shear-induced stress tensor, Σ_s, in the solid phase, which can be written as follows:

$$\Sigma_s = -\mu_l \eta^n (g_s) \dot{\gamma}_{eff} \mathbf{Q} \tag{1}$$

where μ_l is the liquid viscosity, $\dot{\gamma}_{eff} = \sqrt{2\dot{\varepsilon} : \dot{\varepsilon}}$ is the effective shear rate of the suspensions, and $\mathbf{Q}$ represents the anisotropy tensor [18] which in this work has been assumed as being equal to the identity tensor for simplicity. For the dimensionless particle viscosity, $\eta^n(g_s)$, the following expression is employed:

$$\eta^n (g_s) = K_n \left(\frac{g_s}{g_s^p} \right)^2 \left(1 - \frac{g_s}{g_s^p} \right)^{-2} \tag{2}$$

with $K_n = 1$ [19] and the packing solid fraction $g_s^p = 0.68$ [18].

Note that different closure models and parameters are often used for different configurations so that a better fit is achieved with measurements. However, not only this affects the generalizability of the results, it is also important to keep in mind that experimental data are not always possible, particularly in solidification processes. Therefore, in the present work, the same closure models and parameters are kept in all test cases performed so that an objective assessment of the model is carried out. Nevertheless, it is worth noting that, from the set of simulations performed for this paper, it has been found that manipulating the fitting parameters or using the anisotropic tensor, $\mathbf{Q}$, did not produce noteworthy changes in the results.

Contrary to the liquid phase, where the deviatoric stress tensor is given solely by its classical representation for incompressible Newtonian fluids (i.e., $\mathbf{\Sigma}_l^{eff} = \boldsymbol{\tau}_l = 2\mu_l \mathrm{dev}(\dot{\boldsymbol{\varepsilon}}_l)$), in the solid phase, the shear-induced stress tensor contribution is added to the deviatoric stress tensor, yielding:

$$\mathbf{\Sigma}_s^{eff} = \boldsymbol{\tau}_s = 2\mu_s \mathrm{dev}(\dot{\boldsymbol{\varepsilon}}_s) + \Sigma_s \tag{3}$$

where the shear rate is given by $\dot{\boldsymbol{\varepsilon}}_i = \frac{1}{2}\left(\nabla \mathbf{v}_i + \left(\nabla \mathbf{v}_i\right)^T\right)$ and the solid viscosity, μ_s, can be calculated from the power-law viscosity model of Ishii [8] as follows:

$$\mu_s = \mu_\ell / g_s \left(\left(1 - g_s / g_s^p\right)^{-2.5 g_s^p} - (1 - g_s) \right) \tag{4}$$

which assumes the validity of the mixture rule for a suspension as:

$$\mu_{sus} = g_s \mu_s + g_\ell \mu_\ell \tag{5}$$

Introducing equations (1), (2) and (4) into equation (3) gives the deviatoric stress tensor for the solid phase, which is then employed in the momentum conservation equation. Besides momentum, volume-average conservation equations of mass, species, and enthalpy are also solved for both liquid and solid phases. The equations are presented in table 1, where g denotes the volume fraction, ρ the density, **v** the velocity vector, p the shared pressure, Σ_i^{eff} the effective deviatoric stress tensor, **g** the gravity vector, c the species concentration, h the enthalpy per unit mass, **q** the heat flow vector, n the grain number density, and the subscripts i refer to either liquid (l) or solid (s). In each computational cell, the sum of solid and liquid volume fractions (g_s and g_l, respectively) is equal to one. Details on the different interphase exchange terms are given in references [12,15].

Table 1. Volume-Averaged Conservation Equations.

Mass conservation:	$\dfrac{\partial g_i \rho_i}{\partial t} + \nabla \cdot (g_i \rho_i \mathbf{v}_i) = \mp M_{\ell s}$	(6)
Momentum conservation:	$\dfrac{\partial g_i \rho_i \mathbf{v}_i}{\partial t} + \nabla \cdot (g_i \rho_i \mathbf{v}_i \mathbf{v}_i) = -g_i \nabla p + \nabla \cdot g_i \mathbf{\Sigma}_i^{eff} + g_i \rho_i \mathbf{g} \mp \mathbf{U}_{ls}^D \mp \mathbf{U}_{ls}^M$	(7)
Species conservation:	$\dfrac{\partial g_i \rho_i c_i}{\partial t} + \nabla \cdot (g_i \rho_i \mathbf{v}_i c_i) = \nabla \cdot (g_i \rho_i D_i \nabla c_i) \mp C_{\ell s}$	(8)
Enthalpy conservation:	$\dfrac{\partial g_i \rho_i h_i}{\partial t} + \nabla \cdot (g_i \rho_i \mathbf{v}_i h_i) = -\nabla \cdot \mathbf{q}_i \mp H_{\ell s}$	(9)
Grain transport:	$\dfrac{\partial n}{\partial t} + \nabla \cdot (\mathbf{v}_s n) = N_{\ell s}$	(10)

The two-phase Eulerian-Eulerian volume-average model employed in this work is based on the twoPhaseEulerFoam solver from OpenFOAM toolbox [20]. However, the original code has been greatly altered mainly to incorporate the viscoplastic regime (described in detail in reference [12]) and the shear-induced migration phenomenon, which were not supported by the solver. This includes (among other developments) the addition of the bulk viscosity in the viscous stress term, the adaptation of the Navier-Stokes equation to cope with the different rheological material behaviors depending on the regime, the calculation of the effective solid viscosity (equation 5) and the shear-induced stress tensor (equation 1).

The present work is focused on the analysis of the shear-induced migration phenomena and the conditions employed in the different test cases analyzed are such that the viscoplastic regime is most of the time not activated (the onset of the viscoplastic regime occurs at solid fractions above 0.57). In order to

IOP Conf. Series: Materials Science and Engineering **861** (2020) 012066 doi:10.1088/1757-899X/861/1/012066

simplify the analysis of the results and focus our attention solely on the shear-induced migration phenomena, the viscoplastic regime has been neglected from the simulations presented here.

3. Results and discussion

Numerical simulations are performed for both a pressure-driven flow of a suspension in a conduit of rectangular cross-section and a suspension flow in a cylindrical Couette cell, for which experimental validation results are available in the literature. The two setups are illustrated in figure 1.

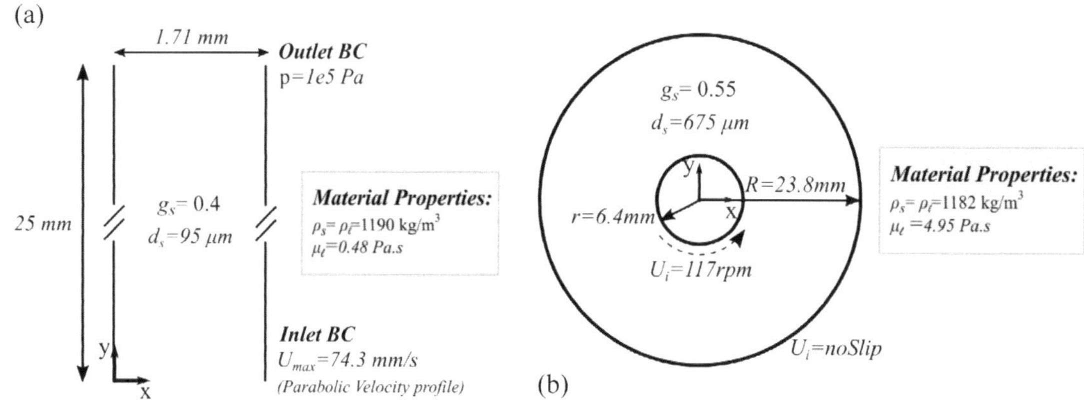

Figure 1. Schematic representation of (a) pressure-driven flow of a suspension in a conduit of rectangular cross-section, and (b) suspension flow in a cylindrical Couette cell.

In both cases, the simulations are performed in a 2D domain and gravity is neglected. Boundary conditions and material properties are defined in figure 1. A two-dimensional orthogonal mesh with 20×100 cells is used in the first case, whereas in the second, the mesh consists of 20 cells in the radial direction and 80 cells in the azimuthal direction (for one complete revolution).

3.1. Validation of results

The numerical results presented in this section are compared against well-known experimental measurements where particle migration has been observed.

Lyon and Leal [2] used a laser-Doppler velocimetry method (LDV) to measure both particle velocity and concentration profiles in a two-dimensional channel flow. The comparison between experimental data and the numerical results are presented in figure 2(a). It can be seen that an inhomogeneous particle distribution is obtained with high solid fraction being obtained along the centerline. The agreement between numerical and experimental results is in line with other comparisons reported in the literature (see e.g. [5] and [7]). Note that Lyon and Leal [2] reported that the particle concentration near the wall was lower than the true value because of the reduced signal-to-noise ratio in this region. This could justify the high discrepancy between numerical and experimental results in this region.

Phillips *et al.* [3] carried out nuclear magnetic resonance (NMR) imaging measurements of particle distributions for a monomodal suspension of polymethyl methacrylate spheres in a Newtonian solvent. Irreversible particle migration in the radial directions was observed. The comparison between experimental data and the numerical results after 200 revolutions of internal cylinder and initial solid fraction of 0.55 are presented in figure 2(b).

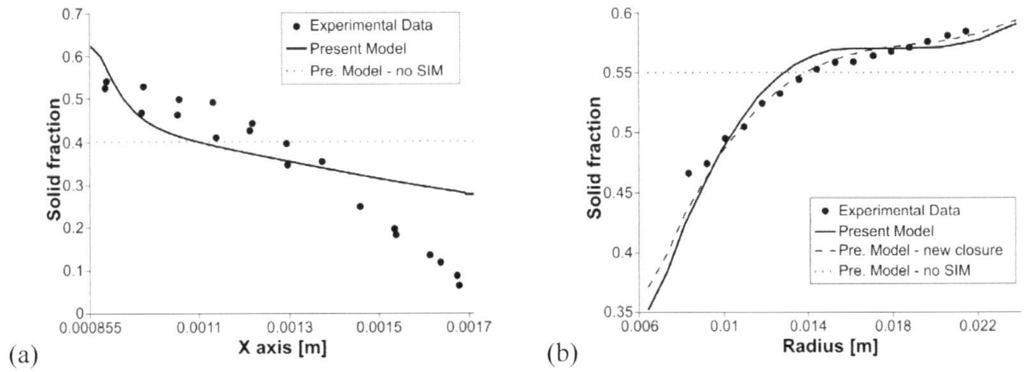

Figure 2. Simulation results for (a) 2D pressure-driven channel flow at y=0.192 for initial solid fraction of 0.4, compared against experimental data of Lyon and Leal [2], and (b) suspension flow in Couette cell after 200 revolutions of internal cylinder and initial solid fraction of 0.55, compared against experimental data of Phillips *et al.* [3]. Results with no shear-induced migration ("Pre. Model - no SIM") are also presented for reference.

The simulation displays a reasonable agreement with experimental data, although an overestimation of the measured solid fraction is observed in the middle of the domain. A better fit can, however, be obtained by assuming that $\mu_{sus} = \mu_s + \mu_l$ [18] – instead of the common mixture law applied in equation (4) – as presented in the dashed line (referred in the legend as "Pre. Model – new closure") of figure 2(b). These results raise some questions regarding the applicability the general rule of mixture for the viscosity employed in many research fields and in a wide range of the phase fraction, Reynolds number and flow regimes [21, 22]. Nevertheless, for the sake of consistency, the original approach is employed in the remainder of the paper.

A closer look at the above results reveals that the jamming effect in the parallel channel flow along the centerline is more significant than in the cylindrical Couette cell. Not only the shear rate is larger in the former configuration, it also keeps a constant value during the simulation since the velocity profile is imposed at the inlet. On the other hand, in the Couette cell, the suspension starts from a rest state which is then affected by the moving inner cylinder. However, as the velocity profile spreads towards the outer cylinder, the velocity gradient fades away, reducing with it the migration flux.

3.2. Effect of shear-induced migration during solidification processes

After having validated the shear-induced migration formulation, the main goal becomes understanding the potential effects of such phenomenon on solidification processes. The investigation of the transport and solidification phenomena occurring during the casting process is essential to avoid internal defects and improve the quality of casting strips. The relative motion between the phases – which might occur during SIM – can be an important source for such defects.

Solidification processes where SIM might be relevant should involve significant and continuous shear rates to the solidifying mush within a narrow and relatively long conduit. High-pressure die casting [14] and thixocasting process are potential examples where such conditions may apply. In the following, a cooling rate is applied on the walls of two parallel plates, in a configuration similar to the one presented in figure 1(a). Such conditions are pertinent in segments of both processes mentioned above, although lower velocity and pressure are assumed here for simplicity. This analysis should provide some insight on how SIM can affect the results, even under simplified conditions. A schematic representation of the domain is illustrated in figure 3 and the simulations are performed for the Al-7wt% Si alloy.

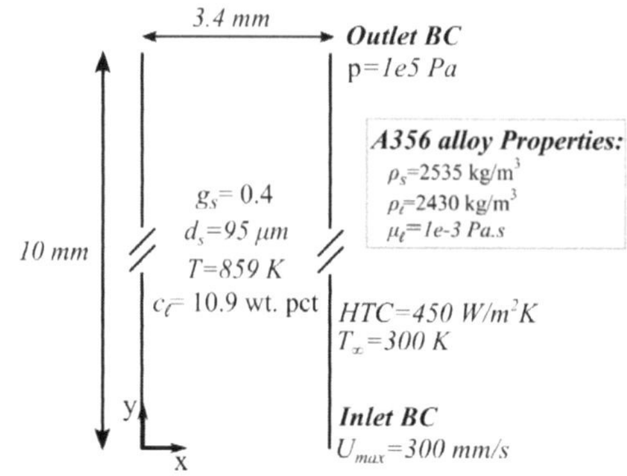

Figure 3. Schematic representation of test case involving shear-induced migration and solidification of A356 alloy.

Figure 4 illustrates the solid fraction profile along a horizontal line located near the outlet. It can be seen that the simulation that considers solidification but neglects SIM ("Pre. model - no SIM") has increased solid fraction values near the walls, while the center maintains a lower value. Such distribution contrasts with the simulation that does not consider solidification of the melt but takes into account SIM ("Pre. model - no Sol"), which presents a solid fraction distribution similar to the pressure-driven channel flow presented above (in figure 2(a)). This means that a jamming of solid particles occurs along the centerline, whereas the solid fraction decreases near the walls. Also, the results of a simulation with no SIM and no solidification are presented in figure 4 for reference ("Pre. model – no SIM + no Sol").

In comparison with the case with no SIM, it can be seen that the solid fraction profile of the complete simulation ("Present model") changes dramatically. The solid fraction distribution maintains some characteristics of the SIM mechanisms, while increasing the general amount of particles along the cross section due to solidification. Due to the larger solidification rate near the walls, the difference in solid fraction between these results is also larger in that region than in the center.

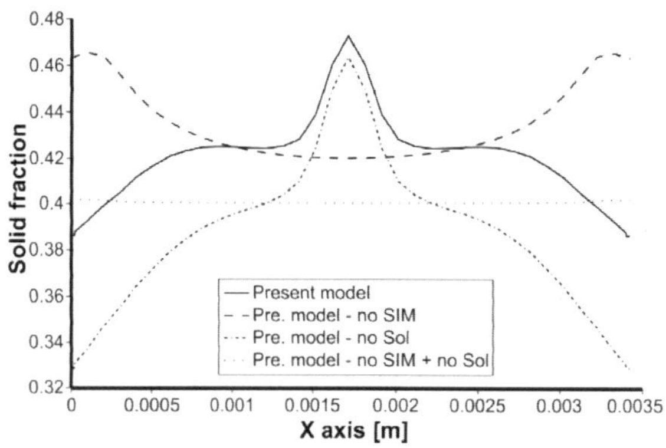

Figure 4. Steady state results of solid fraction distribution along horizontal line located near outlet.

SIM mechanism induces a flux of particles towards the center, with an opposite flux of melt going towards the walls. Solidification growth, in turn, requires an inflow of material predominantly towards the walls to feed solidification. These two mechanisms create a relative motion between the two phases that influence macrosegregation, as observed in figure 5.

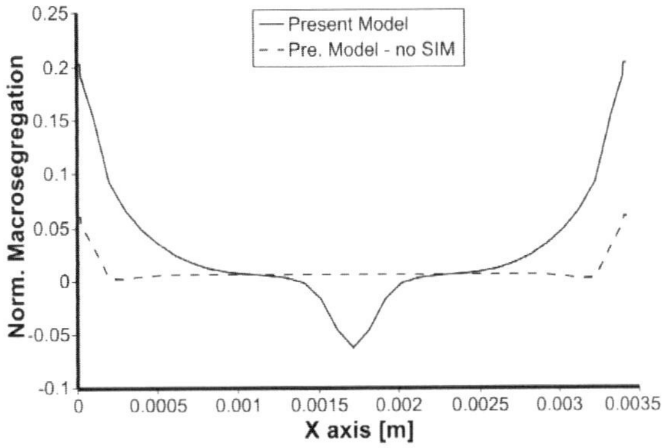

Figure 5. Steady state results of normalized macrosegregation along horizontal line located near the outlet.

Figure 5 illustrate the normalized macrosegregation distribution along the cross section of the conduit (near the outlet). The present results are also compared against the simulation that neglects SIM. In the latter case, a typical macrosegregation profile is observed. A positive composition deviation occurs near the walls, which is due to inverse segregation.

On the other hand, the present results show an increased positive deviation near the walls, while a negative deviation occurs along the centreline. Although inverse segregation plays a role near the wall, the positive deviation is mostly due to the migration effect taking place in this region, which replaces solid particles with solute-enriched melt. In turn, the negative macrosegregation formation along the centreline is the corresponding opposite effect, as the higher solid fraction region is associated with an inflow of crystals and an outflow of melt. The mechanism leading to this macrosegregation formation is similar to the one observed in ingot castings, where negative composition deviations are typically observed at the bottom due to the sedimentation of equiaxed crystals [15].

4. Conclusion

Sophisticated modelling approaches have been successfully employed in the field of granular media to simulate shear bands and, particularly, shear-induced migration. However, despite the fact that partially solidified alloys have been reported to exhibit similar characteristics as granular material, such deformation of the mush has never been addressed in solidification process. Here, the constitutive laws employed for predicting shear-induced migration have been included into a solidification model. The results show that particle migration is actually significant in the configurations where a continuous shear rate is imposed to the solidifying mush within a narrow and relatively long conduit.

Acknowledgments

This work was financially supported mainly by the FWF Austrian Science Fund (P28785-N34), and by the Austrian Federal Ministry of Economy, Family and Youth and the National Foundation for Research, Technology and Development within the framework of the Christian Doppler Laboratory for Metallurgical applications of Magneto-hydrodynamics.

References

[1] Leighton D and Acrivos A 1986 *Chem. Eng. Sci.* **41** 1377–84
[2] Lyon M K and Leal L G 1998 *J. Fluid Mech.* **363** 57–77
[3] Phillips R J, Armstrong R C, Brown R A, Graham A L and Abbott J R 1992 *Phys. Fluids A.* **4** 30–40
[4] Nott P R, Guazzelli E and Pouliquen O 2011 *Phys. Fluids* **23** 043304-13
[5] Dbouk T, Lemaire E, Lobry L and Moukalled F 2013 *Fluid Mech.* **198** 78–95
[6] Drijer I, van de Laar T, Vollebregt H M and Schroën C G P H 2018 *Sep. Purif. Technol* **192** 99–109

[7] Municchi F, Nagrani P P and Christov I C, 2019 *Int. J. Multiph. Flow* **120** 103079-15
[8] Ishii M 1977 ANL-77-47
[9] Dahle A K and StJohn D H 1998 *Acta Mater.* **47** 31–41
[10] Martin C L, Favier D and Suéry M 1999 *Int. J. Plast.* **15** 981–1008
[11] Nguyen T G, Favier D and Suery M 1994 *Int. J. Plast.* **10** 663–93
[12] Rodrigues C M G, Ludwig A, Wu M, Kharicha A and Vakhrushev A 2019 *Metall. Mater. Trans. B* **50** 1334–50
[13] Rodrigues C M G, Ludwig A, Wu M, Kharicha A and Vakhrushev A 2019 *Preprint*
[14] Gourlay C M and Dahle A K 2007 *Nature* **445** 70–73
[15] Ludwig A and Wu M 2002 *Metall. Mater. Trans. A* **33** 3673–83
[16] Fachinotti V D, Le Corre S, Triolet N, Bobadilla M and Bellet M 2006 *Int. J. Numer. Meth. Eng.* **67** 1341–84
[17] Rodrigues C M G R, Ludwig A, Kharicha A, Wu M and Vakhrushev A 2019 *IOP Conf. Mater. Sci. Eng.* **529** 1-7
[18] Morris J F and Boulay F 1999 *J. Rheol.* **43** 1213–37
[19] Boyer F, Guazzelli É and Pouliquen O 2011 *Phys. Rev. Lett.* **107** 1–5
[20] Greenshields C J, 2018 OpenFOAM – The OpenFOAM Foundation – User Guide, http://foam.sourceforge.net/docs/Guides-a4/OpenFOAMUserGuide-A4.pdf
[21] Beckermann C and Viskanta R 1993 *Appl. Mech. Rev.* **46** 1–27
[22] Wu M, Ludwig A and Kharicha A 2019 *Metals* **9** 229 1-43

IOP Conf. Series: Materials Science and Engineering **861** (2020) 012067 doi:10.1088/1757-899X/861/1/012067

Application of multi-phase-field lattice Boltzmann method to semi-solid deformation

N Yamanaka[1], S Sakane[1] and T Takaki[2]

[1] Graduate School of Science and Technology, Kyoto Institute of Technology, Matsugasaki, Sakyo–ku, Kyoto 606–8585, Japan
[2] Faculty of Mechanical Engineering, Kyoto Institute of Technology, Matsugasaki, Sakyo–ku, Kyoto 606–8585, Japan

E-mail: takaki@kit.ac.jp

Abstract. Mechanisms of semi-solid deformation are crucial for reducing solidification defects. In this study, a multi-phase-field lattice Boltzmann (MPF-LB) model, which can simulate growth of multiple dendrites with liquid flow, solid motion and collision, is applied to semi-solid deformation. An elastic contact is introduced to the MPF-LB model to consider collision of two grains. The semi-solid deformation with globular morphology is simulated in the two-dimension by the MPF-LB method. Through the simulations, high potential of the MPF-LB model as the evaluation method of semi-solid deformation is confirmed.

1. Introduction

In the casting process, semi-solid deformation causes serious solidification defects, such as porosity, cracking and macrosegregation [1]. To reduce these defects, it is crucial to elucidate the mechanisms of semi-solid deformation. However, it is very difficult because semi-solid deformation, aside from being a high-temperature and opaque phenomena, is a very complicated multi physics problem including solid-liquid phase transformation, liquid flow, solid motion, solid-solid interaction, solid deformation, and so on.

Owing to recent in-situ observations using synchrotron X-ray radiography, detailed phenomena and mechanisms in semi-solid deformations are gradually being unveiled. It was reported that, for globular materials, rearrangement by translation and rotation of solid grains is a dominant deformation in a wide range of solid fractions, f_S ($f_S < 0.8$) [2-5]. Meanwhile, for globular materials with a high f_S ($f_S > 0.9$) and for equiaxed dendritic morphology [2, 6], deformations of the solid have been observed in addition to the rearrangement of grains. In both solid morphologies, the grain rearrangement causes a dilation which has the greatest effect on producing defects. Indentation experiments of semi-solid materials have also shown transgranular liquation cracking [7] and plastic deformation [8] which results in grain refinement.

Modeling and simulation studies are indispensable for a better understanding of semi-solid deformation under various conditions [9]. The finite element method is a powerful tool to simulate the macroscopic mechanical behavior of a semi-solid material [10, 11], where information on the microstructure is not considered nor introduced through constitutive equations [12]. Finite element simulations have also been conducted for dendritic solid [13] and equiaxed globular grain structures [14]. Moreover, deformation simulations during dendrite growth with an external force have been performed using the phase-field method [15]. However, the liquid flow is not computed in the above simulations. Flow of intergranular liquid has been simulated by a granular model [16, 17]. For globular materials in a range where solids can be considered as a rigid body, a discrete element method was used and could successfully capture some phenomena observed during in-situ observations [18].

As mentioned above, some models succeeded in expressing specific phenomena in semi-solid deformation. However, there is currently no model which can express all phenomena occurred in semi-solid deformation: solid-liquid phase transformation, liquid flow, solid motion, solid-solid interaction, solid deformation, fragmentation of solid and so on. In our previous study [19], we have developed a multi-phase-field lattice Boltzmann (MPF-LB) model which can simulate the growth of multiple dendrites with motion, liquid flow, collision, and coalescence and subsequent grain growth by extending a single-dendrite growth model with motion [20-22]. The model would have great potential to simulate semi-solid deformation. In this study, as a first step of our phase-field study on semi-solid deformation, the MPF-LB model [19] is applied to the semi-solid deformation of a globular material, and the suitability of the MPF-LB model as an evaluation method of semi-solid deformation is investigated.

2. Model

In the MPF-LB model developed in [19], dendrite growth and grain growth after coalescence of multiple grains are expressed by MPF method, liquid flow is computed by LB method, and motion of solids is expressed by solving equations of motion of solids assumed to be a rigid body. Also, a perfectly inelastic collision is assumed for simplicity, when two solids collide with each other. In this study, an elastic collision is introduced in a contact problem of two grains. In addition, the coalescence of grains is not considered, because usually the solid-solid interface energy is larger than twice that of the solid-liquid interface energy [23] and an isothermal condition maintaining a semi-solid condition is assumed in the following simulations.

Multiple grains are expressed by multiple phase-field variables, where phase-field variable ϕ_i is defined as $\phi_i = +1$ in the i-th grain and $\phi_i = -1$ in the others. The time evolution of ϕ_i is expressed by [24]

$$\tau(\theta_i)\left(\frac{\partial \phi_i}{\partial t} + \mathbf{U}_{si}\nabla\phi_i\right) = \nabla\left(W(\theta_i)^2\nabla\phi_i\right) - \frac{\partial}{\partial x}\left(W(\theta_i)\frac{dW(\theta_i)}{d\theta_i}\frac{\partial\phi_i}{\partial y}\right) + \frac{\partial}{\partial y}\left(W(\theta_i)\frac{dW(\theta_i)}{d\theta_i}\frac{\partial\phi_i}{\partial x}\right) - \left(-\phi_i + \phi_i^3\right) - \lambda^*\left(1-\phi_i^2\right)^2 u \tag{1}$$

where $\tau(\theta_i)$ and $W(\theta_i)$ are the phase-field relaxation time and interface thickness, respectively, having anisotropy as $\tau(\theta_i) = \tau_0 a_S(\theta_i)^2$ and $W(\theta_i) = W_0 a_S(\theta_i)$, where θ_i is the crystal orientation, $a_S(\theta_i)$ is the anisotropy function with a form of $a_S(\theta_i) = 1 + \varepsilon_4\cos(4\theta_i)$, and ε_4 is the anisotropy strength. λ^* is a coefficient regarding thermodynamic driving force, and $\mathbf{U}_{Si}$ is the velocity vector of the grain.

The advection-diffusion equation of solute can be expressed as

$$\frac{1}{2}\left[1+k-(1-k)h(\phi_s)\right]\left(\frac{\partial u}{\partial t} + \mathbf{U}\cdot\nabla u\right) = \nabla\left[D_l q(\phi_s)\nabla u - \mathbf{J}_{AT}\right] + \frac{1}{2}\left[1+(1-k)u\right]\frac{\partial h(\phi_s)}{\partial t} - \nabla\cdot\mathbf{J} \tag{2}$$

where ϕ_s is the sum total of phase-field variables expressing the solid, C_l is the solute concentration in the liquid, $u = (C_l - C_l^e)/(C_l^e - C_s^e)$ is the dimensionless supersaturation, C_s is the solute concentration in the solid, C_l^e is the equilibrium concentration in the liquid, C_s^e is the equilibrium concentration in the solid, $k = C_s/C_l = C_s^e/C_l^e$ is the partition coefficient, $q(\phi_s) = [kD_s + D_l + (kD_s - D_l)\phi_s]/(2D_l)$, D_l is the diffusivity of the solute in the liquid, D_s is the diffusivity of the solute in the solid, $\nabla\cdot\mathbf{J}$ is the noise term [25], and $\mathbf{J}_{AT}$ is the anti-trapping current term [24].

The LB equations can be expressed by [20]

$$f_k(\mathbf{x}+\mathbf{c}_k\Delta t, t+\Delta t) = f_k(\mathbf{x},t) - \frac{1}{\tau_{\mathrm{LBM}}}\left[f_k(\mathbf{x},t) - f_k^{\mathrm{eq}}(\mathbf{x},t)\right] + \rho w_k\left[\frac{3(\mathbf{c}_k - \mathbf{U})}{c^2} + \frac{9(\mathbf{c}_k\cdot\mathbf{U})\mathbf{c}_k}{c^4}\right]\cdot\mathbf{G}\Delta t \quad (3)$$

where f_k is the particle velocity distribution in the k-th direction, f_k^{eq} is the equilibrium distribution function, $\mathbf{c}_k$ is the discrete particle velocity vector, Δt is the time increment, τ_{LBM} is the LB relaxation time, $\rho = \sum^Q f_k$ is the liquid density, $\mathbf{U}$ is the liquid velocity vector computed by $\rho\mathbf{U} = \sum^Q \mathbf{c}_k f_k$, Q is the number of discrete velocities, w_k is the weight function, $\mathbf{G} = -2\rho\nu h/W_0((1+\phi_s)/2)^2(\mathbf{U} - \mathbf{U}_{si})$ is the dissipative drag force vector, $\nu = (\tau_{\mathrm{LBM}} - 0.5)\Delta x^2/(3\Delta t)$ is the dynamic viscosity, and $h = 2.757$ [26].

The velocity of i-th rigid body grain, $\mathbf{U}_{si}$, is computed as the sum of the translational velocity $\mathbf{U}_{Ti}$ and the rotational velocity $\boldsymbol{\omega}_i$ as $\mathbf{U}_{si}(\mathbf{x}, t) = \mathbf{U}_{Ti} + \boldsymbol{\omega}_i \times (\mathbf{x} - \mathbf{X}_i)$, where $\mathbf{X}_i$ is the centroid position of the i-th grain. $\mathbf{U}_{Ti}$ and $\boldsymbol{\omega}_i$ are computed by the equations of motion:

$$M_i \frac{d\mathbf{U}_{\mathrm{T}i}}{dt} = \mathbf{F}_i, \quad \mathbf{I}_i \frac{d\boldsymbol{\omega}_i}{dt} = \mathbf{T}_i \quad (4)$$

Here, M_i is the mass and $\mathbf{I}_i$ is the moment of inertia. The total force, $\mathbf{F}_i$, and the total torque, $\mathbf{T}_i$, acting on the i-th grain are expressed by

$$\mathbf{F}_i = -\sum_{\mathbf{x}\in\Omega_i}\{\mathbf{G}(\mathbf{x},t) + \mathbf{F}_{ci}\}\Delta V, \quad \mathbf{T}_i = -\sum_{\mathbf{x}\in\Omega_i}\left[(\mathbf{x}-\mathbf{X}_i)\times(\mathbf{G}(\mathbf{x},t)+\mathbf{F}_{ci})\Delta V\right] \quad (5)$$

where Ω_i is the solid-liquid interface domain of the i-th grain and $\Delta V = \Delta x^2$. $\mathbf{F}_{ci}$ is the collision force acting on the i-th grain by contact with the j-th grain. Here, an elastic contact model [27] is introduced and $\mathbf{F}_{ci}$ is expressed by

$$\mathbf{F}_{ci} = \sum_{j-1}^{n}\left(k_n p_i^2 p_j^2 \mathbf{n}_{ij}\right) \quad (6)$$

where $p_i = (\phi_i + 1)/2$, k_n is a coefficient related to normal spring constant, n is the number of non-zero phase-field variables on a grid point, and $\mathbf{n}_{ij}$ is the unit normal vector expressed by $\mathbf{n}_{ij} = \nabla(p_i - p_j)/|\nabla(p_i - p_j)|$. The overlap distance between the i-th and j-th grains is represented by the overlap of diffuse interface, $p_i^2 p_j^2$.

Table 1. Material and simulation parameters.

Quality	Symbol	Value
Initial concentration	c_0	0.01296 at.frac.
Grid size	Δx	1.0 μm
Interface thickness	W_0	$\Delta x/0.4 = 2.5$ μm
Diffusion coefficient in liquid	D_l	3.0×10^{-9} m²/s
Diffusion coefficient in solid	D_s	3.0×10^{-13} m²/s
Partition coefficient	k	0.14
Anisotropic strength	ε_4	0.02
Gibbs-Thomson constant	Γ	2.4×10^{-7} Km
Liquidus slope	m	–620 K/at.frac.
Temperature	T_0	923.6 K ($u_0 = -0.20$) 921.0 K ($u_0 = -0.40$) 916.7 K ($u_0 = -0.60$)
Time increment	Δt	6.667×10^{-5} s
LB relaxation time	τ_{LBM}	1

3. Computational conditions

The MPF-LB method shown in the previous section is applied to an isothermal shear deformation of semi-solid Al-3wt%Cu alloy with globular morphology [3, 18, 28]. Table 1 shows material and simulation parameters. Figure 1 shows the computational and initial conditions for the solidification simulations to prepare the globular morphology. The square computational domain is divided into 1024 × 1024 grid points with grid size Δx = 1 μm. The wall and push plate are expressed by the 0-th phase-field variable ϕ_0, where the interface between the wall and push plate and liquid phase is formed by one-dimensional equilibrium phase-field profile of $\phi_0 = \tanh(r/(2^{0.5}W_0))$. The diffusion coefficient D_s in the wall and push plate is set to zero. At the initial condition shown in figure 1, a nondimensional supersaturation u_0 is set to u_0 = –0.2, –0.4 and –0.6, and 120 solid seeds are placed in an equilateral triangular array. To prepare the semi-solid mixture with globular morphology, solidification simulations are performed until a quasi-equilibrium state is achieved from the initial condition shown in figure 1. Figure 2 is the computed semi-solid systems with globular morphology. The regular hexagonal array of globular solids is observed except neighboring walls. The solid fractions f_s for figure 2 are f_s = (a) 0.274, (b) 0.458 and (c) 0.634, respectively. These values are slightly larger than the equilibrium values for u_0, because there is the walls and push plate. From those conditions, the push plate is pushed into the semi-solid mixture with a constant velocity of U_{plate} = 6 μm/s; then the right-hand side of equation (1) is set to zero only for ϕ_0. The coefficient representing spring constant is fixed to be k_n = 0.0437 N/mm^3.

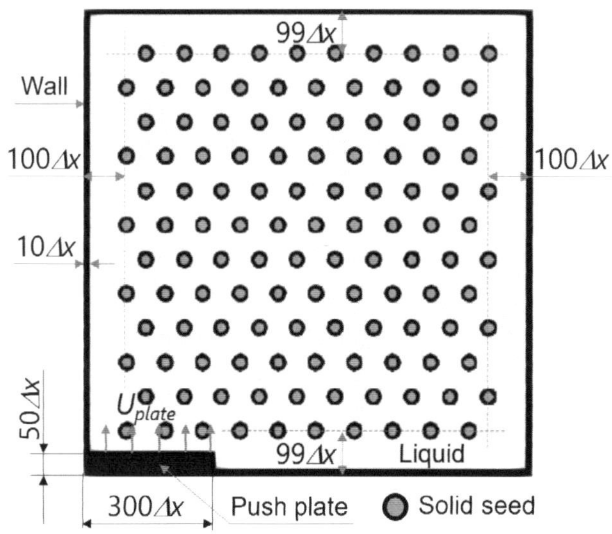

Figure 1. Computational and initial conditions. Black, blue and light blue regions are the wall and push plate ($\phi_0 > -1$), solid and liquid, respectively. Number of solid seeds is set to 120.

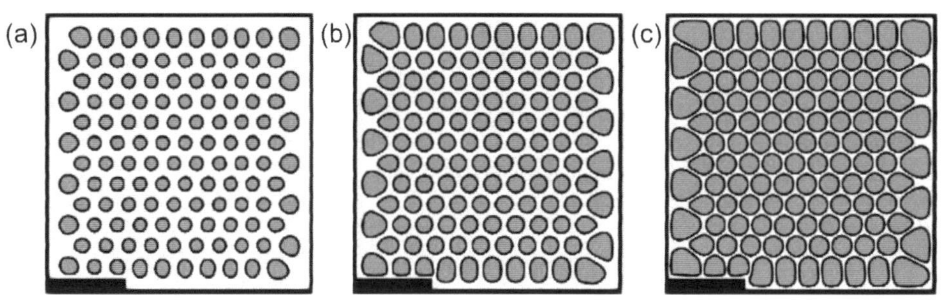

Figure 2. Computed semi-solid structures with globular morphology for u_0 = (a) –0.2 (f_s = 0.274), (b) –0.4 (f_s = 0.458) and (c) –0.6 (f_s = 0.634).

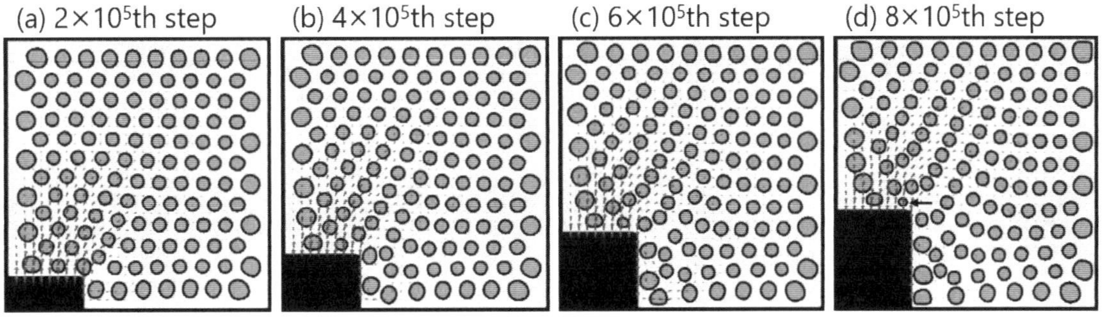

Figure 3. Time slices during semi-solid deformation for $u_0 = -0.2$ ($f_s \approx 0.30$).

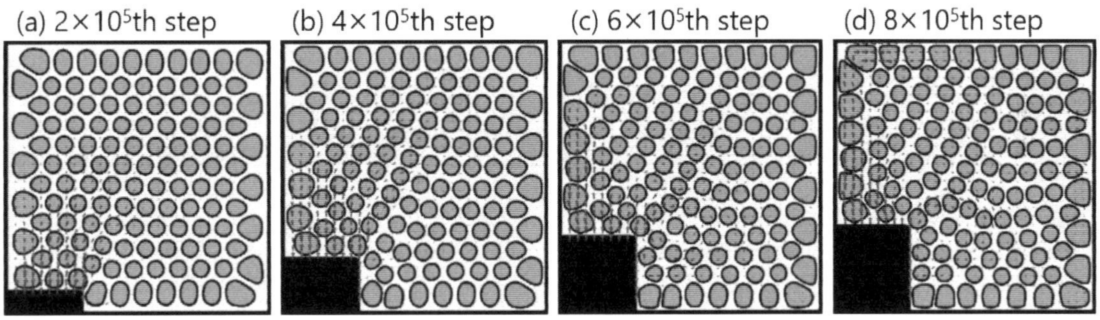

Figure 4. Time slices during semi-solid deformation for $u_0 = -0.4$ ($f_s \approx 0.45$).

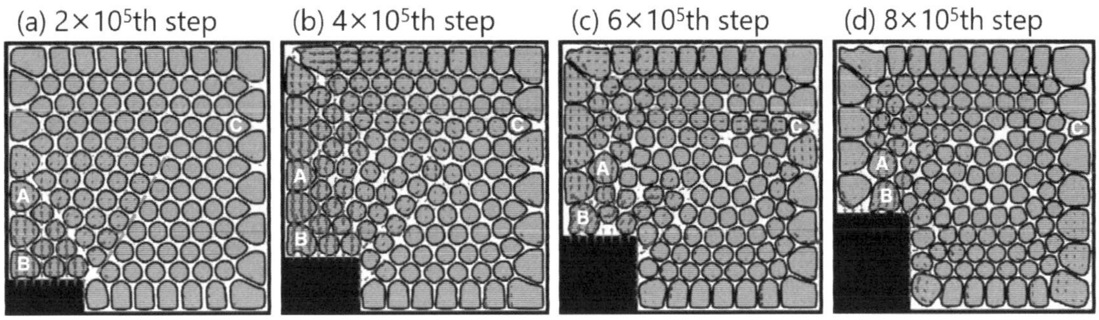

Figure 5. Time slices during semi-solid deformation for $u_0 = -0.6$ ($f_s \approx 0.65$).

4. Results

Figures 3-5 are the time slices during semi-solid deformation by push plate from the structures shown in figure 2. Solid-liquid interfaces are indicated by the black lines with $\sum p_i^2 < 0.85$, and the liquid and solid velocities are expressed by red arrows, of which the length shows the magnitude of velocity. The displacement of the top surface of the push plate are (a) 80 μm, (b) 160 μm, (c) 240 μm, and (d) 320 μm in figures 3-5. Depending on the solid fractions f_s, different rearrangement of solids and velocity patterns are observed.

In figure 3 ($f_s \approx 0.30$), a fan-shaped flow pattern is formed ahead of the push plate and maintained during the simulation. The flow pattern is quite similar to the one observed in in-situ observations [3, 28]. Due to the right upward flow on the push plate, bands of liquid and dense grain are alternatively formed toward the 45° direction after (b) 4 × 10^{5}th step, which result in the formation of dilatancy.

In figure 4 ($f_s \approx 0.45$), similar behavior to figure 3 can be observed in the early stage of figures 4(a)-(b). However, latter stages in figures 4(c)-(d) show different deformation patterns, where the longitudinal force ahead of the push plate is supported by a line of solids in the left side. Due to being pushed upward, some grains located on the top side migrate to the right as shown in figures 4(d).

In figure 5 ($f_s \approx 0.65$), deformation occurred as in a metal solid, where atoms are arranged in a regular pattern. In figure 5(a), the pink dashed line indicates the firstly activated slip surface, where the slip on the surface occurred until around the 1.7×10^5th step. At the step in figure 5(a), slip occurred on the surface indicated by the yellow dashed line. The deformation in figure 5(b) is similar to that in figure 4(d). Meanwhile, due to more dense solid grains, the upward forces are applied overall in front of the push plate compared to figure 4(d). In figure 5(c), grain A was pushed out to the right from the longitudinal solid line on the left. Then, since grain A pushed the other surrounding grains, the grain array drastically changed in the deformation from figure 5(b) to figure 5(c). In figure 5(d), grain B was also pushed out from the left line to the right. The pushing out of grains A and B displaced grain C to the right, into the right solid line.

As shown in figures 3-5, it was confirmed that the semi-solid deformation in the small solid fraction (figure 3) behaves like liquid flow, where the rearrangement of grains is the dominant deformation mechanism, while that in the large solid fraction (figure 5) behaves like a solid deformation, where the solid interaction is the dominant deformation mechanism. In the intermediate condition (figure 4), both deformation behaviors could be observed. In the small solid fraction, the semi-solid deformation is very similar to that observed in the experimental observations [3, 28] and discrete element simulation [18]. In the high solid fraction, plastic deformation of a grain would occur in a real situation. The introduction of the plastic deformation in grains is future work. Also, shrinkage of grains (e.g. a grain indicated by the black arrow in figure 3(d)) and morphological changes of grains due to solidification/melting and liquid flow were observed in figures 3-5. This is another advantage of the present MPF-LB model. On the other hand, mechanical interaction between grains expressed by equation (6) is too strong; thus, liquid phase is always left between grain boundaries. This problem should be solved in next study.

5. Conclusions

The MPF-LB model, which can simulate growth of multiple dendrites with motion and liquid flow, was applied to semi-solid deformation with globular morphology. Here, the coalescence of grains was not considered and an elastic contact model was introduced to the MPF-LB model. By using the model, the simulations of semi-solid deformation by push plate were performed for three different solid fractions. It was observed that, in the small solid fraction, the rearrangement of grains due to the liquid flow was the dominant deformation mechanism, and, in the high solid fraction, solid-solid interaction was the dominant deformation mechanism and it caused the rearrangement of grains. From the results, it was confirmed that the MPF-LB model has high potential as an evaluation model of semi-solid deformation.

In future work, the contact model in a diffuse interface will be improved and the plastic deformation in grains will be considered. Three-dimensional simulations are also expected, though the computational cost is high [29-32].

Acknowledgements

This research was partly supported by KAKENHI, Grant–in–Aid for Scientific Research (A) 17H01237.

References

[1] Flemings M C 1991 *Metall. Trans. B* **22** 269-93
[2] Gourlay C M, Dahle A K, Nagira T, Nakatsuka N, Nogita K, Uesugi K and Yasuda H 2011 *Acta Mater.* **59** 4933-43
[3] Nagira T *et al.* 2011 *Scripta Mater.* **64** 1129-32
[4] Fonseca J, O'Sullivan C, Nagira T, Yasuda H and Gourlay C M 2013 *Acta Mater.* **61** 4169-79
[5] Kareh K M, O'Sullivan C, Nagira T, Yasuda H and Gourlay C M 2017 *Acta Mater.* **125** 187-95
[6] Cai B, Karagadde S, Yuan L, Marrow T J, Connolley T and Lee P D 2014 *Acta Mater.* **76** 371-80
[7] Karagadde S *et al.* 2015 *Nat. Commun.* **6** 8300

[8] Cai B, Lee P D, Karagadde S, Marrow T J and Connolley T 2016 *Acta Mater.* **105** 338-46
[9] Atkinson H V 2005 *Prog. Mater Sci.* **50** 341-412
[10] Zavaliangos A 1998 *Int. J. Mech. Sci.* **40** 1029-41
[11] Kang C G and Jung H K 1999 *Int. J. Mech. Sci.* **41** 1423-45
[12] Gronostajski Z 2000 *J. Mater. Process. Technol.* **106** 40-4
[13] Fuloria D and Lee P D 2009 *Acta Mater.* **57** 5554-62
[14] Phillion A B, Cockcroft S L and Lee P D 2008 *Acta Mater.* **56** 4328-38
[15] Yamaguchi M and Beckermann C 2014 *JOM* **66** 1431–8
[16] Sistaninia M, Phillion A B, Drezet J M and Rappaz M 2012 *Acta Mater.* **60** 3902-11
[17] Sistaninia M, Phillion A B, Drezet J M and Rappaz M 2012 *Acta Mater.* **60** 6793-803
[18] Su T C, O'Sullivan C, Nagira T, Yasuda H and Gourlay C M 2019 *Acta Mater.* **163** 208-25
[19] T. Takaki, R. Sato, R. Rojas, M. Ohno and Shibuta Y 2018 *Comput. Mater. Sci.* **147** 124–31
[20] Rojas R, Takaki T and Ohno M 2015 *J. Comput. Phys.* **298** 29-40
[21] Takaki T, Rojas R, Ohno M, Shimokawabe T and Aoki T 2015 *IOP Conf. Ser. Mater. Sci. Eng.* **84** 012066
[22] Sakane S, Takaki T, Ohno M and Shibuta Y 2019 *Comput. Mater. Sci.* **164** 39-45
[23] Yasuda H, Morishita K, Nakatsuka N, Nishimura T, Yoshiya M, Sugiyama A, Uesugi K and Takeuchi A 2019 *Nat. Commun.* **10** 3183
[24] Ohno M and Matsuura K 2009 *Phys. Rev. E* **79** 031603
[25] Echebarria B, Karma A and Gurevich S 2010 *Phys. Rev. E* **81** 021608
[26] Beckermann C, Diepers H J, Steinbach I, Karma A and Tong X 1999 *J. Comput. Phys.* **154** 468-96
[27] Wang Y U 2007 *Acta Mater.* **55** 3835-44
[28] Nagira T *et al.* 2013 *ISIJ Int.* **53** 1195-201
[29] Takaki T, Sakane S, Ohno M, Shibuta Y and Aoki T 2020 *Comput. Mater. Sci.* **171** 109209
[30] Sakane S, Takaki T, Ohno M, Shibuta Y and Aoki T 2019 *Modell. Simul. Mater. Sci. Eng.* **27** 054004
[31] Sakane S, Takaki T, Ohno M, Shibuta Y, Shimokawabe T and Aoki T 2018 *J. Cryst. Growth* **483** 147-55
[32] Sakane S, Takaki T, Rojas R, Ohno M, Shibuta Y, Shimokawabe T and Aoki T 2017 *J. Cryst. Growth* **474** 154-9

IOP Conf. Series: Materials Science and Engineering **861** (2020) 012068 doi:10.1088/1757-899X/861/1/012068

Concurrent and coupled resolution of fluid flow and solid deformation in solidification processes

S Zhang, G Guillemot, C-A Gandin and M Bellet

MINES Paristech, PSL Research University, CEMEF – Centre de Mise en Forme des Matériaux, CNRS UMR 7635, CS 10207, 1 Rue Claude Daunesse, 06904 Sophia Antipolis cedex, France

E-mail : michel.bellet@mines-paristech.fr

Abstract. Control of macrosegregation phenomena and deformation related defects is the main issue in solidification processes. Numerical modelling of casting provides an answer to industrial needs to master these defects. As a first step, it is essential to achieve a concurrent computation of the fluid flow in the bulk liquid and stress-strain evolution in the already solidified regions. For this specific purpose, a partitioned solution algorithm is developed to model both ingot casting and continuous casting processes: liquid flow induced by natural convection or filling step, solidification shrinkage as well as thermal deformation of solid phase are all taken into consideration. This article is mainly composed of two parts. In the first part, details of the model itself are described. In the second part, applications dedicated to steel solidification are proposed.

1. Introduction

One of the most critical modeling difficulties encountered in solidification processes is the concurrent computation of fluid flow and solid deformation. More precisely, the difficulty lies in the stiffness differences between solid and liquid regions and the explicitly un-describable solid/liquid interface at its microscopic scale. In the current state-of-the-art, there are two types of solutions for this numerical issue: monolithic and partitioned algorithms. In monolithic solution, interactions between the solid and liquid phases are considered as a strong interaction problem. Macroscopic governing equations are developed by using the volume averaging method, thus scaling up physical problems from its dendritic scale to that of the cast product. These equations are then coupled and solved in a single resolution system [1, 2]. The main drawback of this approach is that it requires huge computational resources preventing its application to large industrial casting processes. Typically, a monolithic algorithm consists in the resolution of at least seven unknowns, i.e. velocity vectors in the liquid and solid regions plus pressure, or eventually of nine unknowns if taking into consideration the temperature and solute concentration fields. It is then preferable to consider the interactions between solid and liquid phases as a weak interaction problem, in which the fluid flow computation and the stress-strain analysis are treated as two subproblems. Governing equations respective to each subproblem are solved in two different steps and then coupled [3, 4]. Recently, Zhang *et al.* [4] proposed an original partitioned solution algorithm for the current modeling of fluid flow and solid deformation for solidification processes. However, this algorithm was developed under the specific hypothesis that the movement of the solid phase is negligible compared to that of the fluid flow, which is indeed true for most ingot casting (IC) processes but not for continuous casting (CC). The movement of the solid shell (about several centimeters per second) in industrial CC can be no more neglected compared to that of the fluid flow.

A concurrent and coupled solution algorithm is proposed presently in continuity with this first contribution [4]. The present algorithm extends this work to cover both IC and CC processes. The global non-steady-state resolution strategy is chosen to model CC processes, ensuring continuous increase of the metal volume during the casting process [5]. The Arbitrary Lagrangian-Eulerian (ALE) method is used to update the mesh grid, considering the metal in the fully solid state as Lagrangian [6]. At each time increment, the momentum and mass conservation equations are solved twice at macroscopic scale, respectively in a first solid-oriented resolution step (STEP I) and in a second liquid-oriented resolution step (STEP II). From STEP I, the velocity and pressure fields in the already solidified regions and the associated stress-strain fields are deduced. Subsequently in STEP II, the liquid flow and pressure fields in the bulk liquid and mushy regions are calculated through the resolution of the Navier-Stokes equations based on the volume averaging method. Interactions between the solid and liquid phases in the mushy zone are taken into consideration at this second step through the Darcy-type momentum exchange terms, modeling the resistance of the solid skeleton to the fluid flow. All conservation equations are formulated in a level set framework in order to track the metal/gas interface during the solidification processes. Besides, the current resolution algorithm is coupled with an existing non-linear energy solver to calculate the temperature field [7]. The present paper includes a synthetic presentation of the resolution scheme, followed by an application to a three dimensional (3D) industrial CC process.

2. Numerical model

2.1. Level set method

A gas subdomain is introduced above the metal subdomain within the level set framework. More precisely, the metal/gas interface Γ is represented by the zero-isovalue of the signed distance function $\varphi(\boldsymbol{x}, t)$, defined for any point $\boldsymbol{x}$ and time t in the domain of interest, Ω. The metal and gas parts correspond respectively to Ω^M and Ω^G subdomains in the whole Ω domain. The smoothed Heaviside function $\mathcal{H}^M$ is then defined as:

$$\mathcal{H}^{\mathrm{M}}(\varphi) = \begin{cases} 0 & \text{if } \varphi < -\varepsilon \\ 1 & \text{if } \varphi > \varepsilon \\ \frac{1}{2}\left(1 + \frac{\varphi}{\varepsilon} + \frac{1}{\pi}\sin\left(\frac{\pi\varphi}{\varepsilon}\right)\right) & \text{if } -\varepsilon \leq \varphi \leq \varepsilon \end{cases} \tag{1}$$

where ε is the half thickness of the transition zone introduced in order to distinguish artificially the subdomains. Respectively denoting ψ^G, ψ^M and ψ^l the physical property related to gas subdomain, metal subdomain and liquid phase, the mixed properties, $\hat{\psi}$ and $\hat{\psi}^F$, are given by:

$$\hat{\psi} = \mathcal{H}^M\psi^M + (1 - \mathcal{H}^M)\psi^G; \quad \hat{\psi}^F = \mathcal{H}^M\psi^l + (1 - \mathcal{H}^M)\psi^G \tag{2}$$

Thus, over the thickness [$-\varepsilon$, ε] around the interface Γ, a smooth and continuous transition of properties is defined with $\hat{\psi}^F$ associated to the mixed property of fluid. g^l and g^s correspond hereafter to the volume fraction of liquid and solid phases in the metal subdomain.

2.2. Solution algorithm scheme

The detailed solution algorithm scheme together with the associated governing equations is illustrated hereunder in figure 1. Seven different stages are considered in one time increment Δt:

- 1st stage: **Heat transfer**. The monolithic energy conservation equation, written in the ALE formulation, is solved with the mixing values given in the following table:

Table 1. Mixing properties in the heat transfer stage.

Notation	Definition
$\widehat{\rho h}$	$\mathcal{H}^M\langle\rho h\rangle^M + (1-\mathcal{H}^M)\rho_0^G h^G$
$\widehat{\rho C_p}^F$	$\mathcal{H}^M\langle\rho\rangle^l\langle C_p\rangle^l + (1-\mathcal{H}^M)\rho_0^G \cdot {C_p}_0^G$
$\hat{k}$	$\mathcal{H}^M\langle k\rangle^M + (1-\mathcal{H}^M)k_0^G$

where $\langle\rho h\rangle^M$ is the volume enthalpy in the metal subdomain defined as $g^l\langle\rho\rangle^l\langle h\rangle^l + g^s\langle\rho\rangle^s\langle h\rangle^s$. $\hat{k}$ is the average conductivity equal to $g^l\langle k\rangle^l + g^s\langle k\rangle^s$. $\langle\rho\rangle^\alpha$, $\langle C_p\rangle^\alpha$, $\langle h\rangle^\alpha$, $\langle k\rangle^\alpha$ are respectively the intrinsic density, specific heat, specific enthalpy and thermal conductivity of the α phase in metal. ρ_0^G, ${C_p}_0^G$, k_0^G and h^G are respectively the constant density, constant specific heat, constant thermal conductivity and specific enthalpy of gas. The energy transportation relative to the liquid phase is modelled by a convective term associated to liquid phase velocity. The nodal mesh velocity, v_{msh}, is accounted for in the convective term, considering the mesh updating stage which will be further detailed. From this stage, the global temperature field, in both metal and gas subdomains, is computed.

- 2nd stage: **STEP I** solid-oriented solution. The classical momentum and mass conservation equations of the solid mechanics in solidification processes [6] are extended in the level set framework, with the definition of the following mixed values:

Table 2. Mixing properties in the solid-oriented resolution stage.

Notation	Definition
$\hat{\boldsymbol{s}}$	$\mathcal{H}^M\boldsymbol{s}^M + (1-\mathcal{H}^M)\boldsymbol{s}^G$
$\hat{\rho}$	$\mathcal{H}^M\rho^M + (1-\mathcal{H}^M)\rho_0^G$

where $\boldsymbol{s}^M$, ρ^M, $\boldsymbol{s}^G$, ρ^G are the stress tensor and average density respectively in metal and gas subdomains, considering the cast metal as a homogenized continuum. p is the pressure field and $\boldsymbol{g}$ is the constant gravity vector. Note that the average density of metal is assumed to be the intrinsic density of solid $\langle\rho\rangle^s$ in the entire metal subdomain, to approximate the thermally induced solid deformation without solidification shrinkage effect. Besides, $H(.)$ is the standard Heaviside function, estimated at the temperature difference, $T_C - T$. This function is introduced in the second mass conservation equation as an indicator relative to the use of the thermo-elastic-viscoplastic (TEVP) constitutive model for elements with average temperature lower than the critical temperature T_C. Otherwise the thermo-viscoplastic (TVP) model is used for temperature higher than T_C. In the present work T_C is considered as equal to the solidus temperature to model the TEVP behaviour of material in the solid state. The TVP model is used for metal in the mushy or liquid state. Note that the mushy zone is considered as a homogenized non-Newtonian fluid, in which the strain rate sensitivity is continuously increasing with the liquid fraction, up to a pure Newtonian behaviour in the fully liquid state. Besides, an augmented fluid viscosity (including gas and liquid) is assumed in this stage similarly for numerical stability issues. Nonetheless, such approximation is reasonable as the fluid flow computation will be considered and computed in the next stage. Finally, from this stage, the velocity of the solid phase, v_I, is obtained.

- 3rd stage: **STEP II** fluid-oriented solution. Momentum and mass conservation equations are written in an ALE formulation, under a two phase approach with the use of the volume averaged method [8]. Similarly, expressions of the mixing values involved are resumed in the following table 3. $\langle\rho\rangle_0^l$, $\langle\rho\rangle^l$, $\langle\boldsymbol{s}\rangle^l$, μ^l, $\langle\rho\rangle^s$ are respectively the intrinsic reference liquid density, intrinsic liquid density, intrinsic liquid stress tensor, liquid viscosity and intrinsic solid density. A fixed secondary dendrite arm spacing, λ_2, is considered in the estimation of the permeability value, $\hat{\kappa}^F$. μ_0^G is the constant viscosity of gas, considered as a Newtonian and incompressible fluid. In addition, the Boussinesq approximation is assumed in fluid flow resolution. Note that interactions between solid and liquid phases in the mushy zone are modelled by the Darcy's law with the solid velocity from previous stage $\boldsymbol{v}_I$ as an entry.

Complementary to STEP I, the velocity of the liquid phase, calculated with the nominal liquid viscosity in STEP II, is denoted $\boldsymbol{v}_{II}$.

Table 3. Mixing properties for the fluid-oriented solution.

Notation	Definition
$\hat{\rho}_0^F$	$\mathcal{H}^M\langle\rho\rangle_0^l + (1-\mathcal{H}^M)\rho_0^G$
$\hat{g}^F$	$\mathcal{H}^M g^l + (1-\mathcal{H}^M)$
$\widehat{g\boldsymbol{s}}^F$	$\mathcal{H}^M g^l\langle\boldsymbol{s}\rangle^l + (1-\mathcal{H}^M)\boldsymbol{s}^G$
$\widehat{g\rho}^F$	$\mathcal{H}^M g^l\langle\rho\rangle^l + (1-\mathcal{H}^M)\rho_0^G$
$\hat{\mu}^F$	$\mathcal{H}^M \mu^l + (1-\mathcal{H}^M)\mu_0^G$
$\hat{\kappa}^F$	${\lambda_2}^2{\hat{g}^F}^3/(180(1-\hat{g}^F)^2)$

- 4th stage: **Mesh updating**. The position of each mesh node is updated following an explicit scheme with the mesh velocity, $\boldsymbol{v}_{mesh}$. This velocity is equal to $\boldsymbol{v}_I$ in the fully solidified regions and mushy domain. Based on this consideration, we assume that the energy transportation relative to the solid phase in the solid and mushy zones is achieved through the mesh updating process.
- 5th stage: **Metal/gas interface tracking**. The updating of the level set function permits interface tracking. It is achieved by the convection-reinitialization scheme [9]. The advection velocity is defined according to the node temperature, considering that $\boldsymbol{v}_{II}$ describes better the moving interface of liquid/gas and mush/gas, while $\boldsymbol{v}_I$ describes better the interface solid/gas.
- 6th stage: **Properties mixing.** After the updating of the metal/gas interface, mixed material properties over the artificial transition interface is subsequently updated according to the current level set function.
- 7th stage: **Adaptive remeshing**. The final stage of the algorithm consists in an optional remeshing process when the mesh quality is no longer satisfying.

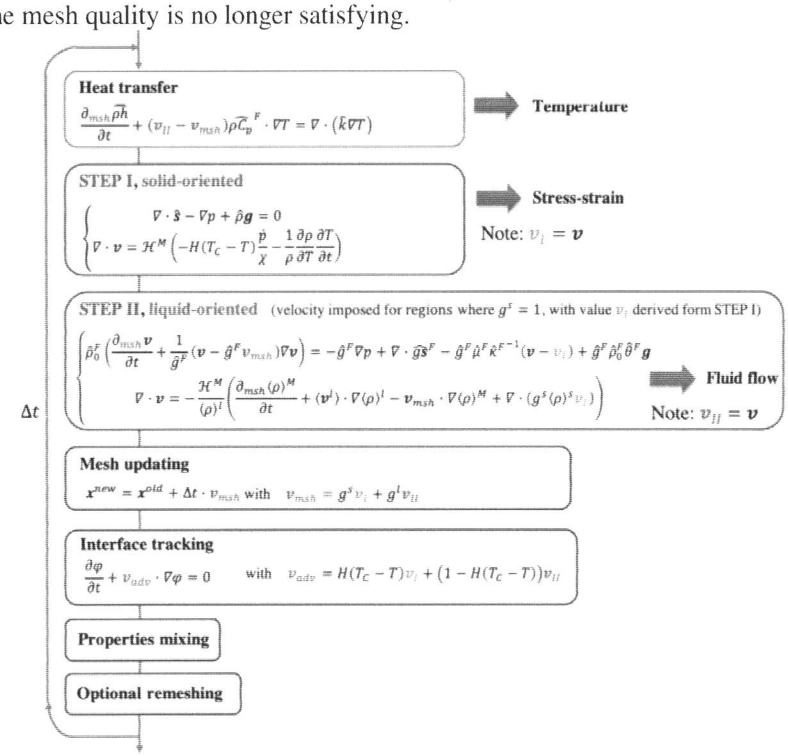

Figure 1. Decomposed solution algorithm in one time increment.

3. Application to steel continuous casting

3.1. Model description

The partitioned algorithm is applied to a 3D CC test case for steel slabs. The initial configuration is given in figure 2(a) with the consideration of a submerged entry nozzle. Only a quarter of the full geometry is modelled considering the symmetry of the system. Heat is extracted from the metal through the lateral vertical surfaces. A constant flux is imposed along surfaces in contact with the mould and a convection type expression is defined along surfaces below the mould exit, in order to model the water cooling effect. An insulation zone is imposed in the vicinity of the meniscus so as to always maintain a liquid-gas interface in contact with the mould. The metal is supposed to slide along the surrounding vertical surfaces in contact with the mould as well as the two symmetry planes. A free velocity is considered on the upper surface in momentum resolution. The surrounding faces of the entry nozzle are defined as sticking. Velocity is imposed for both the liquid inlet surface (vertical velocity $\boldsymbol{v}_{inlet}$ at the top of the entry nozzle) and the bottom surface (vertical casting velocity $\boldsymbol{v}_{cc}$). The same mechanical boundary conditions should be applied for both STEP I and STEP II. Nonetheless, the surrounding vertical surfaces in contact with the mould in STEP II, are assumed to be sticking, in order to reduce numerical instabilities due to the presence of the intensive nozzle jet in contact with the mould.

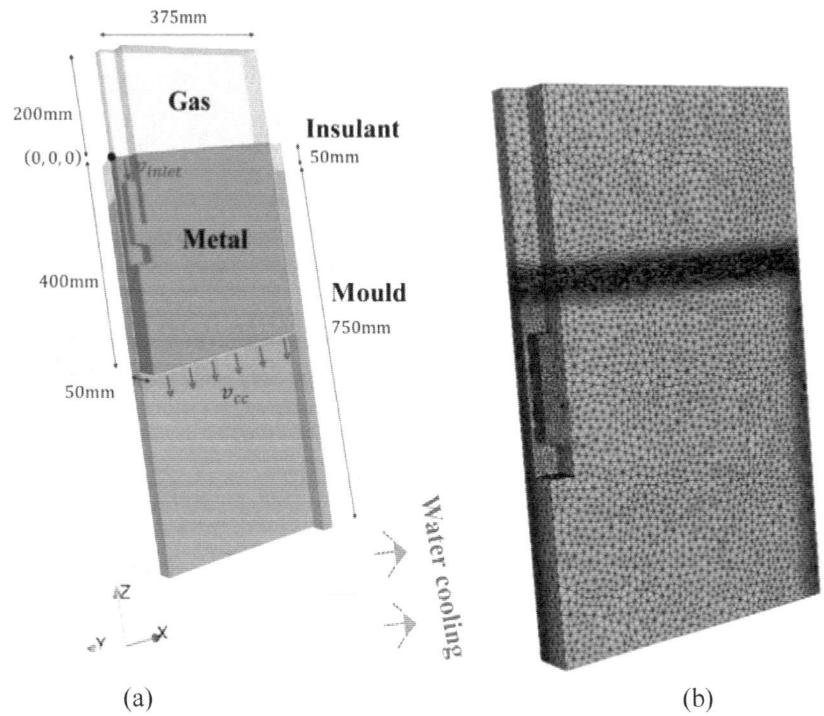

Figure 2. Representation of (a) a schematic for the 3D CC model and (b) the initial mesh configuration.

The initial mesh is defined in figure 2(b). The material properties of the steel grade used in simulation are artificially simplified and are resumed in table 4 together with simulation parameters. Note that solid and liquid densities are taken constant and equal except in the gravity term of STEP II, where the liquid density is governed by the Boussinesq approximation, $\langle\rho\rangle^l = \langle\rho\rangle_0^l\left(1 - \beta_T^l(T - T_{ref})\right)$, and in the thermal dilatation term in STEP I, where the metal density is governed by a linear relation, $\rho = \rho_0\left(1 - \beta_T^s(T - T_{ref})\right)$. Moreover, mechanical properties are detailed in a previous paper [4].

Table 4. Simulation and material parameters in the 3D CC test case.

Material parameters		Symbol	Value	Unit
Solidus temperature			1431	°C
Liquidus temperature			1490.5	°C
Constant specific heat of solid and liquid			800	$J \cdot kg^{-1} \cdot °C^{-1}$
Constant specific heat of gas			1000	$J \cdot kg^{-1} \cdot °C^{-1}$
Latent heat			2.4×10^5	$J \cdot kg^{-1}$
Reference temperature		T_{ref}	1490.5	°C
Reference liquid density		$\langle\rho\rangle_0^l$	7000	$kg \cdot m^{-3}$
Reference metal density		ρ_0	7000	$kg \cdot m^{-3}$
Gas density		ρ_0^G	6000	$kg \cdot m^{-3}$
Liquid thermal dilatation coefficient		β_T^l	1×10^{-4}	$°C^{-1}$
Solid thermal dilatation coefficient		β_T^s	1×10^{-5}	$°C^{-1}$
Secondary dendrite arm spacing		λ_2	5×10^{-4}	m
Gas viscosity	STEP I|STEP II	μ^G	$1 \mid 1 \times 10^{-1}$	$Pa \cdot s$
Liquid viscosity	STEP I|STEP II	μ^l	$1 \mid 5 \times 10^{-3}$	$Pa \cdot s$
Thermal conductivity in gas			0.1	$W \cdot m^{-1} \cdot °C^{-1}$
Thermal conductivity in solid			30	$W \cdot m^{-1} \cdot °C^{-1}$
Thermal conductivity in gas			30	$W \cdot m^{-1} \cdot °C^{-1}$
Simulation parameters				
Casting temperature			1520.5	°C
Critical temperature		T_C	1431	°C
External temperature			25	°C
Time step		Δt	0.1	s
Casting velocity in the vertical direction		v_{cc}	-0.015	$m \cdot s^{-1}$
Liquid inlet velocity in the vertical direction		v_{inlet}	-0.88	$m \cdot s^{-1}$

3.2. Results

Figure 3(a) illustrates the results corresponding to time t = 80 s. The gas subdomain is shown in the upper part with the transparent grey region. The mould exit is shown with the white line. The longitudinal stress component σ_{zz} along the external faces of the product is shown in figure 3(a), revealing that the solid shell is under traction within the mould region while it is in compression below the mould exit, with maximum values reaching respectively 38 MPa and -67 MPa. The fluid flow is concurrently computed, represented by coloured arrows as shown in figure 3(b). It can be noticed that the intensive nozzle jet is mainly influent in the mould region, preventing the formation of the mushy zone due to heat transport. Figure 3(c) shows the temperature evolution at three predefined locations, at positions (373, 0, -200), (372, 0, -400) and (371, 0, -800) with unit in mm respectively for position 1, 2 and 3. All positions are in the mould region, in the vicinity of the narrow face. After 50 s of casting process, the steady-state regime has been reached within the mould region since position 3 is located exactly in the transversal cross section at the mould exit.

Temperature profiles computed over the half thickness of 25 mm (near the wide cooling face) at the narrow symmetry plane are drawn in figure 4(a) at three different times and two different positions respectively at height $z = -0.6$ m above the mould exit (thick lines) and $z = -1.0$ m below the mould exit (thin lines). It can be seen that in the mould region, due to the constant heat flux imposed on the wide cooling face, the cooling rate of the solid shell is higher at the surface than in subsurface. As a consequence, a tensile stress is found in surface. In the secondary cooling region where the cooling is less intense, expressed by the Fourier-type boundary condition, the cooling rate becomes higher in subsurface than in surface. This explains why a compressive stress is found in surface.

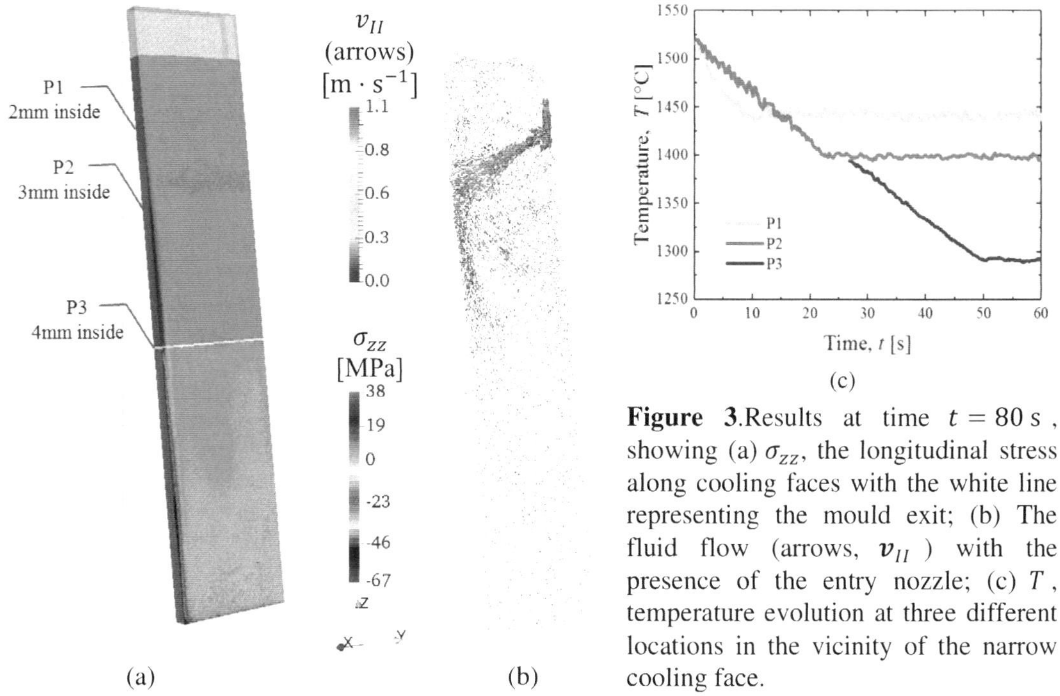

Figure 3.Results at time $t = 80$ s, showing (a) σ_{zz}, the longitudinal stress along cooling faces with the white line representing the mould exit; (b) The fluid flow (arrows, $\boldsymbol{v}_{II}$) with the presence of the entry nozzle; (c) T, temperature evolution at three different locations in the vicinity of the narrow cooling face.

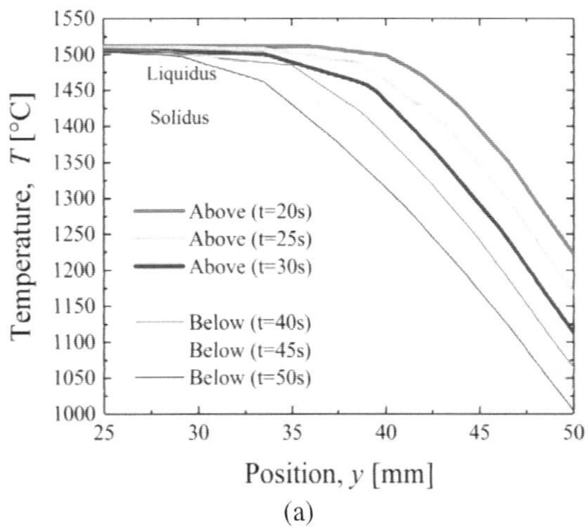

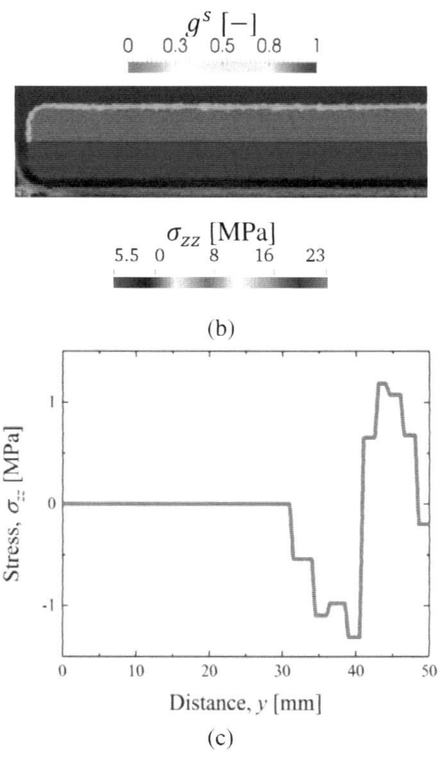

Figure 4.(a) Temperature profiles through the thickness in the transverse symmetry plane (y, z) at three different times and two different positions above (thick lines) and below (thin lines) the mould exit; (b) g^s and σ_{zz}, respectively the solid fraction and the stress component in z direction, shown over the transverse cross section at mould exit at time 80 s; (c) σ_{zz} stress profile through the thickness in the transverse symmetry plane at time 80 s.

Finally, the solid fraction and the stress component in z direction (σ_{zz}) are plotted over the cross section at the mould exit (figure 4(b)), showing a solid shell with thickness between 10 and 15 mm. It can be seen that the external face is submitted to a global tensile stress with the maximum value of 23 MPa located at the corner of the cooling faces while the solidification front undergoes smaller compression stress. It should be mentioned that this maximum predicted value might be unrealistic as the cooling faces are highly constrained (the velocity along faces is imposed to be vertical). The σ_{zz} stress profile through the thickness in the transverse symmetry plane is drawn in figure 4(c), with the position x varying from the centre to the wide cooling face. It can be seen that the vertical stress along the external face has already shifted, from tensile to slightly compressive. Nonetheless, a considerable part of the solid shell near the cooling face is still under traction. Inside the solid shell, compression stresses are found, expressing the necessary mechanical equilibrium along the vertical direction. Finally, it is interesting to note that the predicted stress levels at this specific position are consistent with the flow stress measurement of 1 – 5 MPa, made on solidifying shells by Hiebler *et al.* [10].

4. Conclusions

A new algorithm of concurrent and coupled resolution of fluid flow and solid deformation has been developed for both IC and CC processes. The algorithm consists of a partitioned solution strategy, solving at each time increment a solid-oriented solution and a fluid-oriented solution. A global non steady-state approach is adapted in view of the CC process. All solutions are formulated with the finite element method, using a level set tracking of the moving interface between the solidifying metal subdomain and the gas subdomain. This algorithm has been applied to a 3D CC process. Simulation gives promising results, in which the fluid flow in the bulk liquid or mushy zone and stress formed in the solid regions are concomitantly computed.

The main perspectives of this work are twofold. Firstly, coupling this new algorithm with macrosegregation modelling would allow demonstrating the influence of the solid deformation on the central macrosegregation in CC processes. Secondly, the coupling with equiaxed grain motion would provide access to enhanced predictions of macrosegregation due to thermosolutal convection, shrinkage flow and transport of equiaxed grains.

Acknowledgements

This work is funded by the European Space Agency under project CCEMLCC (AO-2004-017 contract 4200020277) as well as by industrial partners ArcelorMittal Maizières Research, INDUSTEEL France and APERAM ALLOYS IMPHY.

References

[1] Fachinotti V D, Le Corre S, Triolet N, Bobadilla M and Bellet M 2006 *Int. J. Numer. Meth. Eng.* **67** 1341-84
[2] Ludwig A, Vakhrushev A, Wu M, Holzmann T and Kharicha A 2015 *T. Indian I. Metals* **68** 6 1087-94
[3] Kajitani T, Drezet J M and Rappaz M 2001 *Metall. Mater. Trans.* B **32** 1479-91
[4] Zhang S, Guillemot G, Gandin Ch A and Bellet M 2019 *Comput. Method. Appl. M.* **356** 294-324
[5] Bellet M and Heinrich A 2004 *ISIJ Int.* **44** 10 1686-95
[6] Bellet M and Fachinotti V D 2004 *Comput. Method. Appl. M.* **193** 4355-81
[7] Saad A, Gandin Ch A and Bellet M 2015 *Comput. Mater.* **99** 221-31
[8] Ni J and Beckermann C 1991 *Metall. Mater. Trans.* B **22** 349-61
[9] Shakoor M, Scholtes B, Bouchard P O and Bernacki M 2015 *Appl. Math. Model.* **39** 7291-302
[10] Hiebler H, Zirngast J, Bernhard C and Wolf M M 1994 *Steelmaking Conference Proceedings* **77** (Warrendale, PA: Iron and Steel Society, 1994) 405-16

IOP Conf. Series: Materials Science and Engineering **861** (2020) 012069 doi:10.1088/1757-899X/861/1/012069

The ultrasound effects on the formation of the solidification structure of A356 ingots processed via a 2-zone induction melting furnace

A Dong and L Nastac

The University of Alabama, Tuscaloosa, AL, 35487, USA

E-mail: lnastac@eng.ua.edu

Abstract. To study the formation of the solidification structure including the columnar-to-equiaxed transition (CET) under the influence of ultrasound, a 2-zone furnace system and an ultrasound equipment were utilized. The 2-zone furnace system consists of an induction furnace with a top coil and a bottom coil, a graphite crucible, and. a water-cooled chill block located at the bottom of the crucible. By controlling both the top and bottom coil output power independently, the furnace can create various temperature gradients and cooling rates in different regions of the graphite crucible. The ultrasound probe was inserted at the top of the crucible. The top of the crucible was also thermally insulated. Temperature measurements were performed at different locations in the crucible. Optical microscopy was used to characterize the microstructure of the A356 cylindrical ingots. The effects of ultrasound on the microstructure formation during solidification of A356 alloy was studied. A numerical model was developed to compute the temperature gradients and mushy zone evolution in the crucible and to assist in developing of solidification maps.

1. Introduction

The need for lightweight high-performance structural components is expected to continuously increase as automotive, military and aerospace industries are required to further improve the energy efficiency, safety and reliability of their products. Aluminium, magnesium or titanium alloys being much lighter than steel are attractive replacement materials and so there is great interest in enhancing their properties [1-4]. It has been proved that by applying the ultrasonic technology (UST) to molten metal processing columnar dendritic structures can be eliminated and globular non-dendritic structures can be achieved [5-7]. The nonlinear effects like cavitation and acoustic streaming introduced by the ultrasonic field could significantly affect the heat and mass transfer of the solidifying A356 alloy, transforming the grain morphology from dendritic to globular, decreasing the grain size and modifying their properties [8]. UST could also induce degassing during the cavitation process, by creating small hydrogen bubbles that will escape from the molten metal. The pressure oscillations under UST processing can produce large temperature gradients in the liquid, which enhance the nucleation intensity of the bubble. The convection and shock waves produced by UST may promote dendrite fragmentation [9-11]. It is also demonstrated that UST would be more effective when the application temperature is below the liquidus temperature of the alloy [12-13].

In order to create the mushy zone and controlled solidification and cooling conditions, a 2-zone furnace with two separated coils was adopted. By adjusting the top and bottom coil powers of the 2-zone furnace independently, it is possible to create distinct temperature gradients in the crucible with a range of 1 ~ 104 K/m. Cooling rates with a range of 0.005 ~ 0.5 K/s can also be produced in the crucible. Note that the crucible is made of graphite with a water-cooled chill block at the bottom. The 2-zone furnace makes it possible to control the mushy zone size and the grain growth rate during the solidification process.

This study has three main objectives: (1) Compare the solidification process and cooling curves of A356 alloy with and without UST at different cooling rates by controlling the top and bottom coil power levels; (2) Analyse the solidification structure of the A356 alloy at different ingot positions with and without UST; (3) Develop and validate a numerical model to simulate the solidification process and the UST refinement of the A356 microstructure. The results include the temperature profiles and cooling curve analysis, which are also verified by the current experimental work. Microhardness at different positions and for different solidification conditions were compared as well.

2. Experiment description

In the present work, a commercial aluminium alloy A356 was used. A schematic diagram of the two-zone furnace is shown in figure 1(a). The system parameters are as follows: maximum power of the top/bottom zone is 10 kW; height between bottom coil and base is 83 mm; height of top/bottom coil is 152 mm; gap between the top and the bottom zone is 32 mm; total height of the furnace is 425 mm. A graphite crucible opened at both ends with the height, the exterior and the interior diameter of 355, 100, and 75 mm, respectively, was used in the two-zone furnace system. The top of the crucible is covered by a ceramic cap. A water-cooled stainless-steel chill block was applied at the bottom of the furnace as a cooling system. The ultrasonic equipment parameters are as follows: maximum power 2.4 kW, frequency 18 kHz, the diameter of the Nb ultrasonic probe is 40 mm and the probe amplitude is 20 microns. Four K-type thermocouples were fixed at different positions inside the crucible to record the temperature profiles, as shown in figure 1(b). Thermocouples 1, 2, 3 and 4 are located at 260, 190, 120 and 50 mm from the bottom, with wires sealed in ceramic tubes to protect them from the molten alloy.

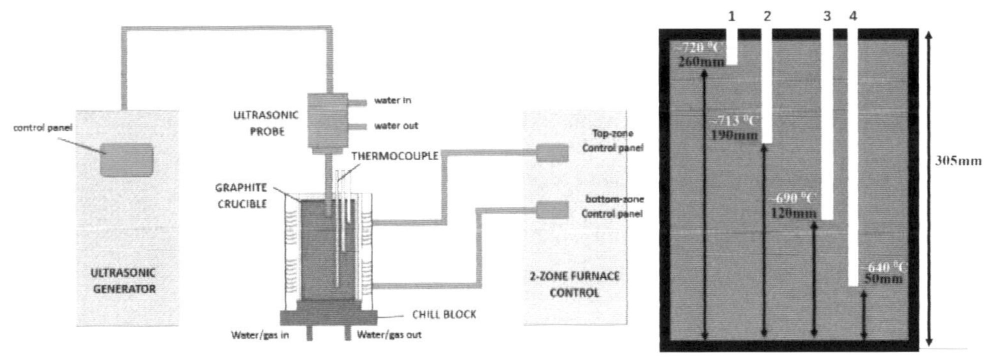

(a) (b)

Figure 1. (a) Sketch of two-zone furnace system and (b) thermocouples positions with initial cooling temperatures

Two different groups of experiments were carried out: (1) Non-UST processing, which is named the control group experiment (N_0 group, top and bottom coils without power during cooling), N_1 group (top 2 kW and bottom 4 kW during cooling), N_2 group (top 3 kW and bottom 6 kW during cooling) and N_3 group (top 4 kW and bottom 7 kW during cooling); (2) UST processing group (UST group, top and bottom coils without power during cooling). In the non-UST groups experiment, 3.3 kg of A356 alloy was melted in the two-zone furnace. Then, four ceramic coated thermocouples were inserted into the designed positions, as shown in figure 1(b), to record the temperature change as a function of time. The top/bottom zone output power values were adjusted to create the desired temperature gradient in the furnace. Firstly, the top and bottom zone output powers were adjusted to the maximum values until the top melt reached about around 730 °C (thermocouple 4), then the melt in the furnace was cooled down by different cooling conditions to the room temperature. In the UST group experiment, the ultrasonic probe was inserted into the melt at about 25 mm below the surface, and the same procedure as explained above was used. After the top melt reached about 730 °C, the furnace was switched off and the ultrasound generator was turned on until the top melt temperature decreased to about 620 °C and then it was shut down and taken out from the melt.

The processed cylindrical cast ingots with a height of 275 mm were cut into 16 disks. Each disk has a height of about 17 mm. Then, these samples were mechanically grounded by #300, 600, 800 and 1200 grinding papers respectively and polished with 3 and 1 μm diamond polishing agents. Finally, the microstructure of the cast samples was characterized via Optical Microscopy by using a Nikon Model Epiphot 200. The microhardness of the specimens was obtained through a Vickers hardness indenter (Buehler Inc.) with a load of 500 gf and dwell time of 15 seconds. Five locations were tested, and the average value was considered as the microhardness.

2.1. Cooling curve comparison

The cooling curves of the all groups are shown in figure 2. The equilibrium liquidus temperature of A356 alloy is about 617 °C. The comparison of non-UST groups from figures 2(a) to 2(d) shows that by adjusting the power of the two coils during solidification, it is possible to control the cooling rates and temperature gradients and create mushy zones with different depths. In addition, due to different cooling rates and temperature gradients, the eutectic solidification process at location 1 is much longer than in location 2, 3 and 4 for the non-UST groups. From figure 2(e), the comparison of N_0 and UST group shows some major differences between these two curves. First, the temperature of the control group drops faster than that of the UST group. Because of the UST application, there is an additional heat source term, which is explained in previous work [14]. Second, the cooling curves of the UST group is smoother during the eutectic stage. The acoustic streaming and cavitation effects due to UST refined the Al matrix and the Si eutectic phase. In addition, by comparing both groups, the heat and fluid flow effects caused by the UST also decreased the temperature gradient in the crucible, which is also indicated by the microstructures shown in figures 3-6.

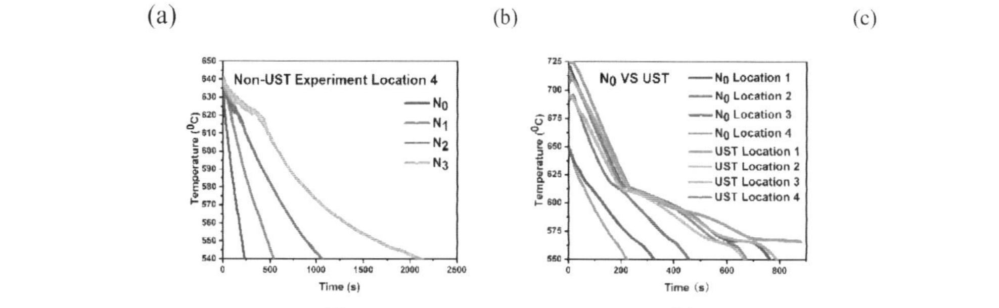

Figure 2. Cooling curves of (a) non-UST groups comparison of location 1; (b) non-UST groups comparison of location 2; (c) non-UST groups comparison of location 3; (d) non-UST groups comparison of location 4; (e) non-UST N_0 group and UST group comparison

From table 1, the largest temperature gradient occurring at location 4 is ~ 10000 K/m and the range of local solidification time is 100 ~ 20000 s, which are both beneficial to create different growth rates and microstructures. In table 1, the G_{TL} is temperature gradient before liquid solidification, V is solidification rate, CR is cooling rate and T_f is local solidification time. By controlling the cooling rates and the temperature gradients in the mushy zone, different microstructures can be achieved (see the microstructures shown in figures 3-6). By comparing the non-UST groups of N_0 to N_3 from the above analysis results of figures 3-6, it can be shown that with the decrease of cooling rates, the grain size becomes larger, and the largest grain size could reach about 1.5 mm. In addition, the grain size decreases from location 1 to 4. The reason is that the temperature gradients in the locations 3 and 4 are much larger than in the locations 1 and 2. Also, the high cooling rates caused by the chill block located at the bottom of the crucible cause the microstructures in the location 4 to be more uniform and finer than in the other regions. By comparing UST with N_0 group, there is an obvious much smaller CET region under the influence of the UST.

Table 1. Analysis results for the non-UST groups and the UST group cooling curves

	Location	UST group	N_0 group	N_1 group	N_2 group	N_3 group
G_{TL} (K/m)	L1	37.1	7.14	4.3	4.3	2.9
	L2	57.1	100	45.7	38.6	29
	L3	485.7	771.4	428.6	214.3	185
	L4	8900	8900	8900	8900	8900
T_f (s)	L1	544	548	1384	4,166	20886
	L2	464	454	1060	2942	14769
	L3	278	274	821	1926	6968
	L4	166	164	328	667	1181
CR (K/s)	L1	1.18E-01	1.17E-01	4.62E-02	1.54E-02	3.06E-03
	L2	1.38E-01	1.41E-01	6.04E-02	2.18E-02	4.33E-03
	L3	2.30E-01	2.34E-01	7.80E-02	3.32E-02	9.18E-03
	L4	3.86E-01	3.90E-01	1.95E-01	9.60E-02	5.42E-02
V (m/s)	L1	3.17E-03	1.64E-02	1.08E-02	3.57E-03	1.06E-03
	L2	2.42E-03	1.41E-03	1.32E-03	5.64E-04	1.49E-04
	L3	4.74E-04	3.03E-04	1.82E-04	1.55E-04	4.96E-05
	L4	4.33E-05	4.38E-05	2.19E-05	1.08E-05	6.09E-06

2.2. Microstructure analysis

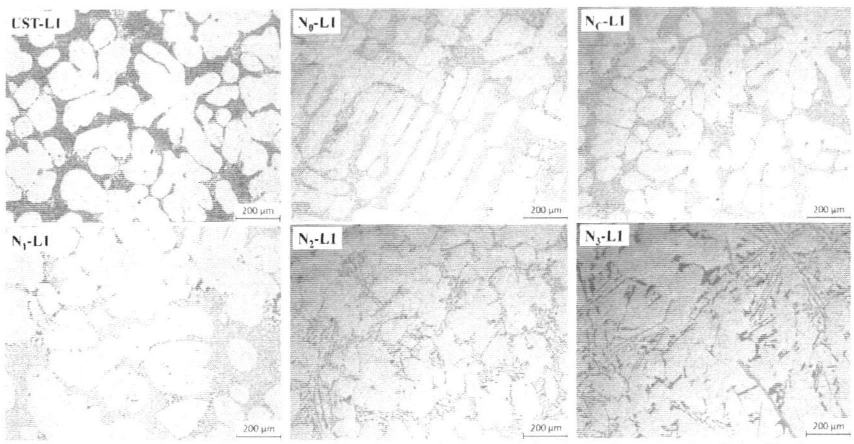

Figure 3. Microstructures of specimens from location 1 of the cast ingots (N-labelled specimens are from the non-UST group, U-labelled specimens from the UST group and N_c specimens from cross section of N_0 group).

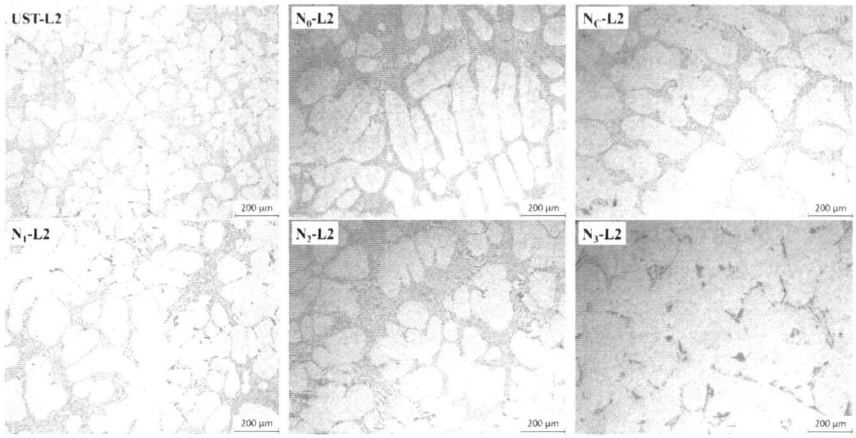

Figure 4. Microstructures of specimens from location 2 of the cast ingots (N labelled specimens are from the non-UST group, U-labelled specimens from the UST group and N_c specimens from cross section of N_0 group).

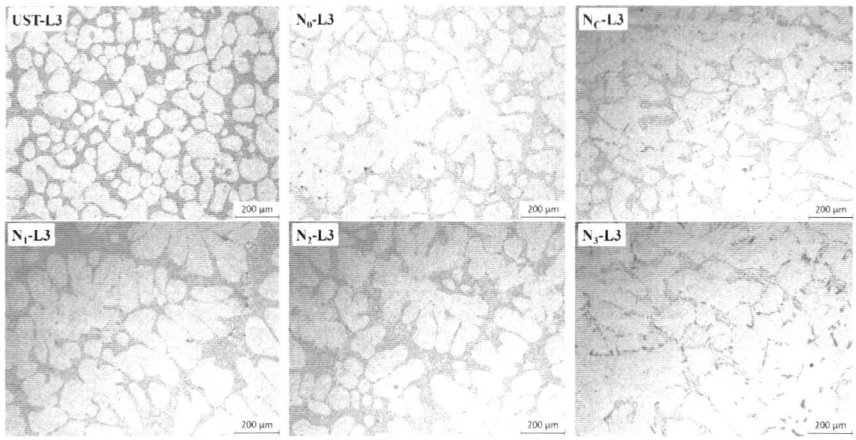

Figure 5. Microstructures of specimens from location 3 of the cast ingots (N-labelled specimens are from the non-UST group, U-labelled specimens from the UST group and N_c specimens from cross section of N_0 group).

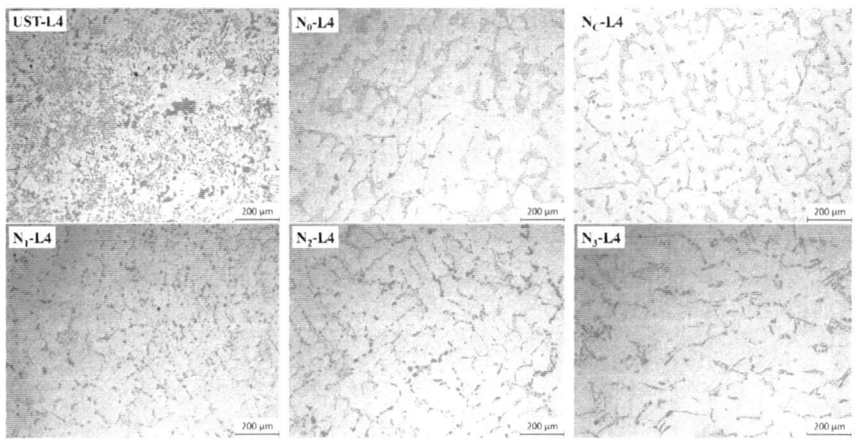

Figure 6. Microstructures of specimens from location 4 of the cast ingots (N labelled specimens are from the non-UST group, U-labelled specimens from the UST group and N_c specimens from cross section of N_0 group).

In addition, it can be seen from figures 3-6 that the eutectic is more refined and modified in the UST case. From location 1 to 3, the UST transforms the morphology of the grains from dendritic to globular. Table 2 presents the detailed solidification analysis data and the microstructural characteristics of different groups. In table 2, the GZ is grain size, GD is grain density, SDAS is secondary dendrite arm spacing, SA is silicon area and SR is silicon roundness in eutectic. The size of the microstructures in table 2 were calculated by averaging at least 5 different measured values. Table 3 shows that the UST group has a smaller CET size than the non-UST N_0 group. The CET size measurements in table 3 include the mixed columnar and equiaxed region. Also, the increase in the top and bottom coil power will decrease the CET size. The microstructure (primary phase and silicon in the eutectic phase) of the UST group is finer than that of the non-UTS groups. Figure 7 further proves that, due to the acoustic streaming and cavitation effect, the UST group has a much refiner microstructure than that of the non-UST groups.

Table 2. Experimental measurements for the non-UST control groups and the UST group

Locations	Group	GZ [μm]	GD [nuclei/m^3]	SDAS [μm]	SA [μm^2]	SR [%]
1	UST	410	1.45E+10	105	29	30.5
	N_0	780	2.11E+09	140	42	18.3
	N_c	720	2.68E+09	147	48	19.2
	N_1	840	1.69E+09	236	36	13.4
	N_2	920	1.28E+09	328	24	8.2
	N_3	1513	2.89E+08	579	14	5.3
2	UST	353	2.27E+10	73	16	53.2
	N_0	634	3.92E+09	95	21	38.4
	N_c	720	2.68E+09	98	20	41.2
	N_1	581	5.10E+09	187	33	28.7
	N_2	565	5.54E+09	249	22	24.3
	N_3	1124	7.04E+08	361	13	11
3	UST	280	4.56E+10	56	8	74.3
	N0	470	9.63E+09	74	11	55.1
	Nc	466	9.88E+09	74	11	54.9
	N1	550	6.01E+09	145	14	41.2
	N2	643	3.76E+09	167	15	35.1
	N3	935	1.22E+09	223	9	10.6
4	UST	212	1.05E+11	53	11	27
	N0	198	1.29E+11	49	15	43.6
	Nc	221	9.26E+10	53	12	41
	N1	521	7.07E+09	186	5	47
	N2	591	4.84E+09	243	6	35
	N3	884	1.43E+09	352	8	19

Table 3. Size of CET (mm) for the non-UST control groups and the UST group

Group	UST	N_0	N_1	N_2	N_3
From ingot top	17	53	22	15	7
From ingot bottom	23	75	29	20	13

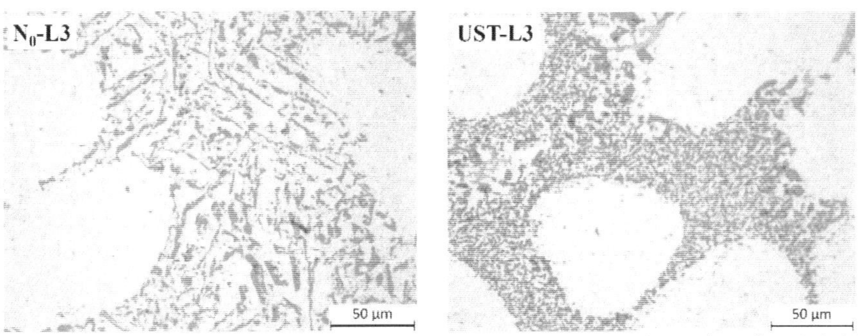

Figure 7. Comparison of microstructures of N_0 and UST samples at location 3 (500×)

2.3. Microhardness results

The Vickers hardness test (figure 8) shows that the micro-hardness is related to the grain size and silicon roundness in eutectic. The smaller grain size and the larger silicon roundness in eutectic, the higher microhardness would be. Due to a good refining of the microstructures by the UST acoustic streaming and cavitation effects, the UST group has the highest microhardness performance. In addition, by comparing with the columnar direction, the cross section of N_0 group has a better performance in microhardness.

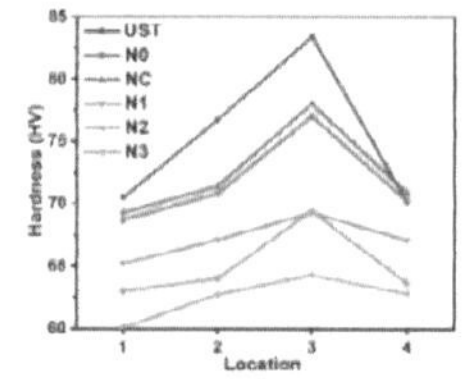

Figure 8. Microhardness results

3. Modelling results

ANSYS's Fluent has been applied to simulate the fluid flow and solidification process in the two-zone furnace. The modelling parameters are as follows: height and diameter of the graphite crucible are 305 and 75 mm respectively, alloy thermal conductivity is 90 W/(m*K) at T_L and 150 W/(m*K) at T_S [15], alloy viscosity is 0.03 kg/(m*s), the solidus temperature (T_S) and liquidus temperature (T_L) are 565.0 °C and 617 °C, respectively. The alloy density is a piece-wise linear relationship with the temperature, varying from 2547 kg m^{-3} at T_S to 2423 kg m^{-3}at T_L. The outlet at the right corner of the top cover of the crucible is set as pressure-outlet. The bottom and top coils are mixed (convention and radiation) wall type with a heat transfer coefficient of 10 W/(m^2*K) and an emissivity coefficient of 0.9. External radiation temperature was measured in these experiments as 150 °C. All other walls are set up as zero heat flux. Simulation time is about 1500 s for N_1 group, 6250 s for N_2 group and 20000 s for N_3 group. The previous work has shown that this model can successfully predicts the solidification process of the N_0 group [16]. Figures 9-11 show the initial temperature contours, which are the same for all groups, *i.e.*, 757 °C at the top and 598 °C at the bottom. The initial large temperature gradient in the liquid is helpful to create a large mushy zone during the solidification/cooling process. The predicted temperature evolution, local cooling rate and solidification time compared well with the experimental data.

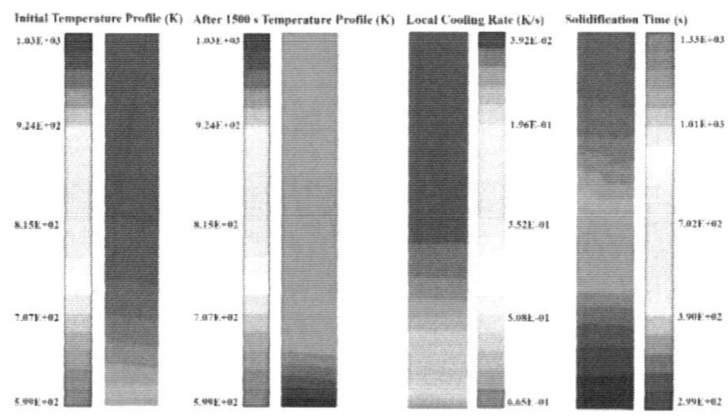

Figure 9. Simulation results of the N_1 group

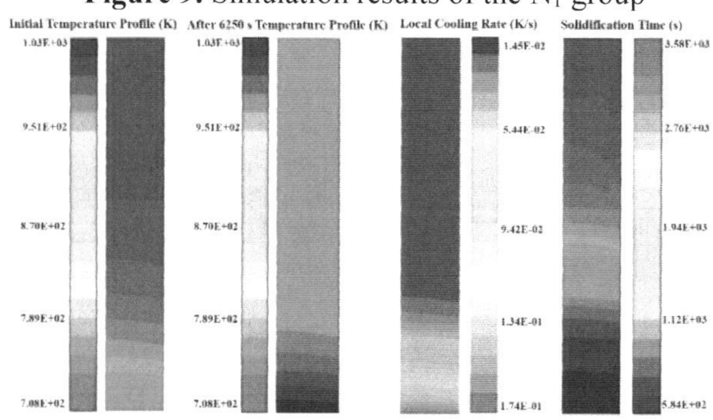

Figure 10. Simulation results of the N_2 group

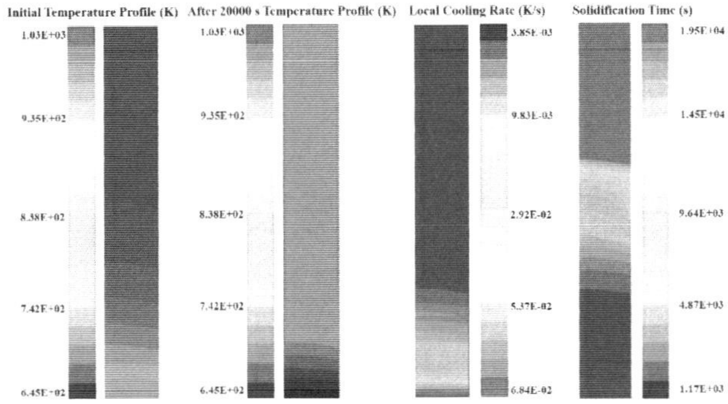

Figure 11. Simulation results of the N_3 group

4. Concluding remarks

An experimental system consisting of a 2-zone furnace with UST has been set up successfully in this work, making possible to refine and modify the microstructure of unidirectionally solidified A356 ingots. The solidification parameters of both the non-UST and the UST cases as well as the microstructures of the processed samples were compared. The results show that the UST is valuable in refinement and modification of the A356 microstructure. The ANSYS's Fluent was used to predict the solidification process of the A356 melt in the two-zone furnace under non-UST conditions. Cooling curve analysis was also performed. The simulation results compare well with the experimental measurements and reveals in more detail the solidification parameters of the carried-out experiments. In a future study, to clearly understand the UST refining mechanism during the solidification process, a CFD model for the 2-zone furnace with UST will be developed and implemented into Fluent.

References

[1] Kaczmar J W, Pietrzak K and Wlosinski W 2000 *J. Mater. Process. Tech.* **106** 58-67
[2] Durisinova K *et al.* 2012 *J. Alloys Compd.* **525** 137-42
[3] William C H 1998 *Mater. Sci. Eng. A* **244** 75–9
[4] Zhang D and Nastac L 2014 *J. Mater. Research Tech.* **3** 296-302
[5] Eskin G I 2002 *Zeitschrift fuer Metallkunde.* **93** 502-7
[6] Eskin D G 2017 *Mater. Sci. Tech.* **33** 636-45
[7] Eskin G I 1998 *Ultrasonic treatment of light alloy melts* (CRC press)
[8] Nastac L 2011 *Metall. Mater. Trans. B* **42** 1297-305
[9] Blundell S J and Blundell K M 2009 *Concepts in thermal physics.* (OUP Oxford)
[10] Hem S L 1967 *Ultrasonics* **5** 202-7
[11] Stefanescu D M 2015 *Science and engineering of casting solidification* (Springer)
[12] Flores A *et al.* 1998 *Intermetallics* **6** 217-27
[13] Puncreobutr C *et al.* 2014 *Acta Mater.* **79** 292-303
[14] Xuan Y, Dong A and Nastac L 2019 *Light Metals* 1545-50
[15] Bakhtiyarov S I, Overfelt R A and Teodorescu S G 2001 *J. Mater. Sci.* **36** 4643-8
[16] Dong A and Nastac L, *"TMS 2020 149th Annual Meeting & Exhibition Supplemental Proceedings,"* https://doi.org/10.1007/978-3-030-36296-6, The Minerals, Metals & Materials Series, 1117-26

Progress in the development of a contactless ultrasonic processing route for alloy grain refinement

K A Pericleous[1], C Beckwith[1], V Bojarevics[1], G Djambazov[1], A Dybalska[2], W D Griffiths[2] and C Tonry[1]

[1] Centre for numerical modelling and process analysis, University of Greenwich, UK
[2] Department of metallurgy and materials, University of Birmingham, UK

E-mail: k.pericleous@gre.ac.uk

Abstract. A high frequency tuned electromagnetic (EM) induction coil can be used to induce ultrasonic pressure waves leading to gas cavitation in alloy melts. This is a useful '*contactless*' approach compared to the usual immersed sonotrode technique. One then expects the same benefits obtained in the traditional ultrasonic treatment (UST) of melts, such as degassing, microstructure refinement and dispersion of particles. However, such an approach avoids melt contamination due to probe erosion prevalent in immersed sonotrodes and it has the potential to be used on higher temperature and reactive alloys. Induction stirring due to the Lorentz force produced by the coil is an added benefit, allowing for the treatment of large melt volumes, a current limitation of UST systems. At ultrasonic frequencies (> 20 kHz), due to the 'skin effect' electromagnetic forces vibrate just a thin volume by the surface of the metal facing the induction source. These vibrations are transmitted as acoustic pressure waves into the bulk and to achieve sufficient fluctuation amplitudes for cavitation, acoustic resonance is sought by carefully adjusting the generator frequency. This is akin to the tuning of a musical instrument, where the geometry and sound properties of the metal, crucible and surrounding structure play an important part. In terms of modelling, this is a multi-physics system, since fluid flow with heat transfer and phase change are coupled to electromagnetic and acoustic fields. The various models used and their coupling are explained in this paper, together with the various complications arising by the physics of cavitation. Experimental validation is obtained on a prototype rig featuring a conical induction coil inserted into the melting crucible containing the various alloys being examined. When resonance is reached, measurements demonstrate strong stirring, evidence of cavitation and finally grain refinement.

1. Introduction

Ultrasonic treatment (UST) of alloys at the liquid state, where a vibrating probe oscillating at ~20 kHz immersed in the melt acts as the US source, has been shown to promote grain refinement, dispersing of immersed particles, breaking up of particle clusters, wetting of surfaces and degassing [1-3]. These beneficial effects are attributed to the onset of cavitation of gases present in the liquid, responding to rapid pressure variations induced by an ultrasonic source. Cavitation involves the formation, growth, oscillation, collapse, and implosion of bubbles in liquids [4]. In the vicinity of collapsing bubbles, extreme temperatures (>10000 K [5]) and pressures (>400 MPa [6]) take place. However, despite these positive attributes, the technique has not been widely adopted in industry [3]. There are many reasons for this: the immersed probe tends to erode in the melt causing contamination and consequently the probe (made of expensive materials, e.g. Nb or W) needs to be replaced quite frequently, leading to unnecessary downtime and cost. Furthermore, the cavitation zone appears to be localised in a small volume surrounding the probe, preventing the US treatment of large quantities of metal, or its use in continuous processes. Increase in the sonotrode power does not necessarily improve efficiency, since as it has been shown [7] the concentrated bubble cloud surrounding the probe leads to pressure wave attenuation. Recent efforts to improve volume efficiency in continuous processes rest on increasing the residence time of liquid in the cavitation zone using baffles, e.g. [8] or alternatively, inducing stirring in the melt using either a mechanical stirrer [9], or electromagnetic induction [10].

Due to these limitations, we proposed [11-13] a contactless electromagnetic vibration device that can excite soundwaves in the melt potent enough to cause cavitation. With this concept, shown schematically in figure 1, a high frequency electromagnetic coil is lowered into the melt. Out of phase currents induced in the melt, which is electrically conducting, result in a repulsive force, which prevents contact between the coil and the metal. Several other events take place at the same time, (i) the time-averaged component of the Lorentz force ($\boldsymbol{F}_L = \boldsymbol{J} \times \boldsymbol{B}$) promotes bulk stirring, and (ii) the time dependent (sinusoidal) component of the force vibrates the melt at a frequency selected to cause cavitation. Although at ultrasonic frequencies the force is concentrated in a thin 'skin layer' region, the vibration is transmitted through the bulk through acoustic waves. In order to maximise the pressure wave amplitude at minimum input power, the frequency needs to be tuned to promote acoustic resonance in the treatment vessel [14].

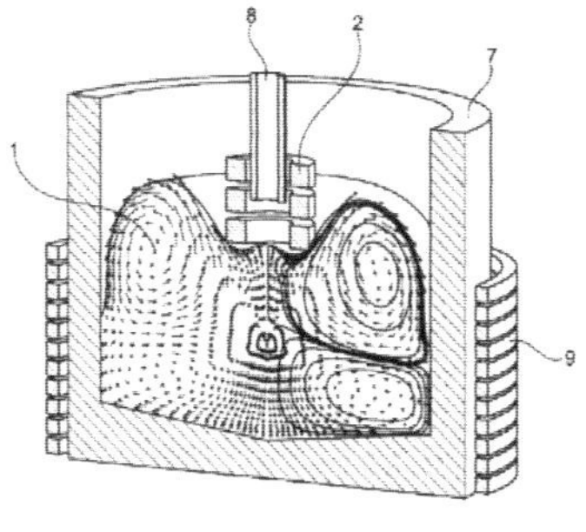

Figure 1. The contactless sonotrode concept [12].

The idea of using electromagnetic vibrations to induce cavitation in liquid metals is not new, but the implementation, that is employing a single induction coil placed in very close proximity to the melt for maximum electromagnetic coupling, plus the use of acoustic resonance to achieve the required wave amplitude is quite novel. In previous studies, Vives [15] coupled AC current running through the melt with an external static magnetic field to promote cavitation. Grants [16] similarly placed an induction heated charge inside the poles of a large DC magnet, to demonstrate cavitation in steel and nano particle dispersion in aluminium for the production of metal matrix composites. Both approaches though demonstrating the principle of contactless EM ultrasonic processing would be difficult to implement in an industrial situation.

In this contribution we demonstrate use of a prototype installation of the contactless sonotrode depicted in figure 1, for grain refinement in aluminium (with or without added grain refiner) and show initial degassing results.

2. Mathematical basis

From the point of view of modelling, this is a truly multi-physics problem, combining magnetically driven flow and heat transfer with acoustics, complicated by the phenomenon of cavitation. Details of the mathematics used have been given in several previous publications and for that reason only a summary will be given here, accompanied by the relevant references.

2.1. MHD flow and heat transfer

The time-dependent conservation equations for mass, momentum and energy are solved, accounting for the effects of electromagnetic interaction and change of phase. An in-house software code 'SPHINX' [17] solves the equations numerically using a spectral colocation technique [18] on a continuously deformable mesh, coincident with the metal volume, as described in [19]. The method enables real-time deformation of the liquid free surface based on the instantaneous balance of forces acting on it. A description of the main equations follows:

Momentum and Continuity

$$\partial_t \mathbf{v} + (\mathbf{v}\cdot\nabla)\mathbf{v} = -\rho^{-1}\nabla p + \nabla\cdot(\nu_e(\nabla\mathbf{v} + \nabla\mathbf{v}^T)) + \rho^{-1}\mathbf{f}_m + S_D \quad (1)$$

$$\nabla.\mathbf{v} = 0 \quad (2)$$

where $\mathbf{v}$ is the velocity, ρ the density, p the pressure, ν_e the effective viscosity (sum of laminar and turbulent contributions), $\mathbf{f}_m$ the volumetric electromagnetic (Lorentz) force and S_D is a Darcy flow resistance term that accounts for the macroscopic effect of phase change on flow in mushy zones. The boundary conditions assume no-slip at solid walls ($\mathbf{v}$=0) or alternatively free surface dynamic and kinematic conditions when the liquid metal is detached from any solid surface.

Energy conservation

$$\frac{\partial(\rho C_p T)}{\partial t} + \nabla(\rho C_p T\mathbf{v}) = \nabla(k_{eff}\nabla(T)) + q_e + q_L \tag{3}$$

where T is the temperature, C_p the specific heat and k_{eff} the effective conductivity. The last two terms in equation (3) represent Joule heating, q_e and latent heat release, q_L

$$q_e = \frac{J^2}{\sigma}\ ;\ q_L = -\frac{\partial}{\partial t}(\rho f_L L) - \nabla.(\rho \mathbf{v} f_L L) \tag{4}$$

where σ is the electrical conductivity of the liquid and L the latent heat constant.

The liquid fraction f_L is defined as a function of temperature T:

$$f_L = \begin{cases} 1 & T > T_L & \text{liquid} \\ \left(\frac{T-T_S}{T_L-T_S}\right) & T_S \le T \le T_L & \text{mushy zone} \\ 0 & T < T_S & \text{solid} \end{cases} \tag{5}$$

Then

$$S_D = \frac{\mu}{K}\boldsymbol{v}\ ;\ K = \frac{f_L^3}{\zeta(1-f_L)^2} \tag{6}$$

Heat transfer boundary conditions are stated for free-surface radiation and wall loss, described in [17, 19]. The effective heat transfer at solid walls is then given by

$$-\rho C_p\, \alpha_e \partial_n T = h(T - T_w) \tag{7}$$

where $h(T)$ is the heat transfer coefficient, with free surface radiation given by:

$$-\rho C_p\, \alpha_e \partial_n T = \varepsilon\sigma_R(T^4 - {T_w}^4) \tag{8}$$

with α_e being the turbulent thermal diffusion coefficient. Turbulence is modeled using the k-ω model of Wilcox [20].

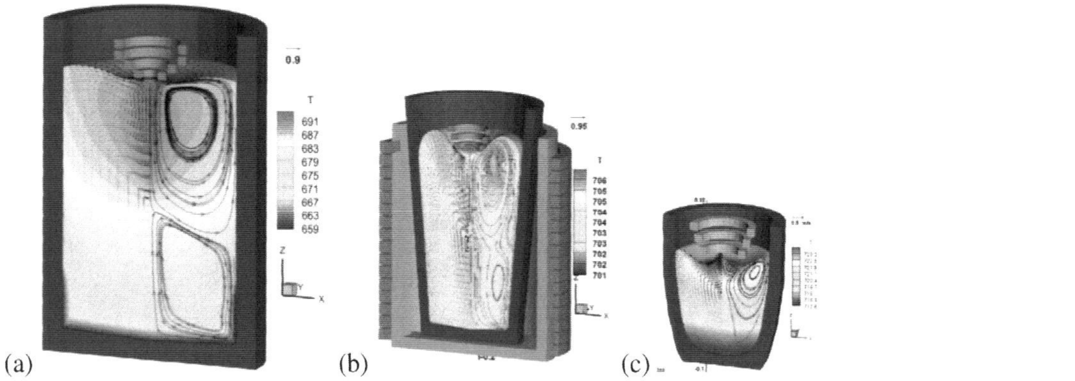

Figure 2. Computed surface deformation, flow patterns and temperature in three crucibles of varying size containing aluminium: (a) 50 kg, (b) 10.5 kg, (c) 5 kg. Note, (a) and (b) include the influence of the furnace coil as well as the top coil. Crucibles (b) and (c) were used in experiments, with (c) used for particle tracking (PEPT) studies.

Typical simulations are shown in figure 2, demonstrating the application of the top coil sonotrode in three different vessel sizes. Ceramic crucibles (b) and (c) correspond to experimental configurations, while the large crucible (a) is a hypothetical application to highlight the scaleup potential of the process.

As the top coil (typically 1700 A at 10 kHz) moves towards the metal free surface, the surface deforms by repulsion. At the same time strong stirring is observed due to combined induction from the furnace and top coils in crucibles 2(a) and 2(b)). In crucible 2(c) the top coil acts alone, but still a deep toroidal vortex is observed due to the time-averaged Lorentz force. In addition to stirring considerable Joule heating is generated by the coil, leading to a hot spot near the surface coinciding with the coil axis. In the same position the free surface is elevated as the Lorentz force decays to zero.

2.2. *Magnetic induction as a source of sound*

The AC magnetic field of angular frequency ω due to the coil, ***B***, and the induced current, ***J***, can be expressed into real and imaginary components (subscripts R and I respectively). The Lorentz force can then be split into time-averaged and time-dependent components:

$$\boldsymbol{F} = \boldsymbol{J} \times \boldsymbol{B} = \bar{F} + \tilde{F} \quad \text{then,}$$
$$\bar{F} = \frac{1}{2}(J_R B_R + J_I B_I) = \frac{1}{2\mu\delta} B_0^2 e^{-2\frac{x}{\delta}}, \qquad \tilde{F} = \frac{1}{2\mu\delta} B_0^2 e^{-2\frac{x}{\delta}} \sqrt{2}\cos(2\omega t - 2\frac{x}{\delta} + \frac{\pi}{4}) \tag{9}$$

where $\delta = \sqrt{\frac{2}{\mu\omega\sigma}}$ is the skin depth indicating the force penetration into the fluid.

The mean value $\bar{F}$, is then responsible for bulk stirring as shown in figure 2, while the sinusoidal part $\tilde{F}$ is the source of vibration. It is important to note (a) that the vibration frequency is double that of the supply current, (b) that the force decays within the skin-depth distance δ from the liquid free surface, hence the importance of acoustic resonance in achieving the required pressure amplitude for cavitation in the bulk volume (see equation (11) below).

2.3. *Sound wave propagation*

The time-dependent force component $\tilde{F}$, is the source of sound waves, computed in the time domain by solving the Euler form of the momentum equations, generating a perturbation velocity field $\tilde{v}$:

$$\frac{\partial p}{\partial t} + \rho c^2 \frac{\partial \tilde{v}}{\partial x} = S; \ \rho \frac{\partial \tilde{v}}{\partial t} + \frac{\partial p}{\partial x} = \tilde{F} \tag{10}$$

A staggered scheme (in space and time) is used to solve equations (10), with details given in Djambazov *et al.* [21]. The source *S* represents pressure contributions due to cavitating bubbles [22]. The solution domain for sound extends beyond the melt to include the crucible and surrounding structures accounting for transmission and reflection of sound through the crucible walls. Constant pressure is assumed at the liquid free surface and a sound-hard boundary (zero flux) is applied at the coil surface. Details of the approach are given in [14].

A characteristic of this process is the appearance of pressure nodes/antinodes deep inside the melt volume during resonance, marking likely cavitation regions (e.g. see figure 3). This contrasts with the immersed sonotrode case, where cavitation activity is restricted to an area surrounding the vibrating probe, leading to shielding effects due to bubble concentration that limit process efficiency.

2.4. *Cavitation*

Cavitation of dissolved gases is considered essential in encouraging nucleation, hence leading to grain refinement in alloys. Seeking the resonant frequencies of the system, ensures pressures greater than the Blake threshold where cavitation may occur can be obtained. The Blake threshold is calculated from

$$P_B = p_0\left(1 + \sqrt{\frac{4}{27}\frac{\xi^3}{1+\xi}}\right), \xi = \frac{2\sigma}{R_0 p_0}, \tag{11}$$

where p_0 is the ambient pressure, σ the interfacial tension and R_0 the bubble radius. For example, bubbles with a radius $R_0 = 5\times10^{-6}$ m, an interfacial tension for aluminium and air of $\sigma = 1.1$ N/m and with $p_0 = 101325.0$ Pa gives a Blake threshold of 254 kPa.

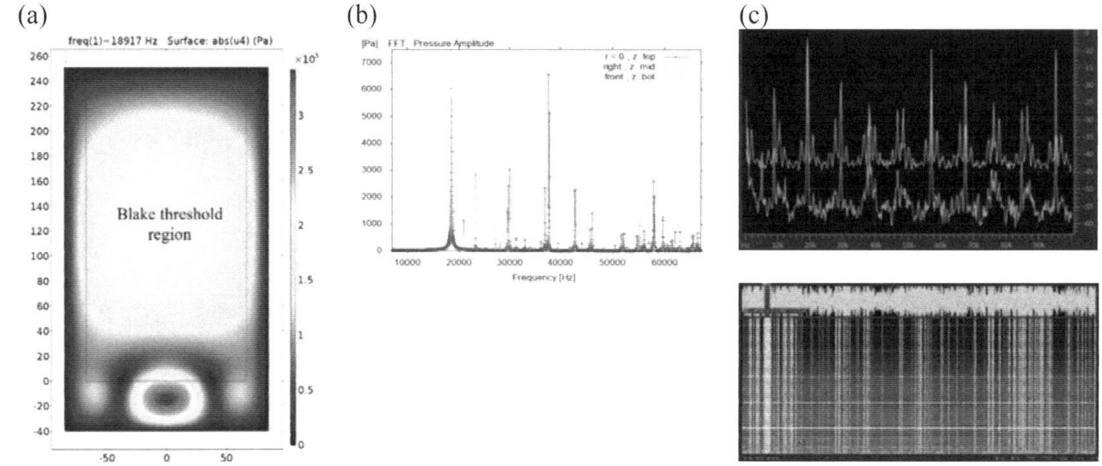

Figure 3. (a) Computed sound-field in a cylindrical crucible with the top coil acting as the sound source (f = 18917 Hz). (b) Predicted sound spectrum obtained with a driving frequency f=18917 Hz, showing several excited harmonics. (c) Experimental sound spectrogram showing intermittent cavitation noise; the accompanying spectrum compares loud and silent regions.

Bubble dynamics due to the pressure change are handled using the Keller-Miksis equation [23]

$$\left(1-\frac{\dot{R}}{C}\right)R\ddot{R}+\frac{3}{2}\dot{R}^2\left(1-\frac{\dot{R}}{3c}\right)=\frac{1}{\rho_l}\left(1+\frac{\dot{R}}{c}+\frac{R}{c}\frac{d}{dt}\right)[p_l-p(t)], \tag{12}$$

which accounts for liquid compressibility and damping. Then p_l represents the liquid pressure at the liquid gas interface defined by

$$p_l=p_g-\frac{2\sigma}{R}-\frac{4\mu\dot{R}}{R}, \tag{13}$$

where σ is the surface tension, μ is the liquid viscosity, and p_g is the pressure in the gas at the interface, calculated from

$$\frac{\partial p_g}{\partial t}=\frac{3}{R}\left[(\gamma-1)\left(k_T\left.\frac{dT}{dR}\right|_{r=R}\right)\gamma p_g\dot{R}\right] \tag{14}$$

γ being the polytropic index, and k_T the thermal conductivity of the gas. The temperature gradient at the bubble surface is calculated using

$$\left.\frac{dT}{dR}\right|_{r=R}=\frac{T-T_\infty}{\sqrt{RD/3(\gamma-1)\dot{R}}} \tag{15}$$

where D is the gas diffusivity, and T_∞ the bulk temperature of the liquid. The Keller-Miksis equation accounts for acoustic radiation terms that address the lack of damping in the original Rayleigh-Plesset equation that overpredicts bubble oscillation amplitudes post collapse. A variable-step variable-order Adams-Bashforth-Moulton method was chosen to solve the Keller-Miksis equation.

To account for the change in void fraction β during a period of pressure oscillation on sound transmission the Caflisch model was used [24]. Based on Trujillo [25] a time harmonic solution of the Caflisch model, introduces dissipation functions $\mathcal{A}$ and $\mathcal{B}$ based on the void fraction $\beta=\frac{4}{3}\pi R^3 N$:

$$\mathcal{A}=-\frac{\rho_l\omega^2}{\pi}\int_0^{2\pi}\frac{\partial\beta}{\partial\tau}\sin\tau\,d\tau,\mathcal{B}=-\frac{\rho_l\omega^2}{\pi}\int_0^{2\pi}\frac{\partial\beta}{\partial\tau}\cos\tau\,d\tau, \tag{16}$$

where N is the assumed number of bubbles. These relations can be used in a Helmholtz equation of the form:

$$(\nabla^2+k^2)p=0 \tag{17}$$

IOP Conf. Series: Materials Science and Engineering **861** (2020) 012070 doi:10.1088/1757-899X/861/1/012070

where k is a modified wave number and p the pressure:

$$k^2 = \left(\frac{\omega}{c}\right)^2 - \frac{\mathcal{A}(p)}{|p|} - i\frac{\mathcal{B}(p)}{|p|} \tag{18}$$

3. Experimental results

The general experimental setup is shown in figure 4. An induction furnace was used to melt and sonicate aluminium in ceramic crucibles. Both cylindrical and tapered crucibles were used with charges ranging from 8 -10 kg. The top coil was lowered into the crucible once the charge was fully melted and then the main furnace coil was switched off to maintain the temperature at around 700 °C.

3.1. Cavitation detection

An ultrasonic microphone was used to detect noise emitted from the crucible and the frequency adjusted until the recorded signal revealed the presence of broad-spectrum noise as a sign of cavitation. Figure 3(c) shows a typical spectrogram, plotting frequency against time. The continuous horizontal lines correspond to the top coil driving frequency, its harmonics and subharmonics. The vertical lines indicate cavitation activity, which is always intermittent. We think this is because the presence of bubbles during cavitation alters the speed of sound locally interrupting resonance. As the pressure amplitude diminishes, cavitation stops, resonance returns, and the cycle repeats. As depicted in figure 3(c) (top) comparing quiet and loud regions in the spectrogram, the spectrum retains its harmonics, but the sound level diminishes by ~ 20 dB.

3.2. Grain refinement

Alloy grain refinement experiments were conducted in a cylindrical clay-graphite crucible with internal and external diameters of 135 and 170 mm respectively, and depth 280 mm. For each experiment the crucible was filled with about 8.5 kg metal, either commercial purity aluminium (CP-Al), or CP-Al with the addition of 0.2 wt.% Al–5Ti–1B grain refiner (100 ppm Ti, 20 ppm B). Figure 4 shows the top coil positioned centrally above the liquid metal surface and above it the Ultramic®200K digital ultrasonic microphone for recording the ambient ultrasonic noise emitted around the crucible during processing. Recorded sound was observed in the form of a FFT (Fast Fourier Transform) sound spectrum in real-time.

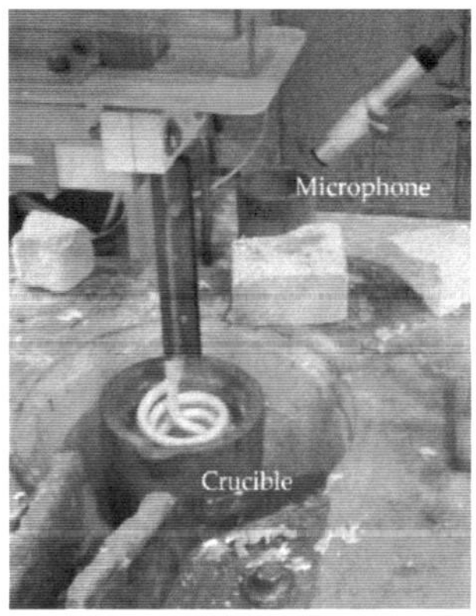

Figure 4. Experimental setup.

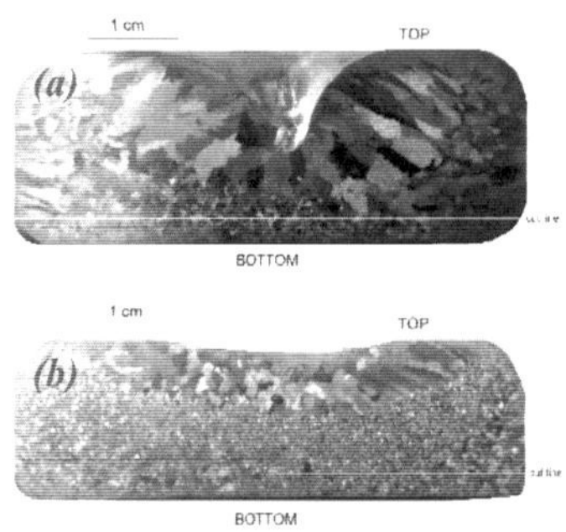

Figure 5. Sectioned KBI ring test samples (a) unprocessed, (b) processed.

Conditions generating resonance accompanied by cavitation noise, (coil frequency and melt temperature), were maintained for about 5 minutes to produce samples for grain structure analysis. The samples were taken using the KBI ring test [25], where liquid metal is poured into a steel ring with an outside diameter of 75 mm, inside diameter 50 mm and height 25 mm, placed on an insulating silica brick and left to solidify. For grain size characterization, the base of the cylindrical samples was removed to about 3 mm above their base and ground (see figure 5), polished and etched with either Poulton's or Keller's solution and the average grain size determined using the mean linear intercept method.

Figures 6 and 7 show the recorded spectrograms and post mortem grain structures of samples obtained following ultrasound treatment. Figure 6 shows the grain of CP-Al at 700°C pre-and post-treatment with 1700 A, 9.35 kHz current through the coil. Control of melt temperature during ultrasound processing found to be important. It must be kept low enough to promote cavitation with this method. In experiments it was maintained at 40 °C above the melting point, which for pure Al was 700 °C, the minimum value for pouring the liquid metal in a casting. For this CP-Al experiment (figure 6), the grain size reduction was about 63%, (a reduction from 256 ± 12 to 95 ± 1 μm). Cavitation is believed to induce heterogeneous nucleation by (i) forced wetting of non-wetted particles present in the melt, resulting in an increased number of nucleation sites, (ii) local undercooling due to pressure changes when bubbles collapse or (iii) undercooling of the melt at the bubble surface when the bubble rapidly expands. In the case of CP-Al, the number of non-wetted particles would be smaller than in alloys with grain refiner addition, making grain refinement more difficult.

The addition of grain refiner (0.2% Al–5Ti–1B) increases the number of active nuclei so that all three mechanisms of cavitation-induced heterogeneous nucleation can take place. Figure 7 shows the grain size reduction achieved for CP-Al with a grain refiner processed for 5 mins at 9.41 kHz. A grain sizes reduction from 223 ± 5 to 121 ± 2 μm was obtained.

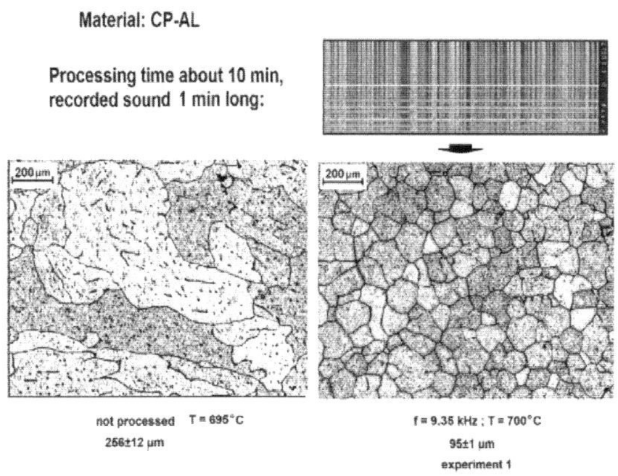

Figure 6. Commercial purity aluminium (CP-Al) showing grain structures before and after processing. Grain size is reduced from 256 μm to 95 μm.

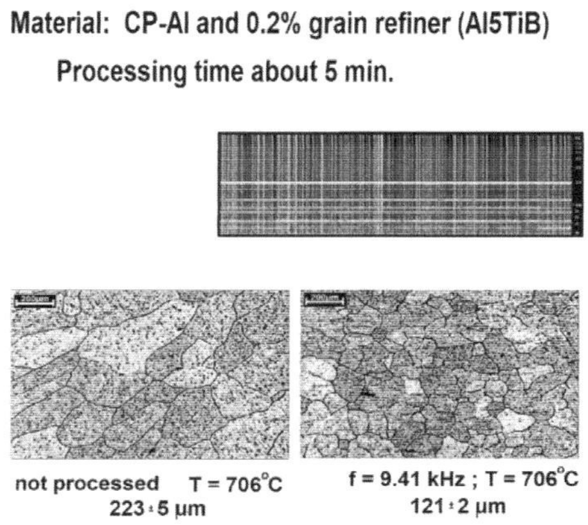

Figure 7. Commercial purity aluminium (CP-Al) showing grain structures before and after processing. Grain size is reduced from 223 μm to 121 μm.

3.3. Correlation between grain refinement and sound frequency spectrum

The basic concept behind the contactless sonotrode relies on the initiation of gas cavitation activity using acoustic resonance as the main driver for grain refinement. Numerical simulations provide an indication of the likely resonant modes, given the sound properties and geometry of the alloy and crucible materials [14]. However, since these properties can vary unpredictably (especially so in ceramic crucibles), the experiment traverses the space about the indicated central frequency value using the spectrogram as an indicator of the most potent value, judged by the frequency of broadband noise bursts. Also evident in all spectrograms is that the cavitation activity as shown by the vertical lines is intermittent. This may be due to local changes in sound velocity in the melt as clouds of bubbles appear and disappear, which would disrupt the resonant conditions.

4. Concluding remarks

This paper shows a contactless electromagnetic processing technique that can generate ultrasonic waves in liquid metals in a crucible, strong enough to produce cavitation. This technique was tested experimentally to treat aluminium alloys in the liquid state with and without commercial grain refiners, achieving substantial grain reduction in both cases. The computational model couples fluid flow, heat transfer, electromagnetics and soundwave generation. Validation was carried out against experimental observations in aspects that are important for the process. These confirm the predicted free surface depression by the coil, induction driven flow leading to strong mixing and the presence of cavitation during resonant conditions. Acoustic resonance was found to be necessary for pressure waves strong enough for cavitation. This means in practice that geometrical details and material sound properties of the setup become important for successful implementation. Using this technique, it has been possible to produce grain-refined samples of both pure aluminium and aluminium with grain refiner added. Future work aims to test the technique in high temperature and reactive alloys, where the traditional immersed sonotrode method cannot be used.

Acknowledgments: The authors acknowledge financial support from the ExoMet Project (EC contract FP7-NMP3-LA-2012-280421), and EPSRC grants EP/P034411/1, EP/R002037/1, EP/R000239/1.

References

[1] Abramov O V 1987 *Ultrasonics* **25** 73
[2] Campbell J Effects of vibration during solidification 1981 *Int Metals Reviews* **26** 71
[3] Eskin G I and Eskin D G 2015 *Ultrasonic treatment of light alloy melts* Taylor & Francis, CRC Press
[4] Leighton T G 1994 *The Acoustic Bubble* (London: Acad. Press)
[5] Flannigan D J and Suslick K S 2005 *Nature* **434** 52–5
[6] Tzanakis I, Eskin D G, Georgoulas A and Fytanidis D 2014 *Ultson* **21** 866–78.
[7] Lebon G S B, Tzanakis I, Pericleous K A and Eskin D 2017 *Ultson* **42** 411-21
[8] Lebon G S B, Pericleous K A, Tzanakis I and Eskin D 2016 *Int. J. Cast Metals Res.* **29** 324
[9] Dieringa H, Katsarou L, Buzolin R, Szakacs G, Horstmann M, Wolff M, Mendis C, Vorozhtsov S and StJon D 2017 *Metals* **7(10)** 388
[10] Xuan Y, Dong A and Nastac L 2019 In: *Light Metals*. Eds Chesonis C *et al.* (Cham: Springer)
[11] Bojarevics V, Djambazov G and Pericleous K A 2015 *Metall. Mater. Trans. A* **46** 2884-92
[12] Jarvis D J, Pericleous K A, Bojarevics V and Lehnert C 2019 *'VibroEM'* Patent US 10, 207, 321 B2
[13] Pericleous K A, Bojarevics V, Djambazov G, Dybalska A, Griffiths W D and Tonry C 2019 *Shape Casting: 7th Int Symp. Celebr. Prof. John Campbell's 80th Birthday*, eds Tiryakioğlu M *et al.* (Cham; Springer) 239-52
[14] Tonry C E H, Djambazov G S, Dybalska A, Bojarevics V, Griffiths W D and Pericleous K A 2019 *Light Metals* TMS Series (Cham: Springer)
[15] Vives C 1996 *J. Crystal Growth* **158** 118
[16] Grants I, Gerbeth G and Bojarevics A 2015 *J. Appl. Phys.* **117** 204
[17] Pericleous K A and Bojarevics V 2007 *Progress in Comp. Fluid Dynamics* **7** Nos. 2/3/4
[18] Canuto C, Hussaini M Y, Quarteroni A, and Zang T A 2006 *Spectral methods. Fundamentals in single domains* (Berlin: Springer-Verlag)
[19] Bojarevics V, Harding RA, Pericleous KA and Wickins M 2004 *Metall. Mat. Trans. B*, **35** 785-803
[20] Wilcox D C 1998 *Turbulence Modelling for CFD*, 2nd ed., (California: DCW Industries)
[21] Djambazov G S, Lai C H and Pericleous K A 2000 *AIAA J.* ***38*** 16–21
[22] Lebon G S B, Tzanakis I, Djambazov G, Pericleous K A and Eskin, D G 2017 *Ultson* **37** 660–8
[23] Keller J B and Miksis M 1980 *J. Acoust. Soc. Am.* **68** 628–33
[24] Caflish R E, Miksis M J, Papanicolaou G C and Ting L 1985 *J. Fluid Mech.* ***153*** 259–73
[25] Trujillo F J 2018 *Ultson* **47** 75-98
[26] Murty B S, Kori S A and Chakraborty M 2002 *Int. Mater. Rev.* ***47*** 3–29

IOP Conf. Series: Materials Science and Engineering **861** (2020) 012071 doi:10.1088/1757-899X/861/1/012071

In situ X-ray observations of transient states in arc weld pools

F Wu[1], K V Falch[1], M Drakopoulos[2] and W U Mirihanage[1]

[1] The University of Manchester, Manchester M13 9PL, UK
[2] Diamond Light Source, Didcot OX11 0DE, UK

E-mail: wajira.mirihanage@manchester.ac.uk

Abstract. Metallic alloys coalesce via extremely rapid melting and subsequent solidification to form fusion welded joints. The melt pool evolution in melting and solidification sequences during the welding process determines the formation of the final weld joint shape, microstructure and defects. The scientific insight on weld pool evolution and related phenomena can be a key contribution to enhance structural integrity and resilience of the welded structures or components. However, inherent complexity with multi-physics phenomena, associated high temperatures and the rapidness of the processes make direct experimental investigation of welding is extremely demanding. Thus, internal flow behaviour during welding or other melt-pool-based metal processing such as additive manufacturing remains unclear and hinders progression to process optimisation. In this contribution we report the observation of melt pool dynamics that take place during electric arc welding, obtained through in situ synchrotron imaging at millisecond scale. The analysis flow patterns along with the quantified weld pool surface dynamics revealed us to how different contributing forces dictate the flow conditions over the distinct durations of the relatively short existence of the liquid phase. Our preliminary results suggest the existence of arc, surface tension and gravity dominant regimes during the evaluation of the weld pool. Further, we present our observations on how different welding parameters influence these regimes and develop into different transient conditions.

1. Introduction

Fusion welding allows two metallic parts to join through a rapid solid – liquid- solid phase transformation under intense heat load. The process can take place with or without intermediate filler metal. The flow behavior of the liquid phase can significantly dominate the heat and mass transfer during the phase transformations. Thus the flow conditions within the weld pool considerably affecting to the microstructures and defects that appear in the final weld joint [1-6]. The surface tension driven forces, the gravity-driven forces of buoyancy gravity and self-weight are commonly affected in the flow conditions within many hot liquids with free surfaces, such as weld pools. In arc welding, an electric arc is employed as the heat source and therefore electromagnetic forces are also expected to play a role to drive the flow within arc weld pools. The arc force is the common term used for the electromagnetic (Lorentz) force generated as a result of the arc and the drag force of the cathode jet on the liquid surface [7, 8]. Recoil pressure as a result of considerable evaporation can be involved with high-density heat inputs such as laser and electron beam welding, but insignificant in arc welding [9-11]. Typically, the negative surface tension gradient (with increasing temperature) creates the outward fluid flow from the weld center towards the pool periphery [1, 5-6]. By having a surface-active element like sulfur in the weld pool, this can be reversed. In such instances, the surface tension driven flow will become inward resulting in a deep and narrow final weld pool [12-14]. The buoyancy force generally causes outward flow while the combined drag and electromagnetic (arc) forces cause inward flow. Magnitudes of the flow caused by the gravity-driven mechanisms are commonly considered as relatively smaller than the surface tension gradient driven force [15-17]. The evolution of melt pools, as a result of multiple driving forces, remains a fundamental consideration of the fusion welding and other melt pool based manufacturing techniques, for example, metal additive manufacturing. Synchrotron X-ray techniques emerged as a very attractive experimental tool to study dynamic processes involved with solidification processing, welding and additive manufacturing of metallic alloys [6, 18-20]. Both direct imaging and X-ray diffraction methods have been employed with X-ray investigations (e.g. [18, 21-22]). This contribution presents in situ X-ray observations that facilitate the identification of the time-evolving dominance of the multiple forces acting simultaneously within the arc weld pools.

2. Experimental

In situ X-ray imaging experiments been carried out at the I12 beamline [23] of the Diamond Light Source, UK. The full-field images over 12 mm × 8 mm field of view (FoV) were collected at 500 Hz frame rate during tungsten inert gas welding.

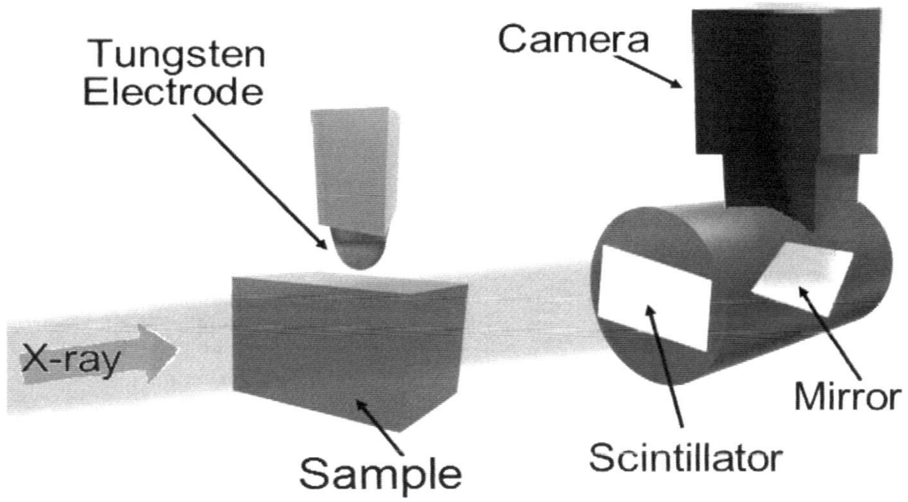

Figure 1. Schematic experimental configuration with X-ray transmission through the weld pool sample size is 25 mm (L) x 12 mm (H) x 6 mm (T).

The samples were binary Al-8%wt. Si alloy with the dimensions of 6 mm × 12 mm × 25 mm (Thickness × Length × Height). Tungsten carbide particles (diameter range 10-40 μm) were included in the samples as tracking particles for velocity measurements. The gas tungsten arc welding (GTAW or commonly referred as TIG) electrode was fixed vertically above the sample. The electrode tip was kept approximately 1.5 mm above the center position of the 6 mm x 25 mm face of the sample. This synchrotron X-ray beam travels through the sample 6 mm thick side towards the imaging detector, as shown in figure 1. 60 keV monochromatic X-rays were used for the experiments while the imaging system consists of a 1 mm thick Nd:YAG Scintillator which is coupled to a Vision Research Inc. MIRO 310M camera through a visible light optics system that yields 10 μm spatial resolution.

3. Results and Discussion

The evolution of a weld pool during a representative arc welding process (with 100 A welding current, 12 V DC voltage and 2 s) is presented in figure 2. In general, once the arc is activated, the surface of the weld pool starts to form beneath the tungsten tip and the free surface level rises rapidly as seen in figure 2(b). With the increase of the weld pool size, the peak point of the pool appears dividing. Once this happen, the height of the center surface of the pool drops slightly. Even though the center of the pool drops (relative to its peak position), it remains above the initial solid-air interface position. A relatively short time after stopping the arc (in this sequence just after 2 s), the next major transition starts. In this transition, the top surface of the entre liquid pool drops below the level of the original solid surface level and forms the final solidified shape of the weld pool.

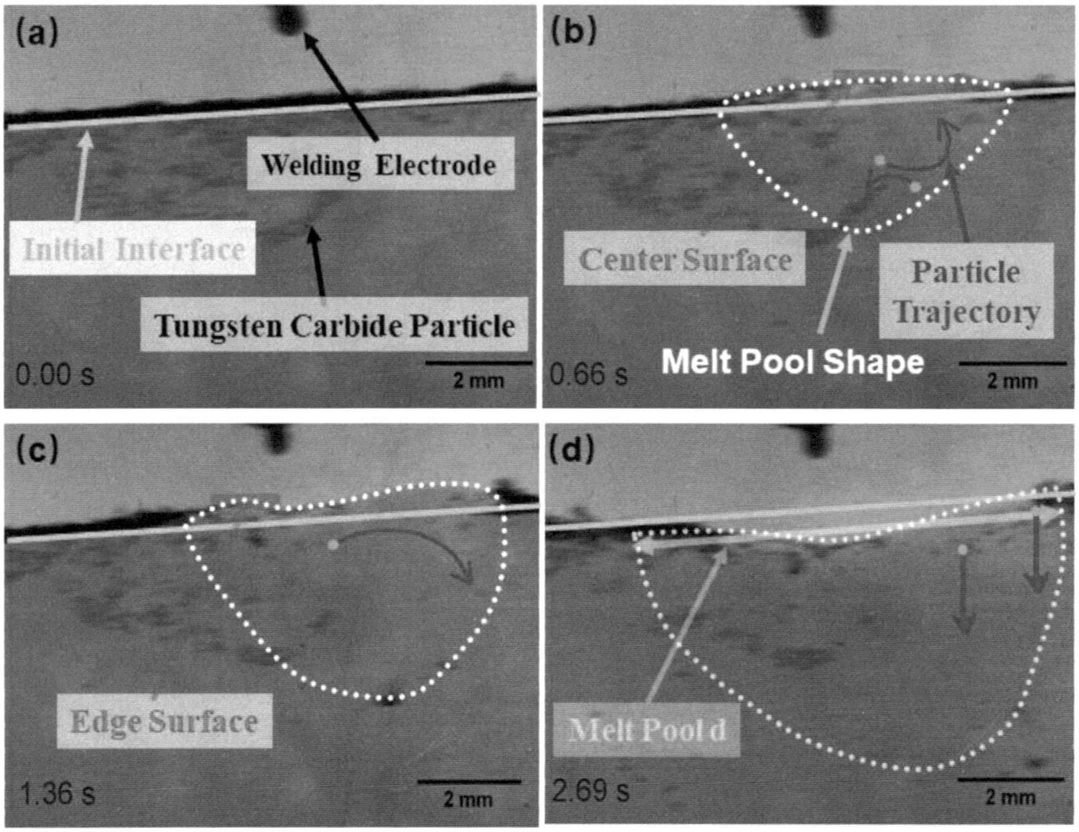

Figure 2. A representative melt pool shape changes with. 100 A DC welding current and 2 s welding time. The top black in the figure is the welding electrode (tungsten), and the dark grey part is the welding material (Al). The orange line represents the initial interface height. Black dots are traced particles (tungsten carbide). The white dotted line is the shape of the melt pool and the blue arrow is the trajectory of particle movement inside the melt pool.

Similar transformations in 4 other comparative welding conditions (with different parameters) are also quantitatively presented in figure 3. The figure quantitatively shows the evolution of the liquid-air interface (free surface) at the weld pool center and the raised interface position close to the circumference of the weld pool.

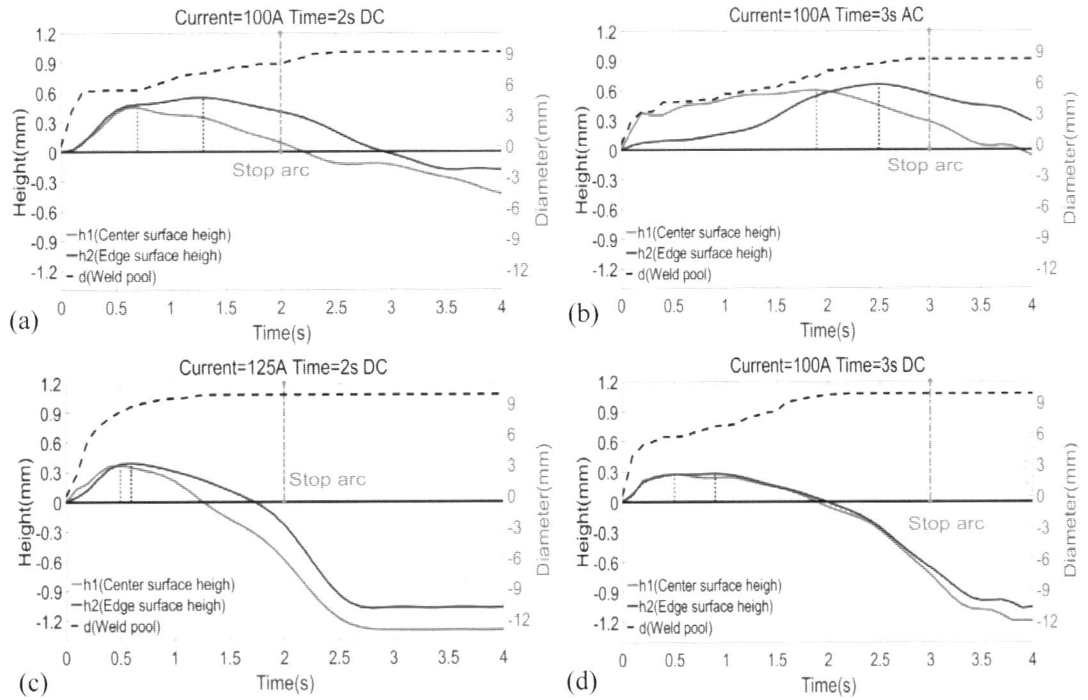

Figure 3. Melt pool height evolution of the centre and boundary with the diameter. Weld current and time: (a) 100 A DC, 2 s (b) 100 A AC, 3 s (c) 125 A DC, 2 s (d) 100 A DC, 3 s.

DC (direct current) power input is compared to same nominal AC (alternating current) power input will take a relatively long time for the center and edge surfaces to reach their peak positions as shown in figures 3(a) and 3(b) (or same welding time but different modes contested in figures 3(b) and 3(d)). This observation is looking more associated with the relatively lower net heating capacity (produced at the sample/substrate). AC mode generates heat nearly half of the time at the sample due to the reversing nature of the polarity. This relatively lower level of heating received at the weld pool is denoted by the slower growth of the pool volume [24]. As a result of higher net heating input at the pool, DC mode appears to enable higher surface tension gradient force driven flow. A higher rate of heating is generated at the center of the pool within a relatively short time compared to the thermal diffusivity in liquid or the heat transferred through the flow [25]. Hence DC input mode appears to support the accumulation of more heat and the resulting increase of the temperature in the center of the melt pool. Thus surface tension force driven flow can be increased as a result of the relatively larger temperature difference across the pool. This is analogous to the conditions between the low and the high welding currents. The diameter of the weld pool with 125 A is expanding faster and the peak values for the free surface levels appear much quicker compared to the case with 100 A (figure 3(c)). Once the pool is big enough, the surface of the melt pool tends to drop below its original solid position, if there no solid boundaries to contain the liquid. This observation indicates the influence of the self-weight of the melt with regard to the behavior of the pool before it solidifies. The dominance of the gravity, on this occasion, over the arc and surface tension forces introduce a major change to the weld pool ape. If the heating is grater, the free surface of the weld pool is observed to plunge and decant much quicker. As indicated in figure 3(c), when the arc force and the current is relatively high, a faster increase in pool size can enable gravity to dominate with the self-weight faster [7]. Once the arc is stopped; gravity-driven self-weight starts to dominate. This phenomenon is mostly indicated through the decrease of the free surface level. However, if the weld pool is grown to a given size, the free surface level can drop even with the presence of arc and power input as indicated in figure 3(c), where the net power input is high or as in figure 3(d), where welding time is hire compared to the conditions shown in figure 3(a).

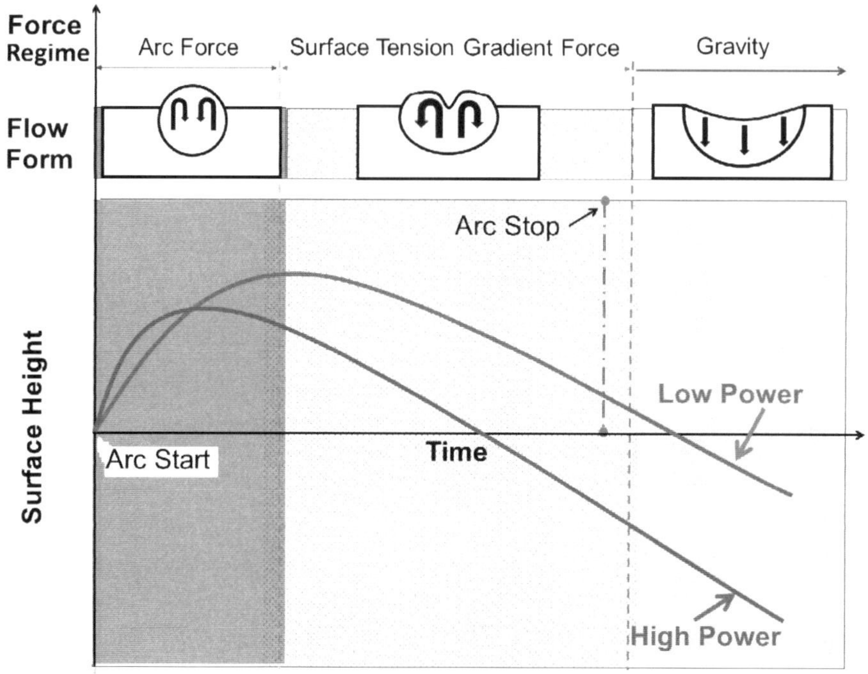

Figure 4. Summarized evolution of the force domination in the arc weld pools and corresponding surface height attributes with respect to the welding power levels.

In overall; at the beginning of weld pool formation, the pool size is relatively small, thus a lower temperature difference across the weld pool results in a relatively week surface tension gradient driven force. Hence, arc force is appeared to dominate the flow initially a short period. Observation of the inward flow inside the melt pool and an apparent increase of free surface provide direct evidence for this. When more and more heat is fed to the center of the pool while the process continues (despite that the excess heat is transported away), the temperature difference between edge and the center of the weld pool can be considerable, and thus a considerable surface tension difference between the center and the edges of the pool can produce. As a result of this considerable difference, increasing Marangoni forces can become dominant in the pool with the flow of outward direction. This can cause the free surface of the pool interface to rise at the edge as seen in figure 2(c). If the net heat input is higher, the Marangoni force can dominate the flow quicker. In the end, the weld pool is large enough and side walls are already molten, the arc and surface tension forces (both upward) struggle to hold the liquid metal against the gravity. As a result, in the end, the liquid metal starts to decant from the sides. If the arc is not stopped and hence both the upward forces exist, the plunging rate of the free can be relatively low. As soon as the arc is distinguished, upward flow diminishes promptly and liquid metal decants much faster. Therefore, the general sequence of the time evolution of the driving forces and the corresponding pool characteristics can be presented as in figure 4. Our observation appeared to be generalized to different metallic materials beyond our observations with aluminum, however, material related physical properties can significantly alter flow condition, see e.g. reference [6]. The outcome presented here can be considered as preliminary guidance to improve numerical modeling of welding. Further analysis with the quantified flow and extended experimental work will provide strong opportunities to validate numerical models and in-depth scientific conclusions.

4. Summary

We have applied X-ray synchrotron imaging to observe the flow in arc weld pools throughout the solid-liquid-solid transformation. The shape evolution of the free surface of the weld pool and the flow patterns suggest the transitions in dominating force regimes that control the flow. Our analysis identifies a general sequence of arc, surface tension, and gravity-driven force domination. Power input of the arc and welding time can alter the dominating period of each type of force, however, the general sequence remains. Weld modeling can be extensively benefited with the real-time information obtained through further studies similar to the one presented here with a wide array of materials and welding techniques.

Acknowledgements
This work is supported by the EPSRC (UK) grant EP/P02680X/1 and EP/R031711/1. Diamond Light Source is acknowledged for granting the beamtime at I12 (EE EE20611-1).

References

[1] Oreper G M and Szekely J 1984 *J. Fluid Mech.* **147** 53
[2] Kim W H and Na S J 1998 *Int. J. Heat Mass Transf.* **41(21)** 3213
[3] Hu J and Tsai H L 2007 *Int. J. Heat Mass Transf.* **50** 808
[4] Hu J and Tsai H L 2007 *Int. J. Heat Mass Transf.* **50** 833
[5] Kou S 2012 *Weld. J.* **91(11)** 287
[6] Aucott L *et al.* 2018 *Nature Comm.* **9** 5414
[7] Tanaka M and Lowke J J 2007 *J. Phys. D: Appl. Phys.* **40(1)** R1
[8] Adonyi Y, Richardson R and Baeslack W 1992 *Weld. J.* **71(9)** 321
[9] He X, Elmer J W and Debroy T 2005 *J. Appl. Phys.* **97** 084909
[10] Raghavan A, Wei H, Palmer T and DebRoy T 2013 *J. Laser Appl.* **25** 052006
[11] Manvatkar V, De A and DebRoy T 2014 *J. Appl. Phys.* **116** 124905
[12] Heiple C, Roper J and Stagner R 1983 *Weld. J.* **62(3)** 72
[13] Wang Y and Tsai H L 2001 *Metall Mat. Trans. B* **32(3)** 501
[14] Zhao Y, Zhou H and Shi Y 2006 *Modell. Simul Mater Sci. Eng.* **14(3)** 331
[15] Wu C S and Dorn L 1994 *Comput. Mater. Sci.* **2** 341
[16] Arivazhagan B, Sundaresan S and Kamaraj M 2008 *Mater. Lett.* **62** 2817
[17] Kim W H, Fan H G and Na S J 1997 *Numer. Heat Transf. A* **32(6)** 633
[18] Mirihanage W U, Di Michiel M, Reiten A, Arnberg L, Dong H B and Mathiesen R H 2014 *Acta Mater.* **68** 159
[19] Cunningham R, Zhao C, Parab N, Kantzos C, Pauza J, Fezzaa K, Sun T and Rollett A D 2019 *Science* **363** 849
[20] Hojjatzadeh S M H *et al.* 2019 *Nature Comm.***10** 3088
[21] Murphy A G, Browne D J, Mirihanage W U and Mathiesen R H 2013 *Acta Mater.* **61** 4559
[22] Sillekens W H, Casari D, Mirihanage W U, Terzi S, Mathiesen R H, Salvo L, Daudin R, Lhuissier P, Guo E and Lee P D *JOM* **68(12)** 3042
[23] Drakopoulos M, Connolley T, Reinhard C, Atwood R, Magdysyuk O, Vo N, Hart M, Connor L, Humphreys B and Howell G 2015 *J. Synchrotron Radiat.* **22** 828
[24] Brumm S and Bürkner G 2015 *Mater. Today: Proc.* **2** S179
[25] Gery D, Long H and Maropoulos P 2005 *J. Mater. Process. Tech.* **167** 393

IOP Conf. Series: Materials Science and Engineering **861** (2020) 012072 doi:10.1088/1757-899X/861/1/012072

Phase-field study on microstructure formation in Mar-M247 during electron beam welding and correlation to hot cracking susceptibility

B Böttger[1], M Apel[1], T Jokisch[2] and A Senger[3]

[1] Access e.V., Aachen, Germany
[2] Siemens AG, Gas and Power, Germany
[3] Institut für Schweißtechnik und Fügetechnik (ISF), RWTH Aachen, Germany

E-mail: b.boettger@access.rwth-aachen.de

Abstract. Ni-base superalloys are frequently used for cast components in the aero-engine and power generation industries. For joining and repair of these components, beam welding is often the method of choice in industrial praxis. However, precipitation-strengthened nickel alloys generally present poor weldability as a consequence of their high weld cracking susceptibility, with high segregating alloys like Mar-M247 even being considered unweldable. Therefore, strong efforts are taken on optimizing techniques and parameters to reduce crack formation during welding of these alloys. Optimization of welding parameters can be assisted by virtual modelling methods through different scales. To be able to focus onto the factors which eventually are responsible for crack formation during welding, comprehensive modelling of the whole process chain is required, starting from a realistic model of the base material and a simulation of the heat source on the macro-scale, and including melting and microstructure formation during welding on the micro-scale. Then, based on the thermal history and the exact microstructure, cracking susceptibilities during solidification can be deduced by hot cracking models adapted to the specific conditions. In this paper, results of microstructure simulations are presented for the technical superalloy MAR-M247 using the phase-field software MICRESS with coupling to Calphad databases. Based on prior phase-field simulations of equiaxed and columnar microstructures of the base material as well as results of a macroscopic simulation of the heat source, melting and subsequent solidification of MAR-M247 has been simulated for two different welding parameter sets. As-weld microstructures are compared to experimental welds, and the virtual hot cracking susceptibility, obtained from the simulation results using a modified Rappaz–Drezet–Gremaud (RDG) hot cracking criterion, is discussed against experimental crack observations.

1. Introduction

Phase-field models have become very popular for simulation of microstructure evolution in material processing. While solidification and phase transformations of alloys have been widely addressed using different approaches [1-3], relatively little simulation work has been done on microstructure development during welding processes. This may be due to specific challenges like the complex thermal conditions which lead to partial melting and subsequent re-solidification, the requirement of defining a realistic initial microstructure with a different typical length scale, the complex conditions how melting leads to formation of new grains, and not to forget the complex alloy chemistry which is involved in may technical welding problems [4].

The software MICRESS® [5] is based on the phase-field concept for multiphase systems which has been extended to multicomponent systems by direct coupling to thermodynamic databases [3, 6]. Since then, the software has been continuously developed not only for advanced scientific applications but also as a simulation tool for microstructure formation in technical processes [7-9].

In this paper, microstructure simulations for welding of Mar-M247 are presented (section 3.4), which have been obtained for two different welding speeds (1 mm/s, 6 mm/s) and two different morphologies (DS, CC) of the base material. While this base material (section 3.3) as well as the welding process at microscale has been simulated using the software MICRESS®, the temperature field and its change in time have been modelled at the macroscopic scale as shown in section 2 using Ansys Version 17.2. The simulated welding microstructures are compared with experimental micrographs obtained under identical conditions. Finally, the hot-cracking susceptibility has been modelled based on the microstructure simulation results using a modified RDG [10-11] criterion and discussed against experimental cracking statistics.

2. Thermal modelling of the welding process

Due to high process complexity, the equivalent heat source (EHS) method is used for the mathematical representation of a location and time-dependant heat effect of an electron beam welding process on Mar-M247. This allows a close representation of temperature distribution in the component. For this purpose, the EHS and the temperature field are adapted and calibrated on the basis of experimental data. The geometric and energetic components of the heat source are set by the solidus line (cross sections) between the base material and the seam and the temperature on the surfaces (measured temperatures via thermocouples during welding process). Due to atypical seam depth to seam width ratio of the electron beam weld at lower welding speeds, a modification of the characteristic beam EHS is necessary [12]. In contrast to the usual high speed beam processes, an extended energy distribution over the seam depth is required. Heat sources used for the phase field simulations are based on a multi-stage approach. The different energy distribution over the depth is varied in the 3D heat source by intensity factors:

$$Q_{TH} = \frac{3\eta_{TH} P_{nom} k_{eff}}{\pi R_{TH}^2 H} step\left(\frac{H-z}{1[mm]}\right) I \exp\left(-\frac{3}{R_{TH}^2}\left((x-x_0)^2+(y-y_0)^2\right)\right) \quad (1)$$

with energy quota η_{TH}, radius of the stepwise deep weld heat source R_{TH}, nominal power P_{nom}, stepwise energy distribution over the depth I and process efficiency k_{eff}.

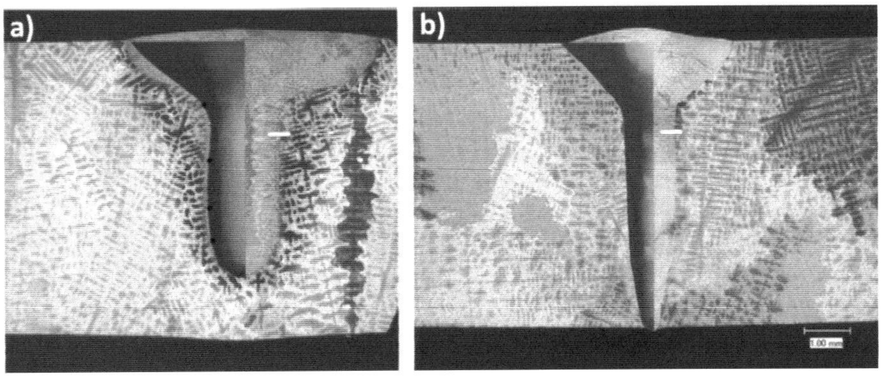

Figure 1. Simulated and observed fusion zones for a welding speed of (a) 1 mm/s, and (b) 6 mm/s. The white lines indicate the position of the results shown in figure 4.

For better temperature distribution near the surface the 3D heat source is superposed with a 2D heat source with exponent 10:

$$Q_{SH} = \beta \frac{\eta_{SH}\, 3 p_{nom}\, k_{eff}\, 0.727}{\pi R_{SH}^2} \exp\left(-\left(\frac{2}{R_{ES}^2} \left((x-x_0)^2 + (y-y_0)^2 \right) \right)^5 \right) \tag{2}$$

with energy quota η_{SH}, radius of the surface heat source R_{ES} and longitudinal energy distribution (degree of asymmetry) β. The good agreement between the simulated and the experimental fusion zone in figure 1 is essentially defined by the intensity factor.

3. Phase-field modelling

3.1. Multiphase-field model and MICRESS® software

The multiphase-field theory describes the evolution of multiple phase-field parameters $\phi_\alpha(\vec{x},t)$ in time and space. They reflect the spatial distribution of multiple grains with different orientation and/or of multiple phases with different thermodynamic properties. At the interfaces, the phase-field variables change continuously over an interface thickness η which can be large compared to the atomic interface thickness but small compared to the microstructure length scale. Their time evolution is calculated by a set of phase-field equations derived by the minimization of the free energy functional [3]:

$$\dot{\phi}_\alpha = \sum_{\beta \neq \alpha}^{v} M_{\alpha\beta} \left[\frac{\sigma_{\alpha\beta}}{v} \left(\frac{1}{2}\left(\nabla^2 \phi_\alpha - \nabla^2 \phi_\beta\right) + \frac{\pi^2}{2\eta_{\alpha\beta}^2}\left(\phi_\beta - \phi_\alpha\right) \right) + \sum_{\gamma \neq \alpha \neq \beta}^{n} J_{\alpha\beta\gamma} + |\nabla \phi| \Delta_{\alpha\beta} \right] \tag{3}$$

with $J_{\alpha\beta\gamma}$ being higher order terms which account for the interaction with additional phases γ in triple or multiple phase regions:

$$J_{\alpha\beta\gamma} = \frac{1}{2}(\sigma_{\beta\gamma} - \sigma_{\alpha\gamma}) \left(\frac{\pi^2}{\eta^2} \phi_\gamma + \nabla^2 \phi_\gamma \right) \tag{4}$$

$$M_{\alpha\beta} = \tilde{M}_{\alpha\beta} \frac{8\eta}{\pi^2} \tag{5}$$

For sake of simplicity, only the isotropic formulation is shown here. The corresponding equations including anisotropy of the interface energy σ and kinetic constants $M_{\alpha\beta}$ are given in [13]. Furthermore, a correction on the interface mobility has been applied to account for the finite interface thickness [14] (sometimes denoted as "quantitative" phase-field model).

In equation (3), $M_{\alpha\beta}$ is the mobility of the interface as a function of the interface orientation, described by the normal vector $\vec{n}$. $\sigma^*_{\alpha\beta}$ is the anisotropic surface stiffness, and $K_{\alpha\beta}$ is related to the local curvature of the interface. The interface, on the one hand, is driven by the curvature contribution $\sigma^*_{\alpha\beta} K_{\alpha\beta}$ and on the other hand by the thermodynamic driving force $\Delta_{\alpha\beta}$. This driving force which is a function of temperature T and local composition $\vec{c}$ couples the phase-field equations to the multi-phase diffusion equations

$$\dot{\vec{c}} = \nabla \sum_{\alpha=1}^{N} \phi_\alpha \vec{D}_\alpha \nabla \vec{c}_\alpha \qquad \text{with } \vec{c} \text{ defined by} \qquad \vec{c} = \sum_{\alpha=1}^{N} \phi_\alpha \vec{c}_\alpha \tag{6}$$

and $\vec{D}_\alpha$ being the multicomponent diffusion coefficient matrix for phase α. $\vec{D}_\alpha$ and $\Delta_{\alpha\beta}$ are calculated online from databases for the given concentration and temperature.

The above equations are implemented in the software package MICRESS® [5] being used for the simulations throughout this paper. Direct coupling to thermodynamic and mobility databases is accomplished via the TQ-interface of Thermo-Calc Software [15]. The thermodynamic driving force ΔG and the solute partitioning are calculated separately using the quasi-equilibrium approach [3], and are introduced into the equation for the multiple phase-fields (equation (3)).

3.2. Model parameters and data

Thermodynamic data were provided by online coupling to the database TTNI7 [16] using the elements given in table 1 and the phases in table 2. In similar way, diffusion data for γ including cross-terms were taken from the MOBNI1 mobility database [15]. As for γ' no diffusion data is given in MOBNI1, values were read from a table as function of temperature also with all cross-terms. This table had been created before by a solidification simulation for Mar-M247 using the databases TCNI7/MOBNI2.

Compositions of both materials are shown in table 1. Phase-related data like nucleation conditions and interface energy data (σ) are given in table 2. Physical interface mobility values μ were taken sufficiently large to ensure diffusion limited growth and corrected for finite interface thickness artefacts using mobility correction [14]. The effects of fluid flow have been neglected.

Table 1: Chemical compositions (in wt. %) of the materials considered in the simulation.

	C	Cr	Co	Mo	W	Ti	Al	B	Ta	Hf
Mar-M247 CC	0.07	8.0	9.0	0.4	9.3	0.6	5.45	0.01	3.1	1.2
Mar-M247 DS	0.1	8.1	9.2	0.5	9.3	0.7	5.6	0.02	3.2	1.4

Table 2: Phases and phase related parameters

No	name in database [15]	identity/ constitution	diffusion data	nucleation in phase	$\mu_{0/x}$ (cm^4/Js)	$\mu_{1/x}$ (cm^4/Js)	$\sigma_{0/x}$ / (J/cm^2)	$\sigma_{1/x}$ / (J/cm^2)
0	LIQUID	melt	10^{-6} cm^2/s	1 (region)	-	0.1	-	2 x 10^{-5}
1	FCC_A1#1	γ	MOBNI1	0	0.1	1 x 10^{-7}	2.0 x 10^{-5}	1.0 x 10^{-5}
2	FCC_A1#2	(Hf,Ta)C	-	0,1	0.1	1 x 10^{-7}	1.2 x 10^{-4}	1.2 x 10^{-4}
3	MB2_C32	$(Ti,Hf)B_2$	-	0,1	0.1	1 x 10^{-7}	1.0 x 10^{-4}	1.0 x 10^{-4}
4	FCC_L12	γ'	MOBNI2	0,1	0.1	1 x 10^{-7}	2.0 x 10^{-5}	4.0 x 10^{-5}
5	M2B_TETR	$(Cr,Mo,W)_2B$	-		0.1	1 x 10^{-7}	1.0 x 10^{-4}	1.0 x 10^{-4}
6	M23C6	$(Cr,MoW)_{23}C_6$	-		0.1	1 x 10^{-7}	1.0 x 10^{-4}	1.0 x 10^{-4}

3.3. Initial microstructure

In view of the complexity of their microstructures, the small size of precipitates, and the difficulty to obtain concentration distributions in consistency with the nominal alloy composition, initial microstructures of the base metal were not taken from metallographic images. Instead, microstructures were obtained from 2D phase-field simulations of solidification with subsequent solution heat treatment according to the material specification. For the equiaxed base material, simulation was done at the scale of two ¼ equiaxed grains including a grain boundary and using the parameter sets given in table 1 and 2. A self-consistent homoenthalpic approach [17] was applied to ensure that microstructure formation was consistent with the macroscopic temperature boundary condition which leads to a realistic initial microstructure.

The columnar initial microstructure was simulated starting with dendrite selection in a temperature gradient of 75 K/cm (domain size 2000 x 2000, Δx=0,5 μm). After stationary growth was reached (figure 3(a)), the lower half of the domain was extracted, and solidification was completed (figure 3(b)) by further cooling with 0.6 K/s without following the dendrite tips (i.e. without co-moving frame).

In both the columnar and the equiaxed case, solidification conditions have been chosen such as to obtain a realistic primary and secondary dendrite arm spacing matching experimental data. A solution heat treatment according to the specifications of the material provider has been added. As visible in figure 2(d), the primary γ'-phase, which forms during solidification due to segregation of γ' forming elements, was not completely dissolved during heat treatment in case of the equiaxed alloy.

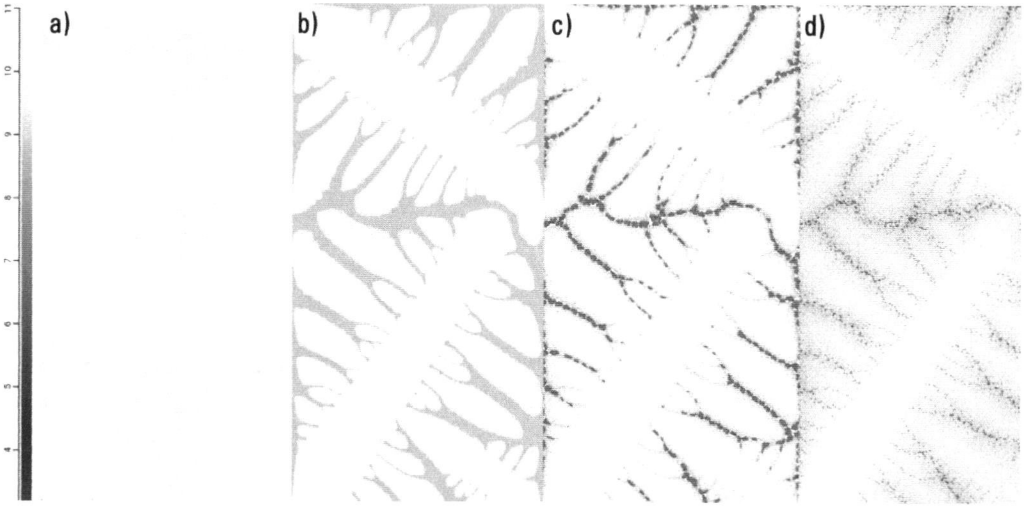

Figure 2. Development of equiaxed microstructure during different processing steps: (a)-(b) during solidification (c) after complete solidification (d) after solution heat treatment. Shown is the concentration distribution of Co. MC carbides and γ'-precipitates appear as dark and red, respectively.

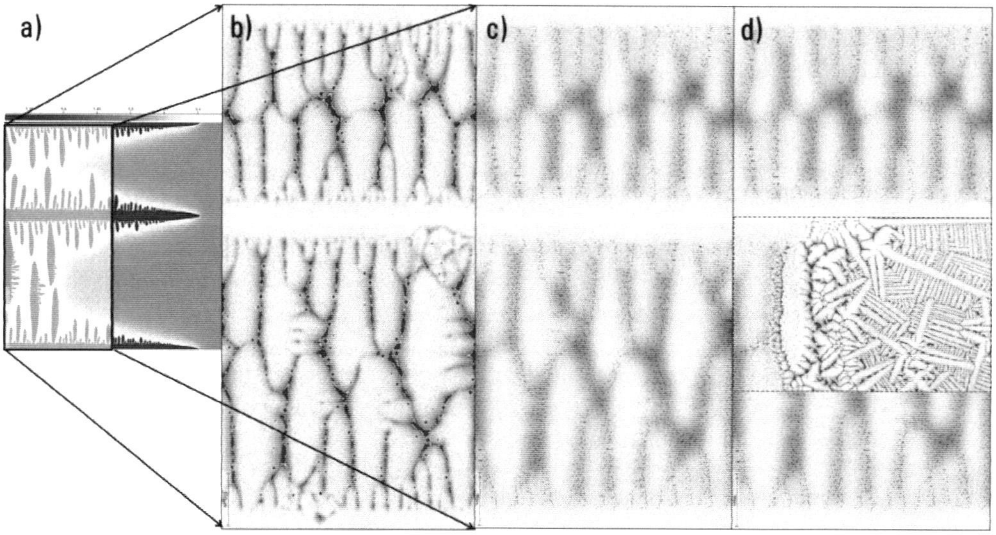

Figure 3. Simulation of directional initial microstructure during different processing steps: (a) Simulation for selection of primary dendrite spacing, (b) after complete solidification of lower half of the domain in (a), (c) after solution heat treatment, (d) subdomain for welding simulation (inserted microstructure is for welding speed 1 mm/s). Shown is the concentration distribution of Co.

3.4. Simulation of welding

As indicated in figure 3(d), a part of the domain has been extracted for the simulation of welding. However, welding microstructures are expected to be substantially finer than the normal solidification microstructures shown in figures 2 and 3. Simulating initial microstructures with a grid resolution sufficient for welding ($\Delta x <\sim 0.2$ μm) would not have been feasible. Therefore, the initial microstructure was interpolated to a three times finer resolution ($\Delta x = 0.167$ μm) using a functionality which refines and reconstructs the phase interfaces by use of the phase-field and diffusion equations.

For simulation of welding, one-dimensional temperature profiles were extracted from the temperature simulations of section 2 (at the positions corresponding to the white lines in figure 1). Simulation comprises melting of the base material, reversal of the front direction shortly before full melting, and re-solidification from the remaining solid (dendrites as well as detached fragments). Melting was initiated by nucleation of liquid phase, "reusing" grain number 0 in order to avoid grain boundaries between different melt "grains". During melting, large numbers of dendrite fragments are formed. In order to allow new grains emerge from these fragments, their complete detachment is detected, a new grain number was assigned, and random rotation of orientation was applied.

Simulation results for the equiaxed base material and two different welding speeds are shown in figure 4 in direct comparison to experimental micrographs obtained under identical conditions. A good qualitative agreement has been found in both cases.

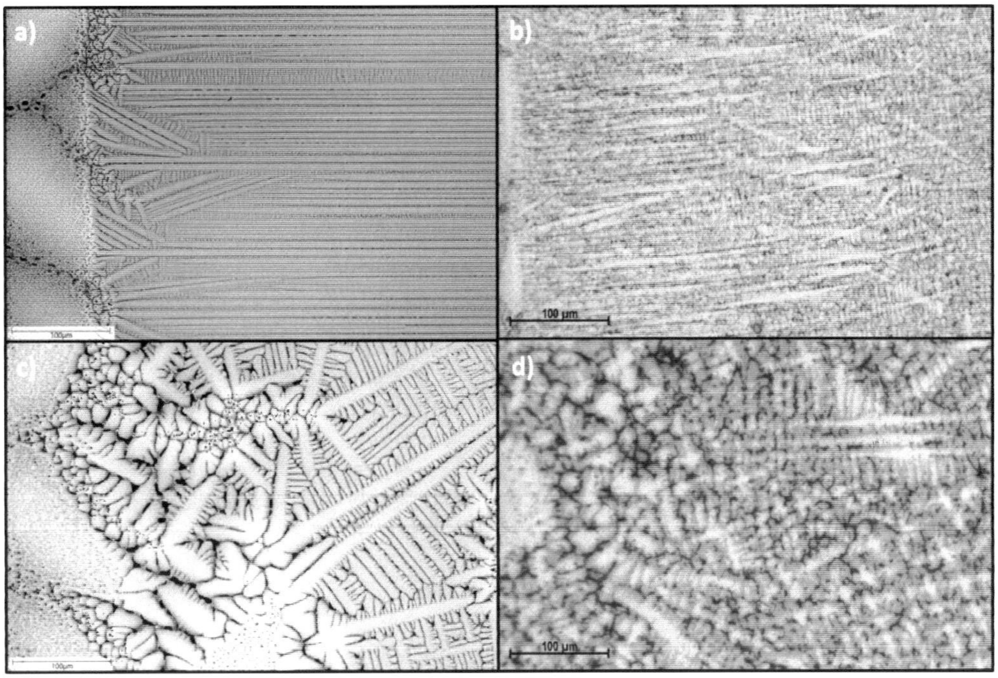

Figure 4. Comparison of simulated microstructures (left) and experimental micrographs (right) for Mar-M247 CC using a welding speed of 6 mm/s (a)-(b) and 1 mm/s (c)-(d)

4. Evaluation of hot-cracking susceptibility

According to Rappaz *et al.* [10] hot-cracking occurs due to insufficient feeding of liquid into strained interdendritic regions and is initiated if a critical strain rate $(d\varepsilon/dt)_{crit}$ is exceeded:

$$(d\varepsilon/dt)_{crit} = \frac{G}{(1+\beta)B}\left[\frac{\lambda_2^2 G\Delta p_{\max}}{180\mu} - v_T\beta A\right] \text{ with } A = \int_{T_c}^{T_l}\frac{(1-f_s)f_s^2}{(1-f_s)^3}dT\ ,\ B = \int_{T_c}^{T_l}\frac{f_s^2}{(1-f_s)^3}\left(\int_{T_c}^{T} f_s dT\right)dT \quad (7)$$

Here, G is the temperature gradient, β the volume shrinkage due to solidification, v_T the solidification speed, µ the melt viscosity, and A and B are integrals over different functions of the fraction solid f_s. The coalescence temperature T_c, which typically is estimated as the temperature for $f_s = 99\%$, and the liquidus temperature T_l are used as integration limits. G, f_s as function of temperature, and the secondary arm spacing λ_2 were statistically evaluated from the simulation results as a function of the distance x from the left border of the simulation domain (table 3). For λ_2, a linear regression was calculated from 3 averaged values for different ranges of x (<200 µm, 200-300 µm,>300 µm). The maximum pressure drop Δp_{max} in equation (7) was estimated as shown below.

$(d\varepsilon/dt)_{crit}$ appears to be of little use comparing the risk of hot-cracking under different conditions, if the strains occurring during the welding process are unknown. Instead, $(d\varepsilon/dT)_{crit} = (d\varepsilon/dt)_{crit}/(dT/dt)$ seems to be more useful in this case [19] as strains are mainly caused by cooling and thus depend on the cooling rate $\dot{T}$. The alternative criterion $(d\varepsilon/dT)_{crit}$ has the units of a thermal expansion coefficient, and its magnitude should be closely related to this material constant. This assumption is very helpful for estimating the remaining unknown Δp_{max}: While the value of Δp_{max} in conventional casting typically is estimated as the sum of the atmospheric and metallostatic pressure [10, 11], in vacuum electron beam melting both quantities are negligibly small. Instead, we assume some pressure must build up in the already interconnected mushy zone at the center of the weld pool, e.g. by thermal contraction or clamping of the specimen. If we further assume that for slow welding (CC, 1 mm/s) there is just no hot-cracking (cf. figure 6), $(d\varepsilon/dT)_{crit}$ at the worst position x should be equal to the thermal expansion coefficient ($d\varepsilon/dT$ (Ni) ~10^{-5} K^{-1}). By this way, Δp_{max} can be estimated to 1.1×10^6 Pa.

Table 3: Parameters for equation (7)

	CC, 1 mm/s	DS, 1 mm/s	CC, 6 mm/s	DS, 6 mm/s
λ_2/μm	12,80 -0,0144x	12,23 -0,0082x	2,85 +0,00074x	3,21 -0,00062x
β	0,13 (Calculated using Thermo-Calc and database TCNI9 [15])			
G/Kcm^{-1}	evaluated at f_s=0.99 for each timestep			
μ/kg(ms)$^{-1}$	0.0032 [18]			
Δp_{max}/Pa	1.1 x 10^6 (calibrated, see text)			

With this knowledge, finally, the different welding conditions can be compared. Figure 5 shows $(d\varepsilon/dT)_{crit}$ as function of the distance (from the left side of the simulation domain, in cm). In case of the equiaxed material and a welding speed of 1 mm/s, $(d\varepsilon/dT)_{crit}$ shows a minimum close to the edge of the weld seam, with a value of ~10^{-5} K^{-1} in accordance with the calibration of Δp_{max} (figure 5(a)). Higher values elsewhere indicate a lower hot-cracking risk at other distances.

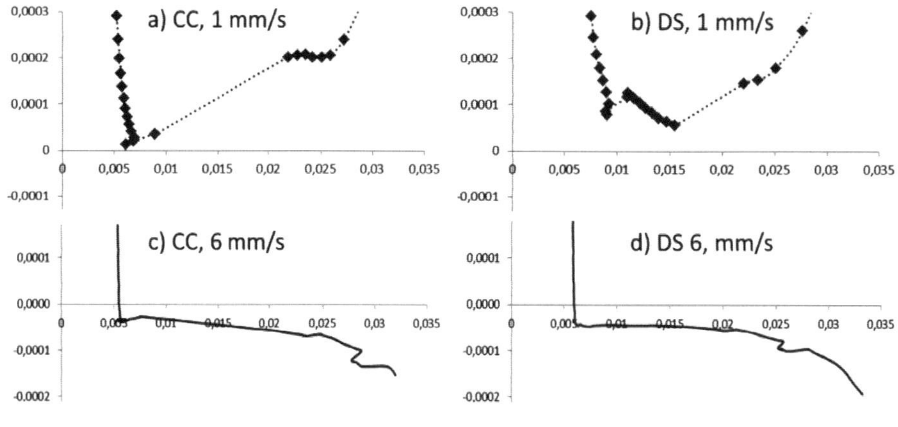

Figure 5. Comparison of $(d\varepsilon/dT)_{crit}$ (as function of the distance from the left side of the simulation domain in cm) for the different welding conditions indicating a low hot cracking risk for a welding speed of 1 mm/s and a high hot cracking risk for 6 mm/s

For the columnar material at 1 mm/s, using the same Δp_{max}, a similar curve of $(d\varepsilon/dT)_{crit}$ is obtained, indicating even a slightly lower risk of hot-cracking (higher minimum value of $(d\varepsilon/dT)_{crit}$, figure 5(b)). However, when changing to a welding speed of 6 mm/s, the behavior changes drastically as the$(d\varepsilon/dT)_{crit}$ curves for both materials go far to the negative for all simulated positions inside the weld pool. This indicates a strongly higher hot-cracking risk predicted by the $(d\varepsilon/dT)_{crit}$ criterion (essentially it means that even negative strain rates would be required to prevent cracking!) and is in good agreement with cracking statistics carried out at Siemens AG for various welding speeds (figure 6). It should be noted that even when we would assume some dynamic pressure build-up at the center of the weld pool and use a proportionally higher value of Δp_{max} (6.6x10^6 Pa) at 6 mm/s, the $(d\varepsilon/dT)_{crit}$ curves would still be widely negative.

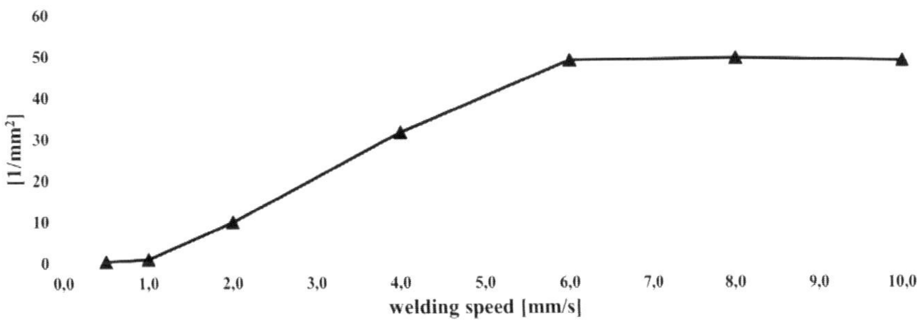

Figure 6. Micro-CT characterisation of total length of the cracks for different welding speeds normalized by the volume of the weld seam

5. Summary and conclusion

Electron beam welding of Mar-M247 has been simulated on the basis of a thermal "heat source" model and of equiaxed as well as columnar initial microstructures obtained by prior microstructure simulations of the solidification and heat treatment process. For two welding experiments with a welding speed of 1 mm/s and 6 mm/s, respectively, a good qualitative agreement between the simulated microstructures and experimental micrographs were found. Furthermore, evaluations of the simulation results using a modified RGB criterion for hot-cracking revealed a significantly higher risk for crack initiation for the higher welding speed for both materials. This is in full agreement with comprehensive experimental cracking statistics for various welding speeds and demonstrates that the presented approach is valid for prediction of hot-cracking under conditions of electron beam welding.

References

[1] Wheeler A A, Boettinger W J and Mc Fadden G B 1993 *Phys. Rev.* E **47** 1893-909
[2] Kim S G, Kim W T and Suzuki T 1999 *Phys. Rev.* E **60** 7186-97
[3] Eiken J, Böttger B and Steinbach I 2006 *Phys. Rev.* E **73** 066122
[4] David S A, Babu S S and Vitek J M 2003 *JOM* **55** 14.
[5] MICRESS software, https://web.micress.de
[6] Böttger B, Grafe U, Ma D and Fries S G 2000 *Mater. Sci. Technol.* **16** 1425-28
[7] Schmitz G J et al 2012 *Int. J. Adv. Eng. Sci. Appl. Math.* **2** 4 126-39
[8] Böttger B, Haberstroh C, Giesselmann N 2016 *JOM* **68(1)** 27–36
[9] Böttger B 2019 *Metall. Mater. Trans.* **50A** 1733
[10] Rappaz M, Drezet J M and Gremaud M 1999 *Metall. Mater. Trans.* **30A** 449
[11] Böttger B, Apel M, Santillana B and Eskin D G 2013 *Metall. Mater. Trans.* **44A** 3765-77
[12] Fang C 2015 *Fusion Engineering and Design*, 2015 357-63
[13] J. Eiken, Dissertation, Faculty of Georesources and Materials Engineering, RWTH Aachen, ISBN 978-8322-9010-8, 2010.
[14] Carré A, Böttger B, and Apel M 2014 *J. Crystal Growth* **380** 5–13
[15] Themo-Calc Software, https://www.thermocalc.com
[16] Thermotech Ltd. https://www.thermotech.co.uk/databases.html
[17] Böttger B, Eiken J and Apel M 2009 *J. Comput. Phys.* **228** 6784-95
[18] Torroba A J et al. 2014 *Integrating Materials and Manufacturing Innovation* **3** 25
[19] Drezet J M and Allehaux D 2008 *Hot Cracking Phenomena in Welds* vol II, ed T Bollinghaus H Herold C Cross and J Lippold (Springer) pp 19-37

IOP Conf. Series: Materials Science and Engineering **861** (2020) 012073 doi:10.1088/1757-899X/861/1/012073

Predicting solidification cracking susceptibility of stainless steels using machine learning

S Feng and H B Dong

Department of Engineering, University of Leicester, UK

E-mail: hd38@le.ac.uk

Abstract. Machine learning, which reveals the complex nonlinear relationship in the archived data, is a powerful complement to theory, experiment, and modeling. In this study we attempted to predict solidification cracking susceptibility of stainless steels as a function of chemistry and processing parameters using machine learning with a data set that contains about 600 longitudinal varestraint test results. Four machine learning models, i.e. decision tree, random forest, shallow neural network and deep neural network, were used to mine the data set. Our results show: deep neural network outperformed other models in prediction accuracy; tree-based models have accepted accuracy and better interpretability than neural network; machine learning models transforms scattered experimental data points into a map in high-dimensional chemistry and processing parameters space. The combination of different machine learning models reveals that the solidification cracking susceptibility of stainless steels was mainly determined by the ratio of Ni content to Cr content, impurity element content and the strain level.

1. Introduction

Solidification cracking is one of the most serious defects which widely occurs in processes involve liquid to solid transformation like casting, welding, and additive manufacturing. It occurs at the last stage of solidification when liquid films exist between dendrites boundaries where local strains cannot be accommodated by liquid feeding and solid deformation [1]. It's a complex problem because of interaction of material factors, mechanical restraint factors, thermal factors in the final crack nucleation and propagation [2]. Metallic material processability e.g. weldability and castability is closely linked with its solidification cracking susceptibility. Due to its complexity and importance, solidification cracking has been researched widely with experiments, theories, and modelling. Studying solidification cracking is time-consuming and expensive. Theories about solidification cracking susceptibility based either on metallurgical features, or strain, or strain rate, or stress criteria are either over-simplified or lack physical basis, thus preventing them from practical application as no existing theory can include all three types of factors mentioned above [3]. Successful computer numeric models which take into consideration of thermal and mechanical interaction, intergranular network geometry and flow field can only predict the position of most susceptible to solidification cracking but not alloys solidification cracking susceptibility [4]. To further study this type multiple variables nonlinear problem, data-based and data-driven machine learning methods, e.g. neural network and tree-based models, increasingly become an alternative to physically based analytical and numerical methods [5]. Based on reliable experimental data, machine learning methods can depict alloy solidification cracking susceptibility as function of alloy composition and processing parameters in a high-dimensional space, like that was shown in our previous work [6].

To explore the relationship between solidification cracking susceptibility, composition and processing parameters, intermediate steps (bridges), i.e. atomic/meso/micro/macro structure analysis, are commonly required in traditional material science route. The relationship between composition, processing parameters and structure and the relationship between structure and solidification cracking susceptibility are researched separately first. A lot of new terminologies and models are proposed like ratio of chromium equivalent to nickel equivalent, primary ferritic/austenitic solidification mode, impurities content, solidification region, etc. to simplify analysis that results in an indirect relationship between solidification cracking susceptibility and three main influential factors. But the main disadvantages of traditional material science methods are the relationship found is always limited to specific composition and processing parameters and most of them are qualitative descriptions, a universal relationship is rarely found, especially for the complex nonlinear phenomena. This shortcoming could be overcome by many data-based methods especially the deep learning by which the hierarchy structure of data can be features engineered automatically [7, 8], thus we can get solidification cracking susceptibility as function of composition and processing parameters, i.e. quantitative mappings from inputs to outputs, directly without the aids of structure analysis.

In this paper we combined different machine learning models including deep neural network of high accuracy and tree-based models of high interpretability to predict solidification cracking susceptibility of stainless steels.

2. Methods

2.1. Data collection

Dozens of test methods exist in order to qualitative and quantitatively evaluate the solidification cracking susceptibility. A controlled strain applied on a geometrically simple specimen and test method of good repetition is preferred for evaluation of cracking tendency such as longitudinal varestraint test, transverse varestraint test, etc. In longitudinal varestraint test a definite bending strain is applied on weld of sheet metals specimen, and total crack length (TCL) and maximum crack length, brittle temperature range are measured as the indicators for solidification cracking susceptibility. A large amount of longitudinal varestraint test results are available in literature. That is why we chose longitudinal varestraint test data in published literature to compile our solidification cracking susceptibility data set. In this study, a data set containing 600 longitudinal varestraint test results of stainless steels was collected from literature. The data set contains information of ferritic, austenitic, duplex, precipitation hardening wrought and casting stainless steels, and comprises 4 types information: stainless steels composition information of 16 elements (contents of C, Si, Mn, P, S, Cr, Ni, Mo, N, Nb, Co, Cu, Al, Ti, V, B, all compositions are expressed in wt% in this paper), samples thickness (Th, mm) and 3 welding parameters (welding current: I, A; voltage: U, V; velocity: Ve, mm/s), restraint information (applied strain: ε, %) and solidification cracking susceptibility indicator total crack length (TCL, mm). The elements contents which were not specified in literature were assumed as follow: Al = 0.02, N = 0.02, Mo = 0, Nb = 0, Cu = 0, V = 0, B = 0. Data missing too much information was discarded, and the final size of training/testing data set is 487.

2.2. Neural network's preprocess and training algorithm

The input variables X were normalized between -0.5 and +0.5, the target variables Z were normalized between -0.99 and +0.99. The neural networks used in this study consisted of 21 input neurons (one neuron corresponds to one input variables), one to several hidden layers, a few hidden neurons (the neurons number is variable) in each hidden layer, and one output neurons for TCL. The transfer function of hidden layer is hyperbolic tangent function tanh(x). The transfer function of output layer is linear function $y = x$.

To achieve unbiased training/testing data division, all data were divided into groups first, and in a group only one or two variables vary, then training data and testing data were randomly chosen in a 2 to 1 ratio, i.e. 2/3 data set (324 samples) was used for training, and the remaining 1/3 data set (163 samples) were kept unseen by neural network in training to test the generalization performance (prediction accuracy on unseen data set) of the trained neural network.

To improve generalization and reduce the possibility of overfitting, a type of regularized performance function mse_{reg}, which combines mean square error with mean square weight, was used in training shallow neural network and fine-tuning deep neural network (after pre-training). The weights and biases (special type of weights with input value of 1) are determined through training neural network which involved minimization of the regularized performance function through gradient descent algorithm (backpropagation algorithm) and auto determination of optimal performance ratio γ using Bayesian regularization proposed by David MacKay. The training and regularization details are described elsewhere [9, 10].

One hidden layer shallow neural network is a mature tool for science and engineering field, including material science. The commonly used ways to increase prediction accuracy of a shallow neural network include: increasing data set size; increasing variables or features of data set; increasing neuron number in hidden layer; changing network's initial weights and biases values; increasing training epoch. And the ways to increase prediction accuracy of the deep neural network include: increasing hidden layer number; changing neuron numbers in hidden layers; pretraining deep neural network using restrict Boltzmann machine and auto-encoder to achieve optimal initiation; changing network structure (from full connected to partially connected); changing training algorithms and using special training strategy (e.g. drop out) [11, 12]. In this study, we tried shallow neural networks of neuron number from 1 to 35 and train every shallow neural network configuration more than 100 times using different initiation values, then the optimal trained shallow neural network was chosen to make prediction. We also attempted 3 hidden layers deep neural networks of 21-(4-3-2)-1 and 21-(5-4-3)-1 structure, 4 hidden layers deep neural networks of 21-(5-4-3-3)-1 and 21-(6-5-4-3)-1 structure, and 5 hidden layers deep neural networks of 21-(6-5-4-3-3)-1 and 21-(5-4-3-3-3)-1 structure. The numbers in bracket represent neuron numbers of hidden layers in sequence. A selection of initiation values for stacked auto-encoder were used to generate different initiation conditions for deep neural networks. Every deep neural network configuration was trained more than 100 times using different initiation conditions, then the optimal deep neural networks was chosen to make prediction. After training Pearson correlation coefficient R of target values and neural network prediction values was calculated as the index of training / testing accuracy.

2.3. Tree-based machine learning models

Tree-based machine learning models are a family of non-parametric, supervised methods. The foundation of tree-based models is the decision tree wherein a series of decision rules (e.g., "If A > B…") are linked. Random forest model is the ensemble of decision trees. A decision tree looks like an upside-down tree, with the first decision rule at the top and subsequent decision rules spreading out below. In a decision tree, every decision rule occurs at a decision node, with the rule creating branches leading to new nodes. A branch without a decision rule at the end is called a leaf. Decision tree regression models attempt to find a decision rule that produces the greatest decrease mean squared error at a node. One benefit of tree-based models is their interpretability. Visualization of decision trees can give researchers an intuitive knowledge of feature importance.

In this work, all machine models are realized on MATLAB 2018a with its statistics and machine learning toolbox and neural network toolbox.

3. Results

3.1. Training and testing accuracy

Table 1 shows the accuracies of four machine learning models on testing data set. Random forest's accuracy (0.88) is better than that of a single decision tree (0.75) and is close to that of shallow neural network. Deep neural network has the highest prediction accuracy in the four machine learning models.

Table 1. The best testing accuracies of four machine learning models

	Decision tree	Random forest	Shallow neural network	Deep neural network
Testing accuracy	0.75	0.88	0.89	0.93

3.2. Neural network structure

The structure schematics of the optimal shallow and deep neural network are shown in figure 1. It shows the optimal 21-(6-5-4-3)-1deep neural network has more compact structure (the hidden layer of deep neural network is narrower than that of shallow neural network) and less parameters (210 vs. 484) than the optimal 21-(21)-1 shallow neural network. The trained neural network can be easily used to estimate solidification cracking susceptibility dependence on compositions and strains (small strains can be utilized to predict solidification cracking susceptibility in casting and large strains can be chosen to predict solidification cracking susceptibility in welding and additive manufacturing) when the time consuming and expensive experimental results are not available. The following sections provide several examples of application, but the application of the machine learning models is not limited to those.

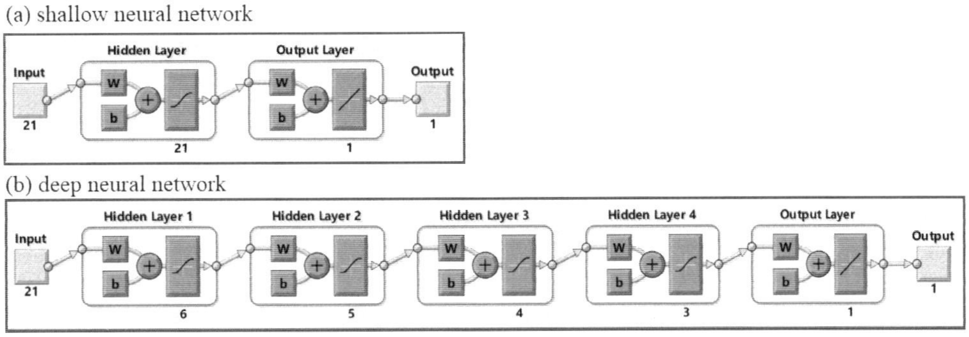

Figure 1. (a) The optimal shallow neural network's structure schematic; (b) the optimal deep neural network's structure schematic.

4. Predicting solidification cracking susceptibility

4.1. Feature importance

Decision tree can give us an intuitive knowledge of how solidification cracking susceptibility is calculated according to a serial of rules. Figure 2 visualizes a decision tree used for predicting solidification cracking susceptibility. The Ni content is used as the first decision rule and the strain is used as the second decision rule, this shows the importance of the Ni content and strain in determining stainless steels' solidification cracking susceptibility.

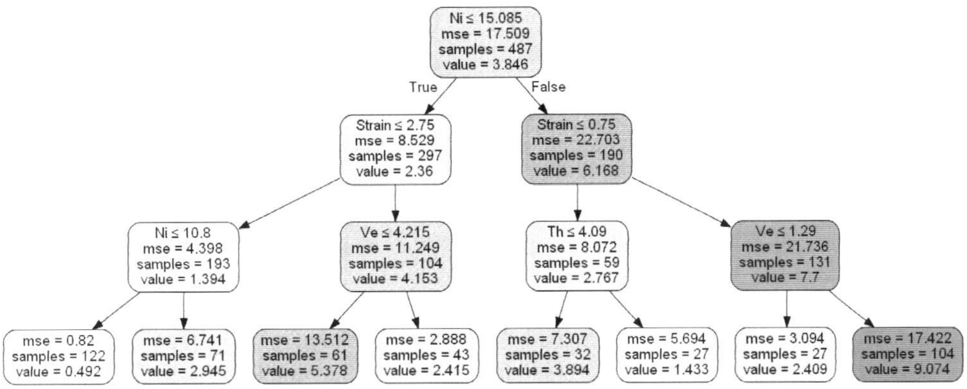

Figure 2. A decision tree for predicting solidification cracking susceptibility.

Besides the intuitive impression of feature importance, tree-based machine learning models can give the value of feature importance. Figure 3 is the feature importance for solidification cracking susceptibility derived from the optimal random forest model. Major alloy elements Ni and Cr, minor alloy elements Mn and Si, impurity elements C, N, P, S, and the strain applied on the specimen play the most important role in determining the solidification cracking susceptibility of stainless steels. The knowledge derived can be used to guide next step predictions. A serial of predictions about solidification cracking susceptibility dependence on compositions and processing parameters in the following part will give us more details and deeper understanding of this complex problem.

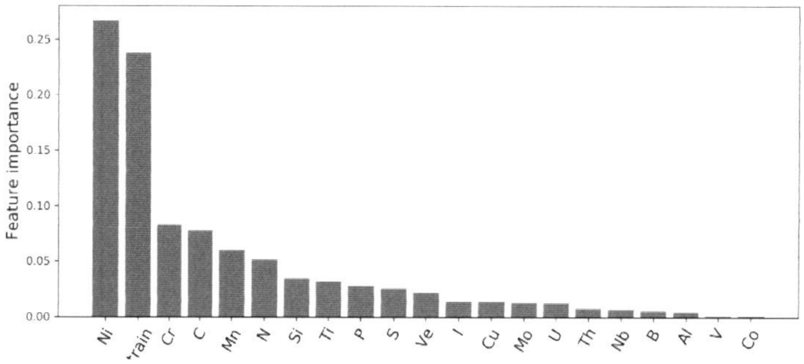

Figure 3. Feature importance for solidification cracking susceptibility from the random forest model.

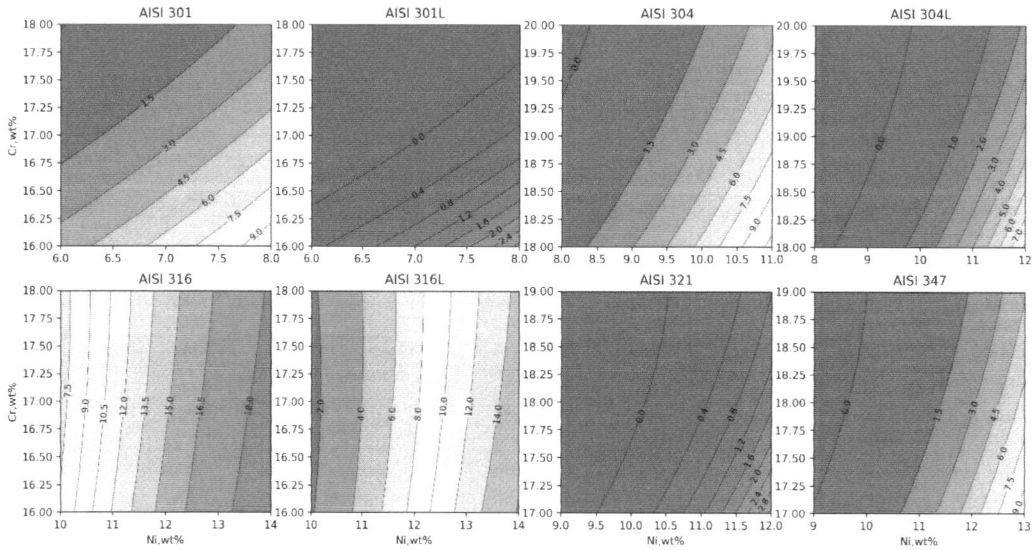

Figure 4. The prediction of 8 stainless steels' solidification cracking susceptibility dependence on Ni and Cr (their Ni and Cr contents are in AISI specification). Compositions are shown in table 2. Processing parameters: Th=3.18 mm, I=100 A, U=12 V, Ve=4.23 mm/s, strain=3%.

4.2. Solidification cracking susceptibility dependence on Ni and Cr

The prediction of 8 stainless steels' solidification cracking susceptibility dependence on the main alloy elements Ni and Cr is shown in figure 4. The predictions are in good accordance with metallurgical experience: fully austenitic stainless steels like AISI 316 is more susceptible to solidification cracking than stainless steels contain a certain amount of ferrite like AISI 304; C which tend to segregate to grains boundaries increase cracking susceptibility (see the contour difference between AISI 316L and AISI 316); machine learning regression has transformed scattered data points from our data set into an expressive high-dimensional map and those contours are some slices of it. The tendency of increasing solidification cracking susceptibility with the increment of Ni is obviously illustrated in the figure 4. High ratio of Cr to Ni (or Cr equivalent to Ni equivalent) is good for solidification cracking resistance by forming a certain amount of ferrite like that in AISI 301 and 304 etc., because ferrite can accept more impurity elements (like S and P) than austenite and irregular ferrite/austenite grain boundary is not in favour of the propagation of cracking cracks.

Table 2. Compositions (wt%) of stainless steels used in predictions in figure 4 (varied Ni and Cr).

Code	C	Si	Mn	P	S	Cr	Ni	Mo	N	Nb	Cu	Al	Ti
301	0.10	1.0	2.0	0.02	0.02	16.0-18.0	6.0-8.0	-	0.10	-	-	-	-
301L	0.03	1.0	2.0	0.02	0.02	16.0-18.0	6.0-8.0	-	0.10	-	-	-	-
304	0.08	1.0	2.0	0.02	0.02	18.0-20.0	8.0-11.0	-	0.10	-	-	-	-
304L	0.03	1.0	2.0	0.02	0.02	18.0-20.0	8.0-12.0	-	0.10	-	-	-	-
316	0.08	0.75	2.0	0.02	0.02	16.0-18.0	10.0-14.0	2.0	0.10	-	-	-	-
316L	0.03	0.75	2.0	0.02	0.02	16.0-18.0	10.0-14.0	2.0	0.10	-	-	-	-
321	0.08	0.75	2.0	0.02	0.02	17.0-19.0	9.0-12.0	-	-	-	-	-	0.5
347	0.08	0.75	2.0	0.02	0.02	17.0-19.0	9.0-13.0	-	-	0.5	-	-	-

4.3. Solidification cracking susceptibility dependence on P and S

The prediction of 8 stainless steels' solidification cracking susceptibility dependence on impurity elements P and S is shown in figure 5. The tendency of increasing solidification cracking susceptibility with the increment of P and S is shown in the prediction. The extent of increasing susceptibility with P and S are different among different steels. The solidification cracking susceptibility of AISI 316 is very sensitive to the P and S contents: small difference in P and S contents can show great difference in solidification cracking behaviour. But that of AISI 316L with less C content is less sensitive to the P and S contents. Figure 5 also shows that impurity elements P and S have different impact on solidification cracking susceptibility (most contours are not symmetry along P=S diagonal), the common assumption of linear superposition P and S used in estimating solidification cracking susceptibility is not always proper.

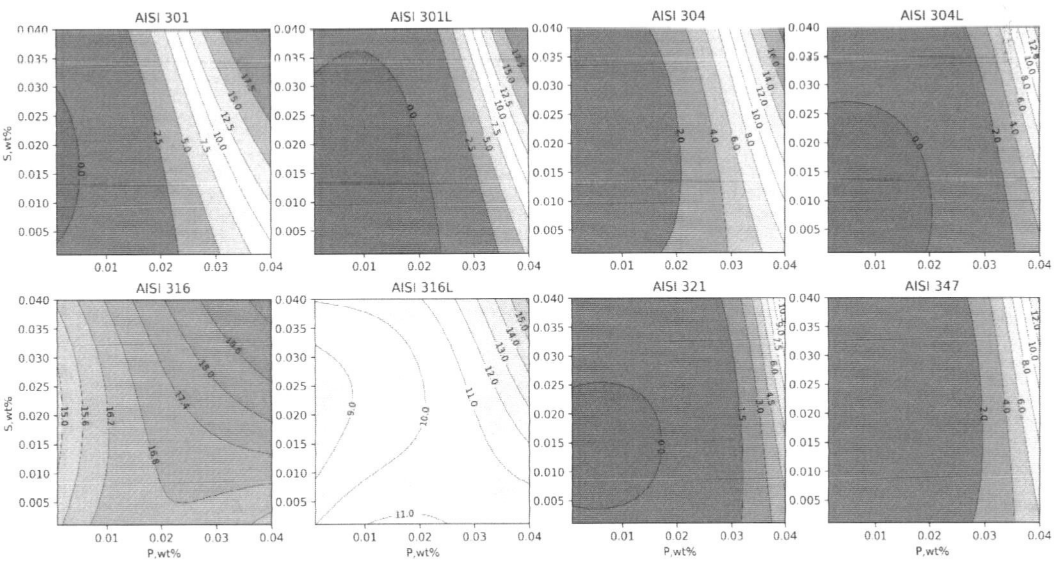

Figure 5. The prediction of 8 stainless steels' solidification cracking susceptibility dependence on P and S. Compositions are shown in table 3. Processing parameters: Th=3.18 mm, I=100 A, U=12 V, Ve=4.23 mm/s, strain=3%.

Table 3. Compositions (wt%) of 8 stainless steels used in predictions in figure 5 (varied P and S).

Code	C	Si	Mn	P	S	Cr	Ni	Mo	N	Nb	Cu	Al	Ti
301	0.10	1.0	2.0	0.001-0.040	0.001-0.040	17.0	7.0	-	0.10	-	-	-	-
301L	0.03	1.0	2.0	0.001-0.040	0.001-0.040	17.0	7.0	-	0.10	-	-	-	-
304	0.08	1.0	2.0	0.001-0.040	0.001-0.040	19.0	9.5	-	0.10	-	-	-	-
304L	0.03	1.0	2.0	0.001-0.040	0.001-0.040	19.0	9.5	-	0.10	-	-	-	-
316	0.08	0.75	2.0	0.001-0.040	0.001-0.040	17.0	12.0	2.0	0.10	-	-	-	-
316L	0.03	0.75	2.0	0.001-0.040	0.001-0.040	17.0	12.0	2.0	0.10	-	-	-	-
321	0.08	0.75	2.0	0.001-0.040	0.001-0.040	18.0	9.5	-	-	-	-	-	0.5
347	0.08	0.75	2.0	0.001-0.040	0.001-0.040	18.0	9.5	-	-	0.5	-	-	-

4.4. Solidification cracking susceptibility dependence on C and N

The prediction of 4 stainless steels' solidification cracking susceptibility dependence on C and N is shown in figure 6. The tendency of increasing solidification cracking susceptibility with the increasement of C and N is revealed in the prediction. The extent of increasing susceptibility with C and N varied among different steels due to the different alloy elements in corresponding steels. The solidification cracking susceptibility of AISI 316 is very sensitive to the C and N contents: small difference in C and N contents can show great difference in solidification cracking behaviour, but cracking susceptibility of AISI 301 and 304 etc. are less sensitive.

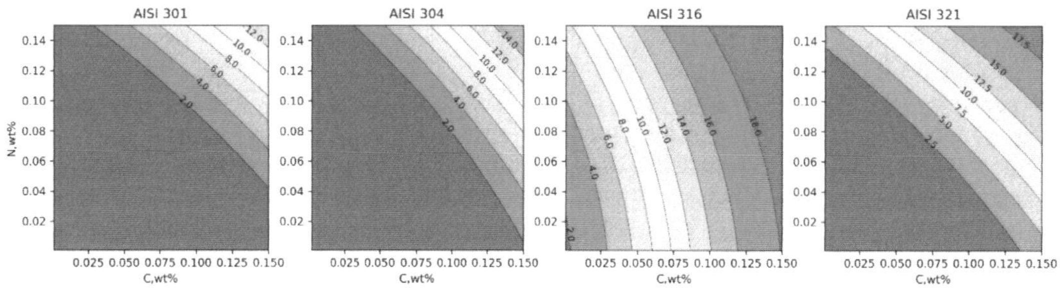

Figure 6. The prediction of 4 stainless steels' solidification cracking susceptibility dependence on C and N. Compositions are shown in table 4. Processing parameters: Th=3.18 mm, I=100 A, U=12 V, Ve=4.23 mm/s, strain=3%.

Table 4. Compositions (wt%) of the 4 stainless steels used in predictions in figure 6 (varied C and N).

Code	C	Si	Mn	P	S	Cr	Ni	Mo	N	Nb	Cu	Al	Ti
301	0.001-0.15	1.0	2.0	0.02	0.02	17.0	7.0	-	0.001-0.15	-	-	-	-
304	0.001-0.15	1.0	2.0	0.02	0.02	19.0	9.5	-	0.001-0.15	-	-	-	-
316	0.001-0.15	0.75	2.0	0.02	0.02	17.0	12.0	2.0	0.001-0.15	-	-	-	-
321	0.001-0.15	0.75	2.0	0.02	0.02	18.0	9.5	-	0.001-0.15	-	-	-	0.5

4.5. Solidification cracking susceptibility dependence on strain

The prediction of AISI 316 stainless steel's solidification cracking susceptibility dependence on strain is shown in figure 7. The tendency of increasing solidification cracking susceptibility with the increment of strain is correctly revealed in the prediction. This tendency is also a nonlinear behaviour, which is very difficult for physical models to depict, but is quite easy for machine learning models.

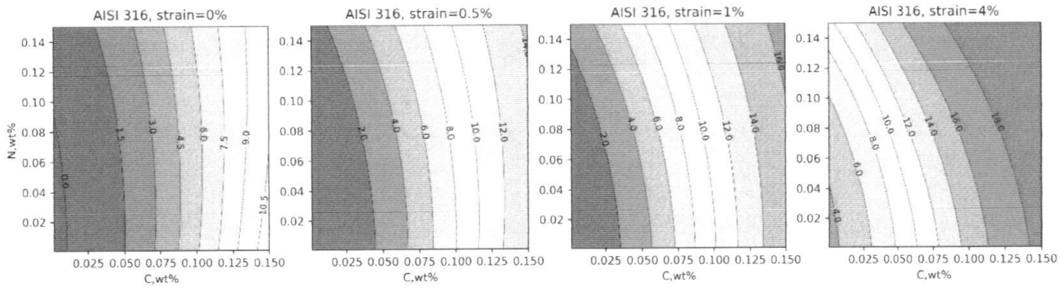

Figure 7. The prediction of AISI 316 stainless steel's solidification cracking susceptibility dependence on strain with variations in C and N. Compositions are shown in table 4. Processing parameters: Th=3.18 mm, I=100 A, U=12 V, Ve=4.23 mm/s.

5. Conclusions

Four machine learning models, i.e. shallow neural network, deep neural network, decision tree and random forest, were used to fit solidification cracking susceptibility of stainless steels as a function of chemistry and processing parameters. Neural network models outperformed tree-based models in prediction accuracy; tree-based models have accepted accuracy and better interpretability than neural networks; the optimal deep neural network has more compact structure and less parameters than the optimal shallow neural network.

Through machine learning regression, vast scattered non-systematic experimental data buried in literature can produce simple quantitative expression: specific material property as function of chemistry composition and processing parameters, etc. The derived models can be used in material property prediction, new alloys development, comparison with experimental results, etc.

The combination of different machine learning models reveals that the solidification cracking susceptibility of stainless steels was mainly determined by Ni content / Cr content (or ferrite content / austenite content), impurity content and the strain level.

Acknowledgements

Shuo Feng wishes to acknowledge EPSRC CDT (Grant No: EP/L016206/1) in Innovative Metal Processing for providing a PhD studentship for this study.

Reference

[1] Aucott L, Huang D, Dong H B, Wen S W, Marsden J, Rack A and Cocks A C F 2018 *Metall. Mater. Trans. A* **49** 1674–82
[2] Eskin D G, Suyitno and Katgerman L 2004 *Prog. Mater. Sci.* **49** 629–711
[3] Kou S 2015 *Acta Mater.* **88** 366–74
[4] Bordreuil C and Niel A 2014 *Comput. Mater. Sci.* **82** 442–50
[5] Mater A C and Coote M L 2019 *J. Chem. Inf. Model.* **59** 2545–59
[6] Feng S, Zhou H and Dong H 2019 *Mater. Des.* **162**
[7] Butler K T, Davies D W, Cartwright H, Isayev O and Walsh A 2018 *Nature* **559** 547–55
[8] LeCun Y, Bengio Y and Hinton G 2015 *Nature* **521** 436–44
[9] MacKay D J C 1992 *Neural Comput.* **4** 448–72
[10] Bhadeshia H K D H, Dimitriu R C, Forsik S, Pak J H and Ryu J H 2009 *Mater. Sci. Technol.* **25** 504–10
[11] Demuth H B, Beale M H, De Jess O and Hagan M T 2014 *Neural network design* (Martin Hagan)
[12] Goodfellow I, Bengio Y and Courville A 2016 *Deep learning* (MIT press)

AUTHOR INDEX

Alapanova, A 510
Andriollo, T 313, 321
Apel, M 289, 578
Atwood, R 118
Baxter, G 118
Bayat, M 72, 88, 124
Beckermann, C 194
Beckwith, C 564
Behnken, H 289, 297
Bellet, M 548
Belouah, S 281
Bilal, M U 233
Bojarevics, V 564
Brachmann, J 297
Budenkova, O 225
Böttger, B 578
Cai, D 357
Chen, Y 104, 118
Chen, Z 257, 502
Cherif, M 273
Clark, S J 104
Clark, S 118
Cockcroft, S 257, 502
Coleman, J 364
Combeau, H 273, 333
Connoley, T 118
Dartfeldt, E 326
De Baere, D 124
De La Torre, U 411
Delannoy, Y 225
Demay, C 281
Djambazov, G 441, 564
Dobravec, T 163, 487
Dong, A 556
Dong, H B 465, 586
Dong, H 357, 387
Dongauzer, K 510
Drakopoulos, M 572
Dubrovskaia, A 510
Dybalska, A 564
Eckert, S 217, 441
Eiken, J 289, 297, 449
Elfsberg, J 326
Elmquist, L 326
Engqvist, J 326
Eriksson, J 305
Esser, P 88, 194
Fainberg, J 155
Falch, K V 572
Fan, P 502
Fautrelle, Y 225
Feng, S 586
Ferrand, M 281
Filho, J E Santos 371
Fujinaga, T 435
Fæster, S 313
Gan, T 96
Gandin, C-A 548
Gandin, Ch-A 427
Gerin, B 273
Gill, S 179
Gomes, L F 457
Gregorc, J 171
Greß, T 471
Griffiths, W D 564
Gu, C 202
Guillemot, G 427, 548
Guo, Z 186
Hackl, G 140
Hall, S 326
Hao, Y Z 379
Hatić, V 241
Hattel, J H 72, 124
Hattel, J 313, 321
He, J C 379
He, S 217
Hepp, E 155
Heringer, R 371
Hort, N 233
Huang, C 110
Hubig, S 132
Ichida, K 419
Inozemtsev, A 510
Irwin, M 305
Isensee, T 403
Jacot, A 494
Jakumeit, J 110, 297
Jarfors, A E W 305
Jing, T 387
Jokisch, T 578
Jolly, M R 305
Jolly, M 265
Jones, C A 305
Jones, M 118
Kao, A 96, 441
Karimi-Sibaki, E 209

Khalajzadeh, V 194
Kharicha, A..... 140, 209, 341, 349, 395, 533
Kouznetsova, V..... 313
Krafthöfer, C..... 155
Krane, M J M..... 364
Krastins, I 96
Kumar, S..... 48
Kunavar, A..... 171
Lacaze, J 411
Laqua, R 110, 297
Larsson, E..... 326
Laschet, G..... 289
Lee, P D..... 80, 104, 118
Lei, H..... 379
Leung, A C L..... 118
Leung, C L A..... 104
Li, J..... 357, 357
Lin, F 321
Ludwig, A..... 140, 209, 341, 349, 395, 533
Luo, A A..... 202
M'Hamdi, M 333
Ma, S..... 387
Maijer, D..... 257, 502
Majkut, M..... 326
Maleev, A 510
Mangelinck-Noël, N 427
Marussi, S..... 104, 118
Mavrič, B..... 163, 241, 487
Mehr, F 502
Mendez, P F..... 25
Miao, M 379
Mirihanage, W U 572
Mitsuyama, Y 249
Mohammadpour, P 64
Mohanty, S..... 72, 124
Morishita, K..... 419
Mramor, K 241
Nadimpalli, V K..... 72
Nakajima, K..... 379
Nakata, T 525
Nardi, V Glück 471
Narumi, T 419, 525
Nastac, L..... 556
Natsume, Y 479
Nenchev, B 179, 465
Nitzl, G..... 140
Ogawa, J 479
Ohno, M..... 249
Ou, J..... 257, 502
Pagone, E..... 265
Pakanati, A..... 333
Papanikolaou, M..... 265
Pereira, C L 457
Pericleous, K A 564
Pericleous, K 96, 441
Phillion, A B 64
Pineau, A..... 427
Poitrault, I 273
Prasad, A 80
Rappaz, M 1
Regordosa, A..... 411
Regula, G 427
Reinhart, G..... 427
Rek, Z..... 241
Robin, V..... 281
Rodrigues, C M G 140, 349, 533
Sabau, A S..... 80
Sakane, S..... 249, 541
Salonitis, K..... 265
Saxena, P..... 265
Schankies, C..... 194
Schleifenbaum, J H 110
Schäfer, W..... 155
Semenov, S 225
Senger, A..... 578
Sertucha, J 411
Shevchenko, N 217, 441
Shibuta, Y 249, 435, 518
Sinclair, L..... 104, 118
Sjögren, T..... 326
Skoglund, P 326
Soar, P..... 441
Spee, J 132
Spinelli, J E 457
Stefan-Kharicha, M 341
Steggo, J..... 305
Steinbach, I 132
StJohn, D..... 80
Strickland, J..... 465
Sturm, J C..... 48
Sturz, L..... 11
Suga, T 419
Svenningsson, R..... 305
Takaki, T 249, 541
Tang, Y 140
Thomas, B G 147
Thorborg, J 48, 88, 313
Tiedje, N S 313
Tiedje, N 321
Tonn, B 471
Tonry, C 96, 564
Tourret, D..... 11, 403
Uddagiri, M 132
Ueno, K 518

Vakhrushev, A ... 140, 349
Vertnik, R ... 163
Viardin, A ... 11
Volk, W ... 471
Wagner, I ... 48
Wang, E ... 225
Wang, Q ... 186, 379
Wang, T ... 225
Wang, Y C ... 379
Watzinger, J ... 140
Wei, C ... 257, 502
Wu, F ... 572
Wu, M ... 140, 209, 341, 349, 395, 533
Xiong, S ... 186
Xu, C L ... 321
Yamanaka, N ... 541
Yasuda, H ... 419, 525
Yuan, L ... 80
Založnik, M ... 11, 273, 333
Zappulla, M L S ... 147
Zhang, A ... 186
Zhang, H W ... 379
Zhang, H ... 209, 395
Zhang, S ... 379, 548
Zhang, Y B ... 321
Zhang, Y ... 257, 313, 502
Zhang, Z ... 209, 395
Zhou, B ... 289
Zhu, Z ... 257, 502
Zielinski, J ... 110
Zong, N ... 387
Šarler, B ... 163, 171, 241, 487